Grundlehren der mathematischen Wissenschaften 293

A Series of Comprehensive Studies in Mathematics

Editors

S. S. Chern B. Eckmann P. de la Harpe
H. Hironaka F. Hirzebruch N. Hitchin
L. Hörmander M.-A. Knus A. Kupiainen
J. Lannes G. Lebeau M. Ratner D. Serre
Ya. G. Sinai N. J. A. Sloane J. Tits
M. Waldschmidt S. Watanabe

Managing Editors

M. Berger J. Coates S. R. S. Varadhan

Daniel Revuz
Marc Yor

Continuous Martingales and Brownian Motion

Corrected Third Printing of the Third Edition

With 8 Figures

Springer

Daniel Revuz
Université Paris VII
Département de Mathématiques
2, place Jussieu
75251 Paris Cedex 05, France
e-mail: revuzd@math.jussieu.fr

Marc Yor
Université Pierre et Marie Curie
Laboratoire de Probabilités
4, place Jussieu, Boîte courrier 188
75252 Paris Cedex 05, France
deaproba@proba.jussieu.fr

3rd edition 1999
Corrected 3rd printing 2005

The Library of Congress has catalogued the original printing as follows:

Revuz, D. Continuous Martingales and Brownian motion / Daniel Revuz, Marc Yor. - 3rd ed. p. cm. - (Grundlehren der mathematischen Wissenschaften; 293) Includes bibliographical references and index. ISBN 3-540-64325-7 (hardcover : alk. paper) 1. Martingales (Mathematics) 2. Brownian motion processes.
I. Yor, Marc. II. Title. III. Series. QA274.5.R48 1999 519.2'87 – dc21

Mathematics Subject Classification (2000): 60G07, 60H05

ISSN 0072-7830

ISBN 3-540-64325-7 Springer Berlin Heidelberg New York

ISBN 3-540-57622-3 2nd edition Springer Berlin Heidelberg New York

Springer is a part of Springer Science+Business Media
springeronline.com

Printed in Germany

Cover design: MetaDesign plus GmbH, Berlin
Typesetting: Typeset by Ingeborg Jebram, Heiligkreuzsteinach, and reformatted by Kurt Mattes, Heidelberg, using a Springer LaTeX macro-package

Printed on acid-free paper 41/3142-5 4 3 2 1 0

Preface

Since the first edition of this book (1991), the interest for Brownian motion and related stochastic processes has not abated in the least. This is probably due to the fact that Brownian motion is in the intersection of many fundamental classes of processes. It is a continuous martingale, a Gaussian process, a Markov process or more specifically a process with independent increments; it can actually be defined, up to simple transformations, as the real-valued, centered process with stationary independent increments and continuous paths. It is therefore no surprise that a vast array of techniques may be successfully applied to its study and we, consequently, chose to organize the book in the following way.

After a first chapter where Brownian motion is introduced, each of the following ones is devoted to a new technique or notion and to some of its applications to Brownian motion. Among these techniques, two are of paramount importance: stochastic calculus, the use of which pervades the whole book, and the powerful excursion theory, both of which are introduced in a self-contained fashion and with a minimum of apparatus. They have made much easier the proofs of many results found in the epoch-making book of Itô and Mc Kean: Diffusion processes and their sample paths, Springer (1965).

These two techniques can both be taught, as we did several times, in a pair of one-semester courses. The first one devoted to Brownian motion and stochastic integration and centered around the famous Itô formula would cover Chapters I through V with possibly the early parts of Chapters VIII and IX. The second course, more advanced, would begin with the local times of Chapter VI and the extension of stochastic calculus to convex functions and work towards such topics as time reversal, Bessel processes and the Ray-Knight theorems which describe the Brownian local times in terms of Bessel processes. Chapter XII on Excursion theory plays a basic role in this second course. Finally, Chapter XIII describes the asymptotic behavior of additive functionals of Brownian motion in dimension 1 and 2 and especially of the winding numbers around a finite number of points for planar Brownian motion.

The text is complemented at the end of each section by a large selection of exercices, the more challenging being marked with the sign $*$ or even $**$. On the one hand, they should enable the reader to improve his understanding of the notions introduced in the text. On the other hand, they deal with many results without which the text might seem a bit "dry" or incomplete; their inclusion in the text however would have increased forbiddingly the size of the book and deprived the reader of the pleasure of working things out by himself. As it is, the text is written with the assumption that the reader will try a good proportion of them, especially those marked with the sign #, and in a few proofs we even indulged in using the results of foregoing exercices.

The text is practically self-contained but for a few results of measure theory. Beside classical calculus, we only ask the reader to have a good knowledge of basic notions of integration and probability theory such as almost-sure and the mean convergences, conditional expectations, independence and the like. Chapter 0 contains a few complements on these topics. Moreover the early chapters include some classical material on which the beginner can hone his skills.

Each chapter ends up with notes and comments where, in particular, references and credits are given. In view of the enormous literature which has been devoted in the past to Brownian motion and related topics, we have in no way tried to draw a historical picture of the subject and apologize in advance to those who may feel slighted.

Likewise our bibliography is not even remotely complete and leaves out the many papers which relate Brownian motion with other fields of Mathematics such as Potential Theory, Harmonic Analysis, Partial Differential Equations and Geometry. A number of excellent books have been written on these subjects, some of which we discuss in the notes and comments.

This leads us to mention some of the manifold offshoots of the Brownian studies which have sprouted since the beginning of the nineties and are bound to be still very much alive in the future:

- the profound relationships between branching processes, random trees and Brownian excursions initiated by Neveu and Pitman and furthered by Aldous, Le Gall, Duquesne, ...
- the important advances in the studies of Lévy processes which benefited from the results found for Brownian motion or more generally diffusions and from the deep understanding of the general theory of processes developed by P. A. Meyer and his "Ecole de Strasbourg". Bertoin's book: Lévy processes (Cambridge Univ. Press, 1996) is a basic reference in these matters; so is the book of Sato: Lévy processes and infinitely divisible distributions (Cambridge Univ. Press, 1999), although it is written in a different spirit and stresses the properties of infinitely divisible laws.
- in a somewhat similar fashion, the deep understanding of Brownian local times has led to intersection local times which serve as a basic tool for the study of multiple points of the three-dimensional Brownian motion. The excellent lecture course of Le Gall (Saint-Flour, 1992) spares us any regret we might have of omitting this subject in our own book. One should also mention the results on the Brownian curve due to Lawler-Schramm-Werner who initiated the study of the Stochastic Loewner Equations.
- stochastic integration and Itô's formula have seen the extension of their domains of validity beyond semimartingales to, for instance, certain Dirichlet processes

i.e. sums of a martingale and of a process with a vanishing quadratic variation (Bertoin, Yamada). Let us also mention the anticipative stochastic calculus (Skorokhod, Nualart, Pardoux). However, a general unifying theory is not yet available; such a research is justified by the interest in fractional Brownian motion (Cheridito, Feyel-De la Pradelle, Valkeila, ...)

Finally it is a pleasure to thank all those, who, along the years, have helped us to improve our successive drafts, J. Jacod, B. Maisonneuve, J. Pitman, A. Adhikari, J. Azéma, M. Emery, H. Föllmer and the late P. A. Meyer to whom we owe so much. Our special thanks go to J. F. Le Gall who put us straight on an inordinate number of points and Shi Zhan who has helped us with the exercises.

Paris, August 2004

Daniel Revuz
Marc Yor

Table of Contents

Chapter 0. Preliminaries 1

§1. Basic Notation 1
§2. Monotone Class Theorem 2
§3. Completion 3
§4. Functions of Finite Variation and Stieltjes Integrals 4
§5. Weak Convergence in Metric Spaces 9
§6. Gaussian and Other Random Variables 11

Chapter I. Introduction 15

§1. Examples of Stochastic Processes. Brownian Motion 15
§2. Local Properties of Brownian Paths 26
§3. Canonical Processes and Gaussian Processes 33
§4. Filtrations and Stopping Times 41
Notes and Comments 48

Chapter II. Martingales 51

§1. Definitions, Maximal Inequalities and Applications 51
§2. Convergence and Regularization Theorems 60
§3. Optional Stopping Theorem 68
Notes and Comments 77

Chapter III. Markov Processes 79

§1. Basic Definitions 79
§2. Feller Processes 88
§3. Strong Markov Property 102
§4. Summary of Results on Lévy Processes 114
Notes and Comments 117

Chapter IV. Stochastic Integration 119

§1. Quadratic Variations 119
§2. Stochastic Integrals 137

§3. Itô's Formula and First Applications 146
§4. Burkholder-Davis-Gundy Inequalities 160
§5. Predictable Processes 171
Notes and Comments 176

Chapter V. Representation of Martingales 179

§1. Continuous Martingales as Time-changed Brownian Motions 179
§2. Conformal Martingales and Planar Brownian Motion 189
§3. Brownian Martingales 198
§4. Integral Representations 209
Notes and Comments 216

Chapter VI. Local Times 221

§1. Definition and First Properties 221
§2. The Local Time of Brownian Motion 239
§3. The Three-Dimensional Bessel Process 251
§4. First Order Calculus 260
§5. The Skorokhod Stopping Problem 269
Notes and Comments 277

Chapter VII. Generators and Time Reversal 281

§1. Infinitesimal Generators 281
§2. Diffusions and Itô Processes 294
§3. Linear Continuous Markov Processes 300
§4. Time Reversal and Applications 313
Notes and Comments 322

Chapter VIII. Girsanov's Theorem and First Applications 325

§1. Girsanov's Theorem 325
§2. Application of Girsanov's Theorem to the Study of Wiener's Space 338
§3. Functionals and Transformations of Diffusion Processes 349
Notes and Comments 362

Chapter IX. Stochastic Differential Equations 365

§1. Formal Definitions and Uniqueness 365
§2. Existence and Uniqueness in the Case of Lipschitz Coefficients 375
§3. The Case of Hölder Coefficients in Dimension One 388
Notes and Comments 399

Chapter X. Additive Functionals of Brownian Motion 401

§1. General Definitions 401

§2. Representation Theorem for Additive Functionals of Linear Brownian Motion 409
§3. Ergodic Theorems for Additive Functionals 422
§4. Asymptotic Results for the Planar Brownian Motion 430
Notes and Comments 436

Chapter XI. Bessel Processes and Ray-Knight Theorems 439

§1. Bessel Processes 439
§2. Ray-Knight Theorems 454
§3. Bessel Bridges 463
Notes and Comments 469

Chapter XII. Excursions 471

§1. Prerequisites on Poisson Point Processes 471
§2. The Excursion Process of Brownian Motion 480
§3. Excursions Straddling a Given Time 488
§4. Descriptions of Itô's Measure and Applications 493
Notes and Comments 511

Chapter XIII. Limit Theorems in Distribution 515

§1. Convergence in Distribution 515
§2. Asymptotic Behavior of Additive Functionals of Brownian Motion 522
§3. Asymptotic Properties of Planar Brownian Motion 531
Notes and Comments 541

Appendix 543

§1. Gronwall's Lemma 543
§2. Distributions 543
§3. Convex Functions 544
§4. Hausdorff Measures and Dimension 547
§5. Ergodic Theory 548
§6. Probabilities on Function Spaces 548
§7. Bessel Functions 549
§8. Sturm-Liouville Equation 550

Bibliography 553

Index of Notation 595

Index of Terms 599

Catalogue 605

Chapter 0. Preliminaries

In this chapter, we review a few basic facts, mainly from integration and classical probability theories, which will be used throughout the book without further ado. Some other prerequisites, usually from calculus, which will be used in some special parts are collected in the Appendix at the end of the book.

§1. Basic Notation

Throughout the sequel, $\mathbb{N}$ will denote the set of integers, namely, $\mathbb{N} = \{0, 1, \dots\}$, $\mathbb{R}$ the set of real numbers, $\mathbb{Q}$ the set of rational numbers, $\mathbb{C}$ the set of complex numbers. Moreover $\mathbb{R}_+ = [0, \infty[$ and $\mathbb{Q}_+ = \mathbb{Q} \cap \mathbb{R}_+$. By positive we will always mean ≥ 0 and say strictly positive for > 0.

Likewise a real-valued function f defined on an interval of $\mathbb{R}$ is increasing (resp. strictly increasing) if $x < y$ entails $f(x) \leq f(y)$ (resp. $f(x) < f(y)$).

If a, b are real numbers, we write:

$$a \wedge b = \min(a, b), \qquad a \vee b = \max(a, b).$$

If E is a set and f a real-valued function on E, we use the notation

$$f^+ = f \vee 0, \qquad f^- = -(f \wedge 0), \qquad |f| = f^+ + f^-,$$

$$\|f\| = \sup_{x \in E} |f(x)|.$$

We will write $a_n \downarrow a$ ($a_n \uparrow a$) if the sequence (a_n) of real numbers decreases (increases) to a.

If $(E, \mathscr{E})$ and $(F, \mathscr{F})$ are measurable spaces, we write $f \in \mathscr{E}/\mathscr{F}$ to say that the function $f : E \to F$ is measurable with respect to $\mathscr{E}$ and $\mathscr{F}$. If $(F, \mathscr{F})$ is the real line endowed with the σ-field of Borel sets, we write simply $f \in \mathscr{E}$ and if, in addition, f is positive, we write $f \in \mathscr{E}_+$. The characteristic function of a set A is written 1_A; thus, the statements $A \in \mathscr{E}$ and $1_A \in \mathscr{E}$ have the same meaning.

If Ω is a set and f_i, $i \in I$, is a collection of maps from Ω to measurable spaces $(E_i, \mathscr{E}_i)$, the smallest σ-field on Ω for which the f_i's are measurable is denoted by $\sigma(f_i, i \in I)$. If $\mathscr{C}$ is a collection of subsets of Ω, then $\sigma(\mathscr{C})$ is the smallest σ-field containing $\mathscr{C}$; we say that $\sigma(\mathscr{C})$ is generated by $\mathscr{C}$. The σ-field $\sigma(f_i, i \in I)$ is generated by the family $\mathscr{C} = \left\{f_i^{-1}(A_i), A_i \in \mathscr{E}_i, i \in I\right\}$. Finally if

$\mathscr{E}_i$, $i \in I$, is a family of σ-fields on Ω, we denote by $\bigvee_i \mathscr{E}_i$ the σ-field generated by $\bigcup_i \mathscr{E}_i$. It is the union of the σ-fields generated by the countable sub-families of $\mathscr{E}_i$, $i \in I$.

A measurable space $(E, \mathscr{E})$ is *separable* if $\mathscr{E}$ is generated by a countable collection of sets. In particular, if E is a LCCB space i.e. a locally compact space with countable basis, the σ-field of its Borel sets is separable; it will often be denoted by $\mathscr{B}(E)$. For instance, $\mathscr{B}(\mathbb{R}^d)$ is the σ-field of Borel subsets of the d-dimensional euclidean space.

For a measure m on $(E, \mathscr{E})$ and $f \in \mathscr{E}$, the integral of f with respect to m, if it makes sense, will be denoted by any of the symbols

$$\int f\,dm, \qquad \int f(x)dm(x), \qquad \int f(x)m(dx), \qquad m(f), \qquad \langle m, f\rangle,$$

and in case E is a subset of a euclidean space and m is the Lebesgue measure, $\int f(x)dx$.

If $(\Omega, \mathscr{F}, P)$ is a probability space, we will as usual use the words random variable and expectation in lieu of measurable function and integral and write

$$E[X] = \int_\Omega X\,dP.$$

We will often write r.v. as shorthand for random variable. The law of the r.v. X, namely the image of P by X will be denoted by P_X or $X(P)$. Two r.v.'s defined on the same space are P-equivalent if they are equal P-a.s.

If $\mathscr{G}$ is a sub-σ-field of $\mathscr{F}$, the conditional expectation of X with respect to $\mathscr{G}$, if it exists, is written $E[X \mid \mathscr{G}]$. If $X = 1_A$, $A \in \mathscr{F}$, we may write $P(A \mid \mathscr{G})$. If $\mathscr{G} = \sigma(X_i, i \in I)$ we also write $E[X \mid X_i, i \in I]$ or $P(A \mid X_i, i \in I)$. As is well-known conditional expectations are defined up to P-equivalence, but we will often omit the qualifying P-a.s. When we apply conditional expectation successively, we shall abbreviate $E\left[E\left[X \mid \mathscr{F}_1\right] \mid \mathscr{F}_2\right]$ to $E\left[X \mid \mathscr{F}_1 \mid \mathscr{F}_2\right]$.

We recall that if Ω is a Polish space (i.e. a metrizable complete topological space with a countable dense subset), $\mathscr{F}$ the σ-field of its Borel subsets and if $\mathscr{G}$ is separable, then there is a regular conditional probability distribution given $\mathscr{G}$.

If μ and ν are two σ-finite measures on $(E, \mathscr{E})$, we write $\mu \perp \nu$ to mean that they are mutually singular, $\mu \ll \nu$ to mean that μ is absolutely continuous with respect to ν and $\mu \sim \nu$ if they are equivalent, namely if $\mu \ll \nu$ and $\nu \ll \mu$. The Radon-Nikodym derivative of the absolutely continuous part of μ with respect to ν is written $\left.\frac{d\mu}{d\nu}\right|_{\mathscr{E}}$ and $\mathscr{E}$ is dropped when there is no risk of confusion.

§2. Monotone Class Theorem

We will use several variants of this theorem which we state here without proof.

(2.1) Theorem. *Let $\mathscr{S}$ be a collection of subsets of Ω such that*

i) $\Omega \in \mathscr{S}$,
ii) *if* $A, B \in \mathscr{S}$ *and* $A \subset B$, *then* $B \backslash A \in \mathscr{S}$,
iii) *if* $\{A_n\}$ *is an increasing sequence of elements of* $\mathscr{S}$ *then* $\bigcup A_n \in \mathscr{S}$.

If $\mathscr{S} \supset \mathscr{F}$ *where* $\mathscr{F}$ *is closed under finite intersections then* $\mathscr{S} \supset \sigma(\mathscr{F})$.

The above version deals with sets. We turn to the functional version.

(2.2) Theorem. *Let* $\mathscr{H}$ *be a vector space of bounded real-valued functions on* Ω *such that*

i) *the constant functions are in* $\mathscr{H}$,
ii) *if* $\{h_n\}$ *is an increasing sequence of positive elements of* $\mathscr{H}$ *such that* $h = \sup_n h_n$ *is bounded, then* $h \in \mathscr{H}$.

If $\mathscr{C}$ *is a subset of* $\mathscr{H}$ *which is stable under pointwise multiplication, then* $\mathscr{H}$ *contains all the bounded* $\sigma(\mathscr{C})$*-measurable functions.*

The above theorems will be used, especially in Chap. III, in the following set-up. We have a family $f_i, i \in I$, of mappings of a set Ω into measurable spaces $(E_i, \mathscr{E}_i)$. We assume that for each $i \in I$ there is a subclass $\mathscr{N}_i$ of $\mathscr{E}_i$, closed under finite intersections and such that $\sigma(\mathscr{N}_i) = \mathscr{E}_i$. We then have the following results.

(2.3) Theorem. *Let* $\mathscr{N}$ *be the family of sets of the form* $\bigcap_{i \in J} f_i^{-1}(A_i)$ *where* A_i *ranges through* $\mathscr{N}_i$ *and* J *ranges through the finite subsets of* I; *then* $\sigma(\mathscr{N}) = \sigma(f_i, i \in I)$.

(2.4) Theorem. *Let* $\mathscr{H}$ *be a vector space of real-valued functions on* Ω, *containing* 1_Ω, *satisfying property ii) of Theorem (2.2) and containing all the functions* 1_Γ *for* $\Gamma \in \mathscr{N}$. *Then,* $\mathscr{H}$ *contains all the bounded, real-valued,* $\sigma(f_i, i \in I)$*-measurable functions.*

§3. Completion

If $(E, \mathscr{E})$ is a measurable space and μ a probability measure on $\mathscr{E}$, the completion $\mathscr{E}^\mu$ of $\mathscr{E}$ with respect to μ is the σ-field of subsets B of E such that there exist B_1 and B_2 in $\mathscr{E}$ with $B_1 \subset B \subset B_2$ and $\mu(B_2 \backslash B_1) = 0$. If γ is a family of probability measures on $\mathscr{E}$, the σ-field

$$\mathscr{E}^\gamma = \bigcap_{\mu \in \gamma} \mathscr{E}^\mu$$

is called the completion of $\mathscr{E}$ with respect to γ. If γ is the family of all probability measures on $\mathscr{E}$, then $\mathscr{E}^\gamma$ is denoted by $\mathscr{E}^*$ and is called the σ-field of *universally measurable sets*.

If $\mathscr{F}$ is a sub-σ-algebra of $\mathscr{E}^\gamma$ we define the *completion of* $\mathscr{F}$ *in* $\mathscr{E}^\gamma$ *with respect to* γ as the family of sets A with the following property: for each $\mu \in \gamma$, there is a set B in $\mathscr{F}$ such that $A \triangle B$ is in $\mathscr{E}^\gamma$ and $\mu(A \triangle B) = 0$. This family will be denoted $\tilde{\mathscr{F}}^\gamma$; the reader will show that it is a σ-field which is larger than $\mathscr{F}^\gamma$. Moreover, it has the following characterization.

(3.1) Proposition. *A set A is in $\tilde{\mathscr{F}}^\gamma$ if and only if for every $\mu \in \gamma$ there is a set B_μ in $\mathscr{F}$ and two μ-negligible sets N_μ and M_μ in $\mathscr{E}$ such that*

$$B_\mu \setminus N_\mu \subset A \subset B_\mu \cup M_\mu.$$

← McCallum & Blumenthal, p.26

Proof. Left to the reader as an exercise. □

The following result gives a means of checking the measurability of functions with respect to σ-algebras of the $\tilde{\mathscr{F}}^\gamma$-type.

(3.2) Proposition. *For $i = 1, 2$, let $(E_i, \mathscr{E}_i)$ be a measurable space, γ_i a family of probability measures on $\mathscr{E}_i$ and $\mathscr{F}_i$ a sub-σ-algebra of $\mathscr{E}_i^{\gamma_i}$. If f is a map which is both in $\mathscr{E}_1/\mathscr{E}_2$ and $\mathscr{F}_1/\mathscr{F}_2$ and if $f(\mu) \in \gamma_2$ for every $\mu \in \gamma_1$ then f is in $\tilde{\mathscr{F}}_1^{\gamma_1}/\tilde{\mathscr{F}}_2^{\gamma_2}$.*

↑ μf^{-1}

Proof. Let A be in $\tilde{\mathscr{F}}_2^{\gamma_2}$. For $\mu \in \gamma_1$, since $\nu = f(\mu)$ is in γ_2, there is a set $B_\nu \in \mathscr{F}_2$ and two ν-negligible sets N_ν and M_ν in $\mathscr{E}_2$ such that

$$B_\nu \setminus N_\nu \subset A \subset B_\nu \cup M_\nu.$$

The set $B_\mu = f^{-1}(B_\nu)$ belongs to $\mathscr{F}_1$, the sets $N_\mu = f^{-1}(N_\nu)$ and $M_\mu = f^{-1}(M_\nu)$ are μ-negligible sets of $\mathscr{E}_1$ and

$$B_\mu \setminus N_\mu \subset f^{-1}(A) \subset B_\mu \cup M_\mu.$$

This entails that $f^{-1}(A) \in \tilde{\mathscr{F}}_1^{\gamma_1}$, which completes the proof. □

§4. Functions of Finite Variation and Stieltjes Integrals

This section is devoted to a set of properties which will be used constantly throughout the book.

We deal with real-valued, right-continuous functions A with domain $[0, \infty[$. The results may be easily extended to the case of $\mathbb{R}$. The value of A in t is denoted A_t or $A(t)$. Let Δ be a subdivision of the interval $[0, t]$ with $0 = t_0 < t_1 < \ldots < t_n = t$; the number $|\Delta| = \sup_i |t_{i+1} - t_i|$ is called the *modulus* or *mesh* of Δ. We consider the sum

$$S_t^\Delta = \sum_i |A_{t_{i+1}} - A_{t_i}|.$$

If Δ' is another subdivision which is a refinement of Δ, that is, every point t_i of Δ is a point of Δ', then plainly $S_t^{\Delta'} \geq S_t^\Delta$.

(4.1) Definition. *The function A is of finite variation if for every t*

$$S_t = \sup_{\Delta} S_t^{\Delta} < +\infty.$$

The function $t \to S_t$ is called the total variation *of A and S_t is the variation of A on $[0, t]$. The function S is obviously positive and increasing and if* $\lim_{t\to\infty} S_t < +\infty$, *the function A is said to be of* bounded variation.

The same notions could be defined on any interval $[a, b]$. We shall say that a function A on the whole line is of *finite variation* if it is of finite variation on any compact interval but not necessarily of bounded variation on the whole of $\mathbb{R}$.

Let us observe that C^1-functions are of finite variation. Monotone finite functions are of finite variation and conversely we have the

(4.2) Proposition. *Any function of finite variation is the difference of two increasing functions.*

Proof. The functions $(S + A)/2$ and $(S - A)/2$ are increasing as the reader can easily show, and A is equal to their difference. □

This decomposition is moreover *minimal* in the sense that if $A = F - G$ where F and G are positive and increasing, then $(S + A)/2 \leq F$ and $(S - A)/2 \leq G$.

As a result, the function A has left limits in any $t \in]0, \infty[$. We write A_{t-} or $A(t\text{-})$ for $\lim_{s\uparrow t} A_s$ and we set $A_{0-} = 0$. We moreover set $\Delta A_t = A_t - A_{t-}$; this is the *jump* of A in t.

The importance of these functions lies in the following

(4.3) Theorem. *There is a one-to-one correspondence between Radon measures μ on $[0, \infty[$ and right-continuous functions A of finite variation given by*

$$A_t = \mu([0, t]).$$

Consequently $A_{t-} = \mu([0, t[)$ and $\Delta A_t = \mu(\{t\})$. Moreover, if $\mu(\{0\}) = 0$, the variation S of A corresponds to the total variation $|\mu|$ of μ and the decomposition in the proof of Proposition (4.2) corresponds to the minimal decomposition of μ into positive and negative parts.

If f is a locally bounded Borel function on $\mathbb{R}_+$, its *Stieltjes integral with respect to A*, denoted

$$\int_0^t f_s dA_s, \qquad \int_0^t f(s)dA(s) \qquad \text{or} \qquad \int_{]0,t]} f(s)dA_s$$

is the integral of f with respect to μ on the interval $]0, t]$. The reader will observe that the jump of A at zero does not come into play and that $\int_0^t dA_s = A_t - A_0$. If we want to consider the integral on $[0, t]$, we will write $\int_{[0,t]} f(s)dA_s$. The integral on $]0, t]$ is also denoted by $(f \cdot A)_t$. We point out that the map $t \to (f \cdot A)_t$ is itself a right-continuous function of finite variation.

A consequence of the Radon-Nikodym theorem applied to μ and to the Lebesgue measure λ is the

(4.4) Theorem. *A function A of finite variation is λ-a.e. differentiable and there exists a function B of finite variation such that $B' = 0$ λ-a.e. and*

$$A_t = B_t + \int_0^t A'_s ds.$$

The function A is said to be *absolutely continuous* if $B = 0$. The corresponding measure μ is then absolutely continuous with respect to λ.

We now turn to a series of notions and properties which are very useful in handling Stieltjes integrals.

(4.5) Proposition (Integration by parts formula). *If A and B are two functions of finite variation, then for any t,*

$$A_t B_t = A_0 B_0 + \int_0^t A_s dB_s + \int_0^t B_{s-} dA_s.$$

Proof. If μ (resp. ν) is associated with A (resp. B) both sides of the equality are equal to $\mu \otimes \nu([0,t]^2)$; indeed $\int_0^t A_s dB_s$ is the measure of the upper triangle including the diagonal, $\int_0^t B_{s-} dA_s$ the measure of the lower triangle excluding the diagonal and $A_0 B_0 = \mu \otimes \nu(\{0,0\})$. □

To reestablish the symmetry, the above formula can also be written

$$A_t B_t = \int_0^t A_{s-} dB_s + \int_0^t B_{s-} dA_s + \sum_{s \le t} \Delta A_s \Delta B_s.$$

The sum on the right is meaningful as A and B have only countably many discontinuities. In fact, A can be written uniquely $A_t = A_t^c + \sum_{s \le t} \Delta A_s$ where A^c is continuous and of finite variation.

The next result is a "chain rule" formula.

(4.6) Proposition. *If F is a C^1-function and A is of finite variation, then $F(A)$ is of finite variation and*

$$F(A_t) = F(A_0) + \int_0^t F'(A_{s-}) dA_s + \sum_{s \le t} \left(F(A_s) - F(A_{s-}) - F'(A_{s-}) \Delta A_s \right).$$

Proof. The result is true for $F(x) = x$, and if it is true for F, it is true for $xF(x)$ as one can deduce from the integration by parts formula; consequently the result is true for polynomials. The proof is completed by approximating a C^1-function by a sequence of polynomials. □

As an application of the notions introduced thus far, let us prove the useful

(4.7) Proposition. *If A is a right continuous function of finite variation, then*

$$Y_t = Y_0 \prod_{s \le t} (1 + \Delta A_s) \exp\left(A_t^c - A_0^c \right)$$

is the only locally bounded solution of the equation

$$Y_t = Y_0 + \int_0^t Y_{s-} dA_s.$$

Proof. By applying the integration by parts formula to $Y_0 \prod_{s\le t}(1+\Delta A_s)$ and $\exp\left(\int_0^t dA_s^c\right)$ which are both of finite variation, it is easily seen that Y is a solution of the above equation.

Let Z be the difference of two locally bounded solutions and $M_t = \sup_{s\le t}|Z_s|$. It follows from the equality $Z_t = \int_0^t Z_{s-}dA_s$ that $|Z_t| \le M_t S_t$ where S is the variation of A; then, thanks to the integration by parts formula

$$|Z_t| \le M_t \int_0^t S_{s-}dS_s \le M_t S_t^2/2,$$

and inductively,

$$|Z_t| \le \frac{M_t}{n!}\int_0^t S_{s-}^n dS_s \le M_t S_t^{n+1}/(n+1)!$$

which proves that $Z = 0$. □

We close this section by a study of the fundamental technique of *time changes*, which allows the explicit computation of some Stieltjes integrals. We consider now an increasing, possibly infinite, right-continuous function A and for $s \ge 0$, we define

$$C_s = \inf\{t : A_t > s\}$$

where, here and below, it is understood that $\inf\{\emptyset\} = +\infty$. We will also say that C is the (right-continuous) *inverse* of A.

To understand what follows, it is useful to draw Figure 1 (see below) showing the graph of A and the way to find C. The function C is obviously increasing so that

$$C_{s-} = \lim_{u\uparrow s} C_u$$

is well-defined for every s. It is easily seen that

$$C_{s-} = \inf\{t : A_t \ge s\}.$$

In particular if A has a constant stretch at level s, then C_s will be at the right end and C_{s-} at the left end of the stretch; moreover $C_{s-} \ne C_s$ only if A has a constant stretch at level s. By convention $C_{0^-} = 0$.

(4.8) Lemma. *The function C is right-continuous. Moreover $A(C_s) \ge s$ and*

$$A_t = \inf\{s : C_s > t\}.$$

Proof. That $A(C_s) \ge s$ is obvious. Moreover, the set $\{A_t > s\}$ is the union of the sets $\{A_t > s+\varepsilon\}$ for $\varepsilon > 0$, which proves the right continuity of C.

If furthermore, $C_s > t$, then $t \notin \{u : A_u > s\}$ and $A_t \le s$. Consequently, $A_t \le \inf\{s : C_s > t\}$. On the other hand, $C(A_t) \ge t$ for every t, hence $C(A_{t+\varepsilon}) \ge t+\varepsilon > t$ which forces

$$A_{t+\varepsilon} \ge \inf\{s : C_s > t\}$$

and because of the right continuity of A

$$A_t \ge \inf\{s : C_s > t\}. \quad □$$

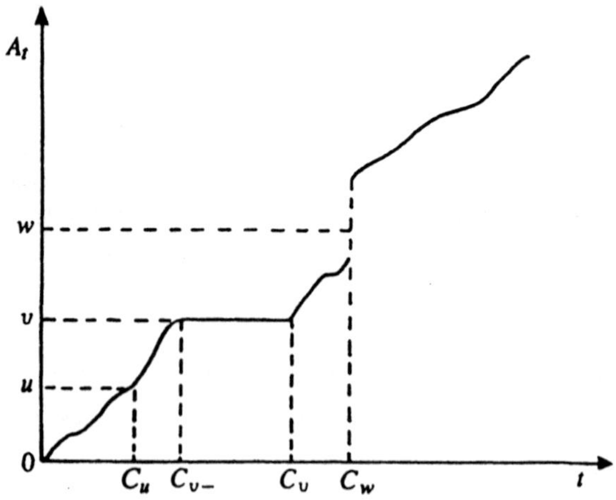

Fig. 1.

Remarks. Thus A and C play symmetric roles. But if A is continuous, C is still only right-continuous in general; in that case, however, $A(C_s) = s$ but $C(A_s) > s$ if s is in an interval of constancy of A. As already observed, the jumps of C correspond to the level stretches of A and vice-versa; thus C is continuous iff A is strictly increasing. The right continuity of C does not stem from the right-continuity of A but from its definition with a strict inequality; likewise, C_{s-} is left continuous.

We now state a "change of variables" formula.

(4.9) Proposition. *If f is a positive Borel function on* $[0, \infty[$,

$$\int_{[0,\infty[} f(u)dA_u = \int_0^\infty f(C_s)1_{(C_s<\infty)}ds.$$

Proof. If $f = 1_{[0,v]}$, the formula reads

$$A_v = \int_0^\infty 1_{(C_s \le v)}ds$$

and is then a consequence of the definition of C. By taking differences, the equality holds for the indicators of sets $]u, v]$, and by the monotone class theorem for any f with compact support. Taking increasing limits yields the result in full generality. □

In the same way, we also have

$$\int_0^\infty f(u)dA_u = \int_0^\infty f(C_s)1_{(0<C_s<\infty)}ds.$$

The right member in the proposition may also be written

$$\int_0^{A_\infty} f(C_s)ds,$$

because $C_s < \infty$ if and only if $A_\infty > s$.

The last result is closely related to time changes.

(4.10) Proposition. *If u is a continuous, non decreasing function on the interval $[a, b]$, then for a non negative Borel function f on $[u(a), u(b)]$*

$$\int_{[a,b]} f\big(u(s)\big)dA_{u(s)} = \int_{[u(a),u(b)]} f(t)dA_t.$$

The integral on the left is with respect to the measure associated with the right continuous increasing function $s \to A\big(u(s)\big)$.

Proof. We define $v_t = \inf\{s : u(s) > t\}$, then $u(v_t) = t$ and v is a measurable mapping from $[u(a), u(b)]$ into $[a, b]$. Let dA be the measure on $[u(a), u(b)]$ associated with A and ν the image of dA by v. Then dA is the image of ν by u and therefore

$$\int_{[a,b]} f\big(u(s)\big)d\nu(s) = \int_{[u(a),u(b)]} f(t)dA_t.$$

In particular, $A\big(u(b)\big) - A\big(u(a)-\big) = \nu([a,b])$ which proves that ν is associated with the increasing function $s \to A\big(u(s)\big)$. The proposition is established. □

§5. Weak Convergence in Metric Spaces

Let E be a metric space with metric d and call $\mathscr{B}$ the Borel σ-algebra on E. We want to recall a few facts about the weak convergence of probability measures on $(E, \mathscr{B})$. If P is such a measure, we say that a subset A of E is a P-continuity set if $P(\partial A) = 0$ where ∂A is the boundary of A.

(5.1) Proposition. *For probability measures P_n, $n \in \mathbb{N}$, and P, the following conditions are equivalent:*

(i) For every bounded continuous function f on E,

$$\lim_n \int f\,dP_n = \int f\,dP;$$

(ii) For every bounded uniformly continuous function f on E,

$$\lim_n \int f\,dP_n = \int f\,dP;$$

(iii) For every closed subset F of E, $\overline{\lim_n} P_n(F) \le P(F)$;

(iv) For every open subset G of E, $\underline{\lim}_n P_n(G) \geq P(G)$;

(v) for every P-continuity set A, $\lim_n P_n(A) = P(A)$.

(5.2) Definition. *If P_n and P satisfy the equivalent conditions of the preceding proposition, we say that (P_n) converges* weakly *to P.*

If π is a family of probability measures on $(E, \mathscr{B})$, we will say that it is weakly relatively compact if every sequence of elements of π contains a weakly convergent subsequence. To prove weak convergence, one needs a criterion for weak compactness which is the *raison d'être* of the following

(5.3) Definition. *A family π is* tight *if for every $\varepsilon \in]0, 1[$, there exists a compact set K_ε such that*

$$P[K_\varepsilon] \geq 1 - \varepsilon, \qquad \textit{for every} \quad P \in \pi.$$

With this definition we have the

(5.4) Theorem (Prokhorov's criterion). *If a family π is tight, then it is weakly relatively compact. If E is a Polish space, then a weakly relatively compact family is tight.*

(5.5) Definition. *If $(X_n)_{n \in \mathbb{N}}$ and X are random variables taking their values in a metric space E, we say that (X_n) converges* in distribution *or* in law *to X if their laws P_{X_n} converge weakly to the law P_X of X. We will then write $X_n \xrightarrow{(d)} X$. We also write $X \overset{(d)}{=} Y$ to mean that X and Y have the same distribution or the same law.*

We stress the fact that the X_n's and X need not be defined on the same probability space. If they are defined on the same probability space $(\Omega, \mathscr{F}, P)$, we may set the

(5.6) Definition. *The sequence (X_n) converges* in probability *to X if, for every $\varepsilon > 0$,*

$$\lim_{n \to \infty} P\left[d(X_n, X) > \varepsilon\right] = 0.$$

We will then write $P\text{-}\lim X_n = X$.

In a Polish space, if $P\text{-}\lim X_n = X$, then $X_n \xrightarrow{(d)} X$. The converse is not true in general nor even meaningful since, as already observed the X_n's need not be defined on the same probability space. However if the X_n's are defined on the same space and converge weakly to a constant c then they converge in probability to the constant r.v. c as the reader can easily check.

The following remarks will be important in Chap. XIII.

(5.7) Lemma. *If (X_n, Y_n) is a sequence of r.v.'s with values in separable metric spaces E and F and such that*

(i) (X_n, Y_n) converges in distribution to (X, Y),

(ii) the law of Y_n does not depend on n,

then, for every Borel function $\varphi : F \to G$ where G is a separable metric space, the sequence $(X_n, \varphi(Y_n))$ converges in distribution to $\big(X, \varphi(Y)\big)$.

Proof. It is enough to prove that if h, k are bounded continuous functions,

$$\lim_n E\left[h(X_n)k\left(\varphi(Y_n)\right)\right] = E\left[h(X)k\big(\varphi(Y)\big)\right].$$

Set $p = k \circ \varphi$; if ν is the common law of the Y_n's, there is a bounded continuous function $\tilde{p}$ in $L^1(\nu)$ such that $\int |p - \tilde{p}| d\nu < \varepsilon$ for any preassigned ε. Then,

$$\begin{aligned}
&|E\left[h(X_n)p(Y_n)\right] - E\left[h(X)p(Y)\right]| \\
&\quad \leq \left|E\left[h(X_n)\left(p(Y_n) - \tilde{p}(Y_n)\right)\right]\right| + \left|E\left[h(X_n)\tilde{p}(Y_n)\right] - E\left[h(X)\tilde{p}(Y)\right]\right| \\
&\qquad + \left|E\left[h(X)\big(p(Y) - \tilde{p}(Y)\big)\right]\right| \\
&\quad \leq 2\|h\|_\infty \varepsilon + \left|E\left[h(X_n)\tilde{p}(Y_n)\right] - E\left[h(X)\tilde{p}(Y)\right]\right|.
\end{aligned}$$

By taking n large, the last term can be made arbitrarily small since h and $\tilde{p}$ are continuous. The proof is complete. □

(5.8) Corollary. *Let $\left(X_n^1, \dots, X_n^k\right)$ be a sequence of k-tuples of r.v.'s with values in separable metric spaces S_j, $j = 1, \dots, k$, which converges in distribution to $\left(X^1, \dots, X^k\right)$. If for each j, the law of X_n^j does not depend on n, then for any Borel functions $\varphi_j : S_j \to U_j$ where U_j is a separable metric space, the sequence $\left(\varphi_1\left(X_n^1\right), \dots, \varphi_k\left(X_n^k\right)\right)$ converges in distribution to $\left(\varphi_1\left(X^1\right), \dots, \varphi_k\left(X^k\right)\right)$.*

Proof. The above lemma applied to $X_n = \left(X_n^1, \dots, X_n^{k-1}\right)$, $Y_n = X_n^k$ permits to replace $\left(X_n^k\right)$ by $\varphi_k\left(X_n^k\right)$; one then takes $X_n = \left(X_n^1, \dots, X_n^{k-2}; \varphi_k\left(X_n^k\right)\right)$, $Y_n = X_n^{k-1}$ and so on and so forth.

§6. Gaussian and Other Random Variables

We will write $X \sim \mathcal{N}(m, \sigma^2)$ to mean that the r.v. X is Gaussian with mean m and variance σ^2. In particular, $X \sim \mathcal{N}(0, 1)$ means that X is a Gaussian *centered* r.v. with unit variance or in other words a reduced Gaussian r.v. In what follows the constant r.v.'s are considered to be a particular case of Gaussian r.v.'s, namely those with $\sigma^2 = 0$. If $X \sim \mathcal{N}(m, \sigma^2)$, $\sigma > 0$, we recall that X has the density $\left(\sqrt{2\pi}\sigma\right)^{-1} \exp\left(-\frac{1}{2}(x - m)^2/\sigma^2\right)$ and that its characteristic function (abbreviated c.f. in the sequel) is given by

$$E\left[e^{itX}\right] = \exp\left(itm - \frac{\sigma^2 t^2}{2}\right).$$

We recall the

(6.1) Proposition. *If (X_n) is a sequence of Gaussian r.v.'s which converges in probability to a r.v. X, then X is a Gaussian r.v., the family $\{|X_n|^p\}$ is uniformly integrable and the convergence holds in L^p for every $p \geq 1$.*

Thus the set of Gaussian r.v.'s defined on a given probability space $(\Omega, \mathscr{F}, P)$ is a closed subset of $L^2(\Omega, \mathscr{F}, P)$. More specifically we set the

(6.2) Definition. *A* Gaussian space *is a closed linear subspace of a space $L^2(\Omega, \mathscr{F}, P)$ consisting only of centered Gaussian r.v.'s.*

If G is a Gaussian space and $X_1, \dots, X_d$ are in G, then the d-dimensional r.v. $X = (X_1, \dots, X_d)$ is a Gaussian r.v., in other words $\alpha(X)$ is a real Gaussian r.v. for every linear form α on $\mathbb{R}^d$. Let us also recall that if K is a symmetric semi-definite positive $d \times d$-matrix (i.e. $\langle x, Kx\rangle \geq 0$ for every $x \in \mathbb{R}^d$), it is the covariance matrix of a d-dimensional centered Gaussian r.v..

We recall the

(6.3) Proposition. *Let G_i, $i \in I$, be a family of closed subspaces of a given Gaussian space; then, the σ-fields $\sigma(G_i)$ are independent if and only if the spaces G_i are pairwise orthogonal.*

In particular, the components of an $\mathbb{R}^d$-valued centered Gaussian variable are independent if and only if they are uncorrelated.

A few probability distributions on the line will occur throughout the book and we recall some of their relationships.

A random variable Y follows the *arcsine* law if it has the density $\left(\pi\sqrt{x(1-x)}\right)^{-1}$ on $[0, 1]$; then $\log Y$ has a characteristic function equal to

$$\Gamma\left(\frac{1}{2} + i\lambda\right) / \sqrt{\pi}\,\Gamma(1 + i\lambda).$$

If $N \sim \mathscr{N}(0, 1)$ then $\log\left(\frac{1}{2}N^2\right)$ has c.f. $\Gamma\left(\frac{1}{2} + i\lambda\right) / \sqrt{\pi}$ and if $\mathbf{e}$ is an exponential r.v. with parameter 1 the c.f. of $\log(\mathbf{e})$ is $\Gamma(1 + i\lambda)$. It follows that

$$N^2 \overset{(d)}{=} 2\mathbf{e}Y$$

with $\mathbf{e}$ and Y independent. This can also be seen directly by writing

$$N^2 = \left(N^2 + N'^2\right)\left(\frac{N^2}{N^2 + N'^2}\right)$$

where $N' \sim \mathscr{N}(0, 1)$ is independent of N and showing that the two factors are independent and have respectively the exponential (mean: 2) and arcsine laws.

The above identity is in fact but a particular case of an identity on Gamma and Beta r.v.'s. Let us call γ_a, $a > 0$, the Gamma r.v. with density $x^{a-1}e^{-x}/\Gamma(a)$ on $\mathbb{R}_+$, and $\beta_{a,b}$, a and $b > 0$, the Beta r.v. with density $x^{a-1}(1-x)^{b-1}/B(a, b)$ on $[0, 1]$. Classical computations show that if γ_a and γ_b are independent, then

i) $\gamma_a + \gamma_b$ and $\gamma_a / (\gamma_a + \gamma_b)$ are independent,

ii) $\gamma_a + \gamma_b \stackrel{(d)}{=} \gamma_{a+b}$, iii) $\gamma_a / (\gamma_a + \gamma_b) \stackrel{(d)}{=} \beta_{a,b}$.

From these properties follows the bi-dimensional equality in law

$$(\gamma_a, \gamma_b) \stackrel{(d)}{=} \gamma_{a+b} \left(\beta_{a,b}, 1 - \beta_{a,b}\right),$$

where, on the right-hand side, γ_{a+b} and $\beta_{a,b}$ are assumed independent.

Further if N and N' are independent reduced Gaussian random variables then N/N' and $N/|N'|$ are *Cauchy* r.v.'s i.e. have the density $\left(\pi(1+x^2)\right)^{-1}$ on the real line. If C is a Cauchy r.v. then $Y \stackrel{(d)}{=} (1+C^2)^{-1}$. Next if $(\mathbf{e}', Y')$ is an independent copy of $(\mathbf{e}, Y)$, then the c.f.'s of $\log(\mathbf{e}/\mathbf{e}')$, $\log(Y/Y')$ and $\log C^2$ are respectively equal to

$$\pi\lambda/\sinh(\pi\lambda), \tanh(\pi\lambda)/\pi\lambda \text{ and } (\cosh(\pi\lambda))^{-1} = \Gamma\left(\frac{1}{2} + i\lambda\right)\Gamma\left(\frac{1}{2} - i\lambda\right)/\pi.$$

Thus, $\log C^2 \stackrel{(d)}{=} \log(\mathbf{e}/\mathbf{e}') + \log(Y/Y')$. Finally, the density of $\log C^2$ is $\left(2\pi\cosh(x/2)\right)^{-1}$.

The above hyperbolic functions occur in many computations. If Φ is such a function, we give below series representations for Φ, as well as for the probability densities ϕ and f defined by

$$\Phi(\lambda) = \int_{-\infty}^{\infty} \exp(i\lambda x)\phi(x)dx = \int_0^{\infty} \exp\left(-\lambda^2 y/2\right) f(y)dy,$$

and the distribution function

$$F(x) = \int_0^x f(y)dy.$$

a) If $\Phi(\lambda) = \tanh\pi\lambda/\pi\lambda$, then also

$$\Phi(\lambda) = \left(2\pi^{-2}\right)\sum_0^{\infty}\left(\lambda^2 + (n - (1/2))^2\right)^{-1},$$

and $\phi(x) = -\pi^{-2}\log\tanh(x/4)$, $f(y) = \pi^{-2}\sum_{n=1}^{\infty}\exp\left(-(n-(1/2))^2 y/2\right)$, from which F is obtained by term by term integration.

b) Likewise if $\Phi(\lambda) = \pi\lambda/\sinh\pi\lambda$, then also

$$\Phi(\lambda) = 1 + 2\lambda^2\sum_{n=1}^{\infty}(-1)^n\left(\lambda^2 + n^2\right)^{-1},$$

and $\phi(x) = (2\cosh(x/2))^{-2}$, $F(x) = \sum_{n=-\infty}^{\infty}(-1)^n\exp(-n^2x/2)$.

c) Furthermore if $\Phi(\lambda) = (\cosh\pi\lambda)^{-1}$, then also

$$\Phi(\lambda) = \pi^{-1}\sum_0^{\infty}(-1)^n(2n+1)\left(\lambda^2 + (n + (1/2))^2\right)^{-1},$$

and $$\phi(x) = (2\pi \cosh(x/2))^{-1},$$

$$f(y) = \pi^{-1} \sum_{0}^{\infty} (-1)^n \,(n + (1/2)) \exp\left(-\,(n + (1/2))^2\, y/2\right).$$

d) Finally, if $\Phi(\lambda) = (\pi\lambda / \sinh \pi\lambda)^2$, then also

$$\Phi(\lambda) = 1 + 2\lambda^2 \sum_{n=1}^{\infty} \left(\lambda^2 - n^2\right)\left(\lambda^2 + n^2\right)^{-2},$$

and

$$\phi(x) = \left((x/2)\cosh(x/2) - 1\right)/2\sinh(x)^2,$$

$$F(x) = \sum_{n=-\infty}^{\infty} \left(1 - n^2 x\right) \exp\left(-n^2 x/2\right).$$

Chapter I. Introduction

§1. Examples of Stochastic Processes. Brownian Motion

A stochastic process is a phenomenon which evolves in time in a random way. Nature, everyday life, science offer us a huge variety of such phenomena or at least of phenomena which can be thought of as a function both of time and of a random factor. Such are for instance the price of certain commodities, the size of some populations, or the number of particles registered by a Geiger counter.

A basic example is the Brownian motion of pollen particles in a liquid. This phenomenon, which owes its name to its discovery by the English botanist R. Brown in 1827, is due to the incessant hitting of pollen by the much smaller molecules of the liquid. The hits occur a large number of times in any small interval of time, independently of each other and the effect of a particular hit is small compared to the total effect. The physical theory of this motion was set up by Einstein in 1905. It suggests that this motion is random, and has the following properties:

i) it has independent increments;
ii) the increments are gaussian random variables;
iii) the motion is continuous.

Property i) means that the displacements of a pollen particle over disjoint time intervals are independent random variables. Property ii) is not surprising in view of the central-limit theorem.

Much of this book will be devoted to the study of a mathematical model of this phenomenon.

The goal of the theory of stochastic processes is to construct and study mathematical models of physical systems which evolve in time according to a random mechanism, as in the above example. Thus, a stochastic process will be a family of random variables indexed by time.

(1.1) Definition. *Let T be a set, $(E, \mathscr{E})$ a measurable space. A* stochastic process *indexed by T, taking its values in $(E, \mathscr{E})$, is a family of measurable mappings X_t, $t \in T$, from a probability space $(\Omega, \mathscr{F}, P)$ into $(E, \mathscr{E})$. The space $(E, \mathscr{E})$ is called the* state space.

The set T may be thought of as "time". The most usual cases are $T = \mathbb{N}$ and $T = \mathbb{R}_+$, but they are by no means the only interesting ones. In this book, we deal mainly with the case $T = \mathbb{R}_+$ and E will usually be $\mathbb{R}^d$ or a Borel subset of $\mathbb{R}^d$ and $\mathscr{E}$ the Borel σ-field on E.

For every $\omega \in \Omega$, the mapping $t \to X_t(\omega)$ is a "curve" in E which is referred to as a *trajectory* or a *path* of X. We may think of a path as a point chosen randomly in the space $\mathscr{F}(T, E)$ of all functions from T into E, or, as we shall see later, in a reasonable subset of this space.

To set up our basic example of a stochastic process, namely the mathematical model of Brownian motion, we will use the following well-known existence result.

(1.2) Theorem. *Given a probability measure μ on $\mathbb{R}$, there exist a probability space $(\Omega, \mathscr{F}, P)$ and a sequence of independent random variables X_n, defined on Ω, such that $X_n(P) = \mu$ for every n.*

As a consequence, we get the

(1.3) Proposition. *Let H be a separable real Hilbert space. There exist a probability space $(\Omega, \mathscr{F}, P)$ and a family $X(h)$, $h \in H$, of random variables on this space, such that*

i) the map $h \to X(h)$ is linear;
ii) for each h, the r.v. $X(h)$ is gaussian centered and

$$E\left[X(h)^2\right] = \|h\|_H^2 .$$

Proof. Pick an orthonormal basis $\{e_n\}$ in H. By Theorem (1.2), there is a probability space $(\Omega, \mathscr{F}, P)$ on which one can define a sequence of independent reduced real Gaussian variables g_n. The series $\sum_0^\infty \langle h, e_n\rangle_H g_n$ converges in $L^2(\Omega, \mathscr{F}, P)$ to a r.v. which we call $X(h)$. The proof is then easily completed. □

We may observe that the above series converges also almost-surely, and $X(h)$ is actually an equivalence class of random variables rather than a random variable. Moreover, the space $\{X(h), h \in H\}$ is a Gaussian subspace of $L^2(\Omega, \mathscr{F}, P)$ which is isomorphic to H. In particular, $E\left[X(h)X(h')\right] = \langle h, h'\rangle_H$. Because in a Gaussian space independence is equivalent to orthogonality, this shows that $X(h)$ and $X(h')$ are independent if and only if h and h' are orthogonal in H.

(1.4) Definition. *Let $(A, \mathscr{A}, \mu)$ be a separable σ-finite measure space. If in Proposition (1.3), we choose $H = L^2(A, \mathscr{A}, \mu)$, the mapping X is called a* Gaussian measure *with intensity μ on $(A, \mathscr{A})$. When $F \in \mathscr{A}$ and $\mu(F) < \infty$, we shall write $X(F)$ instead of $X(1_F)$.*

The term "measure" is warranted by the fact that if $F \in \mathscr{A}$, $\mu(F) < \infty$ and $F = \sum_0^\infty F_n$, then $X(F) = \sum_0^\infty X(F_n)$ a.s. and in $L^2(A, \mathscr{A}, \mu)$; however, the exceptional set depends on F and on the sequence (F_n) and, consequently, there is usually no true measure $m(\omega, \cdot)$ depending on ω such that almost-surely $X(F)(\omega) = m(\omega, F)$ for every $F \in \mathscr{A}$.

Let us also observe that for any two sets F and G, such that $\mu(F) < \infty$, $\mu(G) < \infty$,

$$E\left[X(F)X(G)\right] = \mu(F \cap G);$$

if F and G are disjoint sets, $X(F)$ and $X(G)$ are uncorrelated, hence independent.

We now take a first step towards the construction of Brownian motion. The method we use may be extended to other examples of processes as is shown in Exercises (3.9) and (3.11). We take the space $(A, \mathscr{A}, \mu)$ to be $\mathbb{R}_+ = [0, \infty[$, endowed with the σ-field of Borel sets and the Lebesgue measure. For each $t \in \mathbb{R}_+$, we pick a random variable B_t within the equivalence class $X([0, t])$. We now study the properties of the process B thus defined.

1°) The process B has *independent increments* i.e. for any sequence $0 = t_0 < t_1 < \ldots < t_k$ the random variables $B_{t_i} - B_{t_{i-1}}$, $i = 1, 2, \ldots, k$ are independent. Indeed, $B_{t_i} - B_{t_{i-1}}$ is in the class $X\left(]t_{i-1}, t_i]\right)$ and these classes are independent because the corresponding intervals are pairwise disjoint.

2°) The process B is a *Gaussian process*, that is: for any sequence $0 = t_0 < t_1 < \ldots < t_n$, the vector r.v. $\left(B_{t_0}, \ldots, B_{t_n}\right)$ is a vector Gaussian r.v. This follows from the independence of the increments and the fact that the individual variables are Gaussian.

3°) For each t, we have obviously $E[B_t^2] = t$; in particular, $P[B_0 = 0] = 1$. This implies that for a Borel subset A of the real line and $t > 0$

$$P[B_t \in A] = \int_A g_t(x)dx,$$

where $g_t(x) = (2\pi t)^{-1/2}\exp(-x^2/2t)$. Likewise, the increment $B_t - B_s$ has variance $t - s$. Furthermore, the *covariance* $E[B_s B_t]$ is equal to $\inf(s, t)$; indeed using the independence of increments and the fact that all the B_t's are centered, we have for $s < t$,

$$\begin{aligned} E\left[B_s B_t\right] &= E\left[B_s\left(B_s + B_t - B_s\right)\right] \\ &= E\left[B_s^2\right] + E\left[B_s\left(B_t - B_s\right)\right] = s. \end{aligned}$$

If we refer to our idea of what a model of the physical Brownian motion ought to be, we see that we have got everything but the continuity of paths. There is no reason why an arbitrary choice of B_t within the class $X([0, t])$ will yield continuous maps $t \to B_t(\omega)$. On the other hand, since we can pick any function within the class, we may wonder whether we can do it so as to get a continuous function for almost all ω's. We now address ourselves to this question.

We first need to make a few general observations and give some definitions. From now on, unless otherwise stated, all the processes we consider are indexed by $\mathbb{R}_+$.

(1.5) Definition. *Let E be a topological space and $\mathscr{E}$ the σ-algebra of its Borel subsets. A process X with values in $(E, \mathscr{E})$ is said to be a.s.* continuous *if, for almost all ω's, the function $t \to X_t(\omega)$ is continuous.*

We would like our process B above to have this property; we could then, by discarding a negligible set, get a process with continuous paths. However, whether in discarding the negligible set or in checking that a process is a.s. continuous, we encounter the following problem: there is no reason why the set

$$\{\omega : t \to X_t(\omega) \quad \text{is continuous}\}$$

should be measurable. Since we want to construct a process with state space $\mathbb{R}$, it is tempting, as we hinted at before Theorem (1.2), to use as probability space the set $\mathscr{F}(\mathbb{R}_+, \mathbb{R}) = \mathbb{R}^{\mathbb{R}_+}$ of all possible paths, and as r.v. X_t the coordinate mapping over t, namely $X_t(\omega) = \omega(t)$. The smallest σ-algebra for which the X_t's are measurable is the product σ-algebra, say $\mathscr{F}$. Each set in $\mathscr{F}$ depends only on a countable set of coordinates and therefore the set of continuous ω's is not in $\mathscr{F}$.

This problem of continuity is only one of many similar problems. We will, for instance, want to consider, for a $\mathbb{R}$-valued process X, expressions such as

$$T(\omega) = \inf\{t : X_t(\omega) > 0\}, \quad \lim_{s \downarrow t} X_s, \quad \int_0^t 1_{[-1,1]}(X_s)ds, \quad \sup_{s \leq t} |X_s|,$$

and there is no reason why these expressions should be measurable or even meaningful if the only thing we know about X is that it satisfies Definition (1.1). This difficulty will be overcome by using the following notions.

(1.6) Definition. *Two processes X and X' defined respectively on the probability spaces $(\Omega, \mathscr{F}, P)$ and $(\Omega', \mathscr{F}', P')$, having the same state space $(E, \mathscr{E})$, are said to be* equivalent *if for any finite sequence $t_1, \ldots, t_n$ and sets $A_i \in \mathscr{E}$,*

$$P\left[X_{t_1} \in A_1, X_{t_2} \in A_2, \ldots, X_{t_n} \in A_n\right] = P'\left[X'_{t_1} \in A_1, X'_{t_2} \in A_2, \ldots, X'_{t_n} \in A_n\right].$$

We also say that each one is a version *of the other or that they are* versions *of the same process.*

The image of P by $\left(X_{t_1}, \ldots, X_{t_n}\right)$ is a probability measure on $(E^n, \mathscr{E}^n)$ which we denote by $P_{t_1,\ldots,t_n}$. The family obtained by taking all the possible finite sequences $(t_1, \ldots, t_n)$ is the family of *finite-dimensional distributions* (abbreviated f.d.d.) of X. The processes X and X' are equivalent if they have the same f.d.d.'s. We observe that the f.d.d.'s of X form a projective family, that is, if $(s_1, \ldots, s_k)$ is a subset of $(t_1, \ldots, t_n)$ and if π is the corresponding canonical projection from E^n onto E^k, then

$$P_{s_1,\ldots,s_k} = \pi\left(P_{t_1,\ldots,t_n}\right).$$

This condition appears in the Kolmogorov extension Theorem (3.2).

We shall denote by $\mathscr{M}_X$ the indexed family of f.d.d.'s of the process X. With this notation, X and Y are equivalent if and only if $\mathscr{M}_X = \mathscr{M}_Y$.

It is usually admitted that, most often, when faced with a physical phenomenon, statistical experiments or physical considerations can only give information about the f.d.d.'s of the process. Therefore, when constructing a mathematical model, we may if we can, choose, within the class of equivalent processes, a version for which expressions as those above Definition (1.6) are meaningful. We now work toward this goal in the case of Brownian motion.

(1.7) Definition. *Two processes X and X' defined on the same probability space are said to be* modifications *of each other if for each t*

$$X_t = X'_t \qquad \text{a.s.}$$

They are called indistinguishable *if for almost all ω*

$$X_t(\omega) = X'_t(\omega) \qquad \text{for every} \quad t.$$

Clearly, if X and X' are modifications of each other, they are versions of each other. We may also observe that if X and X' are modifications of each other and are a.s. continuous, they are indistinguishable.

In the next section, we will prove the following

(1.8) Theorem. (Kolmogorov's continuity criterion). *A real-valued process X for which there exist three constants $\alpha, \beta, C > 0$ such that*

$$E\left[|X_{t+h} - X_t|^\alpha\right] \le Ch^{1+\beta}$$

for every t and h, has a modification which is almost-surely continuous.

In the case of the process B above, the r.v. $B_{t+h} - B_t$ is Gaussian centered and has variance h, so that

$$E\left[(B_{t+h} - B_t)^4\right] = 3h^2.$$

The Kolmogorov criterion applies and we get

(1.9) Theorem. *There exists an almost-surely continuous process B with independent increments such that for each t, the random variable B_t is centered, Gaussian, and has variance t.*

Such a process is called a *standard linear Brownian motion* or simply Brownian motion (which we will often abbreviate to BM) and will be our main interest throughout the sequel.

The properties stated in Theorem (1.9) imply those we already know. For instance, for $s < t$, the increments $B_t - B_s$ are Gaussian centered with variance $t - s$; indeed, we can write

$$B_t = B_s + (B_t - B_s)$$

and using the independence of B_s and $B_t - B_s$, we get, taking characteristic functions,

$$\exp\left(-\frac{tu^2}{2}\right) = \exp\left(-\frac{su^2}{2}\right) E\left[\exp\left(iu(B_t - B_s)\right)\right]$$

whence $E\left[\exp\left(iu(B_t - B_s)\right)\right] = \exp\left(-\frac{(t-s)}{2}u^2\right)$ follows. It is then easy to see that B is a Gaussian process (see Definition (3.5)) with covariance $\inf(s,t)$. We leave as an exercise to the reader the task of showing that conversely we could have stated Theorem (1.9) as: there exists an a.s. continuous centered Gaussian process with covariance $\inf(s,t)$.

By discarding a negligible set, we may, and often will, consider that all the paths of B are continuous.

As soon as we have constructed the standard linear BM of Theorem (1.9), we can construct a host of other interesting processes. We begin here with a few.

1°) For any $x \in \mathbb{R}$, the process $X_t^x = x + B_t$ is called the *Brownian motion started* at x, or in abbreviated form a BM(x). Obviously, for any $A \in \mathscr{B}(\mathbb{R})$, $t > 0$,

$$P\left[X_t^x \in A\right] = \frac{1}{\sqrt{2\pi t}} \int_A e^{-(y-x)^2/2t} dy = \int_A g_t(y-x)dy.$$

2°) If $B_t^1, B_t^2, \dots, B_t^d$, are d independent copies of B_t, we define a process X with state space $\mathbb{R}^d$ by stipulating that the i-th component of X_t is B_t^i. This process is called the *d-dimensional Brownian motion*. It is a continuous Gaussian process vanishing at time zero. Again as above by adding x, we can make it start from $x \in \mathbb{R}^d$ and we will use the abbreviation BM$^d(x)$.

3°) The process $X_t = (t, B_t)$ with state space $\mathbb{R}_+ \times \mathbb{R}$ is a continuous process known as the *heat process*. We can replace B_t by the d-dimensional BM to get the heat process in $\mathbb{R}_+ \times \mathbb{R}^d$.

4°) Because of the continuity of paths

$$\sup\{B_s, 0 \le s \le t\} = \sup\{B_s, 0 \le s \le t, s \in \mathbb{Q}\}.$$

Therefore, we can define another process S by setting $S_t = \sup_{s \le t} B_s$. In similar fashion, we can consider the processes $|B_t|$, $B_t^* = \sup_{s \le t} |B_s|$ or $B_t^+ = \sup(0, B_t)$.

5°) Finally, because of the continuity of paths, for any Borel set A, the map

$$(\omega, s) \to 1_A\left(B_s(\omega)\right)$$

is measurable on the product $\left(\Omega \times \mathbb{R}_+, \mathscr{F} \otimes \mathscr{B}(\mathbb{R}_+)\right)$ and therefore

$$X_t = \int_0^t 1_A(B_s)ds$$

is meaningful and defines yet another process, the *occupation time* of A by the Brownian motion B.

We finally close this section by describing a few geometrical invariance properties of BM, which are of paramount importance in the sequel, especially property (iii).

(1.10) Proposition. *Let B be a standard linear BM. Then, the following properties hold:*

(i) *(time-homogeneity). For any $s > 0$, the process $B_{t+s} - B_s$, $t \geq 0$, is a Brownian motion independent of $\sigma(B_u, u \leq s)$;*
(ii) *(symmetry). The process $-B_t$, $t \geq 0$, is a Brownian motion;*
(iii) *(scaling). For every $c > 0$, the process cB_{t/c^2}, $t \geq 0$, is a Brownian motion;*
(iv) *(time-inversion). The process X defined by $X_0 = 0$, $X_t = tB_{1/t}$ for $t > 0$, is a Brownian motion.*

Proof. (i) It is easily seen that $X_t = B_{t+s} - B_s$ is a centered Gaussian process, with continuous paths, independent increments and variance t, hence a Brownian motion. Property (ii) is obvious and (iii) is obtained just as (i).

To prove (iv), one checks that X is a centered Gaussian process with covariance $\inf(s, t)$; thus, it will be a BM if its paths are continuous and, since they are clearly continuous on $]0, \infty[$, it is enough to prove that $\lim_{t\to 0} X_t = 0$. But $X_t, t \in]0, \infty[$ is equivalent to $B_t, t \in]0, \infty[$ and since $\lim_{t\to 0, t\in\mathbb{Q}} B_t = 0$, it follows that $\lim_{t\to 0, t\in\mathbb{Q}} X_t = 0$ a.s. Because X is continuous on $]0, \infty[$, we have $\lim_{t\to 0, t\in\mathbb{R}^+} X_t = 0$ a.s. □

Remarks. 1°) Once we have constructed a BM on a space $(\Omega, \mathscr{A}, P)$, this proposition gives us a host of other versions of the BM on this same space.

2°) A consequence of (iv) is the law of large numbers for the BM, namely $P\left[\lim_{t\to\infty} t^{-1} B_t = 0\right] = 1$.

3°) These properties are translations in terms of BM of invariance properties of the Lebesgue measure as is hinted at in Exercise (1.14) 2°).

(1.11) Exercise. Let B be the standard linear BM on $[0, 1]$, i.e. we consider only $t \in [0, 1]$. Prove that the process $\tilde{B}_t$, $0 \leq t \leq 1$, defined by

$$\tilde{B}_t = B_{1-t} - B_1$$

is another version of B, in other words, a standard BM on $[0, 1]$.

(1.12) Exercise. We denote by H the subspace of $C([0, 1])$ of functions h such that $h(0) = 0$, h is absolutely continuous and its derivative h' (which exists a.e.) satisfies

$$\int_0^1 h'(s)^2 ds < +\infty.$$

1°) Prove that H is a Hilbert space for the scalar product

$$(g, h) = \int_0^1 g'(s)h'(s)ds.$$

2°) For any bounded measure μ on $[0, 1]$, show that there exists an element h in H such that for every $f \in H$

$$\int_0^1 f(x)d\mu(x) = (f, h),$$

and that $h'(s) = \mu(]s, 1])$.

[Hint: The canonical injection of H into $C([0,1])$ is continuous; use Riesz's theorem.]

3°) Let B be a standard linear BM, μ and ν two bounded measures associated as in 2°) with h and g. Prove that

$$X^{\mu}(\omega)=\int_0^1 B_s(\omega)d\mu(s) \qquad \text{and} \qquad X^{\nu}(\omega)=\int_0^1 B_s(\omega)d\nu(s)$$

are random variables, that the pair (X^{μ}, X^{ν}) is Gaussian and that

$$E\left[X^{\mu}X^{\nu}\right]=\int_0^1\int_0^1 \inf(s,t)d\mu(s)d\nu(t)=(h,g).$$

4°) Prove also that with the notation of Exercise (1.14) below

$$X^{\mu}=\int_0^1 \mu(]u,1])dB_u.$$

This will be taken up in Sect. 2 Chap. VIII.

* **(1.13) Exercise.** 1°) Let B be the standard linear BM. Prove that $\overline{\lim}_{t\to\infty}\left(B_t/\sqrt{t}\right)$ is a.s. >0 (it is in fact equal to $+\infty$ as will be seen in Chap. II).

2°) Prove that B is recurrent, namely: for any real x, the set $\{t : B_t = x\}$ is unbounded.

3°) Prove that the Brownian paths are a.s. nowhere locally Hölder continuous of order α if $\alpha > \frac{1}{2}$ (see Sect. 2).

[Hint: Use the invariance properties of Proposition (1.10).]

\# **(1.14) Exercise.** 1°) With the notation of Proposition (1.3) and its sequel, we set for f bounded or more generally in $L^{2+\varepsilon}_{\text{loc}}(\mathbb{R}_+)$ for $\varepsilon > 0$,

$$Y_t=\int_0^t f(s)dB_s = X\left(f1_{[0,t]}\right).$$

Prove that the process Y has a continuous version. This is a particular case of the stochastic integral to be defined in Chapter IV and we will see that the result is true for more general integrands.

2°) For $c>0$ and $f\in L^2(\mathbb{R}_+)$, set $X^c(f)=cX(f^c)$ where $f^c(t)=f(c^2t)$. Prove that X^c is also a Gaussian measure with intensity the Lebesgue measure on $(\mathbb{R}_+, \mathscr{B}(\mathbb{R}_+))$. Derive therefrom another proof of Proposition (1.10)iii). Give similar proofs for properties i) and ii) of Proposition (1.10) as well as for Exercise (1.11).

(1.15) Exercise. Let B be a standard linear BM. Prove that

$$X(\omega)=\int_0^1 B_s^2(\omega)ds$$

is a random variable and compute its first two moments.

\# **(1.16) Exercise.** Let B be a standard linear BM.

1°) Prove that the f.d.d.'s of B are given, for $0 < t_1 < t_2 < \ldots < t_n$, by

$$P\left[B_{t_1} \in A_1, B_{t_2} \in A_2, \ldots, B_{t_n} \in A_n\right]$$
$$= \int_{A_1} g_{t_1}(x_1)dx_1 \int_{A_2} g_{t_2-t_1}(x_2 - x_1)\,dx_2 \ldots \int_{A_n} g_{t_n-t_{n-1}}(x_n - x_{n-1})\,dx_n.$$

2°) Prove that for $t_1 < t_2 < \ldots < t_n < t$,

$$P\left[B_t \in A \middle| B_{t_1}, \ldots, B_{t_n}\right] = \int_A g_{t-t_n}\left(y - B_{t_n}\right)dy.$$

More generally, for $s < t$,

$$P\left[B_t \in A \middle| \sigma\left(B_u, u \le s\right)\right] = \int_A g_{t-s}\left(y - B_s\right)dy.$$

(1.17) Exercise. Let B be the standard BM^2 and let λ be the Lebesgue measure on $\mathbb{R}^2$.

1°) Prove that the sets

$$\gamma_1(\omega) = \{B_t(\omega), 0 \le t \le 1\}, \quad \gamma_2(\omega) = \{B_t(\omega), 0 \le t \le 2\},$$
$$\gamma_3(\omega) = \{B_{1-t}(\omega) - B_1(\omega), 0 \le t \le 1\}, \quad \gamma_4(\omega) = \{B_{1+t}(\omega) - B_1(\omega), 0 \le t \le 1\},$$

are a.s. Borel subsets of $\mathbb{R}^2$ and that the maps $\omega \to \lambda\left(\gamma_i(\omega)\right)$ are random variables.

[Hint: To prove the second point, use the fact that for instance

$$\gamma_1(\omega) = \left\{z : \inf_{u \le 1} |z - B_u(\omega)| = 0\right\}.]$$

2°) Prove that $E[\lambda(\gamma_2)] = 2E[\lambda(\gamma_1)]$, and $E[\lambda(\gamma_1)] = E[\lambda(\gamma_3)] = E[\lambda(\gamma_4)]$.

3°) Deduce from the equality

$$E[\lambda(\gamma_3 \cup \gamma_4)] = E[\lambda(\gamma_3) + \lambda(\gamma_4) - \lambda(\gamma_3 \cap \gamma_4)]$$

that $E[\lambda(\gamma_1)] = 0$, hence that the Brownian curve has a.s. zero Lebesgue measure. One may have to use the fact (Proposition (3.7) Chap. III) that for a BM^1 β and every t, the r.v. $S_t = \sup_{s \le t} \beta_s$ is integrable.

* **(1.18) Exercise.** Let B be the standard linear BM. Using the scaling invariance property, prove that

$$t^{-1/2} \log\left(\int_0^t \exp(B_s)ds\right)$$

converges in law, as t tends to infinity, to $S_1 = \sup_{s \le 1} B_s$.

[Hint: Use the Laplace method, namely $\|f\|_p$ converges to $\|f\|_\infty$ as p tends to $+\infty$ where $\|\ \ \|_p$ is the L^p-norm with respect to the Lebesgue measure on [0, 1].]

The law of S_1 is found in Proposition (3.7) Chap. III.

(1.19) Exercise. 1°) If X is a BM^d, prove that for every $x \in \mathbb{R}^d$ with $\|x\| = 1$, the process $\langle x, X_t \rangle$ is a linear BM.

2°) Prove that the converse is false. One may use the following example: if $B = (B^1, B^2)$ is a BM^2, set

$$X_t^1 = B_{2t/3}^1 - B_{t/3}^2, \qquad X_t^2 = B_{2t/3}^2 + B_{t/3}^1.$$

(1.20) Exercise (Polar functions and points). A continuous function f from $\mathbb{R}_+$ into $\mathbb{R}^2$ is said to be *polar* for BM^2 if for any $x \in \mathbb{R}^2$, $P[\Gamma_x] = 0$ where $\Gamma_x = \{\exists t > 0 : B_t + x = f(t)\}$ (the measurability of the set Γ_x will follow from results in Sect. 4). The set of polar functions is denoted by π.

1°) Prove that f is polar if and only if

$$P[\exists t > 0 : B_t = f(t)] = 0.$$

[Hint: See 1°) in Exercise (3.14).]

2°) Prove that π is left invariant by the transformations:

a) $f \to -f$;
b) $f \to T \circ f$ where T is a rigid motion of $\mathbb{R}^2$;
c) $f \to \big(t \to tf(1/t)\big)$.

3°) Prove that if $f \notin \pi$, then $E[\lambda(\{B_t + f(t), t \geq 0\})] > 0$ where λ is the Lebesgue measure in $\mathbb{R}^2$. Use the result in Exercise (1.17) to show that one-point sets are *polar* i.e. for any $x \in \mathbb{R}^2$,

$$P[\exists t > 0 : B_t = x] = 0.$$

Extend the result to BM^d with $d \geq 3$. Another proof of this important result will be given in Chap. V.

4°) Prove that almost all paths of BM^2 are in π.

[Hint: Use the independent copies B^1 and B^2 of BM^2, consider $B^1 - B^2$ and apply the result in 3°).]

** **(1.21) Exercise.** Let $X = B^+$ or $|B|$ where B is the standard linear BM, p be a real number > 1 and q its conjugate number $\left(p^{-1} + q^{-1} = 1\right)$.

1°) Prove that the r.v. $J_p = \sup_{t \geq 0} \left(X_t - t^{p/2}\right)$ is a.s. strictly positive and finite and has the same law as $\sup_{t \geq 0} \left(X_t / \left(1 + t^{p/2}\right)\right)^q$.

2°) Using time-inversion, show that

$$\sup_{t \geq 1} \left(X_t / \left(1 + t^{p/2}\right)\right) \stackrel{(d)}{=} \sup_{u \leq 1} \left(\frac{1}{1 + u^{p/2}}\right) \left(\frac{X_u}{u^{1-p/2}}\right)$$

and conclude that $E[J_p] < \infty$.

[Hint: Use Theorem (2.1).]

3°) Prove that there exists a constant $C_p(X)$ such that for any positive r.v. L

$$E[X_L] \leq C_p(X) \|L^{1/2}\|_p.$$

[Hint: For $\mu > 0$, write $E[X_L] = E[X_L - \mu L^{p/2}] + \mu E[L^{p/2}]$ and using scaling properties, show that the first term on the right is less than $\mu^{-(q/p)} E[J_p]$.]

4°) Let L_μ be a random time such that

$$X_{L_\mu} - \mu L_\mu^{p/2} = \sup_{t \geq 0} \left(X_t - \mu t^{p/2} \right).$$

Prove that L_μ is a.s. unique and that the constant $C_p(X) = p^{1/p} \left(q E[J_p] \right)^{1/q}$ is the best possible.

5°) Prove that

$$E\left[X_{L_1} | J_p \right] = q J_p, \qquad E\left[L_1^{p/2} | J_p \right] = \frac{q}{p} J_p.$$

** **(1.22) Exercise.** A continuous process X is said to be *self-similar* (of order 1) if for every $\lambda > 0$

$$(X_{\lambda t}, t \geq 0) \stackrel{(d)}{=} (\lambda X_t, t \geq 0)$$

i.e. the two processes have the same law (see Sect. 3).

1°) Prove that if B is the standard BM then $X_t = B_{t^2}$ is self-similar.

2°) Let henceforth X be self-similar and positive and, for $p > 1$, set

$$S_p = \sup_{s \geq 0} \left(X_s - s^p \right) \qquad \text{and} \qquad X_t^* = \sup_{s \leq t} X_s.$$

Prove that there is a constant c_p depending only on p such that for any $a > 0$

$$P\left[c_p \left(X_1^* \right)^q \geq a \right] \leq P\left[S_p \geq a \right] \qquad \text{where} \quad q^{-1} + p^{-1} = 1.$$

[Hint: $S_p = \sup_{t \geq 0} \left(X_t^* - t^p \right)$; apply the self-similarity property to X_t^*.]

3°) Let $k > 1$; prove that for any $a > 0$

$$P\left[S_p \geq a \right] \leq 2P\left[(kX_1^*)^q \geq a \right] + \sum_{n=1}^{\infty} P\left[(kX_1^*)^q \geq k^{np} a \right].$$

[Hint: Observe that $P\left[S_p \geq a \right] \leq \sup \left\{ P\left[X_L^* - L^p \geq a \right];\ L \text{ positive random variable } \right\}$ and write $\Omega = \left\{ L \leq a^{1/p} \right\} \cup \bigcup_{n \geq 0} \left\{ k^n a^{1/p} < L \leq k^{n+1} a^{1/p} \right\}$.]

4°) Prove that if g is a positive convex function on $\mathbb{R}_+$ and $g(0) = 0$, then

$$E\left[g\left(c_p \left(X_1^* \right)^q \right) \right] \leq E\left[g\left(S_p \right) \right] \leq \left(2 + \frac{1}{k^p (k^p - 1)} \right) E\left[g\left((kX_1^*)^q \right) \right].$$

[Hint: If X and Y are two positive r.v.'s such that $P[X \geq a] \leq \sum_n \alpha_n P[Y \geq \beta_n a]$, then

$$E[g(X)] = \int_0^\infty P[X \geq a] dg(a) \leq \sum_n \frac{\alpha_n}{\beta_n} E[g(Y)].]$$

§2. Local Properties of Brownian Paths

Our first task is to figure out what the Brownian paths look like. It is helpful, when reasoning on Brownian paths, to draw a picture of the paths of the heat process or, in other words, of the graph of the mapping $t \to B_t(\omega)$ (Fig. 2 below). This graph should be very "wiggly". How wiggly is the content of this section which deals with the local behavior of Brownian paths.

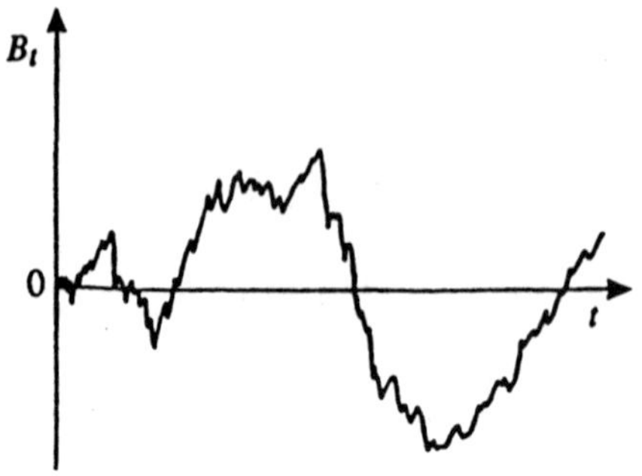

Fig. 2.

We begin with a more general version of the Kolmogorov criterion which was stated in the preceding section. We consider a Banach-valued process X indexed by a d-dimensional parameter. The norm we use on $\mathbb{R}^d$ is $|t| = \sup_i |t_i|$ and we also denote by $| \ |$ the norm of the state space of X. We recall that a Banach-valued function f on $\mathbb{R}^d$ is locally Hölder of order α if, for every $L > 0$,

$$\sup \{|f(t) - f(s)|/|t - s|^\alpha;\ |t|, |s| \leq L, t \neq s\} < \infty.$$

(2.1) Theorem. *Let $X_t, t \in [0, 1[^d$, be a Banach-valued process for which there exist three strictly positive constants γ, c, ε such that*

$$E\left[|X_t - X_s|^\gamma\right] \leq c|t - s|^{d+\varepsilon};$$

then, there is a modification $\tilde{X}$ of X such that

$$E\left[\left(\sup_{s\neq t}\left(|\tilde{X}_t - \tilde{X}_s|/|t - s|^\alpha\right)\right)^\gamma\right] < +\infty$$

for every $\alpha \in [0, \varepsilon/\gamma[$. In particular, the paths of $\tilde{X}$ are Hölder continuous of order α.

Proof. For $m \in \mathbb{N}$, let D_m be the set of d-uples

$$s = \left(2^{-m} i_1, \ldots, 2^{-m} i_d\right)$$

where each i_k is an integer in the interval $[0, 2^m[$ and set $D = \cup_m D_m$. Let further Δ_m be the set of pairs (s, t) in D_m such that $|s - t| = 2^{-m}$; there are fewer than $2^{(m+1)d}$ such pairs. Finally for s and t in D, we say that $s \leq t$ if each component of s is less than or equal to the corresponding component of t.

Let us now set $K_i = \sup_{(s,t)\in\Delta_i} |X_s - X_t|$. The hypothesis entails that for a constant J,

$$E\left[K_i^\gamma\right] \leq \sum_{(s,t)\in\Delta_i} E\left[|X_s - X_t|^\gamma\right] \leq 2^{(i+1)d} \cdot c2^{-i(d+\varepsilon)} = J2^{-i\varepsilon}.$$

For a point s (resp.: t) in D, there is an increasing sequence (s_n) (resp.: (t_n)) of points in D such that s_n (resp.: t_n) is in D_n, $s_n \leq s(t_n \leq t)$ and $s_n = s(t_n = t)$ from some n on.

Let now s and t be in D and $|s - t| \leq 2^{-m}$; either $s_m = t_m$ or $(s_m, t_m) \in \Delta_m$, and in any case

$$X_s - X_t = \sum_{i=m}^{\infty} \left(X_{s_{i+1}} - X_{s_i}\right) + X_{s_m} - X_{t_m} + \sum_{i=m}^{\infty} \left(X_{t_i} - X_{t_{i+1}}\right)$$

where the series are actually finite sums. It follows that

$$|X_s - X_t| \leq K_m + 2\sum_{m+1}^{\infty} K_i \leq 2\sum_{m}^{\infty} K_i.$$

As a result, setting $M_\alpha = \sup\{|X_t - X_s|/|t-s|^\alpha; s, t \in D, s \neq t\}$, we have

$$\begin{aligned} M_\alpha &\leq \sup_{m\in\mathbb{N}} \left\{2^{(m+1)\alpha} \sup_{|t-s|\leq 2^{-m}} |X_t - X_s|; s, t \in D, s \neq t\right\} \\ &\leq \sup_{m\in\mathbb{N}} \left(2.2^{(m+1)\alpha}\left(\sum_{m}^{\infty} K_i\right)\right) \\ &\leq 2^{\alpha+1}\sum_{i=0}^{\infty} 2^{i\alpha} K_i. \end{aligned}$$

For $\gamma \geq 1$ and $\alpha < \varepsilon/\gamma$, we get, with $J' = 2^{\alpha+1}J^{1/\gamma}$,

$$\|M_\alpha\|_\gamma \leq 2^{\alpha+1}\sum_{i=0}^{\infty} 2^{i\alpha}\|K_i\|_\gamma \leq J'\sum_{i=0}^{\infty} 2^{i(\alpha-(\varepsilon/\gamma))} < \infty.$$

For $\gamma < 1$, the same reasoning applies to $E\left[(M_\alpha)^\gamma\right]$ instead of $\|M_\alpha\|_\gamma$.

It follows in particular that for almost every ω, $X_.$ is uniformly continuous on D and it makes sense to set

$$\tilde{X}_t(\omega) = \lim_{\substack{s\to t\\ s\in D}} X_s(\omega).$$

By Fatou's lemma and the hypothesis, $\tilde{X}_t = X_t$ a.s. and $\tilde{X}$ is clearly the desired modification.

Remark. Instead of using the unit cube, we could have used any cube whatsoever.

In the case of Brownian motion, we have

$$E\left[(B_t - B_s)^2\right] = |t - s|$$

and because the increments are Gaussian, for every $p > 0$,

$$E\left[|B_t - B_s|^{2p}\right] = C_p|t - s|^p$$

for some constant C_p. From this result we deduce the

(2.2) Theorem. *The linear Brownian motion is locally Hölder continuous of order α for every $\alpha < 1/2$.*

Proof. As we have already observed, a process has at most one continuous modification (up to indistinguishability). Theorem (2.1) tells us that BM has a modification which is locally Hölder continuous of order α for $\alpha < (p-1)/2p = 1/2 - 1/2p$. Since p can be taken arbitrarily large, the result follows. □

From now on, we may, and will, suppose that all the paths of linear BM are locally Hölder continuous of order α for every $\alpha < 1/2$. We shall prove that the Brownian paths cannot have a Hölder continuity of order α, for $\alpha \geq 1/2$ (see also Exercise (1.13)). We first need a few definitions for which we retain the notation of Sect. 4 Chap. 0.

For a real-valued function X defined on $\mathbb{R}_+$, we set

$$T_t^\Delta = \sum_i \left(X_{t_{i+1}} - X_{t_i}\right)^2.$$

At variance with the S_t^Δ of Chap. 0, it is no longer true that $T_t^{\Delta'} \geq T_t^\Delta$ if Δ' is a refinement of Δ and we set

(2.3) Definition. *A real-valued process X is of* finite quadratic variation *if there exists a finite process $\langle X, X\rangle$ such that for every t and every sequence $\{\Delta_n\}$ of subdivisions of $[0, t]$ such that $|\Delta_n|$ goes to zero,*

$$P\text{-}\lim T_t^{\Delta_n} = \langle X, X\rangle_t.$$

The process $\langle X, X\rangle$ is called the quadratic variation of *X.*

Of course, we may consider intervals $[s, t]$ and, with obvious notation, we will then have

$$P\text{-}\lim T_{s,t}^{\Delta_n} = \langle X, X\rangle_t - \langle X, X\rangle_s;$$

thus, $\langle X, X\rangle$ is an increasing process.

Remark. We stress that a process may be of finite quadratic variation in the sense of Definition (2.3) and its paths be nonetheless a.s. of infinite quadratic variation in the classical sense, i.e. $\sup_\Delta T_t^\Delta = \infty$ for every $t > 0$; this is in particular the case for BM. In this book the words "quadratic variation" will be used only in the sense of Definition (2.3).

(2.4) Theorem. *Brownian motion is of finite quadratic variation and $\langle B, B\rangle_t = t$ a.s. More generally, if X is a Gaussian measure with intensity μ, and F is a set such that $\mu(F) < \infty$, for every sequence $\{F_k^n\}$, $n = 1, 2, \ldots$ of finite partitions of F such that* $\sup_k \mu(F_k^n) \xrightarrow[n\to\infty]{} 0$,

$$\lim_n \sum_k X(F_k^n)^2 = \mu(F)$$

in the L^2-sense.

Proof. Because of the independence of the $X(F_k^n)$'s, and the fact that $E\left[X(F_k^n)^2\right] = \mu(F_k^n)$,

$$\left\| \sum_k X(F_k^n)^2 - \mu(F) \right\|_2^2 = E\left[\left(\sum_k \left(X(F_k^n)^2 - \mu(F_k^n)\right)\right)^2\right],$$

and since for a centered Gaussian r.v. Y, $E[Y^4] = 3E[Y^2]^2$, this is equal to

$$2\sum_k \mu(F_k^n)^2 \le 2\mu(F)\sup_k \mu(F_k^n),$$

which completes the proof. □

Remarks. 1°) This result will be generalized to semimartingales in Chap. IV.

2°) By extraction of a subsequence, one can always choose a sequence (Δ_n) such that the above convergence holds almost-surely; in the case of BM, one can actually show that the a.s. convergence holds for any refining (i.e. $\Delta_n \subset \Delta_{n+1}$) sequence (see Proposition (2.12) in Chap. II and Exercise (2.8) in this section).

(2.5) Corollary. *The Brownian paths are a.s. of infinite variation on any interval.*

Proof. By the foregoing result, there is a set $\Omega_0 \subset \Omega$ such that $P(\Omega_0) = 1$ and for any pair of rationals $p < q$ there exists a sequence (Δ^n) of subdivisions of $[p, q]$ such that $|\Delta^n| \to 0$ and

$$\lim \sum_{t_i \in \Delta^n} \left(B_{t_{i+1}}(\omega) - B_{t_i}(\omega)\right)^2 = q - p$$

for every $\omega \in \Omega_0$.

Let $V(\omega) \le +\infty$ be the variation of $t \to B_t(\omega)$ on $[p, q]$. We have

$$\sum_i \left(B_{t_{i+1}}(\omega) - B_{t_i}(\omega)\right)^2 \le \left(\sup_i \left|B_{t_{i+1}}(\omega) - B_{t_i}(\omega)\right|\right) V(\omega).$$

By the continuity of the Brownian path, the right-hand side would converge to 0 as $n \to \infty$ if $V(\omega)$ were finite. Hence, $V(\omega) = +\infty$ a.s. □

In the following, we will say that a function is nowhere locally Hölder continuous of order α if there is no interval on which it is Hölder continuous of order α.

(2.6) Corollary. *The Brownian paths are a.s. nowhere locally Hölder continuous of order α for $\alpha > 1/2$.*

Proof. It is almost the same as that of Corollary (2.5). If $|B_t(\omega) - B_s(\omega)| \leq k|t-s|^\alpha$ for $p \leq s, t \leq q$ and $\alpha > 1/2$ then

$$\sum_i \left(B_{t_{i+1}}(\omega) - B_{t_i}(\omega)\right)^2 \leq k^2(q-p) \sup_i |t_{i+1} - t_i|^{2\alpha - 1}$$

and we conclude as in the previous proof. □

Theorem (2.2) and Corollary (2.6) leave open the case $\alpha = 1/2$. The next result shows in particular that the Brownian paths are not Hölder continuous of order 1/2 (see also Exercise (2.31) Chap. III).

(2.7) Theorem (Lévy's modulus of continuity). *If $h(t) = (2t \log(1/t))^{1/2}$,*

$$P\left[\overline{\lim_{\varepsilon \to 0}} \left(\sup_{\substack{0 \leq t_1 < t_2 \leq 1 \\ t_2 - t_1 \leq \varepsilon}} \left|B_{t_2} - B_{t_1}\right| \Big/ h(\varepsilon) \right) = 1 \right] = 1.$$

Proof. Pick a number δ in $]0,1[$ and consider the quantity

$$L_n = P\left[\max_{1 \leq k \leq 2^n} \left| B_{k2^{-n}} - B_{(k-1)2^{-n}} \right| \leq (1-\delta) h(2^{-n}) \right].$$

By the independence of the increments, L_n is less than

$$\left[1 - 2\int_{(1-\delta)\sqrt{2\log 2^n}}^{\infty} \frac{e^{-x^2/2}}{\sqrt{2\pi}} dx \right]^{2^n}.$$

By integrating by parts the left side of the inequality

$$\int_a^\infty b^{-2} \exp\left(-\frac{b^2}{2}\right) db < a^{-2} \int_a^\infty \exp\left(-\frac{b^2}{2}\right) db$$

the reader can check that

$$\int_a^\infty \exp\left(-\frac{b^2}{2}\right) db > \frac{a}{a^2+1} \exp\left(-\frac{a^2}{2}\right).$$

Using this and the inequality $1 - s < e^{-s}$, we see that there exists a constant $C > 0$ such that

$$L_n \leq \exp\left(-C 2^{n(1-(1-\delta)^2)} / \sqrt{n}\right).$$

It follows from the Borel-Cantelli lemma that the $\overline{\lim}$ in the statement is a.s. $\geq 1 - \delta$, and it is a.s. ≥ 1 since δ is arbitrary. We shall now prove the reverse inequality.

Again, we pick $\delta \in]0, 1[$ and $\varepsilon > 0$ such that $(1+\varepsilon)^2(1-\delta) > 1+\delta$. Let K be the set of pairs (i, j) of integers such that $0 \leq i < j < 2^n$ and $0 < j - i \leq 2^{n\delta}$ and for such a pair set $k = j - i$. Using the inequality

$$\int_a^\infty \exp\left(-b^2/2\right) db < \int_a^\infty \exp\left(-b^2/2\right)(b/a)db = a^{-1}\exp\left(-a^2/2\right),$$

and setting $L = P\left[\max_K \left(\left|B_{j2^{-n}} - B_{i2^{-n}}\right| / h(k2^{-n})\right) \geq 1+\varepsilon\right]$, we have

$$\begin{aligned} L &\leq \sum_K \frac{2}{\sqrt{2\pi}} \int_{(1+\varepsilon)\log(k^{-1}2^n)}^\infty \exp(-x^2/2)dx \\ &\leq D\sum_K \left(\log(k^{-1}2^n)\right)^{-1/2} \exp\left(-(1+\varepsilon)^2\log(k^{-1}2^n)\right) \end{aligned}$$

where D is a constant which may vary from line to line. Since k^{-1} is always larger than $2^{-n\delta}$, we further have

$$L \leq D2^{-n(1-\delta)(1+\varepsilon)^2} \sum_K \left(\log(k^{-1}2^n)\right)^{-1/2}.$$

Moreover, there are at most $2^{n(1+\delta)}$ points in K and for each of them

$$\log(k^{-1}2^n) > \log 2^{n(1-\delta)}$$

so that finally

$$L \leq Dn^{-1/2}2^{n\left((1+\delta)-(1-\delta)(1+\varepsilon)^2\right)}.$$

By the choice of ε and δ, this is the general term of a convergent series; by the Borel-Cantelli lemma, for almost every ω, there is an integer $n(\omega)$ such that for $n \geq n(\omega)$,

$$\left|B_{j2^{-n}} - B_{i2^{-n}}\right| < (1+\varepsilon)h(k2^{-n})$$

where $(i, j) \in K$ and $k = j - i$. Moreover, as the reader may show, the integer $n(\omega)$ may be chosen so that for $n \geq n(\omega)$,

$$\sum_{m>n} h(2^{-m}) \leq \eta h\left(2^{-(n+1)(1-\delta)}\right)$$

where $\eta > 0$ is any preassigned number. Let ω be a path for which these properties hold; pick $0 \leq t_1 < t_2 \leq 1$ such that $t_2 - t_1 < 2^{-n(\omega)(1-\delta)}$. Next let $n \geq n(\omega)$ be the integer such that

$$2^{-n} < 2^{-(n+1)(1-\delta)} \leq t_2 - t_1 < 2^{-n(1-\delta)}.$$

We may find integers i, j, p_r, q_s such that

$$t_1 = i2^{-n} - 2^{-p_1} - 2^{-p_2} - \dots, \qquad t_2 = j2^{-n} + 2^{-q_1} + 2^{-q_2} + \dots,$$

with $n < p_1 < p_2 < \dots, n < q_1 < q_2 < \dots$ and $0 < j - i \leq (t_2 - t_1) 2^n < 2^{n\delta}$. Since B is continuous, we get

$$\begin{aligned}
\left|B_{t_1}(\omega) - B_{t_2}(\omega)\right| &\leq \left|B_{i2^{-n}}(\omega) - B_{t_1}(\omega)\right| + \left|B_{j2^{-n}}(\omega) - B_{i2^{-n}}(\omega)\right| \\
&\quad + \left|B_{t_2}(\omega) - B_{j2^{-n}}(\omega)\right| \\
&\leq 2(1+\varepsilon) \sum_{m>n} h(2^{-m}) + (1+\varepsilon) h\left((j-i)2^{-n}\right) \\
&\leq 2(1+\varepsilon)\eta h\left(2^{-(n+1)(1-\delta)}\right) + (1+\varepsilon) h\left((j-i)2^{-n}\right).
\end{aligned}$$

Since h is increasing in a neighborhood of 0, for $t_2 - t_1$ sufficiently small, we get

$$\left|B_{t_1}(\omega) - B_{t_2}(\omega)\right| \leq \big(2(1+\varepsilon)\eta + (1+\varepsilon)\big) h(t_2 - t_1).$$

But ε, η and δ can be chosen arbitrarily close to 0, which ends the proof.

\# **(2.8) Exercise.** If B is the BM and Δ_n is the subdivision of $[0, t]$ given by the points $t_j = j2^{-n}t$, $j = 0, 1, \dots, 2^n$, prove the following sharpening of Theorem (2.4):

$$\lim_n T_t^{\Delta_n} = t \qquad \text{almost-surely.}$$

[Hint: Compute the exact variance of $T_t^{\Delta_n} - t$ and apply the Borel-Cantelli lemma.]

This result is proved in greater generality in Proposition (2.12) of Chap. II.

* **(2.9) Exercise (Non-differentiability of Brownian paths).** 1°) If g is a real-valued function on $\mathbb{R}_+$ which is differentiable at t, there exists an integer l such that if $i = [nt] + 1$ then

$$\left|g(j/n) - g\big((j-1)/n\big)\right| \leq 7l/n$$

for $i < j \leq i + 3$ and n sufficiently large.

2°) Let D_t be the set of Brownian paths which are differentiable at $t \geq 0$. Prove that $\bigcup_t D_t$ is contained in the event

$$\Gamma = \bigcup_{l \geq 1} \lim_{n \to \infty} \bigcup_{i=1}^{n+1} \bigcap_{j=i+1}^{i+3} \{|B(j/n) - B((j-1)/n)| \leq l/n\}$$

and finally that $P(\Gamma) = 0$.

\#* **(2.10) Exercise.** Let $\left(X_t^a\right)$ be a family of $\mathbb{R}^k$-valued continuous processes where t ranges through some interval of $\mathbb{R}$ and a is a parameter which lies in $\mathbb{R}^d$. Prove that if there exist three constants $\gamma, c, \varepsilon > 0$ such that

$$E\left[\sup_t \left|X_t^a - X_t^b\right|^\gamma\right] \leq c\, |a - b|^{d+\varepsilon},$$

then, there is a modification of $\left(X_t^a\right)$ which is jointly continuous in a and t and is moreover Hölder continuous in a of order α for $\alpha < \varepsilon/\gamma$, uniformly in t.

(2.11) Exercise (p-variation of BM). 1°) Let B be the standard linear BM and, for every n, let $t_i = i/n, i = 0, \dots, n$. Prove that for every $p > 0$

$$\left(n^{(p/2)-1}\right) \sum_{i=0}^{n-1} \left|B_{t_{i+1}} - B_{t_i}\right|^p$$

converges in probability to a constant v_p as n tends to $+\infty$.

[Hint: Use the scaling invariance properties of BM and the weak law of large numbers.]

2°) Prove moreover that

$$n^{(p-1)/2} \left(\sum_{i=0}^{n-1} \left|B_{t_{i+1}} - B_{t_i}\right|^p - n^{1-p/2} v_p \right)$$

converges in law to a Gaussian r.v.

(2.12) Exercise. For $p > 0$ given, find an example of a right-continuous but discontinuous process X such that

$$E\left[|X_s - X_t|^p\right] \le C|t-s|$$

for some constant $C > 0$, and all $s, t \ge 0$.

[Hint: Consider $X_t = 1_{(Y \le t)}$ for a suitable r.v. Y.]

§3. Canonical Processes and Gaussian Processes

We now come back more systematically to the study of some of the notions introduced in Sect. 1. There, we pointed out that the choice of a path of a process X amounts to the choice of an element of $\mathscr{F}(T, E)$ for appropriate T and E. It is well-known that the set $\mathscr{F}(T, E)$ is the same as the product space E^T. If $w \in \mathscr{F}(T, E)$ it corresponds in E^T to the product of the points $w(t)$ of E. From now on, we will not distinguish between $\mathscr{F}(T, E)$ and E^T. The functions $Y_t, t \in T$, taking their values in E, defined on E^T by $Y_t(w) = w(t)$ are called the *coordinate mappings*. They are random variables, hence form a process indexed by T, if E^T is endowed with the product σ-algebra $\mathscr{E}^T$. This σ-algebra is the smallest for which all the functions Y_t are measurable and it is the union of the σ-algebras generated by the countable sub-families of functions $Y_t, t \in T$. It is also the smallest σ-algebra containing the measurable rectangles $\prod_{t \in T} A_t$ where $A_t \in \mathscr{E}$ for each t and $A_t = E$ but for a finite sub-family $(t_1, \dots, t_n)$ of T.

Let now $X_t, t \in T$, be a process defined on $(\Omega, \mathscr{F}, P)$ with state space $(E, \mathscr{E})$. The mapping ϕ from Ω into E^T defined by

$$\phi(\omega)(t) = X_t(\omega)$$

is measurable with respect to $\mathscr{F}$ and $\mathscr{E}^T$ because $Y_t \circ \phi$ is measurable for each t. Let us call P_X the image of P by ϕ. Plainly, for any finite subset $(t_1, \dots, t_n)$ of T and sets $A_i \in \mathscr{E}$,

$$P\left[X_{t_1} \in A_1, \ldots, X_{t_n} \in A_n\right] = P_X\left[Y_{t_1} \in A_1, \ldots, Y_{t_n} \in A_n\right],$$

that is, the processes X and Y are versions of each other.

(3.1) Definition. *The process Y is called the* canonical version *of the process X. The probability measure P_X is called the* law *of X.*

In particular, if X and X' are two processes, possibly defined on different probability spaces, they are versions of each other if and only if they have the same law, i.e. $P_X = P_{X'}$ on E^T and this we will often write for short

$$X \overset{(d)}{=} X' \qquad \text{or} \qquad (X_t, t \geq 0) \overset{(d)}{=} \left(X'_t, t \geq 0\right).$$

For instance, property iii) in Proposition (1.10) may be stated

$$(B_t, t \geq 0) \overset{(d)}{=} (cB_{c^{-2}t}, t \geq 0).$$

Suppose now that we want to construct a process which models some physical phenomenon. Admittedly, nature (i.e. statistical experiments, physical considerations, ...) gives us a set of f.d.d.'s, and the goal is to construct a process with these given distributions. If we can do that, then by the foregoing we also have a canonical version; actually, we are going to construct this version in a fairly general setting.

We recall from Sect. 1 that the family $\mathscr{M}_X$ of the finite-dimensional distributions of a process X forms a projective family. The existence of the canonical version of a process of given finite-dimensional distributions is ensured by the *Kolmogorov extension theorem* which we recall here without proof.

(3.2) Theorem. *If E is a Polish space and $\mathscr{E}$ the σ-algebra of its Borel subsets, for any set T of indices and any projective family of probability measures on finite products, there exists a unique probability measure on $(E^T, \mathscr{E}^T)$ whose projections on finite products are the given family.*

For our present purposes, this theorem might have been stated: if $(E, \mathscr{E})$ is Polish, given a family $\mathscr{M}$ of finite-dimensional distributions, there exists a unique probability measure on $(E^T, \mathscr{E}^T)$ such that $\mathscr{M}_Y = \mathscr{M}$ for the coordinate process Y.

The above result permits to construct canonical versions of processes. However, for the same reason as those invoked in the first section, one usually does not work with the canonical version. It is only an intermediate step in the construction of some more tractable versions. These versions will usually have some continuity properties. Many processes have a version with paths which are right continuous and have left-hand limits at every point, in other words: $\lim_{s \downarrow t} X_s(\omega) = X_t(\omega)$ and $\lim_{s \uparrow t} X_s(\omega)$ exists; the latter limit is then denoted by $X_{t-}(\omega)$. We denote by $D(\mathbb{R}_+, E)$ or simply D the space of such functions which are called *cadlag*. The space D is a subset of E^T; to say that X has a version which is a.s. cadlag is equivalent to saying that for the canonical version of X the probability measure P_X on $\mathscr{E}^T$ gives the probability 1 to every measurable set which contains D.

In that case, we may use D itself as probability space. Indeed, still calling Y_t the mapping $\omega \to \omega(t)$ on D, let $\mathscr{E}_D^T$ be the σ-algebra $\sigma(Y_t, t \geq 0)$; plainly, $\mathscr{E}_D^T = \mathscr{E}^T \cap D$. The reader will easily check that although D is not measurable in $\mathscr{E}^T$, one defines unambiguously a probability measure Q on $\mathscr{E}_D^T$ by setting

$$Q(\Gamma) = P_X(\bar{\Gamma})$$

where $\bar{\Gamma}$ is any set in $\mathscr{E}^T$ such that $\Gamma = \bar{\Gamma} \cap D$. Obviously, the process Y defined on $(D, \mathscr{E}_D^T, Q)$ is another version of X. This version again is defined on a space of functions and is made up of coordinate mappings and will also be referred to as *canonical*; we will also write P_X instead of Q; this causes no confusion so long as one knows the space we work with.

Finally, if X is a process defined on $(\Omega, \mathscr{F}, P)$ with a.s. continuous paths, we can proceed as above with $C(\mathbb{R}_+, E)$ instead of D and take the image of P by the map ϕ defined on a set of probability 1 by

$$Y_t\big(\phi(\omega)\big) = \phi(\omega)(t) = X_t(\omega).$$

We can do that in particular in the case of Brownian motion and we state

(3.3) Proposition. *There is a unique probability measure W on $C(\mathbb{R}_+, \mathbb{R})$ for which the coordinate process is a Brownian motion. It is called the Wiener measure and the space $C(\mathbb{R}_+, \mathbb{R})$ is called the Wiener space.*

Proof. One gets W as the image by ϕ of the probability measure of the version of BM already constructed. □

It is actually possible to construct directly the Wiener measure, hence a continuous version of BM, without the knowledge of the results of Sect. 1. Let $\mathscr{A}$ be the union of the σ-fields on $C = C(\mathbb{R}_+, \mathbb{R})$ generated by all the finite sets of coordinate mappings; $\mathscr{A}$ is an algebra which generates the σ-field on C. On $\mathscr{A}$, we can define W by the formula in Exercise (1.16) 1°) and then it remains to prove that the set function thus defined extends to a probability measure on the whole σ-field.

For the Brownian motion in $\mathbb{R}^d$, we define similarly the *Wiener space* $C(\mathbb{R}_+, \mathbb{R}^d)$ and the Wiener measure W as the only probability measure on the Wiener space for which the coordinate process is a standard BM^d. The Wiener space will often be denoted by $\mathbf{W}$ or $\mathbf{W}^d$ if we want to stress the dimension. It is interesting to restate the results of Section 2 in terms of Wiener measure, for instance: the Wiener measure is carried by the set of functions which are Hölder continuous of order α for every $\alpha < 1/2$ and nowhere Hölder continuous of order α for $\alpha \geq 1/2$.

We now introduce an important notion. On the three canonical spaces we have defined, namely $E^{\mathbb{R}_+}$, $D(\mathbb{R}_+, E)$, $C(\mathbb{R}_+, E)$, we can define a family of transformations $\theta_t, t \in \mathbb{R}_+$, by

$$Y_s(\theta_t(\omega)) = Y_{t+s}(\omega),$$

where as usual Y is the coordinate process. Plainly, $\theta_t \circ \theta_s = \theta_{t+s}$ and θ_t is measurable with respect to $\sigma(X_s, s \geq t)$ hence, to $\sigma(X_s, s \geq 0)$. The effect of θ_t on a path ω is to cut off the part of the path before t and to shift the remaining part in time. The operators $\theta_t, t \geq 0$, are called the *shift operators*.

(3.4) Definition. *A process X is* stationary *if for every $t_1, \ldots, t_n$ and t and every $A_i \in \mathscr{E}$*

$$P\left[X_{t+t_1} \in A_1, \ldots, X_{t+t_n} \in A_n\right] = P\left[X_{t_1} \in A_1, \ldots, X_{t_n} \in A_n\right].$$

Another way of stating this is to say, using the canonical version, that for every t, $\theta_t(P_X) = P_X$; in other words, the law of X is invariant under the shift operators.

We now illustrate the above notions by studying the case of Gaussian processes.

(3.5) Definition. *A real-valued process $X_t, t \in T$, is a* Gaussian process *if for any finite sub-family $(t_1, \ldots, t_n)$ of T, the vector r.v. $(X_{t_1}, \ldots, X_{t_n})$ is Gaussian. The process X is* centered *if $E[X_t] = 0 \ \forall t \in T$.*

In other words, X is a Gaussian process if the smallest closed subspace of $L^2(\Omega, \mathscr{F}, P)$ containing the r.v.'s $X_t, t \in T$ is a Gaussian subspace. As was already observed, the standard linear BM is a Gaussian process.

(3.6) Definition. *If $X_t, t \in T$, is a Gaussian process, its* covariance Γ *is the function defined on $T \times T$ by*

$$\Gamma(s,t) = \operatorname{cov}(X_s, X_t) = E\left[(X_s - E[X_s])(X_t - E[X_t])\right].$$

Let us recall that a *semi-definite positive function* Γ on T is a function from $T \times T$ into $\mathbb{R}$ such that for any d-uple $(t_1, \ldots, t_d)$ of points in T and any integer d, the $d \times d$-matrix $\left(\Gamma(t_i, t_j)\right)$ is semi-definite positive (Sect. 6, Chap. 0).

(3.7) Proposition. *The covariance of a Gaussian process is a semi-definite positive function. Conversely, any symmetric semi-definite positive function is the covariance of a centered Gaussian process.*

Proof. Let (t_i) be a finite subset of T and a_i some complex numbers; then

$$\sum_{i,j} \Gamma(t_i, t_j) a_i \bar{a}_j = E\left[\left|\sum_i a_i \left(X_{t_i} - E\left[X_{t_i}\right]\right)\right|^2\right] \geq 0$$

which proves the first statement.

Conversely, given a symmetric semi-definite positive function Γ, for every finite subset $t_1, \ldots, t_n$ of T, let $P_{t_1,\ldots,t_n}$ be the centered Gaussian probability measure on $\mathbb{R}^n$ with covariance matrix $\left(\Gamma(t_i, t_j)\right)$ (see Sect. 6 Chap. 0). Plainly, this defines a projective family and under the probability measure given by the Kolmogorov extension theorem, the coordinate process is a Gaussian process with covariance Γ. We stress the fact that the preceding discussion holds for a general set T and not merely for subsets of $\mathbb{R}$. □

Remark. The Gaussian measures of Definition (1.4) may be constructed by applying Proposition (3.7) with $T = \mathscr{A}$ and $\Gamma(A, B) = \mu(A \cap B)$.

We have already seen that the covariance of the standard linear BM is given by $\inf(s, t)$. Another large set of examples is obtained in the following way. If μ is a symmetric probability measure on $\mathbb{R}$, its Fourier transform

$$\phi(t) = \int_{-\infty}^{+\infty} e^{itx} \mu(dx),$$

is real and we get a covariance on $T = \mathbb{R}$ by setting $\Gamma(t, t') = \phi(t - t')$ as the reader will easily show. By Proposition (3.7), the process associated with such a covariance is a stationary process. In particular we know that for $\beta > 0$, the function $\exp(-\beta|t|)$ is the characteristic function of the Cauchy law with parameter β. Consequently, the function $\Gamma(t, t') = c\exp(-\beta|t - t'|)$ with $c > 0$ is the covariance of a stationary Gaussian process called a *stationary Ornstein-Uhlenbeck process* (abbreviated OU in the sequel) with parameter β and size c. If we call this process X, it is easily seen that

$$E\left[(X_t - X_{t'})^2\right] \leq 2c\beta|t - t'|,$$

hence, by Kolmogorov's continuity criterion, X has a continuous modification. Henceforth, we will consider only continuous modifications of this process and the time set will often be restricted to $\mathbb{R}_+$.

Another important example is the *Brownian Bridge*. This is the centered Gaussian process defined on $T = [0, 1]$ and with covariance $\Gamma(s, t) = s(1 - t)$ on $(s \leq t)$. The easiest way to prove that Γ is a covariance is to observe that for the process $X_t = B_t - tB_1$ where B is a BM, $E[X_s X_t] = s(1 - t)$ for $s \leq t$. This gives us also immediately a continuous version of the Brownian Bridge. We observe that $X_1 = 0$ a.s. hence all the paths go a.s. from 0 at time 0 to 0 at time 1; this is the reason for the name given to this process which is also sometimes called the tied-down or pinned Brownian motion. More generally, one may consider the Brownian Bridge X^y between 0 and y which may be realized by setting

$$X_t^y = B_t - t(B_1 - y) = X_t^0 + ty, \qquad 0 \leq t \leq 1,$$

where $X_t^0 = X_t$ is the Brownian Bridge going from 0 to 0. Exercise (3.16) describes how X^y may be viewed as BM conditioned to be in y at time 1. In the sequel, the words "Brownian Bridge" without further qualification will mean the Bridge going from 0 to 0. Naturally, the notion of Bridge may be extended to higher dimensions and to intervals other than [0, 1].

Finally, there is a centered Gaussian process with covariance $\Gamma(s, t) = 1_{[s=t]}$, $s, t \in \mathbb{R}_+$. For such a process, any two r.v.'s X_s, X_t with $s \neq t$ are independent, which can be seen as "great disorder". It is interesting to note that this process does not have a good version; if it had a measurable version, i.e. such that the map $(t, \omega) \to X_t(\omega)$ were measurable (see Definition (1.14) Chap. IV), then for each t, $Y_t = \int_0^t X_s ds$ would be a Gaussian r.v. and, using a Fubini argument, it is easily shown that we would have $E[Y_t^2] = 0$, hence $Y_t = 0$ a.s. Consequently, we would get $X_t(\omega) = 0$ $dt P(d\omega)$-a.s., which is in contradiction with the equality

$$E\left[\int_0^t ds\ X_s^2\right] = t.$$

\# **(3.8) Exercise.** 1°) Let B be a linear BM and λ a positive number; prove that the process

$$X_t = e^{-\lambda t} B_{\exp(2\lambda t)}, \qquad t \in \mathbb{R},$$

is a stationary Ornstein-Uhlenbeck process; compute its parameter and its size. Conclude that the stationary OU process has a continuous version whose paths are nowhere differentiable.

2°) Let X be a continuous OU process with parameter $1/2$ and size 1 and set

$$\beta_t = X_t + (1/2)\int_0^t X_u du \qquad \text{for} \qquad t \geq 0,$$

$$\beta_t = X_t - (1/2)\int_0^t X_u du \qquad \text{for} \qquad t < 0.$$

Prove that β is a Gaussian process with continuous paths. In what sense is β a BM?

[Hint: See Exercise (3.14).]

This is related to the solution of the Langevin equation in Chap. IX.

* **(3.9) Exercise (Fractional Brownian motion).** Let d be a positive integer and p a real number such that $(d/2) - 1 < p < d/2$. We set $\alpha = d - 2p$.

1°) For $x, y \in \mathbb{R}^d$, prove that the function

$$f_y(x) = |x - y|^{-p} - |x|^{-p}$$

is in $L^2(\mathbb{R}^d, m)$ where m is the Lebesgue measure on $\mathbb{R}^d$.

2°) Let X be the Gaussian measure with intensity m and set $Z_y = X(f_y)$. Prove that there is a constant c depending only on d and p such that

$$E\left[Z_y Z_{y'}\right] = c\left\{|y|^\alpha + |y'|^\alpha - |y - y'|^\alpha\right\}.$$

The process Z is called the *fractional Brownian motion* of index α. For $\alpha = 1$, Z is Lévy's Brownian motion with d parameters.

3°) Let X_1 and X_2 be two independent Gaussian measures with intensity m, and set $X = X_1 + iX_2$. For $y \in \mathbb{R}^d$, set $\hat{f}_y(x) = \left(1 - \exp(iy \cdot x)\right)/|x|^{d-p}$, where $\cdot$ indicates the inner product in $\mathbb{R}^d$. Show that

$$\int \hat{f}_y(x) \cdot X(dx) \qquad (y \in \mathbb{R}^d)$$

is a constant multiple of the fractional Brownian motion of index α.

[Hint: The Fourier-Plancherel transform of the function f_y is $\gamma \hat{f}_y$, where γ is a constant independent of y.]

\# **(3.10) Exercise (Brownian Bridge).** Let X be the Brownian Bridge (BB).

1°) Prove that the process X_{1-t}, $0 \leq t \leq 1$ is also a BB and that

$$B_t = (t+1)X_{(t/(t+1))}, \qquad t \geq 0,$$

is a BM. Conversely, if B is a BM, the processes $(1-t)B\big(t/(1-t)\big)$ and $tB(t^{-1}-1)$, $0 \leq t \leq 1$, are BB's.

2°) If we write $\bar{t}$ for t modulo 1, the process

$$Y_t = X_{\overline{t+s}} - X_s, \qquad 0 \leq t \leq 1,$$

where s is a fixed number of $[0, 1]$ is a BB.

3°) (Continuation of Exercise (1.21)). Prove that

$$\sup_{t\geq 0} \frac{1}{t}\left(|B_t| - 1\right) \overset{(d)}{=} \sup_{t \geq 0}\left(|B_t| - t\right) \overset{(d)}{=} \sup_{0\leq t\leq 1} X_t^2.$$

(3.11) Exercise (Brownian sheet). 1°) Prove that the function Γ defined on $\mathbb{R}_+^2 \times \mathbb{R}_+^2$ by

$$\Gamma\big((u,s),(v,t)\big) = \inf(u,v) \times \inf(s,t)$$

is a covariance and that the corresponding Gaussian process has a continuous version. This process is called the *Brownian sheet.*

2°) Prove that the Brownian sheet may be obtained from $L^2(\mathbb{R}_+^2)$ as BM was obtained from $L^2(\mathbb{R}_+)$ in Sect. 1.

Let $\left\{\mathbb{B}(\Gamma);\, \Gamma \in \mathscr{B}(\mathbb{R}_+^2),\, \iint_\Gamma ds\, du < \infty\right\}$ be the Gaussian measure associated with the Brownian sheet $\mathbb{B}$.

Prove that $\{\mathbb{B}(R_t), 0 \leq t \leq 1\}$ is a Brownian bridge, where

$$R_t = \{(s,u);\, 0 \leq s < t < u \leq 1\}, \qquad 0 \leq t \leq 1.$$

3°) If $\mathbb{B}_{(s,t)}$ is a Brownian sheet, prove that the following processes are also Brownian sheets:

a) $\mathbb{B}_{(a^2 s, b^2 t)}/ab$ where a and b are two positive numbers;
b) $s\mathbb{B}_{(s^{-1},t)}$ and $st\mathbb{B}_{(s^{-1},t^{-1})}$;
c) $\mathbb{B}_{(s_0+s,t_0+t)} - \mathbb{B}_{(s_0+s,t_0)} - \mathbb{B}_{(s_0,t_0+t)} + \mathbb{B}_{(s_0,t_0)}$ where s_0, t_0 are fixed positive numbers.

4°) If $\mathbb{B}_{(s,t)}$ is a Brownian sheet, then for a fixed s, $t \to \mathbb{B}_{(s,t)}$ is a multiple of linear BM. The process $t \to \mathbb{B}_{(e^t, e^{-t})}$ is a Ornstein-Uhlenbeck process.

(3.12) Exercise (Reproducing Kernel Hilbert Space). Let Γ be a covariance on $T \times T$, and X_t, $t \in T$, a centered Gaussian process with covariance Γ.

1°) Prove that there exists a unique Hilbert space H of functions on T such that

i) H is the closure of the subspace spanned by $\{\Gamma(t,\cdot), t \in \Gamma\}$;
ii) for every $f \in H$, $\langle f, \Gamma(t,\cdot)\rangle = f(t)$.

2°) Let $\mathscr{H}$ be the Gaussian space generated by the variables $X_t, t \in T$. Prove that H and $\mathscr{H}$ are isomorphic, the isomorphism being given by

$$Z \in \mathscr{H} \to (E[ZX_t], t \in T).$$

3°) In the case of the covariance of BM on [0, 1], prove that H is the space H of Exercise (1.12) in Chap. 1.

* **(3.13) Exercise.** 1°) Let ϕ be a locally bounded measurable function on]0, 1] such that

$$\overline{\lim_{t \downarrow 0}}\, t^{(3/2)-\delta} |\phi(t)| < +\infty$$

for some $\delta > 0$. If B is the BM, prove that the integral

$$X_t^\phi = \int_0^t \phi(s) B_s ds$$

defines a continuous Gaussian process on [0, 1]. (A sharper form is given in Exercise (2.31) of Chap. III). Characterize the functions ϕ for which $B - X^\phi$ is again a BM.

2°) Treat the same questions for the Brownian Bridge.

\# **(3.14) Exercise.** For any $x \in \mathbb{R}^d$, define the translation τ_x on $\mathbf{W} = C(\mathbb{R}_+, \mathbb{R}^d)$ by $\tau_x(w)(t) = x + w(t)$ and call W^x the image of W by τ_x (thus, in particular $W^0 = W$).

1°) Prove that under W^x the coordinate process X is a version of the BM^d started at x. Observe that W^x is carried by the set $\{w : w(0) = x\}$ and conclude that W^x and W^y are mutually singular if $x \neq y$. On the other hand, prove that for any $\varepsilon > 0$, W^x and W^y are equivalent on the σ-algebra $\sigma(X_s, s \geq \varepsilon)$.

2°) For any $\Gamma \in \sigma(X_s, s \geq 0)$, prove that the map $x \to W^x(\Gamma)$ is a Borel function. If μ is a probability measure on the Borel sets of $\mathbb{R}^d$, we define a probability measure W^μ on $\mathbf{W}$ by

$$W^\mu(\Gamma) = \int \mu(dx) W^x(\Gamma).$$

Under which condition is X a Gaussian process for W^μ? Compute its covariance in that case.

* **(3.15) Exercise.** Let us first recall a result from Fourier series theory. Let f be continuous on [0, 1] with $f(0) = f(1) = 0$ and set, for $k \geq 1$,

$$a_k = \sqrt{2} \int_0^1 f(t) \cos 2\pi kt\, dt, \qquad b_k = \sqrt{2} \int_0^1 f(t) \sin 2\pi kt\, dt,$$

then if $\sum_1^\infty a_k$ converges,

$$f(t) = \sqrt{2} \sum_{k=1}^{\infty} \left(a_k(\cos 2\pi kt - 1) + b_k \sin 2\pi kt\right),$$

where the series on the right converges in $L^2([0,1])$. By applying this to the paths of the Brownian Bridge $B_t - tB_1$, prove that there exist two sequences $(\xi_k)_{k=0}^{k=\infty}$ and $(\eta_k)_{k=1}^{k=\infty}$ of independent reduced Gaussian random variables such that almost surely

$$B_t = t\xi_0 + \sqrt{2}\sum_{k=1}^{\infty}\left(\frac{\xi_k}{2\pi k}(\cos 2\pi kt - 1) + \frac{\eta_k}{2\pi k}\sin 2\pi kt\right).$$

\# **(3.16) Exercise. (The Brownian Bridge as conditioned Brownian motion).** We consider $\Omega = C([0,1],\mathbb{R})$ endowed with its Borel σ-field and the Wiener measure W. As Ω is a Polish space and $\mathscr{B}_1 = \sigma(X_1)$ is countably generated, there exists a regular conditional distribution $P(y,\cdot)$ for the conditional expectation with respect to $\mathscr{B}_1$.

1°) Prove that we may take $P(y,\cdot)$ to be the law P^y of the Brownian Bridge between 0 and y. In other words, for any Borel subset Γ of Ω

$$W(\Gamma) = \int_{\mathbb{R}} P^y[\Gamma]\frac{1}{\sqrt{2\pi}}\exp(-y^2/2)dy.$$

2°) The function $y \to P^y$ is continuous in the weak topology (Chap. 0 and Chap. XIII) that is: for any bounded continuous function f on Ω the map $y \to \int_\Omega f(\omega)dP^y(\omega)$ is continuous.

(3.17) Exercise. 1°) Let X be a Gaussian process with covariance Γ on $[0,1]$ and suppose that for each t, X is differentiable at t, i.e., there exists a r.v. X'_t such that

$$\lim_{h\to 0} h^{-1}(X_{t+h} - X_t) = X'_t \qquad \text{a.s.}$$

Prove that $\partial^2\Gamma(s,t)/\partial s\,\partial t$ exists and is equal to the covariance of the process X'.

2°) Let B be the standard linear BM on $[0,1]$ and $T(\omega)$ the derivative in the sense of Schwartz distributions of the continuous map $t \to B_t(\omega)$. If f, g are in $C_K^\infty(]0,1[)$ prove that $\langle T(\omega), f\rangle$ and $\langle T(\omega), g\rangle$ are centered Gaussian random variables with covariance equal to $\langle U, fg\rangle$ where U is the second mixed derivative of $\inf(s,t)$ in the sense of Schwartz distributions, namely a suitable multiple of the Lebesgue measure on the diagonal of the unit square.

§4. Filtrations and Stopping Times

In this section, we introduce some basic notation which will be used constantly in the sequel.

(4.1) Definition. *A* filtration *on the measurable space $(\Omega, \mathscr{F})$ is an increasing family $(\mathscr{F}_t)_{t\geq 0}$, of sub-$\sigma$-algebras of $\mathscr{F}$. In other words, for each t we have a sub-σ-algebra $\mathscr{F}_t$ and $\mathscr{F}_s \subset \mathscr{F}_t$ if $s < t$. A measurable space $(\Omega, \mathscr{F})$ endowed with a filtration $(\mathscr{F}_t)_{t\geq 0}$, is said to be a* filtered space.

(4.2) Definition. *A process X on $(\Omega, \mathscr{F})$ is* adapted *to the filtration $(\mathscr{F}_t)$ if X_t is $\mathscr{F}_t$-measurable for each t.*

Any process X is adapted to its *natural* filtration $\mathscr{F}_t^0 = \sigma(X_s, s \leq t)$ and $(\mathscr{F}_t^0)$ is the minimal filtration to which X is adapted. To say that X is adapted to $(\mathscr{F}_t)$ is to say that $\mathscr{F}_t^0 \subset \mathscr{F}_t$ for each t.

It is the introduction of a filtration which allows for the parameter t to be really thought of as "time". Heuristically speaking, the σ-algebra $\mathscr{F}_t$ is the collection of events which may occur before or at time t or, in other words, the set of possible "pasts" up to time t. In the case of stationary processes, where the law is invariant, it is the measurability with respect to $\mathscr{F}_t$ which places the event in time.

Filtrations are a fundamental feature of the theory of stochastic processes and the definition of the basic objects of our study such as martingales (see the following chapter) or Markov processes, will involve filtrations. We proceed to a study of this notion and introduce some notation and a few definitions.

With a filtration $(\mathscr{F}_t)$, one can associate two other filtrations by setting

$$\mathscr{F}_{t-} = \bigvee_{s<t} \mathscr{F}_s, \qquad \mathscr{F}_{t+} = \bigcap_{s>t} \mathscr{F}_s.$$

We have $\bigvee_{t\geq 0} \mathscr{F}_{t-} = \bigvee_{t\geq 0} \mathscr{F}_t = \bigvee_{t\geq 0} \mathscr{F}_{t+}$ and this σ-algebra will be denoted $\mathscr{F}_\infty$. The σ-algebra $\mathscr{F}_{0-}$ is not defined and, by convention, we put $\mathscr{F}_{0-} = \mathscr{F}_0$.

We always have $\mathscr{F}_{t-} \subseteq \mathscr{F}_t \subseteq \mathscr{F}_{t+}$ and these inclusions may be strict. If for instance $\Omega = D(\mathbb{R}_+, \mathbb{R})$, X is the coordinate process and the filtration is $(\mathscr{F}_t^0)$, the event $\{X_t = a\}$ is in $\mathscr{F}_t^0$ and is not in $\mathscr{F}_{t-}^0$; a little later, we shall see an example of an event in $\mathscr{F}_{t+}^0$ which is not in $\mathscr{F}_t^0$.

We shall encounter other examples of pairs $(\mathscr{F}_t)$, $(\mathscr{G}_t)$ of filtrations such that $\mathscr{F}_t \subset \mathscr{G}_t$ for all t. We can think about this set-up as the knowledge that two different observers can have gained at time t, the $\mathscr{G}_t$-observer being more skillful than the $\mathscr{F}_t$-observer. For instance, the $\mathscr{F}_{t+}^0$-observer above can foresee the immediate future.

This also leads us to remark that, in a given situation, there may be plenty of filtrations to work with and that one may choose the most convenient. That is to say that the introduction of a filtration is no restriction on the problems we may treat; at the extreme, one may choose the constant filtration ($\mathscr{F}_t = \mathscr{F}$ for every t) which amounts to having no filtration.

(4.3) Definition. *If $\mathscr{F}_t = \mathscr{F}_{t+}$ for every t, the filtration is said to be* right-continuous.

For any filtration $(\mathscr{F}_t)$ the filtration $(\mathscr{F}_{t+})$ is right-continuous.

We come now to an important definition.

(4.4) Definition. *A* stopping time *relative to the filtration $(\mathscr{F}_t)$ is a map on Ω with values in* $[0, \infty]$, *such that for every t,*

$$\{T \leq t\} \in \mathscr{F}_t.$$

In particular, T is a positive r.v. taking possibly infinite values. If $(\mathscr{F}_t)$ is right-continuous, it is equivalent to demand that $\{T < t\}$ belongs to $\mathscr{F}_t$ for every t. In that case, the definition is also equivalent to: T is a stopping time if and only if the process $X_t = 1_{]0,T]}(t)$ is adapted (X is then a left-continuous adapted process, a particular case of the predictable processes which will be introduced in Sect. 5 Chap. IV and in Exercise (4.20)).

The class of sets A in $\mathscr{F}_\infty$ such that $A \cap \{T \le t\} \in \mathscr{F}_t$ for all t is a σ-algebra denoted by $\mathscr{F}_T$; the sets in $\mathscr{F}_T$ must be thought of as events which may occur before time T. The constants, i.e. $T(\omega) \equiv s$ for every ω, are stopping times and in that case $\mathscr{F}_T = \mathscr{F}_s$. Stopping times thus appear as generalizations of constant times for which one can define a "past" which is consistent with the "pasts" of constant times.

The proofs of all the above facts are left to the reader whom we also invite to solve Exercises (4.16–19) to become acquainted with stopping times.

A stopping time may be thought of as the first time some physical event occurs. Here are two basic examples.

(4.5) Proposition. *If E is a metric space, A a closed subset of E and X the coordinate process on $\mathbf{W} = C(\mathbb{R}_+, E)$, and if we set*

$$D_A(\omega) = \inf\{t \ge 0, X_t(\omega) \in A\}$$

with the understanding that $\inf(\emptyset) = +\infty$, then D_A is a stopping time with respect to the natural filtration $\mathscr{F}_t^0 = \sigma(X_s, s \le t)$. It is called the entry time of A.

Proof. For a metric d on E, we have

$$\{D_A \le t\} = \left\{\omega : \inf_{s \in \mathbb{Q}, s \le t} d\left(X_s(\omega), A\right) = 0\right\}$$

and the right-hand side set obviously belongs to $\mathscr{F}_t^0$. □

This is one of the rare examples of interesting stopping times with respect to $(\mathscr{F}_t^0)$. If we remove the assumption of path-continuity of X or of closedness of A, we have to use larger filtrations. We have for instance

(4.6) Proposition. *If A is an open subset of E and Ω is the space of right-continuous paths from $\mathbb{R}_+$ to E, the time*

$$T_A = \inf\{t > 0 : X_t \in A\} \qquad \left(\inf(\emptyset) = +\infty\right)$$

is a stopping time with respect to $\mathscr{F}_{t+}^0$. It is called the hitting time of A.

Proof. As already observed, T_A is a $\mathscr{F}_{t+}^0$-stopping time if and only if $\{T_A < t\} \in \mathscr{F}_t^0$ for each t. If A is open and $X_s(\omega) \in A$, by the right-continuity of paths, $X_t(\omega) \in A$ for every $t \in [s, s+\varepsilon[$ for some $\varepsilon > 0$. As a result

$$\{T_A < t\} = \bigcup_{s \in \mathbb{Q}, s < t} \{X_s \in A\} \in \mathscr{F}_t^0.$$

□

It can be seen directly or by means of Exercise (4.21) that T_A is not a $(\mathscr{F}_t^0)$-stopping time and that, in this setting, $\mathscr{F}_{t+}^0$ is strictly larger than $\mathscr{F}_t^0$. This is a general phenomenon; most interesting stopping times will be $(\mathscr{F}_{t+}^0)$-stopping times and we will often have to work with $(\mathscr{F}_{t+}^0)$ rather than $(\mathscr{F}_t^0)$.

We now turn to the use of stopping times in a general setting.

Let $(\mathscr{F}_t)$ be a filtration on $(\Omega, \mathscr{F})$ and T a stopping time. For a process X, we define a new mapping X_T on the set $\{\omega : T(\omega) < \infty\}$ by

$$X_T(\omega) = X_t(\omega) \qquad \text{if} \qquad T(\omega) = t.$$

This is the position of the process X at time T, but it is not clear that X_T is a random variable on $\{T < \infty\}$. Moreover if X is adapted, we would like X_T to be $\mathscr{F}_T$-measurable just as X_t is $\mathscr{F}_t$-measurable. This is why we lay down the following definitions where $\mathscr{B}([0,t])$ is the σ-algebra of Borel subsets of $[0,t]$.

(4.7) Definition. *A process X is* progressively measurable *or simply* progressive *(with respect to the filtration $(\mathscr{F}_t)$) if for every t the map $(s,\omega) \to X_s(\omega)$ from $[0,t] \times \Omega$ into $(E, \mathscr{E})$ is $\mathscr{B}([0,t]) \otimes \mathscr{F}_t$-measurable. A subset Γ of $\mathbb{R}_+ \times \Omega$ is* progressive *if the process $X = 1_\Gamma$ is progressive.*

The family of progressive sets is a σ-field on $\mathbb{R}_+ \times \Omega$ called the progressive *σ-field and denoted by* Prog. *A process X is progressive if and only if the map $(t,\omega) \to X_t(\omega)$ is measurable with respect to* Prog.

Clearly, a progressively measurable process is adapted and we have conversely the

(4.8) Proposition. *An adapted process with right or left continuous paths is progressively measurable.*

Proof. Left to the reader as an exercise.

We now come to the result for which the notion was introduced.

(4.9) Proposition. *If X is progressively measurable and T is a stopping time (with respect to the same filtration $(\mathscr{F}_t)$) then X_T is $\mathscr{F}_T$-measurable on the set $\{T < \infty\}$.*

Proof. The set $\{T < \infty\}$ is itself in $\mathscr{F}_T$. To say that X_T is $\mathscr{F}_T$-measurable on this set is to say that $X_T \cdot 1_{[T \le t]} \in \mathscr{F}_t$ for every t. But the map

$$T : \big((T \le t), (T \le t) \cap \mathscr{F}_t\big) \to ([0,t], \mathscr{B}([0,t]))$$

is measurable because T is a stopping time, hence the map $\omega \to \big(T(\omega), \omega\big)$ from $(\Omega, \mathscr{F}_t)$ into $\big([0,t] \times \Omega, \mathscr{B}([0,t]) \otimes \mathscr{F}_t\big)$ is measurable and X_T is the composition of this map with X which is $\mathscr{B}([0,t]) \otimes \mathscr{F}_t$-measurable by hypothesis. □

With a stopping time T and a process X, we can associate the *stopped process* X^T defined by $X_t^T(\omega) = X_{t \wedge T}(\omega)$. By Exercise (4.16) the family of σ-fields $(\mathscr{F}_{t \wedge T})$ is a filtration and we have the

(4.10) Proposition. *If X is progressive, then X^T is progressive with respect to the filtration $(\mathscr{F}_{t\wedge T})$.*

Proof. Left to the reader as an exercise.

The following remark will be technically important.

(4.11) Proposition. *Every stopping time is the decreasing limit of a sequence of stopping times taking only finitely many values.*

Proof. For a stopping time T one sets

$$\begin{array}{lllll} T_k & = & +\infty & \text{if} & T \geq k, \\ T_k & = & q2^{-k} & \text{if} & (q-1)2^{-k} \leq T < q2^{-k}, q < 2^k k. \end{array}$$

It is easily checked that T_k is a stopping time and that $\{T_k\}$ decreases to T. □

The description we have given so far in this section does not involve any probability measure. We must now see what becomes of the above notions when $(\Omega, \mathscr{F})$ is endowed with a probability measure or rather, as will be necessary in the study of Markov processes, with a family of probability measures.

(4.12) Definition. *If $P_\theta, \theta \in \Theta$, is a family of probability measures on $(\Omega, \mathscr{F})$ a property is said to hold* almost-surely *if it holds P_θ-a.s. for every $\theta \in \Theta$.*

With this notion, two processes are, for instance, indistinguishable if they are indistinguishable for each P_θ. If we have a filtration $(\mathscr{F}_t)$ on $(\Omega, \mathscr{F})$ we will want that a process which is indistinguishable from an adapted process be itself adapted; in particular, we want that a limit (almost-sure, in probability, in the mean) of adapted processes be an adapted process. Another way of putting this is that a process which is indistinguishable from an adapted process can be turned into an adapted process by altering it on a negligible set; this demands that the negligible sets be in $\mathscr{F}_t$ for all t which leads to the following definition.

(4.13) Definition. *If, for each θ, we call $\mathscr{F}_\infty^\theta$ the completion of $\mathscr{F}_\infty$ with respect to P_θ, the filtration $(\mathscr{F}_t)$ is said to be* complete *if $\mathscr{F}_0$, hence every $\mathscr{F}_t$ contains all the negligible sets of $\cap_\theta \mathscr{F}_\infty^\theta$.*

Of course, as follows from Definition (4.12), negligible means negligible for every $P_\theta, \theta \in \Theta$, in other words Γ is negligible if for every θ there exists a set A_θ in $\mathscr{F}_\infty$ such that $\Gamma \subset A_\theta$ and $P_\theta[A_\theta] = 0$.

If $(\mathscr{F}_t)$ is not complete, we can obtain a larger but complete filtration in the following way. For each θ we call $\mathscr{F}_t^{\,\theta}$ the σ-field $\sigma\left(\mathscr{F}_t \cup \mathscr{N}^{\theta}\right)$ where $\mathscr{N}^{\theta}$ is the class of P_θ-negligible, $\mathscr{F}_\infty^\theta$-measurable sets; we then set $\mathscr{G}_t = \cap_\theta \mathscr{F}_t^{\,\theta}$. The filtration $(\mathscr{G}_{t+})$ is complete and right continuous. It is called the *usual augmentation* of $(\mathscr{F}_t)$.

Of course, if we use the usual augmentation of $(\mathscr{F}_t)$ instead of $(\mathscr{F}_t)$ itself, we will have to check that a process with some sort of property relative to $(\mathscr{F}_t)$

retains this property relative to the usual augmentation. This is not always obvious, the completion operation for instance is not an innocuous operation, it can alter significantly the structure of the filtration. Evidence of that will be given later on; in fact, all the canonical processes with the same state spaces have the same uncompleted natural filtrations and we will see that the properties of the completed ones may be widely different.

We close this section with a general result which permits to show that many random variables are in fact stopping times. To this end, we will use a difficult result from measure theory which we now recall.

(4.14) Theorem. *If $(E, \mathscr{E})$ is a LCCB space endowed with its Borel σ-field and $(\Omega, \mathscr{F}, P)$ is a complete probability space, for every set $A \in \mathscr{E} \otimes \mathscr{F}$, the projection $\pi(A)$ of A into Ω belongs to $\mathscr{F}$.*

If Γ is a subset of $\Omega \times \mathbb{R}_+$, we define the *debut* D_Γ of Γ by

$$D_\Gamma(\omega) = \inf\{t \geq 0 : (t, \omega) \in \Gamma\},$$

with the convention that $\inf(\emptyset) = +\infty$.

(4.15) Theorem. *If the filtration $(\mathscr{F}_t)$ is right-continuous and complete, the debut of a progressive set is a stopping time.*

Proof. It is enough to reason when there is only one probability measure involved. Let Γ be a progressive set. We apply Theorem (4.14) above to the set $\Gamma_t = \Gamma \cap ([0, t[\times \Omega)$ which belongs to $\mathscr{B}([0, t]) \otimes \mathscr{F}_t$. As a result $\{D_\Gamma < t\} = \pi(\Gamma_t)$ belongs to $\mathscr{F}_t$.

\# **(4.16) Exercise.** Let $(\mathscr{F}_t)$ be a filtration and S, T be $(\mathscr{F}_t)$-stopping times.

1°) Prove that $S \wedge T$ and $S \vee T$ are stopping times.

2°) Prove that the sets $\{S = T\}$, $\{S \leq T\}$, $\{S < T\}$ are in $\mathscr{F}_S \cap \mathscr{F}_T$.

3°) If $S \leq T$, prove that $\mathscr{F}_S \subset \mathscr{F}_T$.

\# **(4.17) Exercise.** 1°) If (T_n) is a sequence of $(\mathscr{F}_t)$-stopping times, then the r.v. $\sup_n T_n$ is a stopping time.

2°) If moreover $(\mathscr{F}_t)$ is right-continuous, then

$$\inf_n T_n, \qquad \overline{\lim_n} T_n, \qquad \underline{\lim_n} T_n$$

are stopping times. If $T_n \downarrow T$, then $\mathscr{F}_T = \bigcap_n \mathscr{F}_{T_n}$.

\# **(4.18) Exercise.** Let $(\mathscr{F}_t)$ be a filtration. If T is a stopping time, we denote by $\mathscr{F}_{T-}$ the σ-algebra generated by the sets of $\mathscr{F}_0$ and the sets

$$\{T > t\} \cap \Gamma$$

where $\Gamma \in \mathscr{F}_t$.

1°) Prove that $\mathscr{F}_{T-} \subset \mathscr{F}_T$. The first jump time of a Poisson process (Exercise (1.14) Chapter II) affords an example where the inclusion is strict.

2°) If $S \leq T$ prove that $\mathscr{F}_{S^-} \subset \mathscr{F}_{T^-}$. If moreover $S < T$ on $\{S < \infty\} \cap \{T > 0\}$, prove that $\mathscr{F}_S \subset \mathscr{F}_{T^-}$.

3°) Let (T_n) be a sequence of stopping times increasing to a stopping time T and such that $T_n < T$ for every n; prove that $\bigvee_n \mathscr{F}_{T_n} = \mathscr{F}_{T^-}$.

4°) If (T_n) is any increasing sequence of stopping times with limit T, prove that $\bigvee_n \mathscr{F}_{T_n^-} = \mathscr{F}_{T^-}$.

\# **(4.19) Exercise.** Let T be a stopping time and $\Gamma \in \mathscr{F}$. The random variable T_Γ defined by $T_\Gamma = T$ on Γ, $T_\Gamma = +\infty$ on Γ^c is a stopping time if and only if $\Gamma \in \mathscr{F}_T$.

(4.20) Exercise. Let $(\mathscr{F}_t)$ be a right-continuous filtration.

1°) Prove that the σ-fields generated on $\Omega \times \mathbb{R}_+$ by

i) the space of adapted continuous processes,
ii) the space of adapted processes which are left-continuous on $]0, \infty[$,

are equal. (This is solved in Sect. 5 Chap. IV). This σ-field is denoted $\mathscr{P}(\mathscr{F}_t)$ or simply $\mathscr{P}$ and is called the *predictable σ-field* (relative to $(\mathscr{F}_t)$). A process Z on Ω is said to be *predictable* if the map $(\omega, t) \to Z_t(\omega)$ is measurable with respect to $\mathscr{P}(\mathscr{F}_t)$. Prove that the predictable processes are progressively measurable.

2°) If S and T are two $\mathscr{F}_t$-stopping times and $S \leq T$, set

$$]S, T] = \{(\omega, t) : S(\omega) < t \leq T(\omega)\}.$$

Prove that $\mathscr{P}(\mathscr{F}_t)$ is generated by the family of sets $]S, T]$.

3°) If S is a positive r.v., we denote by $\mathscr{F}_{S-}$ the σ-field generated by all the variables Z_S where Z ranges through the predictable processes. Prove that, if S is a stopping time, this σ-field is equal to the σ-field $\mathscr{F}_{S-}$ of Exercise (4.18).

[Hint: For $A \in \mathscr{F}_{S-}$, consider the process $Z_t(\omega) = 1_A(\omega) 1_{]0, S(\omega)]}(t)$.]

Prove that S is $\mathscr{F}_{S-}$-measurable.

4°) In the general situation, it is not true that $S \leq T$ entails $\mathscr{F}_{S-} \subset \mathscr{F}_{T-}$. Give an example of a variable $S \leq 1$ such that $\mathscr{F}_{S-} = \mathscr{F}_\infty$.

*# **(4.21) Exercise (Galmarino's test).** Let $\Omega = D(\mathbb{R}_+, \mathbb{R}^d)$ or $C(\mathbb{R}_+, \mathbb{R}^d)$ and use the notation at the beginning of the section.

1°) Prove that T is a $(\mathscr{F}_t^0)$-stopping time if and only if, for every t, the properties $T(\omega) \leq t$ and $X_s(\omega) = X_s(\omega')$ for every $s \leq t$, imply $T(\omega) = T(\omega')$. Prove that the time T_A of Proposition (4.6) is not a $(\mathscr{F}_t^0)$-stopping time.

2°) If T is a $(\mathscr{F}_t^0)$-stopping time, prove that $A \in \mathscr{F}_T^0$ if and only if $\omega \in A$, $T(\omega) = T(\omega')$ and $X_s(\omega) = X_s(\omega')$ for every $s \leq T(\omega)$ implies $\omega' \in A$.

3°) Let ω_T be the point in Ω defined by $\omega_T(s) = \omega(s \wedge T(\omega))$. Prove that f is $\mathscr{F}_T^0$-measurable if and only if $f(\omega) = f(\omega_T)$ for every ω.

4°) Using the fact that $\mathscr{F}_\infty^0$ is the union of the σ-fields generated by the countable sub-families of coordinate mappings, prove that $\mathscr{F}_T^0 = \sigma\left(X_s^T, s \geq 0\right)$.

5°) Deduce from 4°) that $\mathscr{F}_t^0$ is countably generated.

(4.22) Exercise. A positive Borel function ϕ on $\mathbb{R}_+$ is said to have the property (P) if for every stopping time T in any filtration $(\mathscr{F}_t)$ whatsoever, $\phi(T)$ is a $(\mathscr{F}_t)$-stopping time. Show that ϕ has the property (P) if and only if there is a $t_0 \leq +\infty$ such that $\phi(t) \geq t$ for $t \leq t_0$ and $\phi(t) = t_0$ for $t > t_0$.

Notes and Comments

Sect. 1 There are many rigorous constructions of Brownian motion, some of which are found in the following sections and in the first chapter of the book of Knight [5]. They are usually based on the use of an orthonormal basis of $L^2(\mathbb{R}_+)$ or on a convergence in law, a typical example of which is the result of Donsker described in Chap. XIII. The first construction, historically, was given by Wiener [1] which is the reason why Brownian motion is also often called the Wiener process.

Föllmer [3] and Le Gall [8] give excellent pedagogical presentations of Brownian motion. The approach we have adopted here, by means of Gaussian measures, is a way of unifying these different constructions; we took it from the lecture course of Neveu [2], but it goes back at least to Kakutani. Versions and modifications have long been standard notions in Probability Theory (see Dellacherie-Meyer [1]).

Exercise (1.17) is from Lévy, Exercise (1.18) from Durrett [1] and Exercise (1.19) from Hardin [1]. The first study of polar functions for BM^2 appears in Graversen [1]; this will be taken up more thoroughly in Chap. V. Exercises (1.21) and (1.22) are from Barlow et al. [1], Song-Yor [1] and Yor [14].

Sect. 2 Our proof of Kolmogorov's criterion is borrowed from Meyer [8] (see also Neveu's course [2]). Integral type improvements are found in Ibragimov [1]; we also refer to Weber [1]. A very useful Sobolev type refinement due to Garsia et al. [1] is found in Stroock-Varadhan [1] (see also Dellacherie-Maisonneuve-Meyer [1]): it has been used in manifold contexts, for instance in Barlow-Yor [2], to prove the BDG inequalities in Chap. IV; see also Barlow ([3] and [5]) and Donati-Martin ([1]).

The rest of this section is due to Lévy. The proof of Theorem (2.7) is borrowed from Itô-McKean [1] which contains additional information, namely the "Chung-Erdös-Sirao" test.

Exercise (2.8) is due to Lévy and Exercise (2.9) to Dvoretzky et al. [1].

Sect. 3 The material covered in this section is now the common lore of probabilists (see for instance Dellacherie-Meyer [1] Vol. I). For Gaussian processes, we refer to Neveu [1]. A direct proof of the existence of the Wiener measure is given in Itô [4].

Exercise (3.9) deals with fractional Brownian motion; a number of references about this family of Gaussian processes, together with original results are found in Kahane [1], Chap. 18. The origin of question 2°) in this exercise is found in Albeverio et al. [1], p. 213–216, Stoll [1] and Yor [20]. Fractional Brownian motions were introduced originally in Mandelbrot and Van Ness [1]; they include

Lévy's Brownian motions with several parameters, and arise naturally in limit theorems for intersection local times (Weinryb-Yor [1], Biane [3]).

Exercise (3.11) is due to Aronszajn [1] (see Neveu [1]).

Sect. 4 Filtrations and their associated notions, such as stopping times, have, since the fundamental work of Doob, been a basic feature of Probability Theory. Here too, we refer to Dellacherie-Meyer [1] for the history of the subject as well as for many properties which we have turned into Exercises.

Chapter II. Martingales

Martingales are a very important subject in their own right as well as by their relationship with analysis. Their kinship to BM will make them one of our main subjects of interest as well as one of our foremost tools. In this chapter, we describe some of their basic properties which we shall use throughout the book.

§1. Definitions, Maximal Inequalities and Applications

In what follows, we always have a probability space $(\Omega, \mathscr{F}, P)$, an interval T of $\mathbb{N}$ or $\mathbb{R}_+$ and an increasing family $\mathscr{F}_t$, $t \in T$, of sub-σ-algebras of $\mathscr{F}$. We shall call it a filtration as in the case of $\mathbb{R}_+$ introduced in Sect. 4 Chap. I, the results of which apply as well to this case.

(1.1) Definition. *A real-valued process X_t, $t \in T$, adapted to $(\mathscr{F}_t)$ is a* submartingale *(with respect to $\mathscr{F}_t$) if*

i) $E\left[X_t^+\right] < \infty$ *for every* $t \in T$;

ii) $E\left[X_t \mid \mathscr{F}_s\right] \geq X_s$ *a.s. for every pair* s, t *such that* $s < t$.

A process X such that $-X$ is a submartingale is called a supermartingale *and a process which is both a sub and a supermartingale is a* martingale.

In other words, a martingale is an adapted family of integrable random variables such that

$$\int_A X_s dP = \int_A X_t dP$$

for every pair s, t with $s < t$ and $A \in \mathscr{F}_s$.

A sub(super)martingale such that all the variables X_t are integrable is called an *integrable sub(super)martingale*.

Of course, the filtration and the probability measure P are very important in this definition. When we want to stress this fact, we will speak of $(\mathscr{F}_t)$*-submartingales*, $(\mathscr{F}_t, P)$*-supermartingales*, ... A $(\mathscr{F}_t)$-martingale X is a martingale with respect to its natural filtration $\sigma(X_s, s \leq t)$. Conversely, if $\mathscr{G}_t \supset \mathscr{F}_t$, there is no reason why a $(\mathscr{F}_t)$-martingale should be a $(\mathscr{G}_t)$-martingale. Obviously, the set of martingales with respect to a given filtration is a vector space.

(1.2) Proposition. *Let B be a standard linear BM; then the following processes are martingales with respect to $\sigma(B_s, s \leq t)$:*

i) B_t *itself,* *ii)* $B_t^2 - t$, *iii)* $M_t^\alpha = \exp\left(\alpha B_t - \frac{\alpha^2}{2} t\right)$ *for* $\alpha \in \mathbb{R}$.

Proof. Left to the reader as an exercise. This proposition is generalized in Exercise (1.18). □

These properties will be considerably generalized in Chap. IV. We notice that the martingales in this proposition have continuous paths. The Poisson process of Exercise (1.14) affords an example of a martingale with cadlag paths. Finally, if $(\mathscr{F}_t)$ is a given filtration and Y is an integrable random variable, we can define a martingale Y_t by choosing for each t one random variable in the equivalence class of $E\left[Y \mid \mathscr{F}_t\right]$. Of course, there is no reason why the paths of this martingale should have any good properties and one of our tasks will precisely be to prove the existence of a good version.

Another important remark is that if X_t is a martingale then, because of Jensen's inequality, $|X_t|^p$ is a submartingale for $p \geq 1$ provided $E[|X_t|^p] < \infty$ for every t.

We now turn to the systematic study of martingales and of submartingales. Plainly, by changing X into $-X$ any statement about submartingales may be changed into a statement about supermartingales.

(1.3) Proposition. *Let (X_n), $n = 0, 1, \ldots$ be a (sub)martingale with respect to a discrete filtration $(\mathscr{F}_n)$ and H_n, $n = 1, 2, \ldots$, a positive bounded process such that $H_n \in \mathscr{F}_{n-1}$ for $n > 1$; the process Y defined by*

$$Y_0 = X_0, Y_n = Y_{n-1} + H_n (X_n - X_{n-1})$$

is a (sub)martingale. In particular, if T is a stopping time, the stopped process X^T is a (sub)martingale.

Proof. The first sentence is straightforward. The process Y thus defined is the discrete version of the stochastic integral we will define in Chap. IV. It will be denoted $H \cdot X$.

The second sentence follows from the first since $H_n = 1_{[n \leq T]}$ is $\mathscr{F}_{n-1}$-measurable being equal to $1 - 1_{[T \leq n-1]}$. □

We use this proposition to obtain a first version of the optional stopping theorem which we will prove in Sect. 3. The setting is the same as in Proposition (1.3).

(1.4) Proposition. *If S and T are two bounded stopping times and $S \leq T$, i.e. there is a constant M such that for every ω,*

$$S(\omega) \leq T(\omega) \leq M < \infty,$$

then

$$X_S \leq E\left[X_T | \mathscr{F}_S\right] \quad a.s. ,$$

with equality in case X is a martingale. Moreover, an adapted and integrable process X is a martingale if and only if

$$E[X_S] = E[X_T]$$

for any such pair of stopping times.

Proof. Suppose first that X is a martingale. If $H_n = 1_{[n\leq T]} - 1_{[n\leq S]}$, for $n > M$, we have

$$(H \cdot X)_n - X_0 = X_T - X_S,$$

but since $E\left[(H \cdot X)_n\right] = E[X_0]$ as is easily seen, we get $E[X_S] = E[X_T]$.

If we apply this equality to the stopping times $S^B = S\,1_B + M\,1_{B^c}$ and $T^B = T\,1_B + M\,1_{B^c}$ where $B \in \mathscr{F}_S$ (see Exercise (4.19) Chap. I) we get

$$E\left[X_T 1_B + X_M 1_{B^c}\right] = E\left[X_S 1_B + X_M 1_{B^c}\right],$$

whence it follows that $X_S = E[X_T|\mathscr{F}_S]$ a.s. In particular, the equality $E[X_S] = E[X_T]$ for every pair of bounded stopping times is sufficient to insure that X is a martingale.

If X is a submartingale, $\max(X, a)$ is an integrable submartingale to which we can apply the above reasoning getting inequalities instead of equalities. Letting a tend to $-\infty$, we get the desired result. □

We derive therefrom the following *maximal inequalities.*

(1.5) Proposition. *If (X_n) is an integrable submartingale indexed by the finite set $(0, 1, \ldots, N)$, then for every $\lambda > 0$,*

$$\lambda P\left[\sup_n X_n \geq \lambda\right] \leq E\left[X_N 1_{(\sup_n X_n \geq \lambda)}\right] \leq E\left[|X_N| 1_{(\sup_n X_n \geq \lambda)}\right].$$

Proof. Let $T = \inf\{n : X_n \geq \lambda\}$ if this set is non-empty, $T = N$ otherwise. This is a stopping time and so by the previous result

$$\begin{aligned} E[X_N] &\geq E[X_T] = E\left[X_T 1_{(\sup_n X_n \geq \lambda)}\right] + E\left[X_T 1_{(\sup_n X_n < \lambda)}\right] \\ &\geq \lambda P\left[\sup_n X_n \geq \lambda\right] + E\left[X_N 1_{(\sup_n X_n < \lambda)}\right] \end{aligned}$$

because $X_T \geq \lambda$ on $(\sup_n X_n \geq \lambda)$. By subtracting $E\left[X_N 1_{(\sup_n X_n < \lambda)}\right]$ from the two extreme terms, we get the first inequality, while the second one is obvious. □

(1.6) Corollary. *If X is a martingale or a positive submartingale indexed by the finite set $(0, 1, \ldots, N)$, then for every $p \geq 1$ and $\lambda > 0$*

$$\lambda^p P\left[\sup_n |X_n| \geq \lambda\right] \leq E\left[|X_N|^p\right],$$

and for any $p > 1$,

$$E\left[|X_N|^p\right] \leq E\left[\sup_n |X_n|^p\right] \leq \left(\frac{p}{p-1}\right)^p E\left[|X_N|^p\right].$$

Proof. By Jensen's inequality, if X_N is in L^p, the process $|X_n|^p$ is a submartingale and the first part of the corollary follows from the preceding result.

To prove the second part, we observe that the left-hand side inequality is trivial; to prove the right-hand side inequality we set $X^* = \sup_n |X_n|$. From the inequality $\lambda P[X^* \geq \lambda] \leq E\left[|X_N| 1_{(X^* \geq \lambda)}\right]$, it follows that, for a fixed $k > 0$,

$$\begin{aligned} E[(X^* \wedge k)^p] &= E\left[\int_0^{X^* \wedge k} p\lambda^{p-1} d\lambda\right] = E\left[\int_0^k p\lambda^{p-1} 1_{(X^* \geq \lambda)} d\lambda\right] \\ &= \int_0^k p\lambda^{p-1} P[X^* \geq \lambda] d\lambda \leq \int_0^k p\lambda^{p-2} E\left[|X_N| 1_{(X^* \geq \lambda)}\right] d\lambda \\ &= pE\left[|X_N| \int_0^{X^* \wedge k} \lambda^{p-2} d\lambda\right] = \frac{p}{p-1} E\left[|X_N|(X^* \wedge k)^{p-1}\right]. \end{aligned}$$

Hölder's inequality then yields

$$E\left[(X^* \wedge k)^p\right] \leq \frac{p}{p-1} E\left[(X^* \wedge k)^p\right]^{(p-1)/p} E\left[|X_N|^p\right]^{1/p},$$

and after cancellation

$$E[(X^* \wedge k)^p] \leq \left(\frac{p}{p-1}\right)^p E[|X_N|^p].$$

The proof is completed by making k tend to infinity. □

These results carry over to general index sets. If X is a martingale indexed by an interval T of $\mathbb{R}$, we can look at its restriction to a countable subset D of T. We can then choose an increasing sequence D_n of finite subsets of D such that $\bigcup D_n = D$ and apply the above results to D_n. Since $E[|X_t|^p]$ increases with t, we get, by passing to the limit in n, that

$$\lambda^p P\left[\sup_{t \in D} |X_t| \geq \lambda\right] \leq \sup_t E[|X_t|^p]$$

and for $p > 1$

$$E\left[\sup_{t \in D} |X_t|^p\right] \leq \left(\frac{p}{p-1}\right)^p \sup_t E[|X_t|^p].$$

We insist that this is true without any hypothesis on the filtration $(\mathscr{F}_t)$ which may be neither complete nor right continuous.

(1.7) Theorem (Doob's L^p-inequality). *If X is a right-continuous martingale or positive submartingale indexed by an interval T of $\mathbb{R}$, then if $X^* = \sup_t |X_t|$, for $p \geq 1$,*

$$\lambda^p P[X^* \geq \lambda] \leq \sup_t E[|X_t|^p]$$

and for $p > 1$,

$$\|X^*\|_p \leq \frac{p}{p-1} \sup_t \|X_t\|_p.$$

Proof. If D is a countable dense subset of T, because of the right-continuity $X^* = \sup_{t \in D} |X_t|$ and the results follow from the above remarks. □

If t_d is the point on the right of T, we notice that $\sup_t \|X_t\|_p$ is equal to $\|X_{t_d}\|_p$ if $t_d \in T$ and to $\lim_{t \uparrow t_d} \|X_t\|_p$ if T is open on the right.

For $p = 2$, the second inequality reads

$$\|X^*\|_2 \leq 2 \sup_t \|X_t\|_2$$

and is known as *Doob's L^2-inequality*. Since obviously $\|X_t\|_2 \leq \|X^*\|_2$ for every t, we see that X^* is in L^2 if and only if $\sup_t \|X_t\|_2 < +\infty$, in other words if the martingale is bounded in L^2. We see that in this case the martingale, i.e. the family of variables $\{X_t, t \in T\}$ is uniformly integrable. These remarks are valid for each $p > 1$. They are not for $p = 1$ and a martingale can be bounded in L^1 without being uniformly integrable and a fortiori without X^* being integrable. This has been the subject of many studies. Let us just mention here that

$$E[X^*] \leq \frac{e}{e-1}\left(1 + \sup_t E\left[X_t \log^+ X_t\right]\right)$$

(see Exercise (1.16)).

In the following section, we will apply these inequalities to establish the convergence theorems for martingales. We close this section with some important applications to Brownian motion. The first is known as the *exponential inequality*. It will be considerably generalized in Exercise (3.16) of Chap. IV. We recall that $S_t = \sup_{s \leq t} B_s$.

(1.8) Proposition. *For $a > 0$,*

$$P[S_t \geq at] \leq \exp(-a^2 t/2).$$

Proof. For $\alpha > 0$, we use the maximal inequality for the martingale M^α of Proposition (1.2) restricted to $[0, t]$. Since $\exp\left(\alpha S_t - \frac{\alpha^2 t}{2}\right) \leq \sup_{s \leq t} M_s^\alpha$, we get

$$\begin{aligned} P[S_t \geq at] &\leq P\left[\sup_{s \leq t} M_s^\alpha \geq \exp\left(\alpha a t - \alpha^2 t/2\right)\right] \\ &\leq \exp\left(-\alpha a t + \alpha^2 t/2\right) E[M_t^\alpha]; \end{aligned}$$

but $E[M_t^\alpha] = E[M_0^\alpha] = 1$ and $\inf_{\alpha > 0}\left(-\alpha a t + \frac{\alpha^2}{2} t\right) = -a^2 t/2$, whence the result follows.

Remark. This inequality is also a consequence of the equality in law: $S_t \overset{(d)}{=} |B_t|$ valid for fixed t, which may be derived from the strong Markov property of BM (see Exercise (3.27) in Chap. III).

(1.9) Theorem (Law of the iterated logarithm). *For the linear standard Brownian motion B,*

$$P\left[\overline{\lim_{t\downarrow 0}} \frac{B_t}{\left(2t\log_2(1/t)\right)^{1/2}} = 1\right] = 1$$

where $\log_2 x = \log(\log x)$ *for* $x > 1$.

Proof. Let $h(t) = \sqrt{2t\log_2(1/t)}$ and pick two numbers θ and δ in $]0,1[$. We set

$$\alpha_n = (1+\delta)\theta^{-n}h(\theta^n), \qquad \beta_n = h(\theta^n)/2.$$

By the same reasoning as in the preceding proof

$$P\left[\sup_{s\le 1}(B_s - \alpha_n s/2) \ge \beta_n\right] \le e^{-\alpha_n\beta_n} = Kn^{-1-\delta}$$

for some constant K. Thus, by the Borel-Cantelli lemma, we have

$$P\left[\underline{\lim}_{n\to\infty}\left\{\sup_{s\le 1}(B_s - \alpha_n s/2) < \beta_n\right\}\right] = 1.$$

If we restrict s to $[0,\theta^{n-1}]$ we find that, a fortiori, for almost every ω, there is an integer $n_0(\omega)$ such that for $n > n_0(\omega)$ and $s \in [0,\theta^{n-1}]$

$$B_s(\omega) \le \frac{\alpha_n s}{2} + \beta_n \le \frac{\alpha_n\theta^{n-1}}{2} + \beta_n = \left[\frac{1+\delta}{2\theta} + \frac{1}{2}\right]h(\theta^n).$$

But the function h is increasing on an interval $[0,a[$, with $a > 0$, as can easily be checked. Therefore, for n sufficiently large and s in the interval $]\theta^n,\theta^{n-1}]$, we have, for these ω's,

$$B_s(\omega) \le \left[\frac{1+\delta}{2\theta} + \frac{1}{2}\right]h(s).$$

As a result, $\overline{\lim}_{s\downarrow 0} B_s/h(s) \le (1+\delta)/2\theta + 1/2$ a.s. Letting θ tend to 1 and then δ tend to zero, we get that $\overline{\lim}_{s\downarrow 0} B_s/h(s) \le 1$ a.s.

We now prove the reverse inequality. For $\theta \in]0,1[$, the events

$$\Gamma_n = \left\{B_{\theta^n} - B_{\theta^{n+1}} \ge \left(1-\sqrt{\theta}\right)h(\theta^n)\right\}$$

are independent; moreover (see the proof of Theorem (2.7) Chap. I)

$$\sqrt{2\pi}\,P[\Gamma_n] = \int_a^\infty e^{-u^2/2}du > \frac{a}{1+a^2}e^{-a^2/2}$$

with $a = \left(1-\sqrt{\theta}\right)\left(2\log_2\theta^{-n}/(1-\theta)\right)^{1/2}$. This is of the order of $n^{-(1-2\sqrt{\theta}+\theta)/(1-\theta)}$ $= n^{-\alpha}$ with $\alpha < 1$. As a result, $\sum_1^\infty P[\Gamma_n] = +\infty$ and by the Borel-Cantelli lemma,

$$B_{\theta^n} \ge \left(1-\sqrt{\theta}\right)h\left(\theta^n\right) + B_{\theta^{n+1}} \quad \text{infinitely often a.s.}$$

Since $-B$ is also a Brownian motion, we know from the first part of the proof that $-B_{\theta^{n+1}}(\omega) < 2h(\theta^{n+1})$ from some integer $n_0(\omega)$ on. Putting the last two inequalities together yields, since $h(\theta^{n+1}) \leq 2\sqrt{\theta}h(\theta^n)$ from some n on, that

$$\begin{aligned} B_{\theta^n} &> \left(1-\sqrt{\theta}\right) h\left(\theta^n\right) - 2h\left(\theta^{n+1}\right) \\ &> h(\theta^n)\left(1-\sqrt{\theta}-4\sqrt{\theta}\right) \quad \text{infinitely often,} \end{aligned}$$

and consequently

$$\overline{\lim_{t\downarrow 0}} B_t/h(t) \geq \overline{\lim} B_{\theta^n}/h(\theta^n) > 1 - 5\sqrt{\theta} \quad \text{a.s.}$$

It remains to let θ tend to zero to get the desired inequality. □

Using the various invariance properties of Brownian motion proved at the end of Sect. 1 in Chap. I, we get some useful corollaries of the law of the iterated logarithm.

(1.10) Corollary. $P\left[\underline{\lim}_{t\downarrow 0} B_t \Big/ \sqrt{2t\log_2(1/t)} = -1\right] = 1.$

Proof. This follows from the fact that $-B$ is a BM. □

Since the intersection of two sets of probability 1 is a set of probability 1, we actually have

$$P\left[\overline{\lim_{t\downarrow 0}} B_t/\sqrt{2t\log_2(1/t)} = 1 \quad \text{and} \quad \underline{\lim_{t\downarrow 0}} B_t/\sqrt{2t\log_2(1/t)} = -1\right] = 1$$

which may help to visualize the behavior of B when it leaves zero. We see that in particular 0 is a.s. an accumulation point of zeros of the Brownian motion, in other words B takes a.s. infinitely many times the value 0 in any small interval $[0, a[$.

By translation, the same behavior holds at every fixed time. The reader will compare with the Hölder properties of Sect. 2 in Chap. I.

(1.11) Corollary. *For any fixed s,*

$$P\left[\overline{\lim_{t\downarrow 0}}(B_{t+s} - B_s)/\sqrt{2t\log_2(1/t)} = 1 \quad \text{and} \quad \underline{\lim_{t\downarrow 0}}(B_{t+s} - B_s)/\sqrt{2t\log_2(1/t)} = -1\right] = 1.$$

Proof. $(B_{t+s} - B_s, t \geq 0)$ is also a BM. □

Finally using time inversion, we get

(1.12) Corollary.

$$P\left[\overline{\lim_{t\uparrow\infty}} B_t/\sqrt{2t\log_2 t} = 1 \quad \text{and} \quad \underline{\lim_{t\uparrow\infty}} B_t/\sqrt{2t\log_2 t} = -1\right] = 1.$$

Remark. This corollary entails the *recurrence* property of BM which was proved in Exercise (1.13) of Chap. I, namely, for every $x \in \mathbb{R}$ the set $\{t : B_t = x\}$ is a.s. unbounded.

\# **(1.13) Exercise.** If X is a continuous process vanishing at 0, such that, for every real α, the process $M_t^\alpha = \exp\left\{\alpha X_t - \frac{\alpha^2}{2}t\right\}$ is a martingale with respect to the filtration $(\mathscr{F}_t)$, prove that X is a $(\mathscr{F}_t)$-Brownian motion (see the Definition (2.20) in Chap. III).

[Hint: Use the following two facts:

i) a r.v. X is $\mathscr{N}(0, 1)$ if and only if $E[e^{\lambda X}] = e^{\lambda^2/2}$ for every real λ,

ii) if X is a r.v. and $\mathscr{B}$ is a sub-σ-algebra such that

$$E[e^{\lambda X}|\mathscr{B}] = E[e^{\lambda X}] < +\infty$$

for λ in a neighborhood of 0, then X and $\mathscr{B}$ are independent.]

\# **(1.14) Exercise (The Poisson process).** Let (X_n) be a sequence of independent exponential r.v.'s of parameter c. Set $S_n = \sum_1^n X_k$ and for $t \geq 0$, $N_t = \sum_1^\infty 1_{[S_n \leq t]}$.

1°) Prove that the increments of N_t are independent and have Poisson laws.

2°) Prove that $N_t - ct$ is a martingale with respect to $\sigma(N_s, s \leq t)$.

3°) Prove that $(N_t - ct)^2 - ct$ is a martingale.

(1.15) Exercise (Maximal inequality for positive supermartingales). If X is a right-continuous positive supermartingale, prove that

$$P\left[\sup_t X_t > \lambda\right] \leq \lambda^{-1} E[X_0]$$

(1.16) Exercise (The class $L \log L$). 1°) In the situation of Corollary (1.6), if ϕ is a function on $\mathbb{R}_+$, increasing, right-continuous and vanishing at 0, prove that

$$E\left[\phi(X^*)\right] \leq E\left[|X_N| \int_0^{X^*} \lambda^{-1} d\phi(\lambda)\right].$$

2°) Applying 1°) with $\phi(\lambda) = (\lambda - 1)^+$, prove that there is a constant C such that

$$E[X^*] \leq C\left(1 + \sup_n E\left[|X_n| \log^+(|X_n|)\right]\right)$$

[Hint: For $a, b > 0$, we have $a \log b \leq a \log^+ a + e^{-1} b$.]

The class of martingales for which the right-hand side is finite is called the class $L \log L$. With the notation of the following exercise, we have $L \log L \subset H^1$.

(1.17) Exercise (The Space H^p). The space of continuous martingales indexed by $\mathbb{R}_+$, such that $X^* = \sup_t |X_t|$ is in L^p, $p \geq 1$, is a Banach space for the norm $\|X\|_{H^p} = \|X^*\|_p$.

[Hint: See Proposition (1.22) in Chap. IV.]

Remark. In this book, we focus on continuous processes, which is the reason for limiting ourselves to continuous martingales. In fact, the same result holds for the space $\mathbb{H}^p$ of cadlag martingales such that X^* is in L^p; the space H^p above is a closed subspace of $\mathbb{H}^p$.

\# **(1.18) Exercise.** Retain the notation of Exercise (1.14) of Chap. I and prove that

$$\int_0^t f(s)dB_s, \left(\int_0^t f(s)dB_s\right)^2 - \int_0^t f(s)^2 ds,$$
$$\exp\left(\int_0^t f(s)dB_s - \frac{1}{2}\int_0^t f(s)^2 ds\right)$$

are continuous martingales. This will be considerably generalized in Chap. IV.

(1.19) Exercise. Let X and Y be two positive supermartingales with respect to the same filtration $(\mathscr{F}_n)$ and T a stopping time such that $X_T \geq Y_T$ on $\{T < \infty\}$. Prove that the process Z defined by

$$Z_n(\omega) = X_n(\omega) \quad \text{if} \quad n < T(\omega), \qquad Z_n(\omega) = Y_n(\omega) \quad \text{if} \quad n \geq T(\omega),$$

is a supermartingale.

** **(1.20) Exercise.** 1°) For any measure μ on $\mathbb{R}_+$, prove that

$$P\left[\overline{\lim_{h\downarrow 0}}|B_{s+h} - B_s|/\sqrt{2h\log_2(1/h)} = 1 \quad \text{for} \quad \mu\text{-a.e.} \quad s\right] = 1.$$

The following questions deal with the exceptional set.

2°) Using Lévy's modulus of continuity theorem (Theorem (2.7) of Chap. I) prove that for a.e. ω and any pair $a < b$, one can find s_1 and t_1 such that $a < s_1 < t_1 < b$ and

$$|B_{t_1} - B_{s_1}| > \frac{1}{2}d(t_1 - s_1)$$

where $d(h) = \sqrt{2h\log(1/h)}$.

3°) Having chosen $s_1, \ldots, s_n, t_1, \ldots, t_n$ such that $|B_{t_i} - B_{s_i}| > \frac{i}{i+1}d(t_i - s_i)$, choose $s'_n \in]s_n, t_n[$, $s'_n \leq s_n + 2^{-n}$ and $|B_{t_n} - B_s| > \frac{n}{n+1}d(t_n - s)$ for every $s \in]s_n, s'_n]$. Then, choose s_{n+1}, t_{n+1} in $]s_n, s'_n[$ and so on and so forth. Let $\{s_0\} = \bigcap_n [s_n, t_n]$; prove that

$$\overline{\lim_{h\downarrow 0}}|B_{s_0+h} - B_{s_0}|/d(h) \geq 1.$$

4°) Derive from 3°) that for a.e. ω there is a set of times t dense in $\mathbb{R}_+$ and such that

$$\overline{\lim_{h\downarrow 0}}|B_{t+h} - B_t|/\sqrt{2h\log_2(1/h)} = +\infty.$$

5°) Prove that the above set of times is even uncountable.

[Hint: Remove from $]s_n, s'_n[$ the middle-third part in Cantor set-like fashion and choose two intervals $]s_{n+1}, t_{n+1}[$ in each of the remaining parts.]

* **(1.21) Exercise.** If B is the BM^d, prove that

$$P\left[\overline{\lim_{t\downarrow 0}}\frac{|B_t|}{\sqrt{2t\log_2(1/t)}} = 1\right] = 1.$$

[Hint: Pick a countable subset (e_n) of the unit sphere in $\mathbb{R}^d$ such that $|x| = \sup_n |\langle x, e_n\rangle|$.]

Using the invariance properties, state and prove other laws of the iterated logarithm for BM^d.

(1.22) Exercise. If $\mathbb{B}$ is the Brownian sheet, for fixed s and t,

$$\overline{\lim_{h\downarrow 0}}\left(\mathbb{B}_{(s+h,t)} - \mathbb{B}_{(s,t)}\right)/\sqrt{2h\log_2(1/h)} = \sqrt{t} \quad \text{a.s.}$$

(1.23) Exercise. Prove that if B is the standard BM^d

$$P\left[\sup_{s\le t}|B_s| \ge \delta\right] \le 2d\exp(-\delta^2/2dt).$$

[Hint: Use Proposition (1.8) for $\langle\theta, B_t\rangle$ where θ is a unit vector.]

§2. Convergence and Regularization Theorems

Let us first recall some facts about real-valued functions. Let f be a function which maps a subset T of $\mathbb{R}$ into $\overline{\mathbb{R}}$. Let $t_1 < t_2 < \ldots < t_d$ be a finite subset F of T. For two real numbers a, b with $a < b$, we define inductively

$$s_1 = \inf\{t_i : f(t_i) > b\}, \quad s_2 = \inf\{t_i > s_1 : f(t_i) < a\},$$

$$s_{2n+1} = \inf\{t_i > s_{2n} : f(t_i) > b\}, \quad s_{2n+2} = \inf\{t_i > s_{2n+1} : f(t_i) < a\},$$

where we put $\inf(\emptyset) = t_d$. We set

$$D(f, F, [a,b]) = \sup\{n : s_{2n} < t_d\},$$

and we define the number of *downcrossings* of $[a,b]$ by f as the number

$$D(f, T, [a,b]) = \sup\{D(f, F, [a,b]) : F \quad \text{finite} \quad, F \subset T\}.$$

One could define similarly the number $U(f, T, [a,b])$ of *upcrossings*. The function f has no discontinuity of the second kind, in particular f has a limit at the boundaries of T whenever T is an open interval, if and only if $D(f, T, [a,b])$ (or $U(f, T, [a,b])$) is finite for every pair $[a,b]$ of rational numbers.

We now consider the case where f is the path of a submartingale X; if T is countable, $D(X, T, [a,b])$ is clearly a random variable and we have the

(2.1) Proposition. *If X is a submartingale and T is countable, then for any pair (a,b),*

$$(b-a)E[D(X,T,[a,b])] \leq \sup_{t\in T} E\left[(X_t-b)^+\right].$$

Proof. It is enough to prove the inequality when T is finite and we then use the notation above. The s_k's defined above are now stopping times with respect to the discrete filtration $(\mathscr{F}_{t_i})$. We are in the situation of Proposition (1.4) which we can apply to the stopping time s_k. Set $A_k = \{s_k < t_d\}$; then $A_k \in \mathscr{F}_{s_k}$ and $A_k \supset A_{k+1}$. On A_{2n-1}, we have $X_{s_{2n-1}} > b$, on A_{2n} we have $X_{s_{2n}} < a$ and therefore

$$\begin{aligned} 0 &\leq \int_{A_{2n-1}} \left(X_{s_{2n-1}} - b\right) dP \leq \int_{A_{2n-1}} \left(X_{s_{2n}} - b\right) dP \\ &\leq (a-b)P(A_{2n}) + \int_{A_{2n-1}\backslash A_{2n}} \left(X_{s_{2n}} - b\right) dP. \end{aligned}$$

Consequently, since $s_{2n} = t_d$ on A_{2n}^c,

$$(b-a)P(A_{2n}) \leq \int_{A_{2n-1}\backslash A_{2n}} \left(X_{s_{2n}} - b\right)^+ dP = \int_{A_{2n-1}\backslash A_{2n}} \left(X_{t_d} - b\right)^+ dP.$$

But $P(A_{2n}) = P[D(X,T,[a,b]) \geq n]$ and the sets $A_{2n-1}\backslash A_{2n}$ are pairwise disjoint so that by summing up the above inequalities, we get

$$(b-a)E[D(X,T,[a,b])] \leq E\left[(X_{t_d}-b)^+\right],$$

which is the desired result. □

We now apply this to the convergence theorem for discrete submartingales.

(2.2) Theorem. *If (X_n), $n \in \mathbb{N}$, is a submartingale such that*

$$\sup_n E\left[X_n^+\right] < +\infty,$$

then (X_n) converges almost-surely to a limit which is $< +\infty$ a.s.

Proof. Fatou's lemma ensures that $\underline{\lim} X_n < +\infty$ a.s. So if our claim were false, there would be two real numbers a and b such that $\underline{\lim} X_n < a < b < \overline{\lim} X_n$ with positive probability; thus, we would have $D(X,\mathbb{N},[a,b]) = +\infty$ with positive probability, which, by the foregoing result, is impossible. □

It is also useful to consider decreasing rather than increasing families of σ-algebras, or in other words to "reverse"the time in martingales. Let $(\mathscr{F}_n)_{n\leq 0}$, be a sequence of sub-$\sigma$-fields such that $\mathscr{F}_n \subset \mathscr{F}_m$ if $n \leq m \leq 0$. A submartingale with respect to $(\mathscr{F}_n)$ is an adapted family (X_n) of real-valued r.v.'s such that $E[X_n^+] < \infty$ for every n and $X_n \leq E[X_m|\mathscr{F}_n]$ for $n \leq m \leq 0$. We then get the following

(2.3) Theorem. *If (X_n), $n \in -\mathbb{N}$, is a submartingale, then $\lim_{n\to-\infty} X_n$ exists a.s. If moreover $\sup_n E[|X_n|] < \infty$, then (X_n) is uniformly integrable, the convergence holds in L^1 and, for every n*

$$\lim_{k\to-\infty} X_k \le E\left[X_n | \mathscr{F}_{-\infty}\right]$$

where $\mathscr{F}_{-\infty} = \bigcap_n \mathscr{F}_n$.

Proof. It is easily seen that $\sup_n E[X_n^+] \le E[X_0^+] < +\infty$, so that the first statement is proved as Theorem (2.2). To prove the second, we first observe that the condition $\sup_n E[|X_n|] < +\infty$ is equivalent to $\lim_{n\to-\infty} E[X_n] > -\infty$. Now, for any $c > 0$ and any n, we have

$$\int_{\{|X_n|>c\}} |X_n| dP = \int_{\{X_n \ge -c\}} X_n dP - E[X_n] + \int_{\{X_n > c\}} X_n dP.$$

For $\varepsilon > 0$, there is an integer n_0 such that $E[X_n] > E[X_{n_0}] - \varepsilon$ for $n \le n_0$; using this and the submartingale inequality yields that for $n \le n_0$,

$$\begin{aligned}\int_{\{|X_n|>c\}} |X_n| dP &\le \int_{\{X_n \ge -c\}} X_{n_0} dP - E[X_{n_0}] + \int_{\{X_n > c\}} X_{n_0} dP + \varepsilon \\ (*) \qquad &= \int_{\{|X_n|>c\}} |X_{n_0}| dP + \varepsilon\,.\end{aligned}$$

As $P[|X_n| > c] \le c^{-1} \sup_n E[|X_n|]$, the uniform integrability of the family X_n, $n \in -\mathbb{N}$, now follows readily from $(*)$ and implies that the convergence holds in L^1. Finally, if $\Gamma \in \mathscr{F}_{-\infty}$, for $m < n$,

$$\int_\Gamma X_m dP \le \int_\Gamma X_n dP$$

and we can pass to the limit, thanks to the L^1-convergence, to get

$$\int_\Gamma (\lim_m X_m) dP \le \int_\Gamma X_n dP$$

which ends the proof. □

The following corollary to the above results is often useful.

(2.4) Corollary. *Let X_n be a sequence of r.v.'s converging a.s. to a r.v. X and such that for every n, $|X_n| \le Y$ where Y is integrable. If $(\mathscr{F}_n)$ is an increasing (resp: decreasing) sequence of sub-σ-algebras, then $E[X_n \mid \mathscr{F}_n]$ converges a.s. to $E[X \mid \mathscr{F}]$ where $\mathscr{F} = \sigma(\bigcup \mathscr{F}_n)$ (resp. $\mathscr{F} = \bigcap_n \mathscr{F}_n$).*

Proof. Pick $\varepsilon > 0$ and set

$$U = \inf_{n \geq m} X_n, \qquad V = \sup_{n \geq m} X_n$$

where m is chosen such that $E[V - U] < \varepsilon$. Then, for $n \geq m$ we have

$$E\left[U \mid \mathscr{F}_n\right] \leq E\left[X_n \mid \mathscr{F}_n\right] \leq E\left[V \mid \mathscr{F}_n\right];$$

the left and right-hand sides of these inequalities are martingales which satisfy the conditions of the above theorems and therefore

$$E\left[U \mid \mathscr{F}\right] \leq \underline{\lim} E\left[X_n \mid \mathscr{F}_n\right] \leq \overline{\lim} E\left[X_n \mid \mathscr{F}_n\right] \leq E[V \mid \mathscr{F}].$$

We similarly have

$$E[U \mid \mathscr{F}] \leq E[X \mid \mathscr{F}] \leq E[V \mid \mathscr{F}].$$

It follows that $E\left[\overline{\lim} E\left[X_n \mid \mathscr{F}_n\right] - \underline{\lim} E\left[X_n \mid \mathscr{F}_n\right]\right] \leq \varepsilon$, hence $E\left[X_n \mid \mathscr{F}_n\right]$ converges a.s. and the limit is $E[X \mid \mathscr{F}]$. □

We now turn to the fundamental regularization theorems for continuous time (sub)martingales.

(2.5) Theorem. *If X_t, $t \in \mathbb{R}_+$, is a submartingale, then for almost every ω, for each $t \in]0, \infty[$, $\lim_{r \uparrow t, r \in \mathbb{Q}} X_r(\omega)$ exists and for each $t \in [0, \infty[$, $\lim_{r \downarrow t, r \in \mathbb{Q}} X_r(\omega)$ exists.*

Proof. It is enough to prove the results for t belonging to some compact subinterval I. If t_d is the right-end point of I, then for any $t \in I$,

$$E\left[(X_t - b)^+\right] \leq E\left[X_{t_d}^+\right] + b^-.$$

It follows from Proposition (2.1) that there exists a set $\Omega_0 \subset \Omega$ such that $P(\Omega_0) = 1$ and for $\omega \in \Omega_0$

$$D\left(X(\omega), I \cap \mathbb{Q}, [a, b]\right) < \infty$$

for every pair of rational numbers $a < b$. The same reasoning as in Theorem (2.2) then proves the result. □

We now define, for each $t \in [0, \infty[$,

$$X_{t+} = \overline{\lim_{r \downarrow t, r \in \mathbb{Q}}} X_r$$

and for $t \in]0, \infty[$,

$$X_{t-} = \overline{\lim_{r \uparrow t, r \in \mathbb{Q}}} X_r.$$

By the above result, these upper limits are a.s. equal to the corresponding lower limits. We study the processes thus defined.

(2.6) Proposition. *Suppose that $E[|X_t|] < +\infty$ for every t, then $E[|X_{t+}|] < \infty$ for every t and*

$$X_t \leq E[X_{t+} \mid \mathscr{F}_t] \quad \text{a.s.}$$

This inequality is an equality if the function $t \to E[X_t]$ is right-continuous, in particular if X is a martingale. Finally, (X_{t+}) is a submartingale with respect to $(\mathscr{F}_{t+})$ and it is a martingale if X is a martingale.

Proof. We can restrict ourselves to a compact subinterval. If (t_n) is a sequence of rational numbers decreasing to t, then (X_{t_n}) is a submartingale for which we are in the situation of Theorem (2.3). Thus, it follows immediately that X_{t+} is integrable and that X_{t_n} converges to X_{t+} in L^1. Therefore, we may pass to the limit in the inequality $X_t \leq E\left[X_{t_n} \mid \mathscr{F}_t\right]$ to get

$$X_t \leq E\left[X_{t+} \mid \mathscr{F}_t\right].$$

Also, the L^1-convergence implies that $E[X_{t+}] = \lim_n E[X_{t_n}]$ so that if $t \to E[X_t]$ is right-continuous, $E[X_t] = E[X_{t+}]$ hence $X_t = E[X_{t+} \mid \mathscr{F}_t]$ a.s.

Finally, let $s < t$ and pick a sequence (s_n) of rational numbers smaller than t decreasing to s. By what we have just proved,

$$X_{s_n} \leq E\left[X_t \mid \mathscr{F}_{s_n}\right] \leq E\left[X_{t+} \mid \mathscr{F}_t \mid \mathscr{F}_{s_n}\right] = E\left[X_{t+} \mid \mathscr{F}_{s_n}\right],$$

and applying Theorem (2.3) once again, we get the desired result.

Remark. By considering $X_t \vee a$ instead of X_t, we can remove the assumption that X_t is integrable for each t. The statement has to be changed accordingly.

The analogous result is true for left limits.

(2.7) Proposition. *If $E[|X_t|] < \infty$ for each t, then $E[|X_{t-}|] < +\infty$ for each $t > 0$ and*

$$X_{t-} \leq E\left[X_t \mid \mathscr{F}_{t-}\right] \quad a.s.$$

This inequality is an equality if $t \to E[X_t]$ is left-continuous, in particular if X is a martingale. Finally, $X_{t-}, t > 0$, is a submartingale with respect to $(\mathscr{F}_{t-})$ and a martingale if X is a martingale.

Proof. We leave as an exercise to the reader the task of showing that for every $a \in \mathbb{R}$, $\{X_t \vee a\}, t \in I$, where I is a compact subinterval, is uniformly integrable. The proof then follows the same pattern as for the right limits. □

These results have the following important consequences.

(2.8) Theorem. *If X is a right-continuous submartingale, then*

1) X is a submartingale with respect to $(\mathscr{F}_{t+})$, and also with respect to the completion of $(\mathscr{F}_{t+})$,
2) almost every path of X is cadlag.

Proof. Straightforward. □

(2.9) Theorem. *Let X be a submartingale with respect to a right-continuous and complete filtration $(\mathscr{F}_t)$; if $t \to E[X_t]$ is right-continuous (in particular, if X is a martingale) then X has a cadlag modification which is a $(\mathscr{F}_t)$-submartingale.*

Proof. We go back to the proof of Theorem (2.5) and define

$$\tilde{X}_t(\omega) = X_{t+}(\omega) \quad \text{if} \quad \omega \in \Omega_0, \qquad \tilde{X}_t(\omega) = 0 \quad \text{if} \quad \omega \notin \Omega_0.$$

The process $\tilde{X}$ is a right-continuous modification of X by Proposition (2.6). It is adapted to $(\mathscr{F}_t)$, since this filtration is right-continuous and complete and Ω_0^c is negligible. Thanks again to Proposition (2.6), $\tilde{X}$ is a submartingale with respect to $(\mathscr{F}_t)$ and finally by Theorem (2.5) its paths have left limits. □

These results will be put to use in the following chapter. We already observe that we can now answer a question raised in Sect. 1. If $(\mathscr{F}_t)$ is right-continuous and complete and Y is an integrable random variable, we may choose Y_t within the equivalence class of $E[Y \mid \mathscr{F}_t]$ in such a way that the resulting process is a cadlag martingale. The significance of these particular martingales will be seen in the next section.

From now on, unless otherwise stated, we will consider only right-continuous submartingales. For such a process, the inequality of Proposition (2.1) extends at once to

$$(b-a)E\left[D\left(X, \mathbb{R}_+, [a,b]\right)\right] \le \sup_t E\left[(X_t - b)^+\right]$$

and the same reasoning as in Theorem (2.2) leads to the *convergence theorem*:

(2.10) Theorem. *If $\sup_t E[X_t^+] < \infty$, then $\lim_{t\to\infty} X_t$ exists almost-surely.*

A particular case which is often used is the following

(2.11) Corollary. *A positive supermartingale converges a.s. as t goes to infinity.*

In a fashion similar to Theorem (2.3), there is also a convergence theorem as t goes to zero for submartingales defined on $]0,\infty[$. We leave the details to the reader.

The ideas and results of this section will be used in many places in the sequel. We close this section by a first application to Brownian motion. We retain the notation of Sect. 2 in Chap. I.

(2.12) Proposition. *If $\{\Delta_n\}$ is a sequence of refining (i.e. $\Delta_n \subset \Delta_{n+1}$) subdivisions of $[0,t]$ such that $|\Delta_n| \to 0$, then*

$$\lim_{n\to\infty} \sum_i \left(B_{t_{i+1}} - B_{t_i}\right)^2 = t \qquad \textit{almost-surely} \quad .$$

Proof. We use the Wiener space (see Sect. 3 Chap. I) as probability space and the Wiener measure as probability measure. If $0 = t_0 < t_1 < \ldots < t_k = t$ is a subdivision of $[0, t]$, for each sequence $\varepsilon = (\varepsilon_1, \ldots, \varepsilon_k)$ where $\varepsilon_i = \pm 1$, we define a mapping θ_ε on Ω by

$$\begin{aligned} \theta_\varepsilon\omega(0) &= 0, \\ \theta_\varepsilon\omega(s) &= \theta_\varepsilon\omega(t_{i-1}) + \varepsilon_i\,(\omega(s) - \omega(t_{i-1})) && \text{if } s \in \left[t_{i-1}, t_i\right], \\ \theta_\varepsilon\omega(s) &= \theta_\varepsilon\omega(t_k) + \omega(s) - \omega(t_k) && \text{if } s \geq t_k. \end{aligned}$$

Let $\mathscr{B}$ be the σ-field of events left invariant by all θ_ε's. It is easy to see that W is left invariant by all the θ_ε's as well. For any integrable r.v. Z on $\mathbf{W}$, we consequently have

$$E[Z \mid \mathscr{B}] = 2^{-k} \sum_\varepsilon Z \circ \theta_\varepsilon,$$

hence $E\left[\left(B_{t_i} - B_{t_{i-1}}\right)\left(B_{t_j} - B_{t_{j-1}}\right) \mid \mathscr{B}\right] = 0$ for $i \neq j$. If $\mathscr{B}_n$ is the σ-field corresponding to Δ_n, the family $\mathscr{B}_n$ is decreasing and moreover

$$E\left[B_t^2 \mid \mathscr{B}_n\right] = E\left[\sum_i \left(B_{t_i} - B_{t_{i-1}}\right)^2 \mid \mathscr{B}_n\right] = \sum_i \left(B_{t_i} - B_{t_{i-1}}\right)^2.$$

By Theorem (2.3), $\sum_i \left(B_{t_i} - B_{t_{i-1}}\right)^2$ converges a.s. and, as we already know that it converges to t in L^2, the proof is complete. □

\# **(2.13) Exercise.** 1°) Let $(\Omega, \mathscr{F}, P)$ be a probability space endowed with a filtration $(\mathscr{F}_n)$ such that $\sigma(\bigcup \mathscr{F}_n) = \mathscr{F}$. Let Q be another probability measure on $\mathscr{F}$ and X_n be the Radon-Nikodym derivative of the restriction of Q to $\mathscr{F}_n$ with respect to the restriction of P to $\mathscr{F}_n$.

Prove that (X_n) is a positive $(\mathscr{F}_n, P)$-supermartingale and that its limit X_∞ is the Radon-Nikodym derivative dQ/dP. If $Q \ll P$ on $\mathscr{F}$, then (X_n) is a martingale and $X_n = E[X_\infty \mid \mathscr{F}_n]$.

More on this matter will be said in Sect. 1 Chap. VIII.

2°) Let P be a transition probability (see Sect. 1 Chap. III) on a separable measurable space $(E, \mathscr{E})$ and λ be a probability measure on $\mathscr{E}$. Prove that there is a bimeasurable function f on $E \times E$ and a kernel N on $(E, \mathscr{E})$ such that for each x, the measure $N(x, \cdot)$ is singular with respect to λ and

$$P(x, A) = N(x, A) + \int_A f(x, y)\lambda(dy).$$

* **(2.14) Exercise (Dubins' inequality).** If $(X_n), n = 0, 1, \ldots$ is a positive supermartingale, prove, with the notation of the beginning of the section, that

$$P\left[D(X, \mathbb{N}, [a, b]) \geq p\right] \leq a^{p-1} b^{-p} E[X_0 \wedge b].$$

State and prove a similar result for upcrossings instead of downcrossings.

\# **(2.15) Exercise.** Let $(\Omega, \mathscr{F}, P)$ be a probability space and $(\mathscr{D}_n, n \geq 0)$ be a sequence of sub-σ-fields of $\mathscr{F}$ such that $\mathscr{D}_n \subset \mathscr{D}_m$, if $0 \leq m \leq n$. If $\mathscr{C}$ is another sub-σ-field of $\mathscr{F}$ independent of $\mathscr{D}_0$, prove that

$$\bigcap_n (\mathscr{C} \vee \mathscr{D}_n) = \mathscr{C} \vee \left(\bigcap_n \mathscr{D}_n \right)$$

up to P-negligible sets.

[Hint: Show that, if $C \in \mathscr{C}$, $D \in \mathscr{D}_0$, then $\lim_{n\to\infty} P(CD \mid \mathscr{C} \vee \mathscr{D}_n)$ belongs to $\mathscr{C} \vee \left(\bigcap_n \mathscr{D}_n\right)$.]

(2.16) Exercise. For the standard BM, set $\mathscr{G}_t = \sigma(B_u, u \geq t)$. Prove that for every real λ, the process $\exp\left\{(\lambda B_t/t) - (\lambda^2/2t)\right\}, t > 0$, is a martingale with respect to the decreasing family $(\mathscr{G}_t)$.

[Hint: Observe that $B_s - (s/t)B_t$ is independent of $\mathscr{G}_t$ for $s < t$ or use time-inversion.]

(2.17) Exercise. Suppose that we are given two filtrations $(\mathscr{F}_t^0)$ and $(\mathscr{F}_t)$ such that $\mathscr{F}_t^0 \subseteq \mathscr{F}_t$ for each t and these two σ-fields differ only by negligible sets of $\mathscr{F}_\infty$. Assume further that $(\mathscr{F}_t^0)$ is right-continuous.

1°) Show that every $(\mathscr{F}_t)$-adapted and right-continuous process is indistinguishable from a $(\mathscr{F}_t^0)$-adapted process.

2°) Show that a right-continuous $(\mathscr{F}_t)$-submartingale is indistinguishable from a cadlag $(\mathscr{F}_t^0)$-submartingale.

(2.18) Exercise (Krickeberg decomposition). A process X is said to be L^1-bounded or bounded in L^1 if there is a finite constant K such that for every $t \geq 0$, $E[|X_t|] \leq K$.

1°) If M is a L^1-bounded martingale, prove that for each t the limits

$$M_t^{(\pm)} = \lim_{n\to\infty} E\left[M_n^{\pm} \mid \mathscr{F}_t\right]$$

exist a.s. and the processes $M^{(\pm)}$ thus defined are positive martingales.

2°) If the filtration is right-continuous and complete, prove that a right-continuous martingale M is bounded in L^1 iff it can be written as the difference of two cadlag positive martingales $M^{(+)}$ and $M^{(-)}$.

3°) Prove that $M^{(+)}$ and $M^{(-)}$ may be chosen to satisfy

$$\sup_t E[|M_t|] = E\left[M_0^{(+)}\right] + E\left[M_0^{(-)}\right]$$

in which case the decomposition is unique (up to indistinguishability).

4°) The uniqueness property extends in the following way: if Y and Z are two positive martingales such that $M = Y - Z$, then $Y \geq M^{(+)}$ and $Z \geq M^{(-)}$ where $M^{(\pm)}$ are the martingales of 3°).

§3. Optional Stopping Theorem

We recall that all the (sub, super)martingales we consider henceforth are cadlag. In the sequel, we shall denote by $\mathscr{F}_\infty$ the σ-algebra $\bigvee_t \mathscr{F}_t$. In Theorem (2.9) of last section, the limit variable X_∞ is measurable with respect to $\mathscr{F}_\infty$. We want to know whether the process indexed by $\mathbb{R}_+ \cup \{+\infty\}$ obtained by adjoining X_∞ and $\mathscr{F}_\infty$ is still a (sub)martingale. The corresponding result is especially interesting for martingales and reads as follows.

(3.1) Theorem. *For a martingale X_t, $t \in \mathbb{R}_+$, the following three conditions are equivalent,*

i) $\lim_{t\to\infty} X_t$ *exists in the L^1-sense;*
ii) there exists a random variable X_∞ in L^1, such that $X_t = E\left[X_\infty \mid \mathscr{F}_t\right]$;
iii) the family $\{X_t, t \in \mathbb{R}_+\}$ is uniformly integrable.

If these conditions hold, then $X_\infty = \lim_{t\to\infty} X_t$ a.s. Moreover, if for some $p > 1$, the martingale is bounded in L^p, i.e. $\sup_t E[|X_t|^p] < \infty$, then the equivalent conditions above are satisfied and the convergence holds in the L^p-sense.

Proof. That ii) implies iii) is a classical exercise. Indeed, if we set $\Gamma_t = \left\{\left|E\left[X_\infty \mid \mathscr{F}_t\right]\right| > \alpha\right\}$,

$$a_t \stackrel{\text{def}}{=} \int_{\Gamma_t} \left|E\left[X_\infty \mid \mathscr{F}_t\right]\right| dP \le \int_{\Gamma_t} E\left[|X_\infty| \middle| \mathscr{F}_t\right] dP = \int_{\Gamma_t} |X_\infty| dP.$$

On the other hand, Markov's inequality implies

$$P(\Gamma_t) \le \frac{1}{\alpha} E\left[\left|E[X_\infty | \mathscr{F}_t]\right|\right] \le \frac{E[|X_\infty|]}{\alpha}.$$

It follows that, by taking α large, we can make a_t arbitrarily small independently of t.

If iii) holds, then the condition of Theorem (2.10) is satisfied and X_t converges to a r.v. X_∞ a.s., but since $\{X_t, t \in \mathbb{R}_+\}$ is uniformly integrable, the convergence holds in the L^1-sense so that i) is satisfied.

If i) is satisfied and since the conditional expectation is an L^1-continuous operator, passing to the limit as h goes to infinity in the equality

$$X_t = E\left[X_{t+h} \middle| \mathscr{F}_t\right]$$

yields ii).

Finally, if $\sup_t E[|X_t|^p] < \infty$, by Theorem (1.7), $\sup_t |X_t|$ is in L^p, and consequently the family $\{|X_t|^p, t \in \mathbb{R}_+\}$ is uniformly integrable. □

It is important to notice that, for $p > 1$, a martingale which is bounded in L^p is automatically uniformly integrable and its supremum is in L^p. For $p = 1$, the situation is altogether different. A martingale may be bounded in L^1 without being uniformly integrable, and may be uniformly integrable without belonging to $\mathbb{H}^1$, where $\mathbb{H}^1$ is the space of martingales with an integrable supremum (see Exercise (1.17)). An example of the former is provided by $\exp\{B_t - t/2\}$ where B is the BM; indeed, as B_t takes on negative values for arbitrarily large times, this martingale converges to zero a.s. as t goes to infinity, and thus, by the preceding theorem cannot be uniformly integrable. An example of the latter is given in Exercise (3.15).

The analogous result is true for sub and supermartingales with inequalities in ii); we leave as an exercise to the reader the task of stating and proving them.

We now turn to the *optional stopping theorem*, a first version of which was stated in Proposition (1.4). If X is a uniformly integrable martingale, then X_∞ exists a.s. and if S is a stopping time, we define X_S on $\{S = \infty\}$ by setting $X_S = X_\infty$.

(3.2) Theorem. *If X is a martingale and S, T are two bounded stopping times with $S \leq T$,*

$$X_S = E\left[X_T \middle| \mathscr{F}_S\right] \quad a.s.$$

If X is uniformly integrable, the family $\{X_S\}$ where S runs through the set of all stopping times is uniformly integrable and if $S \leq T$

$$X_S = E\left[X_T \middle| \mathscr{F}_S\right] = E\left[X_\infty \middle| \mathscr{F}_S\right] \quad a.s.$$

Remark. The two statements are actually the same, as a martingale defined on an interval which is closed on the right is uniformly integrable.

Proof. We prove the second statement. We recall from Proposition (1.4) that

$$X_S = E\left[X_T \middle| \mathscr{F}_S\right] = E\left[X_\infty \middle| \mathscr{F}_S\right]$$

if S and T take their values in a finite set and $S \leq T$. It is known that the family U of r.v.'s $E[X_\infty | \mathscr{B}]$ where $\mathscr{B}$ runs through all the sub-σ-fields of $\mathscr{F}$ is uniformly integrable. Its closure $\bar{U}$ in L^1 is still uniformly integrable. If S is any stopping time, there is a sequence S_k of stopping times decreasing to S and taking only finitely many values; by the right-continuity of X, we see that X_S also belongs to $\bar{U}$, which proves that the set $\{X_S,\ S \text{ stopping time}\}$ is uniformly integrable. As a result, we also see that X_{S_k} converges to X_S in L^1. If $\Gamma \in \mathscr{F}_S$, it belongs a fortiori to $\mathscr{F}_{S_k}$ and we have

$$\int_\Gamma X_{S_k} dP = \int_\Gamma X_\infty dP;$$

passing to the limit yields

$$\int_\Gamma X_S dP = \int_\Gamma X_\infty dP ;$$

in other words, $X_S = E[X_\infty \mid \mathscr{F}_S]$ which is the desired result. □

We insist on the importance of uniform integrability in the above theorem. Let X be a positive continuous martingale converging to zero and such that $X_0 = 1$, for instance $X_t = \exp(B_t - t/2)$; if for $\alpha < 1$, $T = \inf\{t : X_t \leq \alpha\}$ we have $X_T = \alpha$, hence $E[X_T] = \alpha$, whereas we should have $E[X_T] = E[X_0] = 1$ if the optional stopping theorem applied. Another interesting example with the same martingale is provided by the stopping times $d_t = \inf\{s > t : B_s = 0\}$. In this situation, all we have is an inequality as is more generally the case with positive supermartingales.

(3.3) Theorem. *If X is a positive right-continuous supermartingale and if we set $X_\infty = 0$, for any pair S, T of stopping times with $S \leq T$,*

$$X_S \geq E\left[X_T | \mathscr{F}_S\right].$$

Proof. Left to the reader as an exercise as well as analogous statements for sub-martingales. □

Before we proceed, let us observe that we have a hierarchy among the processes we have studied which is expressed by the following strict inclusions:

$$\text{supermartingales} \supset \text{martingales} \supset \text{uniformly integrable martingales} \supset \mathbb{H}^1.$$

We now turn to some applications of the optional stopping theorem.

(3.4) Proposition. *If X is a positive right-continuous supermartingale and*

$$T(\omega) = \inf\{t : X_t(\omega) = 0\} \wedge \inf\{t > 0 : X_{t-}(\omega) = 0\}$$

then, for almost every ω, $X_.(\omega)$ vanishes on $[T(\omega), \infty[$.

Proof. Let $T_n = \inf\{t : X_t \leq 1/n\}$; obviously, $T_{n-1} \leq T_n \leq T$. On $\{T_n = \infty\}$, a fortiori $T = \infty$ and there is nothing to prove. On $\{T_n < \infty\}$, we have $X_{T_n} \leq 1/n$. Let $q \in \mathbb{Q}_+$; $T + q$ is a stopping time $> T_n$ and, by the previous result,

$$1/n \geq E\left[X_{T_n} 1_{(T_n < \infty)}\right] \geq E\left[X_{T+q} 1_{(T_n < \infty)}\right].$$

Passing to the limit yields

$$E\left[X_{T+q} 1_{(T_n < \infty, \forall n)}\right] = 0.$$

Since $\{T < \infty\} \subset \{T_n < \infty, \forall n\}$, we finally get $X_{T+q} = 0$ a.s. on $\{T < \infty\}$. The proof is now easily completed. □

(3.5) Proposition. *A cadlag adapted process X is a martingale if and only if for every bounded stopping time T, the r.v. X_T is in L^1 and*

$$E[X_T] = E[X_0].$$

Proof. The "only if" part follows from the optional stopping theorem. Conversely, if $s < t$ and $A \in \mathscr{F}_s$ the r.v. $T = t 1_{A^c} + s 1_A$ is a stopping time and consequently

$$E[X_0] = E[X_T] = E[X_t 1_{A^c}] + E[X_s 1_A].$$

On the other hand, t itself is a stopping time, and

$$E[X_0] = E[X_t] = E[X_t 1_{A^c}] + E[X_t 1_A].$$

Comparing the two equalities yields $X_s = E[X_t \mid \mathscr{F}_s]$.

(3.6) Corollary. *If M is a martingale and T a stopping time, the stopped process M^T is a martingale with respect to $(\mathscr{F}_t)$.*

Proof. The process M^T is obviously cadlag and adapted and if S is a bounded stopping time, so is $S \wedge T$; hence

$$E\left[M_S^T\right] = E\left[M_{S\wedge T}\right] = E\left[M_0\right] = E\left[M_0^T\right].$$

□

Remarks. 1°) By applying the optional stopping theorem directly to M and to the stopping times $T \wedge s$ and $T \wedge t$, we would have found that M^T is a martingale but only with respect to the filtration $(\mathscr{F}_{T\wedge t})$. But actually, a martingale with respect to $(\mathscr{F}_{T\wedge t})$ is automatically a martingale with respect to $\mathscr{F}_t$.

2°) A property which is equivalent to the corollary is that the conditional expectations $E\left[\cdot \mid \mathscr{F}_S\right]$ and $E\left[\cdot \mid \mathscr{F}_T\right]$ commute and that $E\left[\cdot \mid \mathscr{F}_S \mid \mathscr{F}_T\right] = E\left[\cdot \mid \mathscr{F}_{S\wedge T}\right]$. The proof of this fact, which may be obtained also without referring to martingales, is left as an exercise to the reader.

Here again, we close this section with applications to the linear BM which we denote by B. If a is a positive real number, we set

$$T_a = \inf\{t > 0 : B_t = a\}, \qquad \tilde{T}_a = \inf\left\{t > 0 : \left|B_t\right| = a\right\};$$

thanks to the continuity of paths, these times could also be defined as

$$T_a = \inf\{t > 0 : B_t \geq a\}, \qquad \tilde{T}_a = \inf\left\{t > 0 : \left|B_t\right| \geq a\right\};$$

they are stopping times with respect to the natural filtration of B. Because of the recurrence of BM, they are a.s. finite.

(3.7) Proposition. *The Laplace transforms of the laws of T_a and $\tilde{T}_a$ are given by*

$$E\left[\exp\left(-\lambda T_a\right)\right] = \exp\left(-a\sqrt{2\lambda}\right), \quad E\left[\exp\left(-\lambda \tilde{T}_a\right)\right] = \left(\cosh\left(a\sqrt{2\lambda}\right)\right)^{-1}.$$

Proof. For $s \geq 0$, $M_t^s = \exp\left(sB_t - s^2t/2\right)$ is a martingale and consequently, $M_{t\wedge T_a}^s$ is a martingale bounded by e^{sa}. A bounded martingale is obviously uniformly integrable, and therefore, we may apply the optional stopping theorem to the effect that

$$E\left[M_{T_a}^s\right] = E\left[M_0^s\right] = 1,$$

which yields $E\left[\exp\left(-\frac{s^2}{2}T_a\right)\right] = e^{-sa}$, whence the first result follows by taking $\lambda = s^2/2$.

For the second result, the reasoning is the same using the martingale $N_t^s = \left(M_t^s + M_t^{-s}\right)/2 = \cosh(sB_t)\exp\left(-\frac{s^2}{2}t\right)$, as $N_{t\wedge\tilde{T}_a}^s$ is bounded by $\cosh(sa)$. □

Remark. By inverting its Laplace transform, we could prove that T_a has a law given by the density $a(2\pi x^3)^{-1/2}\exp(-a^2/2x)$, but this will be done by another method in the following chapter. We can already observe that

$$(\text{law of } T_a) * (\text{law of } T_b) = \text{law of } T_{a+b}.$$

The reason for that is the independence of T_a and $(T_{a+b} - T_a)$, which follows from the strong Markov property of BM proved in the following chapter.

Here is another application in which we call P_x the law of $x + B$.

(3.8) Proposition. *We have, for $a < x < b$,*

$$P_x\left[T_a < T_b\right] = \frac{b-x}{b-a}, \qquad P_x\left[T_b < T_a\right] = \frac{x-a}{b-a}.$$

Proof. By the recurrence of BM

$$P_x\left[T_a < T_b\right] + P_x\left[T_b < T_a\right] = 1.$$

On the other hand, $B^{T_a\wedge T_b}$ is a bounded martingale to which we can apply the optional stopping theorem to get, since $B_{T_a} = a$, $B_{T_b} = b$,

$$aP_x\left[T_a < T_b\right] + bP_x\left[T_b < T_a\right] = x.$$

We now have a linear system which we solve to get the desired result.

(3.9) Exercise. If X is a positive supermartingale such that $E[\lim_n X_n] = E[X_0] < \infty$, then X is a uniformly integrable martingale.

(3.10) Exercise. Let c and d be two strictly positive numbers, B a standard linear BM and set $T = T_c \wedge T_{-d}$.

1°) Prove that, for every real number s,

$$E\left[e^{-(s^2/2)T}1_{(T=T_c)}\right] = \sinh(sd)/\sinh\left(s(c+d)\right),$$

and derive therefrom another proof of Proposition (3.8). Prove that

$$E\left[\exp\left(-\frac{s^2}{2}T\right)\right] = \cosh\left(s(c-d)/2\right)/\cosh\left(s(c+d)/2\right),$$

and compare with the result in Proposition (3.7).

[Hint: Use the martingale $\exp\left(s\left(B_t - \frac{c-d}{2}\right) - \frac{s^2}{2}t\right)$.]

* 2°) Prove that for $0 \le s < \pi(c+d)^{-1}$,

$$E\left[\exp\left(\frac{s^2}{2}T\right)\right] = \cos\left(s(c-d)/2\right)/\cos\left(s(c+d)/2\right).$$

[Hint: Either use analytic continuation or use the complex martingale $\exp\left(is\left(B_t - \frac{c-d}{2}\right) + \frac{s^2}{2}t\right)$.]

(3.11) Exercise. 1°) With the notation of Proposition (3.7), if B is the standard linear BM, by considering the martingale $B_t^2 - t$, prove that $\tilde{T}_a$ is integrable and compute $E[\tilde{T}_a]$.

[Hint: To prove that $\tilde{T}_a \in L^1$, use the times $\tilde{T}_a \wedge n$.]

2°) Prove that T_a is not integrable.

[Hint: If it were, we would have $a = E\left[B_{T_a}\right] = 0$.]

3°) With the notation of Proposition (3.8), prove that

$$E_x\left[T_a \wedge T_b\right] = (x-a)(b-x).$$

This will be taken up in Exercise (2.8) in Chap. VI.

[Hint: This again can be proved using the martingale $B_t^2 - t$, but can also be derived from Exercise (3.10) 2°).]

\# **(3.12) Exercise.** Let M be a positive continuous martingale converging a.s. to zero as t goes to infinity. Put $M^* = \sup_t M_t$.

1°) For $x > 0$, prove that

$$P\left[M^* \ge x \middle| \mathscr{F}_0\right] = 1 \wedge (M_0/x).$$

[Hint: Stop the martingale when it first becomes larger than x.]

2°) More generally, if X is a positive $\mathscr{F}_0$-measurable r.v. prove that

$$P\left[M^* \ge X \middle| \mathscr{F}_0\right] = 1 \wedge (M_0/X).$$

Conclude that M_0 is the largest $\mathscr{F}_0$-measurable r.v. smaller than M^* and that $M^* \overset{(d)}{=} M_0/U$ where U is independent of M_0 and uniformly distributed on $[0, 1]$.

3°) If B is the BM started at $a > 0$ and $T_0 = \inf\{t : B_t = 0\}$, find the law of the r.v. $Y = \sup_{t<T_0} B_t$.

4°) Let B be the standard linear BM; using $M_t = \exp\left(2\mu\left(B_t - \mu t\right)\right)$, $\mu > 0$, prove that the r.v. $Y = \sup_t\left(B_t - \mu t\right)$ has an exponential density with parameter 2μ. The process $B_t - \mu t$ is called the *Brownian motion with drift* $(-\mu)$ and is further studied in Exercise (3.14).

* 5°) Prove that the r.v. J_2 of Exercise (1.21) Chap. I is integrable and compute the constant $C_2(X)$.

N.B. The questions 3°) through 5°) are independent from one another.

(3.13) Exercise. Let B be the standard linear BM and f be a locally bounded Borel function on $\mathbb{R}$.

1°) If $f(B_t)$ is a right-continuous martingale with respect to the filtration $(\mathscr{F}_t^0) = (\sigma(B_s, s \leq t))$, prove that f is an affine function (one could also make no assumption on f and suppose that $f(B_t)$ is a continuous $(\mathscr{F}_t^0)$-martingale). Observe that the assumption of right-continuity is essential; if f is the indicator function of the set of rational numbers, then $f(B)$ is a martingale.

2°) If we suppose that $f(B_t)$ is a continuous $(\mathscr{F}_t^0)$-submartingale, prove that f has no proper local maximum.

[Hint: For $c > 0$, use the stopping times $T = T_c \wedge T_{-1}$ and

$$S = \inf\{t \geq T : B_t = -1 \quad \text{or} \quad c + \varepsilon \quad \text{or} \quad c - \varepsilon\}.]$$

3°) In the situation of 2°), prove that f is convex.

[Hint: A continuous function is convex if and only if $f(x) + \alpha x + \beta$ has no proper local maximum for any α and β.]

* **(3.14) Exercise.** Let B be the standard linear BM and, for $a > 0$, set

$$\sigma_a = \inf\{t : B_t < t - a\}.$$

1°) Prove that σ_a is an a.s. finite stopping time and that $\lim_{a\to\infty} \sigma_a = +\infty$ a.s.

2°) Prove that $E\left[\exp\left(\frac{1}{2}\sigma_a\right)\right] = \exp(a)$.

[Hint: For $\lambda > 0$, use the martingale $\exp\left(-(\sqrt{1+2\lambda} - 1)(B_t - t) - \lambda t\right)$ stopped at σ_a to prove that $E\left[e^{-\lambda\sigma_a}\right] = \exp\left(-a\left(\sqrt{1+2\lambda} - 1\right)\right)$. Then, use analytic continuation.]

3°) Prove that the martingale $\exp\left(B_t - \frac{1}{2}t\right)$ stopped at σ_a is uniformly integrable.

4°) For $a > 0$ and $b > 0$, define now

$$\sigma_{a,b} = \inf\{t : B_t < bt - a\};$$

in particular, $\sigma_a = \sigma_{a,1}$. Prove that

$$E\left[\exp\left(\frac{1}{2}b^2\sigma_{a,b}\right)\right] = \exp(ab).$$

[Hint: Using the scaling property of BM, prove that $\sigma_{a,b} \overset{(d)}{=} b^{-2}\sigma_{ab,1}$.]

5°) For $b < 1$, prove that $E\left[\exp\left(\frac{1}{2}\sigma_{1,b}\right)\right] = +\infty$.

[Hint: Use 2°).]

* **(3.15) Exercise.** Let $(\Omega, \mathscr{F}, P)$ be $([0,1], \mathscr{B}([0,1]), d\omega)$ where $d\omega$ is the Lebesgue measure. For $0 \leq t < 1$, let $\mathscr{F}_t$ be the smallest sub-σ-field of $\mathscr{F}$ containing the Borel subsets of $[0, t]$ and the negligible sets of $[0, 1]$.

1°) For $f \in L^1([0,1], d\omega)$, give the explicit value of the right-continuous version of the martingale

$$X_t(\omega) = E\left[f \middle| \mathscr{F}_t\right](\omega), 0 \leq t < 1.$$

2°) Set $\tilde{H}f(t) = \frac{1}{1-t}\int_t^1 f(u)du$ and, for $p > 1$, prove *Hardy's L^p-inequality*

$$\left\|\tilde{H}f\right\|_p \leq \frac{p}{p-1}\|f\|_p.$$

[Hint: Use Doob's L^p-inequality.]

3°) Use the above set-up to give an example of a uniformly integrable martingale which is not in H^1.

4°) If $\int_0^1 |f(\omega)|\log^+|f(\omega)|d\omega < \infty$, check directly that $\tilde{H}f$ is integrable. Observe that this would equally follow from the continuous-time version of the result in Exercise (1.16).

** **(3.16) Exercise (BMO-martingales).** 1°) Let Y be a continuous uniformly integrable martingale. Prove that for any $p \in [1, \infty[$, the following two properties are equivalent:

i) there is a constant C such that for any stopping time T

$$E\left[|Y_\infty - Y_T|^p \mid \mathscr{F}_T\right] \leq C^p \quad \text{a.s.};$$

ii) there is a constant C such that for any stopping time T

$$E\left[|Y_\infty - Y_T|^p\right] \leq C^p P[T < \infty].$$

[Hint: Use the stopping time T_Γ of Exercise (4.19) Chap. I.]

The smallest constant for which this is true is the same in both cases and is denoted by $\|Y\|_{\text{BMO}_p}$. The space $\left\{Y : \|Y\|_{\text{BMO}_p} < \infty\right\}$ is called BMO_p and $\| \ \|_{\text{BMO}_p}$ is a semi-norm on this space. Prove that for $p < q$, $\text{BMO}_q \subseteq \text{BMO}_p$. The reverse inclusion will be proved in the following questions, so we will write simply BMO for this space.

2°) The conditions i) and ii) are also equivalent to

iii) There is a constant C such that for any stopping time T there is an $\mathscr{F}_T$-measurable, L^p-r.v. α_T such that

$$E\left[\left|Y_\infty - \alpha_T\right|^p \middle| \mathscr{F}_T\right] < C^p.$$

3°) If $Y_t = E\left[Y_\infty \mid \mathscr{F}_t\right]$ for a bounded r.v. Y_∞, then $Y \in \text{BMO}$ and $\|Y\|_{\text{BMO}_1} \leq 2\|Y_\infty\|_\infty$. Examples of unbounded martingales in BMO will be given in Exercise (3.30) of Chap. III.

4°) If $Y \in \text{BMO}$ and T is a stopping time, $Y^T \in \text{BMO}$ and $\left\|Y^T\right\|_{\text{BMO}_1} \leq \|Y\|_{BMO_1}$.

5°) **(The John-Nirenberg inequality).** Let $Y \in \text{BMO}$ and $\|Y\|_{BMO_1} \leq 1$. Let $a > 1$ and T be a stopping time and define inductively

$$R_0 = T,\ R_n = \inf\left\{t > R_{n-1} : \left|Y_t - Y_{R_{n-1}}\right| > a\right\};$$

prove that $P\left[R_n < \infty\right] \geq aP\left[R_{n+1} < \infty\right]$. Prove that there is a constant C such that for any T

$$P\left[\sup_{t\geq T}|Y_t - Y_T| > \lambda\right] \leq Ce^{-\lambda/e}P[T < \infty];$$

in particular, if $Y^* = \sup_t |Y_t|$,

$$P\left[Y^* \geq \lambda\right] \leq Ce^{-\lambda/e}.$$

As a result, Y^* is in L^p for every p.

[Hint: Apply the inequality $E\left[|Y_S - Y_T|\right] \leq \|Y\|_{BMO_1} P[T < \infty]$ which is valid for $S \geq T$, to the stopping times R_n and R_{n+1}.]

6°) Deduce from 5°) that BMO_p is the same for all p and that all the semi-norms $\|Y\|_{BMO_p}$ are equivalent.

** **(3.17) Exercise (Continuation of Exercise (1.17)). [The dual space of H^1].** 1°) We call *atom* a continuous martingale A for which there is a stopping time T such that

i) $A_t = 0$ for $t \leq T$; ii) $|A_t| \leq P[T < \infty]^{-1}$ for every t. Give examples of atoms and prove that each atom is in the unit ball of H^1.

2°) Let $X \in H^1$ and suppose that $X_0 = 0$; for every $p \in \mathbb{Z}$, define

$$T_p = \inf\left\{t : |X_t| > 2^p\right\}$$

and $C_p = 3 \cdot 2^p P[T_p < \infty]$. Prove that $A^p = \left(X^{T_{p+1}} - X^{T_p}\right)/C_p$ is an atom for each p and that $X = \sum_{-\infty}^{+\infty} C_p A^p$ in H_1. Moreover, $\sum_{-\infty}^{+\infty} \left|C_p\right| \leq 6\,\|X\|_{H^1}$.

3°) Let Y be a uniformly integrable continuous martingale. Prove that

$$\frac{1}{2}\|Y\|_{BMO_1} \leq \sup\{|E[A_\infty Y_\infty]|\,;\, A \quad \text{atom}\} \leq \|Y\|_{BMO_1}$$

and deduce that the dual space $(H^1)^*$ of H^1 is contained in BMO.

[Hint: For the last step, use the fact that the Hilbert space H^2 (Sect. 1 Chap. IV) is dense in H^1.]

4°) If X and Y are in H^2, prove *Fefferman's inequality*

$$|E[(X_\infty Y_\infty)]| \leq 6\,\|X\|_{H^1}\,\|Y\|_{\text{BMO}_1}$$

and deduce that $(H^1)^* = \text{BMO}$.

[Hint: Use 2°) and notice that $\left|\sum_{-N}^{N} C_p A^p\right| \leq 2X^*$.]

The reader will notice that if X is an arbitrary element in H^1 and Y an arbitrary element in BMO_1, we do not know the value taken on X by the linear form associated with Y. This question will be taken up in Exercise (4.24) Chap. IV.

* **(3.18) Exercise (Predictable stopping).** A stopping time T is said to be *predictable* if there exists an increasing sequence (T_n) of stopping times such that

i) $\lim_n T_n = T$

ii) $T_n < T$ for every n on $\{T > 0\}$. (See Sect. 5 Chap. IV)

If $X_t, t \in \mathbb{R}_+$, is a uniformly integrable martingale and if $S \leq T$ are two predictable stopping times prove that

$$X_{S-} = E\left[X_{T-}\middle|\mathscr{F}_{S-}\right] = E\left[X_T\middle|\mathscr{F}_{S-}\right]$$

[Hint: Use Exercise (4.18) 3°) Chap. I and Corollary (2.4).]

Notes and Comments

Sect. 1. The material covered in this section as well as in the following two is classical and goes back mainly to Doob (see Doob [1]). It has found its way in books too numerous to be listed here. Let us merely mention that we have made use of Dellacherie-Meyer [1] and Ikeda-Watanabe [2].

The law of the iterated logarithm is due, in varying contexts, to Khintchine [1], Kolmogorov [1] and Hartman-Wintner [1]. We have borrowed our proof from McKean [1], but the exponential inequality, sometimes called Bernstein's inequality, had been used previously in similar contexts. In connection with the law of the iterated logarithm, let us mention the Kolmogorov and Dvoretzky-Erdös tests which the reader will find in Itô-McKean [1] (see also Exercises (2.32) and (3.31) Chap. III).

Most exercises are classical. The class $L \log L$ was studied by Doob (see Doob [1]). For Exercise (1.20) see Walsh [6] and Orey-Taylor [1].

Sect. 2. The proof of Proposition (2.12) is taken from Neveu [2] and Exercise (2.14) is from Dubins [1]. The result in Exercise (2.13) which is important in some contexts, for instance in the study of Markov chains, comes from Doob [1]; it was one of the first applications of the convergence result for martingales. The relationship between martingales and derivation has been much further studied; the reader is referred to books centered on martingale theory.

Sect. 3. The optional stopping theorem and its applications to Brownian motion have also been well-known for a long time. Exercise (3.10) is taken from Itô-McKean [1] and Lépingle [2].

The series of exercises on H^1 and BMO of this and later sections are copied on Durrett [2] to which we refer for credits and for the history of the subject. The notion of atom appears in the martingale context in Bernard-Maisonneuve [1]. The example of Exercise (3.15) is from Dellacherie et al. [1].

Knight-Maisonneuve [1] show that the optional stopping property for every u.i. martingale characterizes stopping times; a related result is in Williams [14] (See Chaumont-Yor [1], Exercise 6.18).

Chapter III. Markov Processes

This chapter contains an introduction to Markov processes. Its relevance to our discussion stems from the fact that Brownian motion, as well as many processes which arise naturally in its study, are Markov processes; they even have the strong Markov property which is used in many applications. This chapter is also the occasion to introduce the Brownian filtrations which will appear frequently in the sequel.

§1. Basic Definitions

Intuitively speaking, a process X with state space $(E, \mathscr{E})$ is a Markov process if, to make a prediction at time s on what is going to happen in the future, it is useless to know anything more about the whole past up to time s than the present state X_s.

The minimal "past" of X at time s is the σ-algebra $\mathscr{F}_s^0 = \sigma(X_u, u \leq s)$. Let us think about the conditional probability

$$P\left[X_t \in A \mid \sigma\left(X_u, u \leq s\right)\right]$$

where $A \in \mathscr{E}$, $s < t$. If X is Markov in the intuitive sense described above, this should be a function of X_s, that is of the form $g(X_s)$ with g an $\mathscr{E}$-measurable function taking its values in $[0, 1]$. It would better be written $g_{s,t}$ to indicate its dependence on s and t. On the other hand, this conditional expectation depends on A and clearly, as a function of A, it ought to be a probability measure describing what chance there is of being in A at time t, knowing the state of the process at time s. We thus come to the idea that the above conditional expectation may be written $g_{s,t}(X_s, A)$ where, for each A, $x \to g_{s,t}(x, A)$ is measurable and for each x, $A \to g_{s,t}(x, A)$ is a probability measure. We now give precise definitions.

(1.1) Definition. *Let $(E, \mathscr{E})$ be a measurable space. A* kernel *N on E is a map from $E \times \mathscr{E}$ into $\mathbb{R}_+ \bigcup \{+\infty\}$ such that*

i) for every $x \in E$, the map $A \to N(x, A)$ is a positive measure on $\mathscr{E}$;
ii) for every $A \in \mathscr{E}$, the map $x \to N(x, A)$ is $\mathscr{E}$-measurable.

A kernel π is called a *transition probability* if $\pi(x, E) = 1$ for every $x \in E$. In a Markovian context, transition probabilities are often denoted P_i where i ranges through a suitable index set.

If $f \in \mathscr{E}_+$ and N is a kernel, we define a function Nf on E by

$$Nf(x) = \int_E N(x, dy) f(y).$$

It is easy to see that Nf is also in $\mathscr{E}_+$. If M and N are two kernels, then

$$MN(x, A) \stackrel{\text{def}}{=} \int_E M(x, dy) N(y, A)$$

is again a kernel. We leave the proof as an exercise to the reader.

A transition probability π provides the mechanism for a random motion in E which may be described as follows. If, at time zero, one starts from x, the position x_1 at time 1 will be chosen at random according to the probability $\pi(x, \cdot)$, the position x_2 at time 2 according to $\pi(x_1, \cdot)$, and so on and so forth. The process thus obtained is called a *homogeneous Markov chain* and a Markov process is a continuous-time version of this scheme.

Let us now suppose that we have a process X for which, for any $s < t$, there is a transition probability $P_{s,t}$ such that

$$P\left[X_t \in A \middle| \sigma(X_u, u \leq s)\right] = P_{s,t}(X_s, A) \quad \text{a.s.}$$

Then for any $f \in \mathscr{E}_+$, we have $E\left[f(X_t) \middle| \sigma(X_u, u \leq s)\right] = P_{s,t} f(X_s)$ as is proved by the usual arguments of linearity and monotonicity. Let $s < t < v$ be three numbers, then

$$\begin{aligned} P\left[X_v \in A \middle| \sigma(X_u, u \leq s)\right] &= P\left[X_v \in A \middle| \sigma(X_u, u \leq t) \middle| \sigma(X_u, u \leq s)\right] \\ &= E\left[P_{t,v}(X_t, A) \middle| \sigma(X_u, u \leq s)\right] \\ &= \int P_{s,t}(X_s, dy) P_{t,v}(y, A). \end{aligned}$$

But this conditional expectation should also be equal to $P_{s,v}(X_s, A)$. This leads us to the

(1.2) Definition. *A* transition function *(abbreviated t.f.) on $(E, \mathscr{E})$ is a family $P_{s,t}$, $0 \leq s < t$ of transition probabilities on $(E, \mathscr{E})$ such that for every three real numbers $s < t < v$, we have*

$$\int P_{s,t}(x, dy) P_{t,v}(y, A) = P_{s,v}(x, A)$$

for every $x \in E$ and $A \in \mathscr{E}$. This relation is known as the Chapman-Kolmogorov *equation. The t.f. is said to be* homogeneous *if $P_{s,t}$ depends on s and t only through the difference $t - s$. In that case, we write P_t for $P_{0,t}$ and the Chapman-Kolmogorov equation reads*

$$P_{t+s}(x, A) = \int P_s(x, dy) P_t(y, A)$$

for every $s, t \geq 0$; in other words, the family $\{P_t, t \geq 0\}$ forms a semi-group.

The reader will find in the exercises several important examples of transition functions. If we refer to the heuristic description of Markov processes given above, we see that in the case of homogeneous t.f.'s, the random mechanism by which the process evolves stays unchanged as time goes by, whereas in the non homogeneous case, the mechanism itself evolves.

We are now ready for our basic definition.

(1.3) Definition. *Let $(\Omega, \mathscr{F}, (\mathscr{G}_t), Q)$ be a filtered probability space; an adapted process X is a* Markov process *with respect to $(\mathscr{G}_t)$, with transition function $P_{s,t}$ if for any $f \in \mathscr{E}_+$ and any pair (s,t) with $s < t$,*

$$E\left[f(X_t)\middle|\mathscr{G}_s\right] = P_{s,t}f(X_s) \qquad Q\text{-a.s.}$$

The probability measure $X_0(Q)$ is called the initial distribution *of X. The process is said to be* homogeneous *if the t.f. is homogeneous in which case the above equality reads*

$$E\left[f(X_t)\middle|\mathscr{G}_s\right] = P_{t-s}f(X_s).$$

Let us remark that, if X is Markov with respect to $(\mathscr{G}_t)$, it is Markov with respect to the natural filtration $(\mathscr{F}_t^0) = (\sigma(X_u, u \le t))$. If we say that X is Markov without specifying the filtration, it will mean that we use $(\mathscr{F}_t^0)$. Let us also stress the importance of Q in this definition; if we alter Q, there is no reason why X should still be a Markov process. By Exercise (1.16) Chap. I, the *Brownian motion is a Markov process*, which should come as no surprise because of the independence of its increments, but this will be shown as a particular case of a result in Sect. 2.

Our next task is to establish the existence of Markov processes. We will need the following

(1.4) Proposition. *A process X is a Markov process with respect to $(\mathscr{F}_t^0) = \left(\sigma(X_u, u \le t)\right)$ with t.f. $P_{s,t}$ and initial measure ν if and only if for any $0 = t_0 < t_1 < \ldots < t_k$ and $f_i \in \mathscr{E}_+$,*

$$E\left[\prod_{i=0}^{k} f_i\left(X_{t_i}\right)\right] = \int_E \nu(dx_0) f_0(x_0) \int_E P_{0,t_1}(x_0, dx_1) f_1(x_1) \ldots \int_E P_{t_{k-1},t_k}(x_{k-1}, dx_k) f_k(x_k).$$

Proof. Let us first suppose that X is Markov. We can write

$$\begin{aligned} E\left[\prod_{i=0}^{k} f_i\left(X_{t_i}\right)\right] &= E\left[\prod_{i=0}^{k-1} f_i\left(X_{t_i}\right) E\left[f_k\left(X_{t_k}\right)\middle|\mathscr{F}^0_{t_{k-1}}\right]\right] \\ &= E\left[\prod_{i=0}^{k-1} f_i\left(X_{t_i}\right) P_{t_{k-1},t_k} f_k\left(X_{t_{k-1}}\right)\right]; \end{aligned}$$

this expression is the same as the first one, but with one function less and f_{k-1} replaced by $f_{k-1}P_{t_{k-1},t_k}f_k$; proceeding inductively, we get the formula of the statement.

Conversely, to prove that X is Markov, it is enough, by the monotone class theorem, to show that for times $t_1 < t_2 < \ldots < t_k \leq t < v$ and functions $f_1, \ldots, f_k, g$

$$E\left[\prod_{i=0}^{k} f_i\left(X_{t_i}\right) g(X_v)\right] = E\left[\prod_{i=0}^{k} f_i\left(X_{t_i}\right) P_{t,v}g(X_t)\right];$$

but this equality follows readily by applying the equality of the statement to both sides.

Remark. The forbiddingly looking formula in the statement is in fact quite intuitive. It may be written more loosely as

$$Q\left[X_{t_0} \in dx_0, X_{t_1} \in dx_1, \ldots, X_{t_k} \in dx_k\right] = \nu(dx_0)P_{0,t_1}(x_0, dx_1)\ldots P_{t_{k-1},t_k}(x_{k-1}, dx_k)$$

and means that the initial position x_0 of the process is chosen according to the probability measure ν, then the position x_1 at time t_1 according to $P_{0,t_1}(x_0, \cdot)$ and so on and so forth; this is the continuous version of the scheme described after Definition (1.1).

We now construct a canonical version of a Markov process with a given t.f. Indeed, by the above proposition, if we know the t.f. of a Markov process, we know the family of its finite-dimensional distributions to which we can apply the Kolmogorov extension theorem.

From now on, we suppose that $(E, \mathscr{E})$ is a Polish space endowed with the σ-field of Borel subsets. This hypothesis is in fact only used in Theorem (1.5) below and the rest of this section can be done without using it. We set $\Omega = E^{\mathbb{R}_+}$, $\mathscr{F}_\infty^0 = \mathscr{E}^{\mathbb{R}_+}$ and $\mathscr{F}_t^0 = \sigma(X_u, u \leq t)$ where X is the coordinate process.

(1.5) Theorem. *Given a transition function $P_{s,t}$ on $(E, \mathscr{E})$, for any probability measure ν on $(E, \mathscr{E})$, there is a unique probability measure P_ν on $(\Omega, \mathscr{F}_\infty^0)$ such that X is Markov with respect to $(\mathscr{F}_t^0)$ with transition function $P_{s,t}$ and initial measure ν.*

Proof. We define a projective family of measures by setting

$$P_\nu^{t_1,\ldots,t_n}(A_0 \times A_1 \times \ldots \times A_n) = \int_{A_0} \nu(dx_0) \int_{A_1} P_{0,t_1}(x_0, dx_1) \int_{A_2} P_{t_1,t_2}(x_1, dx_2) \ldots \int_{A_n} P_{t_{n-1},t_n}(x_{n-1}, dx_n)$$

and we then apply the Kolmogorov extension theorem. By Proposition (1.4), the coordinate process X is Markov for the resulting probability measure P_ν. □

From now on, unless otherwise stated, we will consider only homogeneous transition functions and processes. In this case, we have

(eq. (1.1)) $$P_\nu\left[X_0 \in A_0, X_{t_1} \in A_1, \dots, X_{t_n} \in A_n\right] =$$

$$\int_{A_0} \nu(dx) \int_{A_1} P_{t_1}(x, dx_1) \int_{A_2} P_{t_2-t_1}(x_1, dx_2) \dots \int_{A_n} P_{t_n-t_{n-1}}(x_{n-1}, dx_n).$$

Thus, for each x, we have a probability measure P_{ε_x} which we will denote simply by P_x. If Z is an $\mathscr{F}_\infty^0$-measurable and positive r.v., its mathematical expectation with respect to P_x (resp. P_ν) will be denoted by $E_x[Z]$ (resp. $E_\nu[Z]$). If, in particular, Z is the indicator function of a rectangle all components of which are equal to E with the exception of the component over t, that is to say, $Z = 1_{\{X_t \in A\}}$ for some $A \in \mathscr{E}$, we get

$$P_x[X_t \in A] = P_t(x, A).$$

This reads: the probability that the process started at x is in A at time t is given by the value $P_t(x, A)$ of the t.f. It proves in particular that $x \to P_x[X_t \in A]$ is measurable. More generally, we have the

(1.6) Proposition. *If Z is $\mathscr{F}_\infty^0$-measurable and positive or bounded, the map $x \to E_x[Z]$ is $\mathscr{E}$-measurable and*

$$E_\nu[Z] = \int_E \nu(dx) E_x[Z].$$

Proof. The collection of sets Γ in $\mathscr{F}_\infty^0$ such that the proposition is true for $Z = 1_\Gamma$ is a monotone class. On the other hand, if $\Gamma = \left\{X_0 \in A_0, X_{t_1} \in A_1, \dots, X_{t_n} \in A_n\right\}$, then $P_x[\Gamma]$ is given by eq.(1.1) with $\nu = \varepsilon_x$ and it is not hard to prove inductively that this is an $\mathscr{E}$-measurable function of x; by the monotone class theorem, the proposition is true for all sets $\Gamma \in \mathscr{F}_\infty^0$. It is then true for simple functions and, by taking increasing limits, for any $Z \in (\mathscr{F}_\infty^0)_+$. □

Remark. In the case of BM^d, the family of probability measures P_ν was already introduced in Exercise (3.14) Chap. I.

In accordance with Definition (4.12) in Chap. I, we shall say that a property of the paths ω holds *almost surely* if the set where it holds has P_ν-probability 1 for every ν; clearly, it is actually enough that it has P_x-probability 1 for every x in E.

Using the translation operators of Sect. 3 Chap. I, we now give a handy form of the Markov property.

(1.7) Proposition (Markov property). *If Z is $\mathscr{F}_\infty^0$-measurable and positive (or bounded), for every $t > 0$ and starting measure ν,*

$$E_\nu\left[Z \circ \theta_t \middle| \mathscr{F}_t^0\right] = E_{X_t}[Z] \qquad P_\nu\text{-a.s.}$$

The right-hand side of this formula is the r.v. obtained by composing the two measurable maps $\omega \to X_t(\omega)$ and $x \to E_x[Z]$, and the formula says that this r.v. is within the equivalence class of the left-hand side. The reader will notice that, by the very definition of θ_t, the r.v. $Z \circ \theta_t$ depends only on the future after time t; its conditional expectation with respect to the past is a function of the present state X_t as it should be. If, in particular, we take $Z = 1_{\{X_s \in A\}}$, the above formula reads

$$P_\nu\left[X_{t+s} \in A \middle| \mathscr{F}_t^0\right] = P_{X_t}[X_s \in A] = P_s(X_t, A)$$

which is the formula of Definition (1.3).

Moreover, it is important to observe that the Markov property as stated in Proposition (1.7) is a property of the *family* of probability measures P_x, $x \in E$.

Proof of Proposition (1.7). We must prove that for any $\mathscr{F}_t^0$-measurable and positive Y,

$$E_\nu[Z \circ \theta_t \cdot Y] = E_\nu\left[E_{X_t}[Z] \cdot Y\right].$$

By the usual extension arguments, it is enough to prove this equality when $Y = \prod_{i=1}^k f_i(X_{t_i})$ with $f_i \in \mathscr{E}_+$ and $t_i \leq t$ and $Z = \prod_{j=1}^n g_j(X_{t_j})$ where $g_j \in \mathscr{E}_+$, but in that case, the equality follows readily from Proposition (1.4). □

We now remove a restriction on P_t. It was assumed so far that $P_t(x, E) = 1$, but there are interesting cases where $P_t(x, E) < 1$ for some x's and t's. We will say that P_t is *Markovian* in the former case, submarkovian in the general case i.e. when $P_t(x, E)$ may be less than one. If we think of a Markov process as describing the random motion of a particle, the submarkovian case corresponds to the possibility of the particle disappearing or dying in a finite time.

There is a simple trick which allows to turn the submarkovian case into the Markovian case studied so far. We adjoin to the state space E a new point Δ called the *cemetery* and we set $E_\Delta = E \cup \{\Delta\}$ and $\mathscr{E}_\Delta = \sigma(\mathscr{E}, \{\Delta\})$. We now define a new t.f. $\tilde{P}$ on $(E_\Delta, \mathscr{E}_\Delta)$ by

$$\begin{aligned} \tilde{P}_t(x, A) &= P_t(x, A) \qquad \text{if} \quad A \subset E, \\ \tilde{P}_t(x, \{\Delta\}) &= 1 - P_t(x, E), \qquad \tilde{P}_t(\Delta, \{\Delta\}) = 1. \end{aligned}$$

In the sequel, we will not distinguish in our notation between P_t and $\tilde{P}_t$ and in the cases of interest for us Δ will be *absorbing*, namely, the process started at Δ will stay in Δ.

By *convention*, all the functions on E will be extended to E_Δ by setting $f(\Delta) = 0$. Accordingly, the Markov property must then be stated

$$E_\nu\left[Z \circ \theta_t \middle| \mathscr{F}_t^0\right] = E_{X_t}(Z) \qquad P_\nu\text{-a.s. on the set} \quad \{X_t \neq \Delta\},$$

because the convention implies that the right-hand side vanishes on $\{X_t = \Delta\}$ and there is no reason for the left-hand side to do so.

Finally, as in Sect. 1 of Chap. I, we must observe that we cannot go much further with the Markov processes thus constructed. Neither the paths of X nor the filtration $(\mathscr{F}_t^0)$ have good enough properties. Therefore, we will devote the following section to a special class of Markov processes for which there exist good versions.

(1.8) Exercise. Prove that the following families of kernels are homogeneous t.f.'s
(i) (Uniform translation to the right at speed v) $E = \mathbb{R}$, $\mathscr{E} = \mathscr{B}(\mathbb{R})$;

$$P_t(x, \cdot) = \varepsilon_{x+vt}.$$

(ii) (Brownian motion) $E = \mathbb{R}$, $\mathscr{E} = \mathscr{B}(\mathbb{R})$; $P_t(x, \cdot)$ is the probability measure with density

$$g_t(y - x) = (2\pi t)^{-1/2} \exp\left(- (y - x)^2/2t\right).$$

(iii) (Poisson process). $E = \mathbb{R}$, $\mathscr{E} = \mathscr{B}(\mathbb{R})$;

$$P_t(x, dy) = \sum_0^\infty \left(e^{-t}t^n/n!\right) \varepsilon_{x+n}(dy).$$

This example can be generalized as follows: Let π be a transition probability on a space $(E, \mathscr{E})$; prove that one can define inductively a transition probability π^n by

$$\pi^n(x, A) = \int \pi(x, dy)\pi^{n-1}(y, A).$$

Then

$$P_t(x, dy) = \sum_0^\infty \left(e^{-t}t^n/n!\right) \pi^n(x, dy)$$

is a transition function. Describe the corresponding motion.

(1.9) Exercise. Show that the following two families of kernels are Markovian transition functions on $(\mathbb{R}_+, \mathscr{B}(\mathbb{R}_+))$:
(i) $P_t f(x) = \exp(-t/x) f(x) + \int_x^\infty ty^{-2} \exp(-t/y) f(y)dy$
(ii) $Q_t f(x) = \left(x/(x + t)\right) f(x + t) + \int_x^\infty t(t + y)^{-2} f(t + y)dy$.

(1.10) Exercise (Space-time Markov processes). If X is an inhomogeneous Markov process, prove that the process (t, X_t) with state space $(\mathbb{R}_+ \times E)$ is a homogeneous Markov process called the "space-time" process associated with X. Write down its t.f. For example, the heat process (see Sect. 1 Chap. I) is the space-time process associated with BM.

(1.11) Exercise. Let X be a Markov process with t.f. (P_t) and f a bounded Borel function. Prove that $(P_{t-s} f(X_s), s \leq t)$ is a P_x-martingale for any x.

(1.12) Exercise. Let X be the linear BM and set $X_t = \int_0^t B_s ds$. Prove that X is not a Markov process but that the pair (B, X) is a Markov process with state space $\mathbb{R}^2$. This exercise is taken up in greater generality in Sect. 1 of Chap. X.

(1.13) Exercise (Gaussian Markov processes). 1°) Prove that a centered Gaussian process X_t, $t \geq 0$, is a Markov process if and only if its covariance satisfies the equality

$$\Gamma(s,u)\Gamma(t,t) = \Gamma(s,t)\Gamma(t,u)$$

for every $s < t < u$.

If $\Gamma(t,t) = 0$, the processes $(X_s, s \leq t)$ and $(X_s, s \geq t)$ are independent. The process $B_t - tB_1$, $t \geq 0$ (the restriction of which to $[0,1]$ is a Brownian Bridge) is an example of such a process for which $\Gamma(t,t)$ vanishes at $t = 1$. The process Y of Exercise (1.14) Chap. I is another example of a centered Gaussian Markov process.

2°) If Γ is continuous on $\mathbb{R}_+^2$ and > 0, prove that $\Gamma(s,t) = a(s)a(t)\rho(\inf(s,t))$ where a is continuous and does not vanish and ρ is continuous, strictly positive and non decreasing. Prove that $(X_t/a(t), t \geq 0)$ is a Gaussian martingale.

3°) If a and ρ are as above, and B is a BM defined on the interval $[\rho(0), \rho(\infty)[$, the process $Y_t = a(t)B_{\rho(t)}$ is a Gaussian process with the covariance Γ of 2°). Prove that the Gaussian space generated by Y is isomorphic to the space $L^2(\mathbb{R}_+, d\rho)$, the r.v. Y_t corresponding to the function $a(t)1_{[0,t]}$.

4°) Prove that the only stationary Gaussian Markov processes are the stationary OU processes of parameter β and size c (see Sect. 3 Chap. I). Prove that their transition functions are given by the densities

$$p_t(x,y) = \left(2\pi c\left(1 - e^{-2\beta t}\right)\right)^{-1/2} \exp\left(-\left(y - e^{-\beta t}x\right)^2 / 2c\left(1 - e^{-2\beta t}\right)\right).$$

Give also the initial measure m and check that it is invariant (Sect. 3 Chap. X) as it should be since the process is stationary. Observe also that $\lim_{t\to\infty} P_t(x,A) = m(A)$.

5°) The OU processes (without the qualifying "stationary") with parameter β and size c are the Markov processes with the above transition functions. Which condition must satisfy the initial measure ν in order that X is still a Gaussian process under P_ν? Compute its covariance in that case.

6°) If u and v are two continuous functions which do not vanish, then

$$\Gamma(s,t) = u\left(\inf(s,t)\right)v\left(\sup(s,t)\right)$$

is a covariance if and only if u/v is strictly positive and non decreasing. This question is independent of the last three.

(1.14) Exercise. 1°) If B is the linear BM, prove that $|B|$ is, for any probability measure P_ν, a homogeneous Markov process on $[0,\infty[$ with transition function given by the density

$$\frac{1}{\sqrt{2\pi t}}\left[\exp\left(-\frac{1}{2t}(y-x)^2\right) + \exp\left(-\frac{1}{2t}(y+x)^2\right)\right].$$

This is the BM *reflected* at 0. See Exercise (1.17) for a more general result.

2°) More generally, prove that, for every integer d, the modulus of BM^d is a Markov process. (This question is solved in Sect. 3 Chap. VI).

* 3°) Define the linear BM reflected at 0 and 1; prove that it is a homogeneous Markov process and compute its transition function.

[Hint: The process may be defined as $X_t = |B_t - 2n|$ on $\{|B_t - 2n| \leq 1\}$.] The questions 2°) and 3°) are independent.

\# **(1.15) Exercise (Killed Brownian motion).** 1°) Prove that the densities

$$\frac{1}{\sqrt{2\pi t}}\left[\exp\left(-\frac{1}{2t}(y-x)^2\right) - \exp\left(-\frac{1}{2t}(y+x)^2\right)\right], \quad x > 0, y > 0,$$

define a submarkovian transition semi-group Q_t on $]0, \infty[$. This is the transition function of the BM *killed* when it reaches 0 as is observed in Exercise (3.29).

2°) Prove that the identity function is invariant under Q_t, in other words, $\int_0^\infty Q_t(x, dy)y = x$. As a result, the operators H_t defined by

$$H_t f(x) = \frac{1}{x}\int_0^\infty Q_t(x, dy)yf(y)$$

also form a semi-group. It may be extended to $[0, \infty[$ by setting

$$H_t(0, dy) = (2/\pi t^3)^{1/2} y^2 \exp(-y^2/2t)dy.$$

This semi-group is that of the Bessel process of dimension 3, which will be studied in Chap. VI and will play an important role in the last parts of this book.

\# **(1.16) Exercise (Transition function of the skew BM).** Let $0 \leq \alpha \leq 1$ and g_t be the transition density (i.e. the density of the t.f. with respect to the Lebesgue measure) of BM. Prove that the following function is a transition density

$$p_t^\alpha(0, y) = 2\alpha g_t(y)1_{(y>0)} + 2(1-\alpha)g_t(y)1_{(y<0)},$$

$$\begin{aligned} p_t^\alpha(x, y) &= 1_{(x>0)}\left[\left(g_t(y-x) + (2\alpha - 1)g_t(y+x)\right)1_{(y>0)} + 2(1-\alpha)g_t(y-x)1_{(y<0)}\right] \\ &+ 1_{(x<0)}\left[\left(g_t(y-x) + (1-2\alpha)g_t(y+x)\right)1_{(y<0)} + 2\alpha g_t(y-x)1_{(y>0)}\right]. \end{aligned}$$

What do we get in the special cases $\alpha = 0$, $\alpha = 1$ and $\alpha = 1/2$?

\# **(1.17) Exercise (Images of Markov processes).** 1°) Let X be a Markov process with t.f. (P_t) and ϕ a Borel function from $(E, \mathscr{E})$ into a space $(E', \mathscr{E}')$ such that $\phi(A) \in \mathscr{E}'$ for every $A \in \mathscr{E}$. If moreover, for every t and every $A' \in \mathscr{E}'$

$$P_t\left(x, \phi^{-1}(A')\right) = P_t\left(y, \phi^{-1}(A')\right) \quad \text{whenever} \quad \phi(x) = \phi(y),$$

then the process $X'_t = \phi(X_t)$ is under P_x, $x \in E$, a Markov process with state space $(E', \mathscr{E}')$. See Exercise (1.14) for the particular case of BM reflected at 0.

2°) Let $X =$ BM^d and ϕ be a rotation in $\mathbb{R}^d$ with center x. For $\omega \in \Omega$, define $\tilde{\phi}(\omega)$ by $X_t(\tilde{\phi}(\omega)) = \phi(X_t(\omega))$. Prove that $\tilde{\phi}$ is measurable and for any $\Gamma \in \mathscr{F}_\infty^0$

$$P_x[\Gamma] = P_x\left[\tilde{\phi}^{-1}(\Gamma)\right].$$

3°) Set $T_r = \inf\{t > 0 : |X_t - X_0| \geq r\}$ and prove that T_r and X_{T_r} are independent. Moreover, under P_x, the law of X_{T_r} is the uniform distribution on the sphere centered at x of radius r.

[Hint: Use the fact that the uniform distribution on the sphere is the only probability distribution on the sphere which is invariant by all rotations.]

These questions will be taken up in Sect. 3 Chap. VIII.

§2. Feller Processes

We recall that all the t.f.'s and processes we consider are time-homogeneous. Let E be a LCCB space and $C_0(E)$ be the space of continuous functions on E which vanish at infinity. We will write simply C_0 when there is no risk of mistake. We recall that a positive operator maps positive functions into positive functions.

(2.1) Definition. *A* Feller semi-group *on $C_0(E)$ is a family T_t, $t \geq 0$, of positive linear operators on $C_0(E)$ such that*

i) $T_0 = Id$ *and* $\|T_t\| \leq 1$ *for every* t;
ii) $T_{t+s} = T_t \circ T_s$ *for any pair* $s, t \geq 0$;
iii) $\lim_{t\downarrow 0} \|T_t f - f\| = 0$ *for every* $f \in C_0(E)$.

The relevance to our discussion of this definition is given by the

(2.2) Proposition. *With each Feller semi-group on E, one can associate a unique homogeneous transition function P_t, $t \geq 0$ on $(E, \mathscr{E})$ such that*

$$T_t f(x) = P_t f(x)$$

for every $f \in C_0$, and every x in E.

Proof. For any $x \in E$, the map $f \to T_t f(x)$ is a positive linear form on C_0; by Riesz's theorem, there exists a measure $P_t(x, \cdot)$ on $\mathscr{E}$ such that

$$T_t f(x) = \int P_t(x, dy) f(y)$$

for every $f \in C_0$. The map $x \to \int P_t(x, dy) f(y)$ is in C_0, hence is Borel, and, by the monotone class theorem, it follows that $x \to P_t(x, A)$ is Borel for any $A \in \mathscr{E}$. Thus we have defined transition probabilities P_t. That they form a t.f. follows from the semi-group property of T_t (Property ii)) and another application of the monotone class theorem. □

(2.3) Definition. *A t.f. associated to a Feller semi-group is called a* Feller transition function.

With the possible exception of the generalized Poisson process, all the t.f.'s of Exercise (1.8) are Feller t.f.'s. To check this, it is easier to have at one's disposal the following proposition which shows that the continuity property iii) in Definition (2.1) is actually equivalent to a seemingly weaker condition.

(2.4) Proposition. *A t.f. is Feller if and only if*

i) $P_t C_0 \subset C_0$ *for each* t*;*
ii) $\forall f \in C_0, \forall x \in E, \lim_{t\downarrow 0} P_t f(x) = f(x)$.

Proof. Of course, only the sufficiency is to be shown. If $f \in C_0$, $P_t f$ is also in C_0 by i) and so $\lim_{s\downarrow 0} P_{t+s} f(x) = P_t f(x)$ for every x by ii). The function $(t, x) \to P_t f(x)$ is thus right-continuous in t and therefore measurable on $\mathbb{R}_+ \times E$. Therefore, the function

$$x \to U_p f(x) = \int_0^\infty e^{-pt} P_t f(x) dt, \qquad p > 0,$$

is measurable and by ii),

$$\lim_{p\to\infty} pU_p f(x) = f(x).$$

Moreover, $U_p f \in C_0$, since one easily checks that whenever $x_n \to x$ (resp. the point at infinity whenever E is not compact), then $U_p f(x_n) \to U_p f(x)$ (resp. 0). The map $f \to U_p f$ is called the *resolvent of order* p *of the semi-group* P_t and satisfies the *resolvent equation*

$$U_p f - U_q f = (q - p) U_p U_q f = (q - p) U_q U_p f$$

as is easily checked. As a result, the image $D = U_p(C_0)$ of U_p does not depend on $p > 0$. Finally $\|pU_p f\| \leq \|f\|$.

We then observe that D is dense in C_0; indeed if μ is a bounded measure vanishing on D, then for any $f \in C_0$, by the dominated convergence theorem,

$$\int f d\mu = \lim_{p\to\infty} \int pU_p f \, d\mu = 0$$

so that $\mu = 0$. Now, an application of Fubini's theorem shows that

$$P_t U_p f(x) = e^{pt} \int_t^\infty e^{-ps} P_s f(x) ds$$

hence

$$\left\| P_t U_p f - U_p f \right\| \leq (e^{pt} - 1) \left\| U_p f \right\| + t \|f\|.$$

It follows that $\lim_{t\downarrow 0} \|P_t f - f\| = 0$ for $f \in D$ and the proof is completed by means of a routine density argument. □

By Fubini's theorem, it is easily seen that the resolvent U_p is given by a *kernel* which will also be denoted by U_p that is, for $f \in C_0$,

$$U_p f(x) = \int U_p(x, dy) f(y).$$

For every $x \in E$, $U_p(x, E) \leq 1/p$ and these kernels satisfy the *resolvent equation*

$$\begin{aligned} U_p(x, A) - U_q(x, A) &= (q-p) \int U_p(x, dy) U_q(y, A) \\ &= (q-p) \int U_q(x, dy) U_p(y, A). \end{aligned}$$

One can also check that for $f \in C_0$, $\lim_{p \to \infty} \|pU_p f - f\| = 0$. Indeed

$$\begin{aligned} \|pU_p f - f\| &= \sup_x |pU_p f(x) - f(x)| \\ &\leq \sup_x \int_0^\infty pe^{-pt} |P_t f(x) - f(x)| dt \\ &\leq \int_0^\infty e^{-s} \|P_{s/p} f - f\| ds \end{aligned}$$

which converges to 0 by the property iii) of Definition (2.1) and Lebesgue's theorem. The resolvent is actually the Laplace transform of the semi-group and therefore properties of the semi-group at 0 translate to properties of the resolvent at infinity.

Basic examples of Feller semi-groups will be given later on in this section and in the exercises.

(2.5) Definition. *A Markov process having a Feller transition function is called a* Feller process.

From now on, we work with the canonical version X of a Feller process for which we will show the existence of a good modification.

(2.6) Proposition. *For any α and any $f \in C_0^+$, the process $e^{-\alpha t} U_\alpha f(X_t)$ is a supermartingale for the filtration $(\mathscr{F}_t^0)$ and any probability measure P_ν.*

Proof. By the Markov property of Proposition (1.7), we have for $s < t$

$$E_\nu \left[e^{-\alpha t} U_\alpha f(X_t) \middle| \mathscr{F}_s^0 \right] = e^{-\alpha t} E_\nu \left[U_\alpha f\left(X_{t-s} \circ \theta_s\right) \middle| \mathscr{F}_s^0 \right] = e^{-\alpha t} P_{t-s} U_\alpha f(X_s).$$

But it is easily seen that $e^{-\alpha(t-s)} P_{t-s} U_\alpha f \leq U_\alpha f$ everywhere so that

$$E_\nu \left[e^{-\alpha t} U_\alpha f(X_t) \middle| \mathscr{F}_t^0 \right] \leq e^{-\alpha s} U_\alpha f(X_s)$$

which is our claim. □

We now come to one of the main results of this section. From now on, we always assume that E_Δ is the one-point compactification of E if E is not compact, the point Δ being the point at infinity and an isolated point in E_Δ if E is compact. We recall (Sect. 3 Chap. I) that an E_Δ-valued cadlag function is a function on $\mathbb{R}_+$ which is right-continuous and has left limits on $]0, \infty[$ with respect to this topology on E_Δ.

(2.7) Theorem. *The process X admits a cadlag modification.*

Since we do not deal with only one probability measure as in Sect. 1 of Chap. I but with the whole family P_ν, it is important to stress the fact that the above statement means that there is a cadlag process $\tilde{X}$ on $(\Omega, \mathscr{F})$ such that $X_t = \tilde{X}_t$ P_ν-a.s. for each t and every probability measure P_ν.

To prove this result, we will need the

(2.8) Lemma. *Let X and Y be two random variables defined on the same space $(\Omega, \mathscr{F}, P)$ taking their values in a LCCB space E. Then, $X = Y$ a.s. if and only if*

$$E\left[f(X)g(Y)\right] = E\left[f(X)g(X)\right]$$

for every pair (f, g) of bounded continuous functions on E.

Proof. Only the sufficiency needs to be proved. By the monotone class theorem, it is easily seen that

$$E\left[f(X, Y)\right] = E\left[f(X, X)\right]$$

for every positive Borel function on $E \times E$. But, since E is metrizable, the indicator function of the set $\{(x, y) : x \neq y\}$ is such a function. As a result, $X = Y$ a.s.

Proof of Theorem (2.7). Let (f_n) be a sequence in C_0^+ which separates points, namely, for any pair (x, y) in E_Δ, there is a function f_n in the sequence such that $f_n(x) \neq f_n(y)$. Since $\alpha U_\alpha f_n$ converges uniformly to f_n as $\alpha \to \infty$, the countable set $\mathscr{H} = \{U_\alpha f_n, \alpha \in \mathbb{N}, n \in \mathbb{N}\}$ also separates points.

Let S be a countable dense subset of $\mathbb{R}_+$. By Proposition (2.6) and Theorem (2.5) in Chap. II, for each $h \in \mathscr{H}$, the process $h(X_t)$ has a.s. right limits along S. Because $\mathscr{H}$ separates points and is countable, it follows that almost-surely the function $t \to X_t(\omega)$ has right limits in E_Δ along S.

For any ω for which these limits exist, we set $\tilde{X}_t(\omega) = \lim_{\substack{s \downarrow t \\ s \in S}} X_s$ and for an ω for which the limits fail to exist, we set $\tilde{X}_.(\omega) \equiv x$ where x is an arbitrary point in E. We claim that for each t, $\tilde{X}_t = X_t$ a.s. Indeed, let g and h be two functions of $C(E_\Delta)$; we have

$$\begin{aligned} E_\nu\left[g(X_t)h(\tilde{X}_t)\right] &= \lim_{\substack{s \downarrow t \\ s \in S}} E_\nu\left[g(X_t)h(X_s)\right] \\ &= \lim_{\substack{s \downarrow t \\ s \in S}} E_\nu\left[g(X_t)P_{s-t}h(X_t)\right] = E_\nu\left[g(X_t)h(X_t)\right] \end{aligned}$$

since $P_{s-t}h$ converges uniformly to h as $s \downarrow t$. Our claim follows from Lemma (2.8) and thus $\tilde{X}$ is a right-continuous modification of X.

This modification has left limits, because for $h \in \mathscr{H}$, the processes $h(\tilde{X}_t)$ are now right-continuous supermartingales which by Theorem (2.8) of Chap. II, have a.s. left limits along $\mathbb{R}_+$. Again, because $\mathscr{H}$ separates points, the process $\tilde{X}$ has a.s. left limits in E_Δ along $\mathbb{R}_+$. □

Remark. In almost the same way we did prove $\tilde{X}_t = X_t$ a.s., we can prove that for each t, $\tilde{X}_t = \tilde{X}_{t-}$ a.s., in other words, $\tilde{X}_{t-}$ is a left continuous modification of X. It can also be said that X has no fixed time of discontinuity i.e. there is no fixed time t such that $P\left[X_{t-} \neq X_t\right] > 0$.

From now on, we consider only cadlag versions of X for which we state

(2.9) Proposition. *If* $\zeta(\omega) = \inf\left\{t \geq 0 : X_{t^-}(\omega) = \Delta \quad \text{or} \quad X_t(\omega) = \Delta\right\}$, *we have almost-surely* $X_. = \Delta$ *on* $[\zeta, \infty[$.

Proof. Let ϕ be a strictly positive function of C_0. The function $g = U_1\phi$ is also strictly positive. The supermartingale $Z_t = e^{-t}g(X_t)$ is cadlag and we see that $Z_{t-} = 0$ if and only if $X_{t-} = \Delta$ and $Z_t = 0$ if and only if $X_t = \Delta$. As a result

$$\zeta(\omega) = \inf\left\{t \geq 0 : Z_{t-}(\omega) = 0 \quad \text{or} \quad Z_t(\omega) = 0\right\};$$

we then conclude by Proposition (3.4) in Chap. II. □

With a slight variation from Sect. 3 in Chap. I, we now call D the space of functions ω from $\mathbb{R}_+$ to E_Δ which are cadlag and such that $\omega(t) = \Delta$ for $t > s$ whenever $\omega(s-) = \Delta$ or $\omega(s) = \Delta$. The space D is contained in the space $\Omega = E_\Delta^{\mathbb{R}_+}$ and, by the same reasoning as in Sect. 3 of Chap. I, we can use it as probability space. We still call X_t the restrictions to D of the coordinate mappings and the image of P_ν by the canonical mapping ϕ will still be denoted P_ν. For each P_ν, X is a cadlag Markov process with transition function P_t, we call it the *canonical cadlag realization* of the semi-group P_t.

For the canonical realization, we obviously have a family θ_t of shift operators and we can apply the Markov property under the form of Proposition (1.7). We will often work with this version but it is not the only version that we shall encounter as will be made clear in the following section. Most often however, a problem can be carried over to the canonical realization where one can use freely the shift operators. The following results, for instance, are true for all cadlag versions. It may nonetheless happen that one has to work with another version; in that case, one will have to make sure that shift operators may be defined and used if the necessity arises.

So far, the filtration we have worked with, e.g. in Proposition (1.7), was the natural filtration $(\mathscr{F}_t^0)$. As we observed in Sect. 4 of Chap. I, this filtration is not right-continuous and neither is it complete; therefore, we must use an augmentation of $(\mathscr{F}_t^0)$.

We shall denote by $\mathscr{F}_\infty^\nu$ the completion of $\mathscr{F}_\infty^0$ with respect to P_ν and by $(\mathscr{F}_t^\nu)$ the filtration obtained by adding to each $\mathscr{F}_t^0$ all the P_ν-negligible sets in $\mathscr{F}_\infty^\nu$. Finally, we will set

$$\mathscr{F}_t = \bigcap_\nu \mathscr{F}_t^\nu, \quad \mathscr{F}_\infty = \bigcap_\nu \mathscr{F}_\infty^\nu.$$

(2.10) Proposition. *The filtrations* $(\mathscr{F}_t^\nu)$ *and* $(\mathscr{F}_t)$ *are right-continuous.*

Proof. Plainly, it is enough to prove that $(\mathscr{F}_t^\nu)$ is right-continuous and, to this end, because $\mathscr{F}_t^\nu$ and $\mathscr{F}_{t+}^\nu$ are P_ν-complete, it is enough to prove that for each $\mathscr{F}_\infty^0$-measurable and positive r.v. Z,

$$E_\nu\left[Z|\mathscr{F}_t^\nu\right] = E_\nu\left[Z|\mathscr{F}_{t+}^\nu\right] \qquad P_\nu\text{-a.s.}$$

By the monotone class theorem, it is enough to prove this equality for $Z = \prod_{i=1}^n f_i(X_{t_i})$ where $f_i \in C_0$ and $t_1 < t_2 < \ldots t_n$. Let us observe that

$$E_\nu\left[Z|\mathscr{F}_t^\nu\right] = E_\nu\left[Z|\mathscr{F}_t^0\right] \qquad P_\nu\text{-a.s. for each } \ t.$$

Let t be a real number; there is an integer k such that $t_{k-1} \leq t < t_k$ and for h sufficiently small

$$E_\nu\left[Z|\mathscr{F}_{t+h}^\nu\right] = \prod_{i=1}^{k-1} f_i(X_{t_i}) g_h(X_{t+h}) \qquad P_\nu\text{-a.s.}$$

where

$$\begin{aligned} g_h(x) &= \int P_{t_k-t-h}(x, dx_k) f_k(x_k) \int P_{t_{k+1}-t_k}(x_k, dx_{k+1}) \ldots \\ &\quad \ldots \int P_{t_n-t_{n-1}}(x_{n-1}, dx_n) f_n(x_n). \end{aligned}$$

If we let h tend to zero, g_h converges uniformly on E to

$$\begin{aligned} g(x) &= \int P_{t_k-t}(x, dx_k) f_k(x_k) \int P_{t_{k+1}-t_k}(x_k, dx_{k+1}) \ldots \\ &\quad \ldots \int P_{t_n-t_{n-1}}(x_{n-1}, dx_n) f_n(x_n). \end{aligned}$$

Moreover, X_{t+h} converges to X_t as h decreases to 0, thanks to the right-continuity of paths and therefore, using Theorem (2.3) in Chap. II,

$$E_\nu\left[Z|\mathscr{F}_{t+}^\nu\right] = \lim_{h\downarrow 0} E_\nu\left[Z|\mathscr{F}_{t+h}^\nu\right] = \prod_{i=1}^{k-1} f_i(X_{t_i}) g(X_t) = E_\nu\left[Z|\mathscr{F}_t^\nu\right] \qquad P_\nu\text{-a.s.,}$$

which completes the proof. □

It follows from this proposition that $(\mathscr{F}_t)$ is the usual augmentation (Sect. 4 Chap. I) of $(\mathscr{F}_t^0)$ and so is $(\mathscr{F}_t^\nu)$ if we want to consider only the probability measure P_ν. It is remarkable that completing the filtration was also enough to make it right-continuous.

The filtrations $(\mathscr{F}_t)$ and $(\mathscr{F}_t^\nu)$ are those which we shall use most often in the sequel; therefore, it is important to decide whether the properties described so far for $(\mathscr{F}_t^0)$ carry over to $(\mathscr{F}_t)$. There are obviously some measurability problems which are solved in the following discussion.

(2.11) Proposition. *If Z is $\mathscr{F}_\infty$-measurable and bounded, the map $x \to E_x[Z]$ is $\mathscr{E}^*$-measurable and*

$$E_\nu[Z] = \int E_x[Z]\nu(dx).$$

Proof. For any ν, there are, by definition of the completed σ-fields, two $\mathscr{F}_\infty^0$-measurable r.v.'s Z_1 and Z_2 such that $Z_1 \leq Z \leq Z_2$ and $E_\nu[Z_2 - Z_1] = 0$. Clearly, $E_x[Z_1] \leq E_x[Z] \leq E_x[Z_2]$ for each x, and since $x \to E_x[Z_i]$, $i = 1, 2$, is $\mathscr{E}$-measurable and $\int (E_x[Z_2] - E_x[Z_1])\, d\nu(x) = E_\nu[Z_2 - Z_1] = 0$, it follows that $E_\cdot[Z]$ is in $\mathscr{E}^\nu$. As ν is arbitrary, the proof is complete.

(2.12) Proposition. *For each t, the r.v. X_t is in $\mathscr{F}_t/\mathscr{E}^*$.*

Proof. This is an immediate consequence of Proposition (3.2) in Chap. 0. □

We next want to extend the Markov property of Proposition (1.7) to the σ-algebras $\mathscr{F}_t$. We first need the

(2.13) Proposition. *For every t and $h > 0$, $\theta_h^{-1}(\mathscr{F}_t) \subset \mathscr{F}_{t+h}$.*

Proof. As $\theta_h \in \mathscr{F}_t^0/\mathscr{F}_{t+h}^0$, the result will follow from Proposition (3.2) in Chap. 0 if we can show that for any starting measure ν, there is a starting measure μ such that $\theta_h(P_\nu) = P_\mu$. Define $\mu = X_h(P_\nu)$; then using the Markov property of Proposition (1.7) we have, for $\Gamma \in \mathscr{F}_\infty^0$

$$P_\nu\left[1_\Gamma \circ \theta_h\right] = E_\nu\left[E_{X_h}[1_\Gamma]\right] = \int \mu(dy) P_y[\Gamma] = P_\mu[\Gamma]$$

which completes the proof. □

We may now state

(2.14) Proposition (Markov property). *If Z is $\mathscr{F}_\infty$-measurable and positive (or bounded), then, for every $t > 0$ and any starting measure ν,*

$$E_\nu\left[Z \circ \theta_t \middle| \mathscr{F}_t\right] = E_{X_t}[Z] \qquad P_\nu\text{-a.s.}$$

on the set $\{X_t \neq \Delta\}$. In particular, X is still a Markov process with respect to $(\mathscr{F}_t)$.

Proof. By Propositions (2.11) and (2.12), the map $E_{X_t(\cdot)}[Z]$ is $\mathscr{F}_t$-measurable, so we need only prove that for any $A \in \mathscr{F}_t$,

$$(*) \qquad E_\nu\left[1_A Z \circ \theta_t\right] = E_\nu\left[1_A E_{X_t}[Z]\right].$$

We may assume that Z is bounded; by definition of $\mathscr{F}_\infty$, there is a $\mathscr{F}_\infty^0$-measurable r.v. Z' such that $\{Z \neq Z'\} \subset \Gamma$ with $\Gamma \in \mathscr{F}_\infty^0$ and $P_\mu[\Gamma] = 0$ where $\mu = X_t(P_\nu)$ as in the preceding proof. We have $\left\{Z \circ \theta_t \neq Z' \circ \theta_t\right\} \subset \theta_t^{-1}(\Gamma)$ and as in the above proof, $P_\nu\left[\theta_t^{-1}(\Gamma)\right] = P_\mu[\Gamma] = 0$. Since it was shown in the last proof that $E_\mu\left[E_{X_t}[\cdot]\right] = E_\nu[\cdot]$, it now follows that

$$E_\nu\left[E_{X_t}[|Z - Z'|]\right] = E_\mu[|Z - Z'|] = 0$$

so that $E_{X_t}[Z] = E_{X_t}[Z']$ P_ν-a.s. Therefore, we may replace Z by Z' on both sides of (*) which is then a straightforward consequence of Proposition (1.7). □

Feller processes are not the only Markov processes possessing good versions, and actually they may be altered in several ways to give rise to Markov processes in the sense of Sect. 1, which still have all the good probabilistic properties of Markov processes but no longer the analytic properties of Feller transition functions. The general theory of Markov processes is not one of the subjects of this book; rather, the Markov theory is more something we have to keep in mind when studying particular classes of processes. As a result, we do not want to go deeper into the remark above, which would lead us to set up axiomatic definitions of "good" Markov processes. In the sequel, if the necessity arises, we will refer to *Markov processes* with values in $(E, \mathscr{E})$ as collections $X = (\Omega, \mathscr{F}, \mathscr{F}_t, P_x, x \in E, \theta_t)$; these symbols will then have the same meaning and can be used in the same manner as for Feller processes. For instance, the maps $t \to X_t$ are supposed to be a.s. cadlag. This may be seen as a sad departure from a rigorous treatment of the subject, but we shall make only a parcimonious use of this liberty, and the reader should not feel uneasiness on this count. Exercise (3.21) gives an example of a Markov process which is not a Feller process.

We proceed to a few consequences of the existence of good versions. The following observation is very important.

(2.15) Theorem (Blumenthal's zero-one law). *For any $x \in E$ and $\Gamma \in \mathscr{F}_0^{\varepsilon_x}$, either $P_x[\Gamma] = 0$ or $P_x[\Gamma] = 1$.*

Proof. If $\Gamma \in \sigma(X_0)$, then $P_x[\Gamma] = 0$ or 1 because $P_x[X_0 = x] = 1$. Since one obtains $\mathscr{F}_0^{\varepsilon_x}$ by adding to $\sigma(X_0)$ sets of P_x-measure zero, the proof is complete. □

(2.16) Corollary. *If T is a $(\mathscr{F}_t^{\varepsilon_x})$-stopping time, then either $P_x[T = 0] = 1$ or $P_x[T > 0] = 1$.*

This corollary has far-reaching consequences, especially in connection with the following result (see Exercise 2.25). If A is a set, we recall from Sect. 4 Chap. I that the *entry and hitting times* of A by X are defined respectively by

$$D_A = \inf\{t \geq 0 : X_t \in A\}, \quad T_A = \inf\{t > 0 : X_t \in A\}$$

where as usual, $\inf(\emptyset) = +\infty$. For any s,

$$s + D_A \circ \theta_s = s + \inf\{t \geq 0 : X_{t+s} \in A\} = \inf\{t \geq s : X_t \in A\}.$$

It follows that $s + D_A \circ \theta_s = D_A$ on $\{D_A \geq s\}$ and also that

$$T_A = \lim_{s \downarrow 0} \downarrow (s + D_A \circ \theta_s).$$

Similarly, one proves that $t + T_A \circ \theta_t = T_A$ on $\{T_A > t\}$.

(2.17) Theorem. *If A is a Borel set, the times D_A and T_A are $(\mathscr{F}_t)$-stopping times.*

Proof. Since X is right-continuous, it is clearly progressively measurable and, since $(\mathscr{F}_t)$ is right-continuous and complete, Theorem (4.15) of Chap. I shows that D_A which is the début of the set $\Gamma = \{(t, \omega) : X_t(\omega) \in A\}$ is a $(\mathscr{F}_t)$-stopping time.

The reader will now check easily (see Proposition (3.3)) that for each s, the time $s + D_A \circ \theta_s$ is a $(\mathscr{F}_t)$-stopping time. As a limit of $(\mathscr{F}_t)$-stopping times, T_A is itself a $(\mathscr{F}_t)$-stopping time. □

We will next illustrate the use of the Markov property with two interesting results. For the first one, let us observe that a basic example of Feller semi-groups is provided by convolution semi-groups i.e. families $(\mu_t, t \geq 0)$ of probability measures on $\mathbb{R}^d$ such that

i) $\mu_t * \mu_s = \mu_{t+s}$ for any pair (s, t);

ii) $\mu_0 = \varepsilon_0$ and $\lim_{t \downarrow 0} \mu_t = \varepsilon_0$ in the vague topology.

If we set

$$P_t(x, A) = \int_{\mathbb{R}^d} 1_A(x + y)\mu_t(dy)$$

we get a Feller t.f. as is easily checked by means of Proposition (2.4) and the well-known properties of convolution. Most of the examples of Exercise (1.8), in particular the t.f. of BM^d, are of this type. A Feller process with such a t.f. has special properties.

(2.18) Proposition. *If the transition function of X is given by a convolution semi-group (μ_t), then X has stationary independent increments. The law of the increment $X_t - X_s$ is μ_{t-s}.*

The word stationary refers to the fact that the law of the increment $X_t - X_s$ depends only on $t - s$, hence is invariant by translation in time. The process X itself is *not* stationary in the sense of Sect. 3 Chap. I.

Proof. For any $f \in \mathscr{E}_+$ and any t we have, since $P_x[X_0 = x] = 1$

$$E_x[f(X_t - X_0)] = E_x[f(X_t - x)] = \mu_t(f)$$

which no longer depends on x. Consequently, by the Markov property, for $s < t$,

$$E_\nu\left[f(X_t - X_s)\middle|\mathscr{F}_s\right] = E_{X_s}\left[f(X_{t-s} - X_0)\right] = \mu_{t-s}(f) \qquad P_\nu\text{-a.s.},$$

which completes the proof. □

Conversely, if a Feller process has stationary independent increments, it is easily checked that its t.f. is given by a convolution semi-group having property ii) above Proposition (2.18). These processes will be called *processes with stationary independent increments or Lévy processes.* Some facts about these processes are collected in Sect. 4.

We now turn to another result which holds for any Markov process with good versions.

(2.19) Proposition. *Let $x \in E$ and $\sigma_x = \inf\{t > 0 : X_t \neq x\}$; there is a constant $a \in [0, \infty]$ depending on x such that*

$$P_x[\sigma_x > t] = e^{-at}.$$

Proof. The time σ_x is the hitting time of the open set $\{x\}^c$ and therefore a stopping time (see Sect. 4 Chap. I). Furthermore $\sigma_x = t + \sigma_x \circ \theta_t$ on $\{\sigma_x > t\}$ as was observed before Theorem (2.17); thus, we may write

$$P_x[\sigma_x > t + s] = P_x[(\sigma_x > t) \cap (\sigma_x > t + s)] = E_x\left[1_{(\sigma_x > t)} 1_{(\sigma_x > s)} \circ \theta_t\right]$$

and by the Markov property, since obviously $X_t \neq \Delta$ on $\{\sigma_x > t\}$, this yields

$$P_x[\sigma_x > t + s] = E_x\left[1_{(\sigma_x > t)} E_{X_t}[\sigma_x > s]\right];$$

but, on $\{\sigma_x > t\}$, we have $X_t = x$, so that finally

$$P_x[\sigma_x > t + s] = P_x[\sigma_x > t]\, P_x[\sigma_x > s]$$

which completes the proof. □

Finally, this proposition leads to a classification of points. If $a = +\infty$, σ_x is P_x-a.s. zero; in other words, the process leaves x at once. This is the case for all points if X is the BM since in that case $P_t(x, \{x\}) = 0$ for every $t > 0$. If $a = 0$, the process never leaves x which can be said to be a *trap* or an *absorbing point*. If $a \in]0, \infty[$, then σ_x has an exponential law with parameter a; we say that x is a *holding point* or that the process stays in x for an *exponential holding time*. This is the case for the Poisson process with $a = 1$ for every x, but, in the general case, a is actually a function of x. Let us further observe that, as will be proved in Proposition (3.13), X can leave a holding point only by a jump; thus, for a process with continuous paths, only the cases $a = 0$ and $a = \infty$ are possible.

We close this section by a few remarks about Brownian motion. We have now two ways to look at it: one as the process constructed in Chap. I which vanishes at time zero and for which we consider only one probability measure; the other one as a Markov process which can be started anywhere so that we have to consider the whole family of probability measures P_ν. The probability measure of the first viewpoint, which is the Wiener measure in the canonical setting, identifies with the probability measure $P_0 = P_{\varepsilon_0}$ of the second viewpoint. Any result proved for P_0 in the Markov process setting will thus be true for the Wiener measure.

In the sequel, the words Brownian motion will refer to one viewpoint or the other. We shall try to make it clear from the context which viewpoint is adopted at a given time; we shall also use the adjective *standard* to mean that we consider only the probability measure for which $B_0 = 0$ a.s., i.e. the Wiener measure.

(2.20) Definition. *If $(\mathscr{G}_t)$ is a filtration, an adapted process B is called a $(\mathscr{G}_t)$-Brownian motion if*

i) it is a Brownian motion,
ii) for each $t \geq 0$, the process $B_{t+s} - B_t$, $s > 0$, is independent of $(\mathscr{G}_t)$.

It is equivalent to say that B is a Markov process with respect to $(\mathscr{G}_t)$ with the t.f. of Exercise (1.8) ii).

In this definition, the notion of independence may refer to one or to a family of probability measures. We want to stress that with the notation of this section, B is a $(\mathscr{F}_t)$-Brownian motion if we consider the whole family of probability measures P_ν or a $(\mathscr{F}_t^\mu)$-Brownian motion if we consider only one probability measure P_μ. Each of these filtrations is, in its context, the smallest right-continuous complete filtration with respect to which B is a BM.

(2.21) Definition. *Let X be a process on a space $(\Omega, \mathscr{F})$ endowed with a family P_θ, $\theta \in \Theta$, of probability measures. We denote by $(\mathscr{F}_t^X)$ the smallest right-continuous and complete filtration with respect to which X is adapted. A $\mathscr{F}_t^X$-stopping time is said to be a* stopping time of X.

In the case of BM, we have $\mathscr{F}_t^B = \mathscr{F}_t$ or $\mathscr{F}_t^\mu$ according to the context. These filtrations will be called the Brownian filtrations.

(2.22) Exercise. Prove that the transition functions exhibited in Exercises (1.8) (with the exception of the generalized Poisson t.f.), (1.14), (1.15) are Feller t.f.'s. Do the same job for the OU processes of Exercise (1.13).

\# **(2.23) Exercise.** Show that the resolvent of the semi-group of linear BM is given by $U_p(x, dy) = u_p(x, y)dy$ where

$$u_p(x, y) = \frac{1}{\sqrt{2p}} \exp\left(-\sqrt{2p}|x - y|\right).$$

(2.24) Exercise. If X is a Markov process, $\mathbf{e}_p$ and $\mathbf{e}_q$ two independent exponential r.v.'s with parameters p and q, independent of X prove that for a positive Borel function f

$$pU_pf(x) = E_x\left[f\left(X_{\mathbf{e}_p}\right)\right], \qquad pqU_pU_qf(x) = E_x\left[f\left(X_{\mathbf{e}_p+\mathbf{e}_q}\right)\right]$$

and derive therefrom the resolvent equation.

* **(2.25) Exercise.** 1°) A subset A of E is called *nearly Borel* if, for every ν, there are two Borel sets A_1, A_2 such that $A_1 \subset A \subset A_2$ and $P_\nu\left[D_{A_2\setminus A_1} < \infty\right] = 0$. Prove that the family of nearly Borel sets is a sub-σ-algebra of the universally measurable sets. Prove that, if A is nearly Borel, then D_A and T_A are $\mathscr{F}_t$-stopping times.

2°) If A is nearly Borel and $x \in E$, prove that either $P_x[T_A = 0] = 1$ or $P_x[T_A = 0] = 0$. In the former (latter) case, the point x is said to be *regular* *(irregular)* for A.

3°) A set O is said to be *finely open* if, for every $x \in O$, there is a nearly Borel set G such that $x \in G \subset O$ and X is irregular for G^c. Prove that the finely open sets are the open sets for a topology which is finer than the locally compact topology of E. This topology is called the *fine topology*.

4°) If a nearly Borel set A is of potential zero, i.e. $\int_0^\infty P_t(\cdot, A)dt = 0$ (see Exercise (2.29)), then A^c is dense for the fine topology.

5°) If f is universally measurable and $t \to f(X_t)$ is right-continuous, then f is finely continuous.

6°) Prove the converse of the property in 5°).

[Hints: Pick $\varepsilon > 0$ and define $T_0 = 0$ and for any ordinal α of the first kind define

$$T_{\alpha+1} = \inf\left\{t > T_\alpha : \left|f(X_t) - f(X_{T_\alpha})\right| > \varepsilon\right\}$$

and if α is a limit ordinal

$$T_\alpha = \sup_{\omega<\alpha} T_\omega.$$

Prove that $T_\alpha < T_{\alpha+1}$ a.s. on $\{T_\alpha < \infty\}$ and that, as a result, there are only countably many finite times T_α.]

* **(2.26) Exercise.** Prove that for a Feller process X, the set $\{X_s(\omega),\ 0 \leq s \leq t,\ t < \zeta(\omega)\}$ is a.s. bounded.

[Hint: Use the quasi-left continuity of Exercise (2.33) applied to exit times of suitable compact sets.]

* **(2.27) Exercise (A criterion for the continuity of paths).** 1°) Let d be a metric on E, and f a function from $[0, 1]$ to E_Δ with left (right) limits on $]0, 1]([0, 1[)$. Then, f is not continuous if and only if there is an $\varepsilon > 0$ such that

$$N_n(f) = \max_{0\leq k\leq n-1} d\big(f(k/n), f((k+1)/n)\big) > \varepsilon$$

for all n sufficiently large.

2°) Let $B(x, \varepsilon) = \{y : d(x, y) \leq \varepsilon\}$. For $\varepsilon > 0$ and a compact set K, define

$$M_n^\varepsilon = \{\omega : N_n(X_.(\omega)) > \varepsilon;\ X_s(\omega) \in K \quad \text{for every} \quad s \in [0, 1]\}.$$

Prove that $P_\nu\left(M_n^\varepsilon\right) \leq n \sup_{x\in K} P_{1/n}\big(x, B(x, \varepsilon)^c\big)$.

3°) Using the result in the preceding exercise, prove that if X satisfies the condition

$$\lim_{t\downarrow 0} \sup_{x\in K} \frac{1}{t} P_t\big(x, B(x, \varepsilon)^c\big) = 0$$

for every $\varepsilon > 0$ and compact set K, then a.s. X has continuous paths.

4°) Check that the condition in 3°) is satisfied for BM. Thus, the results of this section together with 3°) give another construction of BM independent of Chap. I.

* **(2.28) Exercise.** Let B be the BM^d, $\mathscr{A}_t = \sigma(B_s, s \geq t)$ and $\mathscr{A} = \bigcap_t \mathscr{A}_t$ its *asymptotic* σ-field.

1°) Use the time inversion of Sect. 1 Chap. I and Blumenthal's zero-one law to prove that $\mathscr{A}$ is P_0-a.s. trivial i.e. for any $A \in \mathscr{A}$ either $P_0(A) = 0$ or $P_0(A) = 1$.

2°) If A is in $\mathscr{A}$, then for any fixed t, there is an event $B \in \mathscr{A}$ such that $1_A = 1_B \circ \theta_t$. Prove that

$$P_x[A] = \int P_t(x, dy) P_y(B)$$

and conclude that either $P_.[A] \equiv 0$ or $P_.[A] \equiv 1$.

3°) Prove that for any initial distribution ν and $\Gamma \in \mathscr{F}_\infty$

$$\lim_{t\to\infty} \sup_{A \in \mathscr{A}_t} |P_\nu(A \cap \Gamma) - P_\nu(A) P_\nu(\Gamma)| = 0.$$

[Hint: Use Theorem (2.3) of Chap. II.]

4°) If ν_1 and ν_2 are two starting measures, show that

$$\lim_{t\to\infty} \|(\nu_1 - \nu_2)\, P_t\| = 0$$

where the norm is the variation norm on bounded measures.

[Hint: Use a Jordan-Hahn decomposition of $(\nu_1 - \nu_2)$.]

(2.29) Exercise. Let X be a Feller process. For $x \in E$ and $A \in \mathscr{E}$, set

$$U(x, A) = \int_0^\infty P_t(x, A) dt.$$

1°) Prove that this integral is well defined, that U is a kernel on $(E, \mathscr{E})$ and that if $f \in \mathscr{E}_+$, $Uf(x) = E_x\left[\int_0^\infty f(X_t) dt\right]$. The kernel U is called the *potential kernel* of X.

2°) Check that $Uf = \lim_{\lambda \downarrow 0} U_\lambda f$ and that for every $\lambda > 0$

$$U = U_\lambda + \lambda U_\lambda U = U_\lambda + \lambda U U_\lambda.$$

3°) Prove that for $X = \mathrm{BM}^d$, $d \le 2$, the potential kernel takes only the values 0 and $+\infty$ on $\mathscr{E}_+$. This is linked to the recurrence properties of BM in dimensions 1 and 2 (see Sect. 3 Chap. X).

4°) Prove that for BM^d, $d \ge 3$, the potential kernel is the convolution kernel associated with $\left(1/2\pi^{d/2}\right) \Gamma\left((d/2)-1\right) |x|^{2-d}$ i.e. the kernel of *Newtonian potential theory*. In particular, for $d = 3$,

$$Uf(x) = \frac{1}{2\pi} \int \frac{f(y)}{|x-y|} dy.$$

5°) Compute the potential kernel of linear BM killed when it reaches 0 (Exercise (3.29)) and prove that it has the density $2(x \wedge y)$ with respect to the Lebesgue measure on $\mathbb{R}_+$.

6°) Prove that g_t is a density for the potential kernel of the heat process.

(2.30) Exercise. Let A be a Borel set.

1°) Prove that for every $s, t \ge 0$,

$$1_{(T_A > s+t)} = 1_{(T_A > s)} \cdot 1_{(T_A > t)} \circ \theta_s.$$

2°) Let ν be a probability measure such that $\nu(\bar{A}) = 0$. Prove that under P_ν, the process Y defined by

$$Y_t = X_t \quad \text{on} \quad \{t < T_A\}, \qquad Y_t = \Delta \quad \text{on} \quad \{t \ge T_A\}$$

is a Markov process with respect to $(\mathscr{F}_t)$. One says that Y is the process X *killed* when entering A. See Exercise (3.29) for a particular case.

\# **(2.31) Exercise.** Let X be the standard linear BM and set $\psi(t) = t^{-\alpha}$, $\alpha \geq 0$. Prove that the following three properties are equivalent

i) $\lim_{\varepsilon \downarrow 0} \int_\varepsilon^1 B_t \psi(t) dt$ exists on a set of strictly positive probability;
ii) $\alpha < 3/2$;
iii) $\int_0^1 \psi(t)|B_t| dt < \infty$ a.s.

[Hint: Use Blumenthal's zero-one law to prove that i) is equivalent to a stronger property. Then, to prove that i) entails ii) use the fact that for Gaussian r.v.'s almost-sure convergence implies convergence in L^2, hence convergence of the L^2-norms.]

The assertion that for, say, any positive continuous function ψ on $]0, 1]$, the properties i) and iii) are equivalent is false. In fact, it can be shown that if $\psi \in L^1_{\text{loc}}(]0, 1])$, iii) is equivalent to

iv) $\int_0^1 \psi(t) t^{1/2} dt < \infty$,

and there exist functions ψ satisfying i) but not iv).

This subject is taken up in Exercise (3.19) Chap. IV.

* **(2.32) Exercise.** Let B be the standard linear BM and h a continuous function on $]0, 1[$. Let Γ be the event

$$\{\omega : B_t(\omega) < h(t) \quad \text{on some interval} \quad]0, T(\omega)[\subset]0, 1[\}.$$

Prove that either $P(\Gamma) = 0$ or $P(\Gamma) = 1$; in the former (latter) case, h is said to belong to the lower (upper) class. For every $\varepsilon > 0$, $h(t) = (1 + \varepsilon)\sqrt{2t \log_2(1/t)}$ belongs to the upper class and $h(t) = (1 - \varepsilon)\sqrt{2t \log_2(1/t)}$ to the lower class.

* **(2.33) Exercise.** 1°) (Quasi-left continuity). If X is a Feller process and (T_n) a sequence of $(\mathscr{F}_t)$-stopping times increasing to T, prove that

$$\lim_n X_{T_n} = X_T \quad \text{a.s. on} \quad \{T < \infty\}.$$

[Hint: It is enough to prove the result for bounded T. Set $Y = \lim_n X_{T_n}$ (why does it exist?) and prove that for continuous functions f and g

$$E_x[f(Y)g(X_T)] = \lim_{t \downarrow 0} \lim_n E_x\left[f(X_{T_n})g(X_{T_n+t})\right] = E_x[f(Y)g(Y)].].$$

This result is of course totally obvious for processes with continuous paths. For processes with jumps, it shows that if $X_{T_-} \neq X_T$ on $\{0 < T < \infty\}$, then a sequence (T_n) can increase to T only in a trivial way: for a.e. ω, there is an integer $n(\omega)$ such that $T_n(\omega) = T(\omega)$ for $n \geq n(\omega)$. Such a time is said to be totally inaccessible as opposed to the predictable times of Sect. 5 Chap. IV, a typical example being the times of jumps of the Poisson process.

2°) Using only 1°) and Proposition (4.6) in Chap. I, prove that if A is a closed set, then T_A is a $(\mathscr{F}_t)$-stopping time.

§3. Strong Markov Property

Stopping times are of constant use in the study of Markov processes, the reason being that the Markov property extends to them as we now show. We must first introduce some notation.

We shall consider the canonical cadlag version of a Feller process. We use the results and notation of §2. For a $(\mathscr{F}_t)$-stopping time T, we define X_T on the whole space Ω by putting $X_T = \Delta$ on $\{T = \infty\}$. The r.v. X_T is $\mathscr{F}_T$-measurable as follows from Sect. 4 in Chap. I and is the position of the process at time T. We further define a map θ_T from Ω into itself by

$$\theta_T(\omega) = \theta_t(\omega) \quad \text{if} \quad T(\omega) = t, \qquad \theta_T(\omega) = \omega_\Delta \quad \text{if} \quad T(\omega) = +\infty,$$

where ω_Δ is the path identically equal to Δ. Clearly, $X_t \circ \theta_T = X_{T+t}$ so that $\theta_T^{-1}(\mathscr{F}_\infty^0) \subset \sigma(X_{T+t}, t \geq 0)$.

We now prove the *Strong Markov property* of Feller processes.

(3.1) Theorem. *If Z is a $\mathscr{F}_\infty$-measurable and positive (or bounded) random variable and T is a stopping time, for any initial measure ν,*

$$E_\nu\left[Z \circ \theta_T \middle| \mathscr{F}_T\right] = E_{X_T}[Z]$$

P_ν-a.s. on the set $\{X_T \neq \Delta\}$.

Proof. We first prove the formula when T takes its values in a countable set D. We have

$$\begin{aligned} 1_{(X_T \neq \Delta)} E_\nu\left[Z \circ \theta_T \middle| \mathscr{F}_T\right] &= \sum_{d \in D} 1_{(T=d)} 1_{(X_d \neq \Delta)} E_\nu\left[Z \circ \theta_d \middle| \mathscr{F}_d\right] \\ &= \sum_{d \in D} 1_{(T=d)} 1_{(X_d \neq \Delta)} E_{X_d}[Z] \end{aligned}$$

which proves our claim.

To get the general case, let us observe that by setting

$$T_n = \left([2^n T] + 1\right) / 2^n$$

we define a sequence of stopping times taking their values in countable sets and decreasing to T. For functions f_i, $i = 1, 2, \ldots, k$ in C_0^+ and times $t_1 < t_2 < \ldots < t_k$, let

$$g(x) = \int P_{t_1}(x, dx_1) f_1(x_1) \int P_{t_2 - t_1}(x_1, dx_2) \ldots \int P_{t_k - t_{k-1}}(x_{k-1}, dx_k) f_k(x_k).$$

Because X is Feller the function g is in C_0^+ and by the special case,

$$E_\nu\left[\prod_i f_i(X_{t_i}) \circ \theta_{T_n} \mid \mathscr{F}_{T_n}\right] = g(X_{T_n}).$$

Because of the right-continuity of paths, and by Corollary (2.4) Chap. II, we get the result for the special case $\prod_i f_i(X_{t_i})$. By an application of the monotone class Theorem, we get the result for every positive Z in $\mathscr{F}_\infty^0$.

It remains to prove the theorem for $Z \in (\mathscr{F}_\infty)_+$. By working with $P'_\nu = P_\nu\left(\cdot \bigcap (X_T \neq \Delta)\right) / P_\nu(X_T \neq \Delta)$ for which the conditional expectation given $\mathscr{F}_T$ is the same as under P_ν, we may assume that $X_T \neq \Delta$ a.s. and drop the corresponding qualification. Call μ the image of P_ν by X_T i.e. $\mu(A) = P_\nu[X_T \in A]$. By definition of $\mathscr{F}_\infty$, there are two $\mathscr{F}_\infty^0$-measurable r.v.'s Z' and Z'' such that $Z' \leq Z \leq Z''$ and $P_\mu[Z'' - Z' > 0] = 0$. By the first part of the proof

$$P_\nu[Z'' \circ \theta_T - Z' \circ \theta_T > 0] = E_\nu\left[E_{X_T}[Z'' - Z' > 0]\right] = 0.$$

Since ν is arbitrary, it follows that $Z \circ \theta_T$ is $\mathscr{F}_\infty$-measurable.

The conditional expectation $E_\nu[Z \circ \theta_T | \mathscr{F}_T]$ is now meaningful and

$$E_\nu\left[Z' \circ \theta_T | \mathscr{F}_T\right] \leq E_\nu\left[Z \circ \theta_T | \mathscr{F}_T\right] \leq E_\nu\left[Z'' \circ \theta_T | \mathscr{F}_T\right].$$

By the foregoing, the two extreme terms are P_ν-a.s. equal to $E_{X_T}[Z]$, which ends the proof. □

Remark. The qualification $\{X_T \neq \Delta\}$ may be forgotten when $\zeta = +\infty$ a.s. and $T < \infty$ a.s., in which case we will often drop it entirely from the notation.

In the course of the above proof, we saw that θ_T is a $\mathscr{F}_\infty$-measurable mapping. We actually have

(3.2) Lemma. *For any $t > 0$, $T + t$ is a stopping time and $\theta_T^{-1}(\mathscr{F}_t) \subset \mathscr{F}_{T+t}$.*

Proof. By a monotone class argument, it is easily seen that $\theta_T^{-1}(\mathscr{F}_t^0) \subset \mathscr{F}_{T+t}$ and the reasoning in the above proof yields the result. □

We shall use this lemma to prove

(3.3) Proposition. *If S and T are two $(\mathscr{F}_t)$-stopping times, then $S + T \circ \theta_S$ is an $(\mathscr{F}_t)$-stopping time.*

Proof. Since $(\mathscr{F}_t)$ is right-continuous, it is enough to prove that $\{S + T \circ \theta_S < t\} \in \mathscr{F}_t$ for every t. But

$$\{S + T \circ \theta_S < t\} = \bigcup_{q \in \mathbb{Q}} \{S < t - q\} \cap \{T \circ \theta_S < q\}.$$

By the lemma, the set $\{T \circ \theta_S < q\}$ is in $\mathscr{F}_{S+q}$; by definition of $\mathscr{F}_{S+q}$, the set $\{S < t - q\} \cap \{T \circ \theta_S < q\} = \{S + q < t\} \cap \{T \circ \theta_S < q\}$ is in $\mathscr{F}_t$ which proves our claim. □

If we think of a stopping time as the first time some physical event occurs, the stopping time $S + T \circ \theta_S$ is the first time the event linked to T occurs after the event linked to S has occured. For instance, using the notation of §2, if A and B are two sets in $\mathscr{E}$, the stopping time $T_A + T_B \circ \theta_{T_A}$ is the first time the process hits the set B after having hit the set A. This will be used in the sequel of this section.

We now give a first few applications of the strong Markov property. With a stopping time T, we may also associate a kernel P_T on $(E, \mathscr{E})$ by setting

$$P_T(x, A) = P_x[X_T \in A]$$

or more generally for $f \in \mathscr{E}_+$,

$$P_T f(x) = E_x[f(X_T)].$$

The following result tells us how to compose these kernels (see Definition (1.1)).

(3.4) Proposition. *If S and T are two stopping times, then*

$$P_S P_T = P_{S+T\circ\theta_S}.$$

Proof. By definition

$$P_S(P_T f)(x) = E_x\left[E_{X_S}[f(X_T)]\right],$$

so that, using the strong Markov property, we have

$$\begin{aligned} P_S(P_T f)(x) &= E_x\left[1_{(X_S\neq\Delta)} E_x\left[f(X_T)\circ\theta_S \middle| \mathscr{F}_S\right]\right] \\ &= E_x\left[1_{(X_S\neq\Delta)} f(X_T)\circ\theta_S\right]. \end{aligned}$$

Now $f(X_T)\circ\theta_S = f(X_{S+T\circ\theta_S})$ and $f(X_T)\circ\theta_S = 0$ on $\{X_S = \Delta\}$, so that the result follows. □

Remark. We have thus generalized to stopping times the fundamental semi-group property. Indeed if $T = t$ a.s., then $P_T = P_t$ and $S + T\circ\theta_S = S + t$.

We now prove that, if we start to observe a Markov process at a stopping time, the resulting process is still a Markov process with the same t.f.

(3.5) Proposition. *If T is a stopping time, the process $Y_t = X_{T+t}$ is a Markov process with respect to $(\mathscr{F}_{T+t})$ and with the same transition function.*

Proof. Let $f \in \mathscr{E}_+$; for every ν, and every $s \geq 0$,

$$E_\nu\left[f(X_{T+t+s})\middle|\mathscr{F}_{T+t}\right] = E_\nu\left[f(X_s)\circ\theta_{T+t}\middle|\mathscr{F}_{T+t}\right] = P_s f(X_{T+t})$$

on the set $\{X_{T+t} \neq \Delta\}$. But, on the set $\{X_{T+t} = \Delta\}$, the equality holds also so that X_{T+t} satisfies the conditions of Definition (1.3). □

Remarks. 1) The process Y is another version of the Feller process X. This shows that non canonical versions arise naturally even if one starts with the canonical one.

2) The above property is in fact equivalent to the strong Markov property as is stated in Exercise (3.16).

In the case of processes with independent increments, the above proposition can be stated more strikingly.

(3.6) Corollary. *If X has stationary independent increments, the process $(X_{T+t} - X_T, t \geq 0)$ is independent of $\mathscr{F}_T$ and its law under P_ν is the same as that of X under P_0.*

Proof. For $f_i \in \mathscr{E}_+, t_i \in \mathbb{R}_+, i = 1, 2, \ldots, n$,

$$E_\nu\left[\prod_i f_i\left(X_{T+t_i} - X_T\right) \big| \mathscr{F}_T\right] = E_{X_T}\left[\prod_i f_i\left(X_{t_i} - X_0\right)\right]$$

and, as in Proposition (2.18), this is a constant depending only on f_i and t_i. □

In particular, in the case of BM, $B_{T+t} - B_T$ is a $(\mathscr{F}_{T+t})$-Brownian motion independent of $\mathscr{F}_T$. Another proof of this fact is given in Exercise (3.21) Chap. IV. We devote the rest of this section to an application of the Strong Markov property to linear BM.

We recall that the continuous increasing process $S_t = \sup_{s \leq t} B_s$ and the stopping time T_a introduced in Sect. 3 Chap. II are inverses of one another in the sense that

$$T_a = \inf\{t : S_t \geq a\}, \qquad S_t = \inf\{a : T_a \geq t\}.$$

The map $a \to T_a$ is increasing and left-continuous (see Sect. 4 Chap. 0 and Sect. 1 in Chap. V).

In the next result, P is the probability measure for the BM started at 0, i.e. the Wiener measure if we use the canonical version.

(3.7) Proposition (Reflection principle). *For every $a > 0$ and $t \geq 0$,*

$$P[S_t \geq a] = P[T_a \leq t] = 2P[B_t \geq a] = P(|B_t| \geq a).$$

The name for this proposition comes from the following heuristic argument. Among the paths which reach a before time t, "half" will be above a at time t; indeed, if we consider the symmetry with respect to the line $y = a$ (in the usual representation with the heat process paths) for the part of the path between T_a and t, we get a one-to-one correspondence between these paths and those which are under a at time t. Those which are exactly in a at time t have zero probability and therefore

$$\begin{aligned} P[S_t \geq a] &= P[S_t \geq a, B_t > a] + P[S_t \geq a, B_t < a] \\ &= 2P[S_t \geq a, B_t > a] = 2P[B_t > a] \end{aligned}$$

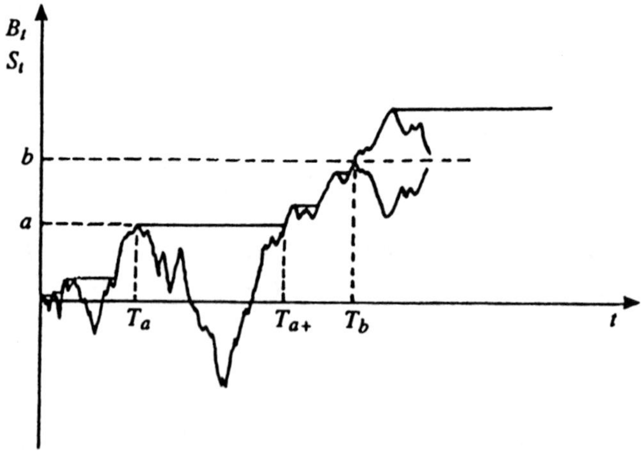

Fig. 3. Reflection in b

since $(B_t > a) \subset (S_t \geq a)$. This argument which is called the reflection principle can be made rigorous but it is easier to use the Strong Markov property. See also Exercise (3.14).

Proof. Indeed, since $B_{T_a} = a$,

$$P\left[S_t \geq a, B_t < a\right] = P\left(T_a \leq t, B_{T_a+(t-T_a)} - B_{T_a} < 0\right],$$

and since $B_{T_a+s} - B_{T_a}$ is a BM independent of $\mathscr{F}_{T_a}$ this is further equal to

$$\frac{1}{2}P[T_a \leq t] = \frac{1}{2}P[S_t \geq a].$$

Since

$$P[S_t \geq a] = P[B_t \geq a] + P[S_t \geq a, B_t < a],$$

the result follows.

Remarks. 1°) For each t, the *random variables* S_t and $|B_t|$ have thus the same law. Of course, the *processes* S *and* $|B|$ do not have the same law (S is increasing and $|B|$ is not). More will be said on this subject in Chap. VI.

2°) As an exercise, the reader may also prove the above result by showing, with the help of the strong Markov property, that the Laplace transforms in t of $P(T_a \leq t)$ and $2P(B_t \geq a)$ are equal.

The preceding result allows to derive from the law of B_t the laws of the other variables involved. As we already pointed out, S_t has the same law as $|B_t|$ namely the density $2(2\pi t)^{-1/2}\exp(-y^2/2t)$ on $[0, \infty[$. As for the law of T_a, it could have been obtained by inverting its Laplace transform $e^{-a\sqrt{2s}}$ found in Proposition (3.7) Chap. II, but we can now observe that

$$P[T_a \leq t] = 2\int_a^\infty \frac{1}{\sqrt{2\pi t}} \exp\left(-y^2/2t\right) dy.$$

Upon differentiation with respect to t, the density f_a of T_a is found to be equal on $[0, \infty[$ to

$$f_a(s) = \frac{1}{\sqrt{2\pi}}\left(-\frac{1}{s^{3/2}}\int_a^\infty \exp\left(-y^2/2s\right) dy + \frac{1}{s^{5/2}}\int_a^\infty y^2 \exp\left(-y^2/2s\right) dy\right)$$

and integrating by parts in the integral on the right yields

$$f_a(s) = a(2\pi s^3)^{-1/2} \exp(-a^2/2s).$$

The reader will find in Proposition (3.10) another proof based on scaling properties. The densities f_a form a convolution semi-group, namely $f_a * f_b = f_{a+b}$; this is an easy consequence of the value of the Laplace transform, but is also a consequence of the following proposition where we look at T_a as a function of a.

(3.8) Proposition. *The process T_a, $a \geq 0$, is a left-continuous increasing process with stationary independent increments and is purely discontinuous, i.e. there is a.s. no interval on which $a \to T_a$ is continuous.*

Proof. It is left-continuous and increasing as already observed. To prove that

$$P\left[\{\omega : a \to T_a(\omega) \quad \text{is continuous on some interval}\}\right] = 0,$$

we only need to prove that for any pair (p, q) of rational numbers with $p < q$

$$P\left[\{\omega : a \to T_a(\omega) \quad \text{is continuous on} \quad [p, q]\}\right] = 0.$$

We remark that $a \to T_a(\omega)$ is continuous on $[p, q]$ iff S is strictly increasing on $[T_p, T_q]$ (see Sect. 4 Chap. 0), but, since $B_{T_p+t} - B_{T_p}$ is a Brownian motion, this is impossible by the law of the iterated logarithm.

To prove the independence of the increments, pick two real numbers $0 < a < b$. Since $T_b > T_a$ a.s. we have $T_b = T_a + T_b \circ \theta_{T_a}$ a.s., hence, for $f \in \mathscr{E}_+$,

$$\begin{aligned} E\left[f\left(T_b - T_a\right) | \mathscr{F}_{T_a}\right] &= E\left[f\left(T_b \circ \theta_{T_a}\right) | \mathscr{F}_{T_a}\right] \\ &= E_{B_{T_a}}\left[f\left(T_b\right)\right] \quad \text{a.s.} \end{aligned}$$

But $B_{T_a} = a$ a.s. and because of the translation invariance of BM the last displayed term is equal to the constant $E\left[f\left(T_{b-a}\right)\right]$ which shows that $T_b - T_a$ is independent of $\mathscr{F}_{T_a}$, thus completing the proof.

Thus, we have proved that T_a, $a \geq 0$, is a process with stationary independent increments, hence a Feller process (see Sect. 2). It is of course not the canonical version of this process since it is defined on the probability space of the Brownian motion. It is in fact not even right-continuous. We get a right-continuous version by setting $T_{a+} = \lim_{b \downarrow a} T_b$ and proving

(3.9) Proposition. *For any fixed a, $T_a = T_{a+}$ P-a.s.*

Proof. We also have $T_{a+} = \inf\{t : S_t > a\}$ (see Sect. 4 Chap. 0 and Sect. 1 Chap. V). The strong Markov property entails that for every $t > 0$, we have $S_{T_a+t} > a$ whence the result follows immediately. □

Remark. This could also have been proved by passing to the limit as h tends to zero in the equality

$$E\left[\exp\left(-\frac{\lambda^2}{2}(T_{a+h} - T_a)\right)\right] = \exp(-\lambda h)$$

which is a consequence of results in Chap. II.

Furthermore, since T_a, $a \geq 0$, is a Feller process, Proposition (3.9) is also a consequence of the remark after Theorem (2.8).

The above results on the law of T_a which can in particular be used to study the Dirichlet problem in a half-space (see Exercise (3.24)) may also be derived from scaling properties of the family T_a which are of intrinsic interest and will be used in Chap. XI. If a is a positive real, we denote by $B_t^{(a)}$ the Brownian motion $a^{-1}B_{a^2t}$ and adorn with the superscript (a) anything which is defined as a function of $B^{(a)}$. For instance

$$T_1^{(a)} = \inf\left\{t : B_t^{(a)} = 1\right\}.$$

With this notation, we obtain

(3.10) Proposition. *We have $T_a = a^2 T_1^{(a)}$ and consequently $T_a \overset{(d)}{=} a^2 T_1$. Moreover, $T_a \overset{(d)}{=} (a/S_1)^2 \overset{(d)}{=} (a/B_1)^2$.*

Proof. By definition

$$\begin{aligned} T_a &= \inf\{t : a^{-1}B_t = 1\} \\ &= \inf\left\{a^2 t : B_t^{(a)} = 1\right\} = a^2 T_1^{(a)}. \end{aligned}$$

As for the second sentence, it is enough to prove that $T_1 \overset{(d)}{=} S_1^{-2}$. This follows from the scaling property of S_t, namely: $S_t \overset{(d)}{=} \sqrt{t} S_1$. Indeed

$$P[T_1 \geq u] = P[S_u \leq 1] = P\left[\sqrt{u} S_1 \leq 1\right] = P\left[S_1^{-2} \geq u\right].$$

□

Knowing that $S_1 \overset{(d)}{=} |B_1|$, it is now easy to derive anew the law of T_1. We will rather use the above results to prove a property of importance in Chap. X (see also Exercise (3.24) in this section, of which it is a particular case).

(3.11) Proposition. *If β is another standard linear BM independent of B, then $\beta_{T_a} \overset{(d)}{=} a \cdot C$ where C is a Cauchy random variable with parameter 1.*

Proof. Because of the independence of β and B and the scaling properties of β,

$$\beta_{T_a} \stackrel{(d)}{=} \sqrt{T_a} \cdot \beta_1 \stackrel{(d)}{=} \frac{a}{S_1} \cdot \beta_1 \stackrel{(d)}{=} \frac{a}{|B_1|} \cdot \beta_1$$

which ends the proof, since $\beta_1/|B_1|$ is known to have the Cauchy distribution. □

Remarks. (i) The reader may look in Sect. 4 for the properties of the *processes* T_{a+} and β_{T_a}, $a \geq 0$.

(ii) Proposition (3.11) may also be seen as giving the distribution of BM^2 when it first hits a straight line.

We now give a first property of the set of zeros of the linear BM.

(3.12) Proposition. *The set* $Z = \{t : B_t = 0\}$ *is a.s. closed, without any isolated point and has zero Lebesgue measure.*

Proof. That Z is closed follows from the continuity of paths. For any x,

$$E_x\left[\int_0^\infty 1_Z(s)ds\right] = \int_0^\infty P_s(x, \{0\})ds = 0,$$

since $P_s(x, \{0\}) = 0$ for each s and x. It follows that Z has a.s. zero Lebesgue measure, hence also empty interior.

The time zero belongs to Z and we already know from the law of the iterated logarithm that it is not isolated in Z. We prove this again with the techniques of this section. Let $T_0 = \inf\{t : B_t = 0\}$; the time $t + T_0 \circ \theta_t$ is the first point in Z after time t; by the Markov property and the explicit value of $E_0[\exp -\alpha T_a]$ we have

$$\begin{aligned} E_0\left[\exp\{-\alpha(t + T_0 \circ \theta_t)\}\right] &= \exp(-\alpha t)E_0\left[E_{B_t}\left[\exp(-\alpha T_0)\right]\right] \\ &= \exp(-\alpha t)E_0\left[\exp\left\{-|B_t|\sqrt{2\alpha}\right\}\right] \end{aligned}$$

and this converges to 1 as t goes to zero. It follows by Fatou's lemma that $P_0\left[\lim_{t\downarrow 0}(t + T_0 \circ \theta_t) = 0\right] = 1$, namely, 0 is a.s. the limit of points in Z. Now for any rational number q, the time $d_q = q + T_0 \circ \theta_q$ is the first point in Z after q; by Corollary (3.6), B_{d_q+t} is a standard linear BM and therefore d_q is a.s. the limit of points in Z. The set $N = \bigcup_{q\in\mathbb{Q}}\left\{d_q \text{ is not a limit of points in } Z\right\}$ is negligible. If $h \in Z(\omega)$ and if we choose a sequence of rational numbers q_n increasing to h, either h is equal to some d_{q_n} or is the limit of the d_{q_n}'s. Thus, if $\omega \notin N$, in either case h is the limit of points in $Z(\omega)$ which establishes the proposition. □

Thus Z is a perfect set looking like the Cantor "middle thirds" set. We will see in Chap. VI that it is the support of a random measure without point masses and singular with respect to the Lebesgue measure and which somehow accounts for the time spent at 0 by the linear BM. Moreover, the complement of Z is a countable union of disjoint intervals I_n called the *excursion intervals*; the restriction of B to such an interval is called an *excursion* of B. Excursions will be studied in great detail in Chap. XII.

Finally, we complete Proposition (2.19) by proving that X can leave a holding point only by a jump, namely, in the notation of Proposition (2.19):

(3.13) Proposition. *If* $0 < a < \infty$, *then* $P_x\left[X_{\sigma_x} = x\right] = 0$.

Proof. If $a < \infty$, then $P_x[\sigma_x < \infty] = 1$. On $\left\{X_{\sigma_x} = x\right\}$ we have $\sigma_x \circ \theta_{\sigma_x} = 0$ and by the strong Markov property

$$\begin{aligned} P_x\left[\sigma_x < \infty; X_{\sigma_x} = x\right] &= P_x\left[\sigma_x < \infty; X_{\sigma_x} = x, \sigma_x \circ \theta_{\sigma_x} = 0\right] \\ &= E_x\left[1_{\{\sigma_x < \infty; X_{\sigma_x} = x\}} P_{X_{\sigma_x}}\left[\sigma_x = 0\right]\right] \\ &= P_x\left[\sigma_x = 0\right] P_x\left[\sigma_x < \infty; X_{\sigma_x} = x\right]. \end{aligned}$$

Thus if $P_x\left[X_{\sigma_x} = x\right] > 0$ we have $P_x\left[\sigma_x = 0\right] = 1$ which completes the proof.

\# **(3.14) Exercise (More on the reflection principle).** We retain the situation and notation of Proposition (3.7). Questions 3°) and 4°) do not depend on each other.

1°) Prove that the process B^a defined by

$$B_t^a = B_t \quad \text{on} \quad \{t < T_a\}, \qquad B_t^a = 2a - B_t \quad \text{on} \quad \{t \geq T_a\}$$

has the same law as B.

2°) For $a \leq b$, $b > 0$, prove that

$$P\left[S_t > b, B_t < a\right] = P\left[B_t < a - 2b\right] = P_{2b}\left[B_t < a\right],$$

and that the density of the pair (B_t, S_t) is given on $\{(a, b); a \leq b, b > 0\}$ by

$$\left(2/\pi t^3\right)^{1/2}(2b - a)\exp\left(-(2b-a)^2/2t\right).$$

This can also be proved by computing the Laplace transform

$$\int_0^\infty e^{-\alpha t} P\left[S_t > b, B_t < a\right] dt.$$

3°) Prove that for each t, the r.v. $S_t - B_t$ has the same law as $|B_t|$ and that $2S_t - B_t$ has the same law as $|\mathrm{BM}_t^3|$; prove further that, conditionally on $2S_t - B_t$, the r.v.'s S_t and $S_t - B_t$ are uniformly distributed on $[0, 2S_t - B_t]$. Much better results will be proved in Chap. VI.

4°) Let $s_t = \inf_{s \leq t} B_s$ and $a > 0$; prove that under the probability measure P_a restricted to $\{s_t > 0\}$, the r.v. B_t has a density equal to

$$(2\pi t)^{-1/2}\left[\exp\left(-\frac{(b-a)^2}{2t}\right) - \exp\left(-\frac{(b+a)^2}{2t}\right)\right], \qquad b > 0.$$

Compare with Exercise (3.29).

* **(3.15) Exercise.** 1°) Let $a < 0 < b$; prove that for $F \subset]-\infty, a]$ and $t > 0$,

$$P[T_b < T_a, B_t \in F] = P[B_t \in \sigma_b F] - P[T_a < T_b, B_t \in \sigma_b F]$$

where $\sigma_b F = \{2b - y, y \in F\}$.

2°) In the notation of Exercise (3.14), prove that for every Borel subset E of $[a, b]$,

$$P[a \leq s_t < S_t \leq b, B_t \in E] = \int_E k(x)dx$$

where

$$k(x) = (2\pi t)^{-1/2} \sum_{k=-\infty}^{+\infty} \left\{ \exp\left(-\frac{1}{2t}(x + 2k(b-a))^2\right) \right.$$
$$\left. - \exp\left(-\frac{1}{2t}(x - 2b + 2k(b-a))^2\right) \right\}.$$

[Hint: $P[T_a < T_b, T_a \leq t, B_t \in E] = P[T_a < T_b, B_t \in \sigma_a E]$. Apply repeatedly the formula of 1°) to the right-hand side.]

3°) Write down the laws of $B_t^* = \sup_{s \leq t} |B_s|$ and $\tilde{T}_a = \inf\{t : |B_t| > a\}$. This can be done also without using 2°)).

(3.16) Exercise. Let X be the canonical version of a Markov process with transition semi-group P_t. If for every stopping time T, every initial distribution ν and every $f \in \mathscr{E}_+$,

$$E_\nu\left[f(X_{T+t}) | \mathscr{F}_T\right] = P_t f(X_T) \qquad P_\nu\text{-a.s.}$$

on $\{X_T \neq \Delta\}$ then X has the strong Markov property.

(3.17) Exercise. Prove that (B_t, S_t) is a Markov process with values in $E = \{(a, b); a \leq b, b > 0\}$ and using 2°) in Exercise (3.14) compute its transition function.

(3.18) Exercise. For the standard linear BM, prove that

$$\lim_{t \to \infty} \sqrt{t} P[B_s \leq 1, \forall s \leq t] = \sqrt{\frac{2}{\pi}}.$$

\# **(3.19) Exercise.** 1°) Let X be a Feller process, T a finite stopping time. Prove that any $\mathscr{F}_\infty$-measurable and positive r.v. Z may be written $\phi(\omega, \theta_T(\omega))$ where ϕ is $\mathscr{F}_T \otimes \mathscr{F}_\infty$-measurable. Then

$$E_\nu\left[Z | \mathscr{F}_T\right](\omega) = \int \phi(\omega, \omega') P_{X_T(\omega)}(d\omega') \qquad P_\nu\text{-a.s.}$$

2°) Let S be ≥ 0 and $\mathscr{F}_T$-measurable. For a positive Borel function f, prove that

$$E_\nu\left[f(X_{T+S}) | \mathscr{F}_T\right](\omega) = E_{X_T(\omega)}\left[f\left(X_{S(\omega)}(\cdot)\right)\right] \qquad P_\nu\text{-a.s.}$$

This can be proved using 1°) or directly from the strong Markov property.

3°) Write down the proof of Proposition (3.7) by using 2°) with $T = T_a$.

(3.20) Exercise (First Arcsine law). The questions 3°) through 5°) may be solved independently of 1°) and 2°).

1°) For a real number u, let $d_u = u + T_0 \circ \theta_u$ as in the proof of Proposition (3.12). Using the BM $(B_{t+u} - B_u, t \geq 0)$, prove that $d_u \overset{(d)}{=} u + B_u^2 \cdot T_1$ where T_1 is independent of B_u. Hence $d_u \overset{(d)}{=} u(1 + C^2)$ where C is a Cauchy variable with parameter 1.

[Hint: $d_u = u + T_{-B_u}$ where $T_.$ refers to $B_{t+u} - B_u$.]

2°) Prove that the r.v. $g_1 = \sup\{t \leq 1 : B_t = 0\}$ has the density $\left(\pi\sqrt{y(1-y)}\right)^{-1}$ on $[0, 1]$.

[Hint: $\{g_1 < u\} = \{d_u > 1\}$.]

3°) Use the strong Markov property and the properties of hitting times recalled before Theorem (2.17) to give another proof of 2°).

4°) Let $d_1 = \inf\{t > 1 : B_t = 0\}$; by the same arguments as in 3°), prove that the pair (g_1, d_1) has the density $\frac{1}{2\pi} y^{-1/2}(z - y)^{-3/2}$ on $0 \leq y \leq 1 \leq z$.

5°) Compute the law of $d_1 - g_1$. In the language of Chap. XII, this is the law of the length of the excursion straddling 1 (see Exercise (3.7) in Chap. XII). For more about g_1 and d_1, see the following Exercise (3.23).

(3.21) Exercise. Let η be a Bernoulli r.v. Define a family X^x of processes on $\mathbb{R} \cup \{\Delta\}$ by

$$\begin{aligned}
X_t^x &= x + t & &\text{if } x < 0 \text{ and } x + t < 0,\\
X_t^x &= x + t & &\text{if } x < 0, x + t \geq 0 \text{ and } \eta = 1,\\
X_t^x &= \Delta & &\text{if } x < 0, x + t \geq 0 \text{ and } \eta = -1,\\
X_t^x &= x + t & &\text{if } x \geq 0.
\end{aligned}$$

Let P_x be the law of X^x. Prove that under $P_x, x \in \mathbb{R}$, the canonical process is a Strong Markov process which is not a Feller process.

(3.22) Exercise. Let E be the union in $\mathbb{R}^2$ of the sets $\{x \leq 0, y = 0\}$, $\{x > 0; y = x\}$ and $\{x > 0; y = -x\}$; define a transition function on E by setting:

$$\begin{aligned}
\text{for } x \leq 0, \quad & P_t((x,0),\cdot) = \varepsilon_{(x+t,0)} \quad \text{if } x + t \leq 0,\\
& P_t((x,0),\cdot) = \tfrac{1}{2}\varepsilon_{(x+t,x+t)} + \tfrac{1}{2}\varepsilon_{(x+t,-x-t)} \quad \text{if } x + t > 0,\\
\text{for } x > 0, \quad & P_t((x,x),\cdot) = \varepsilon_{(x+t,x+t)},\\
& P_t((x,-x),\cdot) = \varepsilon_{(x+t,-x-t)}.
\end{aligned}$$

Construct a Markov process X on E with t.f. P_t and prove that it enjoys neither the Blumenthal zero-one law nor the Strong Markov property.

[Hint: For the latter, consider the time $T = \inf\{t : X_t \in \{x > 0, y = x\}\}$.]

(3.23) Exercise. For the standard BM and $t > 0$, let

$$g_t = \sup\{s < t : B_s = 0\}, \, d_t = \inf\{s > t : B_s = 0\}.$$

1°) By a simple application of the Markov property, prove that the density of the pair (B_t, d_t) is given by

$$(2\pi)^{-1}|x|\left(t(s-t)^3\right)^{-1/2}\exp\left(-sx^2/2t(s-t)\right)1_{(s\geq t)}.$$

2°) By using the time-inversion invariance of Proposition (1.10) in Chap. I, derive from 1°) that the density of the pair (B_t, g_t) is given by

$$(2\pi)^{-1}|x|\left(s(t-s)^3\right)^{-1/2}\exp\left(-x^2/2(t-s)\right)1_{(s\leq t)}.$$

Sharper results along these lines will be given in Sect. 3 Chap. XII.

(3.24) Exercise. 1°) Denote by (X_t, Y_t) the Brownian motion in $\mathbb{R}^n \times \mathbb{R}$ started at $(0, a)$ with $a > 0$. Let $S_a = \inf\{t : Y_t = 0\}$ and prove that the characteristic function of X_{S_a} is $\exp(-a|u|)$. In other words, the law of X_{S_a} is the *Cauchy law* with parameter a, and this generalizes Proposition (3.11); the corresponding density is equal to

$$\Gamma\left((n+1)/2\right)a/\pi\left(|x|^2+a^2\right)^{(n+1)/2}.$$

2°) If (X_t, Y_t) is started at (x, a), write down the density $P_{(x,a)}(z)$ of X_{S_a}. This density is the *Poisson* kernel. If $f \in C_k(\mathbb{R}^n)$, prove that

$$g(x,a) = \int_{\mathbb{R}^n} P_{(x,a)}(z) f(z)dz$$

is a harmonic function in $\mathbb{R}^n \times]0, \infty[$.

(3.25) Exercise. Let (X, Y) be the standard planar BM and for $a > 0$ let $\tilde{T}_a = \inf\{t : |X_t| = a\}$. Show that the r.v. $Y_{\tilde{T}_a}$ has the density

$$(2a\cosh(\pi x/2a))^{-1}.$$

* **(3.26) Exercise (Local extrema of BM).** Let B be the standard linear BM.

1°) Prove that the probability that a fixed real number x be a local extremum of the Brownian path is zero.

2°) For any positive real number r, prove that

$$P\left[\{\omega : S_r(\omega) \quad \text{is a local extremum of} \quad t \to B_t(\omega), t > r\}\right] = 0.$$

3°) Prove that consequently a.e. Brownian path does not have two equal local extrema. In particular, for every r, there is a.s. at most one $s < r$ such that $B_s = S_r$.

4°) Show that the set of local extrema of the Brownian path is a.s. countable.

(3.27) Exercise. Derive the exponential inequality of Proposition (1.8) in Chap. II from the reflection principle.

(3.28) Exercise. 1°) Prove that the stopping time $\sigma_{a,b}$ of Exercise (3.14) in Chap. II has a density equal to

$$a(2\pi t^3)^{-1/2}\exp\left(-(a-bt)^2/2t\right).$$

[Hint: Use the scaling property of Exercise (3.14) in Chap. II and the known forms of the density and Laplace transform for T_a.]

2°) Derive therefrom another proof of 5°) in the above mentioned exercise.

(3.29) Exercise. 1°) Let B be the linear BM and $T = \inf\{t \geq 0 : B_t = 0\}$. Prove that, for any probability measure P_ν such that ν is carried by $]0, \infty[$, the process X defined by

$$X_t = B_t \quad \text{on} \quad \{t < T\}, \qquad X_t = \Delta \quad \text{on} \quad \{t \geq T\}$$

is a Markov process on $]0, \infty[$ with the transition function Q_t of Exercise (1.15). This process can be called the BM *killed* at 0. As a result for $a > 0$, $Q_t\,(a,]0, \infty[) = P_0(T_a < t)$; check this against Proposition (3.10).

[Hint: See Exercise (2.30). To find the transition function, use the joint law of (B_t, S_t) found in Exercise (3.14).]

2°) Treat the same question for the BM absorbed at 0, that is $X_t = B_{t\wedge T}$.

* **(3.30) Exercise. (Examples of unbounded martingales of BMO).** 1°) If B is the standard linear BM, prove that B^1 (i.e. the BM stopped at time 1) is in BMO and that $\|B^1\|_{\mathrm{BMO}_1} = (2/\pi)^{1/2}$. Prove that $E\left[B_1^2 | \mathscr{F}_t\right]$ is not in BMO.

2°) If B is the standard linear BM and S is a positive r.v. independent of B then $X_t = B_{t\wedge S}$ is a martingale of BMO for the filtration $\mathscr{G}_t = \sigma(S, \mathscr{F}_t)$ if and only if S is a.s. bounded.

** **(3.31) Exercise.** If h is a real-valued, continuous and increasing function on $]0, 1[$, if $h(t)/\sqrt{t}$ is decreasing and $\int_{0+} t^{-3/2} h(t) \exp\left(-h^2(t)/2t\right) dt < \infty$, prove that

$$P\big[B_t \geq h(t) \text{ for some } t \in]0, b[\big] \leq \int_{0+}^{b} \frac{h(t)}{\sqrt{2\pi t^3}} \exp\left(-h^2(t)/2t\right) dt.$$

Show that, as a result, the function h belongs to the upper class (Exercise (2.32)).

[Hint: For $0 < a < b$ and a subdivision (t_n) of $]a, b[$,

$$\begin{aligned} P\big[B_t \geq h(t) \text{ for some } t \in]a, b[\big] \;&\leq\; P\left[T_{h(a)} \leq a\right] \\ &+ \sum_k P\left[t_{k-1} < T_{h(t_{k-1})} \leq t_k\right].] \end{aligned}$$

It is also true, but more difficult to prove, that if the integral diverges, h belongs to the lower class. The criterion thus obtained is known as *Kolmogorov's test*.

§4. Summary of Results on Lévy Processes

In Sect. 2, we defined the Lévy processes which include Brownian motion and the Poisson process. In Sect. 3, we found, while studying BM, another example of a Lévy process, namely the process T_{a+}. This is just one of the many examples of Lévy processes cropping up in the study of BM. Thus, it seems worthwhile to pause a little while in order to state without proofs a few facts about Lévy processes which may be used in the sequel. Lévy processes have been widely studied in their own right; if nothing else, their properties often hint at properties of general Markov processes as will be seen in Chap. VII about infinitesimal generators.

In what follows, we deal only with real-valued Lévy processes. We recall that a probability measure μ on $\mathbb{R}$, or a real-valued r.v. Y with law μ, is said to be *infinitely divisible* if, for any $n \geq 1$, there is a probability measure μ_n such that $\mu = \mu_n^{*n}$ or equivalently if Y has the law of the sum of n independent identically distributed random variables. It is easy to see that Gaussian, Poisson or Cauchy variables are infinitely divisible.

Obviously, if X is a Lévy process, then any r.v. X_t is infinitely divisible. Conversely, it was proved by Lévy that any infinitely divisible r.v. Y may be imbedded in a unique convolution semi-group, in other words, there is a Lévy process X such that $Y \stackrel{(d)}{=} X_1$. This can be proved as follows. By analytical methods, one can show that μ is infinitely divisible, if and only if, its Fourier transform $\hat{\mu}$ is equal to $\exp(\psi)$ with

$$\psi(u) = i\beta u - \frac{\sigma^2 u^2}{2} + \int \left(e^{iux} - 1 - \frac{iux}{1+x^2} \right) \nu(dx)$$

where $\beta \in \mathbb{R}, \sigma \geq 0$ and ν is a Radon measure on $\mathbb{R} - \{0\}$ such that

$$\int \frac{x^2}{1+x^2} \nu(dx) < \infty.$$

This formula is known as the *Lévy-Khintchine* formula and the measure ν as the *Lévy measure*. For every $t \in \mathbb{R}_+$, $\exp(t\psi)$ is now clearly the Fourier transform of a probability measure μ_t and plainly $\mu_t * \mu_s = \mu_{t+s}$ and $\lim_{t \downarrow 0} \mu_t = \varepsilon_0$ which proves Lévy's theorem.

The different terms which appear in the Lévy-Khintchine formula have a probabilistic significance which will be further emphasized in Chap. VII. If $\sigma = 0$ and $\nu = 0$, then $\mu = \varepsilon_\beta$ and the corresponding semi-group is that of translation at speed β; if $\beta = 0$ and $\nu = 0$, the semi-group is that of a multiple of BM and the corresponding Lévy process has continuous paths; if $\beta = 0$ and $\sigma = 0$, we get a "pure jump" process as is the case for the process T_{a^+} of the preceding section. Every Lévy process is obtained as a sum of independent processes of the three types above. Thus, the Lévy measure accounts for the jumps of X and the knowledge of ν permits to give a probabilistic construction of X as is hinted at in Exercise (1.18) of Chap. XII.

Among the infinitely divisible r.v.'s, the so-called *stable* r.v.'s form a subclass of particular interest.

(4.1) Definition. *A r.v. Y is stable if, for every k, there are independent r.v.'s $Y_1, \ldots, Y_k$ with the same law as Y and constants $a_k > 0$, b_k such that*

$$Y_1 + \ldots + Y_k \stackrel{(d)}{=} a_k Y + b_k.$$

It can be proved that this equality forces $a_k = k^{1/\alpha}$ where $0 < \alpha \leq 2$. The number α is called the *index* of the stable law. Stable laws are clearly infinitely divisible. For $\alpha = 2$, we get the Gaussian r.v.'s; for $0 < \alpha < 2$ we have the following characterization of the corresponding function ψ.

(4.2) Theorem. *If Y is stable with index $\alpha \in]0, 2[$, then $\sigma = 0$ and the Lévy measure has the density $\left(m_1 1_{(x<0)} + m_2 1_{(x>0)}\right) |x|^{-(1+\alpha)}$ with m_1 and $m_2 \geq 0$.*

With each stable r.v. of index α, we may, as we have already pointed out, associate a Lévy process which will also be called a stable process of index α. The process T_{a^+} of the last section is thus a stable process of index 1/2. If, in the above result, we make $\beta = 0$ and $m_1 = m_2$ we get the *symmetric stable* process of order α. In that case, $\psi(u) = -c|u|^\alpha$ where c is a positive parameter. Among those are the linear BM and the *Cauchy process* which is the symmetric stable process of index 1 such that $\hat{\mu}_t = \exp(-t|u|)$. These processes have a scaling invariance property germane to that of BM, namely, for any $c > 0$, $c^{-1} X_{c^\alpha t}$ has the same law as X_t.

Another interesting subclass of stable processes is that of *stable subordinators*. Those are the non-decreasing stable processes or equivalently the stable processes X such that X_t is a.s. ≥ 0. The corresponding stable law is thus carried by $[0, \infty[$ and may be characterized by its Laplace transform. It turns out that for the index α, this Laplace transform is equal to $\exp(-c\lambda^\alpha)$ for $0 < \alpha < 1$; indeed, for $\alpha \in]1, 2]$ this function is not a Laplace transform and there is no stable subordinator of index α (the case $\alpha = 1$ is obvious). Once again, the process T_{a^+} provides us with an example of a stable subordinator of index 1/2.

If τ_a is a stable subordinator of index $\alpha \in]0, 1[$ vanishing at 0 and X is a Lévy process independent of the process τ_a, the map $a \to X_{\tau_a}$ makes sense and is again a Lévy process as the reader can prove as an exercise. If X is the linear BM, an easy computation shows that X_{τ_a} is a symmetric stable process of index 2α. If, in particular, $\alpha = 1/2$, X_{τ_a} is a symmetric Cauchy process, which generalizes Proposition (3.11).

(4.3) Exercise. If X is a Lévy process, prove that $\exp(iuX_t - t\psi(u))$ is a complex martingale for every real u.

(4.4) Exercise. Derive the scaling invariance property of the process T_{a^+} from the scaling invariance property of BM.

* **(4.5) Exercise.** Let T_a, $a \geq 0$ be a right-continuous stable subordinator of index α vanishing at zero. Since its paths are right-continuous and increasing for every positive Borel function f on $\mathbb{R}_+$, the Stieltjes integral

$$\int_0^t f(a) dT_a$$

makes sense and defines a r.v. $T(f)_t$.

1°) What is the necessary and sufficient condition that f must satisfy in order that $T(f)_t <$ a.s. for every t? What is the law of $T(f)_t$ in that case?

2°) Prove that there is a constant c_α such that the process S defined by

$$S_t = c_\alpha \int_0^{t^{1/(1-\alpha)}} a^{-1} dT_a$$

has the same law as T_a, $a \geq 0$.

(4.6) Exercise (Harnesses and Lévy processes). Let X be an integrable Lévy process i.e. such that $E[|X_1|] < \infty$, and for $0 \leq s < t$ define $\mathscr{F}_{s,t} = \sigma\{X_u, u \leq s;\ X_v, v \geq t\}$.

1°) Prove that for any reals $0 \leq c \leq a < b \leq d$,

$$E\left[(X_b - X_a)/(b-a) \mid \mathscr{F}_{c,d}\right] = (X_d - X_c)/(d-c).$$

A process satisfying this condition is called a *harness*.

2°) For a fixed $T > 0$, prove that if X is a harness, then

$$X_t - \int_0^{t \wedge T} ((X_T - X_s)/(T-s))\, ds, \quad t \leq T,$$

is a $(\mathscr{F}_{t,T})$-martingale. Compare with Exercise (3.18) Chap. IV.

3°) Prove that the process X' defined by $X'_t = tX_{1/t}$, $t > 0$, is also a harness.

Notes and Comments

The material covered in this chapter is classical and is kept to the minimum necessary for the understanding of the sequel. We have used the books by Blumenthal-Getoor [1], Meyer [1] and Chung [2]; the latter is an excellent means of getting more acquainted with Markov processes and their potential theory and complements very nicely our own book. The reader may also use volume 4 of Dellacherie-Meyer [1]. For a more advanced and up-to-date exposition of Markov processes we recommend the book of Sharpe [3].

Most results of this chapter may be found in the above sources. Let us merely mention that Exercise (1.16) is taken from Walsh [3] and that Exercise (1.17) 1°) is from Dynkin [1] whereas the other questions were taken in Chung [2]. In connection with this exercise, let us mention that Pitman-Rogers [1] contains another useful criterion for a function of a Markov process to still be a Markov process.

Kolmogorov's test of Exercise (3.31) may be found in Itô-McKean [1]. The equivalence between iii) and iv) in Exercise (2.31) may be found in Jeulin-Yor [2] and Jeulin [4]. Exercise (3.15) is borrowed from Freedman [1].

For advanced texts on Lévy processes we refer to Bertoin [7] and Sato [1].

The notion of harness of Exercise (4.6) is due to Hammersley, and was discussed by Williams in several papers (see Chaumont-Yor [1], Exercise 6.19). The result in question 2°), in the case of Lévy processes, is found in Jacod-Protter [1]. The result in 3°) is due to Williams (unpublished).

Chapter IV. Stochastic Integration

In this chapter, we introduce some basic techniques and notions which will be used throughout the sequel. Once and for all, we consider below, a filtered probability space $(\Omega, \mathscr{F}, \mathscr{F}_t, P)$ and we suppose that each $\mathscr{F}_t$ contains all the sets of P-measure zero in $\mathscr{F}$. As a result, any limit (almost-sure, in the mean, etc.) of adapted processes is an adapted process; a process which is indistinguishable from an adapted process is adapted.

§1. Quadratic Variations

(1.1) Definition. *A process A is* increasing *(resp. of* finite variation*) if it is adapted and the paths $t \to A_t(\omega)$ are finite, right-continuous and increasing (resp. of finite variation) for almost every ω.*

We will denote by $\mathscr{A}^+$ (resp. $\mathscr{A}$) the space of increasing (resp. of finite variation) processes. Plainly, $\mathscr{A}^+ \subset \mathscr{A}$ and conversely, it is easily seen from Sect. 4 Chap. 0 that any element $A \in \mathscr{A}$ can be written $A_t = A_t^+ - A_t^-$ where A^+ and A^- are in $\mathscr{A}^+$. Moreover, A^+ and A^- can be chosen so that for almost every ω, $A_t^+(\omega) - A_t^-(\omega)$ is the minimal decomposition of $A_t(\omega)$. The process $\int_0^t |dA|_s = A_t^+ + A_t^-$ is in $\mathscr{A}^+$ and for a.e. ω the measure associated with it is the total variation of that which is associated with $A(\omega)$; it is called the *variation* of A.

One can clearly integrate appropriate functions with respect to the measure associated to $A(\omega)$ and thus obtain a "stochastic integral". More precisely, if X is progressively measurable and – for instance – bounded on every interval $[0, t]$ for a.e. ω, one can define for a.e. ω, the Stieltjes integral

$$(X \cdot A)_t(\omega) = \int_0^t X_s(\omega) dA_s(\omega).$$

If ω is in the set where $A_.(\omega)$ is not of finite variation or $X_.(\omega)$ is not locally integrable with respect to $dA(\omega)$, we put $(X \cdot A) = 0$. The reader will have no difficulty in checking that the process $X \cdot A$ thus defined is in $\mathscr{A}$. The hypothesis that X be progressively measurable is precisely made to ensure that $X \cdot A$ is adapted. It is the "stochastic integral" of X with respect to the process A of $\mathscr{A}$.

Our goal is now to define a "stochastic integral" with respect to martingales. A clue to the difficulty, already mentioned in the case of BM, is given by the

(1.2) Proposition. *A continuous martingale M cannot be in $\mathscr{A}$ unless it is constant.*

Proof. We may suppose that $M_0 = 0$ and prove that M is identically zero if it is of finite variation. Let V_t be the variation of M on $[0, t]$ and define

$$S_n = \inf\{s : V_s \geq n\};$$

then the martingale M^{S_n} is of bounded variation. Thus, it is enough to prove the result whenever the variation of M is bounded by a number K.

Let $\Delta = \{t_0 = 0 < t_1 < \ldots < t_k = t\}$ be a subdivision of $[0, t]$; we have

$$\begin{aligned} E\left[M_t^2\right] &= E\left[\sum_{i=0}^{k-1}\left(M_{t_{i+1}}^2 - M_{t_i}^2\right)\right] \\ &= E\left[\sum_{i=0}^{k-1}\left(M_{t_{i+1}} - M_{t_i}\right)^2\right] \end{aligned}$$

since M is a martingale. As a result,

$$E[M_t^2] \leq E\left[V_t\left(\sup_i\left|M_{t_{i+1}} - M_{t_i}\right|\right)\right] \leq K E\left[\sup_i\left|M_{t_{i+1}} - M_{t_i}\right|\right];$$

when the modulus of Δ goes to zero, this quantity goes to zero since M is continuous, hence $M = 0$ a.s. □

Remark. The reader may find more suggestive the proof outlined in Exercise (1.32).

Because of this proposition, we will not be able to define integrals with respect to M by a path by path procedure. We will have to use a global method in which the notions we are about to introduce play a crucial role. We retain the notation of Sect. 2 Chap. I. If $\Delta = \{t_0 = 0 < t_1 < \ldots\}$ is a subdivision of $\mathbb{R}_+$ with only a finite number of points in each interval $[0, t]$ we define, for a process X,

$$T_t^\Delta(X) = \sum_{i=0}^{k-1}\left(X_{t_{i+1}} - X_{t_i}\right)^2 + \left(X_t - X_{t_k}\right)^2$$

where k is such that $t_k \leq t < t_{k+1}$; we will write simply T_t^Δ if there is no risk of confusion. We recall from Sect. 2 Chap. I that X is said to be of *finite quadratic variation* if there exists a process $\langle X, X\rangle$ such that for each t, T_t^Δ converges in probability to $\langle X, X\rangle_t$ as the modulus of Δ on $[0, t]$ goes to zero. The main result of this section is the

(1.3) Theorem. *A continuous and bounded martingale M is of finite quadratic variation and $\langle M, M\rangle$ is the unique continuous increasing adapted process vanishing at zero such that $M^2 - \langle M, M\rangle$ is a martingale.*

Proof. Uniqueness is an easy consequence of Proposition (1.2), since if there were two such processes A and B, then $A - B$ would be a continuous martingale of $\mathscr{A}$ vanishing at zero.

To prove the existence of $\langle M, M\rangle$, we first observe that since for $t_i < s < t_{i+1}$,

$$E\left[\left(M_{t_{i+1}} - M_{t_i}\right)^2 \mid \mathscr{F}_s\right] = E\left[\left(M_{t_{i+1}} - M_s\right)^2 \mid \mathscr{F}_s\right] + \left(M_s - M_{t_i}\right)^2$$

it is easily proved that

$$\begin{aligned} E\left[T_t^{\Delta}(M) - T_s^{\Delta}(M) \mid \mathscr{F}_s\right] &= E\left[(M_t - M_s)^2 \mid \mathscr{F}_s\right] \\ &= E\left[M_t^2 - M_s^2 \mid \mathscr{F}_s\right]. \end{aligned} \tag{1.1}$$

As a result, $M_t^2 - T_t^{\Delta}(M)$ is a continuous martingale. In the sequel, we write T_t^{Δ} instead of $T_t^{\Delta}(M)$.

We now fix $a > 0$ and we are going to prove that if $\{\Delta_n\}$ is a sequence of subdivisions of $[0, a]$ such that $|\Delta_n|$ goes to zero, then $\{T_a^{\Delta_n}\}$ converges in L^2.

If Δ and Δ' are two subdivisions we call $\Delta\Delta'$ the subdivision obtained by taking all the points of Δ and Δ'. By eq. (1.1) the process $X = T^{\Delta} - T^{\Delta'}$ is a martingale and, by eq. (1.1) again, applied to X instead of M, we have

$$E\left[X_a^2\right] = E\left[\left(T_a^{\Delta} - T_a^{\Delta'}\right)^2\right] = E\left[T_a^{\Delta\Delta'}(X)\right].$$

Because $(x + y)^2 \leq 2(x^2 + y^2)$ for any pair (x, y) of real numbers,

$$T_a^{\Delta\Delta'}(X) \leq 2\left\{T_a^{\Delta\Delta'}(T^{\Delta}) + T_a^{\Delta\Delta'}(T^{\Delta'})\right\}$$

and to prove our claim, it is enough to show that $E\left[T_a^{\Delta\Delta'}(T^{\Delta})\right]$ converges to 0 as $|\Delta| + |\Delta'|$ goes to zero.

Let then s_k be in $\Delta\Delta'$ and t_l be the rightmost point of Δ such that $t_l \leq s_k < s_{k+1} \leq t_{l+1}$; we have

$$\begin{aligned} T_{s_{k+1}}^{\Delta} - T_{s_k}^{\Delta} &= \left(M_{s_{k+1}} - M_{t_l}\right)^2 - \left(M_{s_k} - M_{t_l}\right)^2 \\ &= \left(M_{s_{k+1}} - M_{s_k}\right)\left(M_{s_{k+1}} + M_{s_k} - 2M_{t_l}\right), \end{aligned}$$

and consequently,

$$T_a^{\Delta\Delta'}(T^{\Delta}) \leq \left(\sup_k \left|M_{s_{k+1}} + M_{s_k} - 2M_{t_l}\right|^2\right) T_a^{\Delta\Delta'}.$$

By Schwarz's inequality,

$$E\left[T_a^{\Delta\Delta'}(T^{\Delta})\right] \leq E\left[\sup_k \left|M_{s_{k+1}} + M_{s_k} - 2M_{t_l}\right|^4\right]^{1/2} E\left[(T_a^{\Delta\Delta'})^2\right]^{1/2}.$$

Whenever $|\Delta| + |\Delta'|$ tends to zero, the first factor goes to zero because M is continuous; it is therefore enough to prove that the second factor is bounded by a constant independent of Δ and Δ'. To this end, we write with $a = t_n$,

$$
\begin{aligned}
(T_a^\Delta)^2 &= \left(\sum_{k=1}^n \left(M_{t_k} - M_{t_{k-1}}\right)^2\right)^2 \\
&= 2\sum_{k=1}^n \left(T_a^\Delta - T_{t_k}^\Delta\right)\left(T_{t_k}^\Delta - T_{t_{k-1}}^\Delta\right) + \sum_{k=1}^n \left(M_{t_k} - M_{t_{k-1}}\right)^4 .
\end{aligned}
$$

Because of eq. (1.1), we have $E\left[T_a^\Delta - T_{t_k}^\Delta \mid \mathscr{F}_{t_k}\right] = E\left[\left(M_a - M_{t_k}\right)^2 \mid \mathscr{F}_{t_k}\right]$ and consequently

$$
\begin{aligned}
E\left[(T_a^\Delta)^2\right] &= 2\sum_{k=1}^n E\left[\left(M_a - M_{t_k}\right)^2 \left(T_{t_k}^\Delta - T_{t_{k-1}}^\Delta\right)\right] \\
&\quad + \sum_{k=1}^n E\left[\left(M_{t_k} - M_{t_{k-1}}\right)^4\right] \\
&\le E\left[\left(2\sup_k \left|M_a - M_{t_k}\right|^2 + \sup_k \left|M_{t_k} - M_{t_{k-1}}\right|^2\right) T_a^\Delta\right].
\end{aligned}
$$

Let C be a constant such that $|M| \le C$; by eq. (1.1), it is easily seen that $E[T_a^\Delta] \le 4C^2$ and therefore

$$
E\left[(T_a^\Delta)^2\right] \le 12\, C^2 E[T_a^\Delta] \le 48\, C^4 .
$$

We have thus proved that for any sequence $\{\Delta_n\}$ such that $|\Delta_n| \to 0$, the sequence $\{T_a^{\Delta_n}\}$ has a limit $\langle M, M\rangle_a$ in L^2 hence in probability. It remains to prove that $\langle M, M\rangle_a$ may be chosen within its equivalence class in such a way that the resulting process $\langle M, M\rangle$ has the required properties.

Let $\{\Delta_n\}$ be as above; by Doob's inequality applied to the martingale $T^{\Delta_n} - T^{\Delta_m}$,

$$
E\left[\sup_{t\le a}\left|T_t^{\Delta_n} - T_t^{\Delta_m}\right|^2\right] \le 4E\left[\left(T_a^{\Delta_n} - T_a^{\Delta_m}\right)^2\right].
$$

Since, from a sequence converging in L^2, one can extract a subsequence converging a.s., there is a subsequence $\{\Delta_{n_k}\}$ such that $T_t^{\Delta_{n_k}}$ converges a.s. uniformly on $[0, a]$ to a limit $\langle M, M\rangle_t$ which perforce is a.s. continuous. Moreover, the original sequence might have been chosen such that Δ_{n+1} be a refinement of Δ_n and $\bigcup_n \Delta_n$ be dense in $[0, a]$. For any pair (s, t) in $\bigcup_n \Delta_n$ such that $s < t$, there is an n_0 such that s and t belong to Δ_n for any $n \ge n_0$. We then have $T_s^{\Delta_n} \le T_t^{\Delta_n}$ and as a result $\langle M, M\rangle$ is increasing on $\bigcup_n \Delta_n$; as it is continuous, it is increasing everywhere (although the T^{Δ_n} are not necessarily increasing!).

Finally, that $M^2 - \langle M, M\rangle$ is a martingale follows upon passing to the limit in eq. (1.1). The proof is thus complete.

To enlarge the scope of the above result we will need the

(1.4) Proposition. *For every stopping time T,*

$$
\langle M^T, M^T\rangle = \langle M, M\rangle^T .
$$

Proof. By the optional stopping theorem, $(M^T)^2 - \langle M, M\rangle^T$ is a martingale, so that the result is a consequence of the uniqueness in Theorem (1.3).

Much as it is interesting, Theorem (1.3) is not sufficient for our purposes; it does not cover, for instance, the case of the Brownian motion B which is not a bounded martingale. Nonetheless, we have seen that B has a "quadratic variation", namely t, and that $B_t^2 - t$ is a martingale exactly as in Theorem (1.3). We now show how to subsume the case of BM and the case of bounded martingales in a single result by using the fecund idea of *localization*.

(1.5) Definition. *An adapted, right-continuous process X is an $(\mathscr{F}_t, P)$*-local martingale *if there exist stopping times T_n, $n \geq 1$, such that*

i) the sequence $\{T_n\}$ is increasing and $\lim_n T_n = +\infty$ *a.s.;*
ii) for every n, the process $X^{T_n} 1_{[T_n>0]}$ is a uniformly integrable $(\mathscr{F}_t, P)$-martingale.

We will drop $(\mathscr{F}_t, P)$ when there is no risk of ambiguity. In condition ii) we can drop the uniform integrability and ask only that $X^{T_n} 1_{[T_n>0]}$ be a martingale; indeed, one can always replace T_n by $T_n \wedge n$ to obtain a u.i. martingale. Likewise, if X is continuous as will nearly always be in this book, by setting $S_n = \inf\{t : |X_t| = n\}$ and replacing T_n by $T_n \wedge S_n$, we may assume the martingales in ii) to be bounded. This will be used extensively in the sequel. In Sect. 3 we will find a host of examples of continuous local martingales.

We further say that the stopping time T *reduces* X if $X^T 1_{[T>0]}$ is a u.i. martingale. This property can be decomposed in two parts if one introduces the process $Y_t = X_t - X_0$: T reduces X if and only if

i) X_0 is integrable on $\{T > 0\}$; ii) Y^T is a u.i. martingale.

A common situation however is that in which X_0 is constant and in that case one does not have to bother with i). This explains why in the sequel we will often drop the qualifying $1_{[T>0]}$. As an exercise, the reader will show the following simple properties (see also Exercise (1.30)):

i) if T reduces X and $S \leq T$, then S reduces X;
ii) the sum of two local martingales is a local martingale;
iii) if Z is a $\mathscr{F}_0$-measurable r.v. and X is a local martingale then, so is ZX; in particular, the set of local martingales is a vector space;
iv) a stopped local martingale is a local martingale;
v) a positive local martingale is a supermartingale.

Brownian motion or, more generally, any right-continuous martingale is a local martingale as is seen by taking $T_n = n$, but we stress the fact that local martingales are *much more general* than martingales and warn the reader against the common mistaken belief that local martingales need only be integrable in order to be martingales. As will be shown in Exercise (2.13) of Chap. V, there exist local martingales possessing strong integrability properties which nonetheless, are not martingales. However, let us set the

(1.6) Definition. *A real valued adapted process X is said of class (D) if the family of random variables $X_T 1_{(T<\infty)}$ where T ranges through all stopping times is uniformly integrable. It is of class DL if for every $a > 0$, the family of random variables X_T, where T ranges through all stopping times less than a, is uniformly integrable.*

A uniformly integrable martingale is of class (D). Indeed, by Sect. 3 in Chap. II, we then have $X_T 1_{[T<\infty]} = E\left[X_\infty \mid \mathscr{F}_T\right] 1_{[T<\infty]}$ and it is known that if Y is integrable on $(\Omega, \mathscr{A}, P)$, the family of conditional expectations $E[Y \mid \mathscr{B}]$ where $\mathscr{B}$ ranges through the sub-σ-fields of $\mathscr{A}$ is uniformly integrable. But other processes may well be uniformly integrable without being of class (D). For local martingales, we have the

(1.7) Proposition. *A local martingale is a martingale if and only if it is of class (DL).*

Proof. Left to the reader as an exercise. See also Exercise (1.46).

We now state the result for which the notion of local martingale was introduced.

(1.8) Theorem. *If M is a continuous local martingale, there exists a unique increasing continuous process $\langle M, M\rangle$, vanishing at zero, such that $M^2 - \langle M, M\rangle$ is a continuous local martingale. Moreover, for every t and for any sequence $\{\Delta_n\}$ of subdivisions of $[0, t]$ such that $|\Delta_n| \to 0$, the r.v.'s*

$$\sup_{s\le t} \left|T_s^{\Delta_n}(M) - \langle M, M\rangle_s\right|$$

converge to zero in probability.

Proof. Let $\{T_n\}$ be a sequence of stopping times increasing to $+\infty$ and such that $X_n = M^{T_n} 1_{[T_n>0]}$ is a bounded martingale. By Theorem (1.3), there is, for each n, a continuous process A_n in $\mathscr{A}^+$ vanishing at zero and such that $X_n^2 - A_n$ is a martingale. Now, $\left(X_{n+1}^2 - A_{n+1}\right)^{T_n} 1_{[T_n>0]}$ is a martingale and is equal to $X_n^2 - A_{n+1}^{T_n} 1_{[T_n>0]}$. By the uniqueness property in Theorem (1.3), we have $A_{n+1}^{T_n} = A_n$ on $[T_n > 0]$ and we may therefore define unambiguously a process $\langle M, M\rangle$ by setting it equal to A_n on $[T_n > 0]$. Obviously, $(M^{T_n})^2 1_{[T_n>0]} - \langle M, M\rangle^{T_n}$ is a martingale and therefore $\langle M, M\rangle$ is the sought-after process. The uniqueness follows from the uniqueness on each interval $[0, T_n]$.

To prove the second statement, let $\delta, \varepsilon > 0$ and t be fixed. One can find a stopping time S such that $M^S 1_{[S>0]}$ is bounded and $P[S \le t] \le \delta$. Since $T^\Delta(M)$ and $\langle M, M\rangle$ coincide with $T^\Delta(M^S)$ and $\langle M^S, M^S\rangle$ on $[0, S]$, we have

$$P\left[\sup_{s\le t} \left|T_s^\Delta(M) - \langle M, M\rangle_s\right| > \varepsilon\right] \le \delta + P\left[\sup_{s\le t} \left|T_s^\Delta(M^S) - \langle M^S, M^S\rangle_s\right| > \varepsilon\right]$$

and the last term goes to zero as $|\Delta|$ tends to zero. □

Theorem (1.8) may still be further extended by *polarization*.

(1.9) Theorem. *If M and N are two continuous local martingales, there exists a unique continuous process $\langle M, N\rangle$ in $\mathscr{A}$, vanishing at zero and such that $MN - \langle M, N\rangle$ is a local martingale. Moreover, for any t and any sequence $\{\Delta_n\}$ of subdivisions of $[0, t]$ such that $|\Delta_n| \to 0$,*

$$P\text{-}\lim \sup_{s \le t} \left| \tilde{T}_s^{\Delta_n} - \langle M, N\rangle_s \right| = 0,$$

where $\tilde{T}_s^{\Delta_n} = \sum_{t_i \in \Delta_n} \left(M_{t_{i+1}}^s - M_{t_i}^s\right)\left(N_{t_{i+1}}^s - N_{t_i}^s\right)$.

Proof. The uniqueness follows again from Proposition (1.2) after suitable stoppings. Moreover the process

$$\langle M, N\rangle = \frac{1}{4}\left[\langle M + N, M + N\rangle - \langle M - N, M - N\rangle\right]$$

is easily seen to have the desired properties.

(1.10) Definition. *The process $\langle M, N\rangle$ is called the* bracket *of M and N, the process $\langle M, M\rangle$ the* increasing process *associated with M or simply the increasing process of M.*

In the following sections, we will give general examples of computation of brackets; the reader can already look at Exercises (1.36) and (1.44) in this section. In particular, if M and N are independent, the product MN is a local martingale hence $\langle M, N\rangle = 0$ (see Exercise (1.27)).

(1.11) Proposition. *If T is a stopping time,*

$$\langle M^T, N^T\rangle = \langle M, N^T\rangle = \langle M, N\rangle^T.$$

Proof. This is an obvious consequence of the last part of Theorem (1.9). As an exercise, the reader may also observe that $M^T N^T - \langle M, N\rangle^T$ and $M^T(N - N^T)$ are local martingales, hence by difference, so is $M^T N - \langle M, N\rangle^T$. □

The properties of the bracket operation are reminiscent of those of a scalar product. The map $(M, N) \to \langle M, N\rangle$ is bilinear, symmetric and $\langle M, M\rangle \ge 0$; it is also non-degenerate as is shown by the following

(1.12) Proposition. $\langle M, M\rangle = 0$ *if and only if M is constant, that is $M_t = M_0$ a.s. for every t.*

Proof. By Proposition (1.11), it is enough to consider the case of a bounded M and then by Theorem (1.3), $E[(M_t - M_0)^2] = E[\langle M, M\rangle_t]$; the result follows immediately.

This property may be extended in the following way.

(1.13) Proposition. *The intervals of constancy are the same for M and for $\langle M, M\rangle$, that is to say, for almost all ω's, $M_t(\omega) = M_a(\omega)$ for $a \le t \le b$ if and only if $\langle M, M\rangle_b(\omega) = \langle M, M\rangle_a(\omega)$.*

Proof. We first observe that if M is constant on $[a, b]$, its quadratic variation is obviously constant on $[a, b]$. Conversely, for a rational number q, the process $N_t = M_{t+q} - M_q$ is a $(\mathscr{F}_{t+q})$-local martingale with increasing process $\langle N, N\rangle_t = \langle M, M\rangle_{t+q} - \langle M, M\rangle_q$. The random variable

$$T_q = \inf\{s > 0 : \langle N, N\rangle_s > 0\}$$

is a $(\mathscr{F}_{t+q})$-stopping time, and for the stopped local martingale N^{T_q}, we have

$$\langle N^{T_q}, N^{T_q}\rangle = \langle N, N\rangle^{T_q} = \langle M, M\rangle_{q+T_q} - \langle M, M\rangle_q = 0.$$

By Proposition (1.12), M is a.s. constant on the interval $[q, q + T_q]$, hence is a.s. constant on all the intervals $[q, q + T_q]$ where q runs through $\mathbb{Q}_+$. Since any interval of constancy of $\langle M, M\rangle$ is the closure of a countable union of intervals $[q, q + T_q]$, the proof is complete.

The following inequality will be very useful in defining stochastic integrals. It shows in particular that $d\langle M, N\rangle$ is absolutely continuous with respect to $d\langle M, M\rangle$.

(1.14) Definition. *A real-valued process H is said to be* measurable *if the map $(\omega, t) \to H_t(\omega)$ is $\mathscr{F}_\infty \otimes \mathscr{B}(\mathbb{R}_+)$-measurable.*

The class of measurable processes is obviously larger than the class of progressively measurable processes.

(1.15) Proposition. *For any two continuous local martingales M and N and measurable processes H and K, the inequality*

$$\int_0^t |H_s||K_s||d\langle M, N\rangle|_s \le \left(\int_0^t H_s^2 d\langle M, M\rangle_s\right)^{1/2} \left(\int_0^t K_s^2 d\langle N, N\rangle_s\right)^{1/2}$$

holds a.s. for $t \le \infty$.

Proof. By taking increasing limits, it is enough to prove the inequality for $t < \infty$ and for bounded H and K. Moreover, it is enough to prove the inequality where the left-hand side has been replaced by

$$\left|\int_0^t H_s K_s d\langle M, N\rangle_s\right|;$$

indeed, if J_s is a density of $d\langle M, N\rangle_s / |d\langle M, N\rangle|_s$ with values in $\{-1, 1\}$ and we replace H by HJ sgn (HK) in this expression, we get the left-hand side of the statement.

By a density argument, it is enough to prove that for those K's which may be written

$$K = K_0 1_{\{0\}} + K_1 1_{]0,t_1]} + \ldots + K_n 1_{]t_{n-1},t_n]}$$

for a finite subdivision $\{t_0 = 0 < t_1 < \ldots < t_n = t\}$ of $[0, t]$ and bounded measurable r.v. K_i's. By another density argument, we can also take H of the same form and with the same subdivision.

If we now define $\langle M, N\rangle_s^t = \langle M, N\rangle_t - \langle M, N\rangle_s$, we have

$$\left|\langle M, N\rangle_s^t\right| \le \left(\langle M, M\rangle_s^t\right)^{1/2}\left(\langle N, N\rangle_s^t\right)^{1/2} \qquad \text{a.s.}$$

Indeed, almost surely, the quantity

$$\langle M, M\rangle_s^t + 2r\langle M, N\rangle_s^t + r^2\langle N, N\rangle_s^t = \langle M + rN, M + rN\rangle_s^t$$

is non-negative for every $r \in \mathbb{Q}$, hence by continuity for every $r \in \mathbb{R}$, and our claim follows from the usual quadratic form reasoning.

As a result

$$\begin{aligned}\left|\int_0^t H_s K_s d\langle M, N\rangle_s\right| &\le \sum_i |H_i K_i| \left|\langle M, N\rangle_{t_i}^{t_{i+1}}\right| \\ &\le \sum_i |H_i|\,|K_i| \left(\langle M, M\rangle_{t_i}^{t_{i+1}}\right)^{1/2}\left(\langle N, N\rangle_{t_i}^{t_{i+1}}\right)^{1/2}\end{aligned}$$

and using the Cauchy-Schwarz inequality for the summation over i, this is still less than

$$\begin{aligned}&\left(\sum_i H_i^2\langle M, M\rangle_{t_i}^{t_{i+1}}\right)^{1/2}\left(\sum_i K_i^2\langle N, N\rangle_{t_i}^{t_{i+1}}\right)^{1/2} \\ &= \left(\int_0^t H_s^2 d\langle M, M\rangle_s\right)^{1/2}\left(\int_0^t K_s^2 d\langle N, N\rangle_s\right)^{1/2}\end{aligned}$$

which completes the proof.

(1.16) Corollary (Kunita-Watanabe inequality). *For every $p \ge 1$ and $p^{-1} + q^{-1} = 1$,*

$$\begin{aligned}&E\left[\int_0^\infty |H_s|\,|K_s|\,|d\langle M, N\rangle|_s\right] \\ &\le \left\|\left(\int_0^\infty H_s^2 d\langle M, M\rangle_s\right)^{1/2}\right\|_p \left\|\left(\int_0^\infty K_s^2 d\langle N, N\rangle_s\right)^{1/2}\right\|_q .\end{aligned}$$

Proof. Straightforward application of Hölder's inequality.

We now introduce a fundamental class of processes of finite quadratic variation.

(1.17) Definition. *A continuous* $(\mathscr{F}_t, P)$-semimartingale *is a continuous process X which can be written $X = M + A$ where M is a continuous $(\mathscr{F}_t, P)$-local martingale and A a continuous adapted process of finite variation.*

As usual, we will often drop $(\mathscr{F}_t, P)$ and we will use the abbreviation cont. semi. mart. The decomposition into a local martingale and a finite variation process is unique as follows readily from Proposition (1.2); however, if a process X is a continuous semimartingale in two different filtrations $(\mathscr{F}_t)$ and $(\mathscr{G}_t)$, the decompositions may be different even if $\mathscr{F}_t \subset \mathscr{G}_t$ for each t (see Exercise (3.18)).

More generally one can define semimartingales as the sums of local martingales and finite variation processes. It can be proved, but this is outside the scope of this book, that a semimartingale which is a continuous process is a continuous semimartingale in the sense of Definition (1.17), namely that there is a decomposition, necessarily unique, into the sum of a continuous local martingale and a continuous finite variation process.

We shall see many reasons for the introduction of this class of processes, but we may already observe that their definition recalls the decomposition of many physical systems into a signal (the f.v. process) and a noise (the local martingale).

(1.18) Proposition. *A continuous semimartingale $X = M + A$ has a finite quadratic variation and $\langle X, X\rangle = \langle M, M\rangle$.*

Proof. If Δ is a subdivision of $[0, t]$,

$$\left|\sum_i \left(M_{t_{i+1}} - M_{t_i}\right)\left(A_{t_{i+1}} - A_{t_i}\right)\right| \le \left(\sup_i \left|M_{t_{i+1}} - M_{t_i}\right|\right) \operatorname{Var}_t(A)$$

where $\operatorname{Var}_t(A)$ is the variation of A on $[0, t]$, and this converges to zero when $|\Delta|$ tends to zero because of the continuity of M. Likewise

$$\lim_{|\Delta|\to 0} \sum_i \left(A_{t_{i+1}} - A_{t_i}\right)^2 = 0.$$

□

(1.19) Fundamental remark. Since the process $\langle X, X\rangle$ is the limit in probability of the sums $T^{\Delta_n}(X)$, it does not change if we replace $(\mathscr{F}_t)$ by another filtration for which X is still a semimartingale and likewise if we change P for a probability measure Q such that $Q \ll P$ and X is still a Q-semimartingale (see Sect. 1 Chap. VIII).

(1.20) Definition. *If $X = M + A$ and $Y = N + B$ are two continuous semimartingales, we define the* bracket *of X and Y by*

$$\langle X, Y\rangle = \langle M, N\rangle = \frac{1}{4}\left[\langle X + Y, X + Y\rangle - \langle X - Y, X - Y\rangle\right].$$

Obviously, $\langle X, Y\rangle_t$ is the limit in probability of $\sum_i \left(X_{t_{i+1}} - X_{t_i}\right)\left(Y_{t_{i+1}} - Y_{t_i}\right)$, and more generally, if H is left-continuous and adapted,

$$P\text{-}\lim_{|\Delta|\to 0} \sup_{s\le t}\left|\sum_i H_{t_i}\left(X^s_{t_{i+1}} - X^s_{t_i}\right)\left(Y^s_{t_{i+1}} - Y^s_{t_i}\right) - \int_0^s H_u d\langle X, Y\rangle_u\right| = 0,$$

the proof of which is left to the reader as an exercise (see also Exercise (1.33)).

Finally, between the class of local martingales and that of bounded martingales, there are several interesting classes of processes among which the following ones will be particularly important in the next section. We will indulge in the usual confusion between processes and classes of indistinguishable processes in order to get norms and not merely semi-norms in the discussion below.

(1.21) Definition. *We denote by* $\mathbb{H}^2$ *the space of* L^2*-bounded martingales, i.e. the space of* $(\mathscr{F}_t, P)$*-martingales* M *such that*

$$\sup_t E[M_t^2] < +\infty.$$

We denote by H^2 *the subset of* L^2*-bounded continuous martingales, and* H_0^2 *the subset of elements of* H^2 *vanishing at zero.*

An $(\mathscr{F}_t)$-Brownian motion is not in H^2, but it is when suitably stopped, for instance at a constant time. Bounded martingales are in $\mathbb{H}^2$. Moreover, by Doob's inequality (Sect. 1 in Chap. II), $M_\infty^* = \sup_t |M_t|$ is in L^2 if $M \in \mathbb{H}^2$; hence M is u.i. and $M_t = E\left[M_\infty \mid \mathscr{F}_t\right]$ with $M_\infty \in L^2$. This sets up a one to one correspondence between $\mathbb{H}^2$ and $L^2(\Omega, \mathscr{F}_\infty, P)$, and we have the

(1.22) Proposition. *The space* $\mathbb{H}^2$ *is a Hilbert space for the norm*

$$\|M\|_{\mathbb{H}^2} = E\left[M_\infty^2\right]^{1/2} = \lim_{t\to\infty} E\left[M_t^2\right]^{1/2},$$

and the set H^2 *is closed in* $\mathbb{H}^2$.

Proof. The first statement is obvious; to prove the second, we consider a sequence $\{M^n\}$ in H^2 converging to M in $\mathbb{H}^2$. By Doob's inequality,

$$E\left[\left(\sup_t |M_t^n - M_t|\right)^2\right] \leq 4\left\|M^n - M\right\|_{\mathbb{H}^2}^2;$$

as a result, one can extract a subsequence for which $\sup_t \left|M_t^{n_k} - M_t\right|$ converges to zero a.s. which proves that $M \in H^2$. □

The mapping $M \to \left\|M_\infty^*\right\|_2 = E\left[\left(\sup_t |M_t|\right)^2\right]^{1/2}$ is also a norm on $\mathbb{H}^2$; it is equivalent to $\| \quad \|_{\mathbb{H}^2}$ since obviously $\|M\|_{H^2} \leq \|M_\infty^*\|_2$ and by Doob's inequality $\|M_\infty^*\|_2 \leq 2\|M\|_{H^2}$, but it is no longer a Hilbert space norm.

We now study the quadratic variation of the elements of H^2.

(1.23) Proposition. *A continuous local martingale* M *is in* H^2 *if and only if the following two conditions hold*

i) $M_0 \in L^2$;
ii) $\langle M, M\rangle$ *is integrable i.e.* $E\left[\langle M, M\rangle_\infty\right] < \infty$.

In that case, $M^2 - \langle M, M\rangle$ *is uniformly integrable and for any pair* $S \leq T$ *of stopping times*

$$E\left[M_T^2 - M_S^2 \mid \mathscr{F}_S\right] = E\left[(M_T - M_S)^2 \mid \mathscr{F}_S\right] = E\left[\langle M, M\rangle_S^T \mid \mathscr{F}_S\right].$$

Proof. Let $\{T_n\}$ be a sequence of stopping times increasing to $+\infty$ and such that $M^{T_n} 1_{[T_n>0]}$ is bounded; we have

$$E\left[M_{T_n\wedge t}^2 1_{[T_n>0]}\right] - E\left[\langle M, M\rangle_{T_n\wedge t} 1_{[T_n>0]}\right] = E\left[M_0^2 1_{[T_n>0]}\right].$$

If M is in H^2 then obviously i) holds and, since $M_\infty^* \in L^2$, we may also pass to the limit in the above equality to get

$$E[M_\infty^2] - E[\langle M, M\rangle_\infty] = E[M_0^2]$$

which proves that ii) holds.

If, conversely, i) and ii) hold, the same equality yields

$$E\left[M_{T_n \wedge t}^2 1_{[T_n>0]}\right] \le E\left[\langle M, M\rangle_\infty\right] + E\left[M_0^2\right] = K < \infty$$

and by Fatou's lemma

$$E\left[M_t^2\right] \le \varliminf_n E\left[M_{T_n \wedge t}^2 1_{[T_n>0]}\right] \le K$$

which proves that the family of r.v.'s M_t is bounded in L^2. Furthermore, the same inequality shows that the set of r.v.'s $M_{T_n \wedge t} 1_{[T_n>0]}$ is bounded in L^2, hence uniformly integrable, which allows to pass to the limit in the equality

$$E\left[M_{t\wedge T_n} 1_{[T_n>0]} \mid \mathscr{F}_s\right] = M_{s\wedge T_n} 1_{[T_n>0]}$$

to get $E\left[M_t \mid \mathscr{F}_s\right] = M_s$. The process M is a L^2-bounded martingale.

To prove that $M^2 - \langle M, M\rangle$ is u.i., we observe that

$$\sup_t \left|M_t^2 - \langle M, M\rangle_t\right| \le (M_\infty^*)^2 + \langle M, M\rangle_\infty$$

which is an integrable r.v. The last equalities derive immediately from the optional stopping theorem.

(1.24) Corollary. *If $M \in H_0^2$,*

$$\|M\|_{\mathbb{H}^2} = \left\|\langle M, M\rangle_\infty^{1/2}\right\|_2 \equiv E\left[\langle M, M\rangle_\infty\right]^{1/2}.$$

Proof. If $M_0 = 0$, we have $E\left[M_\infty^2\right] = E\left[\langle M, M\rangle_\infty\right]$ as is seen in the last proof.

Remark. The more general comparison between the L^p-norms of M_∞ and $\langle M, M\rangle_\infty^{1/2}$ will be taken up in Sect. 4.

We could have worked in exactly the same way on $[0, t]$ instead of $[0, \infty]$ to get the

(1.25) Corollary. *If M is a continuous local martingale, the following two conditions are equivalent*

i) $M_0 \in L^2$ *and* $E\left[\langle M, M\rangle_t\right] < \infty$;
ii) $\{M_s, s \le t\}$ *is an* L^2*-bounded martingale.*

Remark. It is not true (see Exercise (2.13) Chap. V) that L^2-bounded local martingales are always martingales. Likewise, $E[\langle M, M\rangle_\infty]$ may be infinite for an L^2-bounded cont. loc. mart.

We notice that for $M \in H^2$, simultaneously $\langle M, M\rangle_\infty$ is in L^1 and $\lim_{t\to\infty} M_t$ exists a.s. This is generalized in the following

(1.26) Proposition. *A continuous local martingale M converges a.s. as t goes to infinity, on the set* $\{\langle M, M\rangle_\infty < \infty\}$.

Proof. Without loss of generality, we may assume $M_0 = 0$. Then, if $T_n = \inf\{t : \langle M, M\rangle_t \geq n\}$, the local martingale M^{T_n} is bounded in L^2 as follows from Proposition (1.23). As a result, $\lim_{t\to\infty} M_t^{T_n}$ exists a.s. But on $\{\langle M, M\rangle_\infty < \infty\}$ the stopping times T_n are a.s. infinite from some n on, which completes the proof.

Remark. The converse statement that $\langle M, M\rangle_\infty < \infty$ on the set where M_t converges a.s. will be shown in Chap. V Sect. 1. The reader may also look at Exercise (1.42) in this section.

(1.27) Exercise. 1°) If M and N are two independent continuous local martingales (i.e. the σ-fields $\sigma(M_s, s \geq 0)$ and $\sigma(N_s, s \geq 0)$ are independent), show that $\langle M, N\rangle = 0$. In particular, if $B = (B^1, \dots, B^d)$ is a BM^d, prove that $\langle B^i, B^j\rangle_t = \delta^{ij}t$. This can also be proved by observing that $(B^i + B^j)/\sqrt{2}$ and $(B^i - B^j)/\sqrt{2}$ are linear BM's and applying the polarization formula.

2°) If B is a linear BM and T a stopping time, by considering B^T and $B - B^T$ prove that the converse to the result in 1°) is false.

[Hint: T is measurable with respect to the σ-fields generated by both B^T (observe that $\langle B^T, B^T\rangle_t = t \wedge T$) and $B - B^T$ which thus are not independent if T is not constant a.s.].

If (X, Y) is a BM^2 and T a stopping time, X^T and Y^T provide another example.

(1.28) Exercise. If X is a continuous semimartingale, and T a stopping time, then $\tilde{X}_t = X_{T+t}$ is a $(\mathscr{F}_{T+t})$-semimartingale. Compute $\langle \tilde{X}, \tilde{X}\rangle$ in terms of $\langle X, X\rangle$.

(1.29) Exercise. If $X = (X^1, \dots, X^d)$ is a vector continuous local martingale, there is a unique process $A \in \mathscr{A}$ such that $|X|^2 - A$ is a continuous local martingale.

(1.30) Exercise. 1°) If S and T are two stopping times which reduce M, then $S \vee T$ reduces M.

2°) If (T_n) is a sequence of stopping times increasing a.s. to $+\infty$ and if M^{T_n} is a continuous local martingale for every n, then M is a continuous local martingale. This result may be stated by saying that a process which is locally a cont. loc. mart. is a cont. loc. mart.

3°) If M is a $(\mathscr{F}_t)$-cont. loc. mart., the stopping times $S_k = \inf\{t : |M_t| \geq k\}$ reduce M. Use them to prove that M is also a cont. loc. mart. with respect to $(\mathscr{F}_t^M)$ and even $\sigma(M_s, s \leq t)$. The same result for semimarts. (Stricker's theorem) is more difficult to prove and much more so in the discontinuous case.

4°) In the setting of Exercise (2.17) Chap. II, prove that if X is a $(\mathscr{F}_t)$-cont. semimart., then one can find $(\mathscr{F}_t^0)$-adapted processes M', A', B' such that

i) M' is a $(\mathscr{F}_t^0)$-cont. loc. mart.;

ii) A' and B' are continuous, A' is of finite variation, B' is increasing;
iii) X is indistinguishable from $X' = M' + A'$;
iv) B' is indistinguishable from $\langle X, X\rangle$ and $M'^2 - B'$ is a $(\mathscr{F}_t^0)$-cont. loc. mart..

We will write $\langle X', X'\rangle$ for B'.

5°) Prove that $\langle X', X'\rangle$ is the quadratic variation of X' and that the fundamental Remark (1.19) is still valid. Extend these remarks to the bracket of two cont. semimarts. X and Y.

(1.31) Exercise. If Z is a bounded r.v. and A is a bounded continuous increasing process vanishing at 0, prove that

$$E[ZA_\infty] = E\left[\int_0^\infty E\left[Z \mid \mathscr{F}_t\right] dA_t\right].$$

(1.32) Exercise. (Another proof of Proposition (1.2)). If M is a bounded continuous martingale of finite variation, prove that

$$M_t^2 = M_0^2 + 2\int_0^t M_s dM_s \quad \text{(the integral has then a path by path meaning)}$$

is a martingale. Using the strict convexity of the square function, give another proof of Proposition (1.2).

* **(1.33) Exercise.** All the processes considered below are bounded. For a subdivision $\Delta = (t_i)$ of $[0, t]$ and $\lambda \in [0, 1]$, we set $t_i^\lambda = t_i + \lambda\,(t_{i+1} - t_i)$.

1°) If $X = M + A$, $Y = N + B$ are two cont. semimarts. and H is a continuous adapted process, set

$$K_\Delta^\lambda = \sum_i H_{t_i}\left\{\left(X_{t_i^\lambda} - X_{t_i}\right)\left(Y_{t_i^\lambda} - Y_{t_i}\right) - \langle X, Y\rangle_{t_i}^{t_i^\lambda}\right\}.$$

Prove that

$$\lim_{|\Delta|\to 0} \sup_\lambda E\left[\left(K_\Delta^\lambda\right)^2\right] = 0.$$

2°) If $\lambda = 1$ or if $d\langle X, Y\rangle$ is absolutely continuous with respect to the Lebesgue measure, then

$$\lim_{|\Delta|\to 0} \sum_i H_{t_i}\left(X_{t_i^\lambda} - X_{t_i}\right)\left(Y_{t_i^\lambda} - Y_{t_i}\right) = \lambda \int_0^t H_s d\langle X, Y\rangle_s$$

in the L^2-sense. In the second case, the convergence is uniform in λ.

3°) If F is a C^1-function on $\mathbb{R}$, prove that

$$P\text{-}\lim_{n\to\infty} \sum_\Delta \left(F\left(X_{t_{i+1}}\right) - F\left(X_{t_i}\right)\right)^2 = \int_0^t F'(X_s)^2 d\langle X, X\rangle_s.$$

(1.34) Exercise. If B is the standard linear BM and S and T are two integrable stopping times with $S \le T$, show that

$$E\left[\left(B_T - B_S\right)^2\right] = E\left[B_T^2 - B_S^2\right] = E[T - S].$$

(1.35) Exercise (Gaussian martingales). If M is a continuous martingale and a Gaussian process, prove that $\langle M, M\rangle$ is deterministic i.e. there is a function f on $\mathbb{R}_+$ such that $\langle M, M\rangle_t = f(t)$ a.s. The converse will be proved in Sect. 1 Chap. V.

(1.36) Exercise. 1°) If M is a continuous local martingale, prove that M^2 is of finite quadratic variation and that $\langle M^2, M^2\rangle_t = 4\int_0^t M_s^2 d\langle M, M\rangle_s$.

[Hint: Use 2°) in Exercise (1.33). In the following sections, we will see the profound reasons for which M^2 is of finite quadratic variation as well as a simple way of computing $\langle M^2, M^2\rangle$, but this can already be done here by brute force.]

* 2°) Let R^i, $i = 1, 2, \ldots, r$ be the squares of the moduli of r independent d_i-dimensional BM's and λ_i be r distinct, non zero real numbers. We set

$$X_t^{(k)} = \sum_i \lambda_i^k R_t^i, \qquad k = 1, 2, \ldots.$$

Prove that each $X^{(k)}$ is of finite quadratic variation and that

$$\langle X^{(1)}, X^{(k)}\rangle_t = 4\int_0^t X_s^{(k+1)} ds.$$

Prove that each R^i is adapted to $\left(\mathscr{F}_t^{X^{(1)}}\right)$.

(1.37) Exercise. Let $\mathbf{W} = C(\mathbb{R}_+, \mathbb{R})$, X be the coordinate process and $\mathscr{F}_t^0 = \sigma(X_s, s \le t)$. Prove that the set of probability measures on $(\mathbf{W}, \mathscr{F}^0)$ for which X is a $(\mathscr{F}_t^0)$-local martingale, is a convex set. Is this still true with the space $\mathbf{D}$ of cadlag functions in lieu of $\mathbf{W}$?

* **(1.38) Exercise.** If $X_t(\omega) = f(t)$ for every ω, then X is a continuous semimartingale if and only if f is continuous and of bounded variation on every interval.

[Hint: Write $f(t) = f(0) + M_t + A_t$ for the decomposition of the semimart. f. Pick $a > 0$. Then choose $c > 0$ sufficiently large so that the stopping time

$$T = \inf\left\{t : |M_t| + \int_0^t |dA|_s \ge c\right\}$$

satisfies $P[T \ge a] > 0$ and $|f|$ is bounded by c on $[0, a]$. For a finite subdivision $\Delta = (t_i)$ of $[0, a]$, by comparing $S^\Delta(f)$ with the sum

$$S = \sum_{t_i \le T} |f(t_i) - f(t_{i-1})|$$

prove that $S^\Delta(f)$ is less than $3c/P[T \ge a]$. Another proof is hinted at in Exercise (2.21) and yet another can be based on Stricker's theorem (see the Notes and Comments).]

*# **(1.39) Exercise.** Let $(\Omega, \mathscr{F}_t, P)$, $t \in [0, 1]$ be a filtered probability space. For a right-continuous adapted process X and a subdivision Δ of $[0, 1]$, we set

$$V_\Delta(X) = \sum_{i=1}^n E\left[\left|E\left[X_{t_{i+1}} - X_{t_i} \mid \mathscr{F}_{t_i}\right]\right|\right].$$

1°) If Δ' is a refinement of Δ prove that $V_{\Delta'}(X) \geq V_{\Delta}(X)$. If

$$V(X) = \sup_{\Delta} V_{\Delta}(X) < +\infty,$$

we say that X is a *quasimartingale*. If $M \in H^2$ prove that M^2 is a quasimartingale.

2°) If B is the linear BM, define $(\mathscr{G}_t)$ as $\left(\mathscr{F}_t \vee \sigma(B_1)\right)_+$ where $\mathscr{F}_t$ is the Brownian filtration of Chap. III. Prove that for $0 \leq s \leq t \leq 1$,

$$E\left[B_t - B_s \mid \mathscr{G}_s\right] = \frac{t-s}{1-s}(B_1 - B_s).$$

Prove that B_t, $t \in [0,1]$ is no longer a $(\mathscr{G}_t)$-martingale but that it is a quasimartingale and compute $V(B)$.

This exercise continues in Exercise (3.18).

** **(1.40) Exercise. (Continuation of Exercise (3.16) of Chap. II on BMO).** 1°) If $M \in \text{BMO}$, prove that $\|M\|_{\text{BMO}_2}$ is the smallest constant C such that

$$E\left[\langle M,M\rangle_\infty - \langle M,M\rangle_T \mid \mathscr{F}_T\right] \leq C^2$$

for every stopping time T.

2°) Prove that, for every T and n

$$E\left[(\langle M,M\rangle_\infty - \langle M,M\rangle_T)^n \mid \mathscr{F}_T\right] \leq n!\ \|M\|^{2n}_{\text{BMO}_2}.$$

[Hint: One can use Exercise (1.13) Chap. V.]

3°) If $\|M\|_{\text{BMO}_2} < 1$, then

$$E\left[\exp(\langle M,M\rangle_\infty - \langle M,M\rangle_T) \mid \mathscr{F}_T\right] \leq \left(1 - \|M\|^2_{\text{BMO}_2}\right)^{-1}.$$

* **(1.41) Exercise (A weak definition of BMO).** 1°) Let A be a continuous increasing process for which there exist constants a, b, α, β such that for any stopping time T

$$P\left[A_\infty - A_T > a \mid \mathscr{F}_T\right] \leq \alpha, \quad P\left[A_\infty - A_T > b \mid \mathscr{F}_T\right] \leq \beta.$$

Prove that

$$P\left[A_\infty - A_T > a + b \mid \mathscr{F}_T\right] \leq \alpha\beta.$$

[Hint: Use the stopping time $U = \inf\{t : A_t - A_T > b\}$ and prove that the event $\{A_\infty - A_T > b\}$ is in $\mathscr{F}_U$.]

2°) A continuous local martingale is in BMO if and only if one can find two constants a and $\varepsilon > 0$ such that

$$P\left[\langle M,M\rangle_\infty - \langle M,M\rangle_T > a \mid \mathscr{F}_T\right] \leq 1 - \varepsilon$$

for every stopping time T. When this is the case, prove that for α sufficiently small

$$E\left[\exp(\alpha(\langle M,M\rangle_\infty - \langle M,M\rangle_T)) \mid \mathscr{F}_T\right]$$

is bounded.

* **(1.42) Exercise.** Let $(\Omega, \mathscr{F}, P)$ be a probability space and $(\mathscr{F}_t)$ a complete and right-continuous filtration. A local martingale on $]0, \infty[$ is an adapted process M_t, $t \in]0, \infty[$ such that for every $\varepsilon > 0$, $M_{\varepsilon+t}$, $t \geq 0$, is a continuous local martingale with respect to $(\mathscr{F}_{\varepsilon+t})$. The set of these processes will be denoted by $\mathscr{M}$.

1°) If $A \in \mathscr{F}_0$ and $M \in \mathscr{M}$, then $1_A M \in \mathscr{M}$. If T is a $(\mathscr{F}_t)$-stopping time and $M \in \mathscr{M}$, then $M^T 1_{[T>0]} \in \mathscr{M}$.

2°) For any $M \in \mathscr{M}$, there is a unique random measure $\langle M\rangle(\omega, \cdot)$ on $]0, \infty[$ such that for every $\varepsilon > 0$

$$M_{t+\varepsilon}^2 - \langle M\rangle(\cdot,]\varepsilon, t+\varepsilon])$$

is a $(\mathscr{F}_{t+\varepsilon})$-continuous local martingale.

3°) For any $M \in \mathscr{M}$, prove that the two sets $A = \{\omega : \lim_{t\downarrow 0} M_t(\omega) \text{ exists in } \mathbb{R}\}$ and $B = \{\omega : \langle M\rangle(\omega,]0, 1]) < \infty\}$ are a.s. equal and furthermore, that $1_A M$ is a continuous local martingale in the usual sense.

[Hint: Using 1°) , reduce to the case where M is bounded, then use the continuous time version of Theorem (2.3) of Chap. II.]

This exercise continues in Exercise (3.26).

* **(1.43) Exercise. (Continuation of Exercise (3.15) Chap. II).** Assume that $f \in L^2([0, 1])$ and set

$$[X, X]_t(\omega) = \left(\tilde{H} f(\omega) - f(\omega)\right)^2 1_{[t \geq \omega]}.$$

Prove that $X_t^2 - [X, X]_t$ is a uniformly integrable martingale.

[Hint: Compute $\tilde{H}\left((\tilde{H} f)^2 - 2f(\tilde{H} f)\right)$.]

The reader will notice that albeit we deal with a discontinuous martingale X, the process $[X, X]$ is the quadratic variation process of X and plays a role similar to that of $\langle X, X\rangle$ in the continuous case.

(1.44) Exercise. Let X be a positive r.v. independent of a linear Brownian motion B. Let $M_t = B_{tX}$, $t \geq 0$, and $(\mathscr{F}_t^M)$ be the smallest right-continuous and complete filtration with respect to which M is adapted.

1°) Prove that M is a $(\mathscr{F}_t^M)$-local martingale and that it is a martingale if and only if $E[X^{1/2}] < \infty$.

2°) Find the process $\langle M, M\rangle$.

3°) Generalize the preceding results to $M_t = B_{A_t}$ where A is an increasing continuous process, vanishing at 0 and independent of B.

(1.45) Exercise. Let M and N be two continuous local martingales vanishing at 0 such that $\langle M, N\rangle^2 = \langle M, M\rangle\langle N, N\rangle$. If $R = \inf\{s : \langle M, M\rangle_s > 0\}$, $S = \inf\{s : \langle N, N\rangle_s > 0\}$ prove that a.s. either $R \vee S = \infty$ or $R = S$ and there is a $\mathscr{F}_R \cap \mathscr{F}_S$-measurable r.v. γ vanishing on $\{R \vee S = \infty\}$ and such that $M = \gamma N$.

(1.46) Exercise. Prove that a local martingale X such that for every integer N, the process $(X^-)^N$ is of class (D) is a supermartingale. In particular, a positive local martingale is a supermartingale.

\# **(1.47) Exercise.** Let M and N be two continuous local martingales and T a finite stopping time. Prove that

$$\left|\langle M, M\rangle_T^{1/2} - \langle N, N\rangle_T^{1/2}\right| \leq \langle M-N, M-N\rangle_T^{1/2} \qquad \text{a.s.}$$

[Hint: The map $(M, N) \to \langle M, N\rangle$ has the properties of a scalar product and this follows from the corresponding "Minkowski" inequality.]

* **(1.48) Exercise. (Continuous local martingales on a stochastic interval).** Let T be a stopping time. We define a continuous local martingale M on $[0, T[$ as a process on $[0, T[$ for which there exists a sequence (T_n) of stopping times increasing to T and a sequence of continuous martingales M^n such that $M_t = M_t^n$ on $\{t < T_n\}$. In the sequel, we will always assume $T > 0$ a.s.

1°) Prove that there exists a unique continuous increasing process $\langle M, M\rangle$ on $[0, T[$ such that $M^2 - \langle M, M\rangle$ is a continuous local martingale on $[0, T[$. Prove that M and $\langle M, M\rangle$ have a.s. the same intervals of constancy.

2°) If $E\left[\sup_{t<T} |M_t|\right] < \infty$, then $\lim_{t\uparrow T} M_t = M_{T^-}$ exists and is finite a.s. and if we set $M_t = M_{T^-}$ for $t \geq T$, the process M_t, $t \in \mathbb{R}_+$ is a uniformly integrable martingale.

3°) If $M_0 = 0$ and $E[\langle M, M\rangle_T] < \infty$, prove that M may be continued in a continuous uniformly integrable martingale.

(1.49) Exercise. (Krickeberg decomposition for loc. marts. Continuation of Exercise (2.18) Chap. II). Let X be a loc. mart. and set

$$N_1(X) = \sup_T E\left[|X_T|\right]$$

where T ranges through the family of finite stopping times.

1°) If (T_n) is a sequence of stopping times reducing X and such that $\lim T_n = \infty$ a.s., then $N_1(X) = \sup_n E\left[|X_{T_n}|\right]$. A loc. mart. X is bounded in L^1 iff $N_1(X) < \infty$.

** 2°) If $N_1(X) < \infty$ prove that there is a unique pair $\left(X^{(+)}, X^{(-)}\right)$ of positive loc. marts. such that

i) $X = X^{(+)} - X^{(-)}$,

ii) $N_1(X) = E\left[X_0^{(+)}\right] + E\left[X_0^{(-)}\right]$.

[Hint: See Exercise (2.18) Chap. II and remember that a positive loc. mart. is a supermartingale.]

3°) If $N_1(X) < \infty$, then (X_t) converges a.s. as t tends to infinity, to an integrable r.v.

§2. Stochastic Integrals

For several reasons, one of which is described at length at the end of Sect. 1 of Chap. VII, it is necessary to define an integral with respect to the paths of BM. The natural idea is to consider the "Riemann sums"

$$\sum_i K_{u_i} \left(B_{t_{i+1}} - B_{t_i}\right)$$

where K is the process to integrate and u_i is a point in $[t_i, t_{i+1}]$. But it is known from integration theory that these sums do not converge pathwise because the paths of B are a.s. not of bounded variation (see Exercise (2.21)). We will prove that the convergence holds in probability, but in a first stage we use L^2-convergence and define integration with respect to the elements of H^2. The class of integrands is the object of the following

(2.1) Definition. *If $M \in H^2$, we call $\mathscr{L}^2(M)$ the space of progressively measurable processes K such that*

$$\|K\|_M^2 = E\left[\int_0^\infty K_s^2 d\langle M, M\rangle_s\right] < +\infty.$$

If, for any $\Gamma \in \mathscr{B}(\mathbb{R}_+) \otimes \mathscr{F}_\infty$, we set

$$P_M(\Gamma) = E\left[\int_0^\infty 1_\Gamma(s, \omega) d\langle M, M\rangle_s(\omega)\right]$$

we define a bounded measure P_M on $\mathscr{B}(\mathbb{R}_+) \otimes \mathscr{F}_\infty$ and the space $\mathscr{L}^2(M)$ is nothing else than the space of P_M-square integrable, progressively measurable, functions. As usual, $L^2(M)$ will denote the space of equivalence classes of elements of $\mathscr{L}^2(M)$; it is of course a Hilbert space for the norm $\|\cdot\|_M$.

Since those are the processes we are going to integrate, it is worth recalling that they include all the bounded and left (or right)-continuous adapted processes and, in particular, the bounded continuous adapted processes.

(2.2) Theorem. *Let $M \in H^2$; for each $K \in L^2(M)$, there is a unique element of H_0^2, denoted by $K \cdot M$, such that*

$$\langle K \cdot M, N\rangle = K \cdot \langle M, N\rangle$$

for every $N \in H^2$. The map $K \to K \cdot M$ is an isometry from $L^2(M)$ into H_0^2.

Proof. a) *Uniqueness.* If L and L' are two martingales of H_0^2 such that $\langle L, N\rangle = \langle L', N\rangle$ for every $N \in H^2$, then in particular $\langle L - L', L - L'\rangle = 0$ which by Proposition (1.12) implies that $L - L'$ is constant, hence $L = L'$.

b) *Existence.* Suppose first that M is in H_0^2. By the Kunita-Watanabe inequality (Corollary (1.16)) and Corollary (1.24), for every N in H_0^2 we have

$$\left|E\left[\int_0^\infty K_s d\langle M, N\rangle_s\right]\right| \le \|N\|_{\mathbb{H}^2} \|K\|_M;$$

the map $N \to E[(K\cdot\langle M,N\rangle)_\infty]$ is thus a linear and continuous form on the Hilbert space H_0^2 and, consequently, there is an element $K\cdot M$ in H_0^2 such that

$$(2.1) \qquad E[(K\cdot M)_\infty N_\infty] = E[(K\cdot\langle M,N\rangle)_\infty]$$

for every $N \in H_0^2$. Let T be a stopping time; the martingales of H^2 being u.i., we may write

$$\begin{aligned} E[(K\cdot M)_T N_T] &= E\left[E\left[(K\cdot M)_\infty \mid \mathscr{F}_T\right] N_T\right] = E[(K\cdot M)_\infty N_T] \\ &= E\left[(K\cdot M)_\infty N_\infty^T\right] = E\left[(K\cdot\langle M,N^T\rangle)_\infty\right] \\ &= E\left[(K\cdot\langle M,N\rangle^T)_\infty\right] = E[(K\cdot\langle M,N\rangle)_T] \end{aligned}$$

which proves, by Proposition (3.5) Chap. II, that $(K\cdot M)N - K\cdot\langle M,N\rangle$ is a martingale. Furthermore, by eq. (2.1),

$$\|K\cdot M\|_{\mathbb{H}^2}^2 = E\left[(K\cdot M)_\infty^2\right] = E\left[(K^2\cdot\langle M,M\rangle)_\infty\right] = \|K\|_M^2$$

which proves that the map $K \to K\cdot M$ is an isometry.

If $N \in H^2$ instead of H_0^2, then we still have $\langle K\cdot M,N\rangle = K\cdot\langle M,N\rangle$ because the bracket of a martingale with a constant martingale is zero.

Finally, if $M \in H^2$ we set $K\cdot M = K\cdot(M - M_0)$ and it is easily checked that the properties of the statement carry over to that case.

(2.3) Definition. *The martingale $K\cdot M$ is called the* stochastic integral *of K with respect to M and is also denoted by*

$$\int_0^\cdot K_s dM_s.$$

It is also called the *Itô integral* to distinguish it from other integrals defined in Exercise (2.18). The Itô integral is the only one among them for which the resulting process is a martingale.

We stress the fact that the stochastic integral $K\cdot M$ vanishes at 0. Moreover, as a function of t, the process $(K\cdot M)_t$ may also be seen as an antiderivative (cf. Exercise (2.17)).

The reasons for calling $K\cdot M$ a stochastic integral will become clearer in the sequel; here is one of them. We shall denote by $\mathscr{E}$ the space of *elementary processes* that is the processes which can be written

$$K = K_{-1}1_{\{0\}} + \sum_i K_i 1_{]t_i,t_{i+1}]}$$

where $0 = t_0 < t_1 < t_2 < \ldots$, $\lim_i t_i = +\infty$, and the r.v.'s K_i are $\mathscr{F}_{t_i}$-measurable and uniformly bounded and $K_{-1} \in \mathscr{F}_0$. The space $\mathscr{E}$ is contained in $L^2(M)$. For $K \in \mathscr{E}$, we define the so-called *elementary stochastic integral* $K\cdot M$ by

$$(K\cdot M)_t = \sum_{i=0}^{n-1} K_i\left(M_{t_{i+1}} - M_{t_i}\right) + K_n\left(M_t - M_{t_n}\right)$$

whenever $t_n \leq t < t_{n+1}$. It is easily seen that $K \cdot M \in H_0^2$; moreover, considering subdivisions Δ including the t_i's, it can be proved using the definition of the brackets, that for any $N \in H^2$, we have $\langle K \cdot M, N \rangle = K \cdot \langle M, N \rangle$.

As a result, the elementary stochastic integral coincides with the stochastic integral constructed in Theorem (2.2). This will be important later to prove a property of convergence of Riemann sums which will lead to explicit computations of stochastic integrals.

We now review some properties of the stochastic integral. The first is known as the property of *associativity*.

(2.4) Proposition. *If $K \in L^2(M)$ and $H \in L^2(K \cdot M)$ then $HK \in L^2(M)$ and*

$$(HK) \cdot M = H \cdot (K \cdot M).$$

Proof. Since $\langle K \cdot M, K \cdot M \rangle = K^2 \cdot \langle M, M \rangle$, it is clear that HK belongs to $L^2(M)$. For $N \in H^2$, we further have

$$\langle (HK) \cdot M, N \rangle = HK \cdot \langle M, N \rangle = H \cdot (K \cdot \langle M, N \rangle)$$

because of the obvious associativity of Stieltjes integrals, and this is equal to

$$H \cdot \langle K \cdot M, N \rangle = \langle H \cdot (K \cdot M), N \rangle;$$

the uniqueness in Theorem (2.2) ends the proof. □

The next result shows how stochastic integration behaves with respect to optional stopping; this will be all important to enlarge the scope of its definition to local martingales.

(2.5) Proposition. *If T is a stopping time,*

$$K \cdot M^T = K\, 1_{[0,T]} \cdot M = (K \cdot M)^T.$$

Proof. Let us first observe that $M^T = 1_{[0,T]} \cdot M$; indeed, for $N \in H^2$,

$$\langle M^T, N \rangle = \langle M, N \rangle^T = 1_{[0,T]} \cdot \langle M, N \rangle = \langle 1_{[0,T]} \cdot M, N \rangle.$$

Thus, by the preceding proposition, we have on the one hand

$$K \cdot M^T = K \cdot \left(1_{[0,T]} \cdot M\right) = K\, 1_{[0,T]} \cdot M,$$

and on the other hand

$$(K \cdot M)^T = 1_{[0,T]} \cdot (K \cdot M) = 1_{[0,T]} K \cdot M$$

which completes the proof.

Since the Brownian motion stopped at a fixed time t is in H^2, if K is a process which satisfies

$$E\left[\int_0^t K_s^2 ds\right] < \infty, \qquad \text{for all } t,$$

we can define $\int_0^t K_s dB_s$ for each t hence on the whole positive half-line and the resulting process is a martingale although not an element of H^2. This idea can of course be used for all continuous local martingales.

(2.6) Definition. *If M is a continuous local martingale, we call $L^2_{\mathrm{loc}}(M)$ the space of classes of progressively measurable processes K for which there exists a sequence (T_n) of stopping times increasing to infinity and such that*

$$E\left[\int_0^{T_n} K_s^2 d\langle M, M\rangle_s\right] < +\infty.$$

Observe that $L^2_{\mathrm{loc}}(M)$ consists of all the progressive processes K such that

$$\int_0^t K_s^2 d\langle M, M\rangle_s < \infty \qquad \text{for every } t.$$

(2.7) Proposition. *For any $K \in L^2_{\mathrm{loc}}(M)$, there exists a unique continuous local martingale vanishing at 0 denoted $K \cdot M$ such that for any continuous local martingale N*

$$\langle K \cdot M, N\rangle = K \cdot \langle M, N\rangle.$$

Proof. One can choose stopping times T_n increasing to infinity and such that M^{T_n} is in H^2 and $K^{T_n} \in L^2(M^{T_n})$. Thus, for each n, we can define the stochastic integral $X^{(n)} = K^{T_n} \cdot M^{T_n}$. But, by Proposition (2.5), $X^{(n+1)}$ coincides with $X^{(n)}$ on $[0, T_n]$; therefore, one can define unambiguously a process $K \cdot M$ by stipulating that it is equal to $X^{(n)}$ on $[0, T_n]$. This process is obviously a continuous local martingale and, by localization, it is easily seen that $\langle K \cdot M, N\rangle = K \cdot \langle M, N\rangle$ for every local martingale N.

Remark. To prove that a continuous local martingale L is equal to $K \cdot M$, it is enough to check the equality $\langle L, N\rangle = K \cdot \langle M, N\rangle$ for all bounded N's.

Again, $K \cdot M$ is called the *stochastic integral* of K with respect to M and is alternatively written

$$\int_0^{\cdot} K_s dM_s.$$

Plainly, Propositions (2.4) and (2.5) carry over to the general case after the obvious changes. Also again if $K \in \mathscr{E}$ this stochastic integral will coincide with the elementary stochastic integral. Stieltjes pathwise integrals having been previously mentioned, it is now easy to extend the definition of stochastic integrals to semimartingales.

(2.8) Definition. *A progressively measurable process K is* locally bounded *if there exists a sequence (T_n) of stopping times increasing to infinity and constants C_n such that $|K^{T_n}| \leq C_n$.*

All continuous adapted processes K are seen to be locally bounded by taking $T_n = \inf\{t : |K_t| \geq n\}$. Locally bounded processes are in $L^2_{\mathrm{loc}}(M)$ for every continuous local martingale M.

(2.9) Definition. *If K is locally bounded and $X = M + A$ is a continuous semimartingale, the* stochastic integral of K with respect to X *is the continuous semimartingale*

$$K \cdot X = K \cdot M + K \cdot A$$

where $K \cdot M$ is the integral of Proposition (2.7) and $K \cdot A$ is the pathwise Stieltjes integral with respect to dA. The semimartingale $K \cdot X$ is also written

$$\int_0^{\cdot} K_s dX_s.$$

(2.10) Proposition. *The map $K \to K \cdot X$ enjoys the following properties:*

i) $H \cdot (K \cdot X) = (HK) \cdot X$ *for any pair H, K of locally bounded processes;*
ii) $(K \cdot X)^T = (K 1_{[0,T]}) \cdot X = K \cdot X^T$ *for every stopping time T;*
iii) if X is a local martingale or a process of finite variation, so is $K \cdot X$;
iv) if $K \in \mathscr{E}$, then if $t_n \le t < t_{n+1}$

$$(K \cdot X)_t = \sum_{i=0}^{n} K_i \left(X_{t_{i+1}} - X_{t_i} \right) + K_n \left(X_t - X_{t_n} \right).$$

Proof. Straightforward.

At this juncture, several important remarks are in order. Although we have used Doob's inequality and L^2-convergence in the construction of $K \cdot M$ for $M \in H^2$, the stochastic integral depends on P only through its equivalence class. This is clear from Proposition (2.7) and the fundamental remark (1.19). It is actually true, and we will see a partial result in this direction in Chap. VIII, that if X is a P-semimartingale and $Q \ll P$, then X is also a Q-semimartingale. Since a sequence converging in probability for P converges also in probability for Q, the stochastic integral for P of, say, a bounded process is Q-indistinguishable from its stochastic integral for Q.

Likewise, if we replace the filtration by another one for which X is still a semimartingale, the stochastic integrals of processes which are progressively measurable for both filtrations are the same.

Finally, although we have constructed the stochastic integral by a global procedure, its nature is still somewhat local as is suggested by the following

(2.11) Proposition. *For almost every ω, the function $(K \cdot X)_{.}(\omega)$ is constant on any interval $[a, b]$ on which either $K_{.}(\omega) = 0$ or $X_{.}(\omega) = X_a(\omega)$.*

Proof. Only the case where X is a local martingale has to be proved and it is then an immediate consequence of Proposition (1.13) since $K^2 \cdot \langle X, X \rangle$ hence $K \cdot X$ are then constant on these intervals.

As a result, for K and $\bar{K}$ locally bounded and predictable processes and X and $\bar{X}$ semimartingales we have $(K \cdot X)_t - (K \cdot X)_a = (\bar{K} \cdot \bar{X})_t - (\bar{K} \cdot \bar{X})_a$ a.s. on any interval $[a, b]$ on which $K = \bar{K}$ and $X_{.} - X_a = \bar{X}_{.} - \bar{X}_a$; this follows from the equality

$$K \cdot X - \bar{K} \cdot \bar{X} = K \cdot (X - \bar{X}) + (K - \bar{K}) \cdot \bar{X}.$$

Remark. That stochastic integrals have a path by path significance is also seen in 3°) of Exercise (2.18).

We now turn to a very important property of stochastic integrals, namely the counterpart of the Lebesgue dominated convergence theorem.

(2.12) Theorem. *Let X be a continuous semimartingale. If (K^n) is a sequence of locally bounded processes converging to zero pointwise and if there exists a locally bounded process K such that $|K^n| \leq K$ for every n, then $(K^n \cdot X)$ converges to zero in probability, uniformly on every compact interval.*

Proof. The convergence property which can be stated

$$P\text{-}\lim_{n\to\infty} \sup_{s\leq t} \left|\left(K^n \cdot X\right)_s\right| = 0$$

is clear if X is a process of finite variation. If X is a local martingale and if T reduces X, then $(K^n)^T$ converges to zero in $L^2(X^T)$ and by Theorem (2.2), $(K^n \cdot X)^T$ converges to zero in H^2. The desired convergence is then easily established by the same argument as in Theorem (1.8). □

The next result on "Riemann sums" is crucial in the following section.

(2.13) Proposition. *If K is left-continuous and locally bounded, and (Δ^n) is a sequence of subdivisions of $[0, t]$ such that $|\Delta^n| \to 0$, then*

$$\int_0^t K_s dX_s = P\text{-}\lim_{n\to\infty} \sum_{t_i\in\Delta^n} K_{t_i} \left(X_{t_{i+1}} - X_{t_i}\right).$$

Proof. If K is bounded, the right-hand side sums are the stochastic integrals of the elementary processes $\sum K_{t_i} 1_{]t_i, t_{i+1}]}$ which converge pointwise to K and are bounded by $\|K\|_\infty$; therefore, the result follows from the preceding theorem. The general case is obtained by the use of localization. □

(2.14) Exercise. Let X be a continuous semimartingale and $b\mathscr{F}_0$ the algebra of bounded $\mathscr{F}_0$-measurable random variables. Prove the $b\mathscr{F}_0$-linearity of the map $H \to H \cdot X$ namely, for a and b in $b\mathscr{F}_0$,

$$\int_0^t (aH + bK)dX = a\int_0^t H\, dX + b\int_0^t K\, dX.$$

(2.15) Exercise. Let B^1 and B^2 be two independent linear BM's. For $i = 1, \ldots, 4$, define the following operations on continuous progressively measurable processes K:

$$O_i(K)_t = \int_0^t K_s dB^i(s), \quad i = 1, 2,$$

$$O_3(K)_t = \int_0^t K_s ds, \qquad O_4(K_1, K_2)_t = K_1(t)K_2(t).$$

Let $\mathscr{C}$ be the class of processes obtained from $\{B^1, B^2\}$ by a finite number of operations O_i; we define by induction a real-valued mapping d on $\mathscr{C}$ by

$$d(B^1) = d(B^2) = 1/2,$$

$$d(O_1(K)) = d(O_2(K)) = d(K) + 1/2, \qquad d(O_3(K)) = d(K) + 1$$

$$d(O_4(K_1, K_2)) = d(K_1) + d(K_2).$$

Prove that for every $K \in \mathscr{C}$, there is a constant C_K, which may be zero, such that

$$E[K_t] = C_K t^{d(K)}.$$

(2.16) Exercise. Let f be a locally bounded Borel function on $\mathbb{R}_+$ and B be a BM. Prove that the process

$$Z_t = \int_0^t f(s) dB_s$$

is Gaussian and compute its covariance $\Gamma(s, t)$. Prove that $\exp\left\{Z_t - \frac{1}{2}\Gamma(t, t)\right\}$ is a martingale. This generalizes the example given in Proposition (1.2) iii) of Chap. II.

* **(2.17) Exercise.** 1°) Let B be a BM and H an adapted right-continuous bounded process. Prove that for a fixed t

$$\lim_{h \to 0} (B_{t+h} - B_t)^{-1} \int_t^{t+h} H_s dB_s = H_t \quad \text{in probability.}$$

The result is also true for H unbounded if it is continuous.

[Hint: One may apply Schwarz's inequality to

$$E\left[\left|\frac{1}{B_\varepsilon} \int_0^\varepsilon (H_u - H_0)\, dB_u\right|^{1/4}\right].]$$

2°) Let $B = (B^1, \dots, B^d)$ be a d-dimensional BM and for each j, H^j be a bounded right-continuous adapted process. Prove that for a fixed t,

$$\left(B^1_{t+h} - B^1_t\right)^{-1} \sum_j \int_t^{t+h} H^j_s dB^j_s$$

converges in law as $h \to 0$, to $H^1_t + \sum_{j=2}^d H^j_t \left(N^j / N^1\right)$ where $\left(N^1, \dots, N^d\right)$ is a centered Gaussian r.v. with covariance I_d, independent of $\left(H^1_t, \dots, H^d_t\right)$.

* **(2.18) Exercise.** Let X and Y be two continuous semimartingales. For a subdivision Δ of $[0, t]$, a function $f \in C^1(\mathbb{R})$ and a probability measure μ on $[0, 1]$, we define

$$S^\mu_\Delta = \sum_i \left(Y_{t_{i+1}} - Y_{t_i}\right) \int_0^1 f\left(X_{t_i} + s\left(X_{t_{i+1}} - X_{t_i}\right)\right) d\mu(s).$$

1°) Prove that

$$\lim_{|\Delta|\to 0} S^\mu_\Delta = \int_0^t f(X_s)dY_s + \bar{\mu}\int_0^t f'(X_s)d\langle X, Y\rangle_s$$

in probability, with $\bar{\mu} = \int_0^1 s\, d\mu(s)$. For $\mu = \delta_{1/2}$ and $f(x) = x$, this limit is called the *Stratonovich integral* of X against Y. For $\mu = \delta_0$, we get the Itô stochastic integral of $f(X)$ against Y. Observe that this is the only integral for which the resulting process is a local martingale when Y is a local martingale.

[Hint: Use Exercise (1.33).]

For $\mu = \delta_1$, the limit is called the *backward integral*.

2°) If we set

$$\bar{S}^\mu_\Delta = \sum_i \left(Y_{t_{i+1}} - Y_{t_i}\right)\int_0^1 f\left(X_{t_i+s(t_{i+1}-t_i)}\right)d\mu(s)$$

and if $d\langle X, Y\rangle$ is absolutely continuous with respect to the Lebesgue measure, then $\bar{S}^\mu_\Delta$ has the same limit as S^μ_Δ in probability whenever $|\Delta| \to 0$.

3°) If $\tilde{\omega} = \sum_i f_i(x_1, \dots, x_d)\,dx_i$ is a closed differential form of class C^1 on an open set U of $\mathbb{R}^d$ and $X = \left(X^1, \dots, X^d\right)$ a vector semimartingale with values in U (i.e. $P[\exists t : X_t \notin U] = 0$) then

$$\int_{X(0,t)} \tilde{\omega} = \sum_{i=1}^d \int_0^t f_i(X_s)dX_s^i + \frac{1}{2}\sum_{i,j}\int_0^t \frac{\partial f_i}{\partial x_j}(X_s)d\langle X^i, X^j\rangle_s$$

where $X(0,t)$ is the continuous path $(X_s(\omega), 0 \le s \le t)$. We recall that the integral of a closed form $\tilde{\omega}$ along a continuous but not necessarily differentiable path $\gamma : [0,t] \to \mathbb{R}^d$ is defined as $\pi(\gamma(t)) - \pi(\gamma(0))$ where π is a function such that $d\pi = \tilde{\omega}$ in a string of balls which covers γ.

4°) If B is the planar BM, express as a stochastic integral the area swept by the line segment joining 0 to the point B_s, as s varies from 0 to t.

* **(2.19) Exercise (Fubini-Wiener identity in law).** 1°) Let B and C be two independent standard BM's. If $\phi \in L^2([0,1]^2, ds\,dt)$, by using suitable filtrations, give a meaning to the integrals

$$\int_0^1 dB_u \int_0^1 dC_s\phi(u,s) \qquad \text{and} \qquad \int_0^1 dC_s \int_0^1 dB_u\phi(u,s)$$

and prove that they are almost-surely equal.

2°) Conclude from 1°) that

$$\int_0^1 du\left(\int_0^1 dB_s\phi(u,s)\right)^2 \stackrel{(d)}{=} \int_0^1 du\left(\int_0^1 dB_s\phi(s,u)\right)^2.$$

3°) If f is a C^1-function on $[0,1]$ and $f(1) = 1$, prove that

$$\int_0^1 ds\left(B_s - \int_0^1 f'(t)B_t dt\right)^2 \stackrel{(d)}{=} \int_0^1 ds\left(B_s - f(s)B_1\right)^2.$$

If, in particular, $\tilde{B}$ is a Brownian bridge,

$$\int_0^1 ds \left(B_s - \int_0^1 B_t dt \right)^2 \stackrel{(d)}{=} \int_0^1 ds \tilde{B}_s^2 .$$

[Hint: Take $\phi(s,u) = 1_{(u \le s)} + f(u) - 1$.]

(2.20) Exercise. Prove that in Theorem (2.12) the pointwise convergence of (K_n) may be replaced by suitable convergences in probability.

\# **(2.21) Exercise.** Let X be a real-valued function on $[0, 1]$. For a finite subdivision Δ of $[0, 1]$ and $h \in \mathbf{W} = C([0, 1], \mathbb{R})$, we set

$$S_\Delta(h) = \sum_{t_i \in \Delta} h(t_i) \left(X_{t_{i+1}} - X_{t_i} \right) .$$

1°) Prove that the map $h \to S_\Delta(h)$ is a continuous linear form on $\mathbf{W}$ with norm

$$\|S_\Delta\| = \sum_{t_i \in \Delta} \left| X_{t_{i+1}} - X_{t_i} \right| .$$

2°) If $\{S_{\Delta_n}(h)\}$ converges to a finite limit for every $h \in \mathbf{W}$ and any sequence $\{\Delta_n\}$ such that $|\Delta_n|$ tends to 0, prove that X is of bounded variation. This shows why the stochastic integral with respect to a cont. loc. mart. cannot be defined in the ordinary way.

[Hint: Apply the Banach-Steinhaus theorem.]

3°) Use the same ideas to solve Exercise (1.38).

\# **(2.22) Exercise (Orthogonal martingales).** 1°) Two martingales M and N of H^2, vanishing at 0, are said to be *weakly orthogonal* if $E[M_s N_t] = 0$ for every s and $t \ge 0$. Prove that the following four properties are equivalent:

i) M and N are weakly orthogonal,
ii) $E[M_s N_s] = 0$ for every $s \ge 0$,
iii) $E[\langle M, N \rangle_s] = 0$ for every $s \ge 0$,
iv) $E[M_T N_s] = 0$ for every $s \ge 0$ and every stopping time $T \ge s$.

2°) The two martingales are said to be *orthogonal* (see also Exercise (5.11)) if MN is a martingale. Prove that M and N are orthogonal iff $E[M_T N_s] = 0$ for every $s \ge 0$ and every stopping time $T \le s$.

Prove also that M and N are orthogonal iff $H{\cdot}M$ and N are weakly orthogonal, for every bounded predictable process H.

3°) Give examples of weakly orthogonal martingales which are not orthogonal.

[Hint: One can use stochastic integrals with respect to BM.]

4°) If the two-dimensional process (M, N) is gaussian, and if M and N are weakly orthogonal, prove that they are orthogonal.

Further results relating orthogonality and independence for martingales may be found in Exercise (4.25) Chap. V.

§3. Itô's Formula and First Applications

This section is fundamental. It is devoted to a "change of variables" formula for stochastic integrals which makes them easy to handle and thus leads to explicit computations.

Another way of viewing this formula is to say that we are looking for functions which operate on the class of continuous semimartingales, that is, functions F such that $F(X_t)$ is a continuous semimartingale whatever the continuous semimartingale X is. We begin with the special case $F(x) = x^2$.

(3.1) Proposition (Integration by parts formula). *If X and Y are two continuous semimartingales, then*

$$X_t Y_t = X_0 Y_0 + \int_0^t X_s dY_s + \int_0^t Y_s dX_s + \langle X, Y\rangle_t;$$

In particular,

$$X_t^2 = X_0^2 + 2\int_0^t X_s dX_s + \langle X, X\rangle_t.$$

Proof. It is enough to prove the particular case which implies the general one by polarization. If Δ is a subdivision of $[0, t]$, we have

$$\sum_i \left(X_{t_{i+1}} - X_{t_i}\right)^2 = X_t^2 - X_0^2 - 2\sum_i X_{t_i}\left(X_{t_{i+1}} - X_{t_i}\right);$$

letting $|\Delta|$ tend to zero and using, on one hand the definition of $\langle X, X\rangle$, on the other hand Proposition (2.13), we get the desired result. □

If X and Y are of finite variation, this formula boils down to the ordinary integration by parts formula for Stieltjes integrals. The same will be true for the following change of variables formula. Let us also observe that if M is a local martingale, we have, as a result of the above formula,

$$M_t^2 - \langle M, M\rangle_t = M_0^2 + 2\int_0^t M_s dM_s;$$

we already knew that $M^2 - \langle M, M\rangle$ is a local martingale but the above formula gives us an explicit expression of this local martingale. In the case of BM, we have

$$B_t^2 - t = 2\int_0^t B_s dB_s,$$

which can also be seen as giving us an explicit value for the stochastic integral in the right member. The reader will observe the difference with the ordinary integrals in the appearance of the term t. This is due to the quadratic variation.

All this is generalized in the following theorem. We first lay down the

(3.2) Definition and notation. *A d-dimensional* vector local martingale *(resp.* vector continuous semimartingale*) is a $\mathbb{R}^d$-valued process $X = (X^1, \dots, X^d)$ such that each X^i is a local martingale (resp. cont. semimart.). A complex local martingale (resp. complex cont. semimart.) is a $\mathbb{C}$-valued process whose real and imaginary parts are local martingales (resp. cont. semimarts.).*

(3.3) Theorem (Itô's formula). *Let $X = (X^1, \dots, X^d)$ be a continuous vector semimartingale and $F \in C^2(\mathbb{R}^d, \mathbb{R})$; then, $F(X)$ is a continuous semimartingale and*

$$F(X_t) = F(X_0) + \sum_i \int_0^t \frac{\partial F}{\partial x_i}(X_s)dX_s^i + \frac{1}{2}\sum_{i,j}\int_0^t \frac{\partial^2 F}{\partial x_i \partial x_j}(X_s)d\langle X^i, X^j\rangle_s.$$

Proof. If F is a function for which the result is true, then for any i, the result is true for $G(x_1, \dots, x_d) = x_i F(x_1, \dots, x_d)$; this is a straightforward consequence of the integration by parts formula. The result is thus true for polynomial functions. By stopping, it is enough to prove the result when X takes its values in a compact set K of $\mathbb{R}^d$. But on K, any F in $C^2(\mathbb{R}^d, \mathbb{R})$ is the limit in $C^2(K, \mathbb{R})$ of polynomial functions. By the ordinary and stochastic dominated convergence theorems (Theorem (2.12)), the theorem is established.

Remarks. 1°) The differentiability properties of F may be somewhat relaxed. For instance, if some of the X^i's are of finite variation, F needs only be of class C^1 in the corresponding coordinates; the proof goes through just the same. In particular, if X is a continuous semimartingale and $A \in \mathcal{A}$, and if $\partial^2 F/\partial x^2$ and $\partial F/\partial y$ exist and are continuous, then

$$\begin{aligned} F(X_t, A_t) &= F(X_0, A_0) + \int_0^t \frac{\partial F}{\partial x}(X_s, A_s)dX_s + \int_0^t \frac{\partial F}{\partial y}(X_s, A_s)dA_s \\ &\quad + \frac{1}{2}\int_0^t \frac{\partial^2 F}{\partial x^2}(X_s, A_s)d\langle X, X\rangle_s. \end{aligned}$$

2°) One gets another obvious extension when F is defined only on an open set but X takes a.s. its values in this set. We leave the details to the reader as an exercise.

3°) Itô's formula may be written in "differential" form

$$dF(X_t) = \sum_i \frac{\partial F}{\partial x_i}(X_t)dX_t^i + \frac{1}{2}\sum_{i,j}\frac{\partial^2 F}{\partial x_i \partial x_j}(X_t)d\langle X^i, X^j\rangle_t.$$

More generally, if X is a vector semimartingale, $dY_t = \sum_i H_t^i dX_t^i$ will mean

$$Y_t = Y_0 + \sum_i \int_0^t H_s^i dX_s^i.$$

In this setting, Itô's formula may be read as "the chain rule for stochastic differentials".

4°) Itô's formula shows precisely that the class of semimartingales is invariant under composition with C^2-functions, which gives another reason for the introduction of semimartingales. If M is a local martingale, or even a martingale, $F(M)$ is usually not a local martingale but only a semimartingale.

5°) Let ϕ be a C^1-function with compact support in $]0, 1[$. It is of finite variation, hence may be looked upon as a semimartingale and the integration by parts formula yields

$$X_1\phi(1) = X_0\phi(0) + \int_0^1 \phi(s)dX_s + \int_0^1 X_s\phi'(s)ds + \langle X, \phi\rangle_1$$

which reduces to

$$\int_0^1 \phi(s)dX_s = -\int_0^1 X_s\phi'(s)ds.$$

Stochastic integration thus appears as a random Schwartz distribution, namely the derivative of the continuous function $t \to X_t(\omega)$ in the sense of distributions. The above formula, a special case of the integration by parts formula, was taken as the definition of stochastic integrals in the earliest stage of the theory of stochastic integration and is useful in some cases (cf. Chap. VIII, Exercise (2.14)). Of course, the modern theory is more powerful in that it deals with much more general integrands.

To some extent, the whole sequel of this book is but an unending series of applications of Itô's formula. That is to say that Itô's formula has revolutionized the study of BM and other important classes of processes. We begin here with a few remarks and a fundamental result.

In the following proposition, we introduce the class of *exponential local martingales* which turns out to be very important; they are used in many proofs and play a fundamental role in Chap. VIII. For the time being, they provide us with many new examples of local martingales.

(3.4) Proposition. *If f is a complex valued function, defined on $\mathbb{R} \times \mathbb{R}_+$, and such that $\frac{\partial^2 f}{dx^2}$ and $\frac{\partial f}{dy}$ exist, are continuous and satisfy $\frac{\partial f}{\partial y} + \frac{1}{2}\frac{\partial^2 f}{\partial x^2} = 0$, then for any cont. local mart. M, the process $f(M_t, \langle M, M\rangle_t)$ is a local martingale. In particular for any $\lambda \in \mathbb{C}$, the process*

$$\mathscr{E}^\lambda(M)_t = \exp\left\{\lambda M_t - \frac{\lambda^2}{2}\langle M, M\rangle_t\right\}$$

is a local martingale.

For $\lambda = 1$, we write simply $\mathscr{E}(M)$ and speak of *the* exponential of M.

Proof. This follows at once by making $A = \langle M, M\rangle$ in Remark 1 below Theorem (3.3).

Remarks. 1°) A converse to Proposition (3.4) will be found in Exercise (3.14).

2°) For BM, we already knew that $\exp\{\lambda B_t - \frac{\lambda^2}{2}t\}$ is a martingale. Let us further observe that, for $f \in L^2_{\text{loc}}(\mathbb{R}_+)$, the exponential

$$\mathscr{E}_t^f = \exp\left\{\int_0^t f(u)dB_u - \frac{1}{2}\int_0^t f^2(u)du\right\}$$

is a martingale; this follows easily from the fact that $\int_s^t f(u)dB_u$ is a centered Gaussian r.v. with variance $\int_s^t f^2(u)du$ and is independent of $\mathscr{F}_s$. Likewise for BM^d and a d-uple $f = (f_1, \dots, f_d)$ of functions in $L^2_{\mathrm{loc}}(\mathbb{R}_+)$, and for the same reason,

$$\mathscr{E}_t^f = \exp\left\{\sum_{k=1}^d \int_0^t f_k(s)dB_s^k - \frac{1}{2}\sum_{k=1}^d \int_0^t f_k^2(s)ds\right\}$$

is a martingale. These martingales will be used in the following chapter.

3°) Following the same train of thought as in the preceding remark, one can ask more generally for the circumstances under which exponentials are true martingales (not merely local martingales). This will be studied in great detail in connection with Girsanov's theorem in Chap. VIII. We can already observe the following facts:

a) As already mentioned in Sect. 1, a local martingale which is ≥ 0 is a supermartingale; this can be seen by applying Fatou's lemma in passing to the limit in the equalities

$$E\left[M_{t\wedge T_n} \mid \mathscr{F}_s\right] = M_{s\wedge T_n}.$$

To obtain a martingale, one would have to be able to use Lebesgue's theorem. This will be the case if M is bounded, hence the exponential of a bounded martingale is a martingale.

b) If $M_0 = 0$ then $\mathscr{E}^\lambda(M)$ is a martingale if and only if $E\left[\mathscr{E}^\lambda(M)_t\right] \equiv 1$. The necessity is clear and the sufficiency comes from the fact that a supermartingale with constant expectation is a martingale.

4°) For a cont. semimart. X, we can equally define $\mathscr{E}^\lambda(X)_t = \exp\{\lambda X_t - \frac{\lambda^2}{2}\langle X, X\rangle_t\}$ and we still have

$$\mathscr{E}^\lambda(X)_t = \mathscr{E}^\lambda(X)_0 + \int_0^t \lambda\mathscr{E}^\lambda(X)_s dX_s.$$

This can be stated as: $\mathscr{E}^\lambda(X)$ is a solution to the stochastic differential equation

$$dY_t = \lambda Y_t dX_t.$$

When $X_0 = 0$, it is in fact the unique solution such that $Y_0 = 1$ (see Exercise (3.10)).

In the same spirit as in the above result we may also state the

(3.5) Proposition. *If B is a d-dimensional BM and $f \in C^2(\mathbb{R}_+ \times \mathbb{R}^d)$, then*

$$M_t^f = f(t, B_t) - \int_0^t \left(\frac{1}{2}\Delta f + \frac{\partial f}{\partial t}\right)(s, B_s)ds$$

is a local martingale. In particular if f is harmonic in $\mathbb{R}^d$ then $f(B)$ is a local martingale.

Proof. Because of their independence, the components B^i of B satisfy $\langle B^i, B^j\rangle_t = \delta_{ij}t$ (see Exercise (1.27)). Thus, our claim is a straightforward consequence of Itô's formula. □

The foregoing proposition will be generalized to a large class of Markov processes and will be the starting point of the fundamental method of Martingales problems (see Chap. VII). Roughly speaking, the idea is that Markov processes may be characterized by a set of local martingales of the above type. Actually much less is needed in the case of BM where it is enough to consider M^f for $f(x) = x^i$ and $f(x) = x^i x^j$. This is the content of the fundamental

(3.6) Theorem (P. Lévy's characterization theorem). *For a $(\mathscr{F}_t)$-adapted continuous d-dimensional process X vanishing at 0, the following three conditions are equivalent:*

i) X is an $\mathscr{F}_t$-Brownian motion;
ii) X is a continuous local martingale and $\langle X^i, X^j\rangle_t = \delta_{ij}t$ for every $1 \le i, j \le d$;
iii) X is a continuous local martingale and for every d-uple $f = (f_1, \ldots, f_d)$ of functions in $L^2(\mathbb{R}_+)$, the process

$$\mathscr{E}_t^{if} = \exp\left\{ i \sum_k \int_0^t f_k(s) dX_s^k + \frac{1}{2} \sum_k \int_0^t f_k^2(s) ds \right\}$$

is a complex martingale.

Proof. That i) implies ii) has already been seen (Exercise (1.27)).

Furthermore if ii) holds, Proposition (3.4) applied with $\lambda = i$ and $M_t = \sum_k \int_0^t f_k(s) dX_s^k$ implies that $\mathscr{E}^{if}$ is a local martingale; since it is bounded, it is a complex martingale.

Let us finally assume that iii) holds. Then, if $f = \xi\, 1_{[0,T]}$ for an arbitrary ξ in $\mathbb{R}^d$ and $T > 0$, the process

$$\mathscr{E}_t^{if} = \exp\left\{ i\, (\xi, X_{t\wedge T}) + \frac{1}{2} |\xi|^2 (t \wedge T) \right\}$$

is a martingale. For $A \in \mathscr{F}_s$, $s < t < T$, we get

$$E\left[1_A \exp\{ i\, (\xi, X_t - X_s)\}\right] = P(A) \exp\left(-\frac{|\xi|^2}{2}(t - s) \right).$$

(Here, and below, we use the notation (x, y) for the euclidean scalar product of x and y in $\mathbb{R}^d$, and $|x|^2 = (x, x)$.)

Since this is true for any $\xi \in \mathbb{R}^d$, the increment $X_t - X_s$ is independent of $\mathscr{F}_s$ and has a Gaussian distribution with variance $(t - s)$; hence i) holds. □

(3.7) Corollary. *The linear BM is the only continuous local martingale with t as increasing process.*

Proof. Stated for $d = 1$, the above result says that X is a linear BM if and only if X_t and $X_t^2 - t$ are continuous local martingales.

Remark. The word continuous is essential; for example if N is the Poisson process with parameter $c = 1$, $N_t - t$ and $(N_t - t)^2 - t$ are also martingales. (See Exercise (1.14) Chap. II).

The same principle as in Proposition (3.4) allows in fact to associate many other martingales with a given local martingale M. We begin with a few prerequisites. The *Hermite polynomials* h_n are defined by the identity

$$\sum_{n\geq 0} \frac{u^n}{n!} h_n(x) = \exp\left\{ux - \frac{u^2}{2}\right\}, \qquad u, x \in \mathbb{R},$$

whence it is deduced that

$$h_n(x) = \exp(x^2/2)(-1)^n \frac{d^n}{dx^n}\left(\exp(-x^2/2)\right).$$

For $a > 0$, we also have

$$\begin{aligned}\exp\left(ux - \frac{au^2}{2}\right) &= \exp\left(u\sqrt{a}\left(\frac{x}{\sqrt{a}}\right) - \frac{(u\sqrt{a})^2}{2}\right)\\ &= \sum_{n\geq 0} \frac{u^n}{n!} a^{n/2} h_n(x/\sqrt{a}) = \sum_{n\geq 0} \frac{u^n}{n!} H_n(x, a),\end{aligned}$$

if we set $H_n(x, a) = a^{n/2} h_n(x/\sqrt{a})$; we also set $H_n(x, 0) = x^n$.

(3.8) Proposition. *If M is a local martingale and $M_0 = 0$, the process*

$$L_t^{(n)} = H_n\left(M_t, \langle M, M\rangle_t\right)$$

is, for every n, a local martingale and moreover

$$L_t^{(n)} = n! \int_0^t dM_{s_1} \int_0^{s_1} dM_{s_2} \dots \int_0^{s_{n-1}} dM_{s_n}.$$

Proof. It is easily checked that $\left(\frac{1}{2}\partial^2/\partial x^2 + \partial/\partial a\right) H_n(x, a) = 0$ and that $\partial H_n/\partial x = nH_{n-1}$; thus Itô's formula implies that

$$L_t^{(n)} = n \int_0^t L_s^{(n-1)} dM_s,$$

which entails that $L^{(n)}$ is a loc. mart. and its representation as a multiple stochastic integral is obtained by induction. □

Remark. The reader is invited to compute explicitly $L^{(n)}$ for small n. For $n = 0, 1, 2$, one finds the constant 1, M and $M^2 - \langle M, M\rangle$, but from $n = 3$ on, new examples of local martingales are found.

(3.9) Exercise. Prove the following extension of the integration by parts formula. If $f(t)$ is a right-continuous function of bounded variation on any compact interval and X is a continuous semimartingale

$$f(t)X_t = f(0)X_0 + \int_0^t f(s)dX_s + \int_0^t X_s df(s).$$

\# **(3.10) Exercise.** 1°) If X is a semimartingale and $X_0 = 0$, prove that $\mathscr{E}^\lambda(X)_t$ is the unique solution of $dZ_t = \lambda Z_t dX_t$ such that $Z_0 = 1$.

[Hint: If Y is another solution, compute $Y\,\mathscr{E}^\lambda(X)^{-1}$ using the integration by parts formula and Remark 2 below Itô's formula.]

2°) Let X and Y be two continuous semimartingales. Compute $\mathscr{E}(X+Y)$ and compare it to $\mathscr{E}(X)\mathscr{E}(Y)$. When does the equality occur? This exercise is generalized in Exercise (2.9) of Chap. IX.

(3.11) Exercise. 1°) If $X = M + A$ and $Y = N + B$ are two cont. semimarts., prove that $XY - \langle X, Y\rangle$ is a cont. loc. mart. iff $X \cdot B + Y \cdot A = 0$. In particular, $X^2 - \langle X, X\rangle$ is a cont. loc. mart. iff P-a.s. $X_s = 0$ dA_s-a.e.

2°) If the last condition in 1°) is satisfied, prove that for every C^2-function f,

$$f(X_t) - f(X_0) - f'(0)A_t - (1/2)\int_0^t f''(X_s)d\langle X, X\rangle_s$$

is a cont. loc. mart.

The class Σ of semimartingales X which satisfy the last condition in 1°) is considered again in Definition 4.4, Chap. VI.

\# **(3.12) Exercise. (Another proof and an extension of Itô's formula).** Let X be a continuous semimart. and $g : \mathbb{R} \times (\Omega \times \mathbb{R}_+) \to \mathbb{R}$ a function such that

i) $(x, u) \to g(x, \omega, u)$ is continuous for every ω;
ii) $x \to g(x, \omega, u)$ is C^2 for every (ω, u);
iii) $(\omega, u) \to g(x, \omega, u)$ is adapted for every x.

1°) Prove that, in the notation of Proposition (2.13),

$$P\text{-}\lim_{n\to\infty} \sum_{t_i \in \Delta_n} \left(g\left(X_{t_{i+1}}, t_i\right) - g\left(X_{t_i}, t_i\right)\right) =$$

$$\int_0^t \partial g/\partial x(X_u, u)dX_u + (1/2)\int_0^t \partial^2 g/\partial x^2(X_u, u)d\langle X, X\rangle_u.$$

2°) Prove that if in addition $g(0, \omega, u) \equiv 0$, then

$$P\text{-}\lim_{n\to\infty} \sum_{t_i \in \Delta_n} g\left(X_{t_{i+1}} - X_{t_i}, t_i\right) =$$

$$\int_0^t \partial g/\partial x(0, u)dX_u + (1/2)\int_0^t \partial^2 g/\partial x^2(0, u)d\langle X, X\rangle_u.$$

3°) Resume the situation of 1°) and assume moreover that g satisfies

iv) for every (x, ω), the map $u \to g(x, \omega, u)$ is of class C^1 and the derivative is continuous in the variable x,

then prove the following extension of Itô's formula:

$$\begin{aligned} g(X_t, t) &= g(X_0, 0) + \int_0^t (\partial g/\partial x)(X_u, u) dX_u + \int_0^t (\partial g/\partial u)(X_u, u) du \\ &\quad + (1/2) \int_0^t (\partial^2 g/\partial x^2)(X_u, u) d\langle X, X \rangle_u. \end{aligned}$$

4°) Extend these results to a vector-valued cont. semimart. X.

(3.13) Exercise (Yet another proof of Itô's formula). 1°) Let x be a continuous function which is of finite quadratic variation on $[0, t]$ in the following sense: there exists a sequence (Δ_n) of subdivisions of $[0, t]$ such that $|\Delta_n| \to 0$ and the measures

$$\sum_{t_i \in \Delta_n} \left(x_{t_{i+1}} - x_{t_i}\right)^2 \varepsilon_{t_i}$$

converge vaguely to a bounded measure whose distribution function denoted by $\langle x, x \rangle$ is continuous. Prove that for a C^2-function F,

$$F(x_t) = F(x_0) + \int_0^t F'(x_s) dx_s + \frac{1}{2} \int_0^t F''(x_s) d\langle x, x \rangle_s$$

where $\int_0^t F'(x_s) dx_s = \lim_{n\to\infty} \sum_{t_i \in \Delta_n} F'\left(x_{t_i}\right)\left(x_{t_{i+1}} - x_{t_i}\right)$.

[Hint: Write Taylor's formula up to order two with a remainder r such that $r(a, b) \leq \phi(|a - b|)(b - a)^2$ with ϕ increasing and $\lim_{c\to 0} \phi(c) = 0$.]

2°) Apply the result in 1°) to prove Itô's formula for continuous semimartingales.

\# **(3.14) Exercise.** If M is an adapted continuous process, A is an adapted continuous process of finite variation and if, for every λ, the process $\exp\left\{\lambda M_t - \frac{\lambda^2}{2} A_t\right\}$ is a local martingale, then M is a local martingale and $\langle M, M \rangle = A$.

[Hint: Take derivatives with respect to λ at $\lambda = 0$.]

(3.15) Exercise. If X and Y are two continuous semimartingales, denote by

$$\int_0^t X_s \circ dY_s$$

the Stratonovich integral defined in Exercise (2.18). Prove that if $F \in C^3(\mathbb{R}^d, \mathbb{R})$ and $X = (X^1, \ldots, X^d)$ is a vector semimartingale, then

$$F(X_t) = F(X_0) + \sum_i \int_0^t \frac{\partial F}{\partial x_i}(X_s) \circ dX_s^i.$$

\# **(3.16) Exercise. (Exponential inequality, also called Bernstein's inequality).** If M is a continuous local martingale vanishing at 0, prove that

$$P\left[M_\infty^* \geq x, \langle M, M \rangle_\infty \leq y\right] \leq \exp(-x^2/2y).$$

Derive therefrom that, if there is a constant c such that $\langle M, M\rangle_t \leq ct$ for all t, then

$$P\left[\sup_{s\leq t} M_s \geq at\right] \leq \exp(-a^2t/2c).$$

[Hint: Use the maximal inequality for positive supermartingales to carry through the same proof as for Proposition (1.8) in Chap. II.]

* **(3.17) Exercise.** Let μ be a positive measure on $\mathbb{R}_+$ such that the function

$$f_\mu(x,t) = \int_0^\infty \exp\left(yx - \frac{y^2}{2}t\right) d\mu(y)$$

is not everywhere infinite. For $\varepsilon > 0$, define

$$A(t,\varepsilon) = \inf\left\{x : f_\mu(x,t) \geq \varepsilon\right\}$$

and suppose that this is a continuous function of t.

1°) Prove that for any stopping time T of the linear BM,

$$P\left[\sup_{t\geq T}(B_t - A(t,\varepsilon)) \geq 0 | \mathscr{F}_T\right] = 1 \wedge \frac{f_\mu(B_T, T)}{\varepsilon}.$$

[Hint: Use the result in Exercise (3.12) Chap. II. To prove the necessary convergence to zero, look at times when B vanishes.]

2°) By suitably altering μ, prove that for $h \geq 0$ and $b \in \mathbb{R}$,

$$P\left[\sup_{t\geq T}(B_t - A(t+h,\varepsilon)) \geq -b \mid \mathscr{F}_T\right] = 1 \wedge \frac{f_\mu(b + B_T, h+T)}{\varepsilon}.$$

3°) If μ is a probability measure, prove that

$$\sup_{t\geq 0}(B_t - A(t,1)) \stackrel{(d)}{=} \mathbf{e}/Y$$

where $\mathbf{e}$ is an exponential r.v. with parameter 1, Y is a r.v. with law μ, and $\mathbf{e}$ and Y are independent.

4°) If $\mu(\{0\}) = 0$ and if there is an $N > 0$ such that $\int \exp(-Ny)\mu(dy) < \infty$, then, for every n, the following assertions are equivalent

i) $E\left[\left(\sup_{t\geq 0}(B_t - A(t,\varepsilon))^+\right)^n\right] < \infty$;
ii) $\int_{0+} y^{-n} d\mu(y) < \infty$.

*# **(3.18) Exercise (Brownian Bridges).** 1°) Retain the situation and notation of Exercise (1.39) 2°), and prove that

$$\beta_t = B_t - \int_0^{t\wedge 1} \frac{B_1 - B_s}{1-s} ds$$

is a $\mathscr{G}_t$-Brownian motion, independent of B_1. In particular, B is a $\mathscr{G}_t$-semimartingale.

2°) If $X_t^x = xt + B_t - tB_1$ is the Brownian Bridge of Sect. 3 Chap. I then

$$X_t^x = \beta_t + \int_0^t \frac{x - X_s^x}{1-s} ds.$$

The same equality obtains directly from 1°) by defining X^x as the BM conditioned to be equal to x at time 1.

The following questions, which are independent of 2°) are designed to give another (see Exercise (3.15) Chap. II) probabilistic proof of Hardy's L^2-inequality, namely, if for $f \in L^2([0,1])$ one sets

$$Hf(x) = \frac{1}{x}\int_0^x f(y)dy,$$

then $Hf \in L^2([0,1])$ and $\|Hf\|_2 \leq 2\|f\|_2$.

3°) Prove that if f is in $L^2([0,1])$, there exists a Borel function F on $[0,1[$ such that for any $t < 1$,

$$\int_0^t f(u)\frac{B_1 - B_u}{1-u} du = \int_0^1 F(v \wedge t) dB_v.$$

Then, observe that

$$\int_0^1 f(u) d\beta_u = \int_0^1 \big(f(u) - F(u)\big) dB_u.$$

4°) Prove that

$$\int_0^1 F(v)^2 dv \leq 4\int_0^1 f^2(u)du,$$

then, prove by elementary transformations on the integrals, that this inequality is equivalent to Hardy's L^2-inequality.

* **(3.19) Exercise.** 1°) Let ψ be a Borel function on $]0,1]$ such that for every $\varepsilon > 0$

$$\int_\varepsilon^1 |\psi(u)| du < \infty,$$

and define $\phi(u) = \int_u^1 \psi(s)ds$. If B is the standard linear BM, prove that the limit

$$\lim_{\varepsilon \to 0} \int_\varepsilon^1 \psi(s) B_s ds$$

exists in probability if and only if

$$\int_0^1 \phi^2(u)du < \infty \quad \text{and} \quad \lim_{\varepsilon \to 0} \sqrt{\varepsilon}\phi(\varepsilon) = 0.$$

Compare with Exercise (2.31) of Chap. III.

[Hint: For Gaussian r.v.'s, convergence in probability implies convergence in L^2.]

2°) Deduce from Exercise (3.18) 3°) that for $f \in L^2([0,1])$

$$\lim_{t\to 0}\int_t^1 f(u)u^{-1}B_u du$$

exists a.s. If H^* is the adjoint of the Hardy operator H (see Exercise (3.18)), prove that for every $f \in L^2([0,1])$,

$$\lim_{\varepsilon\to 0}\sqrt{\varepsilon}H^*f(\varepsilon) = 0.$$

3°) Admit the equivalence between iii) and iv) stated in Exercise (2.31) Chap. III. Show that there exists a positive function f in $L^2([0,1])$ such that

$$\lim_{t\to 0}\int_t^1 f(u)u^{-1}B_u du \text{ exists a.s. and } \int_0^1 f(u)u^{-1}|B_u|du = \infty \text{ a.s.}$$

[Hint: Use $f(u) = 1_{[u\le 1/2]}/u^{1/2}(-\log u)^\alpha$, $1/2 < \alpha \le 1$.]

4°) Let X be a stationary OU process. Prove that

$$\lim_{t\to\infty}\int_0^t g(s)X_s ds$$

exists a.s. and in L^2 for every $g \in L^2([0,\infty[)$.

[Hint: Use the representation of X given in Exercise (3.8) of Chap. I and the fact that for $\beta > 0$, the map $g \to (2\beta u)^{-1/2}g\left((2\beta)^{-1}\log(1/u)\right)$ is an isomorphism from $L^2([0,\infty[)$ onto $L^2([0,1[)$.]

Moreover using the same equivalence as in 3°) prove that if g is a positive function of $L^1_{\text{loc}}([0,\infty[)$ then

$$\int_0^\infty g(s)|X_s|ds < \infty \text{ a.s. iff } \int_0^\infty g(s)ds < \infty.$$

5°) For $\mu \in \mathbb{R}$, $\mu \neq 0$, and g locally integrable prove that

$$E\left[\left|\int_0^t g(s)e^{i\mu B_s}ds\right|^2\right] = \mu^{-2}E\left[\left(\int_0^t g(s)X_s ds\right)^2\right]$$

for a suitable stationary OU process X. Conclude that $\lim_{t\to\infty}\int_0^t g(s)e^{i\mu B_s}ds$ exists in L^2 whenever g is in $L^2([0,\infty[)$. Show that the a.s. convergence also holds.

* **(3.20) Exercise.** Let A be a $d\times d$-matrix and B a $BM^d(0)$. Prove that the processes (AB_t, B_t) (where (,) is the scalar product in $\mathbb{R}^d$) and $\int_0^t \left((A + A^t)B_s, dB_s\right)$ have the same filtration.

\# **(3.21) Exercise.** Prove the *strong Markov* property of BM by means of P. Lévy's characterization theorem. More precisely, if B is an $(\mathscr{F}_t)$-BM, prove that for any starting measure ν and any $(\mathscr{F}_t)$-stopping time T, the process $(B_{T+t} - B_T)1_{(T<\infty)}$ is a standard BM for the conditional probability $P_\nu(\cdot \mid T < \infty)$ and is independent of $\mathscr{F}_T$.

(3.22) Exercise. Let B be a BM^d and $O(t) = \left(O^i_j(t)\right)$ be a progressively measurable process taking its values in the set of $d \times d$-orthogonal matrices.

Remark that $\int_0^t \|O(s)\|^2 ds < \infty$ for every t. Prove that the process X defined by

$$X^i_t = \sum_{j=1}^d \int_0^t O^i_j(s) dB^j(s)$$

is a BM^d.

(3.23) Exercise. Let (X, Y) be a standard BM^2 and for $0 < \rho < 1$ put

$$Z_t = \rho X_t + \sqrt{1-\rho^2} Y_t.$$

1°) Prove that Z is a standard linear BM^1 and compute $\langle X, Z\rangle$ and $\langle Y, Z\rangle$.

2°) Prove that X_1 and $\sigma(Z_s, s \geq 0)$ are conditionally independent with respect to Z_1.

(3.24) Exercise. 1°) If M is a continuous process and A an increasing process, then M is a local martingale with increasing process A if and only if, for every $f \in C_b^2$,

$$f(M_t) - f(M_0) - \frac{1}{2}\int_0^t f''(M_s) dA_s$$

is a local martingale.

[Hint: For the sufficiency, use the functions x and x^2 suitably truncated.]

2°) If M and N are two continuous local martingales, the process $\langle M, N\rangle$ is equal to the process of bounded variation C if and only if for every $f \in C_b^2(\mathbb{R}^2, \mathbb{R})$, the process

$$f(M_t, N_t) - f(M_0, N_0) - \frac{1}{2}\int f''_{x^2}(M_s, N_s) d\langle M, M\rangle_s$$
$$-\frac{1}{2}\int f''_{y^2}(M_s, N_s) d\langle N, N\rangle_s - \int f''_{xy}(M_s, N_s) dC_s$$

is a local martingale.

(3.25) Exercise. If M is a continuous local martingale such that $M_0 = 0$, prove that

$$\left\{\mathscr{E}(M)_\infty = 0\right\} = \{\langle M, M\rangle_\infty = \infty\} \quad \text{a.s.}$$

[Hint: $\mathscr{E}(M) = \mathscr{E}\left(\frac{1}{2}M\right)^2 \exp\left(-\frac{1}{4}\langle M, M\rangle\right)$.]

** **(3.26) Exercise (Continuation of Exercise (1.42)).** 1°) Let $X = X_0 + M + A$ be a positive semimartingale such that A is increasing and $X_t \leq c$ a.s. for every t. Prove that $E[A_\infty] \leq c$.

2°) Let $M \in \mathscr{M}$ and suppose that $M_t \leq k$ for every $t > 0$. Prove that

$$V_t = (k+1-M_t)^{-1} - \int_0^t (k+1-M_s)^{-3} d\langle M\rangle_s$$

is in $\mathscr{M}$ and that $\langle V\rangle(\cdot,]0, 1])$ is a.s. finite.

3°) Let $M \in \mathscr{M}$ and suppose that $\overline{\lim}_{t \downarrow 0} M_t < \infty$ a.s. Prove that $\lim_{t \downarrow 0} M_t$ exists a.s.

4°) Prove that for $M \in \mathscr{M}$ and for a.e. ω one of the following three properties holds

i) $\lim_{t \downarrow 0} M_t(\omega)$ exists in $\mathbb{R}$;
ii) $\lim_{t \downarrow 0} |M_t(\omega)| = +\infty$;
iii) $\underline{\lim}_{t \downarrow 0} M_t(\omega) = -\infty$ and $\overline{\lim}_{t \downarrow 0} M_t(\omega) = +\infty$.

(3.27) Exercise. Let $\{\Delta_n\}$ be a sequence of refining (i.e. $\Delta_n \subset \Delta_{n+1}$) finite subdivisions of $[0, T]$ such that $|\Delta_n| \to 0$; write $\Delta_n = \left(0 = t_0 < t_1 < \ldots < t_{p_n} = T\right)$.

1°) If F is a continuous function on $[0, T]$, prove that the sequence of measures

$$\sum_{p=0}^{p_n - 1} \left(t_{p+1} - t_p\right)^{-1} \left(F(t_{p+1}) - F(t_p)\right) 1_{[t_p, t_{p+1}]}(u) du$$

converges to F' in the sense of Schwartz distributions.

2°) Let B be the standard linear BM. Prove that, for $f \in L^2([0, T])$, the sequence of random variables

$$\sum_{p=0}^{p_n - 1} \left(t_{p+1} - t_p\right)^{-1} \int_{t_p}^{t_{p+1}} f(u) du \left(B_{t_{p+1}} - B_{t_p}\right)$$

converges a.s. and in L^2 to a limit which the reader will identify.

3°) Prove that nonetheless the sequence of measures defined by making $F = B_{\cdot}(\omega)$ in 1°) does not converge vaguely to a measure on $[0, T]$.

4°) Prove that if f is a function of bounded variation and θ_p is a point of the interval $\left[t_p, t_{p+1}\right]$, then

$$\sum_{p=0}^{p_n - 1} f(\theta_p) \left(B_{t_{p+1}} - B_{t_p}\right)$$

converges a.s.

* **(3.28) Exercise (Continuation of Exercise (1.48)).** 1°) If M is a continuous local martingale on $[0, T[$, vanishing at 0 and with increasing process $t \wedge T$ and if B is a BM independent of M, then $M_t + B_t - B_{t \wedge T}$ is a BM. We will say that M is a BM on $[0, T[$.

2°) State and prove an Itô formula for continuous local martingales on $[0, T[$. (The two questions are independent).

** **(3.29) Exercise. (Extension of P. Lévy's characterization theorem to signed measures).** Let $(\Omega, \mathscr{F}_t, P)$ be a filtered probability space and Q a bounded signed measure on $\mathscr{F}_\infty$ such that $Q \ll P$. We suppose that $(\mathscr{F}_t)$ is right-continuous and P-complete and call M the cadlag martingale such that for each t, $M_t = (dQ/dP)|_{\mathscr{F}_t}$ a.s. A continuous process X is called a (Q, P)-local martingale if

i) X is a P-semimartingale;
ii) XM is a P-local martingale (see Sect. 1 Chap. VIII for the significance of this condition).

1°) If H is a locally bounded $(\mathscr{F}_t)$-predictable process and $H{\cdot}X$ is the stochastic integral computed under P, then if X is a (Q, P)-local martingale, so is $H \cdot X$.

2°) We assume henceforth that X_t and $X_t^2 - t$ are (Q, P)-local martingales, and $X_0 = 0$, P-a.s. Prove that for any real u the process

$$Y_t = \exp(iuX_t) - 1 + \frac{u^2}{2}\int_0^t \exp(iuX_s)ds$$

is a (Q, P)-local martingale and conclude that $\int Y_t dQ \equiv 0$.

3°) Prove that

$$\int \exp(iuX_t)dQ = Q(1)\exp(-tu^2/2)$$

and, more generally, that for $0 < t_1 < \ldots < t_n$ and real numbers u_k,

$$\int \exp i\left(u_1 X_{t_1} + u_2\left(X_{t_2} - X_{t_1}\right) + \ldots + u_n\left(X_{t_n} - X_{t_{n-1}}\right)\right)dQ$$
$$= Q(1)\exp\left(-\frac{1}{2}\left(t_1u_1^2 + (t_2 - t_1)u_2^2 + \ldots + (t_n - t_{n-1})u_n^2\right)\right).$$

Conclude that

- if $Q(1) = 0$, then $Q = 0$ on $\sigma\{X_s, s \geq 0\}$,
- if $Q(1) \neq 0$, then $Q/Q(1)$ is a probability measure on $\sigma\{X_s, s \geq 0\}$ under which X is a standard BM.

(3.30) Exercise (The Kailath-Segal identity). Let M be a cont. loc. mart. such that $M_0 = 0$, and define the iterated stochastic integrals of M by

$$I_0 = 1, \qquad I_n = \int_0^{\cdot} I_{n-1}(s)dM_s.$$

Prove that for $n \geq 2$,

$$nI_n = I_{n-1}M - I_{n-2}\langle M, M\rangle.$$

Relate this identity to a recurrence formula for Hermite polynomials.

[Hint: See Proposition (3.8).]

* **(3.31) Exercise. (A complement to Exercise (2.17)).** Let B be a $BM^1(0)$ and H and K two $(\mathscr{F}_t^B)$-progressively measurable bounded processes. We set $X_t = \int_0^t H_s dB_s + \int_0^t K_s ds$ and assume in addition that H is right-continuous at 0.

1°) If ϕ is a C^1-function, show that

$$h^{-1/2}\int_0^h \phi(X_s)dX_s$$

converges in law, as h tends to 0, to $\phi(0)H_0B_1$.

2°) Assume now that $\phi(0) = 0$ and that ϕ is C^2, and prove that

$$h^{-1}\int_0^h \phi(X_s)dX_s$$

converges in law to $\phi'(0)H_0^2(B_1^2 - 1)/2$.

3°) State and prove an extension of these results when ϕ is C^{p+1} with $\phi(0) = \phi'(0) = \ldots = \phi^{(p-1)}(0) = 0$.
[Hint: Use Proposition (3.8).]

* **(3.32) Exercise.** Let B and C be two independent BM(0)'s. Prove that

$$\int_0^1 (B_t + C_{1-t})^2 \, dt \stackrel{(d)}{=} \int_0^1 \left(B_t^2 + (B_1 - B_t)^2\right) dt.$$

[Hint: The Laplace transform in $(\lambda^2/2)$ of the right-hand side is the characteristic function in λ of the sum of two stochastic integrals; see Exercise (2.19).]

(3.33) Exercise. 1°) Let B be a BM^1 and H a $(\mathscr{F}_t^B)$-adapted process such that $\int_0^t H_s^2 ds < \infty$ for every t, and $\int_0^\infty H_s^2 ds = \infty$. Set

$$T = \inf\left\{t \geq 0, \int_0^t H_s^2 ds = \sigma^2\right\}$$

where σ^2 is a strictly positive constant. Prove that $\int_0^T H_s dB_s \sim \mathcal{N}(0, \sigma^2)$.

2°) **(Central-limit theorem for stochastic integrals)** Let (B^n) be a sequence of linear BM's defined on the same probability space and (K^n) a sequence of $(\mathscr{F}_t^{B^n})$-adapted processes such that

$$P - \lim_{n\to\infty} \int_0^{T_n} (K_s^n)^2 ds = \sigma^2,$$

for some constants (T_n). Prove that the sequence $\left(\int_0^{T_n} K_s^n \, dB_s^n\right)$ converges in law to $\mathcal{N}(0, \sigma^2)$.

[Hint: Apply 1°) to the processes $H^n = K^n 1_{[0,T_n]} + \sigma 1_{]T_n, T_n+1]}$.]

§4. Burkholder-Davis-Gundy Inequalities

In Sect. 1, we saw that, for an L^2-bounded continuous martingale M vanishing at zero, the norms $\|M_\infty^*\|_2$ and $\|\langle M, M\rangle_\infty^{1/2}\|_2$ are equivalent. We now use the Itô formula to generalize this to other L^p-norms. We recall that if M is a continuous local martingale, we write $M_t^* = \sup_{s\leq t} |M_s|$.

The whole section will be devoted to proofs of the Burkholder-Davis-Gundy inequalities which are the content of

(4.1) Theorem. *For every $p \in]0, \infty[$, there exist two constants c_p and C_p such that, for all continuous local martingales M vanishing at zero,*

$$c_p E\left[\langle M, M\rangle_\infty^{p/2}\right] \leq E\left[(M_\infty^*)^p\right] \leq C_p E\left[\langle M, M\rangle_\infty^{p/2}\right].$$

It is customary to say that the constants c_p and C_p are "universal" because they can be taken the same for all local martingales on any probability space whatsoever. If we call H^p the space of continuous local martingales such that M^*_∞ is in L^p, Theorem (4.1) gives us two equivalent norms on this space. For $p \geq 1$, the elements of H^p are true martingales and, for $p > 1$, the spaces H^p are the spaces of continuous martingales bounded in L^p; this is not true for $p = 1$, the space H^1 is smaller than the space of continuous L^1-bounded martingales and even of uniformly integrable martingales as was observed in Exercise (3.15) of Chap. II.

Let us also observe that, by stopping, the theorem has the obvious, but nonetheless important

(4.2) Corollary. *For any stopping time* T

$$c_p E\left[\langle M, M\rangle_T^{p/2}\right] \leq E\left[(M_T^*)^p\right] \leq C_p E\left[\langle M, M\rangle_T^{p/2}\right].$$

More generally, for any bounded predictable process H

$$\begin{aligned} c_p E\left[\left(\int_0^T H_s^2 d\langle M, M\rangle_s\right)^{p/2}\right] &\leq E\left[\sup_{t\leq T}\left|\int_0^t H_s dM_s\right|^p\right] \\ &\leq C_p E\left[\left(\int_0^T H_s^2 d\langle M, M\rangle_s\right)^{p/2}\right]. \end{aligned}$$

The proof of the theorem is broken up into several steps.

(4.3) Proposition. *For $p \geq 2$, there exists a constant C_p such that for any continuous local martingale M such that $M_0 = 0$,*

$$E\left[(M_\infty^*)^p\right] \leq C_p E\left[\langle M, M\rangle_\infty^{p/2}\right].$$

Proof. By stopping, it is enough to prove the result for bounded M. The function $x \to |x|^p$ being twice differentiable, we may apply Itô's formula to the effect that

$$|M_\infty|^p = \int_0^\infty p|M_s|^{p-1}(\operatorname{sgn} M_s)dM_s + \frac{1}{2}\int_0^\infty p(p-1)|M_s|^{p-2}d\langle M, M\rangle_s.$$

Consequently,

$$\begin{aligned} E\left[|M_\infty|^p\right] &= \frac{p(p-1)}{2} E\left[\int_0^\infty |M_s|^{p-2} d\langle M, M\rangle_s\right] \\ &\leq \frac{p(p-1)}{2} E\left[(M_\infty^*)^{p-2}\langle M, M\rangle_\infty\right] \\ &\leq \frac{p(p-1)}{2}\left\|(M_\infty^*)^{p-2}\right\|_{p/(p-2)} \|\langle M, M\rangle_\infty\|_{p/2}. \end{aligned}$$

On the other hand, by Doob's inequality, we have $\|M_\infty^*\|_p \leq (p/p-1)\|M_\infty\|_p$, and the result follows from straightforward calculations. □

(4.4) Proposition. *For $p \geq 4$, there exists a constant c_p such that*

$$c_p E\left[\langle M, M\rangle_\infty^{p/2}\right] \leq E\left[(M_\infty^*)^p\right].$$

Proof. By stopping, it is enough to prove the result in the case where $\langle M, M\rangle$ is bounded. In what follows, a_p will always designate a universal constant, but this constant may vary from line to line. For instance, for two reals x and y

$$|x+y|^p \le a_p\left(|x|^p + |y|^p\right).$$

From the equality $M_t^2 = 2\int_0^t M_s dM_s + \langle M, M\rangle_t$, it follows that

$$E\left[\langle M, M\rangle_\infty^{p/2}\right] \le a_p\left(E\left[(M_\infty^*)^p\right] + E\left[\left|\int_0^\infty M_s dM_s\right|^{p/2}\right]\right)$$

and applying the inequality of Proposition (4.3) to the local martingale $\int_0^\cdot M_s dM_s$, we get

$$\begin{aligned} E\left[\langle M, M\rangle_\infty^{p/2}\right] &\le a_p\left(E\left[(M_\infty^*)^p\right] + E\left[\left(\int_0^\infty M_s^2 d\langle M, M\rangle_s\right)^{p/4}\right]\right) \\ &\le a_p\left(E\left[(M_\infty^*)^p\right] + \left(E\left[(M_\infty^*)^p\right] E\left[\langle M, M\rangle_\infty^{p/2}\right]\right)^{1/2}\right). \end{aligned}$$

If we set $x = E\left[\langle M, M\rangle_\infty^{p/2}\right]^{1/2}$ and $y = E\left[(M_\infty^*)^p\right]^{1/2}$, the above inequality reads $x^2 - a_p xy - a_p y^2 \le 0$ which entails that x is less than or equal to the positive root of the equation $x^2 - a_p xy - a_p y^2 = 0$, which is of the form $a_p y$. This establishes the proposition.

Theorem (4.1) is a consequence of the two foregoing propositions and of a reduction procedure which we now describe.

(4.5) Definition (Domination relation). *A positive, adapted right-continuous process X is* dominated *by an increasing process A, if*

$$E\left[X_T | \mathscr{F}_0\right] \le E\left[A_T | \mathscr{F}_0\right]$$

for any bounded stopping time T.

(4.6) Lemma. *If X is dominated by A and A is continuous, for x and $y > 0$,*

$$P\left[X_\infty^* > x; A_\infty \le y\right] \le \frac{1}{x} E\left[A_\infty \wedge y\right]$$

where $X_\infty^* = \sup_s X_s$.

Proof. It suffices to prove the inequality in the case where $P(A_0 \le y) > 0$ and, in fact, even $P(A_0 \le y) = 1$, which may be achieved by replacing P by $P' = P(\cdot \mid A_0 \le y)$ under which the domination relation is still satisfied.

Moreover, by Fatou's lemma, it is enough to prove that

$$P\left[X_n^* > x; A_n \le y\right] \le \frac{1}{x} E\left[A_\infty \wedge y\right];$$

but reasoning on $[0, n]$ amounts to reasoning on $[0, \infty]$ and assuming that the r.v. X_∞ exists and the domination relation is true for all stopping times whether bounded or not. We define $R = \inf\{t : A_t > y\}$, $S = \inf\{t : X_t > x\}$, where in both cases the infimum of the empty set is taken equal to $+\infty$. Because A is continuous, we have $\{A_\infty \le y\} = \{R = \infty\}$ and consequently

$$\begin{aligned} P\left[X_\infty^* > x; A_\infty \le y\right] &= P\left[X_\infty^* > x; R = \infty\right] \\ &\le P\left[X_S \ge x; (S < \infty) \cap (R = \infty)\right] \\ &\le P\left[X_{S\wedge R} \ge x\right] \le \frac{1}{x} E\left[X_{S\wedge R}\right] \\ &\le \frac{1}{x} E\left[A_{S\wedge R}\right] \le \frac{1}{x} E\left[A_\infty \wedge y\right], \end{aligned}$$

the last inequality being satisfied since, thanks to the continuity of A, and $A_0 \le y$ a.s., we have $A_{S\wedge R} \le A_\infty \wedge y$. □

(4.7) Proposition. *Under the hypothesis of Lemma (4.6), for any $k \in]0, 1[$,*

$$E\left[\left(X_\infty^*\right)^k\right] \le \frac{2-k}{1-k} E\left[A_\infty^k\right].$$

Proof. Let F be a continuous increasing function from $\mathbb{R}_+$ into $\mathbb{R}_+$ with $F(0) = 0$. By Fubini's theorem and the above lemma

$$\begin{aligned} E\left[F\left(X_\infty^*\right)\right] &= E\left[\int_0^\infty 1_{(X_\infty^* > x)} dF(x)\right] \\ &\le \int_0^\infty \left(P\left[X_\infty^* > x; A_\infty \le x\right] + P\left[A_\infty > x\right]\right) dF(x) \\ &\le \int_0^\infty \left(\frac{1}{x} E\left[A_\infty \wedge x\right] + P\left[A_\infty > x\right]\right) dF(x) \\ &\le \int_0^\infty \left(2P\left[A_\infty > x\right] + \frac{1}{x} E\left[A_\infty \cdot 1_{(A_\infty \le x)}\right]\right) dF(x) \\ &= 2E\left[F(A_\infty)\right] + E\left[A_\infty \int_{A_\infty}^\infty \frac{dF(x)}{x}\right] = E\left[\tilde{F}(A_\infty)\right] \end{aligned}$$

if we set $\tilde{F}(x) = 2F(x) + x\int_x^\infty \frac{dF(u)}{u}$. Taking $F(x) = x^k$, we obtain the desired result.

Remark. For $k \ge 1$ and $f(x) = x^k$, $\tilde{F}$ is identically $+\infty$ and the above reasoning has no longer any interest. Exercise (4.16) shows that it is not possible under the hypothesis of the proposition to find a universal constant c such that $E\left[X_\infty^*\right] \le cE[A_\infty]$. This actually follows also from the case where X is a positive martingale which is not in H^1 as one can then take $A_t = X_0$ for every t.

To finish the proof of Theorem (4.1), it is now enough to use the above result with $X = (M^*)^2$ and $A = C_2\langle M, M\rangle$ for the right-hand side inequality, $X = \langle M, M\rangle^2$ and $A = C_4(M^*)^4$ for the left-hand side inequality. The necessary domination relations follow from Propositions (4.3) and (4.4), by stopping as in Corollary (4.2).

Other proofs of the BDG inequalities in more or less special cases will be found in the exercises. Furthermore, in Sect. 1 of the following chapter, we will see that a method of time-change permits to derive the BDG inequalities from the special case of BM. We will close this section by describing another approach to this special case.

(4.8) Definition. *Let ϕ be a positive real function defined on $]0, a]$, such that $\lim_{x\to 0}\phi(x) = 0$ and β a real number > 1. An ordered pair (X, Y) of positive random variables is said to satisfy the* "good λ inequality" *$I(\phi, \beta)$ if*

$$P[X \geq \beta\lambda; Y < \delta\lambda] \leq \phi(\delta)P[X \geq \lambda]$$

for every $\lambda > 0$ and $\delta \in]0, a]$. We will write $(X, Y) \in I(\phi, \beta)$.

In what follows, F will be a *moderate* function, that is, an increasing, continuous function vanishing at 0 and such that

$$\sup_{x>0} F(\alpha x)/F(x) = \gamma < \infty \quad \text{for some } \alpha > 1.$$

The property then actually holds for every $\alpha > 1$ with γ depending on α. The function $F(x) = x^p$, $0 < p < \infty$ is such a function.

The key to many inequalities is the following

(4.9) Lemma. *There is a constant c depending only on ϕ, β and γ such that if $(X, Y) \in I(\phi, \beta)$, then*

$$E[F(X)] \leq cE[F(Y)].$$

Proof. It is enough to prove the result for bounded F's because the same γ works for F and $F \wedge n$. We have

$$\begin{aligned} E[F(X/\beta)] &= \int_0^\infty P[X \geq \beta\lambda]dF(\lambda) \\ &\leq \int_0^\infty \phi(\delta)P[X \geq \lambda]dF(\lambda) + \int_0^\infty P[Y \geq \delta\lambda]dF(\lambda) \\ &= \phi(\delta)E[F(X)] + E[F(Y/\delta)]. \end{aligned}$$

By hypothesis, there is a γ such that $F(x) \leq \gamma F(x/\beta)$ for every x. Pick $\delta \in]0, a \wedge 1[$ such that $\gamma\phi(\delta) < 1$; then, we can choose γ' such that $F(x/\delta) \leq \gamma' F(x)$ for every x, and it follows that

$$E[F(X)] \leq \gamma' E[F(Y)]/\big(1 - \gamma\phi(\delta)\big).$$

□

The foregoing lemma may be put to use to prove Theorem (4.1) (see Exercise (4.25)). We will presently use it for a result on BM. We consider the canonical BM with the probability measures P_x, $x \in \mathbb{R}$, and translation operators θ_t, $t \geq 0$. We denote by $(\mathscr{F}_t)$ the Brownian filtration of Chap. III. Then, we have the following

(4.10) Theorem. *Let A_t, $t \geq 0$, be an $(\mathscr{F}_t)$-adapted, continuous, increasing process such that*

(i) $\lim_{b\to\infty} \sup_{x,\lambda} P_x[A_{\lambda^2} > b\lambda] = 0$,
(ii) there is a constant K such that for every s and t

$$A_{t+s} - A_s \leq K A_t \circ \theta_s.$$

Then, there exists a constant c_F such that for any stopping time T,

$$E_0[F(A_T)] \leq c_F E_0\left[F(T^{1/2})\right].$$

Proof. It is enough to prove the result for finite T and then it is enough to prove that there exist ϕ and β such that $(A_T, T^{1/2}) \in I(\phi, \beta)$ for every finite T. Pick any $\beta > 1$ and set $S = \inf\{t : A_t > \lambda\}$. Using the strong Markov property of BM at time S, we have

$$\begin{aligned}
P_0\left[A_T \geq \beta\lambda, T^{1/2} < \delta\lambda\right] &= P_0\left[A_T - A_S \geq (\beta-1)\lambda, T < \delta^2\lambda^2, S < T\right] \\
&\leq P_0\left[A_{S+\delta^2\lambda^2} - A_S \geq (\beta-1)\lambda, S < T\right] \\
&\leq P_0\left[A_{\delta^2\lambda^2} \circ \theta_S \geq (\beta-1)\lambda K^{-1}, S < T\right] \\
&\leq E_0\left[E_{B_S}\left[A_{\delta^2\lambda^2} \geq (\beta-1)\lambda K^{-1}\right], S < T\right] \\
&\leq \sup_x P_x\left[A_{\delta^2\lambda^2} \geq (\beta-1)\lambda K^{-1}\right] \cdot P_0[S < T] \\
&\leq \sup_{x,\lambda} P_x\left[A_{\lambda^2} \geq \frac{(\beta-1)}{K\delta}\lambda\right] \cdot P_0[A_T \geq \lambda]
\end{aligned}$$

which ends the proof. □

We may likewise obtain the reverse inequality.

(4.11) Theorem. *If A_t, $t \geq 0$, is an $(\mathscr{F}_t)$-adapted, continuous, increasing process such that*

(i) $\lim_{b\to 0} \sup_{x,\lambda} P_x[A_{\lambda^2} < b\lambda] = 0$,
(ii) there is a constant K such that for every $s < t$

$$A_{t-s} \circ \theta_s \leq K A_t.$$

Then, there is a constant C_F such that for any stopping time T,

$$E_0\left[F(T^{1/2})\right] \leq C_F E_0[F(A_T)].$$

Proof. It follows the same pattern as above. Pick $\beta > 1$, $\delta < 1$; we have

$$\begin{aligned}
P_0\left[T^{1/2} \geq \beta\lambda, A_T < \delta\lambda\right] &\leq P_0\left[T \geq \beta^2\lambda^2, A_{T-\lambda^2} \circ \theta_{\lambda^2} < K\delta\lambda\right] \\
&\leq P_0\left[T \geq \lambda^2, A_{\beta^2\lambda^2-\lambda^2} \circ \theta_{\lambda^2} < K\delta\lambda\right] \\
&= E_0\left[E_{B_{\lambda^2}}\left[A_{\beta^2\lambda^2-\lambda^2} < K\delta\lambda\right], T \geq \lambda^2\right] \\
&\leq \sup_{x,\lambda} P_x\left[A_{(\beta^2-1)\lambda^2} < K\delta\lambda\right] \cdot P_0\left[T^{1/2} \geq \lambda\right]
\end{aligned}$$

which ends the proof. □

The reader will check that these results apply to $A_t = \sup_{s \le t} |B_s - B_0|$, thus yielding the BDG inequalities for B, from which by time-change (see Sect. 1 Chap. V), one gets the general BDG inequalities. This method is actually extremely powerful and, to our knowledge, can be used to prove all the BDG-type inequalities for continuous processes.

\# **(4.12) Exercise.** Let B and B' be two independent standard linear BM's. Prove that for every p, there exist two constants c_p and C_p such that for any locally bounded $(\mathscr{F}_t^B)$-progressively measurable process H,

$$c_p E\left[\left((H \cdot B')^*_\infty\right)^p\right] \le E\left[\left((H \cdot B)^*_\infty\right)^p\right] \le C_p E\left[\left((H \cdot B')^*_\infty\right)^p\right].$$

\# **(4.13) Exercise.** For a continuous semimartingale $X = M + V$ vanishing at 0, we set

$$\|X\|_{\mathscr{S}^p} = \left\| \langle M, M \rangle_\infty^{1/2} + \int_0^\infty |dV|_s \right\|_{L^p}.$$

1°) Check that the set of X's such that $\|X\|_{\mathscr{S}^p} < \infty$ is a vector-space denoted by $\mathscr{S}^p$ and that $X \to \|X\|_{\mathscr{S}^p}$ is a semi-norm on $\mathscr{S}^p$.

2°) Prove that if $X^* = \sup_t |X|_t$, then $\|X^*\|_p \le c_p \|X\|_{\mathscr{S}^p}$ for some universal constant c_p. Is there a constant c'_p such that $\|X\|_{\mathscr{S}^p} \le c'_p \|X^*\|_p$?

3°) For $p > 1$, the quotient of $\mathscr{S}^p$ by the subspace of processes indistinguishable from the zero process is a Banach space and contains the space H^p.

\# **(4.14) Exercise.** 1°) If M is a continuous local martingale, deduce from the BDG inequalities that $\{M^*_\infty < \infty\} = \{\langle M, M \rangle_\infty < \infty\}$ a.s. (A stronger result is proved in Proposition (1.8) Chap. V).

2°) If M^n is a sequence of continuous local martingales, prove that $(M^n)^*_\infty$ converges in probability to zero if and only if $\langle M^n, M^n \rangle_\infty$ does likewise.

[Hint: Observe that it is enough to prove the results when the M^n's are uniformly bounded, then apply Lemma (4.6).]

* **(4.15) Exercise. (A Fourier transform proof of the existence of occupation densities).** 1°) Let M be a continuous local martingale such that $E\left[\langle M, M \rangle_t^2\right] < \infty$; let μ_t be the measure on $\mathbb{R}$ defined by

$$\mu_t(f) = \int_0^t f(M_s) d\langle M, M \rangle_s$$

and $\hat{\mu}_t$ its Fourier transform. Prove that

$$E\left[\int_{-\infty}^{+\infty} |\hat{\mu}_t(\xi)|^2 d\xi\right] < \infty,$$

and conclude that $\mu_t(dx) \ll dx$ a.s.

2°) Prove that for fixed t, there is a family L_t^a of random variables, $\mathscr{B}(\mathbb{R}) \otimes \mathscr{F}_t$-measurable, such that for any positive Borel function f

$$\int_0^t f(M_s)d\langle M, M\rangle_s = \int_{-\infty}^{+\infty} L_t^a f(a)da.$$

This will be taken up much more thoroughly in Chap. VI.

(4.16) Exercise. 1°) If B is a BM, prove that $|B|$ is dominated by $2S$ where $S_t = \sup B_s$.

[Hint: If x and y are two real numbers and $y \geq x^+$, one has $|x| \leq (y-x)+y$.]

2°) By looking at $X_t = |B_{t\wedge T_1}|$ where $T_1 = \inf\{t : B_t = 1\}$, justify the remark following Proposition (4.7).

(4.17) Exercise. Let M be a continuous local martingale with $M_0 = 0$ and define

$$S_t = \sup_{s\leq t} M_s, \qquad s_t = \inf_{s\leq t} M_s.$$

Let A be an increasing adapted continuous process with $A_0 \geq a > 0$.

1°) Remark that

$$\int_0^t (S_s - M_s)dS_s \equiv 0.$$

2°) Suppose M bounded and prove that

$$E\left[A_\infty^{-1}(S_\infty - M_\infty)^2\right] \leq E\left[\int_0^\infty A_s^{-1} d\langle M, M\rangle_s\right].$$

3°) Prove that $(M_t^*)^2 \leq 2\left((S_t - M_t)^2 + (M_t - s_t)^2\right)$ and that

$$\begin{aligned} E\left[\left(M_\infty^*\right)^2 \langle M, M\rangle_\infty^{-1/2}\right] &\leq 4E\left[\int_0^\infty \langle M, M\rangle_s^{-1/2} d\langle M, M\rangle_s\right] \\ &= 8E\left[\langle M, M\rangle_\infty^{1/2}\right] \end{aligned}$$

and extend this result to non bounded M's.

[Hint: To prove the last equality, use the time-change method of Sect. 1 in Chap. V.]

4°) Derive therefrom that

$$E\left[M_\infty^*\right] \leq 2\sqrt{2}E\left[\langle M, M\rangle_\infty^{1/2}\right].$$

5°) Using 2°) , prove also that

$$E\left[(S_\infty - s_\infty)^2\right] \leq 4E\left[M_\infty^2\right].$$

For another proof of this inequality, see Exercise (4.11), Chap. VI.

* **(4.18) Exercise.** 1°) Let M be a continuous local martingale with $M_0 = 0$ and A, B, C three continuous increasing processes such that $B_0 = C_0 = 0$, and $A_0 \geq 0$. If $X = M + B - C$ is ≥ 0, prove that $A^{-1}X$ is dominated by Y where

$$Y_t = \int_0^t A_s^{-1} dB_s.$$

(It is understood that 0/0 is taken equal to 0).

[Hint: Replace A_0 by $A_0 + \varepsilon$, then let ε decrease to 0.]

2°) Prove that for $p > q > 0$, there exists a constant C_{pq} such that

$$E\left[(M^*_\infty)^p \langle M, M\rangle_\infty^{-q/2}\right] \le C_{pq} E\left[(M^*_\infty)^{p-q}\right].$$

(4.19) Exercise. 1°) Let A be an increasing continuous process and X a r.v. in L^1_+ such that for any stopping time S

$$E\left[A_\infty - A_S \, 1_{(S>0)}\right] \le E\left[X \, 1_{(S<\infty)}\right].$$

Prove that for every $\lambda > 0$,

$$E\left[(A_\infty - \lambda) \, 1_{(A_\infty \ge \lambda)}\right] \le E\left[X \, 1_{(A_\infty \ge \lambda)}\right].$$

[Hint: Consider the stopping time $S = \inf\{t : A_t > \lambda\}$.]

2°) Let F be a convex, increasing function vanishing at zero and call f its right derivative. Prove that, under the hypothesis of 1°) ,

$$E\left[F(A_\infty)\right] \le E\left[X f(A_\infty)\right].$$

[Hint: Integrate the inequality of 1°) with respect to $df(\lambda)$.]

3°) If M is a continuous local martingale, show that

$$E\left[\langle M, M\rangle_\infty\right] \le c E\left[\langle M, M\rangle_\infty^{1/2} M^*_\infty\right]$$

for a universal constant c.

4°) For an L^2-bounded martingale M, define

$$S(M)_t = \sup\left(M^*_t, \langle M, M\rangle_t^{1/2}\right),$$
$$I(M)_t = \inf\left(M^*_t, \langle M, M\rangle_t^{1/2}\right).$$

Prove that $E\left[S(M)^2_\infty\right] \le d E\left[I(M)^2_\infty\right]$ for a universal constant d.

(4.20) Exercise. 1°) For $0 < p < 1$ set, in the usual notation,

$$N_t = \int_0^t \langle M, M\rangle_s^{(p-1)/2} dM_s$$

(to prove that this integral is meaningful, use the time-change method of Sect. 1 Chap. V) and prove that, if $E\left[\langle M, M\rangle_t^p\right] < \infty$, then $E\left[\langle M, M\rangle_t^p\right] = pE\left[N_t^2\right]$.

2°) By applying the integration by parts formula to $N_t \langle M, M\rangle_t^{(1-p)/2}$, prove that $|M_t| \le 2N^*_t \langle M, M\rangle_t^{(1-p)/2}$, and conclude that

$$E\left[(M^*_t)^{2p}\right] \le (16/p)^p E\left[\langle M, M\rangle_t^p\right].$$

(4.21) Exercise. Let $M = (M^1, \ldots, M^d)$ be a vector local martingale and set $A = \sum_{i=1}^d \langle M^i, M^i\rangle$. For $\varepsilon, \eta > 0$ and two finite stopping times $S \le T$, prove that

$$P\left[\sup_{S \le t \le T} |M_t - M_S|^2 > \varepsilon\right] \le \frac{\eta}{\varepsilon} + P\left[A_T - A_S \ge \eta\right].$$

* **(4.22) Exercise.** For a continuous local martingale M let (P) be a property of $\langle M, M\rangle_\infty$ such that i) if $\langle N, N\rangle_\infty \leq \langle M, M\rangle_\infty$ and M satisfies (P) then N satisfies (P), ii) if M satisfies (P) then M is a uniformly integrable martingale.

1°) If M satisfies (P), prove that

$$\sup\left\{E\left[\left|\int_0^\infty H_s dM_s\right|\right];\ H \text{ progressively measurable and } |H| \leq 1\right\} < \infty.$$

[Hint: Use the theorem of Banach-Steinhaus.]

2°) By considering $H = \sum \lambda_i 1_{]t_i, t_{i+1}]}$ for a subdivision $\Delta = (t_i)$ prove that (P) entails that $E\left[\langle M, M\rangle_\infty^{1/2}\right] < +\infty$. As a result, the property $E\left[\langle M, M\rangle_\infty^{1/2}\right] < \infty$ is the weakest property for which i) and ii) are satisfied.

[Hint: Prove that $\sup_\Delta E\left[\left(\sum \left(M_{t_{i+1}} - M_{t_i}\right)^2\right)^{1/2}\right] < \infty$.]

*# **(4.23) Exercise.** Let R_t be the modulus of the BM^d, $d \geq 3$, started at $x \neq 0$.

1°) After having developed $\log R_t$ by Itô's formula, prove that

$$\sup_{t \geq 2} E\left[\left(\int_0^t R_s^{-2} ds / \log t\right)^p\right] < +\infty$$

for every $p > 0$.

[Hint: One may use the argument which ends the proof of Proposition (4.4).]

2°) Prove that $(\log R_t / \log t)$ converges in probability as t goes to infinity to a constant c and conclude that $\int_0^t R_s^{-2} ds / \log t$ converges in probability to $1/(d-2)$. The limit holds actually a.s. as is proved in Exercise (3.20) of Chap. X.

3°) Let now x be 0 and study the asymptotic behavior of $\int_\varepsilon^1 R_s^{-2} ds$ as ε tends to zero.

[Hint: Use time-inversion.]

** **(4.24) Exercise. (The duality between H^1 and BMO revisited).** 1°) Prove that $|||X|||_{H^1} = E\left[\langle X, X\rangle_\infty^{1/2}\right]$ is a norm on the space H^1 of Exercise (1.17) of Chap. II which is equivalent to the norm $\|X\|_{H^1}$.

2°) **(Fefferman's inequality).** Let $X \in H^1$ and $Y \in$ BMO; using the result in Exercise (1.40), prove that

$$E\left[\int_0^\infty |d\langle X, Y\rangle|_s\right] \leq 2|||X|||_{H^1}\|Y\|_{\mathrm{BMO}_2}.$$

[Hint: Write $\int_0^\infty |d\langle X, Y\rangle|_s = \int_0^\infty \langle X, X\rangle_s^{-1/4} \langle X, X\rangle_s^{1/4} |d\langle X, Y\rangle|_s$ and apply the Kunita-Watanabe inequality.]

3°) Prove that the dual space of H^1 is BMO and that the canonical bilinear form on $H^1 \times$ BMO is given by

$$(X, Y) \to E\left[\langle X, Y\rangle_\infty\right].$$

* **(4.25) Exercise.** 1°) Let A and B be two continuous adapted increasing processes such that $A_0 = B_0 = 0$ and

$$E\left[(A_T - A_S)^p\right] \leq D\|B_T\|_\infty^p P[S < T]$$

for some positive real numbers p and D and every stopping times S, T with $S \leq T$. Prove that $(A_\infty, B_\infty) \in I(\phi, \beta)$ for every $\beta > 1$ and $\phi(x) = D(\beta - 1)^{-p}x^p$.

[Hint: Set $T = \inf\{t : B_t = \delta\lambda\}$, $S_n = \inf\{t : A_t > \lambda(1 - 1/n)\}$ and prove that the left-hand side in $I(\phi, \beta)$ is less than $P\left[A_T - A_{T\wedge S_n} \geq (\beta - 1 + 1/n)\lambda\right]$.]

2°) If M is a continuous local martingale vanishing at 0, prove, using only the results of Sect. 1, that for $A = \langle M, M\rangle^{1/2}$ and $B = M^*$ or vice-versa, the conditions of 1°) are satisfied with $p = 2$. Conclude that for a moderate function F there are constants c and C such that

$$cE\left[F\left(\langle M, M\rangle_\infty^{1/2}\right)\right] \leq E\left[F\left(M_\infty^*\right)\right] \leq CE\left[F\left(\langle M, M\rangle_\infty^{1/2}\right)\right].$$

3°) Derive another solution to 2°) in Exercise (4.14) from the above inequalities.

[Hint: The function $x/(1+x)$ is moderate increasing.]

* **(4.26) Exercise.** If Z is a positive random variable define

$$\sigma_z = \sup_{\lambda > 0} \lambda P[Z \geq \lambda], \qquad l_z = \overline{\lim_{\lambda \to \infty}} \lambda P[Z \geq \lambda].$$

If $(X, Y) \in I(\phi, \beta)$, prove that there is a constant c depending only on ϕ and β such that

$$\sigma_X \leq c\sigma_Y, \qquad l_X \leq cl_Y.$$

(4.27) Exercise. Apply Theorems (4.10) and (4.11) to

$$A_t = \sup_{0 \leq r \leq s \leq t} \left(|B_s - B_r|/|s - r|^{(1/2)-\varepsilon}\right)^{\varepsilon/2}$$

where $0 < \varepsilon < 1/2$.

* **(4.28) Exercise.** Let A (resp. B) satisfy the assumptions of Theorem (4.10) (resp. (4.11)). For $\alpha > 0$, prove that there is a constant c_F such that for any stopping time

$$E\left[F\left(A_T^{\alpha+1}/B_T^\alpha\right)\right] \leq c_F E\left[F(B_T)\right].$$

* **(4.29) Exercise (Garsia-Neveu lemma).** Retain the situation and notation of Exercise (4.19) and assume further that $\sup_{x>0} xf(x)/F(x) = p < +\infty$.

1°) Prove that if U and V are two positive r.v.'s such that

$$E[Uf(U)] < +\infty, \qquad E[Uf(U)] \leq E[Vf(U)],$$

then

$$E[F(U)] \leq E[F(V)].$$

[Hint: If g is the inverse of f (Sect. 4 Chap. 0) then for $u, v \geq 0$,

$$uf(u) = F(u) + \int_0^{f(u)} g(s)ds, \quad vf(u) = F(v) + \int_0^{f(u)} g(s)ds.\ ]$$

2°) Prove that

$$E\left[F(A_\infty)\right] \le E\left[F(pX)\right] \le p^p E[F(X)].$$

(4.30) Exercise (Improved constants in domination). Let C_k be the smallest constant such that $E\left[\left(X_\infty^*\right)^k\right] \le C_k E\left[A_\infty^k\right]$ for every X and A satisfying the condition of Definition (4.5).

1°) Prove that $C_k \le k^{-k}(1-k)^{-1} \le (2-k)/(1-k)$, for $k \in]0,1[$.

Reverse inequalities are stated in Exercise (4.22) Chap. VI.

[Hint: Follow the proof of Proposition (4.7) using λx instead of x and y/λ instead of y in the inequality of Lemma (4.6).]

2°) Prove that, for $k, k' \in]0,1[$, $C_{kk'} \le C_{k'} \left(C_k\right)^{k'}$.

(4.31) Exercise. 1°) Retain the notation of Exercise (3.14) Chap. II and prove that $E\left[H_n(B_{\sigma_a}, \sigma_a)\right] = 0$ where H_n is the n-th Hermite polynomial.

2°) Prove an induction formula for the moments of σ_a. Compare with the Laplace transform found in Chap. II, Exercise (3.14).

§5. Predictable Processes

Apart from the definition and elementary properties of predictable processes, the notions and results of this section are needed in very few places in the sequel. They may therefore be skipped until their necessity arises.

In what follows, we deal with a filtration $(\mathscr{F}_t)$ supposed to be right-continuous and complete. We shall work with the product space $\Omega \times \mathbb{R}_+$ and think of processes as functions defined on this space. Recall that a σ-field is generated by a set of functions if it is the coarsest σ-field for which these functions are measurable.

(5.1) Proposition. *The σ-fields generated on $\Omega \times \mathbb{R}_+$ by*

i) the space $\mathscr{E}$ of elementary processes,
ii) the space of adapted processes which are left-continuous on $]0, \infty[$,
iii) the space of adapted continuous processes

are equal.

Proof. Let us call τ_i, $i = 1, 2, 3$, the three σ-fields of the statement. Obviously $\tau_3 \subset \tau_2$; moreover $\tau_2 \subset \tau_1$ since a left-continuous process X is the pointwise limit of the processes

$$X_t^n(\omega) = X_0(\omega) 1_{\{0\}}(t) + \sum_{k=0}^{\infty} X_{(k/n)}(\omega) 1_{]k/n,(k+1)/n]}(t).$$

On the other hand, the function $1_{]u,v]}$ is the limit of continuous functions f^n with compact support contained in $]u, v+1/n]$. If $H \in \mathscr{F}_u$, the process Hf^n is continuous and adapted which implies that $\tau_1 \subset \tau_3$.

(5.2) Definition. *The unique σ-field discussed in the preceding proposition is called the* predictable σ-field *and is denoted by $\mathscr{P}$ or $\mathscr{P}(\mathscr{F}_t)$ (when one wants to stress the relevant filtration). A process X with values in $(U, \mathscr{U})$ is* predictable *if the map $(\omega, t) \to X_t(\omega)$ from $(\Omega \times \mathbb{R}_+)$ to $(U, \mathscr{U})$ is measurable with respect to $\mathscr{P}$.*

Observe that if X is predictable and if X_0 is replaced by another $\mathscr{F}_0$-measurable r.v., the altered process is still predictable; predictable processes may be thought of as defined on $]0, \infty[$. It is easily seen that predictable processes are adapted; they are actually $(\mathscr{F}_{t-})$-adapted.

The importance of predictable processes comes from the fact that all stochastic integrals are indistinguishable from the stochastic integrals of predictable processes. Indeed, if we call $L^2_{\mathscr{P}}(M)$ the set of equivalent classes of predictable processes of $\mathscr{L}^2(M)$, it can be proved that the Hilbert spaces $L^2(M)$ and $L^2_{\mathscr{P}}(M)$ are isomorphic, or in other words, that every process of $\mathscr{L}^2(M)$ is equivalent to a predictable process. We may also observe that, since $\mathscr{E}$ is an algebra and a lattice, the monotone class theorem yields that $\mathscr{E}$ is dense in $L^2_{\mathscr{P}}(M)$. Consequently, had we constructed the stochastic integral by continuity starting with elementary stochastic integrals, then $L^2_{\mathscr{P}}(M)$ would have been the class of integrable processes.

We now introduce another important σ-field.

(5.3) Definition. *The σ-field generated on $\Omega \times \mathbb{R}_+$ by the adapted cadlag processes is called the* optional σ-field *and is denoted by $\mathscr{O}$ or $\mathscr{O}(\mathscr{F}_t)$. A process which is measurable with respect to $\mathscr{O}$ is called optional.*

It was already noticed in Sect. 4 Chap. I that, if T is a stopping time, the process $1_{]0,T]}$, namely $(\omega, t) \to 1_{[0<t\leq T(\omega)]}$, is predictable. We can now observe that T is a stopping time if and only if $1_{[0,T[}$ is optional.

Since the continuous processes are cadlag, it is obvious that $\mathscr{O} \supset \mathscr{P}$ and this inclusion is usually strict (however, see Corollary (5.7) below). If we denote by Prog the progressive σ-field, we see that

$$\mathscr{F}_\infty \otimes \mathscr{B}(\mathbb{R}_+) \supset \mathrm{Prog} \supset \mathscr{O} \supset \mathscr{P}.$$

The inclusion Prog $\supset \mathscr{O}$ may also be strict (see Exercise (5.12)).

We proceed to a few properties of predictable and optional processes.

(5.4) Definition. *A stopping time T is said to be* predictable *if there is an increasing sequence (T_n) of stopping times such that almost-surely*

i) $\lim_n T_n = T$,
ii) $T_n < T$ *for every n on* $\{T > 0\}$.

We will now state without proof a result called the *section* theorem. Let us recall that the *graph* $[T]$ of a stopping time is the set $\{(\omega, t) \in \Omega \times \mathbb{R}_+ : T(\omega) = t\}$. If T is predictable, this set is easily seen to be predictable. Let us further call π the canonical projection of $\Omega \times \mathbb{R}_+$ onto Ω.

(5.5) Theorem. *Let A be an optional (resp. predictable) set. For every $\varepsilon > 0$, there is a stopping time (resp. predictable stopping time) such that*

i) $[T] \subset A$,
ii) $P[T < \infty] \geq P(\pi(A)) - \varepsilon$.

This will be used to prove the following *projection* theorem. The σ-field $\mathscr{F}_{T-}$ is defined in Exercise (4.18) of Chap. I. By convention $\mathscr{F}_{0-} = \mathscr{F}_0$.

(5.6) Theorem. *Let X be a measurable process either positive or bounded. There exsists a unique (up to indistinguishability) optional process Y (resp. predictable process Z) such that*

$$E\left[X_T 1_{(T<\infty)} \mid \mathscr{F}_T\right] = Y_T 1_{(T<\infty)} \text{ a.s. } \textit{for every stopping time } T$$

(resp. $E\left[X_T 1_{(T<\infty)} \mid \mathscr{F}_{T-}\right] = Z_T 1_{(T<\infty)}$ *a.s. for every predictable stopping time T).*

The process Y (resp. Z) is called the *optional (predictable) projection* of X.

Proof. The uniqueness follows at once from the section theorem. The space of bounded processes X which admit an optional (predictable) projection is a vector space. Moreover, let X^n be a uniformly bounded increasing sequence of processes with limit X and suppose that they admit projections Y^n and Z^n. The section theorem again shows that the sequences (Y^n) and (Z^n) are a.s. increasing; it is easily checked that $\underline{\lim}\, Y^n$ and $\underline{\lim}\, Z^n$ are projections for X.

By the monotone class theorem, it is now enough to prove the statement for a class of processes closed under pointwise multiplication and generating the σ-field $\mathscr{F}_\infty \otimes \mathscr{B}(\mathbb{R}_+)$. Such a class is provided by the processes

$$X_t(\omega) = 1_{[0,u[}(t) H(\omega), \quad 0 \leq u \leq \infty, \quad H \in L^\infty(\mathscr{F}_\infty).$$

Let H_t be a cadlag version of $E[H \mid \mathscr{F}_t]$ (with the convention that $H_{0-} = H_0$). The optional stopping theorem (resp. the predictable stopping theorem of Exercise (3.18) Chap. II) proves that

$$Y_t = 1_{[0,u[}(t) H_t \quad (\text{resp. } Z_t = 1_{[0,u[}(t) H_{t-})$$

satisfies the condition of the statement. The proof is complete in the bounded case. For the general case we use the processes $X \wedge n$ and pass to the limit.

Remark. The conditions in the theorem might as well have been stated

$E\left[X_T 1_{(T<\infty)}\right] = E\left[Y_T 1_{(T<\infty)}\right]$ for any stopping time T
(resp.
$E\left[X_T 1_{(T<\infty)}\right] = E\left[Z_T 1_{(T<\infty)}\right]$ for any predictable stopping time T).

(5.7) Corollary. *Let $\mathscr{J}$ be the σ-field generated by the processes $M_t - M_{t-}$ where M ranges through the bounded $(\mathscr{F}_t)$-martingales; then*

$$\mathscr{O} = \mathscr{P} \vee \mathscr{J}.$$

In particular, if all $(\mathscr{F}_t)$-martingales are continuous, then $\mathscr{O} = \mathscr{P}$.

Proof. Since every optional process is its own optional projection, it is enough to prove that every optional projection is measurable with respect to $\mathscr{P} \vee \mathscr{J}$, but this is obvious for the processes $1_{[0,u[}H$ considered in the above proof and an application of the monotone class theorem completes the proof.

(5.8) Exercise. Prove that $\mathscr{P}$ is generated by the sets

$$[S, T] = \{(t, \omega) : S(\omega) \leq t \leq T(\omega)\}$$

where S is a predictable stopping time and T an arbitrary stopping time larger than S.

\# **(5.9) Exercise.** Prove that if M_t is a uniformly integrable cadlag martingale, its predictable projection is equal to M_{t-}.

(5.10) Exercise. For any optional process X, there is a predictable process Y such that the set $\{(\omega, t) : X_t(\omega) \neq Y_t(\omega)\}$ is contained in the union of the graphs of countably many stopping times.

[Hint: It is enough to prove it for a bounded cadlag X and then $Y = X_-$ will do. For $\varepsilon > 0$, define $T(\varepsilon) = \inf\{t > 0 : |X_t - X_{t-}| > \varepsilon\}$

$$T_n(\varepsilon) = \inf\{t > T_{n-1}(\varepsilon) : |X_t - X_{t-}| > \varepsilon\}$$

and use $\bigcup_{p,n} T_n(\varepsilon_p)$ with $\varepsilon_p \downarrow 0$.]

\# **(5.11) Exercise.** A closed subspace S of H_0^2 is said to be *stable* if

i) for any $M \in S$ and stopping time T, M^T is in S;
ii) for any $\Gamma \in \mathscr{F}_0$, and $M \in S$, $1_\Gamma M$ is in S.

1°) If S is stable and $\Gamma \in \mathscr{F}_T$, prove that $1_\Gamma(M - M^T) \in S$ for any $M \in S$.

2°) The space S is stable if and only if, for any $M \in S$ and any predictable $H \in L^2(M)$, the martingale $H \cdot M$ is in S.

3°) Let $S(M)$ be the smallest stable subspace of H_0^2 containing the martingale M. Prove that

$$S(M) = \{H \cdot M : H \in L^2(M)\}.$$

4°) Let S be stable and $S^\perp$ be its orthogonal subspace in H_0^2 (for the Hilbert norm). For any $M \in S$ and $N \in S^\perp$, prove that $\langle M, N\rangle = 0$. Show that $S^\perp$ is stable.

5°) If M and N are in H_0^2, prove that the orthogonal projection of N on $S(M)$ is equal to $H \cdot M$ where H is a version of the Radon-Nikodym derivative of $d\langle M, N\rangle$ with respect to $d\langle M, M\rangle$.

* **(5.12) Exercise.** Retain the situation of Proposition (3.12) Chap. III and for $t \geq 0$ set $d_t = t + T_0 \circ \theta_t$.

1°) Prove that the map $t \to d_t$ is a.s. right-continuous.

2°) Let $H = \{(\omega, t) : B_t(\omega) = 0\}$. For each ω, the set $(H(\omega, \cdot))^c$ is a countable union of open intervals and we denote by F the subset of $\Omega \times \mathbb{R}_+$ such that $F(\omega, \cdot)$ is the set of the left-ends of these intervals. Prove that F is progressive with respect to the Brownian filtration.

[Hint: Prove that $H \backslash F = \{(\omega, t) : d_t(\omega) = t\}$.]

3°) Prove that for any stopping time T, one has $P\big[[T] \subset F\big] = 0$.

[Hint: Use the strong Markov property of BM.]

4°) Prove that F is not optional, thus providing an example of a progressive set which is not optional.

[Hint: Compute the optional projection of 1_F.]

(5.13) Exercise. Let $(\mathscr{F}_t) \subset (\mathscr{G}_t)$ be two filtrations. If M is a $(\mathscr{G}_t)$-martingale, prove that its optional projection w.r.t. $(\mathscr{F}_t)$ is a $(\mathscr{F}_t)$-martingale.

(5.14) Exercise. Let X be a measurable process such that, for every t,

$$E\left[\int_0^t |X_s| ds\right] < \infty;$$

and set $Y_t = \int_0^t X_s ds$. If we denote by $\hat{H}$ the optional projection of H w.r.t. a filtration $(\mathscr{F}_t)$, prove that $\hat{Y}_t - \int_0^t \hat{X}_s ds$ is a $(\mathscr{F}_t)$-martingale.

(5.15) Exercise (Filtering). Let $(\mathscr{F}_t)$ be a filtration, B a $(\mathscr{F}_t)$-BM and h a bounded optional process. Let $Y_t = \int_0^t h_s ds + B_t$; if $\hat{h}$ is the optional projection of h w.r.t. $(\mathscr{F}_t^Y)$, prove that the process

$$N_t = Y_t - \int_0^t \hat{h}_s ds$$

is a $\mathscr{F}_t^Y$-BM. In filtering theory, the process N is called the *innovation process*.

[Hint: Prove first that $E[N_t^2] < \infty$ for every t, then compute $E[N_T]$ for bounded stopping times T.]

(5.16) Exercise. Let X and X' be two continuous semimartingales on two filtered probability spaces; let H and H' be two predictable locally bounded processes such that $(H, X) \stackrel{(d)}{=} (H', X')$. Prove that $(H \cdot X, H, X) \stackrel{(d)}{=} (H' \cdot X', H', X')$.

[Hint: Start with the case of elementary H and H'.]

*# **(5.17) Exercise (Fubini's theorem for stochastic integrals).** 1°) Let $(A, \mathscr{A})$ be a measurable space, $(\Omega, \mathscr{F}, P)$ a probability space and $\{X_n(a, \cdot)\}$ a sequence of $\mathscr{A} \otimes \mathscr{F}$-measurable r.v.'s which converges in probability for every a. Prove that there exists a $\mathscr{A} \otimes \mathscr{F}$-measurable r.v., say X, such that for every a, $X(a, \cdot)$ is the limit in probability of $\{X_n(a, \cdot)\}$.

[Hint: Define inductively a sequence $n_k(a)$ by $n_0(a) = 1$ and

$$n_k(a) = \inf\left\{m > n_{k-1}(a) : \sup_{p,q\geq m} P\left[|X_p(a,\cdot)-X_q(a,\cdot)| > 2^{-k}\right] \leq 2^{-k}\right\};$$

prove that $\lim_k X_{n_k(a)}(a,\cdot)$ exists a.s. and answers the question.]

2°) In the setting of Sect. 2, if $H(a, s, \omega)$ is a uniformly bounded $\mathscr{A}\otimes\mathscr{P}$-measurable process and X a continuous semimartingale, there exists a $\mathscr{A}\otimes\mathscr{P}$-measurable process $K(a,\cdot,\cdot)$ such that for each a, $K(a,\cdot,\cdot)$ is indistinguishable from $\int_0^{\cdot} H(a,s,\cdot)dX_s$. Moreover, if ν is a bounded measure on $\mathscr{A}$, then a.s.,

$$\int_A K(a,t,\cdot)\nu(da) = \int_0^t \left(\int_A H(a,s,\cdot)\nu(da)\right) dX_s.$$

[Hint: These properties are easily checked when $H(a,s,\omega) = h(a)g(s,\omega)$ with suitable h and g. Apply the monotone class theorem.]

(5.18) Exercise. 1°) Let $M \in H^2$ and K be a measurable, but not necessarily adapted process, such that

$$E\left[\int_0^\infty K_s^2 d\langle M, M\rangle_s\right] < \infty.$$

Prove that there exists a unique $Z \in H^2$ such that for $N \in H^2$

$$E[Z_\infty N_\infty] = E\left[\int_0^\infty K_s d\langle M, N\rangle_s\right].$$

2°) Let $\tilde{K}$ be the projection (in the Hilbert space sense) of K on $L^2_{\mathscr{P}}(M)$. Prove that $Z = \tilde{K}\cdot M$.

(5.19) Exercise. 1°) If K is a continuous adapted process and X a cont. semimart., prove that

$$P\text{-}\lim_{\varepsilon\to 0}\frac{1}{\varepsilon}\int_0^1 K_s(X_{s+\varepsilon}-X_s)\,ds = \int_0^1 K_s dX_s.$$

[Hint: Use Fubini's theorem for stochastic integrals.]

2°) Under the same hypothesis, prove that

$$P\text{-}\lim_{\varepsilon\to 0}\frac{1}{\varepsilon}\int_0^1 K_s(X_{s+\varepsilon}-X_s)^2\,ds = \int_0^1 K_s d\langle X, X\rangle_s.$$

Notes and Comments

Sect. 1. The notion of local martingale appeared in Itô-Watanabe [1] and that of semimartingale in Doléans-Dade and Meyer [1]. For a detailed study, we direct the reader to the book of Dellacherie and Meyer ([1] vol 2) from which we borrowed some of our proofs, and to Métivier [1].

The proof of Theorem (1.3) is taken from Kunita [4]. The existence of quadratic variation processes is usually shown by using Meyer's decomposition theorem which we wanted to avoid entirely in the present book. Proposition (1.13) is due to Getoor-Sharpe [1]. Let us mention that we do not prove the important result of Stricker [1] asserting that a $(\mathscr{F}_t)$-semimartingale X is still a semimartingale in its own filtration $(\mathscr{F}_t^X)$.

Exercise (1.33) is from Yor [2], Exercise (1.41) from Emery [1] and Exercise (1.42) from Sharpe [1]. Exercise (1.48) is due to Maisonneuve [3] and Exercise (1.49) is taken from Dellacherie-Meyer [1], Vol. II. The method hinted at in Exercise (1.38) was given to us by F. Delbaen (private communication).

Sect. 2. Stochastic integration has a long history which we will not attempt to sketch. It goes back at least to Paley-Wiener-Zygmund [1] in the case of the integral of deterministic functions with respect to Brownian motion. Our method of defining stochastic integrals is that of Kunita and Watanabe [1] and is originally due to Itô [1] in the BM case. Here again, we refer to Dellacherie-Meyer [1] vol. 2 and Métivier [1] for the general theory of stochastic integration with respect to (non continuous) semimartingales.

Exercise (2.17) is due to Isaacson [1] and Yoeurp [2] and Exercise (2.18) is from Yor [2]. Exercise (2.19) is taken from Donati-Martin and Yor [2]; a number of such identities in law have been recently discussed in Chan et al. [1] and Dean-Jansons [1]. The important Exercise (2.21) is taken from a lecture course of Meyer.

Sect. 3. Itô's formula in a general context appears in Kunita-Watanabe [1] and our proof is borrowed from Dellacherie-Meyer [1], Vol. II. Exercise (3.12) is but one of many extensions of Itô's formula; we refer to Kunita ([5], [6]). The proof of Exercise (3.13) is due to Föllmer [2].

Exponentials of semimartingales were studied by Doléans-Dade [2] and are important in several contexts, in particular in connection with Girsanov's theorem of Chap. VIII. We refer to the papers of Kazamaki and Sekiguchi from which some of our exercises are taken, and also to Kazamaki's Lecture Notes [4].

The proof given here of P. Lévy's characterization theorem is that of Kunita-Watanabe [1]; the extension in Exercise (3.29) is due to Ruiz de Chavez [1]. This result plays a central role in Chap. V and its simplicity explains to some extent why the martingale approach to BM is so successful. Moreover, it contains *en germe* the idea that the law of a process may be characterized by a set of martingale properties, as is explained in Chap. VII. This eventually led to the powerful martingale problem method of Stroock and Varadhan (see Chap. VII).

Exercise (3.17) is from Robbins-Siegmund [1]; Exercise (3.19) is from Donati-Yor [1] and Exercise (3.26) from Calais and Génin [1]. Exercise (3.30) comes from Carlen-Krée [1] who obtain BDG type inequalities for multiple stochastic integrals.

Sect. 4. Our proof of the BDG inequalities combines the method of Getoor-Sharpe [1] with the reduction procedure of Lenglart [2] (see also Lenglart et al. [1]). The use of "good λ inequalities" is the original method of Burkholder [2]. The method used in Theorems (4.10) and (4.11), due to Bass [2] (see also Davis [5]), is the most efficient to date and works for all inequalities of BDG type known so far.

For further details about the proof of Fefferman's inequality presented in Exercise (4.24), see e.g. Durrett [2], section 7.2, or Dellacherie-Meyer ([1], Chap. VII).

The scope of Fefferman's inequality may be extended to martingales Y not necessarily in BMO; see, e.g., Yor [26] and Chou [1].

Exercise (4.30) gives an improvement on the constant $(2-k)/(1-k)$ obtained in Proposition (4.7), but the following natural question is still open.

Question 1: For $k \in]0, 1[$, find the best constant C_k for the inequality

$$E\left[\left(X_\infty^*\right)^k\right] \le C_k E\left[A_\infty^k\right]$$

which follows from the domination relation.

Sect. 5. This section contains only the information on predictable processes which we need in the sequel. For a full account, we direct the reader to Dellacherie-Meyer [1].

The reader shall also find some important complements in Chung-Williams [1] (Sect. 3.3) to our discussion, following Definition (5.2), of the various classes of stochastic integrands.

Chapter V. Representation of Martingales

In this chapter, we take up the study of Brownian motion and, more generally, of continuous martingales. We will use the stochastic integration of Chap. IV together with the technique of time changes to be introduced presently.

§1. Continuous Martingales as Time-changed Brownian Motions

It is a natural idea to change the speed at which a process runs through its path; this is the technique of time changes which was described in Sect. 4 Chap. 0 and which we now transpose to a stochastic context.

Let $(\mathscr{F}_t)$ be a right-continuous filtration. Throughout the first part of this section, we consider an increasing, right-continuous, adapted process A with which we associate

$$C_s = \inf\{t : A_t > s\}$$

where, as usual, $\inf(\emptyset) = +\infty$. Since C_s increases with s, the limit $C_{s-} = \lim_{u \uparrow s} C_u$ exists and

$$C_{s-} = \inf\{t : A_t \geq s\}.$$

By convention $C_{0-} = 0$.

(1.1) Proposition. *The family (C_s) is an increasing right-continuous family of stopping times. Moreover for every t, the r.v. A_t is an $(\mathscr{F}_{C_s})$-stopping time.*

Proof. Each C_s is a stopping time, by the reasoning of Proposition (4.6) Chap. I. The right-continuity of the map $s \to C_s$ was shown in Lemma (4.7) Chap. 0, and it follows easily (Exercise (4.17) Chap. I) that the filtration $(\mathscr{F}_{C_s})$ is right-continuous. Finally it was also shown in Lemma (4.7) Chap. 0, that $A_t = \inf\{s : C_s > t\}$ which proves that A_t is an $(\mathscr{F}_{C_s})$-stopping time. □

It was proved in Chap. 0 that $A_{C_t} \geq t$ with equality if t is a point of increase of C, that is $C_{t+\varepsilon} - C_t > 0$ for every $\varepsilon > 0$. If A is strictly increasing, then X is continuous; if A is continuous and strictly increasing, then C is also continuous and strictly increasing and we then have $A_{C_t} = C_{A_t} = t$. While reasoning on this situation, the reader will find it useful to look at Figure 1 in Sect. 4 Chap. 0. He will observe that the jumps of A correspond to level stretches of C and vice-versa. Actually, A and C play symmetric roles as we will see presently.

(1.2) Definition. *A* time-change C *is a family* C_s, $s \geq 0$, *of stopping times such that the maps* $s \to C_s$ *are a.s. increasing and right-continuous.*

Thus, the family C defined in Proposition (1.1) is a time-change. Conversely, given a time change C, we get an increasing right-continuous process by setting

$$A_t = \inf\{s : C_s > t\}.$$

It may happen that A is infinite from some finite time on; this is the case if $C_\infty = \lim_{s\to\infty} C_s < \infty$, but otherwise we see that time-changes are not more general than the inverses of right-continuous increasing adapted processes.

In the sequel, we consider a time-change C and refer to A only when necessary. We set $\hat{\mathscr{F}}_t = \mathscr{F}_{C_t}$. If X is a $(\mathscr{F}_t)$-progressive process then $\hat{X}_t = X_{C_t}$ is a $(\hat{\mathscr{F}}_t)$-adapted process; the process $\hat{X}$ will be called *the time-changed process of* X.

Let us insist once more that, if A is continuous strictly increasing and $A_\infty = \infty$, then C is continuous, strictly increasing, finite and $C_\infty = \infty$. The processes A and C then play totally symmetric roles and for any $(\mathscr{F}_t)$-progressive process X, we have $\hat{X}_A = X$. If $A_\infty < \infty$, the same holds but $\hat{X}$ is only defined for $t < A_\infty$. Finally let us observe that by taking $C_t = t \wedge T$ where T is a stopping time, one gets the stopped processes as a special case of the time-changed processes.

An important property of the class of semimartingales is its invariance under time changes. Since, in this book, we deal only with continuous semimartingales, we content ourselves with some partial results. We will need the following

(1.3) Definition. *If* C *is a time-change, a process* X *is said to be* C-continuous *if* X *is constant on each interval* $[C_{t-}, C_t]$.

If X is increasing and right-continuous, so is $\hat{X}$; thus, if X is right-continuous and of finite variation, so is $\hat{X}$. The Stieltjes integrals with respect to X and $\hat{X}$ are related by the useful

(1.4) Proposition. *If* H *is* $(\mathscr{F}_t)$*-progressive, then* $\hat{H}$ *is* $(\hat{\mathscr{F}}_t)$*-progressive and if* X *is a* C*-continuous process of finite variation, then* $\hat{H}\cdot\hat{X} = \widehat{H\cdot X}$*; in other words*

$$\int_{C_0}^{C_t} H_s dX_s = \int_0^t 1_{(C_u<\infty)} H_{C_u} dX_{C_u} = \int_0^{t\wedge A_\infty} H_{C_u} dX_{C_u}.$$

Proof. The first statement is easy to prove. The second is a slight variation on Proposition (4.10) of Chap. 0. □

In many cases, $C_0 = 0$ and C_t is finite for every t and the above equality reads

$$\int_0^{C_t} H_s dX_s = \int_0^t H_{C_u} dX_{C_u}.$$

We now turn to local martingales. A first problem is that, under time changes, they remain semimartingales but not always local martingales; if for instance, X is a BM and $A_t = S_t$, then $\hat{X}_t = t$. Another problem for us is that C may have jumps so that $\hat{X}$ may be discontinuous even when X is continuous. This is why again, we will have to assume in the next proposition that X is C-continuous.

(1.5) Proposition. *Let C be a.s. finite and X be a continuous $(\mathscr{F}_t)$-local martingale.*

i) If X is C-continuous, then $\hat{X}$ is a continuous $(\hat{\mathscr{F}}_t)$-local martingale and $\langle \hat{X}, \hat{X} \rangle = \widehat{\langle X, X \rangle}$;

ii) If moreover H is $(\mathscr{F}_t)$-progressive and $\int_0^t H_s^2 d\langle X, X\rangle_s < \infty$ a.s. for every t then $\int_0^t \hat{H}_s^2 d\langle \hat{X}, \hat{X}\rangle_s < \infty$ a.s. for every t and $\hat{H} \cdot \hat{X} = \widehat{H \cdot X}$.

Proof. i) Since X is C-continuous the process $\hat{X}$ is continuous. Let T be a $(\mathscr{F}_t)$-stopping time such that X^T is bounded; the time $\hat{T} = \inf\{t; C_t \geq T\}$ is a $(\hat{\mathscr{F}}_t)$-stopping time because $1_{[\hat{T},\infty]} = \widehat{1_{[T,\infty[}}$ is $\hat{\mathscr{F}}_t$-adapted. It is clear that $\hat{X}^{\hat{T}}$ is bounded. Moreover

$$\hat{X}_t^{\hat{T}} = \hat{X}_{\hat{T}\wedge t} = X_{C_{t\wedge \hat{T}}}$$

and because X is C-continuous, X is constant on $[T, C_{\hat{T}}]$ and we have $X_{C_{t\wedge\hat{T}}} = X_{C_t}^T$; it follows from the optional stopping theorem that $\hat{X}^{\hat{T}}$ is a $(\hat{\mathscr{F}}_t)$-martingale. Finally if (T_n) increases to $+\infty$, so does the corresponding sequence $(\hat{T}_n)$; thus, we have proved that $\hat{X}$ is a cont. loc. mart.

By Proposition (1.13) of Chap. IV, the process $\langle X, X\rangle$ is also C-continuous, thus, by the result just proved, $\hat{X}^2 - \widehat{\langle X, X\rangle}$ is a cont. loc. mart. which proves that $\langle \hat{X}, \hat{X}\rangle = \widehat{\langle X, X\rangle}$.

ii) The first part follows from Proposition (1.4). To prove the second part, we need only prove that the increasing process of the local martingale $\widehat{H \cdot X} - \hat{H} \cdot \hat{X}$ vanishes identically and this is a simple consequence of i) and Proposition (1.4). □

Thus, we have proved that suitably time-changed Brownian motions are local martingales. The following converse is the main result of this section.

(1.6) Theorem (Dambis, Dubins-Schwarz). *If M is a $(\mathscr{F}_t, P)$-cont. loc. mart. vanishing at 0 and such that $\langle M, M\rangle_\infty = \infty$ and if we set*

$$T_t = \inf\{s : \langle M, M\rangle_s > t\},$$

then, $B_t = M_{T_t}$ is a $(\mathscr{F}_{T_t})$-Brownian motion and $M_t = B_{\langle M,M\rangle_t}$.

The Brownian motion B will be referred to as the *DDS Brownian motion of M*.

Proof. The family $T = (T_t)$ is a time-change which is a.s. finite because $\langle M, M\rangle_\infty = \infty$ and, by Proposition (1.13) of Chap. IV, the local martingale M is obviously T-continuous. Thus, by the above result, B is a continuous $(\mathscr{F}_{T_t})$-local martingale and $\langle B, B\rangle_t = \langle M, M\rangle_{T_t} = t$. By P. Lévy's characterization theorem, B is a $(\mathscr{F}_{T_t})$-Brownian motion.

To prove that $B_{\langle M,M\rangle} = M$, observe that $B_{\langle M,M\rangle} = M_{T_{\langle M,M\rangle}}$ and although $T_{\langle M,M\rangle_t}$ may be $> t$, it is always true that $M_{T_{\langle M,M\rangle_t}} = M_t$ because of the constancy of M on the level stretches of $\langle M, M\rangle$. □

In the above theorem, the hypothesis $\langle M, M\rangle_\infty = \infty$ ensures in particular that the underlying probability space is rich enough to support a BM. This may be also achieved by enlargement.

We call *enlargement* of the filtered probability space $(\Omega, \mathscr{F}_t, P)$ another filtered probability space $(\tilde{\Omega}, \tilde{\mathscr{F}}_t, \tilde{P})$ together with a map π from $\tilde{\Omega}$ onto Ω, such that $\pi^{-1}(\mathscr{F}_t) \subset \tilde{\mathscr{F}}_t$ for each t and $\pi(\tilde{P}) = P$. A process X defined on Ω may be viewed as defined on $\tilde{\Omega}$ by setting $X(\tilde{\omega}) = X(\omega)$ if $\pi(\tilde{\omega}) = \omega$.

If $\langle M, M\rangle_\infty < \infty$, recall that $M_\infty = \lim_{t\to\infty} M_t$ exists (Proposition (1.26) Chap. IV). Thus we can define a process W by

$$W_t = M_{T_t} \quad \text{for } t < \langle M, M\rangle_\infty, \qquad W_t = M_\infty \quad \text{if } t \geq \langle M, M\rangle_\infty.$$

By Proposition (1.26) Chap. IV, this process is continuous and we have the

(1.7) Theorem. *There exist an enlargement $(\tilde{\Omega}, \tilde{\mathscr{F}}_t, \tilde{P})$ of $(\Omega, \mathscr{F}_{T_t}, P)$ and a BM $\tilde{\beta}$ on $\tilde{\Omega}$ independent of M such that the process*

$$B_t = \begin{cases} M_{T_t} & \text{if } \ t < \langle M, M\rangle_\infty, \\ M_\infty + \tilde{\beta}_{t-\langle M,M\rangle_\infty} & \text{if } \ t \geq \langle M, M\rangle_\infty \end{cases}$$

is a standard linear Brownian motion. The process W is a $(\tilde{\mathscr{F}}_t)$-BM stopped at $\langle M, M\rangle_\infty$.

Proof. Let $(\Omega', \mathscr{F}_t', P')$ be a probability space supporting a BM β and set

$$\tilde{\Omega} = \Omega \times \Omega', \quad \tilde{\mathscr{F}}_t = \mathscr{F}_{T_t} \otimes \mathscr{F}_t', \quad \tilde{P} = P \otimes P', \quad \tilde{\beta}_t(\omega, \omega') = \beta_t(\omega').$$

The process $\tilde{\beta}$ is independent of M and B may as well be written as

$$B_t = M_{T_t} + \int_0^t 1_{(s > \langle M,M\rangle_\infty)} d\tilde{\beta}_s.$$

In the general setting of Proposition (1.5), we had to assume the finiteness of C because M_{C_t} would have been meaningless otherwise. Here where we can define M_{T_t} even for infinite T_t, the reasoning of Proposition (1.5) applies and shows that W is a local martingale (see also Exercise (1.26)).

Being the sum of two $\tilde{\mathscr{F}}_t$-local martingales, B is a local martingale and its increasing process is equal to

$$\langle M_{T_\cdot}, M_{T_\cdot}\rangle_t + \int_0^t 1_{(s > \langle M,M\rangle_\infty)} ds + 2\int_0^t 1_{(s > \langle M,M\rangle_\infty)} d\langle M_{T_\cdot}, \tilde{\beta}\rangle_s.$$

Because of the independence of $M_{T_\cdot}$ and $\tilde{\beta}$, the last term vanishes and it is easily deduced from the last proof that $\langle M_{T_\cdot}, M_{T_\cdot}\rangle_t = t \wedge \langle M, M\rangle_\infty$; it follows that $\langle B, B\rangle_t = t$ which, by P. Lévy's characterization theorem, completes the proof.

Remark. An interesting example is given in Lemma (3.12) Chap. VI.

We may now complete Proposition (1.26) in Chap. IV.

(1.8) Proposition. *For a continuous local martingale M, the sets $\{\langle M, M\rangle_\infty < \infty\}$ and $\{\lim_{t\to\infty} M_t$ exists$\}$ are almost-surely equal. Furthermore, $\overline{\lim}_{t\to\infty} M_t = +\infty$ and $\underline{\lim}_{t\to\infty} M_t = -\infty$ a.s. on the set $\{\langle M, M\rangle_\infty = \infty\}$.*

Proof. We can apply the preceding result to $M - M_0$. Since B is a BM, we have $\overline{\lim}_{t\to\infty} B_t = +\infty$ a.s. On the set $\langle M, M\rangle_\infty = \infty$, we thus have $\overline{\lim}_{t\to\infty} M_{T_t} = +\infty$ a.s. But on this set, T_t converges to infinity as t tends to infinity; as a result, $\overline{\lim}_{t\to\infty} M_t$ is larger than $\overline{\lim}_{t\to\infty} M_{T_t}$, hence is infinite. The same proof works for the inferior limit.

Remarks. 1°) We write $\overline{\lim}\ M_t \geq \overline{\lim}\ M_{T_t}$ because the paths of $M_{T_\cdot}$ could a priori be only a portion of those of M. We leave as an exercise to the reader the task of showing that they are actually the same.

2°) Another event which is also a.s. equal to those in the statements is given in Exercise (1.27) of Chap. VI.

3°) This proposition shows that for a cont. loc. mart., the three following properties are equivalent:

i) $\sup_t M_t = \infty$ a.s., ii) $\inf_t M_t = -\infty$ a.s., iii) $\langle M, M\rangle_\infty = \infty$ a.s.

We now turn to a multi-dimensional analogue of the Dambis, Dubins-Schwarz theorem which says that if $\langle M, N\rangle = 0$, then the BM's associated with M and N are independent.

(1.9) Theorem (Knight). *Let $M = (M^1, \dots, M^d)$ be a continuous vector valued local martingale such that $M_0 = 0$, $\langle M^k, M^k\rangle_\infty = \infty$ for every k and $\langle M^k, M^l\rangle = 0$ for $k \neq l$. If we set*

$$T_t^k = \inf\left\{s : \langle M^k, M^k\rangle_s > t\right\}$$

and $B_t^k = M^k_{T_t^k}$, the process $B = (B^1, \dots, B^d)$ is a d-dimensional BM.

As in the one-dimensional case, the assumption on $\langle M^k, M^k\rangle_\infty$ may be removed at the cost of enlarging the probability space. It is this general version that we will prove below.

(1.10) Theorem. *If $\langle M^k, M^k\rangle_\infty$ is finite for some k's, there is a BM^d β independent of M on an enlargement of the probability space such that the process B defined by*

$$B_t^k = \begin{cases} M^k_{T_t^k} & for \quad t < \langle M^k, M^k\rangle_\infty \\ M^k_\infty + \beta^k_{(t - \langle M^k, M^k\rangle_\infty)} & for \quad t \geq \langle M^k, M^k\rangle_\infty \end{cases}$$

is a d-dimensional BM.

Proof. By the previous results, we know that each B^k separately is a linear BM, so all we have to prove is that they are independent. To this end, we will prove that, with the notation of Theorem (3.6) of Chap. IV, for functions f_k with compact support, $E\left[\mathscr{E}_\infty^{if}\right] = 1$. Indeed, taking $f_k = \sum_{j=1}^p \lambda_k^j 1_{]t_{j-1}, t_j]}$, we will then obtain that the random vectors $\left(B_{t_j}^k - B_{t_{j-1}}^k\right)$, $1 \leq j \leq p$, $1 \leq k \leq d$, have the right characteristic functions, hence the right laws.

In the course of this proof, we write A^k for $\langle M^k, M^k \rangle$. From the equality

$$M_{T_t^k}^k - M_{T_s^k}^k = \int 1_{]s,t]}(A_u^k) dM_u^k$$

which follows from Proposition (1.5), we can derive, by the usual monotone class argument, that

$$\int_0^{A_\infty^k} f_k(s) dM_{T_s}^k = \int_0^\infty f_k(A_u^k) dM_u^k.$$

Consequently,

$$\int_0^\infty f_k(s) dB_s^k = \int_0^\infty f_k(A_u^k) dM_u^k + \int_0^\infty f_k(s + A_\infty^k) d\beta_s^k.$$

Note that the stochastic integral on the right makes sense since A_∞^k is independent of β^k. Passing to the quadratic variations, we get

$$\int_0^\infty f_k^2(s) ds = \int_0^\infty f_k^2(A_u^k) dA_u^k + \int_0^\infty f_k^2(s + A_\infty^k) ds.$$

The process $X = \sum_k \int_0^{\cdot} f_k(A_u^k) dM_u^k$ is a local martingale and, using the hypothesis $\langle M^k, M^l \rangle = 0$ for $k \neq l$, we get

$$\langle X, X \rangle = \sum_k \int_0^{\cdot} f_k^2(A_u^k) dA_u^k.$$

By Itô's formula, $I_t = \exp\left\{iX_t + \frac{1}{2}\langle X, X \rangle_t\right\}$ is a local martingale. Since it is bounded (by $\exp\left\{\sum_k \|f_k\|_2^2/2\right\}$), it is in fact a martingale. Likewise

$$J_t = \exp\left\{\sum_k \left(i \int_0^t f_k\left(s + A_\infty^k\right) d\beta_s^k + \frac{1}{2}\int_0^t f_k^2\left(s + A_\infty^k\right) ds\right)\right\}$$

is a bounded martingale and $E\left[\mathscr{E}_\infty^{if}\right] = E\left[I_\infty J_\infty\right]$.

Now, conditionally on M, J_∞ has the law of $\exp\left\{iZ + \frac{1}{2}\mathrm{Var}(Z)\right\}$ where Z is gaussian and centered. Consequently, $E\left[\mathscr{E}_\infty^{if}\right] = E\left[I_\infty\right]$ and, since I is a bounded martingale, $E\left[\mathscr{E}_\infty^{if}\right] = E\left[I_0\right] = 1$, which ends the proof.

Remarks. 1°) Let us point out that the DDS theorem is both simpler and somewhat more precise than Knight's theorem. The several time-changes of the latter make matters more involved; in particular, there is no counterpart of the filtration $(\mathscr{F}_{T_t})$ of the former theorem with respect to which the time-changed process is a BM.

2°) Another proof of Knight's theorem is given in Exercise (3.18). It relies on a representation of Brownian martingales which is given in Sect. 3.

An important consequence of the DDS and Knight theorems is that, to some extent, a property of continuous local martingales which is invariant by time-changes is little more than a property of Brownian motion and actually many proofs of results on cont. loc. mart. may be obtained by using the associated BM. For instance, the BDG inequalities of Sect. 4 Chap. IV can be proved in that way. Indeed, if $M_t = B_{\langle M,M\rangle_t}$,

$$E\left[(M_t^*)^p\right] = E\left[\left(B^*_{\langle M,M\rangle_t}\right)^p\right]$$

and since $\langle M, M\rangle_t$ is a stopping time for the filtration $(\mathscr{F}_{T_t})$ with respect to which B is a Brownian motion, it is enough to prove that, if $(\mathscr{G}_t)$ is a filtration, for a $(\mathscr{G}_t)$-BM B and a $(\mathscr{G}_t)$-stopping time T,

$$c_p E[T^{p/2}] \le E\left[(B_T^*)^p\right] \le C_p E[T^{p/2}].$$

The proof of this is outlined in Exercise (1.23).

Finally, let us observe that, in Theorem (1.6), the Brownian motion B is measurable with respect to $\mathscr{F}_\infty^M$, where, we recall from Sect. 2 Chap. III, $(\mathscr{F}_t^X)$ is the coarsest right-continuous and complete filtration with respect to which X is adapted; the converse, namely that M is measurable with respect to $\mathscr{F}_\infty^B$ is not always true as will be seen in Exercise (4.16). We now give an important case where it is so (see also Exercise (1.19)).

(1.11) Proposition. *If M is a cont. loc. mart. such that $\langle M, M\rangle_\infty = \infty$ and*

$$M_t = x + \int_0^t \sigma(M_s)d\beta_s$$

for a BM β and a nowhere vanishing function σ, then M is measurable with respect to $\mathscr{F}_\infty^B$ where B is the DDS Brownian motion of M.

Proof. Since σ does not vanish, $\langle M, M\rangle$ is strictly increasing and

$$\langle M, M\rangle_{T_t} = \int_0^{T_t} \sigma^2(M_s)ds = t.$$

Using Proposition (1.4) with the time change T_t, we get $\int_0^t \sigma^2(B_s)dT_s = t$, hence

$$T_t = \int_0^t \sigma^{-2}(B_s)ds.$$

It follows that T_t is $\mathscr{F}_t^B$-measurable, $\langle M, M\rangle_t$ which is the inverse of (T_t) is $\mathscr{F}_\infty^B$-measurable and $M = B_{\langle M,M\rangle}$ is consequently also $\mathscr{F}_\infty^B$-measurable, which completes the proof.

Remark. We stress that, in this proposition, we have *not* proved that M is $(\mathscr{F}_t^B)$-adapted; in fact, it is $\left(\mathscr{F}^B_{\langle M,M\rangle_t}\right)$-adapted (see Exercise (4.16)).

(1.12) Exercise. Let C be a time change and D be a time change relative to $(\hat{\mathscr{F}}_t) = (\mathscr{F}_{C_t})$. Prove that $s \to C_{D_s}$ is a time change and $\hat{\mathscr{F}}_{D_t} = \mathscr{F}_{C_{D_t}}$.

(1.13) Exercise. 1°) Let A be a right-continuous adapted increasing process. If X and Y are two positive measurable processes such that

$$E\left[X_T 1_{(T<\infty)}\right] = E\left[Y_T 1_{(T<\infty)}\right]$$

for every $(\mathscr{F}_t)$-stopping time T, in particular, if Y is the $(\mathscr{F}_t)$-optional projection of X (Theorem (5.6) Chap. IV), then for every $t \le +\infty$,

$$E\left[\int_0^t X_s dA_s\right] = E\left[\int_0^t Y_s dA_s\right].$$

[Hint: Use the same device as in Proposition (1.4) for the time change associated with A.]

2°) If A is continuous, prove that the same conclusion is valid if the assumption holds only for predictable stopping times, hence, in particular, if Y is the predictable projection of X.

[Hint: Use C_{t-} instead of C_t and prove that C_{t-} is a predictable stopping time.]

The result is actually true if A is merely predictable.

3°) If M is a bounded right-continuous positive martingale and A is a right-continuous increasing adapted process such that $A_0 = 0$ and $E[A_t] < \infty$ for every t, then

$$E[M_t A_t] = E\left[\int_0^t M_s dA_s\right]$$

for every $t > 0$. The question 3°) is independent of 2°) .

\# **(1.14) Exercise. (Gaussian martingales. Converse to Exercise (1.35) Chap. IV).**

1°) If M is a cont. loc. mart. vanishing at zero and if $\langle M, M\rangle$ is deterministic, then M is a Gaussian martingale and has independent increments.

[Hint: This can be proved either by applying Theorem (1.6) or by rewriting in that case the proof of P. Lévy's characterization theorem.]

2°) If B is a standard BM^1 and $\beta = H \cdot B$ where H is a $(\mathscr{F}_t^B)$-predictable process such that $|H| = 1$, prove that the two-dimensional process (B, β) is Gaussian iff H is deterministic.

(1.15) Exercise. Let M be a continuous local martingale. Prove that on $\{\langle M, M\rangle_\infty = \infty\}$, one has

$$\overline{\lim_{t\to\infty}} M_t / \left(2\langle M, M\rangle_t \log_2\langle M, M\rangle_t\right)^{1/2} = 1 \quad \text{a.s.}$$

(1.16) Exercise (Law of large numbers for local martingales). Let $A \in \mathscr{A}^+$ be such that $A_0 > 0$ a.s. and M be a continuous local martingale vanishing at 0. We set

$$Z_t = \int_0^t A_s^{-1} dM_s.$$

1°) Prove that

$$M_t = \int_0^t (Z_t - Z_s)\, dA_s + Z_t A_0.$$

2°) It is assumed that $\lim_{t\to\infty} Z_t$ exists a.s. Prove that $\lim_{t\to\infty}(M_t/A_t) = 0$ on the set $\{A_\infty = \infty\}$.

3°) If f is an increasing function from $[0, \infty[$ into $]0, \infty[$ such that $\int^\infty f(t)^{-2} dt < \infty$, then

$$\lim_{t\to\infty} M_t / f(\langle M, M\rangle_t) = 0 \text{ a.s. on } \{\langle M, M\rangle_\infty = \infty\}.$$

In particular $\lim_{t\to\infty} M_t/\langle M, M\rangle_t = 0$ a.s. on $\{\langle M, M\rangle_\infty = \infty\}$.

4°) Prove the result in 3°) directly from Theorem (1.6).

\# **(1.17) Exercise.** 1°) If M is a continuous local martingale, we denote by $\beta(M)$ the DDS BM of M. If C is a finite time-change such that $C_\infty = \infty$ and if M is C-continuous, prove that $\beta(\hat{M}) = \beta(M)$.

2°) If h is > 0, prove that $\beta\left(\frac{1}{h} M\right) = \beta(M)^{(h)}$ where $X_t^{(h)} = \frac{1}{h} X_{h^2 t}$. Conclude that $\beta(M^{(h)}) = \beta(M)^{(h)}$.

* **(1.18) Exercise.** Extend Theorems (1.6) and (1.9) to continuous local martingales defined on a stochastic interval $[0, T[$ (see Exercises (1.48) and (3.28) in Chap. IV).

\# **(1.19) Exercise.** In the situation of Theorem (1.6), prove that $\mathscr{F}_\infty^M$ is equal to the completion of $\sigma\left((B_s, T_s), s \geq 0\right)$. Loosely speaking, if you know B and T, you can recover M.

* **(1.20) Exercise (Hölder condition for semimartingales).** If X is a cont. semimartingale and λ the Lebesgue measure on $\mathbb{R}_+$ prove that

$$\lambda\left(\left\{t \geq 0 : \overline{\lim_{\varepsilon\downarrow 0}}\, \varepsilon^{-\alpha} |X_{t+\varepsilon} - X_t| > 0\right\}\right) = 0 \text{ a.s.},$$

for every $\alpha < 1/2$.

[Hint: Use the DDS and Lebesgue derivation theorems.]

(1.21) Exercise. Let (X, Y) be a BM^2 and H a locally bounded $(\mathscr{F}_t^X)$-predictable process such that $\int_0^\infty H_s^2 ds = \infty$ a.s. Set $M_t = \int_0^t H_s dY_s$ and call T_t the inverse of $\langle M, M\rangle_t$. Prove that the processes M_{T_t} and X_{T_t} are independent.

\# **(1.22) Exercise.** In the notation of the DDS theorem, if there is a strictly positive function f such that

$$\langle M, M\rangle_t = \int_0^t f(M_s) ds \quad \text{and} \quad \langle M, M\rangle_\infty = \infty$$

then

$$T_t = \int_0^t f(B_u)^{-1} du.$$

(1.23) Exercise. Let B be a $(\mathscr{F}_t)$-Brownian motion and T be a bounded $(\mathscr{F}_t)$-stopping time.

1°) Using Exercise (4.25) in Chap. IV and the fact that $E\left[B_T^2\right] = E[T]$, prove that, for $p > 2$, there is a universal constant C_p such that

$$E\left[B_T^{2p}\right] \leq C_p E\left[T^p\right].$$

By the same device as in Sect. 4 Chap. IV, extend the result to all p's.

2°) By the same argument, prove the reverse inequality.

3°) Write down a complete proof of the BDG inequalities via the DDS theorem.

\# **(1.24) Exercise.** This exercise aims at answering in the negative the following question: if M is a cont. loc. mart. and H a predictable process, such that $\int_0^t H_s^2 d\langle M, M\rangle_s < \infty$ a.s. for every t, is it true then that

$$(*) \qquad E\left[\left(\int_0^t H_s dM_s\right)^2\right] = E\left[\int_0^t H_s^2 d\langle M, M\rangle_s\right]$$

whether these quantities are finite or not? The reader will observe that Fatou's lemma entails that an inequality is always true.

1°) Let B be a standard BM and H be a $(\mathscr{F}_t^B)$-predictable process such that

$$\int_0^t H_s^2 ds < \infty \quad \text{for every } t < 1, \quad \text{but} \quad \int_0^1 H_s^2 ds = \infty$$

(the reader will provide simple examples of such H's). Prove that the loc. mart. $M_t = \int_0^t H_s dB_s$ is such that

$$\varliminf_{t \to 1} M_t = -\infty, \qquad \varlimsup_{t \to 1} M_t = \infty \text{ a.s.}$$

2°) For $a \in \mathbb{R}$, $a \neq 0$, give an example of a cont. loc. mart. N vanishing at 0 and such that

i) for every $t_0 < 1$, $(N_t, t \leq t_0)$ is an L^2-bounded martingale.
ii) $N_t = a$ for $t \geq 1$.

Prove furthermore that these conditions force

$$E\left[\langle N, N\rangle_1^{1/2}\right] = \infty,$$

and conclude on the question raised at the beginning of the exercise. This provides another example of a local martingale bounded in L^2 which is nevertheless not a martingale (see Exercise (2.13)).

3°) This raises the question whether the fact that $(*)$ is true for every bounded H characterizes the L^2-bounded martingales. Again the answer is in the negative. For $a > 0$ define

$$S_a = \inf\left\{t \geq 1/2 : \left|\int_{1/2}^{t} (1-s)^{-1} dB_s\right| = a\right\}.$$

By stopping $\int_{1/2}^{\cdot}(1-s)^{-1}dB_s$ at time S_X for a suitable $\mathscr{F}_{1/2}^B$-measurable r.v. X, prove that there exists a cont. loc. mart. M for which (*) obtains for every bounded H and $E\left[M_1^2\right] = E\left[\langle M, M\rangle_1\right] = \infty$. Other examples may be obtained by considering filtrations $(\mathscr{F}_t)$ with non trivial initial σ-field and local martingales, such that $\langle M, M\rangle_t$ is $\mathscr{F}_0$-measurable for every t.

(1.25) Exercise. By using the stopping times T_{a+} of Proposition (3.9) Chap. III, prove that in Proposition (1.5) the C-continuity cannot be omitted.

(1.26) Exercise. A cont. loc. mart. with increasing process $t \wedge T$ where T is a stopping time, is a BM stopped at T.

(1.27) Exercise. A time-changed uniformly integrable martingale is a uniformly integrable martingale, even if the time change takes on infinite values.

§2. Conformal Martingales and Planar Brownian Motion

This section is devoted to the study of a class of two-dimensional local martingales which includes the planar BM. We will use the complex representation of $\mathbb{R}^2$; in particular, the planar BM will be written $B = B^1 + iB^2$ where (B^1, B^2) is a pair of independent linear BM's and we speak of the "*complex Brownian motion*". More generally, we recall from Sect. 3 in Chap. IV that a complex local martingale is a process $Z = X + iY$ where X and Y are real local martingales.

(2.1) Proposition. *If Z is a continuous complex local martingale, there exists a unique continuous complex process of finite variation vanishing at zero denoted by $\langle Z, Z\rangle$ such that $Z^2 - \langle Z, Z\rangle$ is a complex local martingale. Furthermore, the following three properties are equivalent:*

i) Z^2 *is a local martingale;*
ii) $\langle Z, Z\rangle = 0$;
iii) $\langle X, X\rangle = \langle Y, Y\rangle$ *and* $\langle X, Y\rangle = 0$.

Proof. It is enough to define $\langle Z, Z\rangle$ by $\mathbb{C} \times \mathbb{C}$-linearity, that is

$$\langle Z, Z\rangle = \langle X + iY, X + iY\rangle = \langle X, X\rangle - \langle Y, Y\rangle + 2i\langle X, Y\rangle.$$

Plainly, the process thus defined enjoys all the properties of the statement and the uniqueness follows from the usual argument (Proposition (1.12) in Chap. IV) applied to the real and imaginary parts.

(2.2) Definition. *A local martingale satisfying the equivalent properties of the above statement is called a* conformal local martingale *(abbreviated to conf. loc. mart.).*

Obviously, the planar BM is a conf. loc. mart. and if H is a complex-valued locally bounded predictable process and Z a conf. loc. mart., then $U_t = \int_0^t H_s dZ_s$ is a conf. loc. mart.

For a conf. loc. mart. Z, one sees that $\langle \text{Re } Z, \text{Re } Z\rangle = \frac{1}{2}\langle Z, \bar{Z}\rangle$; in particular, $\langle U, \bar{U}\rangle_t = \int_0^t |H_s|^2 d\langle Z, \bar{Z}\rangle_s$. Moreover, Itô's formula takes on a simpler form. Let us recall that $\frac{\partial}{\partial z} = \frac{1}{2}\left(\frac{\partial}{\partial x} - i\frac{\partial}{\partial y}\right)$ and $\frac{\partial}{\partial \bar{z}} = \frac{1}{2}\left(\frac{\partial}{\partial x} + i\frac{\partial}{\partial y}\right)$ and that a function $F : \mathbb{C} \to \mathbb{C}$ which is differentiable as a function of both variables x and y, is holomorphic if and only if $\frac{\partial F}{\partial \bar{z}} = 0$, in which case we set $F' = \frac{\partial F}{\partial z}$.

(2.3) Proposition. *If Z is a conf. loc. mart. and F a complex function on $\mathbb{C}$ which is twice continuously differentiable (as a function of two real variables) then*

$$\begin{aligned} F(Z_t) &= F(Z_0) + \int_0^t \frac{\partial F}{\partial z}(Z_s)dZ_s + \int_0^t \frac{\partial F}{\partial \bar{z}}(Z_s)d\bar{Z}_s \\ &\quad + \frac{1}{4}\int_0^t \Delta F(Z_s)d\langle Z, \bar{Z}\rangle_s. \end{aligned}$$

In particular, if F is harmonic, $F(Z)$ is a local martingale and, if F is holomorphic,

$$F(Z_t) = F(Z_0) + \int_0^t F'(Z_s)dZ_s.$$

Proof. Straightforward computations using Itô's formula. □

Remark. If Z is conformal and F is holomorphic, $F(Z)$ is conformal. We will give shortly a more precise result in the case of BM^2.

We now rewrite Theorem (1.9) in the case of conf. loc. martingales.

(2.4) Theorem. *If Z is a conformal local martingale and $Z_0 = 0$, there exists (possibly on an enlargement of the probability space) a complex Brownian motion B such that*

$$Z_t = B_{\langle X,X\rangle_t}.$$

Proof. Since $\langle X, X\rangle = \langle Y, Y\rangle$ and $\langle X, Y\rangle = 0$, Theorem (1.9), applied to the 2-dimensional local martingale (X, Y), implies the existence of a complex Brownian motion B such that, for $t < \langle X, X\rangle_\infty$,

$$B_t = X_{T_t} + iY_{T_t}$$

where $T_t = \inf\{u : \langle X, X\rangle_u > t\}$. The result follows as in the proof of Theorem (1.6). □

The foregoing theorem has a very important corollary which is known as the *conformal invariance of complex Brownian motion*.

(2.5) Theorem. *If F is an entire and non constant function, $F(B_t)$ is a time-changed BM. More precisely, there exists on the probability space of B a complex Brownian motion $\tilde{B}$ such that*

$$F(B_t) = F(B_0) + \tilde{B}_{\langle X, X \rangle_t}$$

where $\langle X, X\rangle_t = \int_0^t |F'(B_s)|^2 ds$ is strictly increasing and $\langle X, X\rangle_\infty = \infty$.

Proof. If F is an entire function, F^2 is also an entire function and by Proposition (2.3), $F^2(B)$ is a loc. mart. As a result, $F(B)$ is a conformal local martingale to which we may apply Theorem (2.4). By the remarks before Proposition (2.3) and the Proposition itself, for $X = \text{Re } F(B_t)$, we have

$$\langle X, X\rangle_t = \int_0^t |F'(B_s)|^2 ds.$$

As F' is entire and non identically zero, the set Γ of its zeros is countable; therefore $P\left(\int_0^\infty 1_\Gamma(B_s) ds = 0\right) = 1$ and $\langle X, X\rangle$ is strictly increasing.

It remains to prove that $\langle X, X\rangle_\infty = \infty$; the proof of this fact will require some additional information of independent interest. What we have proved so far is that $F(B)$ has the same paths as a complex BM but possibly run at a different speed. The significance of the fact that $\langle X, X\rangle$ is strictly increasing is that these paths are run without gaps and this up to time $\langle X, X\rangle_\infty$. When we prove that $\langle X, X\rangle_\infty = \infty$, we will know that the paths of $F(B)$ are exactly those of a BM. The end of the proof of Theorem (2.5) is postponed until we have proved the recurrence of planar BM in Theorem (2.8). □

We begin with a first result which is important in its own right. We recall from Sect. 2 Chap. III that hitting times T_A are stopping times so that the events $\{T_A < \infty\}$ are measurable.

(2.6) Definition. *For a Markov process with state space E, a Borel set A is said to be* polar *if*

$$P_z[T_A < \infty] = 0 \quad \text{for every } z \in E.$$

(2.7) Proposition. *For the BM in $\mathbb{R}^d$ with $d \geq 2$, the one-point sets are polar sets.*

Proof. Plainly, it is enough to prove the result for $d = 2$, and because of the geometrical invariance properties of BM, it suffices to show that the planar BM started at 0 does not hit the point set $\{(-1, 0)\}$. By what we already know, the process $M_t = \exp(B_t) - 1$ may be written $\tilde{B}_{A_t}$ where $\tilde{B}$ is a planar BM and $A_t = \int_0^t \exp(2X_s) ds$ where X is the real component of B. The process A is clearly strictly increasing. We also claim that $A_\infty = \infty$ a.s. Otherwise, M would converge in $\mathbb{C}$ as t tends to infinity; since $|\exp(B_t)| = \exp(X_t)$ where X, as a linear BM, is recurrent, this is impossible. As a result, the paths of M are exactly the paths of a BM (run at a different speed) and, since $\exp(B_t)$ never vanishes, the result is established.

Remarks. 1°) For BM^1, no non empty set is polar, whereas for BM^2, $d \geq 2$, all one-point sets, hence all countable sets are polar and, for instance, if we call $\mathbb{Q}^2$ the set of points in $\mathbb{R}^2$ with rational coordinates, the Brownian path $\{B_t, t \geq 0\}$ is a.s. contained in $\mathbb{R}^2 \backslash \mathbb{Q}^2$. But, there are also uncountable polar sets even in the case $d = 2$.

2°) Another more elementary proof of Proposition (2.7) was given in Exercise (1.20) of Chap. I and yet another is given in Exercise (2.14).

We may now state the *recurrence* property of the planar BM, another proof of which is given in Exercise (2.14).

(2.8) Theorem. *Almost surely, the set $\{t : B_t \in U\}$ is unbounded for all open subsets U of $\mathbb{R}^2$.*

Proof. By using a countable basis of open balls, it is enough to prove the result whenever U is the ball $B(z, r)$ where $z = x + iy$ and r is > 0.

Since one-point sets are polar, we may consider the process $M_t = \log|B_t - z|$ which by the same reasoning as above is equal to β_{A_t}, where β is a linear BM started at $\log|z|$ and $A_t = \int_0^t |B_u - z|^{-2} du$. But since $\sup_{s \leq t} M_s$ is larger than $\sup_{s \leq t} \log|X_s - x|$ which goes to infinity as t tends to $+\infty$, it follows from Remark 3°) after Proposition (1.8) that A_t converges to infinity. As a result, $\inf_t M_t = +\infty$ a.s., hence M takes on values less than $\log r$ at arbitrary large times which ends the proof.

Remarks. 1°) One can actually prove that for any Borel set A with strictly positive Lebesgue measure, the set $\{t : B_t \in A\}$ is a.s. unbounded and in fact of infinite Lebesgue measure (see Sect. 3 Chap. X).

2°) The above result shows that a.s., the Brownian path, which is of Lebesgue measure zero, is dense in the plane.

We can now turn to the

(2.9) End of the proof of Theorem (2.5). If we had $\langle X, X \rangle_\infty < \infty$, then $F(B_t)$ would have a limit as t tends to infinity. But since F is non constant, one can find two disjoint open sets U_1 and U_2 such that $F(\bar{U}_1) \cap F(\bar{U}_2) = \emptyset$ and as $\{t : B_t \in U_1\}$ and $\{t : B_t \in U_2\}$ are unbounded, $F(B_t)$ cannot have a limit as t tends to infinity. □

We now state one more result about recurrence. We have just seen that the BM in $\mathbb{R}^d$ is recurrent for $d = 1$ and $d = 2$. This is no longer true for $d \geq 3$, in which case the BM is said to be *transient*. More precisely we have the

(2.10) Theorem. *If $d \geq 3$, $\lim_{t \to \infty} |B_t| = +\infty$ almost surely.*

Proof. It is clearly enough to prove this when $d = 3$ and when B is started at $x_0 \neq 0$. Since $\{0\}$ is a polar set, by Itô's formula, $1/|B_t|$ is a positive loc. mart., hence a positive supermartingale which converges a.s. to a r.v. H. By Fatou's lemma, $E_{x_0}[H] \leq \underline{\lim} E_{x_0}[1/|B_t|]$. But, the scaling property shows that

$E_{x_0}[1/|B_t|] = 0(1/\sqrt{t})$. As a result, $H = 0$ P_{x_0}-a.s. and the proof is now easily completed. □

We close this section with a representation result for the complex BM B which we study *under the law* P_a for $a \neq 0$. Since $B_t \neq 0$ for all t P_a-a.s., we may choose a continuous determination $\theta_t(\omega)$ of the argument of $B_t(\omega)$ such that $\theta_0(\omega)$ is a constant and $e^{i\theta_0} = a/|a|$. We then have $B_t = \rho_t \exp(i\theta_t)$ and θ is adapted to the filtration of B_t. The processes ρ_t and θ_t may be analysed in the following way.

(2.11) Theorem. *There is a planar BM* (β, γ) *such that*

$$\rho_t = |a| \exp(\beta_{C_t}), \quad \theta_t = \theta_0 + \gamma_{C_t},$$

where $C_t = \int_0^t \rho_s^{-2} ds$. *Moreover,* $\mathscr{F}_\infty^\beta = \mathscr{F}_\infty^\rho$, *hence* γ *is independent of* ρ.

Proof. Because B almost-surely never vanishes, we may define the conformal local martingale H by

$$H_t = \int_0^t B_s^{-1} dB_s$$

and we have $\langle \operatorname{Re} H, \operatorname{Re} H \rangle_t = C_t$. Applying the integration by parts formula to the product $B_t \exp(-H_t)$, it is easily seen, since $\langle B, B \rangle = 0$, that $B_t = a \exp(H_t)$. By Theorem (2.4), there is a planar BM which we denote by $\beta + i\gamma$ such that $H_t = \beta_{C_t} + i\gamma_{C_t}$ which proves the first half of the statement.

The process β is the DDS Brownian motion of the local martingale $\operatorname{Re} H_t = \log(\rho_t/|a|)$. But with $X = \operatorname{Re} B$, $Y = \operatorname{Im} B$,

$$\operatorname{Re} H_t = \int_0^t \frac{X_s dX_s + Y_s dY_s}{\rho_s^2} = \int_0^t \frac{d\tilde{\beta}_s}{\rho_s}$$

where $\tilde{\beta}$ is a real BM. We may rewrite this as

$$\log \rho_t = \log |a| + \int_0^t \sigma(\log \rho_s) d\tilde{\beta}_s$$

with $\sigma(x) = e^{-x}$; it follows from Proposition (1.11) that $\mathscr{F}_\infty^\beta = \mathscr{F}_\infty^\rho$ which is the second half of the statement. □

This result shows that, as one might expect, the smaller the modulus of B, the more rapidly the argument of B varies. Moreover, as θ_t is a time-changed BM, it is easy to see that

$$\varliminf_{t\to\infty} \theta_t = -\infty, \quad \varlimsup_{t\to\infty} \theta_t = +\infty \quad \text{a.s.}$$

Thus, the planar BM winds itself arbitrarily large numbers of times around 0, then unwinds itself and does this infinitely often. Stronger asymptotic results on θ_t or, more generally, on the behavior of planar BM will be given in Chaps. VII and XIII.

At the cost of losing some information on the modulus, the preceding result may be stated

$$B_t = \rho_t e^{i\gamma_{C_t}}$$

where γ is a linear BM independent of ρ and $C_t = \int_0^t \rho_s^{-2} ds$. This is known as the "*skew-product*" representation of two-dimensional Brownian motion.

It is interesting to stress the fact that we have worked under P_a with $a \neq 0$. For $a = 0$ and for $t > 0$, we may still, by the polarity of $\{0\}$, write a.s. unambiguously $B_t = \rho_t \cdot \mathscr{U}_t$ with $|\mathscr{U}_t| = 1$. But we have no means of choosing a continuous determination of the argument of $\mathscr{U}_t$ adapted to $(\mathscr{F}_t)$. This example hints at the desirability of defining and studying semimartingales on open subsets of $\mathbb{R}_+$.

(2.12) Exercise. Let $A = \{z : |\mathrm{Im}\, z| < \pi/2\}$; compute the law of $B^1_{T_{A^c}}$ for the complex Brownian motion $B^1 + iB^2$ started at 0.

[Hint: The exponential function maps A onto the half-plane H. Use the exit distribution for H.]

\# **(2.13) Exercise (An important counterexample).** 1°) Let B be the BM in $\mathbb{R}^d$ with $d \geq 3$ started at $x \neq 0$. Prove that $|B_t|^{2-d}$ is a local martingale.

[Hint: $|x|^{2-d}$ is harmonic in $\mathbb{R}^d \backslash \{0\}$.]

2°) Use 1°) in the case $d = 3$ to give an example of a local martingale which is bounded in L^2 but is not a true martingale. This gives also an example of a uniformly integrable local martingale which is not a true martingale and an example of a uniformly integrable supermartingale X such that the set $\{X_T;\quad T \text{ stopping time}\}$ is not uniformly integrable. One can also find for every $p \geq 1$ a local martingale bounded in L^p which is not a martingale, as is seen in Exercise (1.16) of Chap. XI.

3°) If B is the complex BM, let $M = \log|B|$ and prove that for $\varepsilon > 0$ and $\alpha < 2$

$$\sup_{\varepsilon \leq t \leq 1} E_x\left[\exp \alpha |M_t|\right] < \infty.$$

This provides an example of a local martingale with exponential moments which is not a martingale. Furthermore, $E_x[\langle M, M\rangle_t] = \infty$, which bears out a remark made after Corollary (1.25) in Chap. IV.

\# **(2.14) Exercise.** 1°) Retain the situation of 1°) in the preceding exercise and let a and b be two numbers such that $0 < a < |x| < b$. Set $R_a = T_{B(0,a)}$ and $S_b = T_{B(0,b)^c}$, and prove that

$$P_x[R_a < S_b] = \left(|x|^{2-d} - b^{2-d}\right) / \left(a^{2-d} - b^{2-d}\right).$$

[Hint: Stop the local martingale $|B_t|^{2-d}$ at $R_a \wedge S_b$ and proceed as in Chap. II.]

Prove that $P_x[R_a < \infty] = (a/|x|)^{d-2}$.

2°) Using the function $\log|x|$ instead of $|x|^{2-d}$, treat the same questions for $d = 2$. Base on this another proof that one-point sets are polar.

3°) Deduce from 2°) that for $d = 2$, any open set U and any z, $P_z[T_U < \infty] = 1$.

4°) Let D_1 and D_2 be two disjoint disks and define inductively a double sequence of stopping times by

$$T_1 = T_{D_1}, \quad U_1 = \inf\{t > T_1, B_t \in D_2\},$$

$$T_n = \inf\{t > U_{n-1}, B_t \in D_1\}, \quad U_n = \inf\{t > T_n, B_t \in D_2\}.$$

Prove that, for any z and every n,

$$P_z[T_n < \infty] = P_z[U_n < \infty] = 1$$

and deduce therefrom another proof of Theorem (2.8).

(2.15) Exercise. In the situation of Theorem (2.11) prove that the filtration generated by θ_t is the same as that of B_t.

[Hint: ρ is $\mathscr{F}^\theta$-adapted.]

(2.16) Exercise. Let B_t be the complex BM and suppose that $B_0 = 1$ a.s. For $r > 0$, let

$$T_r = \inf\{t : |B_t| = r\}.$$

Prove that, if θ_t is the continuous determination of the argument of B_t which vanishes for $t = 0$, then θ_{T_r} is either for $r \leq 1$ or for $r \geq 1$, a process with independent increments and that the law of θ_{T_r} is the Cauchy law with parameter $|\log r|$.

[Hint: Use the results in Proposition (3.11) of Chap. III.]

(2.17) Exercise. 1°) **(Liouville's theorem).** Deduce from the recurrence of BM^2 and the martingale convergence theorem that bounded harmonic functions in the whole plane are constant.

[Hint: See the reasoning in Proposition (3.10) of Chap. X.]

2°) **(D'Alembert's theorem).** Let P be a non constant polynomial with complex coefficients. Use the properties of BM^2 to prove that, for any $\varepsilon > 0$, the compact set $\{z : |P(z)| \leq \varepsilon\}$ is non empty and conclude to the existence of a solution to the equation $P(z) = 0$.

(2.18) Exercise. 1°) Let $Z = X + iY$ be the complex BM started at -1 and

$$T = \inf\{t : Y_t = 0, X_t \geq 0\}.$$

Prove that the law of $\log X_T$ has a density equal to $(2\pi\cosh(x/2))^{-1}$.

[Hint: See Exercise (3.25) in Chap. III.]

2°) As a result, the law of X_T is that of C^2 where C is a Cauchy r.v..

[Hint: See Sect. 6 Chap. 0.]

*# **(2.19) Exercise.** We retain the notation of Exercise (2.18) in Chap. IV.

1°) Let F be a holomorphic function in an open subset U of $\mathbb{C}$ and $\bar{\omega}$ the differential form $F(z)dz$. If Z is a conf. loc. mart. such that $P[\exists t \geq 0 : Z_t \notin U] = 0$, then

$$\int_{Z(0,t)} \bar{\omega} = \int_0^t F(Z_s)dZ_s = \int_0^t F(Z_s) \circ dZ_s \quad \text{a.s.},$$

where $\circ$ stands for the Stratonovich integral.

2°) In the situation of Theorem (2.11), we have

$$\theta_t - \theta_0 = \int_{Z(0,t)} \bar{\omega}$$

with $\bar{\omega} = \left(x^2 + y^2\right)^{-1}(xdy - ydx) = \operatorname{Im}(dz/z)$.

3°) In the same situation, let S_t be the area swept by the segment $[0, Z_u]$, $0 \leq u \leq t$ (see Exercise (2.18) in Chap. IV). Prove that there is a linear BM δ independent of ρ such that $S_t = \delta_{A_t}$ where $A_t = \frac{1}{4}\int_0^t \rho_s^2 ds$. Another proof is given in Exercise (3.10) Chap. IX which removes the condition $a \neq 0$.

(2.20) Exercise. 1°) Let f be a meromorphic function in $\mathbb{C}$ and A be the set of its poles. For $z_0 \notin A$, prove that $f(B)$ is a time-changed BM^2 started at $f(z_0)$.

2°) Let z be a non-zero complex number. If $T = \inf\{t : |B_t| = 1\}$ prove that B_T has the same law under P_z and $P_{1/\bar{z}}$.

* **(2.21) Exercise (Exit time from a cone).** Retain the situation and notation of Theorem (2.11) with $a = 1$ and set $\theta_0 = 0$. For $n, m > 0$, define

$$T = \inf\{u : \theta_u \notin [-n, m]\}, \quad \tau = \inf\{v : \gamma_v \notin [-n, m]\}.$$

1°) For $\delta > 0$, prove that $E\left[\rho_T^{2\delta}\right] = E\left[\exp(2\delta^2\tau)\right]$. Prove that consequently $E[T^\delta] < \infty$ implies $2\delta(n + m) < \pi$.

[Hint: Use the result in Exercise (3.10) 2°) of Chap. II.]

2°) From the equality $T = \int_0^\tau \exp(2\beta_u)du$, deduce that

$$E[T^\delta] \leq 2E\left[\tau^\delta \exp(2\delta^2\tau)\right].$$

Conclude that $E[T^\delta] < \infty$ if and only if $2\delta(n + m) < \pi$.

(2.22) Exercise. Let $C(\omega) = \{B_t(\omega), t \geq 0\}$ where B is the BM^2.

1°) Prove that $C(\omega)$ is a.s. a Borel subset of $\mathbb{R}^2$. Call $\Lambda(\omega)$ its Lebesgue measure.

2°) Prove that

$$\Lambda(\omega) = \int_{\mathbb{R}^2} \left(\sup_{t \geq 0} 1_{\{z\}}(B_t)\right) dx\, dy$$

and conclude that $\Lambda(\omega) = 0$ a.s. Compare with the proof given in Exercise (1.17) Chap. I.

(2.23) Exercise. If Z is the complex BM, a complex C^2-function f on $\mathbb{R}^2$ is holomorphic if and only if

$$f(Z_t) = f(Z_0) + \int_0^t H_s dZ_s$$

for some predictable process H.

* **(2.24) Exercise.** We say that $Z \in \mathscr{C}$ if for every $\varepsilon > 0$, the process $Z_{t+\varepsilon}$, $t \geq 0$ is a $(\mathscr{F}_{t+\varepsilon})$-conformal martingale (see Exercises (1.42) and (3.26) in Chap. IV).

1°) Prove that, if $Z \in \mathscr{C}$, then a.s.,

$$\left\{\omega : \overline{\lim_{t\downarrow 0}} |Z_t(\omega)| < \infty\right\} = \left\{\omega : \lim_{t\downarrow 0} Z_t(\omega) \text{ exists in } \mathbb{C}\right\}.$$

2°) Derive therefrom that for a.e. ω, one of the following three events occurs

i) $\lim_{t\downarrow 0} Z_t(\omega)$ exists in $\mathbb{C}$;
ii) $\lim_{t\downarrow 0} |Z_t(\omega)| = +\infty$;
iii) for every $\delta > 0$, $\{Z_t(\omega), 0 < t < \delta\}$ is dense in $\mathbb{C}$.

[Hint: For $z \in \mathbb{C}$ and $r > 0$ define $T = \inf\{t : |Z_t - z| < r\}$ and look at $V_t = \left(Z_t^T - z\right)^{-1} 1_{(T>0)}$ which is an element of $\mathscr{C}$.]

Describe examples where each of the above possibilities does occur.

\# **(2.25) Exercise.** 1°) Let ρ_t be the modulus of BM^d, $d \geq 2$, say X, and ν a probability measure such that $\nu(\{0\}) = 0$. Prove that, under P_ν,

$$B_t = \rho_t - \rho_0 - \frac{1}{2}(d-1)\int_0^t \rho_s^{-1} ds$$

is a BM. In the language of Chap. IX, ρ_t is a solution to the stochastic differential equation

$$\rho_t = \rho_0 + B_t + \frac{1}{2}(d-1)\int_0^t \rho_s^{-1} ds.$$

The processes ρ_t are *Bessel processes* of dimension d and will be studied in Chap. XI.

2°) We now remove the condition on ν. Prove that under P_0 the r.v. $\int_0^t \rho_s^{-1} ds$ is in L^p for every p and extend 1°) to $\nu = \varepsilon_0$.

[Hint: $\rho_t - \rho_s - \frac{1}{2}(d-1)\int_s^t \rho_u^{-1} du = \int_s^t \langle \operatorname{grad} r(X_u), dX_u\rangle$ where r is the distance to zero. Prove that the right-hand side is bounded in L^2 as s tends to 0.]

* **(2.26) Exercise (On polar functions).** Let $f = (f^1, f^2)$ be a continuous $\mathbb{R}^2$-valued deterministic function of bounded variation such that $f(0) \neq 0$ and $B = (X, Y)$ a planar BM(0). Set $Z_t = B_t + f(t)$ and $T = \inf\{t : Z_t = 0\}$.

1°) Prove that $\beta_t = \int_0^t \frac{(X_s + f^1(s))dX_s + (Y_s + f^2(s))dY_s}{|Z_s|}$ is a linear BM on $[0, T[$ (see Exercise (3.28) Chap. IV).

2°) By first proving that $E\left[\int_{1/n}^t |Z_s|^{-1} |df(s)|\right] < \infty$, show that a.s.

$$\int_0^t |Z_s|^{-1}|df(s)| < \infty, \quad \text{for every } t.$$

[Hint: Use the scaling property of B and the fact that, if G is a two-dimensional Gaussian reduced r.v., then

$$\sup_{m\in\mathbb{C}} E\left[|m+G|^{-1}\right] < \infty.]$$

3°) By considering $\log|Z|$, prove that T is infinite a.s.. In other words, f is a polar function in the sense of Exercise (1.20) of Chap. I.

[Hint: Use Exercise (1.18).]

If f is absolutely continuous, a much simpler proof will be given in Exercise (2.15) Chap. VIII.

(2.27) Exercise (Polarity for the Brownian sheet). Let $X_{(s,t)}$ be a complex Brownian sheet, namely $X_{(s,t)} = W^1_{(s,t)} + iW^2_{(s,t)}$ where W^1 and W^2 are two independent Brownian sheets. If $\gamma(s) = \big(x(s), y(s)\big)$, $s \in [0,1]$, is a continuous path in $]0,\infty[\times]0,\infty[$, write X_γ for the process $s \to X_{\left(x(s),y(s)\right)}$.

1°) If x and y are both increasing (or decreasing), prove that the one-point sets of $\mathbb{C}$ are polar for X_γ.

2°) Treat the same question when x is increasing and y decreasing or the reverse.

[Hint: Use Exercise (1.13) in Chap. III and the above exercise.]

As a result, if γ is a closed path which is piecewise of one of the four kinds described above, the index of any $a \in \mathbb{C}$ with respect to the path of X_γ is a.s. defined.

§3. Brownian Martingales

In this section, we consider the filtration $(\mathscr{F}_t^B)$ but we will write simply $(\mathscr{F}_t)$. It is also the filtration $(\mathscr{F}_t^{\varepsilon_0})$ of Sect. 2 Chap. III, where it was called the Brownian filtration.

We call $\mathscr{T}$ the set of step functions with compact support in $\mathbb{R}_+$, that is, of functions f which can be written

$$f = \sum_{j=1}^n \lambda_j 1_{]t_{j-1},t_j]}.$$

As in Sect. 3 of Chap. IV, we write $\mathscr{E}^f$ for the exponential of $\int_0^\cdot f(s)dB_s$.

(3.1) Lemma. *The set* $\left\{\mathscr{E}^f_\infty, f \in \mathscr{T}\right\}$ *is total in* $L^2(\mathscr{F}_\infty, P)$.

Proof. We show that if $Y \in L^2(\mathscr{F}_\infty, P)$ and Y is orthogonal to every $\mathscr{E}_\infty^f$, then the measure $Y \cdot P$ is the zero measure. To this end, it is enough to prove that it is the zero measure on the σ-field $\sigma\left(B_{t_1}, \dots, B_{t_n}\right)$ for any finite sequence $(t_1, \dots, t_n)$.

The function $\varphi(z_1, \dots, z_n) = E\left[\exp\left(\sum_{j=1}^n z_j\left(B_{t_j} - B_{t_{j-1}}\right)\right) \cdot Y\right]$ is easily seen to be analytic on $\mathbb{C}^n$. Moreover, by the choice of Y, for any $\lambda_i \in \mathbb{R}$, we have

$$\varphi(\lambda_1, \dots, \lambda_n) = E\left[\exp\left(\sum_{j=1}^n \lambda_j\left(B_{t_j} - B_{t_{j-1}}\right)\right) \cdot Y\right] = 0.$$

Consequently, φ vanishes identically and in particular

$$E\left[\exp\left\{i\left(\sum_{j=1}^n \lambda_j\left(B_{t_j} - B_{t_{j-1}}\right)\right)\right\} \cdot Y\right] = 0.$$

The image of $Y \cdot P$ by the map $\omega \to \left(B_{t_1}(\omega), \dots, B_{t_j}(\omega) - B_{t_{j-1}}(\omega), \dots\right)$ is the zero measure since its Fourier transform is zero. The measure vanishes on $\sigma\left(B_{t_i}, \dots, B_{t_{i+1}} - B_{t_i}, \dots\right) = \sigma\left(B_{t_1}, B_{t_2}, \dots, B_{t_n}\right)$ which ends the proof.

(3.2) Proposition. *For any $F \in L^2(\mathscr{F}_\infty, P)$, there exists a unique predictable process H in $L^2(B)$ such that*

$$F = E[F] + \int_0^\infty H_s dB_s.$$

Proof. We call $\mathscr{H}$ the subspace of elements F in $L^2(\mathscr{F}_\infty, P)$ which can be written as stated. For $F \in \mathscr{H}$

$$(*) \qquad E[F^2] = E[F]^2 + E\left[\int_0^\infty H_s^2 ds\right].$$

which implies in particular the uniqueness in the statement. Thus, if $\{F^n\}$ is a Cauchy sequence of elements of $\mathscr{H}$, the corresponding sequence $\{H^n\}$ is a Cauchy sequence in $L^2(B)$, hence converges to a predictable $H \in L^2(B)$; it is clear that $\{F^n\}$ converges in $L^2(\mathscr{F}_\infty, P)$ to

$$\left(\lim_n E[F^n]\right) + \int_0^\infty H_s dB_s$$

which proves that $\mathscr{H}$ is closed.

On the other hand, $\mathscr{H}$ contains all the random variables $\mathscr{E}_\infty^f$ of Lemma (3.1), since, by Itô's formula, we have

$$\mathscr{E}_t^f = 1 + \int_0^t \mathscr{E}_s^f f(s) dB_s, \quad \text{for every } t \leq \infty.$$

This proves the existence of H. The uniqueness in $L^2(B)$ follows from the identity $(*)$. □

Remark. If the condition $H \in L^2(B)$ is removed, there are infinitely many predictable processes H satisfying the conditions of Proposition (3.2); this is proved in Exercise (2.31) Chap. VI, but it may already be observed that by taking $H = 1_{[0,d_T]}$ with $d_T = \inf\{u > T : B_u = 0\}$, one gets $F = 0$.

We now turn to the main result of this section, namely the extension of Proposition (3.2) to local martingales. The reader will observe in particular the following remarkable feature of the filtration $(\mathscr{F}_t)$: *there is no discontinuous* $(\mathscr{F}_t)$-*martingale.* Using Corollary (5.7) of Chap. IV, this entails the

(3.3) Corollary. *For the Brownian filtration, every optional process is predictable.*

The reader who is acquainted with the classification of stopping times will also notice that all the stopping times of the Brownian filtration are predictable.

(3.4) Theorem. *Every $(\mathscr{F}_t)$-local martingale M has a version which may be written*

$$M_t = C + \int_0^t H_s dB_s$$

where C is a constant and H a predictable process which is locally in $L^2(B)$. In particular, any $(\mathscr{F}_t)$-local martingale has a continuous version.

Proof. If M is an L^2-bounded $(\mathscr{F}_t)$-martingale, by the preceding result, there is a process $H \in L^2(B)$ such that

$$\begin{aligned} M_t = E\left[M_\infty \mid \mathscr{F}_t\right] &= E[M_\infty] + E\left[\int_0^\infty H_s dB_s \mid \mathscr{F}_t\right] \\ &= E[M_\infty] + \int_0^t H_s dB_s, \end{aligned}$$

hence the result is true in that case.

Let now M be uniformly integrable. Since $L^2(\mathscr{F}_\infty)$ is dense in $L^1(\mathscr{F}_\infty)$ there is a sequence of L^2-bounded martingales M^n such that $\lim_n E\left[|M_\infty - M^n_\infty|\right] = 0$. By the maximal inequality, for every $\lambda > 0$,

$$P\left[\sup_t |M_t - M^n_t| > \lambda\right] \le \lambda^{-1} E\left[|M_\infty - M^n_\infty|\right].$$

Thanks to the Borel-Cantelli lemma, one can extract a subsequence $\{M^{n_k}\}$ converging a.s. uniformly to M. As a result, M has a continuous version.

If now M is an $(\mathscr{F}_t)$-local martingale, it obviously has a continuous version and thus admits a sequence of stopping times T_n such that M^{T_n} is bounded. By the first part of the proof, the theorem is established. □

It is easy to see that the above reasonings are still valid in a multidimensional context and we have the

(3.5) Theorem. *Every $(\mathscr{F}_t^B)$-local martingale, say M, where B is the d-dimensional BM $(B^1, \dots, B^d)$ has a continuous version and there exist predictable processes H^i, locally in $L^2(B^i)$, such that*

$$M_t = C + \sum_{i=1}^{d} \int_0^t H_s^i dB_s^i.$$

Remarks. 1°) The processes H^i are equal to the Radon-Nikodym derivatives $\frac{d}{dt}\langle M, B^i\rangle_t$ of $\langle M, B^i\rangle$ with respcct to the Lebesgue measure. But, in most concrete examples they can be computed explicitely. A fairly general result to this effect will be given in Sect. 2 Chap. VIII. Exercises (3.13), (3.16) of this section already give some particular cases. When f is harmonic, the representation of the martingale $f(B_t)$ is given by Itô's formula.

2°) It is an interesting, and for a large part unsolved, problem to study the filtration of the general local martingale obtained in the above results. The reader will find some very partial answers in Exercise (3.12).

The above results are, in particular, representation theorems for $L^2(\mathscr{F}_\infty^B)$. We now turn to another representation of this space; for simplicity, we treat the one-dimensional case. We set

$$\Delta_n = \left\{(s_1, \dots, s_n) \in \mathbb{R}_+^n;\ s_1 > s_2 > \dots > s_n\right\}$$

and denote by $L^2(\Delta_n)$ the L^2-space of Lebesgue measure on Δ_n. The subset E_n of $L^2(\Delta_n)$ of functions f which can be written

$$f(s_1, \dots, s_n) = \prod_1^n f_i(s_i)$$

with $f_i \in L^2(\mathbb{R}_+)$ is total in $L^2(\Delta_n)$. For $f \in E_n$, we set

$$J_n(f) = \int_0^\infty f_1(s_1) dB_{s_1} \int_0^{s_1} f_2(s_2) dB_{s_2} \dots \int_0^{s_{n-1}} f_n(s_n) dB_{s_n}.$$

This kind of iterated stochastic integrals has already been encountered in Proposition (3.8) in Chap. IV, and it is easily seen that

$$\|J_n(f)\|_{L^2(\mathscr{F}_\infty)} = \|f\|_{L^2(\Delta_n)}.$$

(3.6) Definition. *For $n \geq 1$, the smallest closed linear subspace of $L^2(\mathscr{F}_\infty^B)$ containing $J_n(E_n)$ is called the n-th* Wiener chaos *and is denoted by K_n.*

The map J_n is extended to $L^2(\Delta_n)$ by linearity and passage to the limit. If f is in the linear space generated by E_n it may have several representations as linear combination of elements of E_n but it is easy to see that $J_n(f)$ is nonetheless defined unambiguously. Moreover J_n is an isometry between $L^2(\Delta_n)$ and K_n. Actually, using Fubini's theorem for stochastic integrals (Exercise (5.17) Chap. IV), the reader may see that $J_n(f)$ could be defined by straightforward multiple stochastic integration.

Obviously, there is a one-to-one correspondence between K_n and $L^2(\Delta_n)$. Moreover, the spaces K_n and K_m are orthogonal if $n \neq m$, the proof of which we leave to the reader as an exercise. We may now state

(3.7) Theorem. *$L^2(\mathscr{F}_\infty^B) = \bigoplus_0^\infty K_n$ where K_0 is the space of constants. In other words, for each $Y \in L^2(\mathscr{F}_\infty^B)$ there exists a sequence (f^n) where $f^n \in L^2(\Delta_n)$ for each n, such that*

$$Y = E[Y] + \sum_{n=1}^{\infty} J_n(f^n)$$

in the L^2-sense.

Proof. By Proposition (3.8) in Chap. IV, the random variables $\mathscr{E}_\infty^f$ of Lemma (3.1) may be written $1 + \sum_1^\infty J_n(f^n)$ pointwise with $f^n(s_1, \ldots, s_n) = f(s_1)f(s_2)\ldots f(s_n)$. As f is bounded and has compact support it is easy to see that this convergence holds in $L^2(\mathscr{F}_\infty)$. Thus, the statement is true for $\mathscr{E}_\infty^f$. It is also true for any linear combination of variables $\mathscr{E}_\infty^f$. Since, by Lemma (3.1), every r.v. $Y \in L^2(\mathscr{F}_\infty)$ is the limit of such combinations, the proof is easily completed.

Remark. The first chaos contains only Gaussian r.v.'s and is in fact the closed Gaussian space generated by the r.v.'s B_t, $t \geq 0$ (see Exercise (3.11)).

We now come to another question. Theorem (3.4) raises the following problem: which martingales can be written as $(H \cdot B)_t$ for a suitable Brownian motion B? We give below a partial answer which will be used in Chap. IX.

(3.8) Proposition. *If M is a continuous local martingale such that the measure $d\langle M, M\rangle_t$ is a.s. equivalent to the Lebesgue measure, there exist an $(\mathscr{F}_t^M)$-predictable process f_t which is strictly positive $dt \otimes dP$ a.s. and an $(\mathscr{F}_t^M)$-Brownian motion B such that*

$$d\langle M, M\rangle_t = f_t dt \quad \text{and} \quad M_t = M_0 + \int_0^t f_s^{1/2} dB_s.$$

Proof. By Lebesgue's derivation theorem, the process

$$f_t = \varliminf_{n\to\infty} n\left(\langle M, M\rangle_t - \langle M, M\rangle_{t-1/n}\right)$$

satisfies the requirements in the statement. Moreover, $(f_t)^{-1/2}$ is clearly in $L^2_{\text{loc}}(M)$ and the process

$$B_t = \int_0^t f_s^{-1/2} dM_s$$

is a continuous local martingale with increasing process t, hence a BM and the proof is easily completed. □

If $d\langle M, M\rangle_t$ is merely absolutely continuous with respect to dt, the above reasoning fails; moreover, the filtration $(\mathscr{F}_t^M)$ is not necessarily rich enough to admit an $(\mathscr{F}_t^M)$-Brownian motion. However, if B' is a BM independent of $\mathscr{F}_\infty^M$ and if we set

$$B_t = \int_0^t 1_{(f_s>0)} f_s^{-1/2} dM_s + \int_0^t 1_{(f_s=0)} dB'_s,$$

then, by Lévy's characterization theorem, B is again a BM and $M_t = M_0 + \int_0^t f_s^{1/2} dB_s$. In other words, the foregoing result is still true provided we enlarge the probability space so as to avail ourselves of an independent BM. Using a little linear algebra and an enlargement $(\tilde{\Omega}, \tilde{\mathscr{F}_t}, \tilde{P})$ of $(\Omega, \mathscr{F}_t, P)$, this can be carried over to the multi-dimensional case. We only sketch the proof, leaving to the reader the task of keeping track of the predictability and integrability properties of the processes involved when altered by algebraic transformations.

(3.9) Theorem. *Let $M = (M^1, \dots, M^d)$ be a cont. vect. loc. mart. such that $d\langle M^i, M^i\rangle_t \ll dt$ for every i. Then there exist, possibly on an enlargement of the probability space, a d-dimensional BM B and a $d \times d$ matrix-valued predictable process α in $L^2_{\mathrm{loc}}(B)$ such that*

$$M_t = M_0 + \int_0^t \alpha_s dB_s.$$

Proof. We may suppose that $M_0 = 0$.

By the inequality of Proposition (1.15) in Chap. IV, we have $d\langle M^i, M^j\rangle_t \ll dt$ for every pair (i, j). The same argument as in the previous proof yields a predictable process γ of symmetric $d \times d$ matrices such that

$$\langle M^i, M^j\rangle_t = \int_0^t \gamma_s^{ij} ds$$

and the matrix γ is $dP \otimes dt$-a.e. semi-definite positive. As a result one can find a predictable process β with values in the set of $d \times d$ orthogonal matrices such that $\rho = \beta^t \gamma \beta$ is diagonal. Setting $\alpha^{ji} = \beta^{ij}(\rho^{jj})^{1/2}$ we get a predictable process such that $\gamma = \alpha^t \alpha = \alpha \alpha^t$. Of course some of the ρ^{jj}'s may vanish and the rank of α, which is equal to the rank of γ, may be less than d. We call ζ_s the predictable process which is equal to the rank of γ_s.

Define a matrix P_ζ by setting $p_\zeta^{ij} = 1$ if $i = j \leq \zeta$ and $p_\zeta^{ij} = 0$ otherwise. There exist a predictable process ϕ such that ϕ_s is a $d \times d$ orthogonal matrix such that $\alpha\phi = \alpha\phi P_\zeta$ and a matrix-valued process λ such that $\lambda\alpha\phi = P_\zeta$. Set $N = \lambda \cdot M$; then N is a cont. vect. loc. mart. and $\langle N^i, N^j\rangle_t = \delta_j^i \int_0^t 1_{[i \leq \zeta_s]} ds$ as follows from the equalities

$$\lambda\gamma\lambda^t = \lambda\alpha\alpha^t\lambda^t = \lambda\alpha\phi\phi^t\alpha^t\lambda^t = P_\zeta.$$

If we set $X = (\alpha\phi) \cdot N$ it is easily seen that $\langle X - M, X - M\rangle = 0$, hence $X = M$.

If we now carry everything over to the enlargement $(\tilde{\Omega}, \tilde{\mathscr{F}_t}, \tilde{P})$, we have at our disposal a BM^d $W = (W^1, W^2, \dots, W^d)$ independent of N and if we define

$$\overline{W}_t^i = N_t^i + \int_0^t 1_{[i>\zeta_s]} dW_s^i,$$

then, by Lévy's theorem, $\overline{W}$ is a BM^d. As β_t is an orthogonal matrix, $B = \beta \cdot \overline{W}$ is again a BM^d (See Exercise (3.22) Chap. IV) and $M = (\alpha\phi\beta^t) \cdot B$. □

Remark. If the matrix γ is $dP \otimes dt$-a.e. of full rank then, as in Proposition (3.8), there is no need to resort to an enlargement of the probability space and the Brownian motion B may be constructed on the space initially given. Actually, one can find a predictable process ψ of invertible matrices such that $d\langle M, M\rangle_s = (\psi_s \psi_s^t) ds$ and $B = \psi^{-1} \cdot M$.

(3.10) Exercise. Prove that Proposition (3.2) and Theorem (3.4) are still true if B is replaced by a continuous Gaussian martingale.

[Hint: See Exercise (1.14).]

(3.11) Exercise. 1°) Prove that the first Wiener chaos K_1 is equal to the Gaussian space generated by B, i.e. the smallest Gaussian space containing the variables B_t, $t \geq 0$.

2°) Prove that a $\mathscr{F}_\infty^B$-measurable r.v. is in K_1 iff the system $(Z, B_t, t \geq 0)$ is Gaussian. As a result there are plenty of $\mathscr{F}_\infty^B$-measurable Gaussian r.v. which are not in K_1.

* **(3.12) Exercise.** Let B be a BM$^1(0)$ and H an $(\mathscr{F}_t^B)$-progressive process such that:

i) $\int_0^t H_s^2 ds < \infty$ a.s. for every t;
ii) $P[\lambda\{s : H_s = 0\} = 0] = 1$ where λ is the Lebesgue measure on $\mathbb{R}_+$. If $\operatorname{sgn} x = 1$ for $x > 0$ and $\operatorname{sgn} x = -1$ for $x \leq 0$, prove that

$$\mathscr{F}_t^{(\operatorname{sgn} H)\cdot B} \subset \mathscr{F}_t^{H\cdot B} \subset \mathscr{F}_t^B$$

for every t. Observe that $(\operatorname{sgn} H) \cdot B$ is itself a Brownian motion.

[Hint: $\operatorname{sgn} H_s = H_s/|H_s|$; replace $|H|$ by a suitable $\mathscr{F}_t^{H\cdot B}$-adapted process.]

\# **(3.13) Exercise.** 1°) Let $t > 0$ and let B be the standard linear BM; if $f \in L^2(\mathbb{R}, g_t(x)dx)$ prove that

$$f(B_t) = P_t f(0) + \int_0^t (P_{t-s} f)'(B_s) dB_s.$$

[Hint: Recall that $\frac{\partial g}{\partial t} + \frac{1}{2}\frac{\partial^2 g}{\partial x^2} = 0$ and look at Exercise (1.11) in Chap. III and Exercise (1.20) in Chap. VII.]

2°) Let B' be an independent copy of B; for $|\rho| < 1$ the process $C_t = \rho B_t + \sqrt{1-\rho^2} B_t'$ is a standard BM^1. Prove that, if $f \in L^2(\mathbb{R}, g_1(x)dx)$, the process

$$E\left[(P_{1-s} f)'(B_s) \mid \mathscr{F}_s^C\right]$$

has a measurable version Z and that if $\int f(x)g_1(x)dx = 0$

$$E[f(B_1) \mid C_1] = \rho \int_0^1 Z_s^{(p)} dC_s$$

where $Z^{(p)}$ is the $\mathscr{F}^C$-predictable projection of Z (see Exercise (1.13)). Instead of $Z^{(p)}$, one can also use a suitable projection in $L^2(ds\, dP)$.

3°) **(Gebelein's Inequality)** If (X, Y) is a centered two-dimensional Gaussian r.v. such that $E[X^2] = E[Y^2] = 1$ and $E[XY] = \rho$, then for any f as in 2°),

$$E\left[(E[f(X) \mid Y])^2\right] \leq \rho^2 E\left[f^2(X)\right].$$

The reader will look at Exercise (3.19) for related results.

** **(3.14) Exercise.** Let B and B' be two independent standard BM[1] and set for s, $t \geq 0$,

$$\mathscr{F}_{s,t} = \sigma\left(B_u, u \leq s;\, B'_v, v \leq t\right).$$

1°) Prove that, as f and g range through $L^2(\mathbb{R}_+)$, the r.v.'s

$$\mathscr{E}\left(\int_0^{\cdot} f(s)dB_s\right)_\infty \mathscr{E}\left(\int_0^{\cdot} g(s)dB'_s\right)_\infty$$

are total in $L^2\left(\mathscr{F}_{\infty,\infty}\right)$.

2°) Define a stochastic integral $\int_0^{\cdot}\int_0^{\cdot} H(s,t)dB_s dB'_t$ of suitably measurable doubly indexed processes H, such that any r.v. X in $L^2\left(\mathscr{F}_{\infty,\infty}\right)$ may be uniquely written

$$X = E[X] + \int_0^\infty h(s)dB_s + \int_0^\infty h'(s)dB'_s + \int_0^\infty \int_0^\infty H(s,t)dB_s dB'_t.$$

where h, resp. h', is predictable w.r.t. the filtration of B, resp. B'.

3°) Let $X_{s,t}$ be a doubly-indexed process, adapted to $\mathscr{F}_{s,t}$ and such that

i) $\sup_{s,t} E\left[X_{s,t}^2\right] < +\infty$;
ii) $E\left[X_{s',t'} \mid \mathscr{F}_{s,t}\right] = X_{s,t}$ a.s. whenever $s \leq s'$ and $t \leq t'$.

Prove that there is a r.v. X in $L^2\left(\mathscr{F}_{\infty,\infty}\right)$ such that $X_{s,t} = E\left[X \mid \mathscr{F}_{s,t}\right]$ a.s. for every pair (s,t) and extend the representation Theorem (3.4) to the present situation.

(3.15) Exercise. 1°) Prove that the family of random variables

$$Z = \prod_{i=1}^n \int_0^\infty e^{-\lambda_i s} f_i(B_s) ds$$

where $\lambda_i \in \mathbb{R}_+$ and the functions f_i are bounded and continuous on $\mathbb{R}$ is total in $L^2(\mathscr{F}_\infty)$.

[Hint: The measures ε_t are the limit in the narrow topology of probability measures whose densities with respect to the Lebesgue measure are linear combinations of exponentials.]

2°) Prove that Z has a representation as in Proposition (3.2) and derive therefrom another proof of this result.

(3.16) Exercise. Let t be fixed and ϕ be a bounded measurable function on $\mathbb{R}$. Find the explicit representation (3.2) for the r.v. $F = \exp\left(\int_0^t \phi(B_s)ds\right)$.

[Hint: Consider the martingale $E\left[F \mid \mathscr{F}_u\right], u \leq t$.]

(3.17) Exercise (Gaussian chaoses). Let G be a centered Gaussian subspace of $L^2(E, \mathscr{E}, P)$ and $\mathscr{G}$ the sub-σ-algebra generated by G. Call K_n the closure in L^2 of the vector space generated by the set

$$\{h_n(X) : X \in G, \|X\|_2 = 1\}$$

where h_n is the n-th Hermite polynomial defined in Sect. 3 Chap. IV.

1°) The map $X \to \exp(X)$ is a continuous map from G into $L^2(E, \mathscr{G}, P)$ and its image is total in $L^2(E, \mathscr{G}, P)$.

2°) If X and Y are in G and $\|X\|_2 = \|Y\|_2 = 1$, prove that

$$\begin{aligned} E\left[h_m(X)h_n(Y)\right] &= n!E[XY]^n \quad \text{if } m = n, \\ &= 0 \quad \text{otherwise.} \end{aligned}$$

3°) Prove that

$$L^2(E, \mathscr{G}, P) = \bigoplus_0^\infty K_n.$$

4°) If H is the Gaussian space of Brownian motion (see Exercise (3.11)), prove that the decomposition of 3°) is one and the same as the decomposition of Theorem (3.7).

\# **(3.18) Exercise (Another proof of Knight's theorem).** 1°) Let M be a continuous $(\mathscr{F}_t)$-local martingale such that $\langle M, M\rangle_\infty = \infty$ and B its DDS Brownian motion. Prove that, for any r.v. $H \in L^2\left(\mathscr{F}_\infty^B\right)$, there is a $(\mathscr{F}_t)$-predictable process K such that $E\left[\int_0^\infty K_s^2 d\langle M, M\rangle_s\right] < \infty$ and $H = E[H] + \int_0^\infty K_s dM_s$. As $\mathscr{F}_\infty^M$ may be strictly larger than $\mathscr{F}_\infty^B$, this does not entail that M has the PRP of the following section.

[Hint: Apply (3.2) to B and use time changes.]

2°) In the situation of Theorem (1.9), prove that for $H \in L^2(\mathscr{F}_\infty^B)$ there exist $(\mathscr{F}_t)$-predictable processes K^i such that

$$H = E[H] + \sum_i \int_0^\infty K_s^i dM_s^i.$$

This can be proved as in 1°) by assuming Theorem (3.4) or by induction from 1°).

3°) Derive Theorem (1.9), i.e. the independence of the B^i's from the above representation property.

*# **(3.19) Exercise (Hypercontractivity).** Let μ be the reduced Gaussian measure of density $(2\pi)^{-1/2}\exp(-x^2/2)$ on $\mathbb{R}$. Let ρ be a real number such that $|\rho| \leq 1$ and set

$$Uf(y) = \int f\left(\rho y + \sqrt{1-\rho^2}x\right)\mu(dx).$$

By making $\rho = e^{-t/2}$, U is the operator given by the transition probability of the Ornstein-Uhlenbeck process.

1°) Prove that for any $p \in [1, \infty[$ the positive operator U maps $L^p(\mu)$ into itself and that its norm is equal to 1. In other words, U is a positive contraction on $L^p(\mu)$.

This exercise aims at proving a stronger result, namely, if $1 < p \leq q \leq q_0 = (p-1)\rho^{-2} + 1$, then U is a positive contraction from $L^p(\mu)$ into $L^q(\mu)$. We call q' the conjugate number of q i.e. $1/q + 1/q' = 1$.

2°) Let (Y_t, Y_t') be a planar standard BM and set $X_t = \rho Y_t + \sqrt{1-\rho^2} Y_t'$. Prove that X is a standard linear BM and that for $f \geq 0$,

$$Uf(Y_1) = E\left[f(X_1) \mid Y_1\right].$$

Observe that (X_1, Y_1) is a pair of reduced Gaussian random variables with correlation coefficient ρ and that U could have been defined using only that.

3°) Let f and g be two bounded Borel functions such that $f \geq \varepsilon$ and $g \geq \varepsilon$ for some $\varepsilon > 0$. Set $M = f^p(X_1)$, $N = g^{q'}(Y_1)$, $a = 1/p$, $b = 1/q'$. Using the representation result (3.2) on $[0, 1]$ instead of $[0, \infty]$, prove that

$$E\left[M^a N^b\right] \leq \|f\|_p \|g\|_{q'}.$$

Since also $E\left[M^a N^b\right] = \int gUf\, d\mu$ derive therefrom the result stated in 1°).

4°) By considering the functions $f(x) = \exp(zx)$ where $z \in \mathbb{R}$, prove that for $q > q_0$, the operator U is unbounded from $L^p(\mu)$ into $L^q(\mu)$.

5°) **(Integrability of Wiener chaoses)** Retaining the notation of Exercise (3.17) prove that $Uh_n = \rho^n h_n$ and derive from the hypercontractivity property of U that if $Z \in K_n$, then for every $q > 2$,

$$\|Z\|_q \leq (q-1)^{n/2} \|Z\|_2.$$

Conclude that there is a constant $\alpha^* > 0$ such that

$$\begin{aligned} E\left[\exp\left(\alpha Z^{2/n}\right)\right] &< \infty \quad \text{if } \alpha < \alpha^*, \\ E\left[\exp\left(\alpha Z^{2/n}\right)\right] &= \infty \quad \text{if } \alpha > \alpha^*. \end{aligned}$$

(3.20) Exercise. Let $(\Omega, \mathscr{G}_t, P)$ be a filtered probability space, B a $\mathscr{G}_t$-BM vanishing at zero and $(\mathscr{F}_t) = (\mathscr{F}_t^B)$.

1°) If M is a $(\mathscr{G}_t)$-martingale bounded in L^2, prove that $X_t \equiv E\left[M_t \mid \mathscr{F}_t\right]$ is a $(\mathscr{F}_t)$-martingale bounded in L^2 which possesses a continuous version. It is the version that we consider henceforth.

2°) Prove that $\langle M, B\rangle_t = \int_0^t a_s ds$ where a is $(\mathscr{G}_t)$-adapted. Prove that the process $t \to E\left[a_t \mid \mathscr{F}_t\right]$ has a $(\mathscr{F}_t)$-progressively measurable version H and that

$$X_t = E[M_0] + \int_0^t H_s dB_s.$$

[Hint: For the last result, compute $E[M_t Y_t]$ in two different ways where Y ranges through the square-integrable $(\mathscr{F}_t)$-martingales.]

3°) If U is $(\mathscr{G}_t)$-progressively measurable and bounded

$$E\left[\int_0^t U_s dB_s \mid \mathscr{F}_t\right] = \int_0^t E\left[U_s \mid \mathscr{F}_s\right] dB_s.$$

If B' is another $(\mathscr{G}_t)$-BM independent of B then

$$E\left[\int_0^t U_s dB'_s \mid \mathscr{F}_t\right] = 0.$$

(3.21) Exercise. 1°) If B is a $\mathrm{BM}^1(0)$ and we set $\mathscr{F}_t = \mathscr{F}_t^B$, prove that if $\mathscr{G}$ is a non-trivial sub-σ-field of $\mathscr{F}_\infty$, there does not exist any $(\mathscr{F}_t)$-BM which is also a $(\mathscr{F}_t \vee \mathscr{G})$-BM.

2°) In contrast, prove that there exist a non-trivial sub-σ-field $\mathscr{G}$ of $\mathscr{F}_\infty^B$ and a $(\mathscr{F}_t)$-martingale which is also a $(\mathscr{F}_t \vee \mathscr{G})$-martingale.

[Hint: Let $\mathscr{G} = \mathscr{F}_\infty^\beta$ where β is the DDS Brownian motion of $1_{(B_\cdot \leq 0)} \cdot B$ and take $M = 1_{(B_\cdot \geq 0)} \cdot B$.]

* **(3.22) Exercise (The Goswami-Rao Brownian filtration).** Let B be the standard BM and define $\mathscr{H}_t$ as the σ-field generated by the variables $\Phi(B_s, s \leq t)$ where Φ is a measurable function on $C([0,t], \mathbb{R})$ such that $\Phi(B_s, s \leq t) = \Phi(-B_s, s \leq t)$.

1°) Prove that the inclusions $\mathscr{F}_t^{|B|} \subset \mathscr{H}_t \subset \mathscr{F}_t^B$ are strict.

[Hint: For $s < t$, the r.v. sgn $(B_s B_t)$ is $\mathscr{H}_t$-measurable but not $\mathscr{F}_t^{|B|}$-measurable.]

2°) Let $Y \in L^2(\mathscr{F}_t^B)$ with $t \leq \infty$. Prove that in the notation of Theorem (3.7),

$$E\left[Y \mid \mathscr{H}_t\right] = E[Y] + \sum_{p=1}^{\infty} J_{2p}(f^{2p})(t)$$

where the notation $J_n(f^n)(t)$ indicates that the first integral in $J_n(f^n)$ is taken only up to time t.

Deduce therefrom that every $(\mathscr{H}_t)$-martingale is a $(\mathscr{F}_t^B)$-martingale.

3°) Prove that consequently every $(\mathscr{H}_t)$-martingale M may be written

$$M_t = c + \int_0^t m(s) d\beta_s$$

where $\beta_t = \int_0^t \mathrm{sgn}\,(B_s) dB_s$ is a $(\mathscr{H}_t)$-BM and m is $(\mathscr{H}_t)$-predictable.

The filtration $\mathscr{F}^{|B|}$ is in fact generated by β (Corollary (2.2) Chap. VI); that $(\mathscr{H}_t)$ is the natural filtration of a BM is the content of the following question.

4°) Let $(t_n)_{n\in\mathbb{Z}}$ be an increasing sequence of positive reals, such that $t_n \underset{n\to-\infty}{\longrightarrow} 0$, and $t_n \underset{n\to+\infty}{\longrightarrow} +\infty$. Prove that $(\mathscr{H}_t)$ is the natural filtration of the BM $\gamma_t = \int_0^t \mu_s dB_s$, where $\mu_s = \mathrm{sgn}\,\left(B_{t_n} - B_{t_{n-1}}\right)$, for $s \in]t_n, t_{n+1}]$.

§4. Integral Representations

(4.1) Definition. *The cont. loc. mart. X has the* predictable representation property *(abbr. PRP) if, for any* $(\mathscr{F}_t^X)$*-local martingale M, there is an* $(\mathscr{F}_t^X)$*-predictable process H such that*

$$M_t = M_0 + \int_0^t H_s dX_s.$$

In the last section, we proved that the Brownian motion has the PRP and, in this section, we investigate the class of cont. loc. martingales which have the PRP.

We need the following lemma which is closely related to Exercise (5.11) in Chap. IV.

(4.2) Lemma. *If X is any cont. loc. mart., every* $(\mathscr{F}_t^X)$*-continuous local martingale M vanishing at 0 may be uniquely written*

$$M = H \cdot X + L$$

where H is predictable and $\langle X, L\rangle = 0$.

Proof. The uniqueness follows from the usual argument.

To prove the existence of the decomposition, let us observe that there is a sequence of stopping times increasing to infinity and reducing both M and X. Let T be one of them. In the Hilbert space H_0^2, the subspace $G = \{H \cdot X^T;\ H \in \mathscr{L}^2_{\mathscr{P}}(X^T)\}$ is easily seen to be closed; thus, we can write uniquely

$$M^T = \bar{H} \cdot X^T + \bar{L}$$

where $\bar{L} \in G^\perp$. For any bounded stopping time S, we have

$$E\left[X_S^T \bar{L}_S\right] = E\left[X_S^T E\left[\bar{L}_\infty \mid \mathscr{F}_S\right]\right] = E\left[X_S^T \bar{L}_\infty\right] = 0$$

since $X^{T\wedge S} \in G$. It follows from Proposition (3.5) Chap. II that $X^T\bar{L}$ is a martingale, hence $\langle X^T, \bar{L}\rangle = \langle X, \bar{L}\rangle^T = 0$.

Because of the uniqueness, the processes $\bar{H}$ and $\bar{L}$ extend to processes H and L which fulfill the requirements of the statement. □

From here on, we will work with the canonical space $\mathbf{W} = C(\mathbb{R}_+, \mathbb{R})$. The coordinate process will be designated by X and we put $\mathscr{F}_t^0 = \sigma(X_s, s \le t)$. Let $\mathscr{H}$ be the set of probability measures on $\mathbf{W}$ such that X is a local martingale (evidently continuous). If $P \in \mathscr{H}$, $(\mathscr{F}_t^P)$ is the smallest right-continuous filtration complete for P and such that $\mathscr{F}_t^0 \subset \mathscr{F}_t^P$. The PRP now appears as a property of P: any $(\mathscr{F}_t^P)$-local martingale M may be written $M = H \cdot X$ where H is $(\mathscr{F}_t^P)$-predictable and the stochastic integration is taken with respect to P.

We will further designate by $\mathscr{K}$ the subset of $\mathscr{H}$ of those probability measures for which X is a martingale. The sets $\mathscr{K}$ and $\mathscr{H}$ are convex sets (see Exercise (1.37) in Chap. IV). We recall the

(4.3) Definition. *A probability measure P of $\mathscr{H}$ (resp. $\mathscr{H}$) is called* extremal *if whenever $P = \alpha P_1 + (1-\alpha)P_2$ with $0 < \alpha < 1$ and $P_1, P_2 \in \mathscr{H}$ (resp. $\mathscr{H}$) then $P = P_1 = P_2$.*

We will now study the extremal probability measures in order to relate extremality to the PRP. We will need the following measure theoretical result.

(4.4) Theorem (Douglas). *Let $(\Omega, \mathscr{F})$ be a measurable space, $\mathscr{L}$ a set of real-valued $\mathscr{F}$-measurable functions and $\mathscr{L}^*$ the vector-space generated by 1 and $\mathscr{L}$. If $\mathscr{H}_{\mathscr{L}}$ is the set of probability measures μ on $(\Omega, \mathscr{F})$ such that $\mathscr{L} \subset L^1(\mu)$ and $\int f \, d\mu = 0$ for every $f \in \mathscr{L}$, then $\mathscr{H}_{\mathscr{L}}$ is convex and μ is an extremal point of $\mathscr{H}_{\mathscr{L}}$ if and only if $\mathscr{L}^*$ is dense in $L^1(\mu)$.*

Proof. That $\mathscr{H}_{\mathscr{L}}$ is convex is clear.

Suppose that $\mathscr{L}^*$ is dense in $L^1(\mu)$ and that $\mu = \alpha\nu_1 + (1-\alpha)\nu_2$ with $0 < \alpha < 1$ and ν_1, ν_2 in $\mathscr{H}_{\mathscr{L}}$. Since $\nu_i = h_i\mu$ for a bounded function h_i, the space $\mathscr{L}^*$ is dense in $L^1(\nu_i)$ for $i = 1, 2$. Since clearly ν_1 and ν_2 agree on $\mathscr{L}^*$, it follows that $\nu_1 = \nu_2$.

Conversely, suppose that $\mathscr{L}^*$ is not dense in $L^1(\mu)$. Then, by the Hahn-Banach theorem there is a non-zero bounded function h such that $\int hf \, d\mu = 0$ for every $f \in \mathscr{L}^*$ and we may assume that $\|h\|_\infty \le 1/2$. The measures $\nu^\pm = (1 \pm h)\mu$ are obviously in $\mathscr{H}_{\mathscr{L}}$ and μ being equal to $(\nu^+ + \nu^-)/2$ is not extremal. □

If we now consider the space $(\mathbf{W}, \mathscr{F}_\infty^0)$ and choose as set $\mathscr{L}$ the set of r.v.'s $1_A(X_t - X_s)$ where $0 \le s < t$ and A ranges through $\mathscr{F}_s^0$, the set $\mathscr{H}_{\mathscr{L}}$ of Theorem (4.4) coincides with the set $\mathscr{H}$ of probability measures for which X is a martingale. We use Theorem (4.4) to prove the

(4.5) Proposition. *If P is extremal in $\mathscr{H}$, then any $(\mathscr{F}_t^P)$-local martingale has a continuous version.*

Proof. Plainly, it is enough to prove that for any r.v. $Y \in L^1(P)$, the cadlag martingale $E_P[Y \mid \mathscr{F}_t^P]$ has in fact a continuous version. Now, this is easily seen to be true whenever Y is in $\mathscr{L}^*$. If $Y \in L^1(P)$, there is, thanks to the preceding result, a sequence (Y_n) in $\mathscr{L}^*$ converging to Y in $L^1(P)$. By Theorem (1.7) in Chap. II, for every $\varepsilon > 0$, every t and every n,

$$P\left[\sup_{s \le t} \left| E_P\left[Y_n \mid \mathscr{F}_s^P\right] - E_P\left[Y \mid \mathscr{F}_s^P\right]\right| \ge \varepsilon\right] \le \varepsilon^{-1} E\left[|Y_n - Y|\right].$$

By the same reasoning as in Proposition (1.22) Chap. IV, the result follows. □

(4.6) Theorem. *The probability measure P is extremal in $\mathscr{H}$ if and only if P has the PRP and $\mathscr{F}_0^P$ is P-a.s. trivial.*

Proof. If P is extremal, then $\mathscr{F}_0^P$ is clearly trivial. Furthermore, if the PRP did not hold, there would, by Lemma (4.2) and the preceding result, exist a non zero continuous local martingale L such that $\langle X, L\rangle = 0$. By stopping, since

$\langle X, L^T\rangle = \langle X, L\rangle^T$, we may assume that L is bounded by a constant k; the probability measures $P_1 = (1 + (L_\infty/2k))\, P$ and $P_2 = (1 - (L_\infty/2k))\, P$ are both in $\mathscr{K}$ as may be seen without difficulty, and $P = (P_1 + P_2)/2$ which contradicts the extremality of P.

Conversely, assume that P has the PRP, that $\mathscr{F}_0^P$ is P-a.s. trivial, and that $P = \alpha P_1 + (1-\alpha)P_2$ with $0 < \alpha < 1$ and $P_i \in \mathscr{K}$. The P-martingale $\frac{dP_1}{dP}\big|_{\mathscr{F}_t}$ has a continuous version L since P has the PRP and XL is also a continuous P-martingale, hence $\langle X, L\rangle = 0$. But since P has the PRP, $L_t = L_0 + \int_0^t H_s dX_s$, hence $\langle X, L\rangle_t = \int_0^t H_s d\langle X, X\rangle_s$; it follows that P-a.s. $H_s = 0$, $d\langle X, X\rangle_s$-a.e., hence $\int_0^t H_s dX_s = 0$ a.s. and L is constant and equal to L_0. Since $\mathscr{F}_0^P$ is P-a.s. trivial, L is equal to 1, hence $P = P_1$ and P is extremal.

Remark. By the results in Sect. 3, the Wiener measure W is obviously extremal, but this can also be proved directly. Indeed, let $Q \in \mathscr{H}$ be such that $Q \ll W$. By the definition of $\langle X, X\rangle$ and the fact that convergence in probability for W implies convergence in probability for Q, $\langle X, X\rangle_t = t$ under Q. By P. Lévy's characterization theorem, X is a BM under Q, hence $Q = W$ which proves that W is extremal. Together with the results of this section, this argument gives another proof of Theorem (3.3).

We will now extend Theorem (4.6) from $\mathscr{K}$ to $\mathscr{H}$. The proof is merely technical and will only be outlined, the details being left to the reader as exercises on measure theory.

(4.7) Theorem. *The probability measure P is extremal in $\mathscr{H}$ if and only if it has the PRP and $\mathscr{F}_0^P$ is P-a.s. trivial.*

Proof. The second part of the proof of Theorem (4.6) works just as well for $\mathscr{H}$ as for $\mathscr{K}$; thus, we need only prove that if P is extremal in $\mathscr{H}$, then $\mathscr{F}_0^P$ is a.s. trivial, which is clear, and that P has the PRP.

Set $T_n = \inf\{t : |X_t| > n\}$; the idea is to prove that if P is extremal in $\mathscr{H}$, then the laws of X^{T_n} under P are extremal in $\mathscr{K}$ for each n and hence the PRP holds up to time T_n, hence for every t.

Let n be fixed; the stopping time T_n is a $(\mathscr{F}_t^0)$-stopping time and the σ-algebra $\mathscr{F}_{T_n}^0$ is countably generated. As a result, there is a regular conditional distribution $Q(\omega, \cdot)$ of P with respect to $\mathscr{F}_{T_n}^0$. One can choose a version of Q such that for every ω, the process $X_t - X_{T_n \wedge t}$ is a $Q(\omega, \cdot)$-local martingale.

Suppose now that the law of X^{T_n} under P, say P^{T_n}, is not extremal. There would exist two different probabilities π_1 and π_2 such that $P^{T_n} = \alpha\pi_1 + (1-\alpha)\pi_2$ for some $\alpha \in]0, 1[$ and these measures may be viewed as probability measures on $\mathscr{F}_{T_n}^0$. Because $\mathscr{F}_{T_n}^0 \vee \theta_{T_n}^{-1}(\mathscr{F}_\infty^0) = \mathscr{F}_\infty^0$, it can be proved that there are two probability measures P_i on $\mathscr{F}_\infty^0$ which are uniquely defined by setting, for two r.v.'s H and K respectively $\mathscr{F}_{T_n}^0$ and $\theta_{T_n}^{-1}(\mathscr{F}_\infty^0)$-measurable,

$$\int HK\, dP_i = \int \pi_i(d\omega) H(\omega) \int Q(\omega, d\omega') K(\omega').$$

Then, under P_i, the canonical process X is a continuous local martingale and $P = \alpha P_1 + (1-\alpha)P_2$ which is a contradiction. This completes the proof.

The equivalent properties of Theorem (4.7) are important in several contexts, but, as a rule-in contrast with the discrete time case (see the Notes and Comments) – it is not an easy task to decide whether a particular local martingale is extremal or not and there is no known characterization – other than the one presented in Theorem (4.7) – of the set of extremal local martingales. The rest of this section will be devoted to a few remarks designed to cope with this problem. As may be surmised from Exercise (3.18), the PRP for X is related to properties of the DDS Brownian motion of X.

(4.8) Definition. *A cont. loc. martingale X adapted to a filtration $(\mathscr{F}_t)$ is said to have the $(\mathscr{F}_t)$-PRP if any $(\mathscr{F}_t)$-local martingale vanishing at zero is equal to $H \cdot X$ for a suitable $(\mathscr{F}_t)$-predictable process H.*

Although $\mathscr{F}_t^X \subset \mathscr{F}_t$, the $(\mathscr{F}_t)$-PRP does not entail the PRP as the reader will easily realize (see Exercise (4.22)). Moreover, if we define $(\mathscr{F}_t)$-extremality as extremality in the set of probability measures Q such that (X_t) is an $(\mathscr{F}_t, Q)$-local martingale, the reader will have no difficulty in extending Theorem (4.7) to this situation.

In what follows, X is a P-continuous local martingale; we suppose that $\langle X, X\rangle_\infty = \infty$ and call B the DDS Brownian motion of X.

(4.9) Theorem. *The following two properties are equivalent*

i) X has the PRP;
ii) B has the $\left(\mathscr{F}_{T_t}^X\right)$-PRP.

Proof. This is left to the reader as an exercise on time-changes.

We turn to another useful way of relating the PRP of X to its DDS Brownian motion.

(4.10) Definition. *A continuous local martingale X such that $\langle X, X\rangle_\infty = \infty$ is said to be* pure *if, calling B its DDS Brownian motion, we have*

$$\mathscr{F}_\infty^X = \mathscr{F}_\infty^B .$$

By Exercise (1.19), X is pure if and only if one of the following equivalent conditions is satisfied

i) the stopping time T_t is $\mathscr{F}_\infty^B$-measurable for every t;
ii) $\langle X, X\rangle_t$ is $\mathscr{F}_\infty^B$-measurable for every t.

One will actually find in Exercise (4.16) still more precise conditions which are equivalent to purity. Proposition (1.11) gives a sufficient condition for X to be pure and it was shown in Theorem (2.11) that if ρ_t is the modulus of BM^2 started at $a \neq 0$ then $\log \rho_t$ is a pure local martingale.

Finally, the reader will show as an exercise, that the pure martingales are those for which the map which sends paths of the martingale into paths of the corresponding DDS Brownian motion is one-to-one.

The introduction of this notion is warranted by the following result, which can also be derived from Exercise (3.18). A local martingale is said to be *extremal* if its law is extremal in $\mathscr{H}$.

(4.11) Proposition. *A pure local martingale is extremal.*

Proof. Let P be a probability measure under which the canonical process X is a pure loc. mart. If $Q \in \mathscr{H}$ and $Q \ll P$, then $\langle X, X\rangle_t$ computed for P is a version of $\langle X, X\rangle_t$ computed for Q and consequently the DDS Brownian motion of X for P, say β, is a version of the DDS Brownian motion of X for Q. As a result, P and Q agree on $\sigma(\beta_s, s \geq 0)$ hence on the completion of this σ-algebra with respect to P. But, since P is pure, this contains $\sigma(X_s, s \geq 0)$ and we get $P = Q$. □

The converse is not true (see Exercise (4.16)) but, as purity is sometimes easier to prove than extremality, this result leads to examples of extremal martingales (see Exercise (3.11) in Chap. IX).

(4.12) Exercise. With the notation of this section a probability measure $P \in \mathscr{H}$ is said to be *standard* if there is no other probability measure Q in $\mathscr{H}$ equivalent to P. Prove that P is standard if and only if P has the PRP, and $\mathscr{F}_0^P$ is P-a.s. trivial.

\# **(4.13) Exercise.** Let (B^1, B^2) be a two-dimensional BM. Prove that $X_t = \int_0^1 B_s^1 dB_s^2$ does not have the PRP.

[Hint: $(B_t^1)^2 - t$ is a $(\mathscr{F}_t^X)$-martingale which cannot be written as a stochastic integral with respect to X.]

* **(4.14) Exercise.** Let X be a cont. loc. mart. with respect to a filtration $(\mathscr{F}_t)$.

1°) Let Δ be the space of real-valued, bounded functions with compact support in $\mathbb{R}_+$. Prove that if, for every t, the set of r.v.'s

$$\mathscr{E}_t^f = \mathscr{E}\left(\int_0^{\cdot} f(s)dX_s\right)_t, \quad f \in \Delta,$$

is total in $L^2(\mathscr{F}_t, P)$, then $\mathscr{F}_0^P$ is P-a.s. trivial and X has the $(\mathscr{F}_t)$-PRP.

2°) Prove the same result with the set of r.v.'s

$$H_n\left(\int_0^t f(s)dX_s, \int_0^t f^2(s)d\langle X, X\rangle_s\right), \quad f \in \Delta, n \in \mathbb{N},$$

where H_n is the Hermite polynomial defined in Sect. 3 Chap. IV.

\# **(4.15) Exercise.** 1°) If B is a standard linear BM, $(\mathscr{F}_t)$ is the Brownian filtration and H is a $(\mathscr{F}_t)$-predictable process a.s. strictly positive with the possible exception of a set of Lebesgue zero measure (depending on ω), then the martingale

$$M_t = \int_0^t H_s dB_s$$

has the PRP.

[Hint: Use ideas of Proposition (3.8) to show that $\mathscr{F}_\infty^M = \mathscr{F}_\infty^B$.]

2°) In particular, the martingales $M_t^n = \int_0^t B_s^n dB_s$, $n \in \mathbb{N}$, are extremal. For n odd, these martingales are actually pure as will be proved in Exercise (3.11) Chap. IX.

3°) Let T be a non constant square-integrable $(\mathscr{F}_t)$-stopping time; prove that $Y_t = B_t - B_t^T$ is not extremal.

[Hint: T is a $\mathscr{F}^Y$-stopping time which cannot be expressed as a constant plus a stochastic integral with respect to Y.]

*# **(4.16) Exercise. (An example of an extremal local martingale which is not pure).** 1°) Let $(\mathscr{F}_t)$ and $(\mathscr{G}_t)$ be two filtrations such that $\mathscr{F}_t \subset \mathscr{G}_t$ for every t. Prove that the following two conditions are equivalent:

i) every $(\mathscr{F}_t)$-martingale is a $(\mathscr{G}_t)$-martingale;
ii) for every t, the σ-algebras $\mathscr{F}_\infty$ and $\mathscr{G}_t$ are conditionally independent with respect to $\mathscr{F}_t$.

If these conditions are in force, prove that $\mathscr{F}_t = \mathscr{G}_t \cap \mathscr{F}_\infty$.

2°) Let M be a continuous $(\mathscr{F}_t)$-loc. mart. having the $(\mathscr{F}_t)$-PRP. If $(\mathscr{G}_t)$ is a filtration such that $\mathscr{F}_t \subset \mathscr{G}_t \subset \mathscr{F}_\infty$ for every t and M is a $(\mathscr{G}_t)$-local martingale, then $(\mathscr{F}_t) = (\mathscr{G}_t)$.

3°) Let M be a cont. loc. mart. with $\langle M, M\rangle_\infty = \infty$ and B its DDS Brownian motion. Prove that M is pure if and only if $\langle M, M\rangle_t$ is, for each t, an $(\mathscr{F}_t^B)$-stopping time.

[Hint: Prove first that, if M is pure, then $\mathscr{F}_t^B = \mathscr{F}_{T_t}^M$.]

Prove further that if M is pure, $\langle M, M\rangle_T$ is a $(\mathscr{F}_t^B)$-stopping time for every $(\mathscr{F}_t^M)$-stopping time T (this result has no bearing on the sequel).

4°) Prove that M is pure if and only if $\mathscr{F}_t^M = \mathscr{F}_{\langle M,M\rangle_t}^B$ for every t.

5°) From now on, β is a standard BM and we set $\tilde{\beta}_t = \int_0^t \operatorname{sgn}(\beta_s) d\beta_s$. In Sect. 2 of Chap. VI it is proved that $\tilde{\beta}$ is a BM and that $\mathscr{F}_t^{\tilde{\beta}} = \mathscr{F}_t^{|\beta|}$ for every t. Prove that $\tilde{\beta}$ has the $(\mathscr{F}_t^\beta)$-PRP.

6°) Set $T_t = \int_0^t (2 + (\beta_s/(1 + |\beta_s|)))\, ds$ and observe that $\mathscr{F}_t^T = \mathscr{F}_t^\beta$ for every t. Let A be the inverse of T and define the $(\mathscr{F}_{A_t}^\beta)$-loc. mart. $M_t = \tilde{\beta}_{A_t}$; prove that M has the PRP, but that M is not pure.

[Hint: Prove that $\mathscr{F}_{T_t}^M = \mathscr{F}_t^\beta$ and use Theorem (4.9).]

(4.17) Exercise. Prove that the Gaussian martingales, in particular αB for any real α, are pure martingales.

(4.18) Exercise. If M is a pure cont. loc. mart. and C an $(\mathscr{F}_t^M)$-time-change such that M is C-continuous, then $\hat{M} = M_C$ is pure.

(4.19) Exercise. Let M be a cont. loc. mart. with the $(\mathscr{F}_t)$-PRP.

1°) Prove that $S = \inf\{t : M_{t+u} = M_t \text{ for every } u > 0\}$ is a $(\mathscr{F}_t)$-stopping time.

2°) If $H(\omega)$ is the largest open subset of $\mathbb{R}_+$ such that $M_t(\omega)$ is constant on each of its connected components, prove that there are two sequences (S_n), (T_n) of $(\mathscr{F}_t)$-stopping times such that $H = \bigcup_n]S_n, T_n[$ and the sets $]S_n, T_n[$ are exactly the connected components of H.

[Hint: For $\varepsilon > 0$, let $]S_n^\varepsilon, T_n^\varepsilon[$ be the n-th interval of H with length $> \varepsilon$. Observe that $S_n^\varepsilon + \varepsilon$ is a stopping time and prove that S_n^ε is a $(\mathscr{F}_t)$-stopping time.]

(4.20) Exercise. 1°) If M is an extremal cont. loc. mart. and if $(\mathscr{F}_t^M)$ is the filtration of a Brownian motion, then $d\langle M, M\rangle_s$ is a.s. equivalent to the Lebesgue measure.

** 2°) Let F be the Cantor middle-fourths set. Set

$$M_t = \int_0^t 1_{F^c}(B_s)dB_s$$

where B is a $\mathrm{BM}^1(0)$. Prove that $(\mathscr{F}_t^M) = (\mathscr{F}_t^B)$ and derive that M has not the PRP.

(4.21) Exercise. Let β be the standard BM^1 and set $M_t = \int_0^t 1_{(\beta_s > 0)} d\beta_s$. Let $\sigma = 1_{(\beta_1 < 0)} + \infty 1_{(\beta_1 \geq 0)}$ and

$$\tau = \sigma + \inf\left\{\varepsilon : \int_1^{1+\varepsilon} 1_{(\beta_s > 0)} ds > 0\right\}.$$

By considering the r.v. $\exp(-\tau)$, prove that M does not have the PRP.

* **(4.22) Exercise.** 1°) Let $(\mathscr{F}_t)$ and $(\mathscr{G}_t)$ be two filtrations such that $\mathscr{G}_t \subset \mathscr{F}_t$ for every t. If M is a continuous $(\mathscr{F}_t)$-local martingale adapted to $(\mathscr{G}_t)$ and has the $(\mathscr{F}_t)$-PRP prove that the following three conditions are equivalent

i) M has the $(\mathscr{G}_t)$-PRP;
ii) every $(\mathscr{G}_t)$-martingale is a $(\mathscr{F}_t)$-martingale;
iii) every $(\mathscr{G}_t)$-martingale is a continuous $(\mathscr{F}_t)$-semimartingale.

2°) Let B be the standard linear BM and $(\mathscr{F}_t)$ be the Brownian filtration. Let t_0 be a strictly positive real and set

$$N_t = \int_0^t \left[1_{(s<t_0)} + \left(2 + \operatorname{sgn}\left(B_{t_0}\right)\right) 1_{(t_0 \leq s)}\right] \operatorname{sgn}(B_s) dB_s.$$

Prove that N has the $(\mathscr{F}_t)$-PRP and has not the PRP.

[Hint: The process $H_t = (\operatorname{sgn}(B_{t_0}))1_{(t_0 \leq t)}$ is a discontinuous $(\mathscr{F}_t^N)$-martingale.]

* **(4.23) Exercise.** Let β be a $(\mathscr{F}_t)$-Brownian motion and suppose that there exists a continuous strictly increasing process A_t, with inverse τ_t, such that $A_t \geq t$ and $\mathscr{F}_t = \mathscr{F}_t^A$. Set $X_t = \beta_{\tau_t}$.

1°) Prove that X is pure if and only if $\mathscr{F}_\infty^B = \mathscr{F}_\infty$.
2°) Prove that X is extremal if and only if β has the $(\mathscr{F}_t)$-PRP.

** **(4.24) Exercise.** 1°) Retain the notation of Exercise (3.29) Chap. IV and prove that the following two conditions are equivalent

i) X has the $(\mathscr{F}_t)$-PRP;
ii) for any Q such that X is a (Q, P)-local martingale there is a constant c such that $Q = cP$.

2°) Use Exercise (3.29) Chap. IV to give another proof of the extremality of the Wiener measure, hence also of Theorem (3.3).

(4.25) Exercise (PRP and independence). 1°) Prove that if two $(\mathscr{F}_t)$-continuous loc. mart. M and N are independent, then they are orthogonal, that is, the product MN is a $(\mathscr{F}_t)$-loc. mart. (see Exercise (2.22) in Chap. IV). Give examples in adequate filtrations of pairs of orthogonal loc. martingales which, nonetheless, are not independent.

2°) Prove that if M and N both have the PRP, then they are orthogonal iff they are independent.

3°) Let B be the standard BM and set

$$M_t = \int_0^t 1_{(B_s>0)} dB_s, \quad N_t = \int_0^t 1_{(B_s<0)} dB_s.$$

Prove that these martingales are not independent and conclude that neither has the PRP. Actually, there exist discontinuous $(\mathscr{F}_t^M)$-martingales.

Notes and Comments

Sect. 1. The technique of time-changes is due to Lebesgue and its application in a stochastic context has a long history which goes back at least to Hunt [1], Volkonski [1] and Itô-McKean [1]. Proposition (1.5) was proved by Kazamaki [1] where the notion of C-continuity is introduced (with a terminology which differs from ours.)

Theorem (1.6) appears in Dubins-Schwarz [1] for martingales with no intervals of constancy and Dambis [1]. The formulation and proof given here borrow from Neveu [2]. Although a nice and powerful result, it says nothing about the distribution of a given continuous martingale M: this hinges on the stochastic dependence between the DDS Brownian motion associated with M and the increasing process of M. Let us mention further that Monroe [1] proves that every semimartingale can be embedded by time change in a Brownian motion, allowing possibly for some extra randomisation. Proposition (1.8) is from Lenglart [1] (see also Doss-Lenglart [1]).

The proof of Knight's theorem (Knight [3]) given in the text is from Cocozza and Yor [1] and the proof in Exercise (3.18) is from Meyer [3]. Knight's theorem has many applications as for instance in Sect. 2 and in Chap. VI where it is used to give a proof of the Arcsine law. Perhaps even more important is its asymptotic version which is discussed and used in Chap. XIII Sect. 2. We refer the reader to Kurtz [1] for an interesting partial converse.

Exercise (1.16) is taken from Lépingle [1], Exercise (1.21) from Barlow [1] and Exercise (1.12) from Bismut [2].

Sect. 2. Conformal martingales were masterfully introduced by Getoor and Sharpe [1] in order to prove that the dual of H^1 is BMO in the martingale setting. The proof which they obtained for continuous martingales uses in particular the fact that if Z is conformal, and $\alpha > 0$, then $|Z|^\alpha$ is a local submartingale. The extension to non-continuous martingales of the duality result was given shortly afterwards by P.A. Meyer. The first results of this section are taken from the paper of Getoor and Sharpe.

The conformal invariance of Brownian motion is a fundamental result of P. Lévy which has many applications to the study of the 2-dimensional Brownian path. The applications we give here are taken from B. Davis [1], McKean [2] and Lyons-McKean [1]. For the interplay between planar BM and complex function theory we refer to the papers of B. Davis ([1] and [3]) with their remarkable proof of Picard's theorems and to the papers by Carne [2] and Atsuji ([1], [2]) on Nevanlinna theory.

For the origin of the skew-product representation we refer to Galmarino [1] and McKean [2]. Extensions may be found in Graversen [2] and Pauwels-Rogers [1]. For some examples involving Bessel processes, see Warren-Yor [1].

The example of Exercise (2.13) was first exhibited by Johnson and Helms [1] and the proof of D'Alembert's theorem in Exercise (2.17) was given by Kono [1]. Exercise (2.19) is from Yor [2]; more general results are found in Ikeda-Manabe [1]. The results and methods of Exercise (2.14) are found in Itô-McKean [1]. Exercise (2.16) is from Williams [4]; they lead to his "pinching" method (see Messulam-Yor [1]).

Exercise (2.24) is from Calais-Génin [1] following previous work by Walsh [1]. Exercise (2.25) originates in McKean [1] and Exercise (2.21) is attributed to H. Sato in Itô-McKean [1]. Exercise (2.21) is from Burkholder [3], but the necessary and sufficient condition of 2°) is already in Spitzer [1].

The subject of polar functions for the planar BM, partially dealt with in Exercise (2.26) was initiated in Graversen [1] from which Exercise (1.20) Chap. I was taken. Graversen has some partial results which have been improved in Le Gall [6]. Despite these results, the following questions remain open

Question 1. What are the polar functions of BM^2?

The result of Exercise (2.26) may be seen as a partial answer to the following

Question 2. Which are the two-dimensional continuous semimartingales for which the one-point sets are polar?

Some partial answers may be found in Bismut [2] and Idrissi-Khamlichi [1]. The answer is not known even for the semimartingales the martingale part of which is a BM^2. The result of Exercise (2.26) is a special case and another is treated in Sznitman-Varadhan [1].

Exercise (2.27) is taken from Yor [3] and Idrissi-Khamlichi [1]. The paper of Yor has several open questions. Here is one of them. With the notation of Exercise (2.27), if a is polar for X_γ the index of X_γ with respect to a is well-defined and it is proved in Yor [3] that its law is supported by the whole set of integers.

Question 3. What is the law of the index of X_γ with respect to a?

This is to be compared to Exercise (2.15) Chap. VIII.

Sect. 3. The first results of this section appeared in Doob [1]. They were one of the first great successes of stochastic integration. They may also be viewed as a consequence of decompositions in chaoses discovered by Wiener [2] in the case of Brownian motion and generalized by Itô to processes with independent increments. The reader may find a more general and abstract version in Neveu [1] (see also Exercise (3.17)).

Theorem (3.5) and the decomposition in chaoses play an important role in Malliavin Calculus as well as in Filtering theory. Those are two major omissions of this book. For the first we refer to Ikeda-Watanabe [2], Nualart [2], and Stroock [3], for the second one to Kallianpur [1]; there is also a short and excellent discussion in Rogers-Williams [1]. A few exercises on Filtering theory are scattered in our book such as Exercise (5.15) Chap. IV and Exercise (3.20) in this section which is taken from Lipster-Shiryaev [1].

Our exposition of Theorem (3.9) follows Jacod [2] and Exercise (3.12) is taken from Lane [1]. Exercise (3.13) is inspired from Chen [1]. Exercise (3.14) is due to Rosen and Yor [1] and Exercise (3.19) to Neveu [3], the last question being taken from Ledoux and Talagrand [1]. The source of Exercise (3.22) is to be found in Goswami and Rao [1]; question 4°) is taken from Attal et al. [1].

Sect. 4. The ideas developed in this section first appeared in Dellacherie [1] in the case of BM and Poisson Process and were expanded in many articles such as Jacod [1], Jacod-Yor [1] and Yor [6] to mention but a few. The method used here to prove Theorem (4.6) is that of Stroock-Yor [1]. Ruiz de Chavez [1] introduces signed measures in order to give another proof (see Exercise (4.24)). The notion of pure martingales was introduced by Dubins and Schwarz [2].

Most of the exercises of this section come from Stroock-Yor ([1] and [2]) and Yor [9] with the exception of Exercise (4.20) taken from Knight [7] and Exercise (4.19) which comes from Stricker [2]. Exercise (4.12) is from Yan and Yoeurp [1]. The results in Exercise (4.25) are completed by Azéma-Rainer [1] who describe all $(\mathscr{F}_t^M)$ martingales.

The unsatisfactory aspect of the results of this section is that they are only of "theoretical" interest as there is no explicit description of extremal martingales (for what can be said in the discrete time case, however, see Dubins-Schwarz [2]). It is even usually difficult to decide whether a particular martingale is extremal or pure or neither. The exercises contain some examples and others may be found in Exercise (4.19) Chap. VI and the exercises of Chap. IX as well as in the papers already quoted. For instance, Knight [7] characterizes the harmonic functions f

in $\mathbb{R}^d$, $d > 1$, such that $f(B_t)$ is pure. However, the subject still offers plenty of open questions, some of which are already found in the two previous editions of this book.

Here, we discuss the state of the matter as it is understood presently (i.e.: in 1998) thanks mainly to the progress initiated by Tsirel'son ([1], [2]), and co-workers.

First, we introduce the following important

Definition. *A filtration* $(\mathscr{F}_t)$ *on the probability space* $(\Omega, \mathscr{F}, P)$ *such that* $\mathscr{F}_0$ *is* P*-a.s. trivial is said to be weakly, resp. strongly, Brownian if there exists a* $(\mathscr{F}_t)$*-* BM^1 β *such that* β *has the* $(\mathscr{F}_t)$*-PRP, resp.* $\mathscr{F}_t = \mathscr{F}_t^{\beta}$.

We will abbreviate these definitions to W.B. and S.B.

Many weakly Brownian filtrations may be obtained through "mild" perturbations of the Brownian filtration. Here are two such examples:

(P1) (Local absolute continuity). If $(\mathscr{F}_t)$ is W.B., and, in the notation of Chapter VIII, $Q \lhd P$, then $(\mathscr{F}_t)$ is also W.B. under Q.

(P2) (Time change). If $(\mathscr{F}_t)$ is W.B., and if (σ_t) is the time-change associated with $A_t = \int_0^t H_s ds$, where H_s is > 0 $dP\,ds$-a.s. and $A_\infty = \infty$ a.s., then $\left(\mathscr{F}_{\sigma_t}\right)$ is also W.B. under P.

A usually difficult, albeit important, question is:

(Q) *Given a W.B. filtration* $(\mathscr{F}_t)$*, is it S.B. and, if so, can one describe explicitly at least one generating* BM^1?

Tsirel'son [1] gives a beautiful, and explicit, example of a W.B. filtration for which a certain Brownian motion is not generating. See Prop. (3.6), Chap. IX, for a proof. Feldman and Smorodinsky [1] give easier examples of the same situation. Emery and Schachermayer [2] prove that the filtration in Tsirel'son's example is S.B. Likewise, Attal et al. [1] prove that the Goswami-Rao filtration is S.B. (see Exercise (3.22)).

A deep and difficult study is made by Dubins et al. [1] to show that the filtration on the canonical space $C(\mathbb{R}_+, \mathbb{R})$ is not S.B. under (many) probabilities Q, equivalent to the Wiener measure, although under such probabilities, $(\mathscr{F}_t)$ is well-known to be W.B. (see (P1) above). The arguments in Dubins et al. [1] have been greatly simplified by Schachermayer [1] and Emery [4].

Tsirel'son [2] has shown that the filtration of Walsh's Brownian motion with at least three rays, which is well-known to be W.B., is not S.B. The arguments in Tsirel'son [2] have been simplified by Barlow, Emery et al. [1]. Based on Tsirel'son's technique, it is shown in this paper that if $\mathscr{F}_t = \mathscr{F}_t^B$, for a BM^d B $(d \geq 1)$, and if L is the end of a $(\mathscr{F}_t)$ predictable set Γ, then $\mathscr{F}_L^+$ differs from $\mathscr{F}_L^-$ by the adjunction of one set A at most, i.e.: $\mathscr{F}_L^+ = \sigma\left\{\mathscr{F}_L^-, A\right\}$, a property of the Brownian filtration which had been conjectured by M. Barlow. Watanabe [6] shows the existence of a 2-dimensional diffusion for which this property does not hold. Warren [2] shows that the filtration generated by sticky Brownian motion and its driving Brownian motion (for the definition of this pair of processes, see Warren [1]) is not S.B. Another simplification of Tsirel'son's work is presented by De Meyer [1]. Emery and Schachermayer [1] show the existence of a pure martingale (M_t) with bracket $\langle M, M\rangle_t$ such that the measure $d\langle M, M\rangle_t$ is equivalent to dt, and nonetheless, the filtration of (M_t) is not S.B., although it is W.B. (see (P2) above).

Let us also recall the question studied by Lane [1].

Question 4. If B is a BM^1 and H a $(\mathscr{F}_t^B)$-predictable process, under which condition on H is the filtration of $M_t = \int_0^t H_s dB_s$ that of a Brownian motion?

Under which conditions are all the $(\mathscr{F}_t^M)$-martingales continuous?

In the case of $H_s = f(B_s)$, Lane [1] has partial and hard to prove results. There are also partial answers in Knight [7] when f is the indicator function of a set (see Exercise (4.20)).

We also list the

Question 5. Which of the martingales of the previous question are extremal or pure?

For $H_s = B_s^n$, Stroock and Yor [2] give a positive answer for n odd (see Exercise (3.11) Chap. IX) and so does Begdhadi-Sakrani [2] for n even. When H is > 0-a.s., then $\int_0^t H_s dB_s$ is extremal (see Exercise (4.15)) but we have the following question, which is a particular case of the previous one:

Question 6. Does there exist a strictly positive predictable process H such that the above stochastic integral is not pure?

Brownian filtrations have been further studied in Tsirel'son [4] who starts a classification of noises, that is, roughly, of families $(\mathscr{F}_{s,t})_{s\leq t}$ of σ-fields such that, for $s < v < t$,

$$\mathscr{F}_{s,v} \vee \mathscr{F}_{v,t} = \mathscr{F}_{s,t}$$

and $\mathscr{F}_{s,v}$ and $\mathscr{F}_{v,t}$ are independent. It is shown that there are many more noises than those generated by the increments of Lévy processes.

Chapter VI. Local Times

§1. Definition and First Properties

With Itô's formula, we saw how C^2-functions operate on continuous semimartingales. We now extend this to convex functions, thus introducing the important notion of local time.

In what follows, f is a convex function. We use the notation and results of Sect. 3 in the Appendix. The following result will lead to a generalization of Itô's formula.

(1.1) Theorem. *If X is a continuous semimartingale, there exists a continuous increasing process A^f such that*

$$f(X_t) = f(X_0) + \int_0^t f'_-(X_s)dX_s + \frac{1}{2}A_t^f$$

where f'_- is the left-hand derivative of f.

Proof. If f is C^2, then this is Itô's formula and $A_t^f = \int_0^t f''(X_s)d\langle X, X\rangle_s$.

Let now j be a positive C^∞-function with compact support in $]-\infty, 0]$ such that $\int_{-\infty}^0 j(y)dy = 1$ and set $f_n(x) = n\int_{-\infty}^0 f(x+y)j(ny)dy$. The function f being convex, hence locally bounded, f_n is well defined for every n and, as n tends to infinity, f_n converges to f pointwise and f'_n increases to f'_-. For each n

$$f_n(X_t) = f_n(X_0) + \int_0^t f'_n(X_s)dX_s + \frac{1}{2}A_t^{f_n},$$

and $f_n(X_t)$ (resp. $f_n(X_0)$) converges to $f(X_t)$ (resp. $f(X_0)$). Moreover, by stopping, we can suppose that X is bounded and then $f'_-(X_s)$ also is bounded. By the dominated convergence theorem (Theorem (2.12) Chap. IV) for stochastic integrals $\int_0^t f'_n(X_s)dX_s$ converges to $\int_0^t f'_-(X_s)dX_s$ in probability uniformly on every bounded interval. As a result, A^{f_n} converges also to a process A^f which, as a limit of increasing processes, is itself an increasing process and

$$f(X_t) = f(X_0) + \int_0^t f'_-(X_s)dX_s + \frac{1}{2}A_t^f.$$

The process A_t^f can now obviously be chosen to be a.s. continuous, which ends the proof. □

The problem is now to compute A^f in an explicit and useful way making clear how it depends on f. We begin with the special cases of $|x|$, $x^+ = x \vee 0$ and $x^- = -(x \wedge 0)$. We define the function sgn by $\operatorname{sgn}(x) = 1$ if $x > 0$ and $\operatorname{sgn}(x) = -1$ if $x \leq 0$. If $f(x) = |x|$, then $f'_-(x) = \operatorname{sgn}(x)$.

(1.2) Theorem (Tanaka formula). *For any real number a, there exists an increasing continuous process L^a called the* local time *of X in a such that,*

$$\begin{aligned}
|X_t - a| &= |X_0 - a| + \int_0^t \operatorname{sgn}(X_s - a)dX_s + L_t^a, \\
(X_t - a)^+ &= (X_0 - a)^+ + \int_0^t 1_{(X_s > a)} dX_s + \frac{1}{2}L_t^a, \\
(X_t - a)^- &= (X_0 - a)^- - \int_0^t 1_{(X_s \leq a)} dX_s + \frac{1}{2}L_t^a.
\end{aligned}$$

In particular, $|X - a|$, $(X - a)^+$ and $(X - a)^-$ are semimartingales.

Proof. The left derivative of $f(x) = (x - a)^+$ is equal to $1_{]a,\infty[}$; by Theorem (1.1), there is a process A^+ such that

$$(X_t - a)^+ = (X_0 - a)^+ + \int_0^t 1_{(X_s > a)} dX_s + \frac{1}{2}A_t^+.$$

In the same way

$$(X_t - a)^- = (X_0 - a)^- - \int_0^1 1_{(X_s \leq a)} dX_s + \frac{1}{2}A_t^-.$$

By subtracting the last identity from the previous one, we get

$$X_t = X_0 + \int_0^t dX_s + \frac{1}{2}\left(A_t^+ - A_t^-\right).$$

It follows that $A_t^+ = A_t^-$ a.s. and we set $L_t^a = A_t^+$.

By adding the same two identities we then get the first formula in the statement. □

We will also write $L_t^a(X)$ for the local time in a of the semimartingale X when there is a risk of ambiguity.

Remark. The lack of symmetry in the last two identities in the statement is due to the fact that we have chosen to work with *left* derivatives. This is also the reason for the choice of the function sgn. See however Exercise (1.25).

With the increasing process L_t^a, we can as usual associate a random measure dL_t^a on $\mathbb{R}_+$. To some extent, it measures the "time" spent at a by the semimartingale X, as is shown in the following result.

(1.3) Proposition. *The measure dL_t^a is a.s. carried by the set $\{t : X_t = a\}$.*

Proof. By applying Itô's formula to the semimartingale $|X - a|$, we get

$$(X_t - a)^2 = (X_0 - a)^2 + 2\int_0^t |X_s - a| d(|X - a|)_s + \langle |X - a|, |X - a| \rangle_t$$

and using the first formula in Theorem (1.2) this is equal to

$$(X_0 - a)^2 + 2\int_0^t |X_s - a| \operatorname{sgn}(X_s - a) dX_s + 2\int_0^t |X_s - a| dL_s^a + \langle X, X \rangle_t.$$

If we compare this with the equality, also given by Itô's formula,

$$(X_t - a)^2 = (X_0 - a)^2 + 2\int_0^t (X_s - a) dX_s + \langle X, X \rangle_t$$

we see that $\int_0^t |X_s - a| dL_s^a = 0$ a.s. which is the result we had to prove. □

Remarks. 1°) This proposition raises the natural question whether $\{t : X_t = a\}$ is exactly the support of dL_t^a. This is true in the case of BM as will be seen in the following section. The general situation is more involved and is described in Exercise (1.26) in the case of martingales.

2°) If $\sigma_a = \sup\{t : X_t = a\}$, then $L_\infty^a = L_{\sigma_a}^a$, a fact which comes in handy in some proofs.

We are now going to study the regularity in the space variable a of the process L_t^a. We need the following

(1.4) Lemma. *There exists a $\mathscr{B}(\mathbb{R}) \otimes \mathscr{P}$-measurable process $\tilde{L}$ such that, for each a, $\tilde{L}(a, \cdot, \cdot)$ is indistinguishable from L^a.*

Proof. Apply Fubini's theorem for stochastic integrals (Exercise (5.17) of Chap. IV) to the process $H(a, s, \cdot) = 1_{(X_s > a)}$.

Consequently, we henceforth suppose that L^a is $\mathscr{B}(\mathbb{R}) \otimes \mathscr{P}$-measurable and we will use the measurability in a to prove the existence of yet another better version. We first prove two important results. We recall that if f is convex, its second derivative f'' in the sense of distributions is a positive measure.

(1.5) Theorem (Itô-Tanaka formula). *If f is the difference of two convex functions and if X is a continuous semimartingale*

$$f(X_t) = f(X_0) + \int_0^t f'_-(X_s) dX_s + \frac{1}{2}\int_{\mathbb{R}} L_t^a f''(da).$$

In particular, $f(X)$ is a semimartingale.

Proof. It is enough to prove the formula for a convex f. On every compact subset of $\mathbb{R}$, f is equal to a convex function g such that g'' has compact support. Thus by stopping X when it first leaves a compact set, it suffices to prove the result when f'' has compact support in which case there are two constants α, β such that

$$f(x) = \alpha x + \beta + \frac{1}{2}\int |x - a| f''(da).$$

Thanks to the previous results we may write

$$\begin{aligned} f(X_t) &= \alpha X_t + \beta + \frac{1}{2}\int |X_t - a| f''(da) \\ &= \alpha(X_t - X_0) + f(X_0) + \int \frac{1}{2}\left(\int_0^t \operatorname{sgn}(X_s - a) dX_s + L_t^a\right) f''(da). \end{aligned}$$

From Sect. 3 in the Appendix and Lemma (1.4), we see that

$$\frac{1}{2}\int_{\mathbb{R}}\int_0^t \operatorname{sgn}(X_s - a) dX_s f''(da) = \int_0^t f'_-(X_s) dX_s - \alpha(X_t - X_0)$$

which completes the proof. □

(1.6) Corollary (Occupation times formula). *There is a P-negligible set outside of which*

$$\int_0^t \Phi(X_s) d\langle X, X\rangle_s = \int_{-\infty}^{\infty} \Phi(a) L_t^a da$$

for every t and every positive Borel function Φ.

Proof. If $\Phi = f''$ with f in C^2, the formula holds for every t, as follows from comparing Itô-Tanaka's and Itô's formulas, outside a P-negligible set Γ_Φ. By considering a countable set (Φ_n) of such functions dense in $C_0(\mathbb{R})$ for the topology of uniform convergence, it is easily seen that outside the P-negligible set $\Gamma = \bigcup_n \Gamma_{\Phi_n}$, the formula holds simultaneously for every t and every Φ in $C_0(\mathbb{R})$. An application of the monotone class theorem ends the proof.

Remarks. 1°) The time t may be replaced by any random time S.

2°) These "occupation times" are defined with respect to $d\langle X, X\rangle_s$ which may be seen as the "natural" time-scale for X. However the name for this formula is particularly apt in the case of Brownian motion where, if $\Phi = 1_A$ for a Borel set A, the left-hand side is exactly the amount of time spent in A by the BM.

3°) A consequence of these formulas is that for a function which is twice differentiable but not necessarily C^2, the Itô formula is still valid in exactly the same form provided f'' is locally integrable; this could also have been proved directly by a monotone class argument.

We now turn to the construction of a regular version of local times with which we will work in the sequel.

(1.7) Theorem. *For any continuous semimartingale X, there exists a modification of the process $\{L_t^a; a \in \mathbb{R}, t \in \mathbb{R}_+\}$ such that the map $(a,t) \to L_t^a$ is a.s. continuous in t and cadlag in a. Moreover, if $X = M + V$, then*

$$L_t^a - L_t^{a-} = 2\int_0^t 1_{(X_s=a)} dV_s = 2\int_0^t 1_{(X_s=a)} dX_s.$$

Thus, in particular, if X is a local martingale, there is a bicontinuous modification of the family L^a of local times.

Proof. By Tanaka's formula

$$L_t^a = 2\left[(X_t - a)^+ - (X_0 - a)^+ - \int_0^t 1_{(X_s>a)} dM_s - \int_0^t 1_{(X_s>a)} dV_s\right].$$

Using Kolmogorov's criterion (Theorem (2.1) Chap. I) with the Banach space $C([0,t],\mathbb{R})$, we first prove that the stochastic integral

$$\hat{M}_t^a = \int_0^t 1_{(X_s>a)} dM_s$$

possesses a bicontinuous modification. Thanks to the BDG-inequalities of Chap. IV and Corollary (1.6), we have, for any $k \geq 1$,

$$\begin{aligned} E\left[\sup_t \left|\hat{M}_t^a - \hat{M}_t^b\right|^{2k}\right] &\leq C_k E\left[\left(\int_0^\infty 1_{(a<X_s\leq b)} d\langle M, M\rangle_s\right)^k\right] \\ &= C_k E\left[\left(\int_a^b L_\infty^x dx\right)^k\right] \\ &= C_k (b-a)^k E\left[\left(\frac{1}{b-a}\int_a^b L_\infty^x dx\right)^k\right] \\ &\leq C_k (b-a)^k E\left[\frac{1}{b-a}\int_a^b (L_\infty^x)^k dx\right]. \end{aligned}$$

By Fubini's theorem, this is less than

$$C_k (b-a)^k \sup_x E\left[(L_\infty^x)^k\right].$$

Now, $L_t^x = 2\left\{(X_t - x)^+ - (X_0 - x)^+ - \int_0^t 1_{(X_s>x)} dX_s\right\}$ and since $|(X_t - x)^+ - (X_0 - x)^+| \leq |X_t - X_0|$, there is a universal constant d_k such that

$$E\left[(L_\infty^x)^k\right] \leq d_k E\left[\sup_t |X_t - X_0|^k + \left(\int_0^\infty |dV|_s\right)^k + \langle M, M\rangle_\infty^{k/2}\right].$$

Remark that the right-hand side no longer depends on x. If it is finite for some $k > 1$, we have proved our claim. If not, we may stop X at the times

$$T_n = \inf\left\{t : \sup_{s\le t} |X_s - X_0|^k + \left(\int_0^t |dV|_s\right)^k + \langle M, M\rangle_t^{k/2} \ge n\right\}.$$

The martingales $(\hat{M}^a)^{T_n}$ have bicontinuous versions, hence also $\hat{M}^a$.

To complete the proof, we must prove that

$$\hat{V}_t^a = \int_0^t 1_{(X_s>a)} dV_s$$

is jointly cadlag in a and continuous in t. But, by Lebesgue's theorem

$$\hat{V}_t^{a-} = \lim_{b\uparrow a} \int_0^t 1_{(X_s>b)} dV_s = \int_0^t 1_{(X_s\ge a)} dV_s.$$

It follows that $L_t^a - L_t^{a-} = 2\left(\hat{V}_t^{a-} - \hat{V}_t^a\right) = 2\int_0^t 1_{(X_s=a)} dV_s$. In the same way

$$\hat{V}_t^{a+} = \lim_{b\downarrow a} \int_0^t 1_{(X_s>b)} dV_s = \int_0^t 1_{(X_s>a)} dV_s$$

so that $L_t^a = L_t^{a+}$.

Finally, the occupation times formula implies that

$$\int_0^t 1_{(X_s=a)} d\langle M, M\rangle_s = \int_0^t 1_{(X_s=a)} d\langle X, X\rangle_s = 0$$

so that $\int_0^t 1_{(X_s=a)} dM_s = 0$ which ends the proof. □

Remark. If X is of finite variation, then $L^a(X) \equiv 0$. Indeed by the occupation times formula, $L_t^a = 0$ for Lebesgue almost every a and, by the right-continuity in a, $L^a(X) = 0$ for every a. However, a semimartingale may have a discontinuous family of local times, or in other words, the above theorem cannot be improved to get continuity in both variables, as is shown in Exercise (1.34).

As a by-product of the use, in the preceding proof, of Kolmogorov's criterion we see that for local martingales we may get Hölder properties in a for the family L^a. This is in particular the case for Brownian motion where

$$E\left[\sup_{s\le t}\left|\hat{B}_s^a - \hat{B}_s^b\right|^{2k}\right] \le C_k |a-b|^k t^{k/2}$$

for any fixed time t and for every k. Thus we may now state the

(1.8) Corollary (Continuity of martingale local times). *The family L^a may be chosen such that almost surely the map $a \to L_t^a$ is Hölder continuous of order α for every $\alpha < 1/2$ and uniformly in t on every compact interval.*

Proof. In the case of BM, only the uniformity in t has to be proved and this follows from Exercise (2.10) Chap. I. The result for local martingales is then a consequence of the DDS Theorem (see Exercise (1.27)). The details are left to the reader. □

From now on, we will of course consider only the version of the local time $L_t^a(X)$ which was exhibited in Theorem (1.7). For this version, we have the following corollary which gives another reason for the name "local time".

(1.9) Corollary. *If X is a continuous semimartingale, then, almost-surely,*

$$L_t^a(X) = \lim_{\varepsilon \downarrow 0} \frac{1}{\varepsilon} \int_0^t 1_{[a,a+\varepsilon[}(X_s) d\langle X, X\rangle_s$$

for every a and t, and if M is a continuous local martingale

$$L_t^a(M) = \lim_{\varepsilon \downarrow 0} \frac{1}{2\varepsilon} \int_0^t 1_{]a-\varepsilon,a+\varepsilon[}(M_s) d\langle M, M\rangle_s.$$

The same result holds with any random time S in place of t.

Proof. This is a straightforward consequence of the occupation times formula and the right-continuity in a of $L_t^a(X)$. □

For BM we have in particular

$$L_t^0 = \lim_{\varepsilon \downarrow 0} \frac{1}{2\varepsilon} \int_0^t 1_{(|B_s|<\varepsilon)} ds,$$

which proves that L_t^0 is adapted to the completion of $\sigma(|B_s|, s \leq t)$. This will be taken up in the following section.

The above corollary is an "approximation" result for the local time. In the case of BM, there are many results of this kind which will be stated in Chap. XII. We begin here with a result which is valid for all the semimartingales of $\mathcal{S}_p$ (Exercise (4.13) Chap. IV). Let X be a continuous semimartingale; for $\varepsilon > 0$, define a double sequence of stopping times by

$$\begin{aligned} \sigma_0^\varepsilon &= 0, \quad \tau_0^\varepsilon = \inf\{t : X_t = \varepsilon\}, \\ \sigma_n^\varepsilon &= \inf\{t > \tau_{n-1}^\varepsilon : X_t = 0\}, \quad \tau_n^\varepsilon = \inf\{t > \sigma_n^\varepsilon : X_t = \varepsilon\}. \end{aligned}$$

We set $d_\varepsilon(t) = \max\{n : \sigma_n^\varepsilon < t\}$; this is the number of "*downcrossings*" of X (see Sect. 2 Chap. II) from level ε to level 0 before time t. On Figure 4, we have $d_\varepsilon(t) = 2$.

For simplicity, we will write only σ_n and τ_n instead of σ_n^ε and τ_n^ε and L_t for L_t^0.

(1.10) Theorem. *If $X = M + V$ is in $\mathcal{S}_p$, $p \geq 1$, i.e.*

$$E\left[\langle M, M\rangle_\infty^{p/2} + \left(\int_0^\infty |dV|_t\right)^p\right] < \infty,$$

then

$$\lim_{\varepsilon \to 0} E\left[\sup_t \left|\varepsilon d_\varepsilon(t) - \frac{1}{2} L_t\right|^p\right] = 0.$$

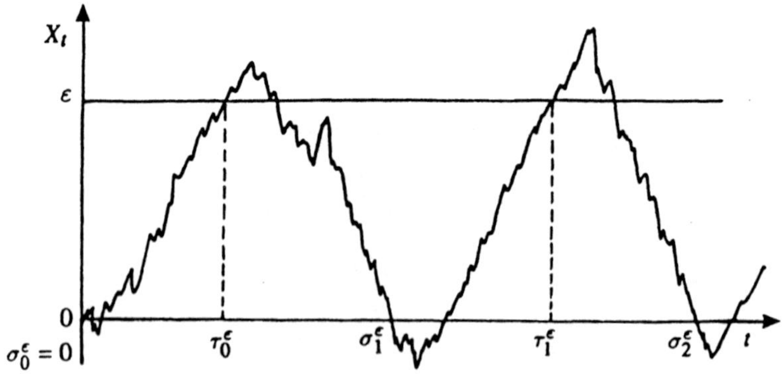

Fig. 4.

Proof. By Tanaka's formula

$$X^+_{\tau_n\wedge t} - X^+_{\sigma_n\wedge t} = \int_{]\sigma_n\wedge t, \tau_n\wedge t]} 1_{(X_s>0)} dX_s + \frac{1}{2}\left(L_{\tau_n\wedge t} - L_{\sigma_n\wedge t}\right).$$

Because X does not vanish on $[\tau_n, \sigma_{n+1}[$, we have $L_{\tau_n\wedge t} - L_{\sigma_n\wedge t} = L_{\sigma_{n+1}\wedge t} - L_{\sigma_n\wedge t}$. As a result

$$\sum_n \left(X^+_{\tau_n\wedge t} - X^+_{\sigma_n\wedge t}\right) = \int_0^t \theta_s^\varepsilon dX_s + \frac{1}{2}L_t$$

where θ_s^ε is the predictable process $\sum_n 1_{]\sigma_n,\tau_n]}(s) 1_{]0,\varepsilon]}(X_s)$. But $X^+_{\tau_n\wedge t} - X^+_{\sigma_n\wedge t} = \varepsilon$ on $\{\tau_n \le t\}$; if $n(t) = \inf\{n : \tau_n > t\}$, the left-hand side of the above equality is equal to $\varepsilon d_\varepsilon(t) + u(\varepsilon)$ where $0 \le u(\varepsilon) = X_t^+ - X_{\sigma_{n(t)}\wedge t} \le \varepsilon$. Thus the proof will be complete if

$$\lim_{\varepsilon\to 0} E\left[\sup_t \left|\int_0^t \theta_s^\varepsilon dX_s\right|^p\right] = 0.$$

But, by the BDG inequalities, this expectation is less than

$$C_p E\left[\left(\int_0^\infty \theta_s^\varepsilon d\langle X, X\rangle_s\right)^{p/2} + \left(\int_0^\infty \theta_s^\varepsilon |dV|_s\right)^p\right]$$

and since θ^ε converges boundedly to zero, it remains to apply Lebesgue's dominated convergence theorem. □

With the same hypothesis and notation, we also have the following proposition which is the key to Exercise (2.13) Chap. XIII.

(1.11) Proposition.

$$\lim_{\varepsilon\downarrow 0} E\left[\sup_{t\ge 0}\left|\varepsilon^{-1}\int_0^t \theta_s^\varepsilon d\langle X, X\rangle_s - \frac{1}{2}L_t\right|^p\right] = 0.$$

Proof. Using Itô's and Tanaka's formulas and taking into account the fact that dL_s does not charge the set $\{s : X_s \neq 0\}$, one can show that

$$(X^+_{\tau_n\wedge t})^2 - (X^+_{\sigma_n\wedge t})^2 = 2\int_{]\sigma_n\wedge t,\tau_n\wedge t]} X_s 1_{(X_s>0)} dX_s + \int_{]\sigma_n\wedge t,\tau_n\wedge t]} 1_{(X_s>0)} d\langle X, X\rangle_s$$

which entails that

$$\varepsilon^2 d_\varepsilon(t) + u^2(\varepsilon) = 2\int_0^t \theta_s^\varepsilon X_s dX_s + \int_0^t \theta_s^\varepsilon d\langle X, X\rangle_s.$$

Therefore

$$\left|\varepsilon^{-1}\int_0^t \theta_s^\varepsilon d\langle X, X\rangle_s - \varepsilon d_\varepsilon(t) - 2\varepsilon^{-1}\int_0^t \theta_s^\varepsilon X_s dX_s\right| \leq \varepsilon,$$

and it is enough to prove that $\varepsilon^{-1}\int_0^t \theta_s^\varepsilon X_s dX_s$ converges to zero uniformly in L^p. But using the BDG inequalities again, this amounts to showing that

$$\varepsilon^{-p} E\left[\left(\int_0^\infty \theta_s^\varepsilon X_s^2 d\langle X, X\rangle_s\right)^{p/2} + \left(\int_0^\infty \theta_s^\varepsilon |X_s||dV|_s\right)^p\right]$$

converges to zero as ε converges to zero, and since $|X_s| \leq \varepsilon$ on $\{\theta_s^\varepsilon > 0\}$, this follows again from Lebesgue's theorem. □

Remark. By stopping, any semimartingale can be turned into an element of $\mathscr{S}_p$. Therefore, the last two results have obvious corollaries for general semimartingales provided one uses convergence in probability on compact sets instead of convergence in L^p.

We close this section with a more thorough study of the dependence of L^a in the space variable a and prove that L^a, as a function of a, has a finite quadratic variation.

For each t, the random function $a \to L_t^a$ is a cadlag function hence admits only countably many discontinuities. We denote by ΔL_t^a the process $L_t^a - L_t^{a-}$. If $X = M + V$,

$$\Delta L_t^a = 2\int_0^t 1_{(X_s=a)} dV_s.$$

Consequently, for $a < b$,

$$\sum_{a<x\leq b} |\Delta L_t^x| \leq 2\int_0^t |dV|_s < \infty$$

and, a fortiori,

$$\sum_{a<x\leq b} |\Delta L_t^x|^2 < \infty.$$

For our later needs, we compute this last sum. We have

$$(\Delta L_t^x)^2 = 4\left(\int_0^t 1_{(X_s=x)} dV_s\right)^2$$

and since, by the continuity of V, the measure $dV \otimes dV$ does not charge the diagonal in $[0, t]^2$,

$$\begin{aligned}(\Delta L_t^x)^2 &= 8\int_0^t 1_{(X_s=x)} dV_s \int_0^s 1_{(X_u=X_s)} dV_u \\ &= 4\int_0^t \Delta L_s^{X_s} 1_{(X_s=x)} dV_s.\end{aligned}$$

Since there are at most countably many $x \in]a, b[$ such that $\Delta L_s^x > 0$ for some $s \in [0, t]$, we finally get

$$\sum_{a<x\leq b} (\Delta L_t^x)^2 = 4\int_0^t \Delta L_s^{X_s} 1_{(a<X_s\leq b)} dV_s.$$

We may now state

(1.12) Theorem. *Let (Δ_n) be a sequence of subdivisions of $[a, b]$ such that $|\Delta_n| \to 0$. For any non negative and finite random variable S*

$$\lim_{n\to\infty} \sum_{\Delta_n} \left(L_S^{a_{i+1}} - L_S^{a_i}\right)^2 = 4\int_a^b L_S^x dx + \sum_{a<x\leq b} (\Delta L_S^x)^2$$

in probability.

Proof. The case of a general S will be covered if we prove that $\sum_{\Delta_n} \left(L_t^{a_{i+1}} - L_t^{a_i}\right)^2$ converges in probability, uniformly for t in a compact subinterval of $\mathbb{R}_+$, to $4\int_a^b L_t^x dx + \sum_{a<x\leq b} (\Delta L_t^x)^2$. To this end, we develop $\frac{1}{4}\left(L_t^{a_{i+1}} - L_t^{a_i}\right)^2$ with the help of Tanaka's formula, namely,

$$\frac{1}{2} L_t^a = (X_t - a)^+ - (X_0 - a)^+ - \hat{M}_t^a - \hat{V}_t^a$$

where we define $\hat{Z}_t^a = \int_0^t 1_{(X_s>a)} dZ_s$ for $Z = M, V, X$.

The function $\phi_t(a) = (X_t - a)^+ - (X_0 - a)^+$ is Lipschitz in a with constant 1 which implies that

$$\lim_{n\to\infty} \sum_{\Delta_n} (\phi_t(a_{i+1}) - \phi_t(a_i))^2 = 0.$$

Likewise, using the continuity of $\hat{M}^a$ as was done in Sect. 1, Chap. IV,

$$\lim_{n\to\infty} \sum_{\Delta_n} (\phi_t(a_{i+1}) - \phi_t(a_i)) \left(\hat{M}_t^{a_{i+1}} - \hat{M}_t^{a_i}\right) = 0.$$

Finally

$$\left|\sum_{\Delta_n} (\phi_t(a_{i+1}) - \phi_t(a_i)) \left(\hat{V}_t^{a_{i+1}} - \hat{V}_t^{a_i}\right)\right| \leq |\Delta_n| \int_0^t |dV|_s$$

which goes to zero as n tends to infinity.

As a result, the limit we are looking for is equal (if it exists) to the limit in probability of

$$L_n = \sum_{\Delta_n} \left(\hat{X}_t^{a_{i+1}} - \hat{X}_t^{a_i}\right)^2.$$

Itô's formula yields

$$\begin{aligned} L_n &= 2\sum_{\Delta_n} \int_0^t \left(\hat{X}_s^{a_i} - \hat{X}_s^{a_{i+1}}\right) 1_{(a_i < X_s \leq a_{i+1})} dX_s \\ &\quad + \sum_{\Delta_n} \int_0^t 1_{(a_i < X_s \leq a_{i+1})} d\langle X, X\rangle_s. \end{aligned}$$

The occupation times formula shows that the second term in the right-hand side is equal to $\int_a^b L_t^x dx$. We shall prove that the first term converges to the desired quantity.

By localization, we may assume that X is bounded; we can then use the dominated convergence theorem for stochastic integrals which proves that this term converges in probability uniformly on every compact interval to

$$J = 2\int_0^t \left(\hat{X}_s^{X_s} - \hat{X}_s^{X_s -}\right) 1_{(a < X_s \leq b)} dX_s = \int_0^t \Delta L_s^{X_s} 1_{(a < X_s \leq b)} dX_s.$$

By the computation preceding the statement, it remains to prove that

$$\int_0^t \Delta L_s^{X_s} 1_{(a < X_s \leq b)} dM_s = 0.$$

But as already observed, there are only countably many $x \in]a, b[$ such that $\Delta L_s^x > 0$ for some $s \in [0, t]$ and moreover P-a.s., $\int_0^t 1_{\{X_s = x\}} d\langle M, M\rangle_s = 0$ for every x. Thus the result follows. □

The following corollary will be used in Chap. XI.

(1.13) Corollary. *If the process $x \to L_S^x$, $x \geq 0$, is a continuous semimartingale (in some appropriate filtration), its bracket is equal to $4 \int_0^x L_S^y dy$.*

\# **(1.14) Exercise.** Let M be a continuous local martingale vanishing at 0 and L its local time at 0.

1°) Prove that $\inf\{t : L_t > 0\} = \inf\{t : \langle M, M\rangle_t > 0\}$ a.s.. In particular $M \equiv 0$ if and only if $L \equiv 0$.

2°) Prove that for $0 < \alpha < 1$, and $M \not\equiv 0$, $|M|^\alpha$ is not a semimartingale.

(1.15) Exercise (Extension of the occupation times formula). If X is a continuous semimart., then almost-surely, for every positive Borel function h on $\mathbb{R}_+ \times \mathbb{R}$,

$$\int_0^t h(s, X_s) d\langle X, X\rangle_s = \int_{-\infty}^{+\infty} da \int_0^t h(s, a) dL_s^a(X).$$

Extend this formula to measurable functions h on $\mathbb{R}_+ \times \Omega \times \mathbb{R}$.

\# **(1.16) Exercise.** 1°) Let X and Y be two continuous semimartingales. Prove that

$$\int_0^t 1_{(X_s = Y_s)} d\langle X, Y\rangle_s = \int_0^t 1_{(X_s = Y_s)} d\langle X, X\rangle_s = \int_0^t 1_{(X_s = Y_s)} d\langle Y, Y\rangle_s.$$

[Hint: Write $\langle X, Y\rangle = \langle X, X\rangle + \langle X, Y - X\rangle$.]

2°) If $X = M + V$ and A is a continuous process of finite variation

$$\int_0^t 1_{(X_s = A_s)} dX_s = \int_0^t 1_{(X_s = A_s)} dV_s.$$

3°) If $X = M + V$ is ≥ 0 and $M_0 = 0$, its local time at 0 is equal to $2\int_0^{\cdot} 1_{(X_s=0)} dV_s = 2\int_0^{\cdot} 1_{(X_s=0)} dX_s$. As a result, if dV_s is carried by the set $\{s : X_s = 0\}$, then V is increasing; moreover if $V_t = \sup_{s \leq t}(-M_s)$, the local time of X at 0 is equal to $2V$.

(1.17) Exercise. 1°) If X is a continuous semimartingale, prove that

$$L_t^a(|X|) = L_t^a(X) + L_t^{(-a)-}(X) \quad \text{if } a \geq 0, \qquad L_t^a(|X|) = 0 \quad \text{if } a < 0,$$

and that $L_t^0(X^+) = L_t^0(X)$.

2°) If X is a continuous semimartingale, prove that

$$L_t^a(-X) = L_t^{(-a)-}(X).$$

3°) Prove a result similar to Theorem (1.10) but with upcrossings instead of downcrossings.

(1.18) Exercise. 1°) If X and Y are two cont. semimarts., prove that $X \vee Y$ and $X \wedge Y$ are cont. semimarts. and that

$$L^0(X \vee Y) + L^0(X \wedge Y) = L^0(X) + L^0(Y).$$

[Hint: By the preceding exercise, it is enough to prove the result for positive X and Y. Use Exercise (1.16), 3°) .]

2°) Prove further that

$$L^0(XY) = X^+ \cdot L^0(Y) + Y^+ \cdot L^0(X) + X^- \cdot L^{0-}(Y) + Y^- \cdot L^{0-}(X).$$

** **(1.19) Exercise.** 1°) If X is the BM, prove that, with the notation of Theorem (1.10),

$$\lim_{\varepsilon \downarrow 0} \varepsilon d_\varepsilon(\cdot) = \frac{1}{2} L \quad \text{a.s.}$$

[Hint: Prove the almost-sure convergence for the sequence $\varepsilon_n = n^{-2}$, then use the fact that $d_\varepsilon(t)$ increases when ε decreases. Another proof is given in Sect. 1 Chap. XII.]

2°) More generally, if for $a < 0 < b$ we denote by $d_{a,b}$ the number of downcrossings from b to a, then

$$\lim_{n\to\infty} (b_n - a_n) d_{a_n,b_n}(t) = \frac{1}{2} L_t, \quad \text{a.s.}$$

if $\sum(b_n - a_n) < \infty$.

* **(1.20) Exercise. (A generalization of P. Lévy's characterization theorem).** Let $f(x,t)$ be a fixed solution to the heat equation i.e. be such that $f'_t + \frac{1}{2} f''_{x^2} = 0$. We recall that such a function is analytic in x for $t > 0$.

1°) If B is a $(\mathscr{G}_t)$-BM and A a $(\mathscr{G}_t)$-adapted continuous increasing process, prove that a.s.

$$\left\{t : f'_t(B_t, A_t) = 0\right\} \subset \left\{t : f''_{xt}(B_t, A_t) = 0\right\}$$

up to m-negligible sets where m is the Lebesgue measure.

[Hint: Apply to the semimartingale $Y_t = f'_t(B_t, A_t)$ the fact that $d\langle Y, X\rangle_t$ does not charge the sets $\{t : Y_t = a\}$.] Arguing inductively prove that if f'_t is not identically zero then $m\{t : f'_t(B_t, A_t) = 0\} = 0$.

[Hint: If not, there would be a point x where all the spatial derivatives of f'_t vanish.]

2°) Let X be a continuous local martingale such that $f(X_t, t)$ is a local martingale and $m\{t : f'_t(X_t, t) = 0\} = 0$; prove that X is a BM.

[Hint: Observe that $\int_0^t f'_s(X_s, s) d\langle X, X\rangle_s = \int_0^t f'_s(X_s, s) ds$.]

(1.21) Exercise. 1°) Let X^1 and X^2 be two continuous semimartingales vanishing at 0. Prove that

$$L^0_t(X^1 \vee X^2) = \int_0^t 1_{(X^2_s \le 0)} dL^0_s(X^1) + L^0_t(X^{2+} - X^{1+}).$$

[Hint: Use Exercise (1.16) 3°) and the equality $(X^1 \vee X^2)^+ = (X^{2+} - X^{1+})^+ + X^{1+}$.]

2°) Suppose henceforth that $L^0(X^2 - X^1) \equiv 0$. Pick a real number α in $]0, 1/2[$ and set $Z^i = X^i - 2\alpha X^{i+}$. After observing that, for $X^2 \ge X^1$,

$$(1 - 2\alpha)(X^2 - X^1) \le Z^2 - Z^1 \le X^2 - X^1,$$

prove that $L^0(Z^2 - Z^1) \equiv 0$.

[Hint: Use Theorem (1.10).]

3°) From the equality $2\alpha(X^{2+} - X^{1+})^+ = (X^2 - X^1)^+ - (Z^2 - Z^1)^+$ derive that

$$L_t^0(X^1 \vee X^2) = \int_0^t 1_{(X_s^2 \leq 0)} dL_s^0(X^1) + \int_0^t 1_{(X_s^1 < 0)} dL_s^0(X^2).$$

(1.22) Exercise. Prove that for the standard BM

$$\sup_{s \leq t} L_s^{B_s} = \sup_a L_t^a.$$

[Hint: One may use the Laplace method.] The process $L_{\cdot}^B$ is further studied in Exercise (2.12) Chap. XI.

(1.23) Exercise. Let f be a strictly increasing function on $\mathbb{R}$, which moreover is the difference of two convex functions. Let X be a cont. semimart.; prove that for every t, the equality

$$L_t^{f(a)}(f(X)) = f'_+(a) L_t^a(X)$$

holds a.s. for every a.

** **(1.24) Exercise.** Let $X = M + V$ be a continuous semimartingale and A a continuous process of bounded variation such that for every t,

$$\text{i)} \int_0^t 1_{(X_s = A_s)} dX_s = 0, \quad \text{ii)} \int_0^t 1_{(X_s = A_s)} dA_s = 0,$$

(see Exercise (1.16) above). If (L_t^x) is the family of local times of X and Λ_t is the local time of $X - A$ at zero, prove that for any sequence (Δ_n) of subdivisions of $[0, t]$ such that $|\Delta_n| \to 0$, and for any continuous adapted process H,

$$\int_0^t H_s d\Lambda_s = \lim_{n \to \infty} \sum_{\Delta_n} H_{t_i} \left(L_{t_{i+1}}^{A_{t_i}} - L_{t_i}^{A_{t_i}} \right)$$

in probability. This applies in particular when $A = V$, in which case we see that the local time of the martingale part of X is the "integral" of the local times of X along its bounded variation part.

(1.25) Exercise (Symmetric local times). Let X be a continuous semimartingale.

1°) If we take $\widetilde{\operatorname{sgn}}(x)$ to be 1 for $x > 0$, -1 for $x < 0$ and 0 for $x = 0$, prove that there is a unique increasing process $\tilde{L}^a$ such that

$$|X_t - a| = |X_0 - a| + \int_0^t \widetilde{\operatorname{sgn}}(X_s - a) dX_s + \tilde{L}_t^a.$$

Give an example in which $\tilde{L}^a$ differs from L^a. Prove that $\tilde{L}^a = (L^a + L^{a-})/2$ and

$$\tilde{L}_t^a = \lim_{\varepsilon \downarrow 0} \frac{1}{2\varepsilon} \int_0^t 1_{]a-\varepsilon, a+\varepsilon[}(X_s) d\langle X, X \rangle_s.$$

2°) Prove the measurability of $\tilde{L}^a$ with respect to a and show that if f is convex

$$f(X_t) = f(X_0) + \int_0^t \frac{1}{2}(f'_+ + f'_-)(X_s)dX_s + \frac{1}{2}\int \tilde{L}_t^a f''(da).$$

Write down the occupation times formula with $\tilde{L}^a$ instead of L^a.

** **(1.26) Exercise (On the support of local times).** Let M be a continuous $(\mathscr{F}_t)$-martingale such that $M_0 = 0$ and set $Z(\omega) = \{t : M_t(\omega) = 0\}$. The set $Z(\omega)$ is closed and $Z(\omega)^c$ is the union of countably many open intervals. We call $R(\omega)$ (resp. $G(\omega)$) the set of right (resp. left) ends of these open intervals.

1°) If T is a $(\mathscr{F}_t)$-stopping time and

$$D_T(\omega) = \inf\{t > T(\omega) : M_t(\omega) = 0\},$$

prove that if M is uniformly integrable, the process

$$Y_t = |M_{T+t}| 1_{[0 \leq t < D_T - T]}$$

is a $(\mathscr{F}_{T+t})$-martingale.

2°) Prove that there exists a sequence of stopping times T_k such that $R(\omega) \subset \bigcup_k \{T_k(\omega)\}$, that for any stopping time T, $P[\omega : T(\omega) \in G(\omega)] = 0$, and finally that $Z(\omega)$ is a.s. a perfect set (i.e. closed and without isolated points).

We now recall that any closed subset Z of $\mathbb{R}$ is the union of a perfect set and a countable set. This perfect set may be characterized as the largest perfect set contained in Z and is called the *perfect core* of Z.

3°) Prove that the support of the measure dL_t is the perfect core of $Z \backslash \overset{\circ}{Z}$ where $\overset{\circ}{Z}$ is the interior of Z. Thus in the case of BM, the support is exactly Z, as is proved in the following section.

[Hint: Use Exercise (1.16).]

4°) Prove the result in 3°) directly from the special case of BM and the DDS theorem of Sect. 1 Chap. V.

(1.27) Exercise. Let τ be a process of time changes and X a τ-continuous semimartingale; prove that

$$L_t^a(X_{\tau_\cdot}) = L_{\tau_t}^a(X).$$

Consequently, if X is a continuous local martingale, and B its DDS BM, prove that $L_t^a(X) = L_{\langle X,X\rangle_t}^a(B)$.

Prove further that if X is a local martingale then for any real a, the sets $\{\lim_{t\to\infty} X_t \text{ exists}\}$ and $\{L_\infty^a < \infty\}$ are a.s. equal.

[This Exercise is partially solved in Sect. 2 Chap. XI.]

* **(1.28) Exercise.** Let X be a continuous semimartingale and L^a the family of its local times.

1°) Let S be a positive r.v. Prove that there exists a unique vector measure on the Borel sets of $\mathbb{R}$ with values in the space of random variables endowed with the topology of convergence in probability, which is equal to

$$\sum f_i \left(L_S^{a_{i+1}} - L_S^{a_i}\right)$$

on the step function $\sum f_i 1_{]a_i, a_{i+1}]}$. We write $\int_{-\infty}^{\infty} f(a) d_a L_S^a$ for the integral with respect to this vector measure.

2°) If f is a bounded Borel function and $F(x) = \int_0^x f(u)du$, prove that

$$F(X_S) = F(X_0) + \int_0^S f(X_u)dX_u - \frac{1}{2}\int_{-\infty}^{\infty} f(a)d_a L_S^a.$$

Extend this formula to locally bounded Borel functions f.

3°) Prove that if $a \to L_S^a$ is a semimartingale (see Sect. 2 Chap. XI) then $\int_{-\infty}^{\infty} f(a) d_a L_S^a$ is equal to the stochastic integral of f with respect to this semimartingale.

* **(1.29) Exercise (Principal values of Brownian local times).** 1°) For the linear BM, prove that for $\alpha > -1$ the process

$$X_t^{(\alpha)} = \int_0^t |B_s|^\alpha ds$$

is finite valued, hence

$$\tilde{X}_t^{(\alpha)} = \int_0^t |B_s|^\alpha (\text{sgn } B_s) ds$$

is also well-defined. Prove that these processes have continuous versions.

2°) Prove that

$$H_t = \lim_{\varepsilon \downarrow 0} \int_0^t B_s^{-1} 1_{(|B_s| \geq \varepsilon)} ds$$

or more generally that for $\alpha > -3/2$

$$H_t^{(\alpha)} = \lim_{\varepsilon \downarrow 0} \int_0^t |B_s|^\alpha (\text{sgn } B_s) 1_{(|B_s| \geq \varepsilon)} ds$$

exists a.s. Prove that these processes have continuous versions.

[Hint: Use the occupation times formula and Corollary (1.8).]

3°) Setting $\hat{B}_t = B_t - H_t$, prove that $\hat{B}_t B_t$ is a local martingale. More generally, if $h : \mathbb{R} \times \mathbb{R}_+ \to \mathbb{R}$ is a solution of the heat equation $\frac{1}{2}\frac{\partial^2 h}{\partial x^2} + \frac{\partial h}{\partial t} = 0$, then $h(\hat{B}_t, t) B_t$ is a local martingale.

4°) If $f = \sum_{k=1}^n \lambda_k 1_{]t_k, t_{k+1}]}$, and $\hat{\mathscr{E}}_t^f = \exp\left\{ i \int_0^t f(s) d\hat{B}_s + \frac{1}{2}\int_0^t f^2(s) ds \right\}$, the process $B_t \hat{\mathscr{E}}_t^f$ is a local martingale. Derive therefrom that

$$E\left[B_t \mid \sigma\left(\hat{B}_s, s \leq t\right)\right] = 0.$$

This discussion is closely related to that of Exercise (3.29), Chap. IV.

* **(1.30) Exercise.** 1°) Using the same notation as in the previous exercise prove that for any C^1-function ϕ we have

$$B_t\phi(H_t) = \int_0^t \phi(H_s)dB_s + \int_0^t \phi'(H_s)ds.$$

2°) For every $a \in \mathbb{R}$, define a process λ^a by the equality

$$B_t 1_{(H_t>a)} = \int_0^t 1_{(H_s>a)}dB_s + \lambda_t^a.$$

Show that there is a jointly measurable version which is again denoted by λ. Then prove that for any positive Borel function f,

$$\int_0^t f(H_s)ds = \int_{-\infty}^{\infty} f(a)\lambda_t^a da.$$

3°) Show that there is a version of λ^a which is increasing in t and such that $\lambda_t^a - B_t 1_{(H_t>a)}$ is jointly continuous.

(1.31) Exercise. If X is a continuous semimartingale, the martingale part of which is a BM, prove that for any locally integrable function f on $\mathbb{R}_+$,

$$\int_0^t ds\ f\left(\int_0^t 1_{(X_u \le X_s)}du\right) = \int_0^t f(v)dv \quad \text{a.s..}$$

More generally, prove that, for every locally integrable function h on $\mathbb{R}_+$,

$$\int_0^t ds\ f\left(\int_0^t du\ 1_{(X_u \le X_s)}\right) h(X_s) = \int_0^t dz\ h\left(\alpha_t(z)\right) f(z),$$

where $\alpha_t(z) = \inf\left\{y : \int_0^t du\ 1_{(X_u \le y)} > z\right\}$.

*# **(1.32) Exercise. (Hölder properties for local times of semimartingales).** Let $X = M + V$ be a continuous semimart. and assume that there is a predictable process v such that

i) $V_t = \int_0^t v_s d\langle M, M\rangle_s$;
ii) there exists a $p > 1$ for which

$$\int_0^t |v_s|^p d\langle M, M\rangle_s < \infty \ \text{a.s.,} \quad \text{for every } t > 0.$$

Call L_t^a, $a \in \mathbb{R}$, $t \geq 0$, the family of the local times of X.

1°) Prove that after a suitable localization of X, there exists, for every $N > 0$, $k \in \mathbb{N}$ and $T > 0$, a constant $C = C(N, k, T)$ such that

$$E\left[\sup_{t \le T} \left|L_t^x - L_t^y\right|^k\right] \le C\left(|x - y|^{k/2} + |x - y|^{k(p-1)/p}\right)$$

for every pair (x, y) for which $|x| \leq N$ and $|y| \leq N$. As a result, there is a bicontinuous modification of the family L_t^a, $a \in \mathbb{R}$, $t \geq 0$.

2°) Prove that, consequently, there exists a r.v. D_T such that

$$\sup_{t \leq T} \left| L_t^x - L_t^y \right| \leq D_T |x - y|^\alpha$$

for every $\alpha < 1/2$ if $p \geq 2$ and every $\alpha < (p-1)/p$ if $p < 2$.

Apply these results to $X_t = B_t + ct^m$ where B is a BM(0), $c \in \mathbb{R}$ and $m < 1$.

The following exercise shows that the conditions stated above are sufficient but not necessary to obtain Hölder properties for local times of semimarts.

\# **(1.33) Exercise.** Let B be a BM(0) and l its local time at 0. Let further $X = B + cl$ where $c \in \mathbb{R}$ and call L^a the family of local times of X. Prove that for every $T > 0$ and $k > 0$, there is a constant $C_{T,k}$ such that

$$E\left[\sup_{t \leq T} \left| L_t^a - L_t^b \right|^k\right] \leq C_{T,k} |a - b|^{k/2},$$

and that consequently the Hölder property of the preceding exercise holds for every $\alpha < 1/2$. Prove the same result for $X = |B| + cl$.

(1.34) Exercise. Let B the standard BM[1] and a and b two different and strictly positive numbers. Prove that the local time for the semimartingale $X = aB^+ - bB^-$ is discontinuous at 0 and compute its jump.

* **(1.35) Exercise.** For the standard BM B prove the following extensions of Theorem (1.12):

$$P\text{-}\lim_{n \to \infty} \sum_{\Delta_n} \left(\int_0^S H_u \, dL_u^{a_{i+1}} - \int_0^S H_u \, dL_u^{a_i} \right)^2 = 4 \int_0^S H_u^2 1_{(a < B_u < b)} du$$

whenever either one of the following conditions applies:

i) H is a $(\mathscr{F}_t^B)$-cont. semimart.;
ii) $H = f(B)$ where f is of bounded variation on every compact interval and S is such that $x \to L_S^x$ is a semimartingale (see Sect. 2, Chap. XI).

Examples of processes H for which the above result fails to be true are found in Exercise (2.13) Chapter XI.

§2. The Local Time of Brownian Motion

In this section, we try to go deeper into the properties of local times for standard linear BM. By the preceding section, it has a bicontinuous family L_t^a of local times. We write L_t instead of L_t^0 as we focus on the local time at zero.

We first introduce some notation, which is valid for a general semimartingale X and will be used also in Sect. 4 and in Chap. XII. We denote by $Z(\omega)$ the random set $\{s : X_s(\omega) = 0\}$. The set Z^c is open and therefore is a countable union of open intervals.

For $t \geq 0$, we define

$$d_t = \inf\{s > t : X_s = 0\}$$

with the convention $\inf(\emptyset) = +\infty$. Those are stopping times which are easily seen to be predictable and for $t \geq 0$, we have $L_t = L_{d_t}$ because as was proved in the last section, dL_t is carried by Z. In the case of BM we will prove shortly a more precise result (see also Exercise (1.26)).

Our first goal is to find the law of the process L. We shall use Itô-Tanaka's formula

$$|B_t| = \int_0^t \operatorname{sgn}(B_s) dB_s + L_t.$$

We saw in Exercise (1.14) of Chap. III that $|B_t|$ is a Markov process (see also Exercise (2.18) in this section). It is also clearly a semimartingale and its local time at 0 is equal to $2L$ (see Exercises (1.17) and (2.14)). A process having the law of $|B|$ is called a *reflecting Brownian motion* (we shall add "at zero" if there is a risk of ambiguity).

To analyse L, we need the following lemma which is useful in other contexts as well (see Exercise (2.14) Chap. IX).

(2.1) Lemma (Skorokhod). *Let y be a real-valued continuous function on $[0, \infty[$ such that $y(0) \geq 0$. There exists a unique pair (z, a) of functions on $[0, \infty[$ such that*

i) $z = y + a$,
ii) *z is positive,*
iii) *a is increasing, continuous, vanishing at zero and the corresponding measure da_s is carried by $\{s : z(s) = 0\}$.*

The function a is moreover given by

$$a(t) = \sup_{s \leq t} \big(-y(s) \vee 0 \big).$$

Proof. We first remark that the pair (a, z) defined by

$$a(t) = \sup_{s \le t} \big(- y(s) \vee 0 \big), \quad z = y + a$$

satisfies properties i) through iii).

To prove the uniqueness of the pair (a, z), we remark that if $(\tilde{a}, \tilde{z})$ is another pair which satisfies i) through iii), then $z - \tilde{z} = a - \tilde{a}$ is a process of bounded variation, and we can use the integration by parts formula to obtain

$$0 \le (z - \tilde{z})^2(t) = 2 \int_0^t \big(z(s) - \tilde{z}(s)\big) d\big(a(s) - \tilde{a}(s)\big).$$

Thanks to iii) this is further equal to

$$-2 \int_0^t \tilde{z}(s) da(s) - 2 \int_0^t z(s) d\tilde{a}(s)$$

which by ii) and iii) is ≤ 0. □

(2.2) Corollary. *The process* $\beta_t = \int_0^t \operatorname{sgn}(B_s) dB_s$ *is a standard BM and* $\mathscr{F}_t^\beta = \mathscr{F}_t^{|B|}$. *Moreover,* $L_t = \sup_{s \le t}(-\beta_s)$.

Proof. That β is a BM is a straightforward consequence of P. Lévy's characterization theorem. The second sentence follows at once from the previous lemma and Tanaka's formula

$$|B|_t = \beta_t + L_t.$$

It is now obvious that $\mathscr{F}_t^{|B|} \subset \mathscr{F}_t^\beta$ and, since it follows from Corollary (1.9) that $\mathscr{F}_t^L \subset \mathscr{F}_t^{|B|}$, we have $\mathscr{F}_t^\beta \subset \mathscr{F}_t^{|B|}$; the proof is complete.

Remark. Another proof of the equality $\mathscr{F}_t^{|B|} = \mathscr{F}_t^\beta$ will be given in Exercise (3.16) Chap. IX.

This corollary entails that the processes L_t and S_t have the same law; in particular, by Proposition (3.8) Chap. III, L_t is the inverse of a stable subordinator of index 1/2. Another proof of this is given in Exercise (1.11) Chap. X. The equality of the laws of L_t and S_t can be still further improved.

(2.3) Theorem (Lévy). *The two-dimensional processes* $(S_t - B_t, S_t)$ *and* $(|B_t|, L_t)$ *have the same law.*

Proof. On one hand, we have by Tanaka's formula, $|B_t| = \beta_t + L_t$; on the other hand, we may trivially write $S_t - B_t = -B_t + S_t$. Thus, Lemma (2.1) shows that one gets S and $S - B$ (resp. L and $|B|$) from $-B$ (resp. β) by the same deterministic procedure. Since $-B$ and β have the same law, the proof is finished.

Remark. The filtration of $S - B$ is actually that of B (see Exercise (2.12)) hence, the filtration of $(S - B, S)$ is also $(\mathscr{F}_t^B)$ whereas the filtration of $(|B|, L)$, which is $(\mathscr{F}_t^{|B|})$, is strictly coarser.

The following corollary will be important in later sections.

(2.4) Corollary. *$P\left[L_\infty^a = \infty\right] = 1$ for every a.*

Proof. By the recurrence properties of BM, we have obviously $P[S_\infty = \infty] = 1$; thus $P\left[L_\infty^0 = \infty\right] = 1$ follows from Theorem (2.3). For every $a \neq 0$, we then get $P\left[L_\infty^a = \infty\right] = 1$ from the fact that $B_{T_a+t} - a$ is a standard BM. □

We now turn to the result on the support of the measure dL_s which was announced earlier. We call (τ_t) the time-change associated with L_s i.e.

$$\tau_t = \inf\{s > 0 : L_s > t\}.$$

By the above corollary, the stopping times τ_t are a.s. finite. We set

$$\mathscr{O}(\omega) = \bigcup_{s \geq 0}]\tau_{s^-}(\omega), \tau_s(\omega)[.$$

The sets $]\tau_{s^-}, \tau_s[$ are empty unless the local time L has a constant stretch at level s and this stretch is then precisely equal to $[\tau_{s^-}, \tau_s]$. The sets $]\tau_{s^-}, \tau_s[$ are therefore pairwise disjoint and $\mathscr{O}(\omega)$ is in fact a countable union. We will prove that this set is the complement of $Z(\omega)$. We recall from Sect. 3 in Chap. III that Z has almost surely an empty interior and no isolated points; the sets $]\tau_{s^-}, \tau_s[$ are precisely the excursion intervals defined in Chap. III.

(2.5) Proposition. *The following three sets*
i) $Z(\omega)$, ii) $\mathscr{O}(\omega)^c$, iii) the support $\Sigma(\omega)$ of the measure $dL_t(\omega)$,
are equal for almost every ω.

Proof. An open set has zero dL_t-measure if and only if L is constant on each of its connected components. Thus, the set $\mathscr{O}(\omega)$ is the largest open set of zero measure and $\Sigma(\omega) = \mathscr{O}(\omega)^c$.

We already know from Proposition (1.3) that $\Sigma(\omega) \subset Z(\omega)$ a.s. To prove the reverse inclusion, we first observe that $L_t > 0$ a.s. for any $t > 0$ or in other words that $\tau_0 = 0$ a.s. Furthermore, since d_t is a stopping time and $B_{d_t} = 0$, Tanaka's formula, for instance, implies that $L_{d_t+s} - L_{d_t}$, $s \geq 0$, is the local time at zero of the BM B_{d_t+s}, $s \geq 0$ and therefore $L_{d_t+s} - L_{d_t} > 0$ a.s. for every $s > 0$. We conclude that for any fixed t the point $d_t(\omega)$ is in $\Sigma(\omega)$ for a.e. ω and, consequently, for a.e. ω the point $d_r(\omega)$ is in $\Sigma(\omega)$ for every $r \in \mathbb{Q}_+$.

Pick now s in $Z(\omega)$ and an interval $I \ni s$. Since $Z(\omega)$ is a.s. closed and has empty interior, one may find r such that $r < s$, $r \in \mathbb{Q}_+ \cap I$ and $r \notin Z(\omega)$; plainly $d_r \leq s$, thus s is the limit of points of the closed set $\Sigma(\omega)$, hence belongs to $\Sigma(\omega)$ and we are done.

Remarks. 1°) The fact that $\tau_0 = 0$ a.s. is worth recording and will be generalized in Chap. X.

2°) The equality between $Z(\omega)$ and $\Sigma(\omega)$ is also a consequence of Exercise (1.26) and Proposition (3.12) in Chap. III.

(2.6) Corollary. *$P(\forall s \geq 0, B_{\tau_s} = B_{\tau_{s-}} = 0) = 1$. Conversely, for any $u \in Z$, either $u = \tau_s$ or $u = \tau_{s-}$ for some s.*

Proof. The first statement is obvious. To prove the second let $u > 0$ be a point in $Z = \Sigma$; then, either $L_{u+\varepsilon} - L_u > 0$ for every $\varepsilon > 0$, hence $u = \inf\{t : L_t > L_u\}$ and $u = \tau_s$ for $s = L_u$, or L is constant on some interval $[u, u+\varepsilon]$, hence $L_u - L_{u-\eta} > 0$ for every $\eta > 0$ and u is equal to τ_{s-} for $s = L_u$.

Remark. We have just proved that the points of Z which are not left-ends of intervals of $\mathcal{O}$ are points of right-increase of L.

We close this section with P. Lévy's *Arcsine law* which we prove by using the above ideas in a slightly more intricate context. The following set-up will be used again in an essential way in Sect. 3 Chap. XIII.

We set

$$A_t^+ = \int_0^t 1_{(B_s > 0)} ds, \quad A_t^- = \int_0^t 1_{(B_s < 0)} ds,$$

and call α_t^+ and α_t^- the associated time-changes. Our aim is to find the law of the r.v. A_1^+ and since $\{A_1^+ > t\} = \{\alpha_t^+ < 1\}$, this amounts to finding the law of α_t^+. But since $u = A_u^+ + A_u^-$ entails $\alpha_t^+ = t + A^-(\alpha_t^+)$, we will look for the law of $A^-(\alpha_t^+)$. The following considerations serve this particular goal.

The processes $A^\pm$ are the increasing processes associated with the martingales

$$M_t^+ = \int_0^t 1_{(B_s > 0)} dB_s, \quad M_t^- = \int_0^t 1_{(B_s < 0)} dB_s.$$

Obviously $\langle M^+, M^- \rangle = 0$, and so, by Knight's theorem (Theorem (1.9) Chap. V), since $A_\infty^+ = A_\infty^- = \infty$, there exist two independent Brownian motions, say δ^+ and δ^-, such that $M^\pm(\alpha_\cdot^\pm) = \delta^\pm$ and by the definition of the local time L of B,

$$B^+(\alpha_\cdot^+) = \delta^+ + \frac{1}{2}L(\alpha_\cdot^+), \quad B^-(\alpha_\cdot^-) = -\delta^- + \frac{1}{2}L(\alpha_\cdot^-),$$

where B^+ and B^- are simply the positive and negative parts of B (and not other processes following the $\pm$ notational pattern).

Using Lemma (2.1) once more, we see that $\left(B^\pm(\alpha_\cdot^\pm), \frac{1}{2}L(\alpha_\cdot^\pm)\right)$ has the same law as $(|B|, L)$ and moreover that

$$\frac{1}{2}L(\alpha_t^+) = \sup_{s \leq t}(-\delta_s^+), \quad \frac{1}{2}L(\alpha_t^-) = \sup_{s \leq t}(\delta_s^-).$$

We now prove the

(2.7) Theorem. *The law of A_1^+ is the Arcsine law on* $[0, 1]$.

Proof. For $a > 0$ let us put

$$T^{\delta^-}(a) \equiv T_a^{\delta^-} = \inf\left\{t : \delta_t^- > a\right\}.$$

We claim that for every t, we have $A_{\tau_t}^- = T^{\delta^-}(t/2)$. Indeed, by definition of α_t^-,

$$B_t^- = -\delta^-(A_t^-) + L_t/2$$

whence $B^-(\tau_t) = 0 = -\delta^-(A_{\tau_t}^-)+t/2$. Moreover, τ is a.s. a point of right increase of L (see Corollary (2.6)), hence there is a sequence (s_n) decreasing to τ_t such that $B_{s_n}^- = 0$ and $L_{s_n} > L_{\tau_t} = t$ and consequently $\delta^-(A_{s_n}^-) > t/2$. It follows that $A_{\tau_t}^- \geq T^{\delta^-}(t/2)$. Now, if $u < A_{\tau_t}^-$, then $u = A_v^-$ for some $v < \tau_t$. If $v < \tau_{t^-}$ then

$$\delta^-(A_v^-) = \frac{1}{2}L_v - B_v^- < \frac{t}{2}.$$

If $\tau_{t^-} \leq v < \tau_t$, then A^- increases between τ_{t^-} and τ_t which implies that B is negative on this interval, hence B^- is > 0 and again $\delta^-(A_v^-)$ is less than $t/2$, which proves the reverse inequality $A_{\tau_t}^- \leq T^{\delta^-}(t/2)$.

Moreover $A^-(\alpha^+) = A^-\left(\tau_{L(\alpha^+)}\right)$; indeed $\tau_{L_v} = v$ if v is a point of increase of L and if v belongs to a level stretch of L and $v = \alpha_t^+$ for some t, then B is positive on this stretch and $A^-(\tau_{L_v}) = A_v^-$.

Combining these two remarks we may write that for each fixed t,

$$A^-(\alpha_t^+) = T^{\delta^-}\left(\frac{1}{2}L(\alpha_t^+)\right).$$

Now, $\frac{1}{2}L(\alpha_t^+)$ is independent of δ^- since it is equal to $\sup_{s\leq t}(-\delta_s^+)$. Thus we have proved that $A^-(\alpha_t^+) = T^{\delta^-}(S_t)$ where S_t is the supremum process of a BM β independent of δ^-. By the scaling properties of the family $T_a^{\delta^-}$ and the reflection principle seen in Sect. 3 of Chap. III we get

$$A^-(\alpha_t^+) \overset{(d)}{=} S_t^2 \cdot T_1^{\delta^-} \overset{(d)}{=} t\beta_1^2 \cdot (\delta_1^-)^{-2} \overset{(d)}{=} t \cdot C^2$$

where C is a Cauchy variable with parameter 1.

By the discussion before the statement, $\alpha_t^+ \overset{(d)}{=} t(1 + C^2)$ and

$$P\left[A_1^+ > t\right] = P\left[(1 + C^2)^{-1} > t\right].$$

Hence, the r.v. A_1^+ follows the law of $(1 + C^2)^{-1}$ which is the Arcsine law as can be checked by elementary computations (see Sect. 6 Chap. 0).

The reader ought to ponder what the Arcsine law intuitively means. Although the BM is recurrent and comes back infinitely often to zero, the chances are that at a given time, it will have spent much more time on one side of zero than on the other.

(2.8) Exercise. 1°) For the linear BM and $a \leq x \leq y \leq b$ prove that

$$E_x\left[L^y_{T_a\wedge T_b}\right] = 2u(x, y)$$

where $u(x, y) = (x-a)(b-y)/(b-a)$.

2°) For $a \leq y \leq x \leq b$ set $u(x, y) = u(y, x)$ and prove that for any positive Borel function f

$$E_x\left[\int_0^{T_a\wedge T_b} f(B_s)ds\right] = 2\int_a^b u(x, y)f(y)dy.$$

This gives the potential of BM killed when it first hits either a or b.

(2.9) Exercise. Let (P_t) be the semi-group of BM and put $f(x) = |x|$. Prove that

$$L_t = \lim_{h\to 0}\frac{1}{h}\int_0^t (P_h f(B_s) - f(B_s))\,ds \quad \text{a.s.}$$

[Hint: Write the occupation times formula for the function $P_h f - f$.]

(2.10) Exercise. 1°) Prove that for $\alpha > 0$, the processes

$$\left(S_t - B_t + \alpha^{-1}\right)\exp(-\alpha S_t) \text{ and } \left(|B_t| + \alpha^{-1}\right)\exp(-\alpha L_t)$$

are local martingales.

2°) Let $U_x = \inf\{t : S_t - B_t > x\}$ and $\tilde{T}_x = \inf\{t : |B_t| > x\}$. Prove that both S_{U_x} and $L_{\tilde{T}_x}$ follow the exponential law of parameter x^{-1}. This can also be proved by the methods of Sect. 4.

(2.11) Exercise (Invariance under scaling). Let $0 < c < \infty$.

1°) Prove that the doubly-indexed processes (B_t, L^a_t) and $\left(B_{ct}, L^{a\sqrt{c}}_{ct}\right)/\sqrt{c}$, $a \in \mathbb{R}$, $t \geq 0$, have the same law.

2°) Prove that the processes (τ_t) and $\left(c^{-1}\tau_{\sqrt{c}t}\right)$ have the same law.

3°) If as usual $T_a = \inf\{t : B_t = a\}$ prove that the doubly-indexed processes $(L^x_{T_a})$ and $\left(c^{-1}L^{cx}_{T_{ca}}\right)$, $x \in \mathbb{R}$, $a \geq 0$, have the same law.

(2.12) Exercise. Prove that $\mathscr{F}^B_t = \mathscr{F}^{S-B}_t$. In other words, if you know $S - B$ up to time t you can recover B up to time t.

(2.13) Exercise. If $B = (B^1, B^2)$ is a standard planar BM and τ_t is the inverse of the local time of B^1 at zero, prove that $X_t = B^2_{\tau_t}$ is a symmetric Cauchy process. Compare with Exercise (3.25) of Chap. III.

(2.14) Exercise. 1°) Prove that the two-dimensional process $\left(|B_t|, \frac{1}{2}L(|B|)_t\right)$ has the same law as the processes of Theorem (2.3).

2°) Conclude that the local time of $|B_t|$ (resp. $S_t - B_t$) is equal to $2L_t$ (resp. $2S_t$). See also Exercise (1.17).

(2.15) Exercise. 1°) Fix $t > 0$. Prove that for the standard linear BM, there is a.s. exactly one $s < t$ such that $B_s = S_t$, in other words

$$P[\exists (r, s) : r < s \leq t \text{ and } B_r = B_s = S_t] = 0.$$

[Hint: $2S$ is the local time at 0 of the reflected BM $S - B$. This result can actually be proved by more elementary means as is hinted at in Exercise (3.26) of Chap. III.]

2°) Prove that $G_1 = \sup\{s < 1 : B_s = S_1\}$ has also the Arcsine law; thus G_1 and A_1 have the same law.

[Hint: Use Exercise (3.20) Chapter III.]

(2.16) Exercise. Let X be the standard BM reflected at 0 and 1 (see Exercise (1.14) of Chap. III).

1°) Prove that $X_t = \beta_t + \tilde{L}_t^0 - \tilde{L}_t^1$ where β is a standard linear BM and $\tilde{L}^a$ the symmetric local time (Exercise (1.25)) of X at a.

2°) By extending Lemma (2.1) to this situation prove that

$$\tilde{L}_t^0 = \sup_{s \leq t} \left(-\beta_s + \tilde{L}_s^1\right)^+, \quad \tilde{L}_t^1 = \sup_{s \leq t} \left(\beta_s + \tilde{L}_s^0 - 1\right)^+.$$

(2.17) Exercise. Prove that the filtration $(\mathscr{F}_t^X)$ of the martingale $X_t = \int_0^t B_s^1 dB_s^2$ introduced in Exercise (4.13) of Chap. V, is the filtration of a BM^2.

[Hint: Compute $\langle X, X \rangle$.]

(2.18) Exercise. 1°) Prove that the joint law of $(|B_t|, L_t)$ has a density given by

$$(2/\pi t^3)^{1/2}(a + b)\exp\left(-(a + b)^2/2t\right), \quad a, b \geq 0.$$

Give also the law of (B_t, L_t).

2°) Prove that the 2-dimensional process $(|B_t|, L_t)$ is a Markov process with respect to $(\mathscr{F}_t)$ and find its transition function.

The reader will find a more general result in Exercise (1.13) of Chap. X and may also compare with Exercise (3.17) in Chap. III.

(2.19) Exercise. 1°) Prove that almost-surely the random measure ν on $\mathbb{R}$ defined by

$$\nu(f) = \int_0^1 \int_0^1 f(B_t - B_s) ds\, dt$$

has a continuous density with respect to the Lebesgue measure. Prove that this density is Hölder continuous of order β for every $\beta < 1$.

[Hint: This last result can be proved by using the same kind of devices as in the proof of Corollary (1.8); it is also a consequence of the fact that the convolution of two Hölder continuous functions of order β is Hölder continuous of order 2β.]

2°) More generally, for every Borel subset Γ of $[0, 1]^2$, the measure

$$\nu(f) = \iint_\Gamma f(B_t - B_s) ds\, dt$$

has a continuous density $\alpha(x, \Gamma)$ with respect to the Lebesgue measure. The map α is then a kernel on $\mathbb{R} \times [0, 1]^2$.

[Hint: Use Exercise (1.15).]

3°) Prove that, if $f \in L^1(\mathbb{R})$, a.s.

$$\lim_{n\to\infty} n \int_0^1 \int_0^1 f(n(B_t - B_s))\,dt\,ds = \left(\int f(a)da\right)\left(\int \left(L_1^b\right)^2 db\right).$$

(2.20) Exercise. With the notation used in the proof of the Arcsine law, prove that B is a deterministic function of δ^+ and δ^-, namely, there is a function f on $C(\mathbb{R}_+, \mathbb{R})^2$ such that $B = f(\delta^+, \delta^-)$.

[This exercise is solved in Chap. XIII, Proposition (3.5).]

(2.21) Exercise. Prove the result of Exercise (1.26) on the support of dL_t by means of Proposition (2.5) and the DDS theorem of Sect. 1 in Chap. V.

(2.22) Exercise. Let f be a locally bounded odd function on $\mathbb{R}$ with a constant sign on each side of 0 and such that the set $\{x : f(x) = 0\}$ is of zero Lebesgue measure. Prove that the filtration generated by $M_t = \int_0^t f(B_s)dB_s$ is that of a Brownian motion.

[Hint: Use Exercise (3.12) Chap. V.]

(2.23) Exercise. Let B be the standard linear BM. Prove that $f(B_t)$ is a $(\mathscr{F}_t)$-local submartingale if and only if f is a convex function.

[Hint: A function f is convex if and only if $f + l$ admits no proper local maximum for any affine function l whatsoever.]

* **(2.24) Exercise.** Let X be a continuous semimartingale, if it exists, such that

$$(*) \qquad X_t = x + B_t + \int_0^t \alpha(s)dL_s$$

where B is a BM, α is a deterministic Borel function on $\mathbb{R}_+$ and $L = L^0(X)$.

1°) Prove that if $\alpha < 1$ the law of the process L is uniquely determined by α.

[Hint: Write the expression of $|X|$ and use Lemma (2.1).]

2°) Let $g_t(\lambda) = E\left[\exp(i\lambda X_t)\right]$ and prove that

$$g_t(\lambda) = \exp(i\lambda x) - \frac{\lambda^2}{2}\int_0^t g_s(\lambda)ds + i\lambda E\left[\int_0^t \alpha(s)dL_s\right].$$

As a result the law of the r.v. X_t is also determined by α. Using the same device for conditional laws, prove that all continuous semimartingales satisfying equation $(*)$ have the same law. In the language of Chap. IX, there is uniqueness in law for the solution to $(*)$. The skew BM of Exercise (2.24) Chap. X is obtained in the special case where α is constant.

3°) Prove that $L_t^{0-} = \int_0^t \left(1-2\alpha(s)\right)dL_s$ and that as a result there is no solution X to $(*)$ if α is a constant $> 1/2$.

(2.25) Exercise. Let X be a continuous process and L^a the family of local times of BM. Prove that for each t the process

$$Y_a = \int_0^t X_u dL_u^a$$

is continuous.

* **(2.26) Exercise (Hausdorff dimension of the set of zeros of BM).** 1°) For every $n \geq 1$, define inductively two sequences of stopping times by

$$U_1^n = 0, \quad V_k^n = \inf\{t \geq U_k^n : |B_t| = 2^{-n}\}, \quad U_{k+1}^n = \inf\{t \geq V_k^n : |B_t| = 0\}.$$

For an integer K, prove that if $\alpha > \frac{1}{2}$, then almost-surely

$$\sup\{(V_k^n - U_k^n)^\alpha, \quad 1 \leq k \leq K2^n\} \leq 2^{-n}$$

for n sufficiently large.

[Hint: The r.v.'s $(V_k^n - U_k^n)$ are i.i.d. with the law of $2^{-2n}\tilde{T}_1$.]

2°) Using the approximation result of Exercise (1.19), prove that, a.s. on the set $\{L_1 < K\}$, for n sufficiently large

$$\sum_1^{N_n} (V_k^n - U_k^n)^\alpha \leq K$$

where $N_n = \sup\{k \geq 1 : U_k^n \leq 1\}$. Conclude that the Hausdorff dimension of the set $Z = \{t \leq 1 : B_t = 0\}$ is a.s. $\leq \frac{1}{2}$.

The reverse inequality is the subject of the following exercise.

* **(2.27) Exercise.** 1°) Let ν be a measure on $\mathbb{R}$, C and α two constants > 0, such that in the notation of Appendix 4, $\nu(I) \leq C|I|^\alpha$ for every interval. If A is a Borel set of strictly positive ν-measure, prove that the Hausdorff dimension of A is $\geq \alpha$.

2°) By applying 1°) to the measure dL_t, prove that the Hausdorff dimension of Z is $\geq 1/2$ a.s. Together with the preceding exercise, this shows that the Hausdorff dimension of Z is almost-surely equal to 1/2.

** **(2.28) Exercise (Local times of BM as stochastic integrals).**

1°) For $f \in C_K(\mathbb{R})$, prove that for the standard linear BM,

$$\int_0^t f(B_s)ds = \int_0^t E[f(B_s)]\,ds + \lim_{\varepsilon \downarrow 0} \int_\varepsilon^t ds \int_0^{s-\varepsilon} (P_{s-v}f)'(B_v)dB_v.$$

[Hint: Use the representation of Exercise (3.13) Chap. V.]

2°) If we put $q(x) = 2\int_{|x|}^\infty \exp(-u^2/2)du$ prove that

$$\int_0^t f(B_s)ds = \int_0^t E[f(B_s)]\,ds + \lim_{\varepsilon \downarrow 0} \frac{1}{\sqrt{2\pi}} \int_{-\infty}^{+\infty} f(y)dy \ldots$$

$$\ldots \int_0^{t-\varepsilon} \operatorname{sgn}(B_v - y)\left[q\left(\frac{B_v - y}{\sqrt{t-v}}\right) - q\left(\frac{B_v - y}{\sqrt{\varepsilon}}\right)\right] dB_v.$$

[Hint: Use Fubini's theorem for stochastic integrals extended to suitably integrable processes.]

3°) Conclude that $\int_0^t f(B_s)ds = \int_{-\infty}^{+\infty} f(y)L_t^y dy$ where

$$L_t^y = \int_0^t g_s(y)ds - \frac{1}{\sqrt{2\pi}}\int_0^t \operatorname{sgn}(B_v - y)q\left(\frac{B_v - y}{\sqrt{t-v}}\right)dB_v.$$

** **(2.29) Exercise (Pseudo-Brownian Bridge).** 1°) Let h be a bounded Borel function on $\mathbb{R}_+$ and set

$$M_t = \mathscr{E}\left(\int_0^{\cdot} h(s)dB_s\right)_t.$$

If ϕ is a continuous function with compact support, prove that the process

$$Y_a = \int_0^\infty \phi(t)M_t dL_t^a, \quad a \in \mathbb{R},$$

is continuous (see Exercise (2.25)).

2°) Let Q_x^u be the law of the Brownian Bridge from 0 to x over the interval $[0, u]$ and write simply Q^u for Q_0^u (see Exercise (3.16) Chap. I for the properties of the family (Q_x^u)). If Z is a positive predictable process prove that

$$\int_0^\infty E[Z_{\tau_s}]ds = \int_0^\infty Q^u[Z_u]\frac{du}{\sqrt{2\pi u}}$$

where $Q^u[Z_u] = \int Z_u dQ^u$.

[Hint: Compute in two different ways $E\left[\int_0^\infty f_n(B_t)\phi(t)M_t dt\right]$ where (f_n) is an approximation of the Dirac mass δ_0, then let n tend to $+\infty$.]

3°) If F is a positive Borel function on $C([0,1],\mathbb{R})$ and g a positive Borel function on $\mathbb{R}_+$ prove that

$$\int_0^\infty E\left[F\left(\frac{1}{\sqrt{\tau_s}}B_{v\tau_s}; v \le 1\right)g(\tau_s)\right]ds = \int_0^\infty \frac{g(u)}{\sqrt{2\pi u}}E\left[F\left(\frac{1}{\sqrt{u}}\beta_{vu}^u, v \le 1\right)\right]du$$

where β^u is the Brownian Bridge from 0 to 0 over the time interval $[0, u]$.

4°) Let X be the process defined on $[0, 1]$ by

$$X_u = \frac{1}{\sqrt{\tau_1}}B_{u\tau_1}, \quad 0 \le u \le 1$$

which may be called the *Pseudo-Brownian Bridge*. Prove that

$$E\left[F\left(X_u, u \le 1\right)\right] = E\left[F\left(\beta_u^1, u \le 1\right)\sqrt{\frac{2}{\pi}}\frac{1}{\lambda}\right]$$

where λ is the local time of β^1 at level 0 and at time 1.

[Hint: Use the scaling invariance properties to transform the equality of 3°). Then observe that $1/\sqrt{\tau_1}$ is the local time of X at level 0 and time 1 and that

$$\frac{1}{\sqrt{\tau_1}} = \lim_{\varepsilon \to 0} \frac{1}{2\varepsilon} \int_0^1 1_{]-\varepsilon,\varepsilon[}(X_s)ds \quad .]$$

5°) Prove that the processes $(B_t; t \leq \tau_1)$ and $\left(B_{\tau_1 - t}; t \leq \tau_1\right)$ have the same law.

6°) Prove that λ has the law of $\sqrt{2\mathbf{e}}$ where $\mathbf{e}$ is exponential with parameter 1. This is taken up in Exercise (3.8) of Chap. XII. This question can be solved independently of 5°) .

(2.30) Exercise. In the notation of Exercise (3.23) Chap. III, prove that the process $\{g_1^{-1/2} B_{ug_1},\ 0 \leq u \leq 1\}$ is a Brownian Bridge which is independent of $\sigma(g_1, B_{g_1+u}, u \geq 0)$.

[Hint: Use time-inversion and Exercise (3.10) Chap. I.]

\# **(2.31) Exercise.** In the situation of Proposition (3.2) Chap. V, prove that there are infinitely many predictable processes H such that $\int_0^\infty H_s^2 ds < \infty$, and

$$F = E[F] + \int_0^\infty H_s dB_s.$$

[Hint: In the notation of this section, think about $1_{[0,\tau_t]}$.]

(2.32) Exercise. Let X be a cont. loc. mart. vanishing at 0 and set

$$S_t = \sup_{s \leq t} X_s \quad \text{and} \quad \hat{X}_t = \int_0^t \operatorname{sgn}(X_s) dX_s.$$

1°) Prove that the following two properties are equivalent:

i) the processes $S - X$ and $|X|$ have the same law,
ii) the processes X and $-\hat{X}$ have the same law.

2°) Let X and Y be two cont. loc. mart. vanishing at zero and call β and γ their DDS Brownian motions. Prove that X and Y have the same law iff the 2-dimensional processes $(\beta, \langle X, X\rangle)$ and $(\gamma, \langle Y, Y\rangle)$ have the same law. In particular, if $Y = \hat{X}$, then X and Y have the same law iff conditionally w.r.t. $\mathscr{F}_\infty^{\langle X,X\rangle}$, the processes β and γ have the same law.

3°) If $\mathscr{F}_\infty^{\langle X,X\rangle} = \mathscr{F}_\infty^{|\beta|}$, then the laws of X and $\hat{X}$ are not equal. Let ϕ be a continuous, one-to-one and onto function from $\mathbb{R}_+$ to $[a, b[$ with $0 < a < b < \infty$, β a BM^1 and A the time-change associated with $\int_0^t \phi(|\beta_s|)ds$, then if $X_t = \beta_{A_t}$, the laws of X and $\hat{X}$ are not equal. Notice in addition that there is a BM denoted by B such that $X_t = \int_0^t \phi(|X_s|)^{-1/2} dB_s$.

4°) Likewise prove that if $\mathscr{F}_\infty^{\langle X,X\rangle} = \mathscr{F}_\infty^{\beta}$, then the laws of X and $\hat{X}$ are not equal. Change suitably the second part of 3°) to obtain an example of this situation.

** **(2.33) Exercise (Complements to the Arcsine law).** 1°) In the notation of Theorem (2.7), prove that for every positive Borel function F on $C([0,1],\mathbb{R})$,

$$E\left[F(B_u, u \le 1)\, 1_{(B_1>0)}\right] = E\left[\left(1/\alpha_1^+\right) F\left(B_{u\alpha_1^+}/\sqrt{\alpha_1^+}, u \le 1\right)\right].$$

[Hint: It may be helpful to consider the quantity

$$E\left[\int_0^\infty F\left(t^{-1/2}B_{st}, s \le 1\right)\phi(t)\,dA_t^+\right]$$

where ϕ is a positive Borel function on $\mathbb{R}_+$.]

2°) Prove that for every positive Borel function f on $[0,1]\times\mathbb{R}_+$, one has

$$E\left[f\left(A_1^+, L_1^2\right) 1_{(B_1>0)}\right] = E\left[\left(A_{\tau_1}^+/\tau_1\right) f\left(A_{\tau_1}^+/\tau_1, 1/\tau_1\right)\right].$$

3°) Prove that the law of the triple $T^{-1}\left(A_T^+, A_T^-, L_T^2\right)$ is the same for all the following random times:

i) $T = t$ (a constant time), ii) $T = \alpha_s^+$, iii) $T = \tau_u$.

(2.34) Exercise. 1°) Let B be the standard BM^1 and L its local time at 0; for $h > 0$, prove that there is a loc. mart. M such that

$$\log(1 + h|B_t|) = M_t - (1/2)\langle M, M\rangle_t + hL_t.$$

2°) Let γ be the DDS BM of $-M$; set $\beta_t = \gamma_t + (1/2)t$ and $\sigma_t = \sup_{s\le t}\beta_s$. Using Skorokhod's lemma (2.1), prove that for every $t \ge 0$,

$$\log(1 + h|B_t|) = \sigma_{V_t} - \beta_{V_t} \quad \text{a.s.}$$

where $V_t = \int_0^t \left(h^{-1} + |B_s|\right)^{-2} ds$.

3°) Define as usual $\tilde{T}_a = \inf\{t : |B_t| = a\}$ and $\tau_s = \inf\{u : L_u > s\}$ and prove that

$$V_{\tilde{T}_a} = \inf\{u : \sigma_u - \beta_u = \log(1+h)\} \text{ and } V_{\tau_s} = \inf\{u : \beta_u = hs\}.$$

The laws of $V_{\tilde{T}_a}$ and V_{τ_s} are obtained at the end of Exercise (3.18) Chapter VIII.

(2.35) Exercise (Local times of the Brownian bridge). 1°) Let $x > 0$. In the notation of this section, by considering the BM $B_{t+T_x} - x$, and using Theorem (2.3) prove that

$$P\left[L_1^x \ge y, B_1 > b\right] = (1/2)P\left[S_1 > x + y, S_1 - B_1 > |b - x|\right].$$

Consequently, prove that, conditionally on $B_1 = b$, the law of L_1^x, for $0 \le x \le b$, does not depend on x.

2°) Using the result in Exercise (2.18) and conditioning with respect to $\{B_1 = 0\}$, prove that if (l^x) is the family of local times of the standard BB, then

$$l_1^x \overset{(d)}{=} (R - 2x)^+,$$

where R has the density $1_{(r>0)} r \exp\left(-(r^2/2)\right)$.

§3. The Three-Dimensional Bessel Process

In Chap. XI we will make a systematic study of the one-parameter family of the so-called Bessel processes using some notions which have yet to be introduced. In the present section, we will make a first study of the 3-dimensional Bessel process, which crops up quite often in the description of linear BM, using only the tools we have introduced so far.

We first take up the study of the euclidean norm of BM^δ which was begun in Chap. V for $\delta = 2$ and in the preceding section for $\delta = 1$. Let us suppose that δ is an integer ≥ 1 and let ρ_t be the modulus of BM^δ. As usual, we denote by P_x the probability measure of the BM^δ started at x and $(\mathscr{F}_t)$ is the complete Brownian filtration introduced in Sect. 2 Chap. III. For Bessel functions see Appendix 7.

(3.1) Proposition. *For every $\delta \geq 1$, the process ρ_t, $t \geq 0$, is a homogeneous $(\mathscr{F}_t)$-Markov process with respect to each P_x, $x \in \mathbb{R}^\delta$. For $\delta \geq 2$, its semi-group P_t^δ is given on $[0, \infty[$ by the densities*

$$p_t^\delta(a, b) = (a/t)(b/a)^{\delta/2} I_{\delta/2-1}(ab/t) \exp\left(-(a^2 + b^2)/2t\right) \quad \text{for } a, b > 0,$$

where I_ν is the modified Bessel function of index ν, and

$$p_t^\delta(0, b) = \Gamma(\delta/2) 2^{(\delta/2)-1} t^{-\delta/2} b^{\delta-1} \exp(-b^2/2t).$$

Proof. Let f be a positive Borel function on $[0, \infty[$. For $s < t$,

$$E_x\left[f(\rho_t) \mid \mathscr{F}_s\right] = E_{B_s}\left[f(|B_{t-s}|)\right] = P_{t-s}\tilde{f}(B_s) \quad P_x - \text{a.s.},$$

where $\tilde{f}(x) = f(|x|)$ and P_t is the semi-group of BM^δ.

For $\delta \geq 2$, we have

$$P_t\tilde{f}(x) = (2\pi t)^{-\delta/2} \int \exp\left(-|x - y|^2/2t\right) f(|y|) dy$$

and using polar coordinates

$$P_t\tilde{f}(x) = (2\pi t)^{-\delta/2} \int \exp\left(-(|x|^2 + \rho^2)/2t\right) \exp\left(-|x|\rho\cos\theta/t\right) f(\rho)\rho^{\delta-1} d\rho\, \sigma(d\eta)$$

where η is the generic element of the unit sphere and θ the angle between x and η. It turns out that $P_t\tilde{f}(x)$ depends only on $|x|$ which proves the first part of the result (the case $\delta = 1$ was studied in Exercise (1.14) of Chap. III). Moreover, setting $P_t^\delta f(a) = P_t\tilde{f}(x)$ where x is any point such that $|x| = a$, we see that P_t^δ has a density given by

$$(2\pi t)^{-\delta/2} b^{\delta-1} \exp\left(-(a^2 + b^2)/2t\right) \int_{S^{\delta-1}} \exp\left(-ab\cos\theta/t\right) \sigma(d\eta)$$

which entails the desired result.

(3.2) Definition. *A Markov process with semi-group P_t^δ is called a* δ-dimensional Bessel process.

Bessel processes are obviously Feller processes. We will write for short BES^δ and $\text{BES}^\delta(x)$ will designate a δ-dimensional Bessel process started at $x \geq 0$. The above result says that the modulus of a BM^δ is a realization of BES^δ. From the results obtained for BM^δ, we thus deduce that a BES^δ never reaches 0 after time 0 if $\delta \geq 2$. Moreover for $\delta \geq 3$, it is a transient process, that is, it converges a.s. to infinity.

From now on, we will focus on the 3-dimensional process BES^3, which we will designate by ρ_t. The semi-group P_t^3 has a particularly simple form which can be seen from the expression of $I_{1/2}$. We call Q_t the semi-group of the linear BM on $]0, \infty[$ killed when it hits zero. It was seen in Exercise (1.15) of Chap. III that Q_t is given by the density

$$q_t(x, y) = g_t(x - y) - g_t(x + y), \quad x > 0 \text{ and } y > 0.$$

If we set $h(x) = x$ on $]0, \infty[$, it is readily checked that $Q_t h = h$. The semi-group P_t^3 is what will be termed in Chap. VIII as the h-transform of Q_t, namely

$$P_t^3 f(x) = h(x)^{-1} Q_t(fh)(x), \quad x > 0;$$

in other words, P_t^3 is given by the density $x^{-1} q_t(x, y) y$. For $x = 0$, we have

$$P_t^3 f(0) = \int_0^\infty (2/\pi t^3)^{1/2} \exp(-y^2/2t) y^2 f(y) dy.$$

We will also need the following

(3.3) Proposition. *If (ρ_t) is a* $\text{BES}^3(x)$ *with $x \geq 0$, there is a Brownian motion β such that*

$$\rho_t = x + \beta_t + \int_0^t \rho_s^{-1} ds.$$

Moreover, ρ_t^{-1} is a local martingale (in the case $x = 0$, the time-set is restricted to $]0, \infty[$ *).*

Proof. We know that ρ_t may be realized as the modulus of BM^3; using the fact that $\beta_t = \sum_1^3 \int_0^t \rho_s^{-1} B_s^i dB_s^i$ is a BM^1 (P. Lévy's characterization theorem) the result follows easily from Itô's formula, and the fact that ρ_t never visits 0 after time 0.

Remark. The first result says that ρ is a solution to the stochastic differential equation $d\rho_s = d\beta_s + \rho_s^{-1} ds$ (see Chap. IX).

Another proof of the fact that ρ_t^{-1} is a local martingale is hinted at in Exercise (2.13) of Chap. V where it is used to give an important counter-example. It will now be put to use to prove the

(3.4) Corollary. *Let P_x^3 be the probability measure governing* $\mathrm{BES}^3(x)$ *with* $x > 0$ *and* T_a *be the hitting time of* $a > 0$. *For* $0 < a < x < b$,

$$P_x^3[T_a < T_b] = \left(b^{-1} - x^{-1}\right) / \left(b^{-1} - a^{-1}\right)$$

and $P_x^3[T_a < \infty] = a/x$. *Moreover,* $J_0 = \inf_{s \geq 0} \rho_s$ *is uniformly distributed on* $[0, x]$.

Proof. The local martingale ρ_t^{-1} stopped at T_a is bounded, hence is a martingale to which we may apply the optional stopping theorem. The proof then follows exactly the same pattern as in Proposition (3.8) of Chap. II. We then let b go to infinity to get $P_x^3[T_a < \infty]$. Finally

$$P_x^3[J_0 \leq a] = P_x^3[T_a < \infty] = a/x$$

which ends the proof.

We now turn to our first important result which complements Theorem (2.3), the notation of which we keep below, namely B is a $\mathrm{BM}^1(0)$ and $S_t = \sup_{s \leq t} B_s$.

(3.5) Theorem (Pitman). *The process* $\rho_t = 2S_t - B_t$ *is a* $\mathrm{BES}^3(0)$. *More precisely, if* ρ_t *is a* $\mathrm{BES}^3(0)$ *and* $J_t = \inf_{s \geq t} \rho_s$, *then the processes* $(2S_t - B_t, S_t)$ *and* (ρ_t, J_t) *have the same law.*

Proof. Let ρ be a $\mathrm{BES}^3(0)$. If we put $X_t = 2J_t - \rho_t$, we shall prove that for each t, $J_t = \sup_{s \leq t} X_s$. Indeed, if $J_t = \rho_t$, then $X_t = J_t$ and for $s \leq t$, since $J_s \leq \rho_s$, we get $X_s = 2J_s - \rho_s \leq J_s \leq J_t = X_t$ which proves our claim in this case; if $\rho_t \neq J_t$, then $\rho_t > J_t$ and $X_t > J_t = J_{g_t}$ where $g_t = \sup\{s < t : J_s = \rho_s\}$. Since by the first part $J_{g_t} = \sup_{s \leq g_t} X_s$, we get the result in all cases.

We have thus proved that $(\rho_t, J_t) = (2J_t - X_t, J_t)$ with $J_t = \sup_{s \leq t} X_s$ and consequently, it remains to prove that X is a BM. To this end it is enough by P. Lévy's characterization theorem (Sect. 3 Chap. IV) to prove that X is a martingale, since plainly $\langle X, X\rangle_t = \langle \rho, \rho\rangle_t = t$.

We first notice that $J_s = J_t \wedge \inf_{s \leq u \leq t} \rho_u$ for $s < t$ and therefore the knowledge of J_t and of ρ_s, $s \leq t$, entails the knowledge of J_s, $s \leq t$. As a result $\mathscr{F}_t^X \subset \mathscr{F}_t^\rho \vee \sigma(J_t)$. On the other hand $\sigma(J_t) \subset \mathscr{F}_t^X$ since $J_t = \sup_{s \leq t} X_s$ and $\mathscr{F}_t^\rho \subset \mathscr{F}_t^X$ since $\rho_t = 2J_t - X_t$. Consequently $\mathscr{F}_t^X = \mathscr{F}_t^\rho \vee \sigma(J_t)$. Since $X_t \leq \rho_t$ and $-X_t \leq \rho_t$, each r.v. X_t is integrable; thus, to prove that X is a martingale, it suffices to show that for any $a > 0$ and $s \leq t$,

$$(*) \qquad E_0^3\left[X_t 1_{(J_s > a)} \mid \mathscr{F}_s^\rho\right] = E_0^3\left[X_s 1_{(J_s > a)} \mid \mathscr{F}_s^\rho\right].$$

Call K the right-hand side of $(*)$. Corollary (3.4) implies that

$$P_z^3[J_0 > a] = P_z^3[T_a = \infty] = \begin{cases} 1 - az^{-1} & \text{if } a < z, \\ 0 & \text{if } a \geq z, \end{cases}$$

and

$$E_z^3\left[J_0 1_{(J_0>a)}\right] = \begin{cases} \left(z - a^2 z^{-1}\right)/2 & \text{if } a < z, \\ 0 & \text{if } a \geq z. \end{cases}$$

Using this and the Markov property for ρ, it is not difficult to prove that $K = \left(a - a^2\rho_s^{-1}\right) 1_{(\rho_s>a)}$.

Now, call k the left-hand side of $(*)$; we have

$$\begin{aligned} k &= E_0^3\left[X_t 1_{(J_t>a)} 1_{(\inf_{s\leq u\leq t}\rho_u>a)} | \mathscr{F}_t^{\rho} | \mathscr{F}_s^{\rho}\right] \\ &= E_0^3\left[E_0^3\left[X_t 1_{(J_t>a)} \mid \mathscr{F}_t^{\rho}\right] 1_{(\inf_{s\leq u\leq t}\rho_u>a)} \mid \mathscr{F}_s^{\rho}\right], \end{aligned}$$

and using the above computation of K with t instead of s, we obtain

$$\begin{aligned} k &= E_0^3\left[\left(a - a^2\rho_t^{-1}\right) 1_{(\inf_{s\leq u\leq t}\rho_u>a)} \mid \mathscr{F}_s^{\rho}\right] \\ &= E_0^3\left[\left(a - a^2\rho_t^{-1}\right) 1_{(\rho_s>a)} 1_{(s+T_a\circ\theta_s>t)} \mid \mathscr{F}_s^{\rho}\right] \\ &= E_0^3\left[\left(a - a^2\rho_{(t-s)\wedge T_a}^{-1} \circ \theta_s\right) 1_{(\rho_s>a)} \mid \mathscr{F}_s^{\rho}\right] \\ &= E_{\rho_s}^3\left[a - a^2\rho_{(t-s)\wedge T_a}^{-1}\right] 1_{(\rho_s>a)} = \left(a - a^2\rho_s^{-1}\right) 1_{(\rho_s>a)}, \end{aligned}$$

since $\left(\rho_{t\wedge T_a}^{-1}\right)$ is a bounded martingale. It follows that $k = K$ which ends the proof. □

It is interesting to observe that although $2S - B$ is a Markov process and is $(\mathscr{F}_t^B)$-adapted, it is not a Markov process with respect to $(\mathscr{F}_t^B)$; indeed by the early part of the preceding proof, $\mathscr{F}_t^B$ contains some information on the future of $2S - B$ after time t. This is at variance with the case of $S - B$ studied in the last section where $\mathscr{F}_t^{S-B} = \mathscr{F}_t^B$ (see Exercise (2.12)). Here $\mathscr{F}_t^{2S-B}$ is strictly contained in $\mathscr{F}_t^B$ as is made plain by the following

(3.6) Corollary. *The conditional distribution of S_t, hence also of $S_t - B_t$, with respect to $\mathscr{F}_t^{2S-B}$ is the uniform distribution on $[0, 2S_t - B_t]$.*

Proof. By the preceding result it is also the conditional distribution of J_t with respect to $\sigma(\rho_s, s \leq t)$, but because of the Markov property of ρ, this is the distribution of $J_0 = \inf_{s\geq 0}\tilde{\rho}_s$ where $\tilde{\rho}$ is a BES^3 started at ρ_t. Our claim follows from Corollary (3.4).

It was also shown in the proof of Theorem (3.5) that

(3.7) Corollary. *If ρ is a* $\mathrm{BES}^3(x)$, *then $B_t = 2J_t - \rho_t$ is a Brownian motion started at $2J_0 - x$ and $\mathscr{F}_t^B = \sigma\left(J_t, \mathscr{F}_t^{\rho}\right)$.*

Finally Theorems (2.3) and (3.5) brought together yield the

(3.8) Corollary. *If B is* BM(0) *and L is its local time at* 0, *then $|B| + L$ is a* $\mathrm{BES}^3(0)$.

We will now strive to say more about J_0 which is the absolute minimum of ρ.

(3.9) Proposition. *Let ρ be a* BES^3. *If T is a stopping time of the bivariate process (ρ, J) such that $\rho_T = J_T$, then $\rho_{T+t} - \rho_T$ is a* $\mathrm{BES}^3(0)$ *independent of* $\{\rho_t, t < T\}$.

Proof. Let us first suppose that ρ starts at 0. Using the notation and result of Corollary (3.7), we see that T is also a stopping time of B. Consequently by the strong Markov property or by Exercise (3.21) of Chap. IV, $\tilde{B}_t = B_{T+t} - B_T$ is a BM(0) independent of $\mathscr{F}_T^B$. By the hypothesis made on T

$$B_T = 2J_T - \rho_T = \rho_T.$$

As a result, the equality $J_t = \sup_{s \leq t} B_s$ proved at the beginning of the proof of Theorem (3.5) shows that, with obvious notation,

$$J_{T+t} - J_T = \sup_{s \leq T+t} B_s - B_T = \tilde{S}_t.$$

The definition of B implies that

$$\begin{aligned} \rho_{T+t} - \rho_T &= 2J_{T+t} - B_{T+t} - \rho_T \\ &= 2\left(J_{T+t} - J_T\right) - \left(B_{T+t} - B_T\right) \\ &= 2\tilde{S}_t - \tilde{B}_t. \end{aligned}$$

Pitman's theorem then implies that $\rho_{T+t} - \rho_T$ is a $\mathrm{BES}^3(0)$. We know moreover that it is independent of $\mathscr{F}_T^B$ hence of $\{\rho_t, t < T\}$.

If ρ starts at $x > 0$, by the strong Markov property, it has the same law as $\tilde{\rho}_{T_x+t}$, $t \geq 0$, where $\tilde{\rho}$ is a $\mathrm{BES}^3(0)$. It suffices to apply the above result to $\tilde{\rho}$; the details are left to the reader. □

The first time at which ρ attains its absolute minimum

$$\tau = \inf\{t : \rho_t = J_0\}$$

obviously satisfies the conditions of the above result. Therefore, since a BES^3 never reaches 0, it follows that τ is the *only* time at which ρ is equal to J_0. We recall moreover that $\rho_\tau = J_0$ is uniformly distributed on $[0, x]$ and we state

(3.10) Proposition. *Let ρ be a* $\mathrm{BES}^3(x)$ *with $x > 0$; the process $(\rho_t, t < \tau)$ is equivalent to $\left(B_t, t < T_\gamma\right)$ where B is a* BM(x) *and T_γ is the hitting time by B of an independent random point γ uniformly distributed on* $[0, x]$.

Proof. By Corollary (3.7), $B_t = 2J_t - \rho_t$ is a BM started at $2J_0 - x$. For $t < \tau$, we have $J_t = J_0$, hence $B_t = 2J_0 - \rho_t$; as a result, for $t < \tau$, we have $\rho_t = 2J_0 - B_t = \beta_t$ where β is a BM(x). Moreover $\beta_t = x - (B_t - B_0)$ is independent of B_0 hence of $J_0 = \rho_\tau$ and $\tau = \inf\{t : \beta_t = J_0\}$. Since J_0 is uniformly distributed on $[0, x]$, the proof is complete.

The above results lead to an important decomposition of the path of $\mathrm{BES}^3(0)$.

(3.11) Theorem (Williams). *Pick $c > 0$ and the following four independent elements*

i) *a r.v.* α *uniformly distributed on* $[0, c]$;
ii) *a* BM(c) *called* B;
iii) *two* $\mathrm{BES}^3(0)$ *called* ρ *and* $\tilde{\rho}$.

Put $R_c = \inf\{t : \rho_t = c\}$; $T_\alpha = \inf\{t : B_t = \alpha\}$. *Then, the process* X *defined by*

$$X_t = \begin{cases} \rho_t, & \text{if } t < R_c \\ B_{t-R_c}, & \text{if } R_c \le t < R_c + T_\alpha \\ \alpha + \tilde{\rho}_{t-R_c-T_\alpha} & \text{if } t \ge R_c + T_\alpha \end{cases}$$

is a $\mathrm{BES}^3(0)$.

Proof. If we look at a $\mathrm{BES}^3(0)$, say ρ, the strong Markov property (Proposition (3.5) in Chap. III) entails easily that the processes $\{\rho_t, t < R_c\}$ and $\{\rho_{t+R_c}, t \ge 0\}$ are independent and the second one is a $\mathrm{BES}^3(c)$. Thus, the theorem follows from the preceding results. □

This theorem, as any path decomposition theorem, is awkward to state but is easily described by a picture such as Figure 5 in which α is uniformly distributed on $[0, c]$.

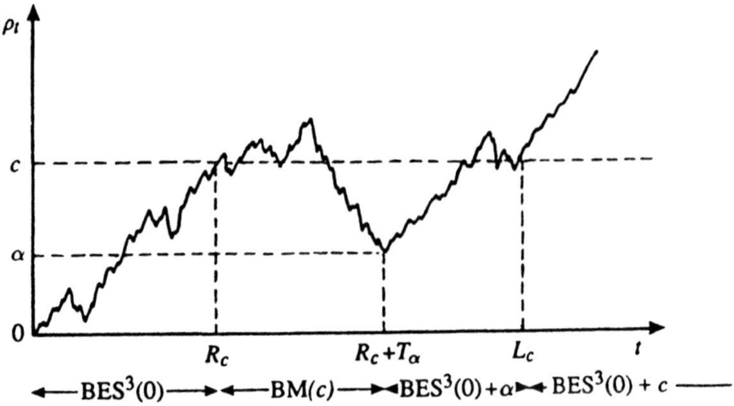

Fig. 5.

According to Proposition (3.9), the last part can be further split up in two independent parts at time L_c or, for that matter, at any time L_d with $d > c$. Indeed, since the BES^3 converges a.s. to infinity, for every $c > 0$, the time

$$L_c = \sup\{t \ge 0 : \rho_t = c\},$$

where we agree that $\sup(\emptyset) = 0$, is a.s. finite. Since we have also

$$L_c = \inf\{t : \rho_t = J_t = c\},$$

this is also a stopping time as considered in Proposition (3.9).

In Sect. 4 of Chap. VII, the foregoing decomposition will be turned into a decomposition of the Brownian path. We close this section with another application to BM. We begin with a lemma which complements Proposition (3.3).

(3.12) Lemma. *If ρ is a* $\mathrm{BES}^3(x)$, $x > 0$, ρ^{-1} *is a time-changed* $\mathrm{BM}(x^{-1})$ *restricted to* $[0, T_0[$.

Proof. By Proposition (3.3) and the DDS theorem of Sect. 1 Chap. V, we have $\rho_t^{-1} = \beta(A_t)$ where $A_t = \int_0^t \rho_s^{-4} ds$ and β is a $\mathrm{BM}(x^{-1})$. Since $\rho > 0$ and $\lim_{t\to\infty} \rho_t^{-1} = 0$, we have $A_\infty = \inf\{t : \beta_t = 0\}$ and the result follows.

We may now state

(3.13) Proposition. *Let B be a* $\mathrm{BM}(a)$, $a > 0$, *and* $M = \max\{B_t, t < T_0\}$; *then, the following properties hold:*

(i) *the r.v. M has the density ax^{-2} on $[a, \infty[$;*
(ii) *there is a.s. a unique time $\nu < T_0$ for which $B_\nu = M$.*

Furthermore, conditionally on $M = m$,

(iii) *the processes $X^1 = (B_t, t < \nu)$ and $X^2 = (B_{\nu+t}, 0 \le t < T_0 - \nu)$ are independent;*
(iv) *the process X^1 is a* $\mathrm{BES}^3(a)$ *run until it hits m;*
(v) *the process $m - X^2$ is a* $\mathrm{BES}^3(0)$ *run until it hits m.*

Proof. Using the notation of the preceding proof, we have

$$(B_t, t < T_0) \stackrel{(d)}{=} \left(\rho_{C_t}^{-1}, t < A_\infty\right)$$

where C is the inverse of A. Thus properties i) and ii) are straightforward consequences of Propositions (3.9) and (3.10). Property iii) follows equally from Proposition (3.9) applied at time τ when the $\mathrm{BES}^3(a^{-1})$ process ρ reaches its absolute minimum.

To prove iv) let us observe that

$$X^1 = \left(\rho_{C_t}^{-1}, t < A_\tau \mid \rho_\tau = 1/m\right).$$

By Proposition (3.10) there is a $\mathrm{BM}(a^{-1})$, say γ, such that

$$\left(\rho_t^{-1}, t < \tau \mid \rho_\tau = 1/m\right) \stackrel{(d)}{=} \left(\gamma_t^{-1}, t < T_{1/m}\right).$$

As a result,

$$X^1 \stackrel{(d)}{=} \left(\gamma_{\tilde{C}_t}^{-1}, t < \tilde{A}_{T_{1/m}}\right)$$

where $\tilde{A}_t = \int_0^t \gamma_s^{-4} ds$ and $\tilde{C}$ is the inverse of $\tilde{A}$. But by Itô's formula,

$$\gamma_t^{-1} = a - \int_0^t \gamma_s^{-2} d\gamma_s + \int_0^t \gamma_s^{-3} ds$$

which entails

$$\gamma_{\bar{C}_t}^{-1} = a + \bar{\beta}_t + \int_0^t \gamma_{\bar{C}_s}^{-1} ds$$

with $\bar{\beta}$ another BM. As a result, the process $\gamma_{\bar{C}_t}^{-1}$ satisfies the same stochastic differential equation as BES^3 and, by the uniqueness result for solutions of SDE's which we will see in Sect. 3 of Chap. IX, this process is a $\mathrm{BES}^3(a)$. Property (iv) now follows easily. Property (v) has a similar proof which we leave as an exercise to the reader.

Remark. The law of M was already derived in Exercise (3.12) of Chap. II.

(3.14) Exercise. Extend Corollary (3.6) to the stopping times of the filtration $(\mathscr{F}_t^{2S-B})$.

(3.15) Exercise. Let X be a $\mathrm{BES}^3(0)$ and put $L = \sup\{t : X_t = 1\}$. Prove that the law of L is the same as the law of $T_1 = \inf\{t : B_t = 1\}$ where B is a $\mathrm{BM}(0)$.
[Hint: Use the fact that $2S_t - B_t$ is a $\mathrm{BES}^3(0)$.]
This will also follow directly from a time-reversal result in Sect. 4 Chap. VII.

** **(3.16) Exercise (A Markov proof of Theorem (3.11)).** Pick $b > 0$ and the following three independent elements:

(i) a r.v. γ uniformly distributed on $[0, b]$;
(ii) a $\mathrm{BES}^3(0)$-process ρ,
(iii) a $\mathrm{BM}(b)$-process B,

and define $T_\gamma = \inf\{t : B_t = \gamma\}$,

$$X_t = B_t \text{ on } \{t < T_\gamma\}, \quad X_t = \rho(t - T_\gamma) + \gamma \text{ on } \{t \geq T_\gamma\}.$$

1°) Prove that for $t > 0$, the conditional probability distribution $P[X_t \in \cdot; T_\gamma < t \mid \gamma]$ has a density equal to

$$1_{(y>\gamma)}(\gamma - y)\frac{\partial}{\partial\gamma}q_t(b - \gamma, y - \gamma)$$

with respect to the Lebesgue measure dy.

2°) Prove that consequently, if $t_0 = 0 < t_1 < t_2 < \ldots < t_n$, the conditional probability distribution of the restriction of $(X_{t_1}, \ldots, X_{t_n})$ to $\{t_{i-1} < T_\gamma < t_i\}$ with respect to γ has a density equal to

$$1_{(\gamma<\inf(x_i))}\left(-\frac{\partial}{\partial\gamma}q_{t_i-t_{i-1}}(x_{i-1} - \gamma, x_i - \gamma)\right)$$
$$\left(\prod_{k\neq i} q_{t_k-t_{k-1}}(x_{k-1} - \gamma, x_k - \gamma)\right)(x_n - \gamma)$$

with respect to the Lebesgue measure $dx_1 \ldots dx_n$.

3°) For $0 \leq c < b$, prove that X conditioned by the event $\{\gamma > c\}$ is equivalent to $\tilde{\rho} + c$ where $\tilde{\rho}$ is a $\mathrm{BES}^3(b - c)$. In particular, X is a $\mathrm{BES}^3(0)$.

4°) Use 3°) to give another proof of Theorem (3.11).

* **(3.17) Exercise.** 1°) Let B be the standard linear BM and set $X_t = B_t + t$. Prove that the process $\exp(-2X_t)$ is a time-changed BM(1) killed when it first hits 0.

2°) Let $\gamma = \inf_t X_t$; prove that $(-\gamma)$ is exponentially distributed (hence a.s. finite) with parameter 2 and that there exists a unique time ρ such that $X_\rho = \gamma$. The law of $(-\gamma)$ was already found in Exercise (3.12) of Chap. II; the point here is to derive it from the results in the present section.

** **(3.18) Exercise.** 1°) With the usual notation, put $X = |B| + L$. Write the canonical decompositions of the semimartingale X in the filtrations $(\mathscr{F}_t^X)$ and $(\mathscr{F}_t^{|B|})$ and deduce therefrom that the inclusion $\mathscr{F}_t^X \subset \mathscr{F}_t^{|B|}$ is strict. This, by the way, gives an example of a semimartingale $X = M + A$ such that $(\mathscr{F}_t^X) \subset (\mathscr{F}_t^M)$ strictly.

2°) Derive also the conclusion of 1°) from the equality

$$L_t = \inf_{s \geq t} (|B_s| + L_s) .$$

3°) Let now $c \neq 1$ and put $X = |B| + cL$. The following is designed to prove that conversely $\mathscr{F}^X = \mathscr{F}^{|B|}$ in that case.

Let D be the set of (ω, t)'s such that

$$\lim_n \frac{1}{n} \sum_{k \leq n} 1_{\{X(t-2^{-k},\omega) - X(t-2^{-k-1},\omega) \geq 0\}} = 1/2.$$

Prove that

a) D is $\mathscr{F}^X$-predictable;
b) for fixed $t > 0$, $P[\{\omega : (\omega, t) \in D\}] = 1$;
c) for fixed $t > 0$, $P[\{\omega : (\omega, \tau_t(\omega)) \in D\}] = 0$ where as usual τ_t is the inverse of L.

Working in $\mathscr{F}^B$, prove that

$$\int_0^t 1_D(s) dX_s = \int_0^t \operatorname{sgn}(B_s) dB_s$$

and derive the sought-after conclusion.

(3.19) Exercise. If ρ is the $\mathrm{BES}^3(0)$ prove that for any t,

$$P\left[\overline{\lim_{h \downarrow 0}} |\rho_{t+h} - \rho_t| / \sqrt{2h \log_2 1/h} = 1 \right] = 1.$$

[Hint: Use the result in Exercise (1.21) Chap. II and the Markov property.]

§4. First Order Calculus

If M is a martingale, K a predictable process such that $\int_0^t K_s^2 d\langle M, M\rangle_s < \infty$, we saw how stochastic integration allows to construct a new local martingale, namely $K{\cdot}M$, with increasing process $K^2{\cdot}\langle M, M\rangle$. We want to study the analogous problem for the local time at zero. More precisely, if L is the local time of M at 0 and if K is a predictable process such that $|K|{\cdot}L$ is finite, can we find a local martingale with local time $|K| \cdot L$ at 0? The answer to this question will lead to a first-order calculus (see Proposition (4.5)) as opposed to the second order calculus of Itô's formula.

Throughout this section, we consider a fixed continuous semimartingale X with local time L at 0. We use a slight variation on the notation of Sect. 2, namely $Z = \{t : X_t = 0\}$ and for each t

$$g_t = \sup\{s < t : X_s = 0\}, \quad d_t = \inf\{s > t : X_s = 0\}.$$

(4.1) Lemma. *If K is a locally bounded predictable process, the process $K_{g_\cdot}$ is locally bounded and predictable.*

Proof. Let T be a stopping time; since $g_T \le T$, the process $K_{g_\cdot}$ is bounded on $[0, T]$ if K is and therefore $K_{g_\cdot}$ is locally bounded if K is. It is enough to prove the second property for a bounded K and by Exercise (4.20) Chap. I and the monotone class theorem, for $K = 1_{[0,T]}$. But in that case $K_{g_\cdot} = 1_{[0,d_T]}$ and one easily checks that d_T is a stopping time, which completes the proof.

The following result supplies an answer to the question raised above.

(4.2) Theorem. *i) If Y is another cont. semimart. such that $Y_{d_t} = 0$ for every $t \ge 0$, then $K_{g_t} Y_t$ is a continuous semimartingale and more precisely*

$$K_{g_t} Y_t = K_0 Y_0 + \int_0^t K_{g_s} dY_s.$$

In particular, $K_{g_\cdot} X$ is a continuous semimartingale.

ii) If Y is a local martingale with local time Λ at zero, $K_{g_\cdot} Y$ is also a local martingale and its local time at zero is equal to

$$\int_0^t |K_{g_s}| d\Lambda_s.$$

In particular, if X is a local martingale, then $K_{g_\cdot} X$ is a local martingale with local time at 0 equal to $\int_0^t |K_{g_s}| dL_s = \int_0^t |K_s| dL_s$.

Proof. By the dominated convergence theorem for stochastic integrals, the class of predictable processes K for which the result is true is closed under pointwise bounded convergence. Thus, by the monotone class theorem, to prove i), it is once again enough to consider the case of $K = 1_{[0,T]}$. Then, because $Y_{d_T} = 0$ and $K_{g_t} = 1_{[g_t \le T]} = 1_{[t \le d_T]}$,

$$K_{g_t}Y_t = Y_{t\wedge d_T} = K_0Y_0 + \int_0^{t\wedge d_T} dY_s$$
$$= K_0Y_0 + \int_0^t 1_{[0,d_T]}(s)dY_s = K_0Y_0 + \int_0^t K_{g_s}dY_s$$

which proves our claim.

To prove ii) , we apply i) to the semimartingale $|Y|$ which clearly satisfies the hypothesis; this yields

$$|K_{g_t}Y_t| = |K_{g_t}|\,|Y_t| = |K_0Y_0| + \int_0^t |K_{g_s}|d|Y|_s$$
$$= |K_0Y_0| + \int_0^t |K_{g_s}|\mathrm{sgn}(Y_s)dY_s + \int_0^t |K_{g_s}|d\Lambda_s$$
$$= |K_0Y_0| + \int_0^t \mathrm{sgn}(K_{g_s}Y_s)d(K_{g_s}Y_s) + \int_0^t |K_{g_s}|d\Lambda_s$$

which is the desired result.

We may obviously apply this result to X as $X_{d_\cdot} \equiv 0$. The local time of $K_{g_\cdot}X$ is thus equal to

$$\int_0^t |K_{g_s}|dL_s.$$

But the measure dL_s is carried by Z and $g_s = s$ for any s which is the limit from the left of points of Z; the only points in Z for which $g_s \neq s$ are therefore the right-end points of the intervals of Z^c, the set of which is countable. Since L is continuous, the measure dL_s has no point masses so that $|K_{g_s}| = |K_s|$ for dL-almost all s which completes the proof. □

If f is the difference of two convex functions, we know that $f(X)$ is a semimartingale, and if $f(0) = 0$, then $f(X)_{d_\cdot} \equiv 0$ and consequently

$$f(X_t)K_{g_t} = f(X_0)K_0 + \int_0^t K_{g_s}df(X)_s.$$

In this setting, we moreover have the

(4.3) Proposition. *If $\phi : \mathbb{R}_+ \to \mathbb{R}_+$ is locally bounded,*

$$f(X_t)\phi(L_t) = f(X_0)\phi(0) + \int_0^t \phi(L_s)df(X)_s.$$

Proof. We apply the above formula with $K_t = \phi(L_t)$ and take into account that $L_{g_t} = L_t$ as was already observed. □

We now apply the above results to a special class of semimartingales.

(4.4) Definition. *We call Σ the class of semimartingales $X = N + V$ such that the measure dV_t is a.s. carried by $Z = \{t : X_t = 0\}$.*

If, for instance, M is a local martingale, then the semimartingales $|M|$ and M^+ are in Σ with respectively $V = L$ and $V = \frac{1}{2}L$. This will be used in the

(4.5) Proposition. *If $X \in \Sigma$, the process*

$$X_t K_{g_t} - \int_0^t K_s dV_s = X_0 K_0 + \int_0^t K_{g_s} dN_s$$

is a local martingale. If M is a local martingale, the processes

$$|M_t| K_{g_t} - \int_0^t K_s dL_s, \quad M_t^+ K_{g_t} - \frac{1}{2}\int_0^t K_s dL_s,$$

are local martingales. Finally, if φ is a locally bounded Borel function and $\phi(x) = \int_0^x \varphi(u)du$, then

$$\phi(L_t) - |M_t|\varphi(L_t), \quad \frac{1}{2}\phi(L_t) - M_t^+\varphi(L_t),$$

are local martingales.

Proof. The first two statements are straightforward consequences of Theorem (4.2) and its proof. The third one follows from the second by making $K_t = \varphi(L_t)$ and using the fact that $\int_0^t \varphi(L_s)dL_s = \phi(L_t)$ which is a simple consequence of time-change formulas. □

In Theorem (2.3), we saw that the processes $(S - B, S)$ and $(|B|, L)$ have the same law; if for a local martingale M we put $S_t = \sup_{s \le t} M_s$, it should not be surprising that one can replace in the above formulas L_t by S_t and $|M_t|$ by $S_t - M_t$, although, in this generality, the equality in law of Theorem (2.3) will not necessarily obtain (see Exercise (2.32)). In fact, the semimartingale $X = S - M$ is in Σ, with $V = S$, since clearly S increases only on the set $\{S = M\} = \{X = 0\}$. As a result, the processes

$$(S_t - M_t)K_{\gamma_t} - \int_0^t K_s dS_s, \quad \phi(S_t) - (S_t - M_t)\varphi(S_t)$$

where $\gamma_t = \sup\{s < t : S_s = M_s\}$, are local martingales.

Observe that Proposition (4.3) can also be deduced from the integration by parts formula if ϕ is C^1 and likewise the last formula of Proposition (4.5) if ϕ is C^2. For instance, if $\phi \in C^2$, then $\varphi(S_t)$ is a continuous semimartingale and

$$\begin{aligned}\phi(S_t) - (S_t - M_t)\varphi(S_t) &= \phi(S_t) - \int_0^t \varphi(S_s)dS_s + \int_0^t \varphi(S_s)dM_s \\ &\quad + \int_0^t (S_s - M_s)\varphi'(S_s)dS_s.\end{aligned}$$

But the last integral is zero, because dS is carried by $\{S - M = 0\}$. Moreover $\int_0^t \varphi(S_s)dS_s = \phi(S_t)$ so that

$$\phi(S_t) - (S_t - M_t)\varphi(S_t) = \int_0^t \varphi(S_s)dM_s$$

which is a local martingale. We can thus measure what we have gained by the methods of this section which can also be used to give yet another proof of Itô's formula (see Notes and Comments). Finally, an elementary proof of the above result is to be found in Exercise (4.16).

Itô's and Tanaka's formulas may also be used to extend the above results to functions $F(M_t, S_t, \langle M, M\rangle_t)$ or $F(M_t, L_t, \langle M, M\rangle_t)$ as is seen in Exercise (4.9). These results then allow to compute explicitly a great variety of laws of random variables related to Brownian motion; here again we refer to the exercises after giving a sample of these computations.

(4.6) Proposition. *If B is a standard* BM *and $T_b = \inf\{t : B_t = b\}$, then for $0 \leq a \leq b$ and $\lambda \geq 0$,*

$$E\left[\exp\left(-\lambda L^a_{T_b}\right)\right] = \left(1 + 2\lambda(b-a)\right)^{-1}$$

that is to say, $L^a_{T_b}$ has an exponential law with parameter $\left(2(b-a)\right)^{-1}$.

Proof. By Proposition (4.5), $\left((1/2)+\lambda(B_t - a)^+\right)\exp(-\lambda L^a_t)$ is a local martingale for any $\lambda \geq 0$. Stopped at T_b, it becomes a bounded martingale and therefore

$$E\left[\left(\frac{1}{2} + \lambda(b-a)\right)\exp\left(-\lambda L^a_{T_b}\right)\right] = \frac{1}{2}.$$

□

We now state a partial converse to Proposition (4.5).

(4.7) Proposition. *Let $(l, x) \to F(l, x)$ be a real-valued $C^{1,2}$-function on $\mathbb{R}^2_+$. If M is a cont. loc. mart. and $\langle M, M\rangle_\infty = \infty$ a.s., and if $F(L_t, |M_t|)$ is a local martingale, then there exists a C^2-function f such that*

$$F(l, x) = f(l) - xf'(l).$$

Proof. If $F(L_t, |M_t|)$ is a local martingale, Itô's and Tanaka's formulae imply that

$$\left(F'_l(L_s, |M_s|) + F'_x(L_s, |M_s|)\right)dL_s + \frac{1}{2}F''_{x^2}(L_s, |M_s|)d\langle M, M\rangle_s = 0 \ \text{ a.s.}$$

The occupation times formula entails that the measures dL_s and $d\langle M, M\rangle_s$ are a.s. mutually singular; therefore

$$\left(F'_l(L_s, 0) + F'_x(L_s, 0)\right)dL_s = 0, \quad F'_{x^2}(L_s, |M_s|)d\langle M, M\rangle_s = 0 \ \text{ a.s.}$$

Now since $\langle M, M\rangle_\infty = \infty$, the local time L is the local time of a time-changed BM and consequently $L_\infty = \infty$ a.s. By the change of variables formula for Stieltjes integrals, the first equality then implies that

$$F_l'(l,0) + F_x'(l,0) = 0$$

for Lebesgue-almost every l, hence for every l by the assumption of continuity of the derivatives. By using the time-change associated with $\langle M, M\rangle$, the second equality yields $P\left[F''_{x^2}(l_t, |B_t|) = 0\ dt\text{-a.e.}\right] = 1$ where B is a BM and l its local time at 0 and because of the continuity

$$P\left[F''_{x^2}(l_t, |B_t|) = 0 \text{ for every } t\right] = 1.$$

For every $b > 0$, it follows that

$$F''_{x^2}\left(l_{T_b}, b\right) = 0 \quad \text{a.s.}$$

But by Proposition (4.6) the law of l_{T_b} is absolutely continuous with respect to the Lebesgue measure; using the continuity of F''_{x^2}, it follows that $F''_{x^2}(\cdot, b) = 0$ for every b. As a result

$$F(l,x) = g(l)x + f(l)$$

and since F is continuously differentiable in l for every x, it follows that f and g are C^1-functions; furthermore the equality $F_l'(l,0) + F_x'(l,0) = 0$ yields $g(l) = -f'(l)$ which entails that f is in C^2 and completes the proof.

\# **(4.8) Exercise.** Let M be a uniformly integrable continuous martingale and L its local time at 0. Set $G = \sup\{s : M_s = 0\}$.

1°) Prove that for any bounded predictable process K,

$$E\left[\int_0^\infty K_s dL_s\right] = E\left[1_{(G>0)} K_G |M_\infty|\right].$$

2°) Assume now first that $P(M_\infty = 0) = P(G = 0) = 0$ and deduce from 1°) that

$$E\left[K_{\tau_u}; \tau_u < \infty\right] du = E\left[K_G |M_\infty| \,\Big|\, L_\infty = u\right] P(L_\infty \in du),$$

where $\tau_u = \inf\{t : L_t > u\}$. In particular,

$$P(L_\infty \in du) = \left(E\left(|M_\infty| \,\Big|\, L_\infty = u\right)\right)^{-1} P(L_\infty > u) du;$$

check this formula whenever M is a stopped BM, say B^T, for some particular stopping times such as $T = t$, $T = \tilde{T}_a$, ...

Finally, prove that in the general setting $dP\, du$-a.s.,

$$E\left[K_{\tau_u}; \tau_u < \infty\right] = E\left[K_G |M_\infty| \,\Big|\, L_\infty = u\right] \Big/ E\left[|M_\infty| \,\Big|\, L_\infty = u\right].$$

\# **(4.9) Exercise.** Let M be a local martingale, L its local time, S its supremum.

1°) Prove that the measures $d\langle M, M\rangle$ and dS are mutually singular.

2°) If $F : (x,y,z) \to F(x,y,z)$ is defined on $\mathbb{R} \times \mathbb{R}_+^2$, and is sufficiently smooth and if

$$\frac{1}{2} F''_{x^2} + F'_z \equiv 0, \quad F'_y(y,y,z) = 0 \quad \text{for every } x, y \text{ and } z,$$

then $F(M_t, S_t, \langle M, M\rangle_t)$ is a local martingale. Find the corresponding sufficient condition for $F(M_t, L_t, \langle M, M\rangle_t)$ to be a local martingale.

3°) Prove that $(S_t - M_t)^2 - \langle M, M\rangle_t$ is a local martingale and that for any reals α, β, the process

$$Z_t^{\alpha,\beta} = [\beta \cosh \beta(S_t - M_t) - \alpha \sinh \beta(S_t - M_t)] \exp\left\{\alpha S_t - \frac{\beta^2}{2}\langle M, M\rangle_t\right\}$$

is a local martingale. Prove the same results when $S - M$ is replaced by $|M|$ and S by L.

4°) If B is a BM and $\tilde{T}_a = \inf\{t : |B_t| = a\}$, then for $\alpha > 0$, $\beta \neq 0$,

$$E\left[\exp\left\{-\alpha L_{\tilde{T}_a} - \frac{\beta^2}{2}\tilde{T}_a\right\}\right] = \beta\,[\beta \cosh a\beta + \alpha \sinh a\beta]^{-1}.$$

5°) Prove an analogous formula for S and $R_a = \inf\{t : S_t - B_t = a\}$.

6°) Again for the BM and with the notation of Sect. 2, prove that

$$E\left[\exp\left(-\frac{\alpha^2}{2}\tau_t\right)\right] = e^{-\alpha t}.$$

* **(4.10) Exercise.** 1°) Let M be a martingale and L its local time at 0. For any $p \geq 1$ prove that $\|L_t\|_p \leq p\|M_t\|_p$. For $p = 1$ and $M_0 = 0$, prove that $\|M_t\|_1 = \|L_t\|_1$.

[Hint: Localize so as to deal with bounded M and L; then apply Proposition (4.5).] For $p > 1$, prove that $\|S_t\|_p \leq (p/p-1)\|M_t\|_p$.

2°) Show that there is no converse inequality, that is, for $p > 1$, there is no universal constant C_p such that

$$\|M_t\|_p \leq C_p\|L_t\|_p$$

for every M locally bounded in L^p.

(4.11) Exercise. Let M be a square-integrable martingale vanishing at zero and set $s_t = \inf_{s \leq t} M_s$. Prove the following reinforcement of Doob's inequality

$$E\left[(S_t - s_t)^2\right] \leq 4E\left[M_t^2\right].$$

[Hint: Use 3°) in Exercise (4.9).] Prove that this inequality cannot be an equality unless M vanishes identically on $[0, t]$.

(4.12) Exercise. For the BM and $b > 0$ set $\tilde{T}_b = \inf\{t : |B_t| = b\}$. Prove that $L_{\tilde{T}_b}$ has an exponential law with parameter $1/b$. (In another guise, this is already in Exercise (2.10)).

(4.13) Exercise. For the BM call μ the law of the r.v. s_{T_1} where $s_t = \inf_{s \leq t} B_s$.

1°) Using the analogue for s of the local martingale $\phi(S_t) - (S_t - B_t)\phi'(S_t)$, prove that $\mu = \big(-(1-x)\mu\big)'$ where the derivative is taken in the sense of distributions.

Prove that consequently, μ has the density $(1-x)^{-2}$ on $]-\infty, 0[$.

2°) Using this result, prove that the result of Exercise (4.11) cannot be extended to $p \neq 2$, in the form

$$E\left[(S_t - s_t)^p\right] \leq C_p E\left[|M_t|^p\right]$$

with $C_p = (p/p-1)^p$.

[Hint: Take $M_t = B_{t\wedge \tilde{T}_1}$, where $\tilde{T}_1 = \inf\{t : |B_t| = 1\}$.]

\# **(4.14) Exercise.** 1°) Let M be a continuous local martingale vanishing at 0, and F a C^2-function. For $a > 0$, prove that

$$\frac{1}{2}F(L_t^a) + \left[(M_t^+ - a)\wedge 0\right]F'(L_t^a) - \frac{1}{2}\int_0^t F'(L_s^a)dL_s^0$$

is a local martingale.

2°) For the Brownian motion, compute the Laplace transform of the law of $L_{\tau_t}^a$ where $\tau_t = \inf\{s : L_s^0 > t\}$.

(4.15) Exercise. Let M be a positive continuous martingale such that $M_\infty = 0$. Using the local martingale $\phi(S_t) - (S_t - M_t)\varphi(S_t)$ for a suitably chosen φ, find a new derivation of the law of the r.v. S_∞ conditioned on $\mathscr{F}_0$, which was found in Exercise (3.12) of Chap. II.

\# **(4.16) Exercise.** Following the hint below, give an elementary proof of the fact that, in the notation of Proposition (4.5), the process $\phi(S_t) - (S_t - M_t)\varphi(S_t)$ is a local martingale.

[Hint: Assume first that M is a bounded martingale, then derive from the equality

$$E\left[M_T 1_{(T>T_a)}\right] = E\left[M_{T_a} 1_{(T>T_a)}\right]$$

that

$$E\left[(S_T - a)^+ - (S_T - M_T)1_{(S_T>a)}\right]$$

does not depend on the stopping time T. This is the result for $\varphi(x) = 1_{(x>a)}$; extend to all functions φ by monotone class arguments.]

* **(4.17) Exercise.** 1°) Let X be a cont. loc. mart. and (L^a) the family of its local times. For any C^∞ function f on $\mathbb{R}_+^n$ and $a_1 < a_2 < \ldots < a_n$, prove that

$$f\left(L_t^{a_1}, L_t^{a_2}, \ldots, L_t^{a_n}\right) - \sum_i 2(X_t - a_i)^+ \frac{\partial f}{\partial x_i}\left(L_t^{a_1}, \ldots, L_t^{a_n}\right)$$

$$+ \sum_{d=2}^n (-2)^d \sum_{i_1<i_2<\ldots<i_d\leq n} (a_{i_1} - a_{i_2})\ldots$$

$$\ldots(a_{i_{d-1}} - a_{i_d})(X_t - a_{i_d})^+ \frac{\partial^d f}{\partial x_{i_1}\ldots\partial x_{i_d}}(L_t^{a_1}, \ldots, L_t^{a_n})$$

is a local martingale.

2°) For $(\gamma_1, \dots, \gamma_n) \in \mathbb{R}^n$ and $a_n \leq 1$, prove that for the BM

$$E\left[\exp\left(-\sum_1^n \gamma_i L_{T_1}^{a_i}\right)\right] = \phi(0,\gamma)/\phi(1,\gamma)$$

where

$$\phi(x,\gamma) = 1 + \sum_i 2(x-a_i)^+\gamma_i$$
$$+ \sum_{d=2}^n 2^d \sum_{i_1<i_2\dots<i_d\leq n} (a_{i_2}-a_{i_1})\dots(a_{i_d}-a_{i_{d-1}})(x-a_{i_d})^+\gamma_{i_1}\dots\gamma_{i_d}.$$

This generalizes Proposition (4.6) and gives the law of the process $a \to L_{T_1}^a$ which will be further identified in the Ray-Knight theorem of Sect. 2 Chap. XI. (See Exercise (2.11) in that chapter).

* **(4.18) Exercise.** Let ϕ be an increasing C^1-function on $\mathbb{R}_+$ such that $\phi(0) = 0$ and $0 \leq \phi(x) \leq x$ for every x. Let f be a C^1-function such that

$$f(x)\phi'(x) = f'(x)(x - \phi(x))$$

and set

$$F(x) = \int_0^x f(y)dy \equiv f(x)(x - \phi(x)).$$

1°) In the notation of this chapter, prove that

$$\int_0^{T_a} 1_{(B_s > \phi(S_s))}ds \stackrel{(d)}{=} \int_0^{T_{F(a)}} \left(1/f^2 \circ F^{-1}(S_s)\right) 1_{(B_s>0)}ds.$$

[Hint: Using Theorem (2.3), prove first that the left-hand side has the same law as $\int_0^{\tau_a} 1_{(f(L_s)|B_s|<F(L_s))}ds$, then apply Proposition (4.5) to $f(L_s)B_s$.] The same equality holds with $<$ in place of $>$.

2°) If l is the local time at 0 of the semimartingale $B - \phi(S)$ prove that

$$l_{T_a} \stackrel{(d)}{=} \int_0^{T_{F(a)}} \left(f \circ F^{-1}(S_s)\right)^{-1} dL_s.$$

[Hint: Write l as the limit of $\varepsilon^{-1}\int_0^\cdot 1_{(0<B_s-\phi(S_s)<\varepsilon)}ds$, then follow the same pattern as in 1°) .]

3°) Carry out the computations for $\phi(x) = ax$, $a < 1$.

* **(4.19) Exercise.** Let B be the standard linear BM and L its local time at 0.

1°) If K is a strictly positive, locally bounded $(\mathscr{F}_t^B)$-predictable process, prove that $M_t = K_{g_t}B_t$, $t \geq 0$, has the same filtration as B.

2°) Prove that $N_t = \int_0^t K_{g_s}dL_s - K_{g_t}|B_t|$ has the same filtration as $|B|$.

3°) For $p > 0$, prove that the local martingale $M = L^{p-1}B$ is pure.

[Hint: If τ_t is the time-change associated with $\langle M, M\rangle$, express τ_t as a function of $\left(L_{\tau_t}^p/p\right)$ which is the local time of the DDS Brownian motion of M.]

4°) Prove that, for $p > 0$, the local martingale $L_t^p - p|B_t|L_t^{p-1}$, $t \geq 0$, is also pure.

[Hint: Use Lemma (2.1).]

* **(4.20) Exercise. (More on the class $L \log L$ of Exercise (1.16) Chap. II).** We consider a cont. loc. mart. X such that $X_0 = 0$ and write $X_t^* = \sup_{s \leq t} |X_s|$ and $L_t = L_t^0(X)$.

1°) Prove that

$$|X_t| \log^+ X_t^* - (X_t^* - 1)^+ - \int_0^t \log^+ X_s^* dL_s$$

is a local martingale.

2°) Prove that

$$E\left[X_\infty^*\right] \leq (e/(e-1))\left(1 + \sup_T E\left[|X_T| \log^+ |X_T|\right]\right)$$

where T ranges through the family of all finite stopping times and that

$$\sup_T E\left[|X_T| \log^+ X_T^*\right] \leq ((e+1)/e) E\left[X_\infty^*\right] + E\left[L_\infty \log^+ L_\infty\right].$$

3°) Prove likewise that $L_t \log^+(L_t) - (L_t - 1)^+ - |X_t| \log^+(L_t)$ is a local martingale and derive that

$$E\left[L_\infty \log^+(L_\infty)\right] \leq \sup_T \left\{((e+1)/e) E\left[|X_T|\right] + E\left[|X_T| \log^+ |X_T|\right]\right\}.$$

4°) Conclude that the following two conditions are equivalent:

i) $\sup_T E\left[|X_T| \log^+ |X_T|\right] < \infty$;
ii) $E\left[X_\infty^*\right] < \infty$ and $E\left[L_\infty \log^+ L_\infty\right] < \infty$.

\# **(4.21) Exercise (Bachelier's equation).** 1°) Let M be a cont. loc. mart. vanishing at 0. Prove that there exists a unique strictly positive cont. loc. mart. $\tilde{M}$ such that $\tilde{M}_0 = 1$ and, in the notation of this section,

$$S_t - M_t = \left(\tilde{M}_t / \tilde{s}_t\right) - 1,$$

where $\tilde{s}_t = \inf_{u \leq t} \tilde{M}_u$.

[Hint: Try for $\tilde{s}_t = \exp(-S_t)$.]

2°) Let h be a function on $\mathbb{R}_+$ which is strictly decreasing, of class C^1 with non vanishing derivative and such that $h(0) = 1$, $h(\infty) = 0$. Given a loc. mart. $\tilde{M}$ satisfying the conditions of 1°) , prove that there exists a unique cont. loc. mart. M such that

$$-h'\left(h^{-1}(\tilde{s}_t)\right)(S_t - M_t) = \tilde{M}_t - \tilde{s}_t.$$

3°) Prove that the local martingale M of 2°) satisfies the *Bachelier equation*

$$dM_t = d\tilde{M}_t / h'(S_t).$$

Give a sufficient condition on h in order that this equation have a unique solution and then express it as a function of $\tilde{M}$.

(4.22) Exercise (Improved constants in domination). 1°) For the BM B prove, in the usual notation, that for $k < 1$,

$$E\left[(-s_{T_1})^k\right] \le E\left[\sup_{s \le T_1}(S_s - B_s)^k\right] \le C_k,$$

where C_k is the constant defined in Exercise (4.30) Chap. IV. Conclude that

$$(\pi k/\sin \pi k) \le \Gamma(1-k) \le C_k.$$

[Hint: Use the results in Exercises (4.12) and (4.13).]

2°) Combining 1°) with Exercise (4.30) Chap. IV, prove that

$$\lim_{k \to 1}(1-k)C_k = 1.$$

(4.23) Exercise. 1°) In the usual notation, set $T = \inf\{t : |L_t B_t| = 1\}$ and prove that the law of L_T is that of $\sqrt{\mathbf{e}}$ where $\mathbf{e}$ is an exponential r.v. with parameter 1.

[Hint: Express the local time at 0 of the loc. mart. LB as a function of L and follow the pattern of the proof of Proposition (4.6).]

2°) Prove that consequently, $L^T B^T$ is a bounded martingale such that the process H of Proposition (3.2) Chap. V is unbounded.

(4.24) Exercise (An extension of Pitman's theorem). Retain the notation of Proposition (4.5) and the remarks thereafter and suppose that φ is positive and that $M_0 = 0$.

1°) Prove that

$$\phi(S_t) = \sup_{s \le t} \int_0^s \varphi(S_u) dM_u.$$

[Hint: Apply Lemma (2.1).]

2°) Prove that $(S_t - M_t)\varphi(S_t)$ is a time-changed reflected BM and identify its local time at 0.

3°) Prove that $\phi(S_t) + (S_t - M_t)\varphi(S_t)$ is a time-changed $\mathrm{BES}^3(0)$.

§5. The Skorokhod Stopping Problem

Let μ be a probability measure on $\mathbb{R}$; we wish to find a stopping time T of the standard linear BM such that the law of B_T is μ. In this generality the problem has a trivial solution which is given in Exercise (5.7); unfortunately this solution is uninteresting in applications as T is too large, namely $E[T] = \infty$. We will therefore amend the problem by demanding that $E[T]$ be finite. This however imposes restrictions on μ. If $E[T] < \infty$, the martingale B^T is in M_0^2 which implies $E[B_T] = 0$ and furthermore $(B_t^T)^2 - T \wedge t$ is uniformly integrable so that $E[B_T^2] = E[T]$. The conditions

$$\int x^2 d\mu(x) < \infty, \quad \int x \, d\mu(x) = 0,$$

are therefore necessary. They are also sufficient, and indeed, the problem thus amended has several known solutions one of which we now describe. We actually treat the case of cont. loc. martingales which by the DDS theorem is equivalent to the case of BM.

In what follows all the probability measures μ we consider will be *centered*, i.e., will satisfy

$$\int |x| d\mu(x) < \infty, \quad \int x\, d\mu(x) = 0.$$

For such a μ, we define $\bar{\mu}(x) = \mu([x, \infty[)$ and

$$\begin{aligned} \psi_\mu(x) &= \bar{\mu}(x)^{-1} \int_{[x,\infty[} t\, d\mu(t) \text{ if } x < b = \inf\{x : \bar{\mu}(x) = 0\}, \\ \psi_\mu(x) &= x \text{ if } x \geq b. \end{aligned}$$

The functions $\bar{\mu}$ and ψ_μ are left-continuous; ψ_μ is increasing and converges to b as $x \to b$. Moreover, $\psi_\mu(x) > x$ on $]-\infty, b[$ and $b = \inf\{x : \psi_\mu(x) = x\}$.

(5.1) Lemma. *For every $x \in]-\infty, b[$ and every $a \leq 0$,*

$$\frac{\bar{\mu}(x)}{\bar{\mu}(a)} = \frac{\psi_\mu(a) - a}{\psi_\mu(x) - x} \exp\left(-\int_a^x \frac{ds}{\psi_\mu(s) - s}\right).$$

Proof. By definition

$$\bar{\mu}(x)\psi_\mu(x) = \int_{[x,\infty[} s\, d\mu(s),$$

and taking regularizations to the right,

$$(*) \qquad \bar{\mu}(x+)\psi_\mu(x+) = \int_{]x,\infty[} s\, d\mu(s).$$

Let us call ν the measure associated (Sect. 4 Chap. 0) with the right-continuous function of finite variation $\psi_\mu(x+)$; using the integration by parts formula of Sect. 4 Chap. 0, the above equality reads

$$(+) \qquad \bar{\mu}(x) d\nu(x) - \left(\psi_\mu(x+) - x\right) d\mu(x) = 0,$$

this equality being valid in $]-\infty, b[$ where $\psi_\mu(x) > x$.

By the reasoning in Proposition (4.7) Chap. 0, there exists only one locally bounded solution $\bar{\mu}$ to (+) with a given value in a given point. But for $a < 0$, the function

$$\phi(x) = \left(\psi_\mu(x) - x\right)^{-1} \exp\left(-\int_a^x \frac{ds}{\psi_\mu(s) - s}\right)$$

is a solution to (+). This is seen by writing

$$\phi(x+)\left(\psi_\mu(x+) - x\right) = \exp\left(-\int_a^x \frac{ds}{\psi_\mu(s) - s}\right)$$

and applying again the integration by parts formula which yields

$$\left(\psi_\mu(x+) - x\right) d\phi(x) + \phi(x) d\nu(x) = 0.$$

The proof is then easily completed.

The preceding lemma shows that the map $\mu \to \psi_\mu$ is one-to-one; the next lemma is a step in the direction of showing that it is onto, namely, that every function possessing the properties listed above Lemma (5.1) is equal to ψ_μ for some μ.

(5.2) Lemma. *Let ψ be a left-continuous increasing function and $a < 0 < b$ two numbers such that $\psi(x) = 0$ for $x \leq a$, $\psi(x) > x$ for $a < x < b$, and $\psi(x) = x$ for $x \geq b$. Then, there is a unique centered probability measure μ with support in $[a, b]$ and such that $\psi_\mu = \psi$.*

Proof. We set $\bar{\mu}(x) = 1$ for $x \leq a$,

$$\bar{\mu}(x) = \big(-a/(\psi(x) - x)\big) \exp\left(-\int_a^x \big(\psi(s) - s\big)^{-1} ds\right) \quad \text{for } a < x \leq b.$$

This function is left-continuous on $]-\infty, b[$; furthermore, it is decreasing. Indeed, this is easy to see if ψ is C^1; if not, we use a C^∞ function $j > 0$ with support in $]0, 1]$ and such that $\int j(y)dy = 1$, and we set $\psi_n(x) = n \int f(x+y) j(ny)dy$ as in the proof of Theorem (1.1). Then $\psi_n \geq \psi$ and $\lim_n \psi_n(x) = \psi(x+)$, whence the property of $\bar{\mu}$ follows. We complete the definition of $\bar{\mu}$ by setting $\bar{\mu}(b) = \lim_{x \to b, x < b} \bar{\mu}(x)$, and $\bar{\mu}(x) = 0$ for $x > b$. There is then a unique probability measure μ such that $\bar{\mu}(x) = \mu([x, \infty[)$. By differentiating the equality

$$\bar{\mu}(x+)\big(\psi(x+) - x\big) = -a \exp\left(-\int_a^x \big(\psi(s) - s\big)^{-1} ds\right)$$

we get

$$\big(\psi(x+) - x\big) d\bar{\mu}(x) = -\bar{\mu}(x) d\psi(x)$$

which may be written $d(\psi\bar{\mu}) = x\, d\bar{\mu}$. As a result

$$\psi(x)\bar{\mu}(x) = \int_{[x,\infty[} t\, d\mu(t).$$

Taking $x < a$ this equality shows that μ is centered and the proof is complete.

We will also need the following

(5.3) Lemma. *If $\int x^2 d\mu(x) < +\infty$, then*

$$\int x^2 d\mu(x) = \int \psi_\mu(x)\bar{\mu}(x)dx.$$

Proof. By eq($*$) we have

$$\int x^2 d\mu(x) = -\int x d\left(\bar{\mu}(x+)\psi_\mu(x+)\right)$$

and integrating by parts

$$\int x^2 d\mu(x) = \int \bar{\mu}(x)\psi_\mu(x)dx - \left[x\bar{\mu}(x+)\psi_\mu(x+)\right]_{-\infty}^{+\infty}.$$

But

$$\lim_{x\to\infty} x\bar{\mu}(x+)\psi_\mu(x+) = \lim_{x\to\infty} x\int_{]x,\infty[} t\, d\mu(t) \leq \lim_{x\to\infty}\int_{]x,\infty[} t^2 d\mu(t) = 0$$

whereas, because μ is centered,

$$\begin{aligned}\lim_{x\to-\infty}(-x)\bar{\mu}(x+)\psi_\mu(x+) &= \lim_{x\to-\infty} x\int_{]-\infty,x]} t\, d\mu(t)\\ &\leq \lim_{x\to-\infty}\int_{]-\infty,x[} t^2 d\mu(t) = 0.\end{aligned}$$

□

From now on we consider a cont. loc. mart. M such that $M_0 = 0$ and $\langle M, M\rangle_\infty = \infty$. We set $T_x = \inf\{t : M_t = x\}$ and $S_t = \sup_{s\leq t} M_s$. The main result of this section is the following

(5.4) Theorem. *If μ is a centered probability measure, the stopping time*

$$T_\mu = \inf\left\{t \geq 0 : S_t \geq \psi_\mu(M_t)\right\}$$

is a.s. finite, the law of M_T is equal to μ and, moreover,

i) M^{T_μ} *is a uniformly integrable martingale,*
ii) $E\left[\langle M, M\rangle_{T_\mu}\right] = \int x^2 d\mu(x)$.

We illustrate the theorem with the next

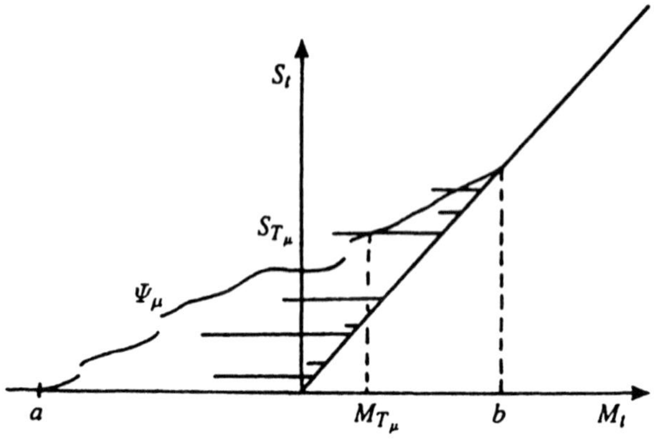

Fig. 6.

Proof of the Theorem. The time T_μ is a.s. finite because for any $x > 0$ we have

$$T_\mu \leq \inf\left\{t > T_x : M_t = \psi_\mu(x)\right\}$$

and this is finite since $\langle M, M\rangle_\infty = \infty$ (Proposition (1.8) Chap. V). To complete the proof of the theorem we use an approximation procedure which is broken into several steps.

(5.5) Proposition. *Let (μ_n) be a sequence of centered probability measures such that (ψ_{μ_n}) increases pointwise to ψ_μ; then*

i) (μ_n) *converges weakly to* μ;
ii) $\lim_n \int |x| d\mu_n(x) = \int |x| d\mu(x)$;
iii) $(M_{T_{\mu_n}})$ *converges a.s. to* M_{T_μ}.

Proof. The sequence (μ_n) is weakly relatively compact because for any $K > 0$,

$$\begin{aligned}\mu_n([K,\infty[) &\leq \frac{1}{K}\int_{[K,\infty[} t\, d\mu_n(t) \leq \frac{1}{K}\int_{[0,\infty[} t\, d\mu_n(t)\\ &= \frac{1}{K}\psi_{\mu_n}(0)\bar{\mu}_n(0) \leq \frac{1}{K}\psi_\mu(0),\end{aligned}$$

and, recalling that μ is centered,

$$\begin{aligned}\mu_n(]-\infty,-K]) &\leq \frac{1}{K}\int_{]-\infty,-K]} (-t) d\mu_n(t) \leq \frac{1}{K}\int_{]-\infty,0]} (-t) d\mu_n(t)\\ &= \frac{1}{K}\psi_{\mu_n}(0)\bar{\mu}_n(0) \leq \frac{1}{K}\psi_\mu(0).\end{aligned}$$

Moreover, since ψ_{μ_n} increases to ψ_μ it is easily seen that $b_n = \inf\{t : \psi_{\mu_n}(t) = t\}$ increases to b. From Lemma (5.1) it follows that $\bar{\mu}_n(x)/\bar{\mu}_n(0)$ converges to $\bar{\mu}(x)/\bar{\mu}(0)$ on $]-\infty, b[$ and $\bar{\mu}_n = \bar{\mu} = 0$ on $]b,\infty[$. Let (μ_{n_k}) be a subsequence converging weakly to a probability measure ν. If x is a point of continuity of $\bar{\nu}$, then

$$\lim_k \bar{\mu}_{n_k}(x) = \bar{\nu}(x).$$

If we choose $x < b$, it follows that $\bar{\mu}_{n_k}(0)$ has a limit l, otherwise $\bar{\mu}_{n_k}(x)/\bar{\mu}_{n_k}(0)$ could not have a limit. Moreover

$$\bar{\nu}(x)/l = \bar{\mu}(x)/\bar{\mu}(0).$$

By taking a sequence x_n of continuity points of $\bar{\nu}$ increasing to zero, it follows that $l = \bar{\mu}(0)$ and $\bar{\nu} = \bar{\mu}$ which proves i).

Using the lower semi-continuity of $\mu \to \int_{[0,\infty[} t\, d\mu(t)$, we have

$$\begin{aligned}\int_{[0,\infty[} t\, d\mu(t) &\leq \varliminf_n \int_{[0,\infty[} t\, d\mu_n(t) \leq \varlimsup_n \int_{[0,\infty[} t\, d\mu_n(t)\\ &= \lim_n \bar{\mu}_n(0)\psi_{\mu_n}(0) = \bar{\mu}(0)\psi_\mu(0) = \int_{[0,\infty[} t\, d\mu(t);\end{aligned}$$

using the fact that μ_n and μ are centered we also have

$$\lim_n \int_{]-\infty,0]} t\, d\mu_n(t) = \int_{]-\infty,0]} t\, d\mu(t)$$

which establishes ii).

Finally, the sequence $T_n = T_{\mu_n}$ increases to a stopping time $R \leq T_\mu$. But for $p \leq n$, we have

$$\psi_{\mu_p}(M_{T_n}) \leq \psi_{\mu_n}(M_{T_n}) \leq S_{T_n} \leq S_R;$$

passing to the limit, we get

$$\psi_{\mu_p}(M_R) \leq \underline{\lim}\, \psi_{\mu_n}(M_{T_n}) \leq S_R$$

hence $\psi_\mu(M_R) \leq S_R$ and $R = T_\mu$. By the continuity of M the proof is complete.

(5.6) Lemma. *There exists a sequence of centered probability measures μ_n such that*

i) *μ_n has a compact support contained in an interval $[a_n, b_n]$;*
ii) *ψ_{μ_n} is continuous on $\mathbb{R}$ and strictly increasing on $[a_n, \infty[$;*
iii) *the sequence (ψ_{μ_n}) increases to ψ_μ everywhere on $\mathbb{R}$.*

Proof. Let $a = \sup\{x : \psi_\mu(x) = 0\}$, and (a_n) (resp: (b_n)) a sequence of strictly negative (resp: positive) numbers decreasing (resp: increasing) to a (resp: b). The number

$$\delta_n = \inf\{\psi(x) - x; -\infty < x \leq b_n\}$$

is finite and > 0. Let j be the function in the proof of Theorem (1.1), and for each n, pick an integer k_n such that

$$\int y\, j(y)\, dy > -\delta_n k_n.$$

For $-\infty < x \leq b_n$, set

$$\psi_n(x) = \left((x - a_{n+1})1_{(a_{n+1} \leq x < a_n)}/(a_n - a_{n+1}) + 1_{(a_n \leq x)}\right) \int \psi_\mu(x+y) k_n j(k_n y)\, dy.$$

The function ψ_n enjoys the following properties:

i) $\psi_n \leq \psi_\mu$ and ψ_n is continuous on $]-\infty, b_n]$,
ii) $\psi_n > 0$ on $]a_{n+1}, b_n]$,
iii) $\psi_n(x) > x$ on $]-\infty, b_n]$,

(indeed, for $x \in [0, b_n]$,

$$\psi_n(x) - x = k_n \int (\psi_\mu(x+y) - (x+y))\, j(k_n y)\, dy + k_n \int y\, j(k_n y)\, dy$$

which is > 0 by the choice of k_n).

We now define ψ_n on $]b_n, \infty[$ in the following manner. Since $b_n < \psi_n(b_n) \leq \psi_\mu(b_n)$, we let ψ_n be affine on $[b_n, \psi_\mu(b_n)]$ and set $\psi_n(x) = x$ for $x \geq \psi_\mu(b_n)$ in such a way that ψ_n is continuous. Finally, we set $\tilde{\psi}_n = \psi_1 \vee \psi_2 \vee \cdots \vee \psi_n$; by Lemma (5.2), the sequence $(\tilde{\psi}_n)$ enjoys the properties of the statement.

End of the Proof of Theorem (5.4). By Proposition (5.5) and Lemma (5.6) it is enough to prove the first sentence when the support of μ is contained in the compact interval $[a, b]$ and ψ_μ is continuous and strictly increasing on $[a, \infty[$. Since $\psi_\mu(x) = 0$ for $x < a$ we have $T_\mu < T_a$ and since $\psi(x) = x$ for $x > b$ we also have $T_\mu \leq T_b$, hence M^{T_μ} is bounded.

Let γ be the inverse of the restriction of ψ_μ to $[a, \infty[$; for $\phi \in C_K$ we set $g = \phi \circ \gamma$ and $G(x) = \int_0^x g(u)du$. By the remarks following Proposition (4.5) the process $X_t = G(S_t) - (S_t - M_t)g(S_t)$ is a local martingale. The functions ϕ and G being bounded, X^{T_μ} is a bounded martingale and consequently

$$E\left[G\left(S_{T_\mu}\right) - \left(S_{T_\mu} - M_{T_\mu}\right) g\left(S_{T_\mu}\right)\right] = 0.$$

By the definitions of g and T_μ this may be written

$$E\left[\int_0^{\psi_\mu(M_{T_\mu})} \phi \circ \gamma(u)du - \left(\psi_\mu\left(M_{T_\mu}\right) - M_{T_\mu}\right) \phi\left(M_{T_\mu}\right)\right] = 0.$$

If ν is the law of M_{T_μ} we have

$$\int \nu(dx) \int_{-\infty}^x \phi(v) d\psi_\mu(v) + \int \left(x - \psi_\mu(x)\right) \phi(x) \nu(dx) = 0,$$

and after integrating by parts

$$\int \phi(x) \left[-\bar{\nu}(x) d\psi_\mu(x) + \left(x - \psi_\mu(x)\right) d\bar{\nu}(x)\right] = 0.$$

Since ϕ is arbitrary in C_K it follows from Lemma (5.1) and its proof that $\nu = \mu$.

To prove i) choose a sequence (ψ_n) according to Lemma (5.6). For each n, the process $M^{T_{\mu_n}}$ is a bounded martingale. Moreover $|M_{T_{\mu_n}}|$ converges a.s. to $|M_{T_\mu}|$ and by Proposition (5.5) ii), $E\left[|M_{T_{\mu_n}}|\right]$ to $E\left[|M_{T_\mu}|\right]$; it follows that $|M_{T_{\mu_n}}|$ converges to $|M_{T_\mu}|$ in L^1. The proof of i) is then easily completed.

It remains to prove ii). When μ has compact support, M^{T_μ} is bounded and by Proposition (1.23) Chap. IV, $E\left[\langle M, M\rangle_{T_\mu}\right] = \int x^2 d\mu(x)$. To get the general case we use again an approximating sequence. Set, with $\lambda(x) = x$,

$$\psi_n = \psi_\mu 1_{]-n,n]} + \psi_\mu(n) 1_{]n, \psi_\mu(n)]} + \lambda 1_{]\psi_\mu(n), \infty[}.$$

By Lemma (5.2), ψ_n corresponds to a measure μ_n and by Lemma (5.3) if $\int x^2 d\mu_n(x) < \infty$,

$$\int x^2 d\mu_n(x) = \int_{-n}^{\psi_\mu(n)} \psi_{\mu_n}(x) \bar{\mu}_n(x)\, dx.$$

By Lemma (5.1), for $-n < x < n$, we have $\bar{\mu}_n(x) = C_n \bar{\mu}(x)$ where $\lim_n C_n = 1$. Therefore

$$\int x^2 d\mu_n(x) = C_n \int_{-n}^{n} \psi_\mu(x) \bar{\mu}(x)\, dx + \int_{-n}^{\psi_\mu(n)} \psi_\mu(x) \bar{\mu}_n(x)\, dx.$$

We will prove that the last integral on the right, say I_n, goes to 0 as n tends to infinity. Indeed, $\bar{\mu}_n$ is also constant on $[n, \psi_\mu(n)]$ and is equal to $C_n\bar{\mu}(n)$. Hence

$$I_n = C_n\psi_\mu(n)\bar{\mu}(n)(\psi_\mu(n) - n).$$

If X is a r.v. with law μ, then:

$$\begin{aligned} I_n/C_n &= E[X\,1_{(X\geq n)}]\,E[(X-n)\,1_{(X\geq n)}]/P(X\geq n) \\ &\leq E[X^2]^{1/2}\,E[((X-n)^+)^2]^{1/2}, \end{aligned}$$

which proves our claim.

As a result, $\lim_n \int x^2 d\mu_n(x) < \infty$. By the proof of Proposition (5.5) the sequence $\{T_{\mu_n}\}$ increases to T_μ so that

$$E\left[\langle M, M\rangle_{T_\mu}\right] = \lim_n E\left[\langle M, M\rangle_{T_{\mu_n}}\right] = \int x^2 d\mu(x).$$

□

\# **(5.7) Exercise.** Let B be the standard linear BM.

1°) For any probability measure μ on $\mathbb{R}$ prove that there is a $\mathscr{F}^B_{1/2}$ measurable r.v. say Z, such that $Z(P) = \mu$.

2°) Define a $(\mathscr{F}_t^B)$-stopping time T by

$$T = \inf\{t \geq 1 : B_t = Z\}.$$

Prove that the law of B_T is μ and that $E[T] = \infty$.

* **(5.8) Exercise (A uniqueness result).** 1°) Let g be a continuous strictly increasing function such that $\lim_{x\to-\infty} g(x) = 0$, $g(x) \geq x$ and $g(x) = x$ for all $x \geq \inf\{u : g(u) = u\}$. If for $T = \inf\{t : S_t \geq g(M_t)\}$ the process M^T is a uniformly integrable martingale and if M_T has law μ, prove that

$$g(x) = -\frac{1}{\bar{\mu}(x)}\int_{-\infty}^x t\, d\mu(t).$$

In particular if μ is centered, then $g = \psi_\mu$.

[Hint: Use the first part of the proof of Theorem (5.4).]

2°) Extend the result to the case where g is merely left-continuous and increasing.

(5.9) Exercise. In the situation of Theorem (5.4) prove that the law of S_{T_μ} is given by

$$P\left[S_T \geq x\right] = \exp\left(-\int_0^x \frac{ds}{s - \gamma(s)}\right),$$

where $\gamma(s) = \inf\left\{t : \psi_\mu(t) \geq s\right\}$.

(5.10) Exercise. Let B be the standard linear BM and μ a centered probability measure. Prove that there is an increasing sequence of finite stopping times T_n such that the random variables $B_{T_{n+1}} - B_{T_n}$ are independent, identically distributed with law μ and $E\left[T_{n+1} - T_n\right] = \int x^2 d\mu(x)$.

* **(5.11) Exercise.** Prove that the time T_μ of Theorem (5.4) has the following minimality property: if R is a stopping time such that $R \leq T_\mu$ and $M_R \overset{(d)}{=} M_{T_\mu}$, then $R = T_\mu$.

(5.12) Exercise. 1°) For $p, q \in \mathbb{R}_+$, call $\nu(p,q)$ the probability measure $(q\varepsilon_p + p\varepsilon_{-q})/(p+q)$. If μ is a probability measure satisfying the conditions of Theorem (5.4), prove that there is a probability measure α on $\mathbb{R}_+^2$ such that

$$\mu = \int \nu(p,q)\alpha(dp, dq).$$

[Hint: Start with μ of finite support and use weak convergence.]

2°) Let (P, Q) be a r.v. independent of B with law α. The r.v.

$$T = \inf\{t \geq 0 : B_t \notin]-Q, P[\}$$

is such that the law of B_T is μ. Observe that it is a stopping time but for a filtration strictly larger than the Brownian filtration.

Notes and Comments

Sect. 1. The concept and construction of local time in the case of Brownian motion are due to P. Lévy [2]. The theory expanded in at least three directions. The first to appear was the theory of local times for Markov processes which is described in Blumenthal-Getoor [1] (see also Sharpe [1]) and will be taken up in the Notes and Comments of Chap. X. A second approach is that of occupation densities (Geman and Horowitz [1]). The point there is to show that the measure $A \to \int_0^t 1_A(X_s)ds$ is absolutely continuous with respect to a given deterministic measure which usually is the Lebesgue measure on $\mathbb{R}$. This is often done by Fourier transform methods and generalizes in the theory of intersection local times which has known much progress in recent years and for which the reader may consult Geman et al. [1], Rosen [1], Le Gall [4] and Yor [18]; the Markov view-point on this question being thoroughly developed in Dynkin [2]. These two approaches to local times are fleetingly alluded to in some exercises e.g. Exercise (2.19).

The third and possibly most useful line of attack stems from the desire to enlarge the scope of Itô's formula; this is the semimartingale point of view which first appeared in Meyer [5] after earlier results of Tanaka [1] for Brownian motion and Millar [2] for processes with independent increments. The case of continuous semimartingales is the subject of this section. The reader can find another exposition based on the general theory of processes in Azéma-Yor [1] and the extension to local times of regenerative sets in Dellacherie-Meyer [1] vol 4.

Theorem (1.7) which extends or parallels early results of Trotter [1], Boylan [1] and Ray [1] is taken from Yor [4] and Theorem (1.12) from Bouleau-Yor [1] as well as Exercise (1.29). The approximation results of Corollary (1.9), Theorem (1.10) and Exercise (1.20) as well as some others to be found in Chap. XII, were, in the case of Brownian motion, originally due to Lévy (see Itô-McKean [1] and for semimartingales see El Karoui [1]). Ouknine-Rutkowski [1] give many interesting "algebraic" formulae for the computation of local times, some of which we have turned into exercises.

Exercise (1.21) is from Weinryb [1], Exercise (1.20) from McGill et al. [1] and Exercise (1.26) from Pratelli [1]. Exercise (1.17) is due to Yoeurp [1], Exercises (1.14) and (1.22) are respectively from Yor [5] and [12]. Exercise (1.29) is from Biane-Yor [1]; it extends a result which is in Itô-McKean [1] (Problem 1, p. 72). Principal values of Brownian local times have been studied in depth by Yamada ([2], [3], [4], [6]) and also by Bertoin [2] to whom Exercise (1.30) is due; they have been investigated for physical purposes by Ezawa et al. ([1], [2], [3]).

Tanaka's formula expresses the Doob-Meyer decomposition of the absolute value of a martingale. Conversely, Gilat [1] has shown that the law of a positive submart. is the law of the absolute value of a martingale. A pathwise construction of this martingale (involving a possible extra-randomization) has been given, among others, by Protter-Sharpe [1], Barlow [7], Barlow-Yor [3] and Maisonneuve [8].

Sect. 2. The results of the first half of this section are due to Lévy but the proofs are totally different. Actually Lévy's study of Brownian local time was based on the equivalence theorem (2.3), whereas we go the other way round, thanks to Lemma (2.1) which is due to Skorokhod [2] (see also El Karoui and Chaleyat-Maurel [1]). Among other things, Theorem (2.3) shows that the Brownian local time is not after all such an exotic object since it is nothing else than the supremum process of another BM.

Corollary (2.8) gives a precise labeling of the excursions of BM away from zero which will be essential in Chap. XII.

The first proof of the Arcsine law appears in Lévy ([4], [5]). The proof presented here is found in Pitman-Yor [5] and Karatzas-Shreve [1] but the original ideas are due to Williams [1] and McKean [3]. There are other proofs of the Arcsine law, especially by the time-honoured Feynman-Kac's approach which may be found in Itô-McKean [1]. Another proof relying on excursion theory is found in Barlow-Pitman-Yor [1] (see Exercise (2.17) Chap. XII).

Exercise (2.13) is due to Spitzer [1] and Exercise (2.15) is in Itô-McKean [1]. The equality in law of Exercise (2.15), 2°) is a particular case of a general result on Lévy processes (see Bertoin [7]) which is an extension to continuous time of a combinatorial result of Sparre Andersen (see Feller [4]).

Exercise (2.22) is taken from Lane [1] and Exercise (2.24) from Weinryb [1]. Exercise (2.27) is taken from Le Gall [3]; another method as well as Exercise (2.26) may be found in Itô-McKean [1]; the results are originally due to Besicovitch and

Taylor [1] and Taylor [1]. Exercise (2.28) is from Yor [18] and Exercise (2.29) from Biane et al. [1].

Pushing the ideas of Exercise (2.32) a little further we are led to the following open

Question 1. Is the map $\beta \to \hat{\beta}$ ergodic?

Dubins and Smorodinsky [1] have given a positive answer to a discrete analogue of Question 1 for the standard random walk on $\mathbb{Z}$. Furthermore Dubins et al. [1] (1993) give another interesting question equivalent to Question 1 along the lines of Exercise (2.32). Let us note that Question 1 may be extended to functions of modulus 1 other than sgn and also to a d-dimensional setting using the result in Exercise (3.22) Chapter IV. Finally, Malric [2] obtains density results for the sets of zeros of the iterates of $\hat{\beta}$ which may help to solve question 1.

Exercise (2.33) presents one of many absolute continuity relationships (see, e.g., Yor [25]) which follow from scaling invariance properties; for a different example, see Exercise (2.29) relating the Brownian and Pseudo-Brownian bridges. The method used in Exercise (2.33) may be extended to yield Petit's generalization of Lévy's Arcsine law (see Petit [1] and Carmona et al. [1]).

Sect. 3. Most of the results of this section are from Pitman [1], but we have borrowed our proof of Theorem (3.5) from Ikeda-Watanabe [2]. The original proof of Pitman uses a limiting procedure from the discrete time case to the continuous time case. This can be used successfully in many contexts as for instance in Le Gall [5] or to prove some of the results in Sect. 3 Chap. VII as in Breiman [1]. For other proofs of Theorem (3.5) see Pitman-Rogers [1], Rogers [2] and Jeulin [1], as well as Exercise (4.15) in Chapter VII. A simple proof has been given by Imhof [3].

Theorem (3.11) is due to Williams [3] as well as several exercises. Exercise (3.18) is taken from Emery-Perkins [1].

Sect. 4. The better part of this section comes from Azéma-Yor [1] and Yor [8]. Earlier work may be found in Azéma [2] and extensions to random closed sets in Azéma [3]. As mentioned below Proposition (4.5) another proof of Itô's formula may be based on the results of this section (see Azéma-Yor [1]) and thus it is possible to give a different exposition of many of the results in Chaps. IV and VI.

Exercise (4.9) is due in part to Kennedy [1]. Exercise (4.11) is borrowed from Dubins-Schwarz [3] and Pitman [2] and is continued in Exercise (4.13). Further computations of best constants for Doob-like inequalities are found in Jacka [1].

Exercise (4.17) is in Itô-McKean [1] (see also Azéma-Yor [2]) and Exercise (4.20) in Brossard and Chevalier [1]. Exercise (4.21) is due to L. Carraro (private communication). Exercise (4.25) comes from Pitman [7].

Let us mention the

Question 2. Is it possible to relax the hypothesis on F in Proposition (4.7)?

Sect. 5. The problem dealt with in this section goes back to Skorokhod [2] and has received a great many solutions such as in Dubins [3], Root [1], Chacon-Walsh [1] to mention a few. In discrete time the subject has been investigated by Rost (see Revuz [3]) and has close connections with Ergodic theory.

The solution presented in this section is taken from Azéma-Yor [2] with a proof which was simplified by Pierre [1] (see also Meilijson ([1] and [2]) and Zaremba [1]). It is only one of the many solutions given over the years, each of which has drawbacks and advantages. A complete discussion is given by Obloj [1].

Further Skorokhod type problems are solved in Obloj-Yor [1] using either stochastic calculus or excursion theory.

Chapter VII. Generators and Time Reversal

In this chapter, we take up the study of Markov processes. We assume that the reader has read Sect. 1 and 2 in Chap. III.

§1. Infinitesimal Generators

The importance of the theory of Markov processes is due to several facts. On the one hand, Markov processes provide models for many a natural phenomenon; that the present contains all the information needed on the past to make a prediction on the future is a natural, if somewhat overly simplifying idea and it can at least often be taken as a first approximation. On the other hand, Markov processes arise naturally in connection with mathematical and physical theories.

However, the usefulness of the theory will be limited by the number of processes that can be constructed and studied. We have seen how to construct a Markov process starting from a t.f., but the snag is that there aren't many t.f.'s which are explicitly known; moreover, in most phenomena which can be modeled by a Markov process, what is grasped by intuition is not the t.f. but the way in which the process moves from point to point. For these reasons, the following notions are very important.

(1.1) Definition. *Let X be a Feller process; a function f in C_0 is said to belong to the domain $\mathscr{D}_A$ of the infinitesimal generator of X if the limit*

$$Af = \lim_{t \downarrow 0} \frac{1}{t} (P_t f - f)$$

exists in C_0. The operator $A : \mathscr{D}_A \to C_0$ thus defined is called the infinitesimal generator *of the process X or of the semi-group P_t.*

By the very definition of a Markov process with semi-group (P_t), if f is a bounded Borel function

$$E\left[f(X_{t+h}) - f(X_t) \mid \mathscr{F}_t\right] = P_h f(X_t) - f(X_t).$$

As a result, if $f \in \mathscr{D}_A$, we may write

$$E\left[f(X_{t+h}) - f(X_t) \mid \mathscr{F}_t\right] = hAf(X_t) + o(h).$$

Thus A appears as a means of describing how the process moves from point to point in an infinitesimally small time interval.

We now give a few properties of A.

(1.2) Proposition. *If* $f \in \mathscr{D}_A$, *then*

i) $P_t f \in \mathscr{D}_A$ *for every* t;

ii) the function $t \to P_t f$ *is strongly differentiable in* C_0 *and*

$$\frac{d}{dt} P_t f = A P_t f = P_t A f;$$

iii) $P_t f - f = \int_0^t P_s A f\, ds = \int_0^t A P_s f\, ds.$

Proof. For fixed t, we have, using the semi-group property,

$$\lim_{s \to 0} \frac{1}{s} [P_s(P_t f) - P_t f] = \lim_{s \to 0} P_t \left[\frac{1}{s} (P_s f - f) \right] = P_t A f$$

which proves i) and $A P_t f = P_t A f$. Also, $t \to P_t f$ has a right-hand derivative which is equal to $P_t A f$.

Consider now the function $t \to \int_0^t P_s A f\, ds$. This function is differentiable and its derivative is equal to $P_t A f$. Since two continuous functions which have the same right derivatives differ by a constant, we have $P_t f = \int_0^t P_s A f\, ds + g$ for some g, which completes the proof of ii); by making $t = 0$, it follows that $g = f$ which proves iii). □

Remark. The equation $\frac{d}{dt} P_t f = P_t A f$ may be written in a formal way

$$\frac{d}{dt} P_t(x, \cdot) = A^* P_t(x, \cdot)$$

where A^* is the formal adjoint of A for the duality between functions and measures. It is then called the *forward* or *Fokker-Planck* equation. The reason for the word forward is that the equation is obtained by perturbing the final position, namely, $P_t A f$ is the limit of $P_t \left(\frac{1}{\varepsilon} (P_\varepsilon f - f) \right)$ as $\varepsilon \to 0$. Likewise, the equation $\frac{d}{dt} P_t f = A P_t f$ which is obtained by perturbing the initial position i.e. $A P_t f = \lim_{\varepsilon \to 0} \frac{1}{\varepsilon} (P_\varepsilon - I) P_t f$, is called the *backward* equation. These names are especially apt in the non-homogeneous case where the forward (resp. backward) equation is obtained by differentiating $P_{s,t}$ with respect to t (resp. s).

(1.3) Proposition. *The space* $\mathscr{D}_A$ *is dense in* C_0 *and* A *is a closed operator.*

Proof. Set $A_h f = \frac{1}{h} (P_h f - f)$ and $B_s f = \frac{1}{s} \int_0^s P_t f\, dt$. The operators A_h and B_s are bounded on C_0 and moreover

$$A_h B_s = B_s A_h = A_s B_h = B_h A_s.$$

For every $s > 0$ and $f \in C_0$,

$$\lim_{h\to 0} A_h B_s f = \lim_{h\to 0} A_s(B_h f) = A_s f;$$

therefore $B_s f \in \mathscr{D}_A$ and since $\lim_{s\to 0} B_s f = f$, $\mathscr{D}_A$ is dense in C_0.

Let now (f_n) be a sequence in $\mathscr{D}_A$, converging to f and suppose that (Af_n) converges to g. Then

$$\begin{aligned} B_s g &= \lim_n B_s A f_n = \lim_n B_s \left(\lim_h A_h f_n\right) \\ &= \lim_n \lim_h A_s\,(B_h f_n) = \lim_n A_s f_n = A_s f. \end{aligned}$$

It follows that $f \in \mathscr{D}_A$ and $Af = \lim_{s\to o} A_s f = g$ which proves that A is a closed operator. □

The resolvent U_p, which was defined in Sect. 2 Chap. III, is the resolvent of the operator A as is shown in the next

(1.4) Proposition. *For every $p > 0$, the map $f \to pf - Af$ from $\mathscr{D}_A$ to C_0 is one-to-one and onto and its inverse is U_p.*

Proof. If $f \in \mathscr{D}_A$, then

$$\begin{aligned} U_p(pf - Af) &= \int_0^\infty e^{-pt} P_t(pf - Af)dt \\ &= p\int_0^\infty e^{-pt} P_t f\, dt - \int_0^\infty e^{-pt}\left(\frac{d}{dt}P_t f\right)dt; \end{aligned}$$

integrating by parts in the last integral, one gets $U_p(pf - Af) = f$. Conversely, if $f \in C_0$, then, with the notation of the last proposition

$$\lim_{h\to 0} A_h U_p f = \lim_{h\to 0} U_p A_h f = \lim_{h\to 0}\int_0^\infty e^{-pt} P_t\left(\frac{P_h f - f}{h}\right)dt$$

which is easily seen to be equal to $pU_p f - f$. As a result, $(pI - A)U_p f = f$ and the proof is complete. □

The last three propositions are actually valid for any strongly continuous semi-group of contractions on a Banach space. Our next result is more specific.

(1.5) Proposition. *The generator A of a Feller semi-group satisfies the following* positive maximum principle*: if $f \in \mathscr{D}_A$, and if x_0 is such that $0 \le f(x_0) = \sup\{f(x), x \in E\}$, then*

$$Af(x_0) \le 0.$$

Proof. We have $Af(x_0) = \lim_{t\downarrow 0} \frac{1}{t}\left(P_t f(x_0) - f(x_0)\right)$ and

$$P_t f(x_0) - f(x_0) \le f(x_0)\left(P_t(x_0, E) - 1\right) \le 0.$$

□

The probabilistic significance of generators which was explained below Definition (1.1) is also embodied in the following proposition where X is a Feller process with transition function (P_t).

(1.6) Proposition. *If $f \in \mathscr{D}_A$, then the process*

$$M_t^f = f(X_t) - f(X_0) - \int_0^t Af(X_s)ds$$

is a $\left(\mathscr{F}_t^0, P_\nu\right)$-martingale for every ν. If, in particular, $Af = 0$, then $f(X_t)$ is a martingale.

Proof. Since f and Af are bounded, M_t^f is integrable for each t. Moreover

$$E_\nu\left[M_t^f \mid \mathscr{F}_s^0\right] = M_s^f + E_\nu\left[f(X_t) - f(X_s) - \int_s^t Af(X_u)du \,\middle|\, \mathscr{F}_s^0\right].$$

By the Markov property, the conditional expectation on the right is equal to

$$E_{X_s}\left[f\left(X_{t-s}\right) - f(X_0) - \int_0^{t-s} Af(X_u)du\right].$$

But for any $y \in E$,

$$\begin{aligned} E_y\left[f\left(X_{t-s}\right) - f(X_0) - \int_0^{t-s} Af(X_u)du\right] \\ = P_{t-s}f(y) - f(y) - \int_0^{t-s} P_u Af(y)du \end{aligned}$$

which we know to be zero by Proposition (1.2). This completes the proof. We observe that in lieu of $(\mathscr{F}_t^0)$, we could use any filtration $(\mathscr{G}_t)$ with respect to which X is a Markov process. □

Remark. This proposition may be seen as a special case of Exercise (1.8) in Chap. X. We may also observe that, if $f \in \mathscr{D}_A$, then $f(X_t)$ is a semimartingale; in the case of BM, a converse will be found in Exercise (2.23) of Chap. X.

Conversely, we have the

(1.7) Proposition. *If $f \in C_0$ and, if there exists a function $g \in C_0$, such that*

$$f(X_t) - f(X_0) - \int_0^t g(X_s)ds$$

is a $(\mathscr{F}_t, P_x)$-martingale for every x, then $f \in \mathscr{D}_A$ and $Af = g$.

Proof. For every x we have, upon integrating,

$$P_t f(x) - f(x) - \int_0^t P_s g(x) ds = 0,$$

hence

$$\left\| \frac{1}{t}(P_t f - f) - g \right\| = \left\| \frac{1}{t} \int_0^t (P_s g - g) ds \right\| \leq \frac{1}{t} \int_0^t \| P_s g - g \| \, ds$$

which goes to zero as t goes to zero. □

The two foregoing results lead to the following

(1.8) Definition. *If X is a Markov process, a Borel function f is said to belong to the domain $\mathbb{D}_A$ of the* extended infinitesimal generator *if there exists a Borel function g such that, a.s., $\int_0^t |g(X_s)| ds < +\infty$ for every t, and*

$$f(X_t) - f(X_0) - \int_0^t g(X_s) ds$$

is a $(\mathscr{F}_t, P_x)$-right-continuous martingale for every x.

Of course $\mathbb{D}_A \supset \mathscr{D}_A$; moreover we still write $g = Af$ and call the "operator" A thus defined the *extended infinitesimal generator*. This definition makes also perfect sense for Markov processes which are not Feller processes. Actually, most of the above theory can be extended to this more general case (see Exercise (1.16)) and the probabilistic significance is the same. Let us observe however that g may be altered on a set of potential zero (Exercise (2.25) Chap. III) without altering the martingale property so that the map $f \to g$ is actually multi-valued and only "almost" linear.

The remainder of this section is devoted to a few fundamental examples. Some of the points we will cover are not technically needed in the sequel but are useful for a better understanding of some of the topics we will treat.

There are actually few cases where $\mathscr{D}_A$ and A are completely known and one has generally to be content with subspaces of $\mathscr{D}_A$. We start with the case of independent increment processes for which we use the notation of Sect. 4 in Chap. III. Let $\mathscr{S}$ be the Schwartz space of infinitely differentiable functions f on the line such that $\lim_{|x|\to\infty} f^{(k)}(x)P(x) = 0$ for any polynomial P and any integer k. The Fourier transform is a one-to-one map from $\mathscr{S}$ onto itself.

(1.9) Proposition. *Let X be a real-valued process with stationary independent increments; the space $\mathscr{S}$ is contained in $\mathscr{D}_A$ and for $f \in \mathscr{S}$*

$$Af(x) = \beta f'(x) + \frac{\sigma^2}{2} f''(x) + \int \left[f(x+y) - f(x) - \frac{y}{1+y^2} f'(x) \right] \nu(dy).$$

Proof. We first observe that $|\psi|$ increases at most like $|u|^2$ at infinity. Indeed

$$\left|\int_{[-1,1]^c}\left(e^{iux}-1-\frac{iux}{1+x^2}\right)\nu(dx)\right| \leq 2\nu\left([-1,1]^c\right)+|u|\int_{[-1,1]^c}\frac{|x|}{1+x^2}\nu(dx)$$

and

$$\begin{aligned}\left|\int_{-1}^{1}\left(e^{iux}-1-\frac{iux}{1+x^2}\right)\nu(dx)\right| &\leq |u|\int_{-1}^{1}\left|\frac{x}{1+x^2}-x\right|\nu(dx) \\ &\quad+\int_{-1}^{1}\left|e^{iux}-1-iux\right|\nu(dx);\end{aligned}$$

it remains to observe that the last integrand is majorized by $c|x|^2|u|^2$ for a constant c.

Let then f be in $\mathscr{S}$; there exists a unique $g\in\mathscr{S}$ such that $f(y)=\int e^{iyv}g(v)dv$. If we set $g_x(v)=e^{ixv}g(v)$, we have $P_tf(x)=\langle\hat{\mu}_t,g_x\rangle$ as is proved by the following string of equalities:

$$\begin{aligned}\langle\hat{\mu}_t,g_x\rangle &= \int e^{ixv}g(v)\hat{\mu}_t(v)dv=\int e^{ixv}g(v)\left(\int e^{iyv}\mu_t(dy)\right)dv \\ &= \int\mu_t(dy)\int e^{i(x+y)v}g(v)dv=\int f(x+y)\mu_t(dy)=P_tf(x).\end{aligned}$$

As a result

$$P_tf(x)=\int e^{t\psi(u)}g_x(u)du=\langle 1,g_x\rangle+t\langle\psi,g_x\rangle+\frac{t^2}{2}H(t,x)$$

where, because $e^{s\psi}$ is the Fourier transform of a probability measure,

$$|H(t,x)|\leq\sup_{0\leq s\leq t}|\langle\psi^2e^{s\psi},g_x\rangle|\leq\langle|\psi|^2,|g_x|\rangle;$$

by the above remark, $\langle|\psi|,|g_x|\rangle$ and $\langle|\psi|^2,|g_x|\rangle$ are finite so that $\frac{1}{t}\left(P_tf(x)-f(x)\right)$ converges uniformly to $\langle\psi,g_x\rangle$. As $f'(y)=i\int vg_y(v)dv$ and $f''(y)=i^2\int v^2g_y(v)dv$, we get for A the announced formula. □

The three following particular cases are fundamental. To some extent, they provide the "building blocks" of large classes of Markov processes.

If X is the linear BM, then obviously $Af(x)=\frac{1}{2}f''(x)$ for every $f\in\mathscr{S}$, and if $X=\sigma B$ where B is the linear BM, then $Af(x)=\frac{\sigma^2}{2}f''(x)$, $f\in\mathscr{S}$. In this case we can actually characterize the space $\mathscr{D}_A$. We call C_0^2 the space of twice continuously differentiable functions f on $\mathbb{R}^d$ $(d\geq 1)$ such that f and its first and second order derivatives are in C_0.

(1.10) Proposition. *For the linear* BM, *the space* $\mathscr{D}_A$ *is exactly equal to the space* C_0^2 *and* $Af(x)=\frac{1}{2}f''$ *on this space.*

Proof. From Proposition (1.4), we know that $\mathscr{D}_A = U_p(C_0)$ for any $p > 0$, and that $AU_pf = pU_pf - f$. We leave as an exercise to the reader the task of showing, by means of the explicit expression of U_p computed in Exercise (2.23) of Chap. III, that if $f \in C_0$ then $U_pf \in C_0^2$ and $pU_pf - f = \frac{1}{2}(U_pf)''$.

If, conversely, g is in C_0^2 and we define a function f by

$$f = pg - \frac{1}{2}g''$$

the function $g - U_pf$ satisfies the differential equation $y'' - 2py = 0$ whose only bounded solution is the zero function. It follows that $g = U_pf$, hence $g \in \mathscr{D}_A$ and $Ag = \frac{1}{2}g''$. □

The other particular cases are: i) the translation at speed β for which $\mathscr{D}_A$ is the space of absolutely continuous functions in C_0 such that the derivative is in C_0 and $Af(x) = \beta f'(x)$, ii) the Poisson process with parameter λ for which $\mathscr{D}_A = C_0$ (see Exercise (1.14)) and

$$Af(x) = \lambda\big(f(x+1) - f(x)\big).$$

In all these cases, we can describe the whole space $\mathscr{D}_A$, but this is a rather unusual situation, and, as a rule, one can only describe subspaces of $\mathscr{D}_A$.

We turn to the case of BM^d.

(1.11) Proposition. *For $d \geq 2$, the infinitesimal generator of* BM^d *is equal to $\frac{1}{2}\Delta$ on the space C_0^2.*

Proof. For $f \in C_0$, we may write

$$P_tf(x) = (2\pi)^{-d/2}\int_{\mathbb{R}^d} e^{-|z|^2/2} f\left(x + z\sqrt{t}\right) dz.$$

If $f \in C_0^2$, using Taylor's formula, we get

$$P_tf(x) = f(x) + \frac{1}{2}t\Delta f(x) + (2\pi)^{-d/2}\frac{t}{2}J(t,x)$$

where

$$J(t,x) = \int e^{-|z|^2/2}\left(\sum_{i,j=1}^{d}\left[\frac{\partial^2 f}{\partial x_i \partial x_j}(\theta) - \frac{\partial^2 f}{\partial x_i \partial x_j}(x)\right] z_i z_j\right) dz$$

with θ some point on the segment $\left[x, x + z\sqrt{t}\right]$. Set

$$F(x,z,t) = \max_{i,j}\left|\frac{\partial^2 f}{\partial x_i \partial x_j}(\theta) - \frac{\partial^2 f}{\partial x_i \partial x_j}(x)\right|.$$

For any $R > 0$, we have

$$|J(t,x)| \leq \int_{|z|\leq R} F(x,z,t)e^{-|z|^2/2}\left(\sum_{i,j}|z_i|\,|z_j|\right)dz$$
$$+2\int_{|z|>R}\max_{i,j}\left\|\frac{\partial^2 f}{\partial x_i\partial x_j}\right\| e^{-|z|^2/2}\left(\sum_{i,j}|z_i|\,|z_j|\right)dz.$$

As t goes to zero, the uniform continuity of second partial derivatives entails that the first half of the sum above goes to zero uniformly in x; consequently

$$\overline{\lim_{t\downarrow 0}}\sup_{x\in\mathbb{R}^d}|J(t,x)| \leq 2\max_{i,j}\left\|\frac{\partial^2 f}{\partial x_i\partial x_j}\right\|\int_{|z|>R} e^{-|z|^2/2}\left(\sum_{i,j}|z_i|\,|z_j|\right)dz.$$

By taking R large, we may make this last expression arbitrarily small which implies that

$$\lim_{t\downarrow 0}\left\|\frac{1}{t}(P_t f - f) - \frac{1}{2}\Delta f\right\| = 0$$

for every $f \in C_0^2$. □

Remarks. 1) At variance with the case $d = 1$, for $d > 1$, the space C_0^2 is not equal to $\mathscr{D}_A$. Using the closedness of the operator A, one can show without too much difficulty that $\mathscr{D}_A$ is the subspace (of C_0) of functions f such that Δf taken in the sense of distributions is in C_0 and A is then equal to $\frac{1}{2}\Delta f$.

2) In the case of BM, it follows from Proposition (1.2) that for $f \in C_0^2$

$$\frac{d}{dt}P_t f = \frac{1}{2}\Delta P_t f = \frac{1}{2}P_t\Delta f.$$

Actually, this can be seen directly since elementary computations prove that

$$\frac{\partial}{\partial t}g_t + \frac{1}{2}\frac{\partial^2}{\partial x^2}g_t = 0,$$

and the similar formula in dimension d. It turns out that the equality $\frac{d}{dt}P_t f = \frac{1}{2}\Delta P_t f$ is valid for any bounded Borel function f and $t > 0$. In the language of PDE's, g_t and its multidimensional analogues are fundamental solutions of the heat equation $\frac{\partial f}{\partial t} + \frac{1}{2}\Delta f = 0$.

If B is a $\mathrm{BM}^d(0)$ and σ a $d \times d$-matrix, one defines a $\mathbb{R}^d$-valued Markov process X by stipulating that if $X_0 = x$ a.s., then $X_t = x + \sigma B_t$. We then have the

(1.12) Corollary. *The infinitesimal generator of X is given on C_0^2 by*

$$Af(x) = \frac{1}{2}\sum_{i,j}\gamma_{ij}\frac{\partial^2 f}{\partial x_i\partial x_j}(x),$$

where $\gamma = \sigma\sigma'$, with σ' being the transpose of σ.

Proof. We have to find the limit, for $f \in C_0^2$, of

$$\frac{1}{t}E[f(x+\sigma B_t) - f(x)]$$

where E is the expectation associated with B; it is plainly equal to $\Delta g(0)/2$, where $g(y) = f(x+\sigma y)$, whence the result follows by straightforward computations. □

Remark. The matrix γ has a straightforward interpretation, namely, $t\gamma$ is the covariance matrix of X_t.

Going back to Proposition (1.9), we now see that, heuristically speaking, it tells us that a process with stationary independent increments is a mixture of a translation term, a diffusion term corresponding to $\frac{\sigma^2}{2}f''$ and a jump term, the jumps being described by the Lévy measure ν. The same description is valid for a general Markov process in $\mathbb{R}^d$ as long as $\mathscr{D}_A \supset C_K^2$; but since these processes are no longer, as was the case with independent increments, translation-invariant in space, the translation, diffusion and jump terms will vary with the position of the process in space.

(1.13) Theorem. *If P_t is a Feller semi-group on $\mathbb{R}^d$ and $C_K^\infty \subset \mathscr{D}_A$, then*

i) $C_K^2 \subset \mathscr{D}_A$;

ii) For every relatively compact open set U, there exist functions a_{ij}, b_i, c on U and a kernel N such that for $f \in C_K^2$ and $x \in U$

$$\begin{aligned} Af(x) &= c(x)f(x) + \sum_i b_i(x)\frac{\partial f}{\partial x_i}(x) + \sum_{i,j} a_{ij}(x)\frac{\partial^2 f}{\partial x_i \partial x_j}(x) \\ &\quad + \int_{\mathbb{R}^d\setminus\{x\}} \left[f(y) - f(x) - 1_U(y)\sum_i (y_i - x_i)\frac{\partial f}{\partial x_i}(x)\right] N(x,dy) \end{aligned}$$

where $N(x,\cdot)$ is a Radon measure on $\mathbb{R}^d\setminus\{x\}$, the matrix $a(x) = \|a_{ij}(x)\|$ is symmetric and non-negative, c is ≤ 0. Moreover, a and c do not depend on U.

Fuller information can be given about the different terms involved in the description of A, but we shall not go into this and neither shall we prove this result (see however Exercise (1.19)) which lies outside of out main concerns. We only want to retain the idea that a process with the above infinitesimal generator will move "infinitesimally" from a position x by adding a translation of vector $b(x)$, a gaussian process with covariance $a(x)$ and jumps given by $N(x,\cdot)$; the term $c(x)f(x)$ corresponds to the possibility for the process of being "killed" (see Exercise (1.26)). If the process has continuous paths, then its infinitesimal generator is given on C_K^2 by

$$Af(x) = c(x)f(x) + \sum_i b_i(x)\frac{\partial f}{\partial x_i}(x) + \sum_{i,j} a_{ij}(x)\frac{\partial^2 f}{\partial x_i \partial x_j}(x)$$

where the matrix $a(x)$ is symmetric and non-negative. Such an operator is said to be a *semi-elliptic second order differential operator.*

As was noted at the beginning of this section, a major problem is to go the other way round, that is, given an operator satisfying the positive maximum principle, to construct a Feller process whose generator is an extension of the given operator.

Let us consider a semi-elliptic second order differential operator $\sum a_{ij}(x)\frac{\partial^2}{\partial x_i \partial x_j}$ without terms of order 0 or 1. If the generator A of a Feller process X is equal to this operator on C_K^2, we may say, referring to Corollary (1.12), that between times t and $t+h$, the process X moves like $\sigma(x)B$ where $\sigma(x)$ is a square root of $a(x)$, i.e., $a(x) = \sigma(x)\sigma'(x)$, and B a BM^d; in symbols

$$X_{t+h} = X_t + \sigma(X_t)(B_{t+h} - B_t) + 0(h).$$

The idea to construct such a process is then to see it as an integral with respect to B,

$$X_t = \int_0^t \sigma(X_s)dB_s,$$

or, to use a terminology soon to be introduced, as a solution to the stochastic differential equation $dX = \sigma(X)dB$. As the paths of B are not of bounded variation, the integral above is meaningless in the Stieltjes-Lebesgue sense and this was one of the main motivations for the introduction of stochastic integrals. These ideas will be developed in the following section and in Chap. IX.

(1.14) Exercise (Bounded generators). 1°) Let π be a transition probability on E such that $\pi(C_0(E)) \subset C_0(E)$ and I be the identity on $C_0(E)$. Prove that $P_t = \exp\left(t(\pi - I)\right)$ is a Feller semi-group such that $\mathscr{D}_A = C_0(E)$ and $A = \pi - I$. Describe heuristically the behavior of the corresponding process of which the Poisson process is a particular case.

[Hint: See the last example in Exercise (1.8), Chap. III.]

2°) More generally, if A is a bounded operator in a Banach space, then $T_t = \exp(tA)$ is a uniformly continuous semi-group (i.e. $\lim_{t\downarrow 0}\|T_{t+s} - T_s\| = 0$) of bounded operators (not necessarily contractions) with infinitesimal generator A.

3°) Prove that actually the three following conditions are equivalent

i) (T_t) is uniformly continuous;
ii) $\mathscr{D}_A$ is the whole space;
iii) A is a bounded operator.

If these conditions are in force, then $T_t = \exp(tA)$.

[Hint: Use the closed graph and Banach-Steinhaus theorems.]

(1.15) Exercise. A strongly continuous resolvent on C_0 is a family (U_λ), $\lambda > 0$, of kernels such that

i) $\|\lambda U_\lambda\| \le 1$ for every $\lambda > 0$;
ii) $U_\lambda - U_\mu = (\mu - \lambda)U_\lambda U_\mu = (\mu - \lambda)U_\mu U_\lambda$ for every pair (λ, μ);
iii) for every $f \in C_0$, $\lim_{\lambda\to\infty}\|\lambda U_\lambda f - f\| = 0$.

It was shown in Sect. 2 Chap. III that the resolvent of a Feller semi-group is a strongly continuous resolvent.

1°) If (U_λ), $\lambda > 0$, is a strongly continuous resolvent, prove that each operator U_λ is one-to-one and that if the operator A is defined by $\lambda I - A = U_\lambda^{-1}$, then A does not depend on λ. If (U_λ) is the resolvent of a Feller semi-group, A is the corresponding generator.

2°) Prove that $f \in \mathscr{D}_A$ if and only if $\lim_{\lambda\to\infty} \lambda(\lambda U_\lambda f - f)$ exists and the limit is then equal to Af.

(1.16) Exercise. For a homogeneous transition function P_t, define B_0 as the set of bounded Borel functions f such that $\lim_{t\to 0} \|P_t f - f\| = 0$, where $\|f\| = \sup_x |f(x)|$. Define $\mathscr{D}_A$ as the set of those functions f for which there exists a function Af such that

$$\lim_{t\to 0} \left\| \frac{1}{t}(P_t f - f) - Af \right\| = 0.$$

Prove that $\mathscr{D}_A \subset B_0$ and extend the results of the present section to this general situation by letting B_0 play the role held by C_0 in the text.

(1.17) Exercise. If (P_t) is a Feller semi-group and f a function in C_0 such that $t^{-1}(P_t f - f)$ is uniformly bounded and converges pointwise to a function g of C_0, then $f \in \mathscr{D}_A$ and $Af = g$.

[Hint: Prove that $f \in U_\lambda(C_0)$.]

\# **(1.18) Exercise.** Let P_t and Q_t be two Feller semi-groups on the same space with infinitesimal generators A and B. If $\mathscr{D}_A \subset \mathscr{D}_B$ and $B = A$ on $\mathscr{D}_A$, prove that $P_t = Q_t$. Consequently, the map $P_t \to A$ is one-to-one and no strict continuation of an infinitesimal generator can be an infinitesimal generator.

[Hint: For $f \in \mathscr{D}_A$, differentiate the function $s \to Q_s P_{t-s} f$.]

* **(1.19) Exercise.** Let A be a linear map from $C^\infty(\mathbb{R}^d)$ into $C(\mathbb{R}^d)$, satisfying the positive maximum principle and such that $A1 = 0$. We assume moreover that A is a *local* operator, namely, if $f \equiv 0$ on some neighborhood of x, then $Af(x) = 0$.

1°) Prove that A satisfies the local maximum principle: if f has a local maximum at x, then $Af(x) \le 0$.

2°) If, for some x, the function f is such that $|f(y) - f(x)| = 0(|y - x|^2)$ as $y \to x$, prove that $Af(x) = 0$.

[Hint: Apply A to the function $f(y) + \alpha|y - x|^2$ for suitably chosen α.]

3°) Call χ^i the coordinate mappings and set $b_i(x) = A\chi^i(x)$, $a_{ij}(x) = A(\chi^i\chi^j)(x) - b_i(x)\chi^j(x) - b_j(x)\chi^i(x)$. Prove that for every x, the matrix $\left(a_{ij}(x)\right)$ is non-negative.

[Hint: Use the functions $f(y) = \left|\sum_{i=1}^d \theta_i \left(\chi^i(y) - \chi^i(x)\right)\right|^2$ where $\theta \in \mathbb{R}^d$.]

4°) Prove that

$$Af(x) = \frac{1}{2}\sum_{i,j} a_{ij}(x)\frac{\partial^2 f}{\partial x_i \partial x_j}(x) + \sum_i b_i(x)\frac{\partial f}{\partial x_i}(x).$$

[Hint: Use Taylor's formula.]

\# **(1.20) Exercise.** Let γ be the Gaussian measure in $\mathbb{R}^d$ with mean 0 and covariance matrix I_d. Let f be a C^∞-function in $\mathbb{R}^d$, bounded as well as its derivatives and B the $\mathrm{BM}^d(0)$.

1°) **(Chernoff's inequality).** Prove that

$$f(B_1) - E[f(B_1)] = \int_0^1 \nabla(P_{1-t}f)(B_t)dB_t$$

and derive that

$$\int \left(f - \int f\,d\gamma\right)^2 d\gamma \leq \int |\nabla f|^2 d\gamma.$$

2°) Suppose further that $\|\nabla f\| \leq 1$ and prove that there exists a BM^1 say X and a r.v. $\tau \leq 1$ such that $f(B_1) - E[f(B_1)] = X_\tau$.

[Hint: $\|\nabla(P_{1-t}f)\| \leq 1$.]

3°) Prove that for every $u > 0$,

$$\gamma\left(\left\{x \in \mathbb{R} : f(x) > \int f\,d\gamma + u\right\}\right) \leq \sqrt{\frac{2}{\pi}} \int_u^\infty \exp(-x^2/2)dx.$$

4°) Extend the result of 1°) to functions f in $L^2(\gamma)$ such that ∇f is in $L^2(\gamma)$ and prove that the equality obtains if and only if f is an affine function.

* **(1.21) Exercise.** In the case of BM^1 let F be a r.v. of $L^2(\mathscr{F}_\infty^B)$. Assume that there is a function ϕ on $\Omega \times \mathbb{R}_+ \times \Omega$ such that for each t, one has $F(\omega) = \phi(\omega, t, \theta_t(\omega))$ (see Exercise (3.19) Chap. III) and $(x, t) \to \Phi(\omega, x, t) = E_x[\phi(\omega, t, \cdot)]$ is for a.e. ω a function of $C^{2,1}_{x,t}$. Prove then that the representation of F given in Sect. 3 Chap. V is equal to

$$E[F] + \int_0^t \Phi'_x(\omega, B_s(\omega), s)\,dB_s(\omega).$$

Give examples of variables F for which the above conditions are satisfied.

[Hint: Use Exercise (3.12) Chap. IV.]

(1.22) Exercise. Prove that the infinitesimal generator of the BM killed at 0 is equal to the operator $\frac{1}{2}\frac{d^2}{dx^2}$ on $C^2_K(]0, \infty[)$.

\# **(1.23) Exercise (Skew Brownian motion).** Prove that the infinitesimal generator of the semi-group defined in Exercise (1.16) of Chap. III is equal to $\frac{1}{2}\frac{d^2}{dx^2}$ on the space $\{f \in C_0 : f'' \text{ exists in } \mathbb{R}\backslash\{0\}, f''(0-) = f''(0+) \text{ and } (1-\alpha)f'(0-) = \alpha f'(0+)\}$.

\# **(1.24) Exercise.** If X is a homogeneous Markov process with generator A, prove that the generator of the space-time process associated with X (Exercise (1.10) of Chap. III) is equal to $\frac{\partial}{\partial t} + A$ on a suitable space of functions on $\mathbb{R}_+ \times E$.

(1.25) Exercise. 1°) Let X be a Feller process and U the potential kernel of Exercise (2.29) Chap. III. If $f \in C_0$ is such that $Uf \in C_0$, then $Uf \in \mathscr{D}_A$ and $-AUf = f$. Thus, the potential kernel appears as providing an inverse for A.

[Hint: Use Exercise (1.17).]

2°) Check that for BM^3 the conditions of 1°) are satisfied for every $f \in C_K$. In the language of PDE's, $-1/|x|$ is a fundamental solution for A, that is: $-A(1/|x|) = \delta_0$ in the sense of distributions.

(1.26) Exercise. Let X be a Feller process and c a positive Borel function.

1°) Prove that one defines a homogeneous transition function Q_t by setting

$$Q_t(x, A) = E_x\left[1_A(X_t)\exp\left(-\int_0^t c(X_s)ds\right)\right].$$

Although we are not going into this, Q_t corresponds to the curtailment or "killing" of the trajectories of X performed at the "rate" $c(X)$.

2°) If f is in the domain of the generator of X and, if c is continuous, prove that

$$\lim_{t \downarrow 0} t^{-1}(Q_t f - f) = Af - cf$$

pointwise. The reader is invited to look at Proposition (3.10) in the following chapter.

(1.27) Exercise. 1°) Let Z be a strictly positive r.v. and define a family of kernels on $\mathbb{R}_+$ by

$$P_t f(x) = E\left[f\left((tZ) \vee x\right)\right].$$

Prove that (P_t) is a semi-group iff Z^{-1} is an exponential r.v., i.e. $Z^{-1} \overset{(d)}{=} \lambda \mathbf{e}$, $\lambda > 0$.

2°) If this is the case, write down the analytical form of (P_t), then prove that it is a Feller semi-group, that $\mathscr{D}_A = C_0(\mathbb{R}_+)$ and that

$$Af(x) = \lambda \int_x^\infty \left(f(y) - f(x)\right)y^{-2}dy.$$

(1.28) Exercise. In the notation of this section, call A^n the n-th iterate of A and $\mathscr{D}_{A^n}$ its domain.

1°) If $\varphi \in C_K^\infty(]0, \infty[)$ prove that for $f \in C_0$, the function $P_\varphi f$ (see above Lemma (4.3)) is in $\mathscr{D}_{A^n}$ for every n. Prove then that $\bigcap_n \mathscr{D}_{A^n}$ is dense in C_0.

2°) If the paths of X are continuous and if $f \in \mathscr{D}_{A^n}$ prove that

$$\sum_{k=0}^{n-1}\left((-t)^k/k!\right)A^k f(X_t) + (-1)^n \int_0^t \left(u^{n-1}/(n-1)!\right)A^n f(X_u)du$$

is a $\left(\mathscr{F}_t^0, P_\nu\right)$-martingale for every ν. This question does not depend on the first one; furthermore the continuity of paths is needed only because of the limitations of this book.

§2. Diffusions and Itô Processes

In the foregoing section, we have seen, in a heuristic way, that some Markov processes ought to be solutions to "stochastic differential equations". We now take this up and put it in a rigorous and systematic form thus preparing for the discussion in Chap. IX and establishing a bridge between the theory of the infinitesimal generator and stochastic calculus.

In the sequel, a and b will denote a matrix field and a vector field on $\mathbb{R}^d$ subject to the conditions

i) the maps $x \to a(x)$ and $x \to b(x)$ are Borel measurable and locally bounded,
ii) for each x, the matrix $a(x)$ is symmetric and non-negative i.e. for any $\lambda \in \mathbb{R}^d$, $\sum_{i,j} a_{ij}(x)\lambda_i\lambda_j \geq 0$.

With such a pair (a, b), we associate the second order differential operator

$$L = \frac{1}{2}\sum_{i,j=1}^{d} a_{ij}(\cdot)\frac{\partial^2}{\partial x_i \partial x_j} + \sum_{i=1}^{d} b_i(\cdot)\frac{\partial}{\partial x_i}.$$

In Sect. 1, we have mentioned that some Markov processes have infinitesimal generators which are extensions of such operators. It is an important problem to know if conversely, given such an operator L, we can find a Markov process whose generator coincides with L on C_K^2.

(2.1) Definition. *A Markov process $X = (\Omega, \mathscr{F}, \mathscr{F}_t, X_t, P_x)$ with state space $\mathbb{R}^d$ is said to be a* diffusion process with generator L *if*

i) it has continuous paths,
ii) for any $x \in \mathbb{R}^d$ and any $f \in C_K^\infty$,

$$E_x[f(X_t)] = f(x) + E_x\left[\int_0^t Lf(X_s)ds\right].$$

We further say that X has *covariance* or *diffusion coefficient* a and *drift* b. This is justified by the considerations in Sect. 1.

Let us stress that the hypothesis of continuity of paths includes that $\zeta = \infty$ a.s. As a result, if $\{K_n\}$ is an increasing sequence of compact sets such that $K_n \subset \mathring{K}_{n+1}$ and $\bigcup_n K_n = \mathbb{R}^d$, then setting $\sigma_n = T_{K_n^c}$, we have $\lim_n \sigma_n = +\infty$. Furthermore, the necessity of the non-negativity of a follows from Theorem (1.13) and Exercise (1.19) but is also easily explained by Exercise (2.8).

Observe also that one could let a and b depend on the time s, and get for each s a second-order differential operator L_s equal to

$$\frac{1}{2}\sum_{i,j} a_{ij}(s,\cdot)\frac{\partial^2}{\partial x_i \partial x_j} + \sum_i b_i(s,\cdot)\frac{\partial}{\partial x_i}.$$

The notion of diffusion would have to be extended to that of *non homogeneous diffusion*. In that case, one would have probability measures $P_{s,x}$ corresponding to the process started at x at time s and demand that for any $f \in C_K^\infty$, $s < t$,

$$E_{s,x}[f(X_t)] = f(x) + E_{s,x}\left[\int_s^t L_u f(X_u)du\right].$$

In the sequel, we will deal mainly with homogeneous diffusions and write for $f \in C^2$,

$$M_t^f = f(X_t) - f(X_0) - \int_0^t Lf(X_s)ds.$$

The process M^f is continuous; it is moreover locally bounded, since, by the hypothesis made on a and b, it is clearly bounded on $[0, \sigma_n \wedge n]$. Likewise, if $f \in C_K^\infty$, M^f is bounded on every interval $[0, t]$ and the integrals in ii) are finite.

(2.2) Proposition. *The property ii) above is equivalent to each of the following:*

iii) for any $f \in C_K^\infty$, M^f is a martingale for any P_x;
iv) for any $f \in C^2$, M^f is a local martingale for any P_x.

Proof. If iii) holds, then, since $M_0^f = 0$,

$$P_t f(x) - f(x) - E_x\left[\int_0^t Lf(X_s)ds\right] = E_x[M_t^f] = 0,$$

and ii) holds.

Conversely, if ii) holds, then $E_{X_s}[M_t^f] = 0$ for every s and t. By the Markov property, we consequently have

$$\begin{aligned} E_x\left[M_t^f \mid \mathscr{F}_s\right] &= M_s^f + E_x\left[f(X_t) - f(X_s) - \int_s^t Lf(X_u)du \middle| \mathscr{F}_s\right] \\ &= M_s^f + E_{X_s}\left[M_{t-s}^f\right] = M_s^f, \end{aligned}$$

which shows that ii) implies iii).

If M^f is a local martingale and is bounded on $[0, t]$ for each t, then it is a martingale; thus, iv) implies iii).

To prove that iii) implies iv), let us begin with f in C_K^2. There is a compact set H and a sequence $\{f_p\}$ of functions in C_K^∞ vanishing on H^c and such that $\{f_p\}$ converges uniformly to f on H as well as the first and second order derivatives. For every t, the process $M_t^{f_p} - M_t^f$ is bounded on $[0, t]$ by a constant c_p which goes to zero as $p \to \infty$. By passing to the limit in the right-hand side of the inequality

$$\left|E_x\left[M_t^f \mid \mathscr{F}_s\right] - M_s^f\right| \le \left|E_x\left[M_t^f \mid \mathscr{F}_s\right] - E_x\left[M_t^{f_p} \mid \mathscr{F}_s\right]\right| + \left|M_s^{f_p} - M_s^f\right|,$$

we see that M_t^f is a martingale.

Let now f be in C^2; we may find a sequence $\{g_n\}$ of functions in C_K^2 such that $g_n = f$ on K_n. The processes M^f and M^{g_n} coincide up to time σ_n. Since M^{g_n} is a martingale by what we have just seen, the proof is complete.

Remarks. 1°) The local martingales M_t^f are local martingales with respect to the uncompleted σ-fields $\mathscr{F}_t^0 = \sigma(X_s, s \leq t)$ and with respect to the usual augmentation of $(\mathscr{F}_t^0)$.

2°) Proposition (2.2) says that any function in C_b^2 is in the domain of the extended infinitesimal generator of X. If X is Feller, by arguing as in Proposition (1.7), we see that $C_K^2 \subset \mathscr{D}_A$ and $A = L$ on C_K^2. In the same vein, if $Lf = 0$, then $f(X_t)$ is a local martingale, which generalizes what is known for BM (Proposition (3.4) Chap. IV). By making $f(x) = x$, we also see that X is a local martingale if and only if L has no first order terms.

If we think of the canonical version of a diffusion where the probability space is $\mathbf{W} = C(\mathbb{R}_+, \mathbb{R}^d)$ and X is the coordinate process, the above result leads to the

(2.3) Definition. *A probability measure P on $\mathbf{W}$ is a solution to the* Martingale problem *$\pi(x, a, b)$ if*

i) $P[X_0 = x] = 1$;
ii) for any $f \in C_K^\infty$, the process

$$M_t^f = f(X_t) - f(X_0) - \int_0^t Lf(X_s)ds$$

is a P-martingale with respect to the filtration $(\sigma(X_s, s \leq t)) = (\mathscr{F}_t^0)$.

The idea is that if (Ω, X_t, P_x) is a diffusion with generator L, then $X(P_x)$ is a solution to the martingale problem $\pi(x, a, b)$. Therefore, if one wants to construct a diffusion with generator L, we can try in a first step to solve the corresponding martingale problem; then in a second step if we have a solution for each x, to see if these solutions relate in such a way that the canonical process is a diffusion with L as its generator. This will be discussed in Chap. IX.

For the time being, we prove that the conditions in Proposition (2.2) are equivalent to another set of conditions. We do it in a slightly more general setting which covers the case of non-homogeneous diffusions. Let a and b be two progressively measurable, locally bounded processes taking values in the spaces of non-negative symmetric $d \times d$-matrices and d-vectors. For $f \in C^2(\mathbb{R}^d)$, we set

$$L_s(\omega)f(x) = \frac{1}{2}\sum_{i,j} a_{ij}(s, \omega)\frac{\partial^2 f}{\partial x_i \partial x_j}(x) + \sum_i b_i(s, \omega)\frac{\partial f}{\partial x_i}(x).$$

(2.4) Proposition. *Let X be a continuous, adapted, $\mathbb{R}^d$-valued process; the three following statements are equivalent:*

i) for any $f \in C^2$, the process $M_t^f = f(X_t) - f(X_0) - \int_0^t L_s f(X_s)ds$ is a local martingale;

ii) for any $\theta \in \mathbb{R}^d$, the process $M_t^\theta = \langle \theta, X_t - X_0 - \int_0^t b(s)ds \rangle$ is a local martingale and

$$\langle M^\theta, M^\theta \rangle_t = \int_0^t \langle \theta, a(s)\theta \rangle ds;$$

iii) for any $\theta \in \mathbb{R}^d$,

$$\mathscr{E}_t^\theta = \exp\left(\langle \theta, X_t - X_0 - \int_0^t b(s)ds \rangle - \frac{1}{2}\int_0^t \langle \theta, a(s)\theta \rangle ds\right)$$

is a local martingale.

Proof. i) $\Rightarrow$ ii). For $f(x) = \langle \theta, x \rangle$, we get $M^f = M^\theta$ which is thus a local martingale. Making $f(x) = \langle \theta, x \rangle^2$ in i), we get that

$$H_t^\theta \equiv \langle \theta, X_t \rangle^2 - \langle \theta, X_0 \rangle^2 - 2\int_0^t \langle \theta, X_s \rangle \langle \theta, b(s) \rangle ds - \int_0^t \langle \theta, a(s)\theta \rangle ds$$

is a local martingale. Writing $X \sim Y$ if $X - Y$ is a local martingale, we have, since $2\langle \theta, X_0 \rangle M_t^\theta$ is a local martingale, that

$$(M_t^\theta)^2 - \int_0^t \langle \theta, a(s)\theta \rangle ds \sim \left(M_t^\theta + \langle \theta, X_0 \rangle\right)^2 - \int_0^t \langle \theta, a(s)\theta \rangle ds.$$

Setting $A_t = \langle \theta, \int_0^t b(s)ds \rangle$, we further have

$$\begin{aligned}(M_t^\theta)^2 - \int_0^t \langle \theta, a(s)\theta \rangle ds &\sim \left(M_t^\theta + \langle \theta, X_0 \rangle\right)^2 - \langle \theta, X_0 \rangle^2 - \int_0^t \langle \theta, a(s)\theta \rangle ds - H_t^\theta \\ &= (\langle \theta, X_t \rangle - A_t)^2 - \langle \theta, X_t \rangle^2 + 2\int_0^t \langle \theta, X_s \rangle dA_s \\ &= -2\langle \theta, X_t \rangle A_t + A_t^2 + 2\int_0^t \langle \theta, X_s \rangle dA_s.\end{aligned}$$

As $\langle \theta, X_t \rangle = M_t^\theta + \langle \theta, X_0 \rangle + A_t$ is a semimartingale, we may apply the integration by parts formula to the effect that

$$(M_t^\theta)^2 - \int_0^t \langle \theta, a(s)\theta \rangle ds \sim A_t^2 - 2\int_0^t A_s dA_s = 0$$

which completes the proof.

ii) $\Rightarrow$ iii). By Proposition (3.4) in Chap. IV, there is nothing to prove as $\mathscr{E}_t^\theta = \mathscr{E}(M^\theta)_t$.

iii) $\Rightarrow$ i). We assume that iii) holds and first prove i) for $f(y) = \exp(\langle \theta, y \rangle)$. The process

$$V_t = \exp\left[\langle \theta, \int_0^t b(s)ds \rangle + \frac{1}{2}\int_0^t \langle \theta, a(s)\theta \rangle ds\right]$$

is of bounded variation; integrating by parts, we obtain that the process

$$\begin{aligned}&\mathscr{E}_t^\theta V_t - \int_0^t \mathscr{E}_s^\theta dV_s \\ &= \exp(\langle \theta, X_t - X_0 \rangle) - \int_0^t \exp\left(\langle \theta, X_s - X_0 \rangle\right)\left(\langle \theta, b(s) + \frac{1}{2}a(s)\theta \rangle\right) ds\end{aligned}$$

is a local martingale. Since $L_s f(x) = \exp(\langle\theta, X_s\rangle)\left(\langle\theta, b(s) + \frac{1}{2}a(s)\theta\rangle\right)$, we have proved that

$$f(X_0)\left(f(X_t) - \int_0^t L_s f(X_s)ds\right)$$

is a local martingale. The class of local martingales being invariant under multiplication by, or addition of, $\mathscr{F}_0$-measurable variables, our claim is proved in that case. To get the general case, it is enough to observe that exponentials are dense in C^2 for the topology of uniform convergence on compact sets of the functions and their two first order derivatives. □

Remarks. 1°) Taking Proposition (2.2) into account, the implication ii) ⇒ i) above is a generalization of P. Lévy's characterization theorem (Theorem (3.6) of Chapter IV) as is seen by making $a = Id$ and $b = 0$.

2°) Another equivalent condition is given in Exercise (2.11).

(2.5) Definition. *A process X which satisfies the conditions of Proposition (2.4) is called an* Itô process *with* covariance *or* diffusion coefficient *a and* drift *b.*

Obviously Itô processes, hence diffusions, are continuous semimartingales. We now show that they coincide with the solutions of some "stochastic differential equations" which we shall introduce and study in Chap. IX. We will use the following notation. If $X = (X^1, \dots, X^d)$ is a vector semimartingale and $K = (K_{ij})$ a process taking its values in the space of $r \times d$-matrices, such that each K_{ij} is progressive and locally bounded we will write $K \cdot X$ or $\int_0^t K_s dX_s$ for the r-dimensional process whose i-th component is equal to $\sum_j \int_0^t K_{ij}(s)dX^j(s)$.

(2.6) Proposition. *Let β be an $(\mathscr{F}_t)$-BM^r defined on a probability space $(\Omega, \mathscr{F}, \mathscr{F}_t, P)$ and σ (resp. b) a locally bounded predictable process with values in the $d \times r$-matrices (resp. $\mathbb{R}^d$); if the adapted continuous process X satisfies the equation*

$$(*) \qquad X_t = X_0 + \int_0^t \sigma(s)d\beta_s + \int_0^t b(s)ds,$$

it is an Itô process with covariance $\sigma\sigma^t$ and drift b. If, in particular, $\sigma(s) = \sigma(X_s)$ and $b(s) = b(X_s)$ for two fields σ and b defined on $\mathbb{R}^d$, and if $X_0 = x$ a.s., then $X(P)$ is a solution to the martingale problem $\pi(x, \sigma\sigma^t, b)$.

Proof. A straightforward application of Itô formula shows that condition i) of Proposition (2.4) holds. □

We now want to prove a converse to this proposition, namely that given an Itô process X, and in particular a diffusion, there exists a Brownian motion β such that X satisfies $(*)$ for suitable σ and b. The snag is that the space on which X is defined may be too poor to carry a BM; this is for instance the case if X is the translation on the real line. We will therefore have to enlarge the probability space unless we make an assumption of non degeneracy on the covariance a.

(2.7) Theorem. *If X is an Itô process with covariance a and drift b, there exist a predictable process σ and a Brownian motion B on an enlargement of the probability space such that*

$$X_t = X_0 + \int_0^t \sigma(s)dB_s + \int_0^t b(s)ds.$$

Proof. By ii) of Proposition (2.4), the continuous vector local martingale $M_t = X_t - X_0 - \int_0^t b(s)ds$ satisfies $\langle M^i, M^j\rangle_t = \int_0^t a_{ij}(s)ds$. The result follows immediately from Proposition (3.8) in Chap. V. □

Remark. By the remark after Proposition (3.8) in Chap. V, we see that if a is $dP\otimes dt$ a.e. strictly non-negative, then σ and B may be chosen such that $a = \sigma\sigma^t$. If in particular $a(s) = c(X_s)$ where c is a measurable field of symmetric strictly non-negative matrices on $\mathbb{R}^d$, we can pick a measurable field γ of matrices such that $\gamma\gamma^t = c$ and take $\sigma(s) = \gamma(X_s)$.

\# **(2.8) Exercise.** 1°) In the situation of Proposition (2.2) and for $f, g \in C^2$ prove that

$$\langle M^f, M^g\rangle_t = \int_0^t \left(L(fg) - fLg - gLf\right)(X_s)ds = \int_0^t \langle\nabla f, a\nabla g\rangle(X_s)ds.$$

[This exercise is solved in Sect. 3 Chap. VIII].

2°) Deduce from 1°) the necessity for the matrices $a(x)$ to be non-negative.

\# **(2.9) Exercise.** In the situation of Proposition (2.2), prove that if f is a strictly positive C^2-function, then

$$\left(f(X_t)/f(X_0)\right)\exp\left(-\int_0^t (Lf/f)(X_s)ds\right)$$

is a P_x-local martingale for every x.

(2.10) Exercise. If X is a d-dimensional Itô process with covariance a and drift 0, vanishing at 0, prove that for $2 \le p < \infty$, there is a constant C depending only on p and d such that

$$E\left[\sup_{0\le s\le t} |X_s|^p\right] \le CE\left[\left(\int_0^t \text{Trace } a(s)\, ds\right)^{p/2}\right].$$

(2.11) Exercise. Prove that the conditions in Proposition (2.4) are also equivalent to

iv) for any f on $[0,\infty[\times\mathbb{R}^d$ which is once (twice) differentiable in the first (second) variable, the process

$$f(t, X_t) - \int_0^t \left(\frac{\partial f}{\partial s} + L_s f\right)(X_s)ds$$

is a local martingale. Compare with Exercise (1.24).

(2.12) Exercise. In the situation of Proposition (2.4), suppose that a and b do not depend on ω. If u is a function on $[0, \infty[\times\mathbb{R}^d$, which is sufficiently differentiable and such that $\frac{\partial u}{\partial t} = L_t u + g$ in $]0, \infty[\times\mathbb{R}^d$, prove that

$$u(t-s, X_s) + \int_0^s g(t-r, X_r)dr$$

is a local martingale on $[0, t[$.

§3. Linear Continuous Markov Processes

Beside the linear BM itself, many Markov processes, with continuous paths, defined on subsets of $\mathbb{R}$ such as the BES^3 or the reflected BM have cropped up in our study. The particular case of Bessel processes will be studied in Chap. XI. This is the reason why, in this section, we make a systematic study of this situation and compute the corresponding generators.

We will therefore deal with a Markov process X whose state space E is an interval (l, r) of $\mathbb{R}$ which may be closed, open or semi-open, bounded or unbounded. The death-time is as usual denoted by ζ. We assume, throughout the section, that the following assumptions are in force:

i) the paths of X are continuous on $[0, \zeta[$;
ii) X enjoys the strong Markov property;
iii) if $\zeta < \infty$ with strictly positive probability then at least one of the points l and r does not belong to E and $\lim_{t\uparrow\zeta} X_t \notin E$ a.s. on $\{\zeta < \infty\}$; in other words X can be "killed" only at the end-points of E which do not belong to E.

Property i) entails that the process started at x cannot hit a point y without hitting all the points located between x and y. The hitting time of the one-point set $\{x\}$ is denoted by T_x; we have

$$T_x = \inf\{t > 0 : X_t = x\}$$

where as usual $\inf(\emptyset) = +\infty$. Naturally, $X_{T_x} = x$ on $\{T_x < \infty\}$.

Finally, we will make one more assumption, namely, that X is *regular*: for any $x \in \overset{\circ}{E} =]l, r[$ and $y \in E$, $P_x[T_y < \infty] > 0$. This last hypothesis means that E cannot be decomposed into smaller sets from which X could not exit (see Exercise (3.22)).

From now on, we work with the foregoing set of hypotheses.

For any interval $I =]a, b[$ such that $[a, b] \subset E$, we denote by σ_I the exit time of I. For $x \in I$, we have $\sigma_I = T_a \wedge T_b$ P_x-a.s. and for $x \notin I$, $\sigma_I = 0$ P_x-a.s. We also put $m_I(x) = E_x[\sigma_I]$.

(3.1) Proposition. *If I is bounded, the function m_I is bounded on I. In particular σ_I is almost-surely finite.*

Proof. Let y be a fixed point in I. Because of the regularity of X, we may pick $\alpha < 1$ and $t > 0$ such that

$$\max\left(P_y[T_a > t], P_y[T_b > t]\right) = \alpha.$$

If now $y < x < b$, then

$$P_x[\sigma_I > t] \leq P_x[T_b > t] \leq P_y[T_b > t] \leq \alpha;$$

the same reasoning applies to $a < x < y$ and consequently

$$\sup_{x \in I} P_x[\sigma_I > t] \leq \alpha < 1.$$

Now, since $\sigma_I = u + \sigma_I \circ \theta_u$ on $\{\sigma_I > u\}$, we have

$$P_x[\sigma_I > nt] = P_x[(\sigma_I > (n-1)t) \cap \left((n-1)t + \sigma_I \circ \theta_{(n-1)t} > nt\right)],$$

and using the Markov property

$$P_x[\sigma_I > nt] = E_x\left[1_{(\sigma_I > (n-1)t)} E_{X_{(n-1)t}}[1_{(\sigma_I > t)}]\right].$$

On $\{\sigma_I > (n-1)t\}$, we have $X_{(n-1)t} \in I$ P_x-a.s. and therefore

$$P_x[\sigma_I > nt] \leq \alpha P[\sigma_I > (n-1)t].$$

It follows inductively that $P_x[\sigma_I > nt] \leq \alpha^n$ for every $x \in I$, and therefore

$$\sup_{x \in I} E_x[\sigma_I] \leq \sup_{x \in I} \sum_{n=0}^{\infty} t P_x[\sigma_I > nt] \leq t(1-\alpha)^{-1},$$

which is the desired result. □

For a and b in E and $l \leq a < x < b \leq r$, the probability $P_x[T_b < T_a]$ is the probability that the process started at x exits $]a, b[$ by its right-end. Because of the preceding proposition, we have

$$P_x[T_a < T_b] + P_x[T_b < T_a] = 1.$$

(3.2) Proposition. *There exists a continuous, strictly increasing function s on E such that for any a, b, x in E with $l \leq a < x < b \leq r$*

$$P_x[T_b < T_a] = \left(s(x) - s(a)\right)/\left(s(b) - s(a)\right).$$

If $\tilde{s}$ is another function with the same properties, then $\tilde{s} = \alpha s + \beta$ with $\alpha > 0$ and $\beta \in \mathbb{R}$.

Proof. Suppose first that E is the closed bounded interval $[l, r]$. The event $\{T_r < T_l\}$ is equal to the disjoint union

$$\{T_r < T_l, T_a < T_b\} \cup \{T_r < T_l, T_b < T_a\}.$$

Now $T_l = T_a + T_l \circ \theta_{T_a}$ and $T_r = T_a + T_r \circ \theta_{T_a}$ on the set $\{T_a < T_b\}$. Thus

$$P_x[T_r < T_l; T_a < T_b] = E_x\left[1_{(T_a<T_b)} 1_{(T_r<T_l)} \circ \theta_{T_a}\right]$$

and since $\{T_a < T_b\} \in \mathscr{F}_{T_a}$, the strong Markov property yields

$$P_x[T_r < T_l; T_a < T_b] = E_x\left[1_{(T_a<T_b)} E_{X_{T_a}}[1_{(T_r<T_l)}]\right]$$

and because $X_{T_a} = a$ a.s.,

$$P_x[T_r < T_l; T_a < T_b] = P_x[T_a < T_b]\, P_a[T_r < T_l].$$

We finally get

$$P_x[T_r < T_l] = P_x[T_a < T_b]\, P_a[T_r < T_l] + P_x[T_b < T_a]\, P_b[T_r < T_l].$$

Setting $s(x) = P_x[T_r < T_l]$ and solving for $P_x[T_b < T_a]$ we get the formula in the statement.

To prove that s is strictly increasing, suppose there exist $x < y$ such that $s(x) = s(y)$. Then, for any $b > y$, the formula just proved yields $P_y[T_b < T_x] = 0$ which contradicts the regularity of the process.

Suppose now that E is an arbitrary interval. If $l_2 < l_1 < r_1 < r_2$ are four points in E, we may apply the foregoing discussion to $[l_1, r_1]$ and $[l_2, r_2]$; the functions s_1 and s_2 thus defined obviously coincide on $]l_1, r_1[$ up to an affine transformation. As a result, a function s may be defined which satisfies the equality in the statement for any three points in E. It remains to prove that it is continuous.

If $a < x$ and $\{a_n\}$ is a sequence of real numbers smaller than x and decreasing to a, then $T_{a_n} \uparrow T_a$ P_x-a.s.; indeed, because of the continuity of paths, $X_{\lim_n T_{a_n}} = a$ P_x-a.s., so that $\lim_n T_{a_n} \geq T_a$, and the reverse inequality is obvious. Consequently, $\{T_a < T_b\} = \lim_n \{T_{a_n} < T_b\}$ P_x-a.s. since obviously $P_x[T_a = T_b] = 0$. It follows that s is right-continuous in a and the left-continuity is shown in exactly the same way. The proof is complete. □

(3.3) Definition. *The function s of the preceding result is called the* scale function *of X.*

We speak of *the* scale function although it is defined up to an affine transformation. If $s(x)$ may be taken equal to x, the process is said to be on its *natural scale*. A process on its natural scale has as much tendency to move to the right as to the left as is shown in Exercise (3.15). The linear BM, the reflected and absorbed linear BM's are on their natural scale as was proved in Proposition (3.8) of Chap. II. Finally, a simple transformation of the state space turns X into a process on its natural scale.

(3.4) Proposition. *The process $\tilde{X}_t = s(X_t)$ satisfies the hypotheses of this section and is on its natural scale.*

Proof. Straightforward. □

The scale function was also computed for BES^3 in Chap. VI. In that case, since 0 is not reached from the other points, to stay in the setting of this section, we must look upon BES^3 as defined only on $]0, \infty[$. The point 0 will be an entrance boundary as defined in Definition (3.9). In this setting, it was proved in Chap. VI that $s(x) = -1/x$ (see also Exercise (3.20) and the generalizations to other Bessel processes in Chap. XI). The proof used the fact that the process $\tilde{X}$ of Proposition (3.4) is a local martingale; we now extend this to the general case. We put $R = \zeta \wedge T_l \wedge T_r$. Let us observe that if ζ is finite with positive probability and $\lim_{t\uparrow\zeta} X_t = l$ (say), then $\lim_{x\to l} s(x)$ is finite and we will by extension write $s(l)$ for this limit. Accordingly, we will say that $s(X_\zeta) = s(l)$ in that case.

(3.5) Proposition. *A locally bounded Borel function f is a scale function if and only if $f(X_t)^R$ is a local martingale. In particular, X is on its natural scale if and only if X^R is a local martingale.*

Proof. If $f(X_t)^R$ is a local martingale, then for $a < x < b$, the process $f(X_t)^{T_a\wedge T_b}$ is a bounded $(\mathscr{F}_t, P_x)$-martingale and the optional stopping theorem yields

$$f(x) = f(a)P_x[T_a < T_b] + f(b)P_x[T_b < T_a].$$

On the other hand, as already observed,

$$1 = P_x[T_a < T_b] + P_x[T_b < T_a].$$

Solving this system of linear equations in $P_x[T_a < T_b]$ shows that f is a scale function.

Conversely, by the reasoning in the proof of Proposition (3.2), if f is a scale function, then it is continuous. As a result, for $[a, b] \subset \overset{\circ}{E}$ and $\varepsilon > 0$, we may find a finite increasing sequence $A = (a_k)_{k=0,\dots,K}$ of numbers such that $a_0 = a$, $a_K = b$ and $|f(a_{k+1}) - f(a_k)| < \varepsilon$ for each $k \leq K - 1$. We will write S for $T_a \wedge T_b$.

We define a sequence of stopping times T_n by $T_0 = 0$, $T_1 = T_A$ and

$$T_n = \inf\left(t > T_{n-1} : X_t \in A\backslash\left\{X_{T_{n-1}}\right\}\right) \wedge S.$$

Clearly, $\{T_n\}$ increases to S and if f is a scale function, the strong Markov property implies, for $x \in]a, b[$,

$$\begin{aligned} E_x\left[f(X^S_{T_n}) \mid \mathscr{F}_{T_{n-1}}\right] &= E_x\left[f(X^S_{T_1}) \circ \theta_{T_{n-1}} \mid \mathscr{F}_{T_{n-1}}\right] \\ &= E_{X_{T_{n-1}}}\left[f(X^S_{T_1})\right] = f(X^S_{T_{n-1}}) \quad P_x\text{-a.s.}; \end{aligned}$$

in other words, $\left\{f(X^S_{T_n}), \mathscr{F}_{T_n}\right\}$ is a bounded P_x-martingale.

For $t > 0$, let $N = \inf\{n \geq 1 : T_n \geq t\}$ where as usual $\inf(\emptyset) = \infty$. The r.v. N is a stopping time with respect to $(\mathscr{F}_{T_n})$, so that, by the optional stopping theorem

$$f(x) = E_x[f(X^S_{T_N})].$$

But, on the set $\{t < S\}$, N is finite and $T_{N-1} < t \leq T_N \leq S$ which by the choice of A implies that

$$E_x\left[|f(X_t^S) - f(X_{T_N}^S)|\right] < \varepsilon.$$

Since ε is arbitrary, it follows that $f(x) = E_x[f(X_t^S)]$ and another application of the Markov property shows that $f(X_t^S)$ is a martingale. The proof is now easily completed.

Remarks. We have thus proved that, up to an affine transformation, there is at most one locally bounded Borel function f such that $f(X_t)$ is a local martingale. This was stated for BM in Exercise (3.13) of Chap. II. If $R = \infty$, we see that s belongs to the domain of the extended infinitesimal generator A and that $As = 0$, a fact which agrees well with Theorem (3.12) below.

We now introduce another notion, linked to the speed at which X runs through its paths. We will see in Chap. X how to time-change a BM so as to preserve the Markov property. Such a time-changed BM is on its natural scale and the converse may also be shown. Thus, the transformation of Proposition (3.4) by means of the scale function turns the process into a time-changed BM. The time-change which will further turn it into a BM may be found through the speed-measure which we are about to define now. These questions will be taken up in Sect. 2 of Chap. X.

Let now $J =]c, d[$ be an open subinterval of I. By the strong Markov property and the definition of s, one easily sees that for $a < c < x < d < b$,

$$(*)\qquad \begin{aligned} m_I(x) &= E_x\left[\sigma_J + \sigma_I \circ \theta_{\sigma_J}\right] \\ &= m_J(x) + \frac{s(d) - s(x)}{s(d) - s(c)} m_I(c) + \frac{s(x) - s(c)}{s(d) - s(c)} m_I(d). \end{aligned}$$

Since $m_J(x) > 0$, it follows that m_I is an s-concave function. Taking our cue from Sect. 3 in the Appendix and Exercise (2.8) in Chap. VI, we define a function G_I on $E \times E$ by

$$G_I(x, y) = \begin{cases} \dfrac{(s(x) - s(a))(s(b) - s(y))}{s(b) - s(a)} & \text{if } a \le x \le y \le b, \\ \dfrac{(s(y) - s(a))(s(b) - s(x))}{s(b) - s(a)} & \text{if } a \le y \le x \le b, \\ 0 & \text{otherwise.} \end{cases}$$

(3.6) Theorem. *There is a unique Radon measure m on the interior $\mathring{E}$ of E such that for any open subinterval $I =]a, b[$ with $[a, b] \subset E$,*

$$m_I(x) = \int G_I(x, y) m(dy)$$

for any x in I.

Proof. By Sect. 3 in the Appendix, for any I, there is a measure ν_I on I for which

$$m_I(x) = \int G_I(x, y) \nu_I(dy).$$

Thus, we need only prove that if $J \subset I$ as in $(*)$, ν_J coincides with the restriction of ν_I to J. But, this is a simple consequence of the definition of ν_I; indeed, if we take the s-derivatives in $(*)$, we see that the derivatives of m_I and m_J differ by a constant and, consequently, the associated measures are equal. □

(3.7) Definition. *The measure m is called the* speed measure *of the process X.*

For example, using the result in Exercise (3.11) of Chap. II or Exercise (2.8) of Chap. VI, one easily checks that, if we take $s(x) = x$, the speed measure of Brownian motion is twice the Lebesgue measure. The reader will also observe that, under m, every open sub-interval of E has a strictly positive measure.

We will see, in Theorem (3.12), that the knowledge of the scale function and of the speed measure entails the knowledge of the infinitesimal generator. It is almost equivalent to say that they determine the potential operator of X killed when it exits an interval, which is the content of the

(3.8) Corollary. *For any $I =]a, b[$, $x \in I$ and $f \in \mathscr{B}(\mathbb{R})_+$,*

$$E_x\left[\int_0^{T_a \wedge T_b} f(X_s)ds\right] = \int G_I(x, y)f(y)m(dy).$$

Proof. Pick c such that $a < c < b$. The function $v(x) = E_x\left[\int_0^{T_a \wedge T_b} 1_{]c,b[}(X_u)du\right]$ is s-concave on $]a, b[$ and $v(a) = v(b) = 0$; therefore

$$v(x) = -\int G_I(x, y)v''(dy)$$

where v'' is the measure which is associated with the s-derivative of v (see Appendix 3). On the other hand, by the Markov property,

$$\begin{aligned} v(x) &= E_x[T_c \wedge T_b] + \frac{s(b) - s(x)}{s(b) - s(c)} v(c) \quad \text{on }]c, b[, \\ v(x) &= \frac{s(x) - s(a)}{s(c) - s(a)} v(c) \quad \text{on }]a, c\}. \end{aligned}$$

Since the function $E_.[T_c \wedge T_b]$ is equal to $\int G_{]c,b[}(., y)m(dy)$, the measure associated to its second s-derivative is equal to $-1_{]c,b[}m$; the measures associated with the other terms are obviously 0 since their s-derivatives are constant. Therefore

$$v(x) = \int G_I(x, y)1_{]c,b[}(y)m(dy),$$

which is the result stated in the case where $f = 1_{]c,b[}$. The proof is completed by means of the monotone class theorem and the usual extensions arguments. □

From now on, we will specialize to the case of importance for us, namely when E is either $]0, \infty[$ or $[0, \infty[$ and we will investigate the behavior of the process at the boundary $\{0\}$. The reader may easily carry over the notions and results to the other cases, in particular to that of a compact subinterval. If $0 \notin E$ and if $\zeta = \infty$ a.s., we introduce the following classification.

(3.9) Definition. *If $E =]0, \infty[$, the point 0 is said to be a* natural *boundary if for all $t > 0$ and $y > 0$,*

$$\lim_{x \downarrow 0} P_x[T_y < t] = 0.$$

It is called an entrance *boundary if there are $t > 0$ and $y > 0$ such that*

$$\lim_{x \downarrow 0} P_x[T_y < t] > 0.$$

An example where 0 is an entrance boundary is given by the BES^3 process described in Sect. 3 of Chap. VI (see more generally Bessel processes of dimension ≥ 2 in Chap. XI). Indeed, the limit in Definition (3.9) is monotone, hence is equal, in the case of BES^3 to the probability that $\mathrm{BES}^3(0)$ has reached y before time t which is strictly positive. In this case, we see that the term "entrance boundary" is very apt as BES^3 is a process on $[0, \infty[$ (see Sect. 3 Chap. VI) which does not come back to zero after it has left it.

At this juncture, we will further illustrate the previous results by computing the speed measure of BES^3. Since the scale function of BES^3 is equal to $(-1/x)$, by passing to the limit in Corollary (3.8), we find that the potential operator of BES^3 is given, for $x > 0$, by

$$Uf(x) = E_x\left[\int_0^\infty f(X_s)ds\right] = \int_0^\infty u(x, y) f(y) m(dy)$$

where m is the speed measure and $u(x, y) = \inf(1/x, 1/y)$. To compute the potential kernel $U(0, \cdot)$ of $\mathrm{BES}^3(0)$, we may pass to the limit; indeed for $\varepsilon > 0$, if f vanishes on $[0, \varepsilon]$, the strong Markov property shows that

$$Uf(0) = Uf(\varepsilon) = \int_\varepsilon^\infty y^{-1} f(y) m(dy).$$

Passing to the limit yields that $U(0, \cdot)$ is the measure with density $y^{-1} 1_{(y>0)}$ with respect to m. On the other hand, since the modulus of $\mathrm{BM}^3(0)$ is a $\mathrm{BES}^3(0)$, $U(0, \cdot)$ may be computed from the potential kernel of BM^3 (Exercise (2.29) Chap. III); it follows, using polar coordinates, that

$$\begin{aligned} Uf(0) &= \frac{1}{2\pi}\int_{\mathbb{R}^3} \frac{f(|x|)}{|x|} dx = \frac{1}{2\pi}\int_0^{2\pi}\int_{-\pi/2}^{\pi/2}\int_0^\infty \rho^{-1} f(\rho)\rho^2 \cos\phi \, d\theta \, d\phi \, d\rho \\ &= 2\int_0^\infty f(\rho)\rho \, d\rho. \end{aligned}$$

Comparing the two formulas for $Uf(0)$ shows that the speed measure for BES^3 is the measure with density $2y^2 \, 1_{(y>0)}$ with respect to the Lebesgue measure.

We now turn to the case $E = [0, \infty[$. In that situation, $s(0)$ is a finite number and we then always make $s(0) = 0$. The hypothesis of regularity implies that 0 is visited with positive probability; on the other hand, since regularity again bars the possibility that 0 be absorbing, by the remarks below Proposition (2.19) Chap. III, the process started at 0 leaves $\{0\}$ immediately.

The speed measure is defined so far only on $]0, \infty[$ and the formula in Theorem (3.6) gives for $b > 0$, the mean value of $T_0 \wedge T_b$ for the process started at $x \in]0, b[$. We will show that the definition of G may be extended and that $m(\{0\})$ may be defined so as to give the mean value of T_b for the process started at $x \in [0, b[$. We first define a function $\tilde{s}$ on $]-\infty, \infty[$ by setting $\tilde{s}(x) = s(x)$ for $x \geq 0$ and $\tilde{s}(x) = -s(-x)$ for $x < 0$. For $J = [0, b[$, we define $\tilde{G}_J$ as the function G_I defined for $I = [-b, b]$ by means of $\tilde{s}$ in lieu of s. For $x, y \geq 0$, we next define a function G_J by

$$G_J(x, y) = \tilde{G}_J(x, y) + \tilde{G}_J(x, -y).$$

(3.10) Proposition. *One can choose $m(\{0\})$ in order that for any $x \in J$ and any positive Borel function f,*

$$E_x\left[\int_0^{T_b} f(X_s)ds\right] = \int_J G_J(x, y)f(y)m(dy).$$

Proof. Thinking of the case of reflected BM, we define a process $\tilde{X}$ on $]-\infty, \infty[$ from which X is obtained by reflection at 0. This may be put on a firm basis by using the excursion theory of Chap. XII and especially the ideas of Proposition (2.5) in that chapter. We will content ourselves here to observe that we can define the semi-group of $\tilde{X}$ by setting for $x \geq 0$ and $A \subset \mathbb{R}_+$

$$\tilde{P}_t(x, A) = E_x\left[1_A(X_t)1_{(t \leq T_0)}\right] + \frac{1}{2}E_x\left[1_A(X_t)1_{(t \geq T_0)}\right]$$

and for $A \subset \mathbb{R}_-$,

$$\tilde{P}_t(x, A) = \frac{1}{2}E_x\left[1_{\tilde{A}}(X_t)1_{(t \geq T_0)}\right]$$

where $\tilde{A} = -A$. For $x < 0$, we set $\tilde{P}_t(x, A) = \tilde{P}_t(-x, \tilde{A})$. Using the fact that $T_0 = t + T_0 \circ \theta_t$ on $\{T_0 > t\}$, it is an exercise on the Markov property to show that this defines a transition semi-group. Moreover, excursion theory would insure that $\tilde{X}$ has all the properties we demanded of X at the start of the section. It is easy to see that $\tilde{s}$ is a scale function for $\tilde{X}$ and the corresponding speed measure $\tilde{m}$ then coincides with m on $]-\infty, 0[$ and $]0, \infty[$.

Let now $\tilde{f}$ be the function equal to f on $[0, \infty[$ and defined by $\tilde{f}(x) = -f(-x)$ on $]-\infty, 0[$. By applying Corollary (3.8) to $\tilde{X}$, we have

$$\begin{aligned} E_x\left[\int_0^{T_b} f(X_s)ds\right] &= \tilde{E}_x\left[\int_0^{T_b \wedge T_{-b}} \tilde{f}(\tilde{X}_s)ds\right] = \int_{-b}^{b} \tilde{G}_J(x, y)f(y)\tilde{m}(dy) \\ &= \int_{]0,b]} G_J(x, y)f(y)m(dy) + \frac{1}{2}G_J(x, 0)f(0)\tilde{m}(\{0\}). \end{aligned}$$

It remains to set $m(\{0\}) = \tilde{m}(\{0\})/2$ to get the result.

We observe that $m(\{0\}) < \infty$.

(3.11) Definition. *The point* 0 *is said to be* slowly reflecting *if* $m(\{0\}) > 0$ *and* instantaneously reflecting *if* $m(\{0\}) = 0$.

If absorbing points were not excluded by regularity, it would be consistent to set $m(\{0\}) = \infty$ for 0 absorbing.

For the reflected BM, the point 0 is instantaneously reflecting; the Lebesgue measure of the set $\{t : X_t = 0\}$ is zero which is typical of instantaneously reflecting points. For slowly reflecting points, the same set has positive Lebesgue measure as is seen by taking $f = 1_{\{0\}}$ in the above result. An example of a slowly reflecting point will be given in Exercise (2.29) Chap. X.

We next turn to the description of the extended infinitesimal generator A of X; its domain (see Sect. 1) is denoted by $\mathbb{D}_A$. Here E is any subinterval of $\mathbb{R}$. We recall that the s-derivative of a function f at a point x is the limit, if it exists, of the ratios $\big(f(y) - f(x)\big)/\big(s(y) - s(x)\big)$ as y tends to x. The notions of right and left derivatives extend similarly.

(3.12) Theorem. *For a bounded function* f *of* $\mathbb{D}_A$ *and* $x \in \overset{\circ}{E}$,

$$Af(x) = \frac{d}{dm}\frac{d}{ds}f(x)$$

in the sense that

i) the s*-derivative* $\frac{df}{ds}$ *exists except possibly on the set* $\{x : m(\{x\}) > 0\}$,
ii) if x_1 *and* x_2 *are two points for which this* s*-derivative exists*

$$\frac{df}{ds}(x_2) - \frac{df}{ds}(x_1) = \int_{x_1}^{x_2} Af(y)m(dy).$$

Proof. If $f \in \mathbb{D}_A$, by definition

$$M_t^f = f(X_t) - f(X_0) - \int_0^t Af(X_s)ds$$

is a martingale. Moreover, $|M_t^f| \leq 2\|f\| + t\|Af\|$ so that if T is a stopping time such that $E_x[T] < \infty$, then $M_{t\wedge T}^f$ is uniformly integrable under P_x and therefore

$$E_x[f(X_T)] - f(x) = E_x\left[\int_0^T Af(X_s)ds\right].$$

For $I =]a, b[\subset \overset{\circ}{E}$ and $a < x < b$, we may apply this to $T = T_a \wedge T_b$, and, by Corollary (3.8), it follows that

$$(\#)\quad \begin{aligned} &f(a)\big(s(b) - s(x)\big) + f(b)\big(s(x) - s(a)\big) - f(x)\big(s(b) - s(a)\big) \\ &= \big(s(b) - s(a)\big)\int_I G_I(x, y)Af(y)m(dy). \end{aligned}$$

By straightforward computations, this may be rewritten

$$\frac{f(b)-f(x)}{s(b)-s(x)} - \frac{f(x)-f(a)}{s(x)-s(a)} = \int H_I(x,y)Af(y)m(dy)$$

where

$$H_I(x,y) = \begin{cases} \dfrac{s(y)-s(a)}{s(x)-s(a)} & \text{if } a < y \le x, \\ \dfrac{s(b)-s(y)}{s(b)-s(x)} & \text{if } x \le y < b, \\ 0 & \text{otherwise.} \end{cases}$$

If we let b decrease to x, the integrand $\frac{s(b)-s(y)}{s(b)-s(x)} 1_{(x \le y < b)}$ tends to $1_{\{x\}}$; by an application of Lebesgue's theorem, we see that the right s-derivative of f exists. Similarly, the left s-derivative exists. If we let simultaneously a and b tend to x, we find, applying again Lebesgue's theorem, that

$$\frac{df^+}{ds}(x) - \frac{df^-}{ds}(x) = 2m(\{x\})Af(x)$$

which yields the part (i) of the statement.

To prove the second part, we pick h such that $a < x+h < b$. Applying (#) to x and $x+h$ and subtracting, we get

$$\begin{aligned} f(b) - f(a) - \big(s(b)-s(a)\big)\frac{f(x+h)-f(x)}{s(x+h)-s(x)} \\ = \big(s(b)-s(a)\big)\int \frac{G_I(x+h,y)-G_I(x,y)}{s(x+h)-s(x)} Af(y)m(dy). \end{aligned}$$

If the s-derivative of f in x exists, letting h go to zero and applying once more Lebesgue's theorem yields

$$\begin{aligned} f(b) &- f(a) - \big(s(b)-s(a)\big)\frac{df}{ds}(x) \\ &= -\int_a^x \big(s(y)-s(a)\big)Af(y)m(dy) + \int_x^b \big(s(b)-s(y)\big)Af(y)m(dy) \\ &= -\int_a^b s(y)Af(y)m(dy) + s(a)\int_a^x Af(y)m(dy) + s(b)\int_x^b Af(y)m(dy). \end{aligned}$$

Let $x_1 < x_2$ be two such points; by subtraction, we obtain

$$\frac{df}{ds}(x_2) - \frac{df}{ds}(x_1) = \int_{x_1}^{x_2} Af(y)m(dy)$$

which is the desired result.

Remarks. 1°) The reader will observe that, if s is multiplied by a constant, since by its very definition m is divided by the same constant, the generator is unchanged, as it should be.

2°) For the linear BM, we get $A = \frac{1}{2}\frac{\partial^2}{\partial x^2}$ as we ought to. The reader can further check, using the values for s and m found above, that for the BES^3, we have $A = \frac{1}{2}\frac{\partial^2}{\partial x^2} + \frac{1}{x}\frac{\partial}{\partial x}$; this jibes with the SDE satisfied by BES^3 which was given in Sect. 3 of Chap. VI.

3°) The fact that $s(X)^R$ is a local martingale agrees with the form of the infinitesimal generator and Proposition (1.6).

We now investigate what happens at the boundary point when $E = [0, \infty[$. The positive maximum principle shows that the functions of $\mathbb{D}_A$ must satisfy some condition on their first derivative. More precisely, we have the

(3.13) Proposition. *If $E = [0, \infty[$, for every bounded $f \in \mathbb{D}_A$*

$$\frac{df^+}{ds}(0) = m(\{0\})Af(0).$$

Proof. Using Proposition (3.10) instead of Corollary (3.8), the proof follows the same patterns as above and is left to the reader as an exercise. □

For the reflected BM, we see that, by continuity, $Af(0) = \frac{1}{2}f''(0)$ and that $f'(0) = 0$ which is consistent with the positive maximum principle; indeed, a function such that $f'(0) < 0$, could have a maximum in 0 with $f''(0) > 0$. It is also interesting to observe that the infinitesimal generators of the reflected BM and the absorbed BM coincide on $C_K^2(]0, \infty[)$ (see Exercise (3.16)).

As may be surmised, the map $X \to (s, m)$ is one-to-one or in other words, the pair (s, m) is characteristic of the process X. We prove this in a special case which will be useful in Chap. XI.

(3.14) Proposition. *If X and $\tilde{X}$ are two Feller processes on $[0, \infty[$ such that $s = \tilde{s}$ and $m = \tilde{m}$, then they are equivalent.*

Proof. By Propositions (1.6) and (1.7) on the one hand, and Theorem (3.12) and Proposition (3.13) on the other hand, the spaces $\mathscr{D}_A$ and $\mathscr{D}_{\tilde{A}}$ are equal and so are the generators A and $\tilde{A}$. It follows from Exercise (1.18) that the semi-groups of X and $\tilde{X}$ are equal whence the result follows. □

(3.15) Exercise. Prove that X is on its natural scale if and only if for any $a < b$ and $x_0 = (a + b)/2$,

$$P_{x_0}[T_a < T_b] = 1/2.$$

(3.16) Exercise. Prove that the domain of the infinitesimal generator of the reflected BM is exactly $\{f \in C_0^2([0, \infty[) : f'(0) = 0\}$.

* **(3.17) Exercise.** 1°) **(Dynkin's operator).** For $x \in \overset{\circ}{E}$ and h sufficiently small, call $I(h)$ the interval $]x - h, x + h[$. For $f \in \mathbb{D}_A$, prove that

$$Af(x) = \lim_{h \downarrow 0} \left(E_x\left[f(X_{\sigma_{I(h)}})\right] - f(x)\right) / m_{I(h)}(x).$$

The limit on the right may exist even for functions f which are not in $\mathbb{D}_A$. We will then still call the limit Af, thus defining a further extension of the operator A.

2°) For $I =]a, b[\subset \overset{\circ}{E}$, define $p_I(x) = P_x[T_b < T_a]$. Prove that $Ap_I = 0$ on I and $Am_I = -1$ on I.

(3.18) Exercise. If ϕ is a homeomorphism of an interval E onto an interval $\tilde{E}$ and if X is a process on E satisfying the hypothesis of this section, then $\tilde{X} = \phi(X)$ is a process on $\tilde{E}$ satisfying the same hypothesis. Prove that $\tilde{s} = s \circ \phi^{-1}$ and that $\tilde{m}$ is the image of m by ϕ.

(3.19) Exercise. Suppose that X is on its natural scale and that $E = [0, \infty[$; prove that for any $\varepsilon > 0$, $\int_{]0,\varepsilon[} ym(dy) < \infty$.

\# **(3.20) Exercise.** Let X be a diffusion on $\mathbb{R}$ with infinitesimal generator

$$L = \frac{1}{2}\sigma^2(x)\frac{d^2}{dx^2} + b(x)\frac{d}{dx},$$

where σ and b are locally bounded Borel functions and σ does not vanish. We assume that X satisfies the hypothesis of this section (see Exercise (2.10) Chap. IX).

1°) Prove that the *scale function* is given by

$$s(x) = \int_c^x \exp\left(-\int_c^y 2b(z)\sigma^{-2}(z)dz\right)dy$$

where c is an arbitrary point in $\mathbb{R}$. In particular, if $b = 0$, the process is on natural scale.

2°) Prove that the *speed measure* is the measure with density $(2/s'\sigma^2)$ with respect to the Lebesgue measure where s' is the derivative of s.

[Hint: Use Exercise (3.18).]

(3.21) Exercise. 1°) If $E =]l, r[$ and there exist x, b with $l < x < b < r$ such that $P_x[T_b < \infty] = 1$ (for instance in the case of BES^d, $d > 2$), then $s(l+) = \lim_{a \downarrow l} s(a) = -\infty$. Conversely, if $s(l+) = -\infty$, prove that $P_x[T_b < \infty] = 1$ for every $l < x < b < r$.

2°) If, moreover, $s(r-) = \lim_{b \uparrow r} s(b) = \infty$, prove that X is recurrent.

3°) If instead of the condition in 2°), we have $s(r-) < \infty$, prove that

$$P_x\left[\lim_{t \uparrow \zeta} X_t = r\right] = P_x\left[\inf_t X_t > l\right] = 1$$

and find the law of $\gamma = \inf_t X_t$ under P_x. As a result, X is recurrent if and only if $s(l+) = -\infty$ and $s(r-) = \infty$.

4°) Under the conditions of 3°), i.e. $s(l+) = -\infty$, $s(r-) < \infty$, prove that there is a unique time ρ such that $X_\rho = \gamma$.

[Hint: use the same method as in Exercise (3.17) of Chap. VI, the hypotheses of which, the reader may observe, are but a particular case of those in 3°).]

(3.22) Exercise. Let X be a strong Markov process on $\mathbb{R}$ with continuous paths. For $a \in \mathbb{R}$, set

$$D_{a+} = T_{]a,\infty[}, \qquad D_{a-} = T_{]-\infty,a[}.$$

1°) Prove that either $P_a[D_{a+} = \infty] = 1$ or $P_a[D_{a+} = 0] = 1$ and similarly with D_{a-}. The point a is said to be regular (resp.: a left shunt, a right shunt, a trap) if $P_a[D_{a+} = 0, D_{a-} = 0] = 1$ (resp.: $P_a[D_{a+} = \infty, D_{a-} = 0] = 1$, $P_a[D_{a+} = 0, D_{a-} = \infty] = 1$, $P_a[D_{a+} = \infty, D_{a-} = \infty] = 1$). Find examples of the four kinds of points. Prove that a regular point is regular for itself in the sense of Exercise (2.25) in Chap. III, in other words $P_a[T_a = 0] = 1$. For a regular process in the hypothesis of this section, prove that all the points in the interior of I are regular.

2°) Prove that the set of regular points is an open set, hence a union of intervals. Assume that $\zeta = \infty$ a.s. and show that the process can be restricted to any of these intervals so as to obtain a process satisfying the hypothesis of this section.

(3.23) Exercise. If $E =]0, \infty[$, $\zeta = \infty$ a.s. and X is on its natural scale, then 0 is a natural boundary.

(3.24) Exercise. Let $E = \mathbb{R}$ and X be recurrent (Exercise (3.21), 3°)). Choose s such that $s(0) = 0$. If μ is a probability measure on $\mathbb{R}$ such that

$$\int |s(x)| d\mu(x) < \infty, \qquad \int s(x) d\mu(x) = 0,$$

prove that there exists a stopping time T such that the law of X_T under P_0 is μ.

* **(3.25) Exercise (Asymptotic study of a particular diffusion).** In the notation of Exercise (3.20) let $\sigma(x) = 2(1 + x^2)$ and $b(x) = 2x$.

1° Prove that $X_t = \tan(\gamma_{H_t})$ where γ is a BM^1 and

$$H_t = 2\int_0^t \left(1 + X_s^2\right)^{-1} ds = \inf\left\{u : \int_0^u \left(1 + \tan^2(\gamma_s)\right) ds > 2t\right\}.$$

[Hint: Use Exercise (3.20) and Proposition (3.4).]

2°) Show that as t tends to ∞,

$$\lim_{t\to\infty} H_t = \inf\{u : |\gamma_u| = \pi/2\} \quad \text{a.s.},$$

and $\lim_{t\to\infty} |X_t| = \infty$ a.s.

3°) Show that for t_0 sufficiently large (for instance $t_0 = \sup\{t : |X_t| = 1\}$), the following formula holds for every $t > t_0$,

$$\log|X_t| = \log|X_{t_0}| + \int_{t_0}^t X_s^{-1} \left(2(1 + X_s^2)\right)^{1/2} d\beta_s + t - t_0 - \int_{t_0}^t X_s^{-2} ds$$

where β is a BM.

4°) Show that $t^{-1/2}(\log|X_t| - t)$ converges in law as t tends to ∞, to $2G$ where G is a standard Gaussian r.v.

§4. Time Reversal and Applications

In this section, we consider a Markov process with general state space and continuous paths on $[0, \zeta[$ and assume that $X_{\zeta-}$ exists a.s. on $\{\zeta < \infty\}$ as is the case for Feller processes (Theorem (2.7) of Chap. III). Our goal is to show that, under suitable analytic conditions, one can get another Markov process by running the paths of X in the reverse direction starting from a special class of random times which we now define.

(4.1) Definition. *A positive r.v. L on Ω is a* cooptional time *if*

i) $\{L < \infty\} \subset \{L \leq \zeta\}$*;*
ii) For every $t \geq 0$, $L \circ \theta_t = (L - t)^+$.

The reader will check that ζ is a cooptional time and so is the *last exit time* of a Borel set A defined by

$$L_A(\omega) = \sup\{t : X_t(\omega) \in A\}$$

where $\sup(\emptyset) = 0$. We also have the

(4.2) Proposition. *If L is cooptional, then for any $s \geq 0$, the r.v. $(L - s)^+$ is also cooptional.*

Proof. Condition i) of (4.1) is obviously satisfied by $(L - s)^+$ and moreover

$$(L-s)^+ \circ \theta_t = (L \circ \theta_t - s)^+ = \left((L-t)^+ - s\right)^+ = (L - t - s)^+ = \left((L-s)^+ - t\right)^+$$

which is condition ii). □

In what follows, L is a fixed, a.s. finite and strictly positive cooptional time and we define a new process $\tilde{X}$ taking its values in E_Δ by setting for $t > 0$

$$\tilde{X}_t(\omega) = \begin{cases} \Delta & \text{if } \ L(\omega) \leq t \ \text{ or } \ L(\omega) = \infty, \\ X_{L(\omega)-t}(\omega) & \text{if } \ 0 < t < L(\omega), \end{cases}$$

and $\tilde{X}_0 = X_{L-}$ if $0 < L < \infty$ and $\tilde{X}_0 = \Delta$ otherwise.

We will set $\tilde{\mathscr{F}}_t = \sigma(\tilde{X}_s, s \leq t)$. On $\{L > t + u\}$, we have, using Property ii) of Definition (4.1), $L(\theta_u) = L - u > t$, hence

$$\tilde{X}_t(\theta_u) = X_{L(\theta_u)-t}(\theta_u) = X_{L-u-t+u} = \tilde{X}_t.$$

It follows from the monotone class theorem that, if Γ is in $\tilde{\mathscr{F}}_t$, then for every $u \geq 0$

$$\theta_u^{-1}(\Gamma) \cap \{t + u < L\} = \Gamma \cap \{t + u < L\}.$$

We now introduce the set-up in which we will show that $\tilde{X}$ is a Markov process. We assume that:

i) there is a probability measure μ such that the potential $\nu = \mu U$ where U is the potential kernel of X (Exercise (2.29) Chap. III), is a Radon measure.

ii) there is a second semi-group on E, denoted by $(\hat{P}_t)$, such that
   a) if $f \in C_K(E)$, then $\hat{P}_t f$ is right-continuous in t;
   b) the resolvents $(\hat{U}_p)$ and (U_p) are in *duality with respect* to ν, namely

$$\int U_p f \cdot g \, d\nu = \int f \cdot \hat{U}_p g \, d\nu$$

   for every $p > 0$ and every positive Borel functions f and g.

Examples will be given later in this section. The last equality will also be written

$$\langle U_p f, g\rangle_\nu = \langle f, \hat{U}_p g\rangle_\nu.$$

If $\hat{X}$ is another Markov process with $\hat{P}_t$ as transition semi-group, we say that X and $\hat{X}$ are in duality with respect to ν. Using the Stone-Weierstrass and monotone class theorems, it is not difficult to see that this relationship entails that for any positive Borel function ϕ on $\mathbb{R}_+$

$$\langle P_\phi f, g\rangle_\nu = \langle f, \hat{P}_\phi g\rangle_\nu,$$

where $P_\phi f(x) = \int_0^\infty \phi(t) P_t f(x) dt$.

Our goal is to prove that $\tilde{X}$ is a Markov process with transition semi-group $(\hat{P}_t)$. We will use the following lemmas.

(4.3) Lemma. *Given $r > 0$, ϕ a positive Borel function on $\mathbb{R}_+$ and H a positive $\tilde{\mathscr{F}}_r$-measurable r.v., then for any positive Borel function f on E,*

$$\int_0^\infty \phi(t) E_\mu \left[f(\tilde{X}_{t+r}) H \right] dt = \int f h_\phi d\nu$$

where $h_\phi(x) = E_x\left[H\phi(L-r)1_{\{r<L\}}\right]$. Moreover, for $s > r$,

$$\int_0^\infty \phi(t) E_\mu \left[f(\tilde{X}_{t+s}) H \right] dt = \int f P_{s-r} h_\phi d\nu.$$

Proof. By considering $(L-r)^+$ instead of L, we may make $r = 0$ in the equality to be proven. The left-hand side is then equal to

$$\begin{aligned} \int_0^\infty E_\mu\left[f(X_{L-t})1_{(L>t)}H\right]\phi(t)dt &= \int_0^\infty E_\mu\left[Hf(X_u)\phi(L-u)1_{(L>u)}\right]du \\ &= \int_0^\infty E_\mu\left[(H\phi(L)1_{(L>0)})\circ\theta_u f(X_u)\right]du \end{aligned}$$

since, as a consequence of the definition of $\tilde{\mathscr{F}}_0$, we have $H = H \circ \theta_u$ on $(L > u)$. Furthermore, by the Markov property of X, the last expression is equal to

$$\int_0^\infty du E_\mu\left[h_\phi(X_u) f(X_u)\right] = \int f h_\phi d\nu,$$

which proves the first part of the lemma.

To prove the second part, observe that since $\tilde{\mathscr{F}}_r \subset \tilde{\mathscr{F}}_s$, the r.v. H is also in $\tilde{\mathscr{F}}_s$ so that, by the first part, we may write

$$\int_0^\infty \phi(t) E_\mu \left[f(\tilde{X}_{t+s}) H \right] dt = \int f(x) E_x \left[\phi(L-s) H \ 1_{(s<L)} \right] \nu(dx).$$

But since $1_{(r<L)} \circ \theta_{s-r} = 1_{\{s<L\}}$ and $H \circ \theta_{r-s} = H$ on $\{L > s\}$, we have $\phi(L-s)H \ 1_{(s<L)} = \left(\phi(L-r)H \ 1_{\{r<L\}}\right) \circ \theta_{r-s}$ and the second part follows from the Markov property. □

(4.4) Lemma. *Keeping the notation of Lemma (4.3), if ψ is another positive Borel function on $\mathbb{R}_+$, then*

$$P_\psi h_\phi = h_{\psi * \phi} = P_\phi h_\psi .$$

Proof. From the above proof, it follows that

$$P_{s-r} h_\phi = E_\cdot \left[\phi(L-s) H \ 1_{(s<L)} \right].$$

As a result

$$\begin{aligned} P_\psi h_\phi &= \int_r^\infty \psi(s-r) P_{s-r} h_\phi ds = E_\cdot \left[\int_r^\infty \psi(s-r)\phi(L-s) H \ 1_{(s<L)} ds \right] \\ &= E_\cdot \left[1_{(r<L)} H \int_r^L \psi(s-r)\phi(L-s) ds \right] = h_{\psi*\phi}, \end{aligned}$$

which is the first equality. The second one follows by symmetry. □

We may now turn to the main result of this section.

(4.5) Theorem. *Under P_μ, the process $\tilde{X}$ is a Markov process with respect to $(\tilde{\mathscr{F}}_t)$ with transition function $(\hat{P}_t)$.*

Proof. We want to prove that $E_\mu \left[f(\tilde{X}_{t+s}) \mid \tilde{\mathscr{F}}_s \right] = \hat{P}_t f(\tilde{X}_s)$ P_μ-a.s. But using the notation and result in Lemma (4.3) we have

$$\int_0^\infty \phi(t) E_\mu \left[f(\tilde{X}_{t+s}) H \right] dt = \langle f, P_{s-r} h_\phi \rangle_\nu$$

and on the other hand, Fubini's theorem yields

$$\int_0^\infty \phi(t) E_\mu \left[\hat{P}_t f(\tilde{X}_s) H \right] dt = E_\mu \left[\hat{P}_\phi f(\tilde{X}_s) H \right].$$

Let us compare the right members of these identities. Using the above lemmas and the duality property, we get

$$\begin{aligned} \int_r^\infty \psi(s-r) \langle f, P_{s-r} h_\phi \rangle_\nu ds &= \langle f, P_\psi h_\phi \rangle_\nu = \langle f, P_\phi h_\psi \rangle_\nu \\ &= \langle \hat{P}_\phi f, h_\psi \rangle_\nu = \int_r^\infty \psi(s-r) E_\mu \left[\hat{P}_\phi f(\tilde{X}_s) H \right] ds. \end{aligned}$$

It follows that there is a Lebesgue-negligible set $N(f, r, H, \phi) \subset \mathbb{R}_+$ such that for $s > r$, $s \notin N(f, r, H, \phi)$,

$$E_\mu\left[\hat{P}_\phi f(\tilde{X}_s) \cdot H\right] = \langle f, P_{s-r} h_\phi \rangle_\nu.$$

Let N be the union of the sets $N(f, r, H, \phi)$ where f runs through a dense sequence in $C_K(E)$, ϕ runs through a sequence which is dense in $C_K(\mathbb{R}_+)$, r runs through rational numbers > 0 and for each r, H runs through a countable algebra of bounded functions generating $\sigma(\tilde{X}_u, u \leq r)$; then, for $s \notin N$, the last displayed equality holds simultaneously for every $f \in C_K(E)$, $\phi \in C_K(\mathbb{R}_+)$, $r \in \mathbb{Q}_+$, $s > r$ and every H which is $\sigma(\tilde{X}_u, u \leq r)$-measurable and bounded. As a result, under the same conditions

$$E_\mu\left[f(\tilde{X}_{s+t})H\right] = E_\mu\left[\hat{P}_t f(\tilde{X}_s)H\right]$$

for almost every t. But, by the property ii) a) of $\hat{P}_t$, both sides are right continuous in t and the equality holds for every t.

Next, because of the continuity of paths, the filtration $(\tilde{\mathscr{F}}_t)$ is right and left continuous (i.e. $\tilde{\mathscr{F}}_{t-} = \tilde{\mathscr{F}}_t = \tilde{\mathscr{F}}_{t+}$ for each t) up to sets of P_μ-measure zero; it follows first that the last displayed equality is valid with H in $\tilde{\mathscr{F}}_s$ and, finally, since each s is a limit of points in N^c, that this equality holds without restriction. As a result, for $f \in C_K(E)$,

$$E_\mu\left[f(\tilde{X}_{s+t}) \mid \tilde{\mathscr{F}}_s\right] = \hat{P}_t f(\tilde{X}_s)$$

for every s and t, which is the desired result.

Remarks. 1°) This result does not show that $\tilde{X}$ has good properties such as the Feller or Strong Markov property. However, in many applications the semi-group $(\hat{P}_t)$ is a Feller semi-group, which insures that $\tilde{X}$ has good properties (see also Exercise (4.13)).

2°) If $\hat{P}_t$ is already known to be the semi-group of a process $\hat{X}$, then $\tilde{X}$ under P_μ has the same law as $\hat{X}$ under $\hat{P}_{\hat{\mu}}$ where $\hat{\mu} = X_{L-}(P_\mu)$ is the law of X_{L-} under P_μ.

We will now give two applications of Theorem (4.5) to the Bessel process of dimension 3 which was introduced and studied in Sect. 3 Chap. VI. This process and the BM killed at 0 (which we will denote by B^0) are in duality with respect to the measure $\nu(dx) = 2x\,dx$. Indeed, in the notation of Chap. VI, using the fact that Q_t is in duality with itself with respect to Lebesgue's measure, we have, for $f, g \geq 0$ and $h(x) = x$,

$$\int_0^\infty f(x) P_t^3 g(x) x\,dx = \int_0^\infty f(x) \frac{1}{x} Q_t(gh)(x) x\,dx = \int_0^\infty Q_t f(x) g(x) x\,dx$$

which proves our claim. This is a particular instance of a more general result; the process BES^3 is the h-process of BM^0 as defined in Sect. 3 Chap. VIII and as such is in duality with BM^0 with respect to the measure $h(x)dx$ (see Exercise (3.17) in Chap. VIII).

Furthermore, it was shown in the last section that ν is precisely equal to the potential measure $U(0, \cdot)$ of $\mathrm{BES}^3(0)$. Thus, we are exactly in the setting of Theorem (4.5) with $\mu = \varepsilon_0$, and we may state the

(4.6) Corollary (Williams). *Let X be a* $\mathrm{BES}^3(0)$ *and B a* $\mathrm{BM}(b)$ *with $b > 0$, then if $L_b = \sup\{t : X_t = b\}$, the processes $\{X_{L_b - t}, 0 \le t \le L_b\}$ and $\{B_t, 0 \le t \le T_0\}$ have the same law.*

Remarks. 1°) Another proof of this result relying on excursion theory will be given in Chap. XII.

2°) This corollary implies that the law of L_b for $\mathrm{BES}^3(0)$ is the same as the law of T_0 for $\mathrm{BM}(b)$, which was computed in Chap. II, Proposition (3.7) and Chap. III, Proposition (3.7).

Our second application deals with the process BES^3 killed when it first hits a point $b > 0$. More precisely, if X is a BES^3, we consider the process X^b defined by

$$X_t^b = X_t \quad \text{if } t < T_b \text{ and } X_0 \in [0, b[, \quad X_t^b = \Delta \quad \text{otherwise}$$

where as usual $T_b = \inf\{t > 0 : X_t = b\}$. It was shown in Exercise (2.30) of Chap. III that this is a Markov process on $[0, b[$ and clearly T_b is the deathtime, hence a cooptional time, for X^b .

(4.7) Lemma. *The processes X^b and $b - X^b$ are in duality with respect to the measure $\xi(dx) = x(b - x)dx$ on $[0, b]$.*

Proof. We have already used the fact, that the potential U of X has the density $u(x, y) = \inf(1/x, 1/y)$ with respect to the measure $2y^2dy$. By a simple application of the strong Markov property, we see that the potential V of X^b is given by, for $x < b$,

$$\begin{aligned} Vf(x) &= E_x\left[\int_0^{T_b} f(X_t)dt\right] = Uf(x) - P_{T_b}Uf(x) \\ &= 2\int_0^b \big(u(x, y) - u(b, y)\big)y^2dy; \end{aligned}$$

in other words, V has the density $v(x, y) = \inf(1/x, 1/y) - 1/b$ with respect to the measure $2y^2 1_{(0 \le y \le b)}dy$. Clearly, the potential $\hat{V}$ of the process $b - X^b$ has the density $v(b - x, b - y)$ with respect to the measure $2(b - y)^2 1_{(0 \le y \le b)}dy$. It is then a tedious but elementary computation to check that for $f, g \ge 0$,

$$\int Vf \cdot g\, d\xi = \int f \cdot \hat{V}g\, d\xi.$$

Now the mapping $f \to Vf$ (resp. $f \to \hat{V}f$) is bounded on the space of bounded functions on $[0, b]$ so that the result follows from Exercise (4.17). □

(4.8) Proposition. *If X is a* $\mathrm{BES}^3(0)$ *and b is strictly positive, the processes* $\left(X_{T_b-t}, 0 \le t \le T_b\right)$ *and* $(b - X_t, 0 \le t \le T_b)$ *are equivalent.*

Proof. The potential measure $V(0, dy)$ is equal, by what we have just seen, to $2(1/y - 1/b)y^2dy = b\xi(dy)$. Thus the result follows at once from Theorem (4.5) and the above lemma. □

Bringing together Corollary (4.6), Proposition (4.8) and Theorem (3.11) of Chap. VI we obtain

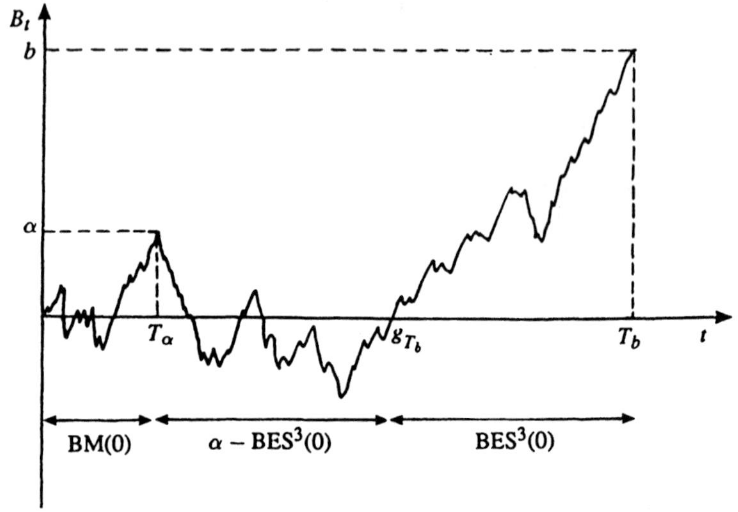

Fig. 7.

(4.9) Theorem (Williams' Brownian path decomposition). *For $b > 0$, let be given the four following independent elements:*

i) *a r.v. α uniformly distributed on* $[0, b]$*;*
ii) *a standard* BM *B;*
iii) *two* $\mathrm{BES}^3(0)$ *processes ρ and ρ',*

and define

$$T_\alpha = \inf\{t : B_t = \alpha\}, \quad g_{T_b} = T_\alpha + \sup\{t : \alpha - \rho(t) = 0\},$$
$$T_b = g_{T_b} + \inf\left\{t : \rho'(t) = b\right\},$$

then, the process $\tilde{X}$ defined for $0 \le t \le T_b$ *by*

$$\tilde{X}_t = \begin{cases} B_t, & 0 \le t \le T_\alpha, \\ \alpha - \rho(t - T_\alpha), & T_\alpha \le t \le g_{T_b}, \\ \rho'(t - g_{T_b}), & g_{T_b} \le t \le T_b, \end{cases}$$

is a BM(0) *killed when it first hits b.*

Proof. By Corollary (4.6), a BM killed at time T_b is a time-reversed $BES^3(0)$ to which we apply the decomposition Theorem (3.11) of Chap. VI. The time-reversed parts are easily identified by means of Corollary (4.6) and Proposition (4.8). Here again, the result is best described by Figure 7; it is merely Figure 5 of Chap. VI put "upside down".

Remark. There are actually other proofs of the fact that BM taken between g_{T_b} and T_b (if g_{T_b} is the last zero before T_b) is a BES^3. If this result were known, then the above decomposition theorem might be deduced from (4.6) and Theorem (3.11) of Chap. VI without having to resort to Proposition (4.8) above.

(4.10) Exercise. If L and L' are two cooptional times, then $L \vee L'$ and $L \wedge L'$ are cooptional times.

* **(4.11) Exercise.** Let L be a cooptional time and $\mathscr{G}_L$ be the family of sets $\Gamma \in \mathscr{F}$ such that for every $u \geq 0$,

$$\Gamma \cap \{L > u\} = \theta_u^{-1}(\Gamma) \cap \{L > u\}.$$

1°) Prove that $\mathscr{G}_L$ is a σ-algebra (see also Exercise (4.13) below) and that L and X_L are $\mathscr{G}_L$-measurable.

2°) If $A \in \mathscr{G}_L$, prove that the r.v. L^A defined by

$$L^A = L \quad \text{on } A, \qquad L^A = 0 \quad \text{on } A^c,$$

is cooptional.

* **(4.12) Exercise.** 1°) Let ρ_t be the modulus of BM^2 and suppose that $\rho_0 = r$ with $0 < r < 1$. Prove that there exists a BM^1 γ started at $(-\log r)$ such that $-\log \rho_t = \gamma_{C_t}$ where $C_t = \inf\left\{u : \int_0^u \exp(-2\gamma_s)ds > t\right\}$.

[Hint: Use the ideas of Sect. 2 Chap. V.]

2°) Let X be a $BES^2(0)$ and $T_1 = \inf\{t : X_t = 1\}$. Prove that there exists a $BES^3(0)$, say Y, such that

$$(-\log X_t, 0 < t \leq T_1) = \left(Y_{A_t}, 0 < t \leq \int_0^\infty \exp(-2Y_s)ds\right),$$

where $A_t = \sup\left\{u : \int_u^\infty \exp(-2Y_s)ds > t\right\}$.

[Hint: Apply Corollary (4.6) to the BM γ of 1°), then let r converge to 0.]

3°) Extend the result of 1°) and 2°) to $\rho = |BM^d|$ with $d \geq 3$. More precisely prove that if X is a $BES^d(0)$

$$\left((X_t)^{2-d}, t > 0\right) = \left(Y_{A_t}, t > 0\right)$$

where $A_t = \sup\left\{u : (d-2)^{-2} \int_u^\infty Y_s^{-\alpha} ds > t\right\}$, and $\alpha = 2(d-1)/(d-2)$.

* **(4.13) Exercise.** With the notation of this section, let $\tilde{\mathscr{G}}_t$ be the σ-algebra of sets Γ in $\mathscr{F}$ such that for every $u \geq 0$

$$\theta_u^{-1}(\Gamma) \cap \{t + u < L\} = \Gamma \cap \{t + u < L\}.$$

1°) Prove that $(\tilde{\mathscr{G}}_t)$ is a right-continuous filtration which is larger than $(\tilde{\mathscr{F}}_t)$. Check that Lemmas (4.3) and (4.4) are still valid with $(\tilde{\mathscr{G}}_t)$ instead of $(\tilde{\mathscr{F}}_t)$.

2°) Prove that if T is a $(\tilde{\mathscr{G}}_t)$-stopping time, then $(L-T)^+$ is a cooptional time.

3°) Prove that in Theorem (4.5), one can replace $(\tilde{\mathscr{F}}_t)$ by $(\tilde{\mathscr{G}}_t)$; then using 2°), prove that $\tilde{X}$ has the strong Markov property.

(4.14) Exercise. Let L be a cooptional time and set $\phi(x) = P_x[L > 0]$.

1°) Prove that ϕ is an excessive function (see Definition (3.1) of Chap. X).

2°) If f is excessive and finite, prove that one defines a new transition semi-group P^f by setting

$$\begin{aligned} P_t^f(x, dy) &= f^{-1}(x) P_t(x, dy) f(y) \qquad \text{if } f(x) \neq 0 \\ &= 0 \qquad \text{otherwise.} \end{aligned}$$

(See also Proposition (3.9) in Chap. VIII).

* 3°) Let $Y_t(\omega) = X_t(\omega)$ if $t < L(\omega)$ and $Y_t(\omega) = \Delta$ if $t \geq L(\omega)$, and prove that for any probability measure μ, the process Y is a Markov process with transition semi-group P_t^f.

* **(4.15) Exercise (Another proof of Pitman's theorem).** Let B be the standard linear BM, L its local time at 0 and as usual $\tau_1 = \inf\{t : L_t > 1\}$. We call (T) the following property which is proved in Exercise (2.29) of Chap. VI and in Exercise (4.17) of Chap. XII: the processes $(|B_t|, t \leq \tau_1)$ and $(|B_{\tau_1 - t}|, t \leq \tau_1)$ are equivalent. Call (P) the property proved in Pitman's theorem (Sect. 3 Chap. VI) namely

$$(2S_t - B_t, S_t, t \geq 0) \stackrel{(d)}{=} (Z_t, J_t, t \geq 0)$$

where Z is a $\text{BES}^3(0)$ and $J_t = \inf_{s \geq t} Z_s$. Call further (R) the time-reversal property of Corollary (4.6). The aim of this exercise is to show that together with the Lévy equivalence $(S_t - B_t, S_t, t \geq 0) \stackrel{(d)}{=} (|B_t|, L_t, t \geq 0)$ proved in Sect. 2 of Chap. VI and which we shall call (L), any two of the properties (T), (P), (R) imply the third one.

1°) Let as usual $T_1 = \inf\{t : B_t = 1\}$; deduce from (L) that

$$\left(|B_{\tau_1 - u}|, u \leq \tau_1\right) \stackrel{(d)}{=} \left(-1 + S_{T_1 - u} + (1 - B_{T_1 - u}), u \leq T_1\right)$$

and conclude that (R) and (P) imply (T).

2°) Using (L) (or Tanaka's formula) prove that

$$(B_u, u \leq T_1) \stackrel{(d)}{=} (L_u - |B_u|, u \leq \tau_1)$$

and conclude that (T) and (P) imply (R).

[Hint: If (L) is known, (P) is equivalent to (P'), namely

$$(|B_u| + L_u, u \geq 0) \stackrel{(d)}{=} (Z_u, u \geq 0).]$$

3°) Use (T), then (L), to prove that

$$(|B_u| + L_u, u \leq \tau_1) \overset{(d)}{=} \left(1 - B_{T_1 - u}, u \leq T_1\right).$$

Use the scaling invariance properties to deduce that for any $a > 0$,

$$(|B_u| + L_u, u \leq \tau_a) \overset{(d)}{=} (Z_u, u \leq L_a)$$

and conclude that (T) and (R) imply (P'), hence (P).

* **(4.16) Exercise (On last passage times).** Let L be a cooptional time.

1°) In the notation of Exercise (4.14) prove that the supermartingale $Z_t = \phi(X_t)$ (see Proposition (3.2) Chap. X) is equal to $P_x[L > t \mid \mathscr{F}_t]$ P_x-a.s.

2°) Suppose that X is a Feller process on $]0, \infty[$ and that the scale function s is such that $s(0+) = -\infty$ and $s(\infty) = 0$ (see Exercise (3.21)). For $a > 0$, let $L = L_a = \sup\{t : X_t = a\}$ and Λ^x be the family of local times of the local martingale $s(X)$. Prove that

$$Z_t + \frac{-1}{2s(a)} \Lambda_t^{s(a)}$$

is a local martingale (a particular instance of Meyer's decomposition theorem).

3°) Prove that for every positive predictable process H,

$$E_x[H_L] = \frac{-1}{2s(a)} E_x\left[\int_0^\infty H_u d\Lambda_u^{s(a)}\right].$$

This may be stated: $\frac{-1}{2s(a)}\Lambda^{s(a)}$ is the dual predictable projection of $1_{[0<L\leq \cdot]}$.

4°) We now assume in addition that there exists a measure μ such that

i) there exists a continuous function $p^{\cdot}$ on $]0, \infty[^3$ such that

$$P_t(x, dy) = p_t^{\cdot}(x, y)\mu(dy);$$

ii) for every positive Borel function f on $\mathbb{R}_+$,

$$\int_0^t f(X_s)ds = \int f(y)\Lambda_t^{s(y)}\mu(dy)$$

(see Exercise (2.32) Chap. X and Sect. 1 Chap. XI). Prove that

$$\frac{\partial}{\partial t} E_x[\Lambda_t^{s(a)}] = p_t^{\cdot}(x, a).$$

5°) Show that

$$P_x(L_a \in dt) = \frac{-1}{2s(a)} p_t^{\cdot}(x, a)dt.$$

An important complement to this exercise is Exercise (1.16) in Chap. X where it is shown how conditioning with respect to L is related to the distribution of the bridges of X.

(4.17) Exercise. 1°) Let V^p and $\hat{V}^p$ be two resolvents on $(E, \mathscr{E})$ such that the kernels $V = V^0$ and $\hat{V} = V^0$ are bounded on the space $b\mathscr{E}$ of bounded Borel functions. If ξ is a measure such that

$$\int Vf \cdot g\, d\xi = \int f \cdot \hat{V}g\, d\xi$$

for every pair of positive Borel functions, prove that the two resolvents are in duality with respect to ξ.

[Hint: For p sufficiently small, $V^p = \sum_0^\infty p^n V^{n+1}$.]

2°) If the two resolvents are the resolvents of two right- or left-continuous processes with semi-groups P_t and $\hat{P}_t$, prove that for every t,

$$\int P_t f \cdot g\, d\xi = \int f \cdot \hat{P}_t g\, d\xi.$$

(4.18) Exercise. Let X be a Feller process on $[0, \infty[$ such that 0 is not reached from the other points and such that the restriction of X to $]0, \infty[$ satisfies the hypothesis of Sect. 3. We call s and m the corresponding scale function and speed measure and assume that $s(0+) = -\infty$, $s(\infty) = 0$ (see Exercise (3.21)).

1°) Compute the potential kernel of X and prove that X is in duality with itself with respect to m.

2°) Prove that for every $b > 0$, the process $Y_t = \left\{X_{L_b - t}, t < L_b\right\}$ is under P_0 a Markov process on $]0, b[$ with initial measure ε_b and semi-group Q_t given by

$$Q_t f(x) = P_t(fs)(x)/s(x).$$

As a result the law of L_b under P_0 is the same as the law of T_0 for the process Y.

3°) Prove that $(-1/s)$ is a scale function for the process with semi-group Q_t and that the corresponding speed measure is $s^2 m$.

Notes and Comments

Sect. 1. This section is devoted to the minimum of semi-group theory which is necessary (for intuition more than for technical needs) in the sequel. For a detailed account, we recommend the book of Pazy [1]; another exposition designed for probabilists is that of Dellacherie-Meyer [1] vol. IV.

We are uncertain about the origin of Theorem (1.13) but we can mention that it is a special case of a much more general result of Kunita [1] and Roth [1]; the same is true of Exercise (1.19). Exercise (1.20) comes from Chen [1] and Ledoux [1].

In relation to Exercise (1.21), the reader shall find more extensions of Itô's formula in Kunita [5].

Most exercises of this section just record classical properties of semi-groups and may be found in the textbooks on the subject.

Sect. 2. The bulk of this section is taken from Stroock-Varadhan [1] and Priouret [1]. The systematic use of martingale problems in the construction of diffusions is due to Stroock and Varadhan. Their ideas were carried over to other contexts and still play a great role in present-day research although it is only of marginal interest in our own exposition which favors the SDE aspect.

The exercises of this section have the same origin as the text. Exercise (2.8) is taken from a lecture course by Meyer.

Sect. 3. The material covered in this section appeared in a series of papers of Feller. There are many extensive – much more so than ours – expositions of the subject, e.g. in Dynkin [1], Itô-McKean [1], Freedman [1] and Mandl [1]. Our exposition is borrowed from Breiman [1] with other proofs, however, where he uses approximation by discrete time processes. The exercises are very classical.

Sect. 4. The main result of the section, namely Theorem (4.5), is due to Nagasawa [1]. In our exposition, and in some of the exercises, we borrowed from Meyer [2] and Meyer et al. [1]. Corollary (4.6) is stated in Williams [3]. The remarkable Theorem (4.9), which was the first of this kind, is from Williams [2] and [3].

Exercise (4.12) is from Williams [3] and from Yor [16]. Exercise (4.16) comes from Pitman-Yor [1]; see Getoor [2] for the particular case of transient Bessel processes.

In connection with Exercise (4.15) let us mention that Pitman's theorem has now been extended to other processes (see Tanaka ([2], [3]) for random walks, Bertoin [5] for Lévy processes, and Saisho-Tanemura ([1]) for certain diffusions; see in particular Exercise (1.29) Chap. XI).

Chapter VIII. Girsanov's Theorem and First Applications

In this chapter we study the effect on the space of continuous semimartingales of an absolutely continuous change of probability measure. The results we describe have far-reaching consequences from the theoretical point of view as is hinted at in Sect. 2; they also permit many explicit computations as is seen in Sect. 3.

§1. Girsanov's Theorem

The class of semimartingales is invariant under many operations such as composition with C^2-functions or more generally differences of convex functions. We have also mentioned the invariance under time changes. It is also invariant under an absolutely continuous change of probability measures. This is the content of *Girsanov's theorem*: If Q is a probability measure on $(\Omega, \mathscr{F})$ which is absolutely continuous with respect to P, then every semimartingale with respect to P is a semimartingale with respect to Q.

The above theorem is far from intuitive; clearly, a process of finite variation under P is also a process of finite variation under Q but local martingales may lose the martingale property. They however remain semimartingales and one of our goals is precisely to describe their decomposition into the sum of a local martingale and a process with finite variation.

We will work in the following setting. Let $(\mathscr{F}_t^0)$, $t \geq 0$, be a right-continuous filtration with terminal σ-field $\mathscr{F}_\infty^0$ and P and Q two probability measures on $\mathscr{F}_\infty^0$. We assume that for each $t \geq 0$, the restriction of Q to $\mathscr{F}_t^0$ is absolutely continuous with respect to the restriction of P to $\mathscr{F}_t^0$, which will be denoted by $Q \lhd P$. We stress the fact that we may have $Q \lhd P$ without having $Q \ll P$. Furthermore, we call D_t the Radon-Nikodym derivative of Q with respect to P on $\mathscr{F}_t^0$. These (classes of) random variables form a $(\mathscr{F}_t^0, P)$-martingale and since $(\mathscr{F}_t^0)$ is right-continuous, we may choose D_t within its P-equivalence class so that the resulting process D is a $(\mathscr{F}_t^0)$-adapted martingale almost every path of which is cadlag (Theorem (2.5) and Proposition (2.6) Chap. II). In the sequel we always consider such a version.

(1.1) Proposition. *The following two properties are equivalent:*

i) the martingale D is uniformly integrable;
ii) $Q \ll P$ on $\mathscr{F}_\infty^0$.

Proof. See Exercise (2.13) Chap. II.

As was observed in Sect. 4 of Chap. I and at the beginning of Chap. IV, when dealing with stochastic processes defined on $(\Omega, \mathscr{F}_\infty^0, P)$, one has most often to consider the usual augmentation of $(\mathscr{F}_t^0)$ with respect to P, or in other words to consider the σ-fields $\mathscr{F}_t$ obtained by adding to $\mathscr{F}_t^0$, $0 \le t \le \infty$, the P-negligible sets of the completion $\mathscr{F}_\infty$ of $\mathscr{F}_\infty^0$ with respect to P. If $Q \lhd P$ but Q is not absolutely continuous with respect to P on $\mathscr{F}_\infty^0$, then Q cannot be extended to $\mathscr{F}_\infty$. Indeed, if $P(A) = 0$ and $Q(A) > 0$, then all the subsets of A belong to each $\mathscr{F}_t$ and there is no reason why Q could be consistently defined on $\mathscr{P}(A)$. In contrast, if $Q \ll P$ on $\mathscr{F}_\infty^0$ we have the following complement to Proposition (1.1).

(1.1') Proposition. *The conditions of Proposition (1.1) are equivalent to:*

iii) Q may be extended to a probability measure $\bar{Q}$ on $\mathscr{F}_\infty$ such that $\bar{Q} \ll P$ in restriction to each of the completed σ-fields $(\mathscr{F}_t)$, $0 \le t \le \infty$.

If these conditions hold, $D_t = E\left[dQ/dP \mid \mathscr{F}_t^0\right]$ P-a.s.

Proof. It is easy to see that ii) implies iii). If iii) holds and if $A \in \mathscr{F}_\infty^0$ and $P(A) = 0$, then $A \in \mathscr{F}_t$, hence $\bar{Q}(A) = 0$ since $\bar{Q} \ll P$ on $\mathscr{F}_t$ and therefore $Q \ll P$ on $\mathscr{F}_\infty^0$. □

In the sequel, whenever $Q \ll P$ on $\mathscr{F}_\infty^0$, we will not distinguish between Q and $\bar{Q}$ in the notation and we will work with $(\mathscr{F}_t)$-adapted processes without always recalling the distinction. *But if we have merely* $Q \lhd P$, it will be understood that we work only with $(\mathscr{F}_t^0)$-adapted processes, which in view of Exercise (1.30) Chap. IV is not too severe a restriction. The following two results deal with the general case where $Q \lhd P$ only.

The martingale D is positive, but still more important is the

(1.2) Proposition. *The martingale D is strictly positive Q-a.s.*

Proof. Let $T = \inf\{t : D_t = 0 \text{ or } D_{t-} = 0\}$; by Proposition (3.4) in Chap. II, D vanishes P-a.s. on $[T, \infty[$, hence $D_t = 0$ on $\{T < t\}$ for every t. But since $Q = D_t \cdot P$ on $\mathscr{F}_t^0$, it follows that $Q(\{T < t\}) = 0$ which entails the desired conclusion. □

We will further need the following remark.

(1.3) Proposition. *If $Q \lhd P$, then for every $(\mathscr{F}_t^0)$-stopping time T,*

$$Q = D_T \cdot P \quad \text{on } \mathscr{F}_T^0 \cap \{T < \infty\}.$$

If $Q \ll P$, then $Q = D_T \cdot P$ on $\mathscr{F}_T^0$.

Proof. The particular case $Q \ll P$ follows from the optional stopping theorem. For the general case, we use the fact that D is uniformly integrable on each $[0, t]$. Let $A \in \mathscr{F}_T^0$; for every t, the event $A \cap (T \leq t)$ is in $\mathscr{F}_{T\wedge t}^0$ and therefore

$$\begin{aligned} Q\big(A \cap (T \leq t)\big) &= \int_{A\cap(T\leq t)} E\left[D_t \mid \mathscr{F}_{T\wedge t}^0\right] dP = \int_{A\cap(T\leq t)} D_{T\wedge t} dP \\ &= \int_{A\cap(T\leq t)} D_T dP. \end{aligned}$$

Letting t tend to infinity, we get

$$Q\big(A \cap (T < \infty)\big) = \int_{A\cap(T<\infty)} D_T dP$$

which completes the proof. □

The martingale D plays a prominent role in the discussion to follow. But, since the *parti-pris* of this book is to deal only with continuous semimartingales, we have not developed the techniques needed to deal with the discontinuous case; as a result, we are compelled to prove Girsanov's theorem only in the case of continuous densities. Before we turn to the proof, we must point out that if X and Y are two continuous semimartingales for both P and Q and if $Q \lhd P$, any version of the process $\langle X, Y\rangle$ computed for P is a version of the process $\langle X, Y\rangle$ computed for Q and this version may be chosen $\big(\mathscr{F}_t^0\big)$-adapted (the reader is referred to Remark (1.19) and Exercise (1.30) in Chap. IV). In the sequel, it will always be understood that $\langle X, Y\rangle$ is such a version even if we are dealing with Q.

The following theorem, which is the main result of this section, is Girsanov's theorem in our restricted framework; it specifies the decomposition of P-martingales as continuous semimartingales under Q.

(1.4) Theorem (Girsanov's theorem). *If $Q \lhd P$ and if D is continuous, every continuous $\big(\mathscr{F}_t^0, P\big)$-semimartingale is a continuous $\big(\mathscr{F}_t^0, Q\big)$-semimartingale. More precisely, if M is a continuous $\big(\mathscr{F}_t^0, P\big)$-local martingale, then*

$$\widetilde{M} = M - D^{-1} \cdot \langle M, D\rangle$$

is a continuous $\big(\mathscr{F}_t^0, Q\big)$-local martingale. Moreover, if N is another continuous P-local martingale

$$\langle \widetilde{M}, \widetilde{N}\rangle = \langle \widetilde{M}, N\rangle = \langle M, N\rangle.$$

Proof. If X is a cadlag process and if XD is a $\big(\mathscr{F}_t^0, P\big)$-loc. mart., then X is a $\big(\mathscr{F}_t^0, Q\big)$-loc. mart. Indeed, by Proposition (1.3), D_t^T is a density of Q with respect to P on $\mathscr{F}_{T\wedge t}^0$ and if $(XD)^T$ is a P-martingale, it is easily seen that X^T is a Q-martingale (see the first remark after Corollary (3.6) Chap. II). Moreover a sequence of $\big(\mathscr{F}_t^0\big)$-stopping times increasing to $+\infty$ P-a.s. increases also to $+\infty$ Q-a.s., as the reader will easily check.

Let $T_n = \inf\{t : D_t \leq 1/n\}$; it is easy to see that the process $\left(D^{-1} \cdot \langle M, D\rangle\right)^{T_n}$ is P-a.s. finite and consequently, $\left(\widetilde{M} D\right)^{T_n}$ is the product of two semimarts. By the integration by parts formula, it follows that

$$\begin{aligned}
\left(\widetilde{M} D\right)_t^{T_n} &= M_0 D_0 + \int_0^{T_n \wedge t} \widetilde{M}_s dD_s + \int_0^{T_n \wedge t} D_s d\widetilde{M}_s + \langle \widetilde{M}, D\rangle_{T_n \wedge t} \\
&= M_0 D_0 + \int_0^{T_n \wedge t} \widetilde{M}_s dD_s + \int_0^{T_n \wedge t} D_s dM_s - \langle M, D\rangle_{T_n \wedge t} \\
&\quad + \langle \widetilde{M}, D\rangle_{T_n \wedge t} \\
&= M_0 D_0 + \int_0^{T_n \wedge t} \widetilde{M}_s dD_s + \int_0^{T_n \wedge t} D_s dM_s,
\end{aligned}$$

which proves that $(\widetilde{M} D)^{T_n}$ is a P-loc. mart. By the first paragraph of the proof $\widetilde{M}^{T_n}$ is thus a Q-loc. mart.; but (T_n) increases Q-a.s. to $+\infty$ by Proposition (1.2) and a process which is locally a loc. mart. is a loc. mart. (Exercise (1.30) Chap. IV) which completes the proof of the second statement. The last statement follows readily from the fact that the bracket of a process of finite variation with any semimart. vanishes identically. □

Furthermore it is important to note that stochastic integration commutes with the transformation $M \to \widetilde{M}$. The hypothesis is the same as for Theorem (1.4) and here again the processes involved are $(\mathscr{F}_t^0)$-adapted.

(1.5) Proposition. *If H is a predictable process in $L^2_{\text{loc}}(M)$, then it is also in $L^2_{\text{loc}}(\widetilde{M})$ and $\widetilde{H \cdot M} = H \cdot \widetilde{M}$.*

Proof. The first assertion follows at once from the equality $\langle M, M\rangle = \langle \widetilde{M}, \widetilde{M}\rangle$.

Moreover, if H is locally bounded,

$$H \cdot \widetilde{M} = H \cdot M - H D^{-1} \cdot \langle M, D\rangle = H \cdot M - D^{-1} \cdot \langle H \cdot M, D\rangle = \widetilde{H \cdot M}.$$

We leave to the reader the task of checking that the expressions above are still meaningful for $H \in L^2_{\text{loc}}(M)$, hence that $\widetilde{H \cdot M} = H \cdot \widetilde{M}$ in that case too.

With a few exceptions (see for instance Exercise (1.22) in Chap. XI), Q is usually actually equivalent to P on each $\mathscr{F}_t^0$, in which case the above results take on an even more pleasant and useful form. In this situation, D is also strictly positive P-a.s. and may be represented in an exponential form as is shown in

(1.6) Proposition. *If D is a strictly positive continuous local martingale, there exists a unique continuous local martingale L such that*

$$D_t = \exp\left\{L_t - \frac{1}{2}\langle L, L\rangle_t\right\} = \mathscr{E}(L)_t;$$

L is given by the formula

$$L_t = \log D_0 + \int_0^t D_s^{-1} dD_s.$$

Proof. Once again, the uniqueness is a consequence of Proposition (1.2) in Chap. IV.

As to the existence of L, we can, since D is > 0 a.s., apply Itô's formula to the process $\log D$, with the result that

$$\begin{aligned}\log D_t &= \log D_0 + \int_0^t D_s^{-1} dD_s - \frac{1}{2}\int_0^t D_s^{-2} d\langle D, D\rangle_s \\ &= L_t - \frac{1}{2}\langle L, L\rangle_t.\end{aligned}$$

□

If P and Q are equivalent on each $\mathscr{F}_t^0$, we then have $Q = \mathscr{E}(L)_t \cdot P$ on $\mathscr{F}_t^0$ for every t, which we write simply as $Q = \mathscr{E}(L) \cdot P$. Let us restate Girsanov's theorem in this situation.

(1.7) Theorem. *If $Q = \mathscr{E}(L) \cdot P$ and M is a continuous P-local martingale, then*

$$\widetilde{M} = M - D^{-1} \cdot \langle M, D\rangle = M - \langle M, L\rangle$$

is a continuous Q-local martingale. Moreover, $P = \mathscr{E}(-\widetilde{L})Q$.

Proof. To prove the first statement, we need only show the identity $D^{-1} \cdot \langle M, D\rangle = \langle M, L\rangle$ which follows from the equality $L_t = \log D_0 + \int_0^t D_s^{-1} dD_s$.

For the second statement, observe that, because of the first, $-\widetilde{L} = -L + \langle L, L\rangle$ is a Q-local martingale with $\langle \widetilde{L}, \widetilde{L}\rangle = \langle L, L\rangle$. As a result

$$\mathscr{E}\left(-\widetilde{L}\right)_t = \exp\left\{-L_t + \langle L, L\rangle_t - \frac{1}{2}\langle L, L\rangle_t\right\} = \mathscr{E}(L)_t^{-1};$$

consequently, $P = \mathscr{E}(-\widetilde{L}) \cdot Q$. □

We now particularize the situation still further so that P and Q can play totally symmetric roles.

(1.8) Definition. *We call* Girsanov's pair *a pair (P, Q) of probability measures such that $Q \sim P$ on $\mathscr{F}_\infty^0$ and the P-martingale D is continuous.*

If (P, Q) is a Girsanov pair, then the completion of $\left(\mathscr{F}_t^0\right)$ for Q is the same as the completion for P and we may deal with the filtration $\left(\mathscr{F}_t\right)$ and forget about $\left(\mathscr{F}_t^0\right)$. Moreover, by the above results, we see that (Q, P) is also a Girsanov pair and that the class of continuous semimartingales is the same for P and Q.

(1.9) Definition. *If (P, Q) is a Girsanov pair, we denote the map $M \to \widetilde{M}$ from the space of continuous P-local martingales into the space of continuous Q-local martingales by G_P^Q and we call it the* Girsanov transformation *from P to Q.*

This map is actually one-to-one and onto as a consequence of the following proposition; it moreover shows that the relation $P \sim Q$ if (P, Q) is a Girsanov pair is an equivalence relation on the set of probability measures on $\mathscr{F}_\infty$.

(1.10) Proposition. *If (P, Q) and (Q, R) are two Girsanov pairs, then (P, R) is a Girsanov pair and $G_Q^R \circ G_P^Q = G_P^R$. In particular,*

$$G_Q^P \circ G_P^Q = G_P^Q \circ G_Q^P = I.$$

Proof. Since $\frac{dR}{dP}\big|_{\mathscr{F}_t} = \frac{dR}{dQ}\big|_{\mathscr{F}_t} \times \frac{dQ}{dP}\big|_{\mathscr{F}_t}$, the pair (P, R) is obviously a Girsanov pair.

By definition, $G_P^R(M)$ is the local martingale part of M under R, but as M and $G_P^Q(M)$ differ only by a finite variation process, $G_P^R(M)$ is also the local martingale part of $G_P^Q(M)$ under R which proves the result.

Remark. This can also be proved by using the explicit expressions of Theorem (1.7).

In this context, Propositon (1.5) may be restated as

(1.11) Proposition. *The Girsanov transformation G_P^Q commutes with stochastic integration.*

Before we turn to other considerations, let us sketch an alternative proof of Theorem (1.7) in the case of a Girsanov pair, based on Exercises (3.11) and (3.14) in Chap. IV. The point of this proof is that the process A which has to be subtracted from M to get a Q-local martingale appears naturally.

By Exercise (3.14), Chap. IV, it is enough to prove that for any λ, the process $\exp\left\{\lambda(M - A) - \frac{\lambda^2}{2}\langle M, M\rangle\right\}$ is a Q-local martingale, which amounts to showing that

$$\exp\left\{\lambda(M - A) - \frac{\lambda^2}{2}\langle M, M\rangle\right\}\exp\left\{L - \frac{1}{2}\langle L, L\rangle\right\}$$

is a P-martingale. But this product is equal to

$$\exp\left\{\lambda M + L - \frac{\lambda^2}{2}\langle M, M\rangle - \frac{1}{2}\langle L, L\rangle - \lambda A\right\}$$

which, by Exercise (3.11), Chap. IV, is equal to $\mathscr{E}(\lambda M + L)$ provided that $A = \langle M, L\rangle$.

Remark. We observe that if Q is equivalent to P on each $\mathscr{F}_t$ and D is continuous, (P, Q) becomes a Girsanov pair if we restrict the time interval to $[0, t]$ and the above results apply. However, one should pay attention to the fact that the filtrations to consider may not be complete with respect to P (or Q); when working on $[0, t]$, we complete with the negligible sets in $\mathscr{F}_t$ in order to recover the above set-up.

To illustrate the above results, we treat the case of BM which will be taken up more thoroughly in the following section.

(1.12) Theorem. *If $Q \lhd P$ and if B is a $(\mathscr{F}_t^0, P)$-BM, then $\widetilde{B} = B - \langle B, L\rangle$ is a $(\mathscr{F}_t^0, Q)$-BM.*

Proof. The increasing process of $\widetilde{B}$ is equal to that of B, namely t, so that P. Lévy's characterization theorem (Sect. 3, Chap. IV) applies.

Remarks. 1°) The same proof is valid for BM^d.

2°) If $(\mathscr{F}_t)$ is the Brownian filtration, then for any Q equivalent to P on $\mathscr{F}_\infty$, the pair (P, Q) is a Girsanov pair as results from Sect. 3 in Chap. V. Even in that case the filtration of $\widetilde{B}$ may be strictly smaller than the filtration of B as will be seen in Exercise (3.15) in Chap. IX.

3°) The same result is true for any Gaussian martingale as follows from Exercise (1.35) in Chap. IV.

In this and the following chapters, we will give many applications of Girsanov's theorem, some of them under the heading "Cameron-Martin formula" (see the notes and comments at the end of this chapter). In the usual setting D, or rather L, is given and one constructs the corresponding Q by setting $Q = D_t \cdot P$ on $\mathscr{F}_t$. This demands that L being given, the exponential $\mathscr{E}(L)$ be a "true" P-martingale (which is equivalent to $E\left[\mathscr{E}(L)_t\right] \equiv 1$) and not merely a local martingale (otherwise, see Exercise (1.38)). Sufficient criterions ensuring this property will be given below.

When $E\left[\mathscr{E}(L)_t\right] \equiv 1$ obtains, the formula $Q = \mathscr{E}(L)\cdot P$ defines a set function Q on the algebra $\bigcup_t \mathscr{F}_t$ which has to be extended to a probability measure on $\mathscr{F}_\infty$. To this end, we need Q to be σ-additive on $\bigcup_t \mathscr{F}_t$.

(1.13) Proposition. *Let X be the canonical process on $\Omega = \mathbf{W} = C(\mathbb{R}_+, \mathbb{R}^d)$ and $\mathscr{F}_t^0 = \sigma(X_s, s \leq t)_+$; if $E\left[\mathscr{E}(L)_t\right] \equiv 1$, there is a unique probability measure Q on $(\Omega, \mathscr{F}_\infty^0)$ such that $Q = \mathscr{E}(L)\cdot P$.*

This is a consequence of Theorem (6.1) in the Appendix.

The above discussion stresses the desirability of a criterion ensuring that $\mathscr{E}(L)$ is a martingale. Of course, if L is bounded or if $E[\exp(L_\infty^*)] < \infty$ this property obtains easily, but we will close this section with two criterions which can be more widely applied. Once again, we recall that $\mathscr{E}(L)$ is a supermartingale and is a martingale if $E\left[\mathscr{E}(L)_t\right] \equiv 1$ as L is always assumed to vanish at 0. The first of these two criterions is known as *Kazamaki's criterion*. We return to the setting of a general filtered space $(\Omega, \mathscr{F}, \mathscr{F}_t, P)$.

(1.14) Proposition. *If L is a local martingale such that $\exp\left(\frac{1}{2}L\right)$ is a uniformly integrable submartingale, then $\mathscr{E}(L)$ is a uniformly integrable martingale.*

Proof. Pick a in $]0, 1[$. Straightforward computations yield

$$\mathscr{E}(aL)_t = \left(\mathscr{E}(L)_t\right)^{a^2}\left(Z_t^{(a)}\right)^{1-a^2}$$

with $Z_t^{(a)} = \exp(aL_t/(1+a))$. By the optional stopping theorem for uniformly integrable submartingales (Sect. 3 Chap. II), it follows from the hypothesis that the family $\left\{Z_T^{(a)},\ T \text{ stopping time}\right\}$ is uniformly integrable. If Γ is a set in $\mathscr{F}$ and T a stopping time, Hölder's inequality yields

$$E\left[1_\Gamma \mathscr{E}(aL)_T\right] \leq E\left[\mathscr{E}(L)_T\right]^{a^2} E\left[1_\Gamma Z_T^{(a)}\right]^{1-a^2} \leq E\left[1_\Gamma Z_T^{(a)}\right]^{1-a^2}$$

since $E\left[\mathscr{E}(L)_T\right] \leq 1$. It follows that $\left\{\mathscr{E}(aL)_T,\ T \text{ stopping time}\right\}$ is uniformly integrable, hence $\mathscr{E}(aL)$ is a uniformly integrable martingale. As a result

$$1 = E\left[\mathscr{E}(aL)_\infty\right] \leq E\left[\mathscr{E}(L)_\infty\right]^{a^2} E\left[Z_\infty^{(a)}\right]^{1-a^2}.$$

The hypothesis implies also that L_∞ exists a.s. and that $\exp\left(\frac{1}{2}L_\infty\right)$ is integrable. Since $Z_\infty^{(a)} \leq 1_{(L_\infty \leq 0)} + \exp\left(\frac{1}{2}L_\infty\right) 1_{(L_\infty > 0)}$, Lebesgue's dominated convergence theorem shows that $\lim_{a\uparrow 1} E\left[Z_\infty^{(a)}\right]^{1-a^2} = 1$. By letting a tend to 1 in the last displayed inequality, we get $E\left[\mathscr{E}(L)_\infty\right] \geq 1$, hence $E\left[\mathscr{E}(L)_\infty\right] = 1$ and the proof is complete. □

Remarks. If L is a u.i. martingale and $E\left[\exp\left(\frac{1}{2}L_\infty\right)\right] < \infty$, then $\exp\left(\frac{1}{2}L_t\right)$ is a u.i. submartingale and Kazamaki's criterion applies; this will be used in the next proof. Let us also observe that if M is a local martingale, the condition $E[\exp M_t] < \infty$ for every t does not entail that $\exp(M)$ is a submartingale; indeed, if Z is the planar BM and $0 < \alpha < 2$ then $\exp(-\alpha \log|Z_t|) = |Z_t|^{-\alpha}$ has an expectation decreasing in t and in fact is a supermartingale (see also Exercise (1.24) in Chap. XI).

From the above proposition, we may derive *Novikov's criterion*, often easier to apply, but of narrower scope as is shown in Exercises (1.30) and (1.34).

(1.15) Proposition. *If L is a continuous local martingale such that*

$$E\left[\exp\left(\frac{1}{2}\langle L, L\rangle_\infty\right)\right] < \infty,$$

then $\mathscr{E}(L)$ is a u.i. martingale. Furthermore $E\left[\exp\left(\frac{1}{2}L_\infty^*\right)\right] < \infty$, and as a result, L is in H^p for every $p \in [1, \infty[$.

Proof. The hypothesis entails that $\langle L, L\rangle_\infty$ has moments of all orders, therefore, by the BDG-inequalities, so has L_∞^*; in particular L is a u.i. martingale. Moreover

$$\exp\left(\frac{1}{2}L_\infty\right) = \mathscr{E}(L)_\infty^{1/2} \exp\left(\frac{1}{4}\langle L, L\rangle_\infty\right),$$

so that, by Cauchy-Schwarz inequality,

$$E\left[\exp\left(\frac{1}{2}L_\infty\right)\right] \leq E\left[\mathscr{E}(L)_\infty\right]^{1/2} E\left[\exp\left(\frac{1}{2}\langle L, L\rangle_\infty\right)\right]^{1/2}$$

and since $E\left[\mathscr{E}(L)_\infty\right] \leq 1$, it follows that $E\left[\exp\left(\frac{1}{2}L_\infty\right)\right] < \infty$. By the above remarks, $\mathscr{E}(L)$ is a u.i. martingale.

To prove that $E\left[\exp\left(\frac{1}{2}L^*_\infty\right)\right] < \infty$, let us observe that we now know that for $c < 1/2$,

$$\exp(cL_t) = \mathscr{E}^c(L)_t \exp\left(\frac{c^2}{2}\langle L, L\rangle_t\right)$$

is a positive submartingale. Applying Doob's inequality with $p = 1/2c$, yields

$$E\left[\sup_t \exp\left(\frac{1}{2}L_t\right)\right] \leq C_p E\left[\mathscr{E}^c(L)_\infty^{1/2} \exp\left(\frac{c}{4}\langle L, L\rangle_\infty\right)\right].$$

Then, by Hölder's inequality, we get

$$E\left[\sup_t \exp\left(\frac{1}{2}L_t\right)\right] \leq C_p E\left[\mathscr{E}^c(L)_\infty^{(2-c)/4c}\right]^{2/(2-c)} E\left[\exp\left(\frac{1}{2}\langle L, L\rangle_\infty\right)\right]^{c/2},$$

and since $(2-c)/4c < 1$ for $c > 2/5$, the left-hand side is finite. The same reasoning applies to $-L$, thus the proof is complete.

Remark. It is shown in Exercise (1.31) that $1/2$ cannot be replaced by $(1/2) - \delta$, with $\delta > 0$, in the above hypothesis.

The above results may be stated on the intervals $[0, t]$; by using an increasing homeomorphism from $[0, \infty]$ to $[0, t]$, we get

(1.16) Corollary. *If L is a local martingale such that either $\exp\left(\frac{1}{2}L\right)$ is a submartingale or $E\left[\exp\left(\frac{1}{2}\langle L, L\rangle_t\right)\right] < \infty$ for every t, then $\mathscr{E}(L)$ is a martingale.*

Let us close this section with a few comments about the above two criterions. Their main difference is that in Kazamaki's criterion, one has to assume that $\exp\left(\frac{1}{2}L\right)$ is a submartingale, whereas it is part of the conclusion in Novikov's criterion. Another difference is that Novikov's criterion is "two-sided"; it works for L if and only if it works for $-L$, whereas Kazamaki's criterion may work for L without working for $-L$. Finally we stress the fact that Kazamaki's criterion is not a necessary condition for $\mathscr{E}(L)$ to be a martingale as will be seen from examples given in Exercise (2.10) of Chap. IX.

\# **(1.17) Exercise.** Suppose that $Q = D_t \cdot P$ on $\mathscr{F}_t$ for a positive continuous martingale D. Prove the following improvement of Proposition (1.2): the r.v. $Y = \inf_t D_t$ is > 0 Q-a.s. If moreover P and Q are mutually singular on $\mathscr{F}_\infty$ then, under Q, Y is uniformly distributed on $[0, 1]$.

(1.18) Exercise. Let P_i, $i = 1, 2, 3$ be three probability measures such that any two of them form a Girsanov pair and call $D^i_j = \mathscr{E}\left(L^i_j\right)$ the martingale such that $P_i = D^i_j \cdot P_j$. There is a P_1-martingale M such that $D^3_2 = \mathscr{E}\left(G^{P_2}_{P_1}(M)\right)$. Prove that $L^3_1 = M + L^2_1$.

(1.19) Exercise. Call $\mathcal{M}(P)$ the space of cont. loc. mart. with respect to P. Let (P, Q) be a Girsanov pair and Γ a map from $\mathcal{M}(P)$ into $\mathcal{M}(Q)$.

1°) If $\langle \Gamma(M), N \rangle = \langle M, N \rangle$ for every $M \in \mathcal{M}(P)$ and $N \in \mathcal{M}(Q)$ prove that $\Gamma = G_P^Q$.

2°) If $\langle \Gamma(M), \Gamma(N) \rangle = \langle M, N \rangle$ for every $M, N \in \mathcal{M}(P)$ there exists a map J from $\mathcal{M}(P)$ into itself such that $\langle J(M), J(N) \rangle = \langle M, N \rangle$ and $\Gamma = G_P^Q \circ J$.

(1.20) Exercise. If (P, Q) is a Girsanov pair with density D and if M is a P-martingale, then MD^{-1} is a Q-martingale. Express it as a Girsanov transform. In relation with the above exercise, observe that the map $M \to MD^{-1}$ does not leave brackets invariant and does not commute with stochastic integration.

\# **(1.21) Exercise.** 1°) Let B be the standard linear BM and for $a > 0$ and $b > 0$ set $\sigma_{a,b} = \inf\{t : B_t + bt = a\}$. Use Girsanov's theorem to prove that the density of $\sigma_{a,b}$ is equal to $a(2\pi t^3)^{-1/2} \exp\left(-(a - bt)^2/2t\right)$. This was already found by other means in Exercise (3.28) of Chap. III; compare the two proofs.

2°) Prove Novikov's criterion directly from the DDS theorem and the above result.

(1.22) Exercise. If for some $\varepsilon > 0$, $E\left[\exp\left(\left(\frac{1}{2} + \varepsilon\right) \langle M, M \rangle_t\right)\right] < \infty$ for every t, prove, using only Hölder's inequality and elementary computations, that $\mathcal{E}(M)$ is a martingale.

(1.23) Exercise. Let B be the standard linear BM. For any stopping time T such that $E\left[\exp\left(\frac{1}{2}T\right)\right] < \infty$, prove that

$$E\left[\exp\left(B_T - \frac{1}{2}T\right)\right] = 1.$$

(1.24) Exercise. 1°) Let B be the standard linear BM and prove that

$$T = \inf\left\{t : B_t^2 = 1 - t\right\}$$

is a stopping time such that $P[0 < T < 1] = 1$.

2°) Set $H_s = -2B_s \cdot 1_{(T \geq s)}/(1 - s)^2$ and prove that for every t,

$$\int_0^t H_s^2 ds < \infty \quad \text{a.s.}$$

3°) If $M_t = \int_0^t H_s dB_s$, compute $M_t - \frac{1}{2}\langle M, M \rangle_t + (1 - t \wedge T)^{-2} B_{T \wedge t}$.

4°) Prove that $E\left[\mathcal{E}(M)_1\right] < 1$ and hence that $\mathcal{E}(M)_t$, $t \in [0, 1]$, is not a martingale.

\# **(1.25) Exercise.** Let $(\mathscr{F}_t)$ be a filtration such that every $(\mathscr{F}_t)$-martingale is continuous (see Sect. 3 Chap. V). If H_n is a sequence of predictable processes converging a.s. to a process H and such that $|H_n| \leq K$ where K is a locally bounded predictable process, prove that, for every t and for every continuous $(\mathscr{F}_t)$-semimartingale X,

$$P\text{-}\lim_{n\to\infty}\int_0^t H_n dX = \int_0^t H\, dX$$

[Hint: Use the probability measure $Q = P(\cdot\cap\Gamma)/P(\Gamma)$ where Γ is a suitable set on which the processes H_n are uniformly bounded.]

(1.26) Exercise. (Continuation of Exercise (5.15) of Chap. IV). Prove that N has the $(\mathscr{F}_t^Y)$-PRP. As a result every $(\mathscr{F}_t^Y)$-local martingale is continuous.

[Hint: Start with a bounded h and change the law in order that Y become a BM.]

* **(1.27) Exercise.** 1°) Let (P, Q) be a Girsanov pair relative to a filtration $(\mathscr{F}_t)$. Prove that if M is a P-cont. loc. mart. which has the $(\mathscr{F}_t)$-PRP, then $G_P^Q(M)$ has also the $(\mathscr{F}_t)$-PRP. It is shown in Exercise (3.12) Chap. IX that this does not extend to the purity property.

2°) Let $B = (B^1, B^2)$ be a BM^2 and set

$$X_t = B_t^1 + \int_0^t B_s^2 ds.$$

Prove that B^1 is not adapted to $(\mathscr{F}_t^X)$.

[Hint: Use $Q = \mathscr{E}\left(-\int_0^\cdot B_s^2 dB_s^1\right)\cdot P$ to prove that there is a BM^1 which has the $(\mathscr{F}_t^X)$-PRP.]

(1.28) Exercise. In the notation of this section, assume that $Q \ll P$ on $\mathscr{F}_\infty$ and that D is continuous. If dQ/dP is in $L^2(P)$ prove that (in the notation of Exercise (4.13) of Chap. IV) any semimartingale of $\mathscr{S}_2(P)$ belongs to $\mathscr{S}_1(Q)$.

(1.29) Exercise. Prove that if $\langle M, M\rangle_\infty = \infty$, then $\mathscr{E}(M)_t$ converges a.s. to 0 as t tends to $+\infty$, hence cannot be uniformly integrable.

(1.30) Exercise. Let B be the standard linear BM, $T_1 = \inf\{t : B_t = 1\}$, and set

$$\tau_t = \frac{t}{1-t}\wedge T_1 \quad \text{if } t < 1, \qquad \tau_t = T_1 \quad \text{if } t \geq 1.$$

Prove that $M_t = B_{\tau_t}$ is a continuous martingale for which Kazamaki's criterion applies and Novikov's does not.

[Hint: Prove that $\mathscr{E}(-M)$ is not a martingale and observe that Novikov's criterion applies to M if and only if it applies to $-M$.]

* **(1.31) Exercise.** Retain the situation and notation of Exercises (3.14) in Chap. II and (3.28) in Chap. III (see also Exercise (1.21) above).

1°) Prove that

$$E\left[\exp\left(B_{\sigma_{1,b}} - \frac{1}{2}\sigma_{1,b}\right)\right] < 1.$$

2°) Derive therefrom that, for any $\delta > 0$, there exists a continuous martingale M such that $E\left[\exp\left(\frac{1}{2}-\delta\right)\langle M, M\rangle_\infty\right] < +\infty$ and $\mathscr{E}(M)$ is not a uniformly integrable martingale.

* **(1.32) Exercise.** Let $M \in \mathrm{BMO}$ and $M_0 = 0$; using Exercise (1.40) in Chap. IV prove that for any stopping time T,

$$E\left[\mathscr{E}(M)_\infty \mathscr{E}(M)_T^{-1} \mid \mathscr{F}_T\right] \geq \exp\left(-\frac{1}{2}\|M\|^2_{\mathrm{BMO}_2}\right).$$

Prove that consequently $\mathscr{E}(M)$ is a uniformly integrable martingale.

* **(1.33) Exercise.** For a continuous local martingale M vanishing at 0 and a real number α, we set

$$G_t^\alpha = \exp\left\{\alpha M_t + \left(\frac{1}{2} - \alpha\right)\langle M, M\rangle_t\right\},$$

$$g(\alpha) = \sup\left\{E\left[G_T^\alpha\right];\ T \text{ stopping time}\right\}.$$

1°) For $\alpha \leq \beta < 1$, prove that $g(\beta) \leq g(\alpha)^{(1-\beta)/(1-\alpha)}$, and that for $1 < \alpha \leq \beta$, $g(\alpha) \leq g(\beta)^{(\alpha-1)/(\beta-1)}$.

2°) If $a \neq 0$ and $T_t = \inf\{s : \langle M, M\rangle_s > t\}$, then $\mathscr{E}(aM)$ is a uniformly integrable martingale if and only if

$$\lim_{t\to\infty} E\left[\mathscr{E}(aM)_{T_t} 1_{(T_t<\infty)}\right] = 0.$$

3°) If $\alpha \neq 1$ and $g(\alpha) < \infty$, then $\mathscr{E}(\alpha M)$ and $\mathscr{E}(M)$ are uniformly integrable martingales.

4°) If $M \in \mathrm{BMO}$, then $g(\alpha) < \infty$ for some $\alpha \neq 1$.

(1.34) Exercise. Let ρ_t be the modulus of the planar BM started at $a \neq 0$. Prove that $L_t = \log(\rho_t/|a|)$ is a local martingale for which Kazamaki's criterion applies (on the interval $[0, t]$) and Novikov's criterion does not.

* **(1.35) Exercise.** Assume that the filtration $(\mathscr{F}_t)$ is such that all $(\mathscr{F}_t)$-martingales are continuous. Let X and Y be two continuous semimartingales. Prove that their martingale parts are equal on any set $\Gamma \in \mathscr{F}_\infty$ on which $X - Y$ is of finite variation. (This is actually true in full generality but cannot be proved with the methods of this book).

[Hint: Use $Q = P(\cdot \cap \Gamma)/P(\Gamma)$.]

* **(1.36) Exercise.** Let P be the Wiener measure on $\Omega = C([0,1], \mathbb{R})$, $\mathscr{F}_t = \sigma(\omega(s), s \leq t)$ and b be a bounded predictable process. We set

$$Q = \exp\left\{\int_0^1 b(s,\omega)d\omega(s) - \frac{1}{2}\int_0^1 b^2(s,\omega)ds\right\} \cdot P$$

and $\theta(\omega)_t = \omega(t) - \int_0^t b(s,\omega)ds$.

Prove that if $(M_t, t \leq 1)$ is a $(\mathscr{F}_t, P)$-martingale then $(M_t \circ \theta, t \leq 1)$ is a $(\mathscr{F}_t, Q)$-martingale. For instance if h is a function of class $C^{2,1}$ such that $\frac{1}{2}\frac{\partial^2 h}{\partial x^2} + \frac{\partial h}{\partial t} = 0$ then $h(\theta(\omega)_t, t)$ is a $(\mathscr{F}_t, Q)$-martingale.

[Hint: If $\hat{\mathscr{F}}_t = \sigma(\theta(\omega)_s, s \leq t)$, one can use the representation theorem to prove that every $(\hat{\mathscr{F}}_t, Q)$-martingale is a $(\mathscr{F}_t, Q)$-martingale.]

* **(1.37) Exercise.** Let $(\Omega, \mathscr{F}_t, P)$ be a filtered space such that every $(\mathscr{F}_t, P)$-martingale is continuous. Let X be a $(\mathscr{F}_t)$-adapted continuous process and π the set of probability measures Q such that

i) $Q|_{\mathscr{F}_t} \sim P|_{\mathscr{F}_t}$ for every t,
ii) X is a $(\mathscr{F}_t, Q)$-local martingale.

1°) Show that if π is non-empty, then X is a P-semimartingale with canonical decomposition $X = M + A$ such that $dA \ll d\langle M, M\rangle$ a.s.

2°) Conversely if under P the condition of 1°) is satisfied we call h a good version of $dA/d\langle M, M\rangle$ (see Sect. 3 Chap. V). Assume that one can define a probability measure $\widetilde{Q}$ on $\mathscr{F}_\infty$ by setting

$$\widetilde{Q}|_{\mathscr{F}_t} = \mathscr{E}\left(\int_0^{\cdot} h_s dM_s\right)_t \cdot P|_{\mathscr{F}_t};$$

prove then that $\widetilde{Q}$ is in π.

3°) If the conditions in 2°) are satisfied, describe the set π. Prove that $\widetilde{Q}$ is the only element of π if and only if M has the PRP under P.

* **(1.38) Exercise.** Let P and Q be two probability measures on a space with a filtration $(\mathscr{F}_t)$ which is right-continuous and complete for both P and Q. Let D_t be the martingale density of Q with respect to $2^{-1}(P+Q)$. Remark that $0 \le D_t \le 2$, $P+Q$-a.s. and set $T = \inf\{t : D_t = 2\}$.

1°) Prove that $P[T = \infty] = 1$ and that the Lebesgue decomposition of Q with respect to P on $\mathscr{F}_t$ may be written

$$Q(B) = \int_B Z_t dP + Q(B \cap (T \le t))$$

where $Z_t = D_t/(2-D_t)$ on $\{t < T\}$, $Z_t = 0$ on $\{t \ge T\}$. Prove that Z is a positive P-supermartingale and if (T', Z') is another pair with the same properties, then $T' = T$ Q-a.s. and $Z' = Z$ up to P-equivalence.

2°) Assume that D is continuous and prove Girsanov theorem for P and Q and the filtration $(\mathscr{F}_{t\wedge T})$.

(1.39) Exercise. Let H be a predictable process with respect to the Brownian filtration such that $0 < c \le H \le C < \infty$ for two constants c and C. For any f in $L^2_{\text{loc}}(\mathbb{R}_+, ds)$, prove that

$$\exp\left((c^2/2)\int_0^t f(s)^2 ds\right) \le E\left[\exp\left(\int_0^t f(s)H_s dB_s\right)\right] \le \exp\left((C^2/2)\int_0^t f(s)^2 ds\right).$$

[Hint: Apply Novikov's criterion.]

* **(1.40) Exercise (Another Novikov's type criterion).** Let B be a $(\mathscr{F}_t)$-BM and H an adapted process such that

$$E\left[\exp\left(aH_s^2\right)\right] \leq c,$$

for every $s \leq t$ and two constants a and $c > 0$.

Prove that for $r \leq s \leq t$ and $s - r$ sufficiently small,

$$E\left[\left(\mathscr{E}(H\cdot B)_s/\mathscr{E}(H\cdot B)_r\right) \mid \mathscr{F}_r\right] = 1,$$

and conclude that $E\left[\mathscr{E}(H\cdot B)_t\right] = 1$. This applies in particular if H is a Gaussian process.

[Hint: Use a truncation of H and pass to the limit using Novikov's criterion and Jensen's inequality.]

(1.41) Exercise (A converse to Theorem (1.12)). Let P be a probability measure on **W** such that, in the notation of this section, the law of $\tilde{X}$ under Q is the same as the law of X under P, for any Q equivalent to P. Prove that P is the law of a Gaussian martingale.

[Hint: For any positive functional F,

$$E_P[F(\langle X, X\rangle)(dQ/dP)] = E_P[F(\langle X, X\rangle)]\ .]$$

§2. Application of Girsanov's Theorem to the Study of Wiener Space

This section is a collection of results on Wiener space which may seem to be loosely related to one another but are actually linked by the use of Girsanov's transformation and the ubiquitous part played by the Cameron-Martin space (reproducing kernel Hilbert space of Exercise (3.12) in Chap. I) and the so-called *action functional* of BM.

We restrict the time interval to $[0, T]$ for a positive real T and we will consider the Wiener space **W** of $\mathbb{R}^d$-valued continuous functions on $[0, T]$ vanishing at 0. We endow **W** with the topology of uniform convergence; the corresponding norm is denoted by $\|\ \|_\infty$ or simply $\|\ \|$. The Wiener measure will be denoted by W and the coordinate mappings by β_t, $0 \leq t \leq T$. As **W** is a vector space, we can perform translations in **W**; for $h \in \mathbf{W}$, we call τ_h the map defined by

$$\beta_t(\tau_h(\omega)) = \beta_t(\omega) + h(t).$$

Let W_h be the image of W under τ_h. By definition, for any finite set of reals $t_i \leq T$ and Borel sets $A_i \subset \mathbb{R}^d$, we have

$$W_h\left(\bigcap_i \left\{\omega : \beta_{t_i}(\omega) \in A_i\right\}\right) = W\left(\bigcap_i \left\{\omega : \beta_{t_i}(\omega) + h(t_i) \in A_i\right\}\right);$$

thus a probability measure Q is equal to W_h if and only if, under Q, $\beta = B + h$ where B is a standard BM^d.

We are going to investigate the conditions under which W_h is equivalent to W. For this purpose, we need the following

(2.1) Definition. *The space H of functions h defined on $[0, T]$ with values in $\mathbb{R}^d$, such that each component h_i is absolutely continuous, $h_i(0) = 0$, and*

$$\int_0^T |h'(s)|^2 ds = \sum_{i=1}^d \int_0^T h_i'(s)^2 ds < +\infty$$

is called the Cameron-Martin *space.*

For $d = 1$, this space was introduced in Exercise (1.12) of Chap. I (see also Exercise (3.12) in the same chapter). The space H is a Hilbert space for the scalar product

$$\langle h, g \rangle = \sum_i \int_0^T h_i'(s) g_i'(s) ds,$$

and we will denote by $\|h\|_H$ the Hilbert norm $\langle h, h \rangle^{1/2}$. It is easy to prove, by taking linear approximations, that H is dense in $\mathbf{W}$ for the topology of uniform convergence.

If h is in H, the martingale

$$\left(h' \cdot \beta\right)_t = \sum_i \int_0^t h_i'(s) d\beta_s^i$$

where the stochastic integrals are taken under W, satisfies Novikov's criterion (1.15) and therefore $\mathscr{E}(h' \cdot \beta)$ is a martingale. This can also be derived directly from the fact that $h' \cdot \beta$ is a Gaussian martingale, and has actually been widely used in Chap. V.

(2.2) Theorem. *The probability measure W_h is equivalent to W if and only if $h \in H$ and then $W_h = \mathscr{E}(h' \cdot \beta)_T \cdot W$.*

This may also be stated: W is *quasi-invariant* with respect to H and the Radon-Nikodym derivative is equal to $\mathscr{E}(h' \cdot \beta)_T$.

Proof. If W_h is equivalent to W, then by the results in Sect. 3 Chap. V and the last section, there is a continuous martingale M such that $W_h = \mathscr{E}(M)_T \cdot W$; by Girsanov's theorem, we have $\beta^i = \widetilde{B}^i + \langle \beta^i, M \rangle$ where $\widetilde{B}$ is a W_h-Brownian motion. But by definition of W_h, we also have $\beta = B + h$ where B is a W_h-BM. Under W_h, the function h is therefore a deterministic semimartingale; by Exercise (1.38) (see also Exercise (2.19)) Chap. IV, its variation is bounded. Thus, under W_h, we have two decompositions of the semimartingale β in its own filtration; by the uniqueness of such decompositions, it follows that $h_i(t) = \langle \beta^i, M \rangle_t$ a.s.

Now by the results in Sect. 3 of Chap. V, we know that there is a predictable process ϕ such that $\int_0^T |\phi_s|^2 ds < \infty$ W-a.s. and $M_t = \sum_i \int_0^t \phi_s^i d\beta_s^i$. Consequently, we get $h_i(t) = \int_0^t \phi_s^i ds$. This proves on the one hand that $h \in H$, on the other hand that ϕ^i can be taken equal to the deterministic function h_i' whence the last equality in the statement follows.

Conversely, if $h \in H$, the measure $Q = \mathscr{E}(h' \cdot \beta)_T \cdot W$ is equivalent to W and under Q we have

$$\beta^i = B^i + \int_0^{\cdot} h_i'(s)ds = B^i + h_i$$

where B is a BM, which shows that $Q = W_h$. □

Remark. Whatever h may be, the process β is, under W_h, a Gaussian process and therefore, by a general result, W_h and W are either equivalent or mutually singular. Thus if $h \notin H$ then W and W_h are mutually singular.

Let us now recall (see Sect. 1 Chap. XIII) that the σ-algebra $\mathscr{F}_T$ generated by β is the Borel σ-algebra of $\mathbf{W}$ and that the support of a measure is the smallest closed subset which carries this measure.

(2.3) Corollary. *The support of the measure W is the whole space* $\mathbf{W}$.

Proof. The support contains at least one point; by the previous results, it contains all the translates of that point by elements of H. One concludes from the density of H in $\mathbf{W}$. □

We next supplement the representation Theorem (3.4) of Chap. V, which we just used in the proof of Theorem (2.2). Let $(\mathscr{F}_t)$ be the W-complete filtration generated by β; if X is in $L^2(\mathscr{F}_T, W)$, then

$$X = E[X] + \int_0^T \phi_s d\beta_s$$

for a $\phi \in L^2_{\mathscr{P}}(\beta^T)$. On the other hand, X is a.s. equal to a function, say F, of the path $\omega \in \mathbf{W}$ such that $E\left[F(\omega)^2\right] < \infty$. Our aim is to compute ϕ as a function of F.

If $\psi \in \mathbf{W}$, $\omega + \psi$ is also in $\mathbf{W}$ and $F(\omega + \psi)$ makes perfect sense. We will also sometimes write $F(\beta)$ instead of $F(\omega)$. We will henceforth assume that F enjoys the following properties:

i) there exists a constant K such that

$$|F(\beta + \psi) - F(\beta)| \leq K \|\psi\|$$

for every $\psi \in \mathbf{W}$;

ii) there exists a kernel F' from Ω to $[0, T]$ such that for every $\psi \in H$

$$\lim_{\varepsilon \to 0} \varepsilon^{-1}\big(F(\beta + \varepsilon\psi) - F(\beta)\big) = \int_0^T F'(\beta, dt)\psi(t), \qquad \text{for a.e. } \beta.$$

It is worth recording that condition i) implies the integrability condition $E[F(\beta)^2] < \infty$. Moreover, if F is a differentiable function with bounded derivative on the Banach space $\mathbf{W}$, then, by the mean value theorem, condition i) is in force and since a continuous linear form on $\mathbf{W}$ is a bounded measure on $[0, T]$, ii) is also satisfied.

Under conditions i) and ii), we have *Clark's formula*:

(2.4) Theorem. *The process ϕ is the predictable projection of $F'(\beta,]t, T])$.*

Proof. Let u be a bounded $(\mathscr{F}_t)$-predictable process and set

$$\psi_t = \int_0^t u_s ds, \qquad \widetilde{\psi}_t = \int_0^t u_s d\beta_s;$$

for each ω, the map $t \to \psi_t(\omega)$ belongs to $\mathbf{W}$.

For $\varepsilon > 0$, $\mathscr{E}(\varepsilon\widetilde{\psi})_t$, $t \leq T$, is, by Novikov's criterion, a uniformly integrable martingale and if $Q = \mathscr{E}(\varepsilon\widetilde{\psi})_T \cdot W$, then, under Q, the process $\beta - \varepsilon\psi$ is a BM. As a result $E[F(\beta)] = \int_{\mathbf{W}} F(\beta - \varepsilon\psi) dQ$, where, we recall, E is the expectation with respect to W and consequently

$$E[F(\beta)] = E\left[F(\beta - \varepsilon\psi)\mathscr{E}(\varepsilon\widetilde{\psi})_T\right].$$

This may be rewritten

$$E\left[\left(F(\beta - \varepsilon\psi) - F(\beta)\right)\left(\mathscr{E}(\varepsilon\widetilde{\psi})_T - 1\right)\right] + E[F(\beta - \varepsilon\psi) - F(\beta)]$$
$$+E\left[F(\beta)\left(\mathscr{E}(\varepsilon\widetilde{\psi})_T - 1\right)\right] = 0.$$

Let us divide by ε and let ε tend to zero. By condition i), the modulus of the first part is majorized by $KT\|u\| E\left[\left|\mathscr{E}(\varepsilon\widetilde{\psi})_T - 1\right|\right]$ and this converges to zero as ε goes to zero; indeed, $\mathscr{E}(\varepsilon\widetilde{\psi})_T$ converges to 1 pointwise and $\mathscr{E}(\varepsilon\widetilde{\psi})_T \leq \exp\left(\varepsilon\left|\widetilde{\psi}_T\right|\right)$. Moreover we will show at the end of the proof that, for α sufficiently small, $E\left[\exp\left(\alpha\widetilde{\psi}_T^*\right)\right] < \infty$. Thus, the necessary domination condition obtains and one may pass to the limit in the expectation. In the second term, conditions i) and ii) allow us to use the dominated convergence theorem. Finally

$$\varepsilon^{-1}\left(\mathscr{E}(\varepsilon\widetilde{\psi})_T - 1\right) = \int_0^T \mathscr{E}(\varepsilon\widetilde{\psi})_s u_s d\beta_s$$

and $\mathscr{E}(\varepsilon\widetilde{\psi})u$ converges to u in $L^2(dW \otimes ds)$. Indeed, by Doob's inequality,

$$E\left[\sup_{s\leq T}\left(\left(\mathscr{E}(\varepsilon\widetilde{\psi})_s - 1\right)^2 u_s^2\right)\right] \leq 4\|u\|^2 E\left[\left(\mathscr{E}(\varepsilon\widetilde{\psi})_T - 1\right)^2\right]$$

and this last term converges to zero as ε tends to zero because the integrand is dominated by $\left(\exp\left(\alpha\widetilde{\psi}_T^*\right) + 1\right)^2$ for $\varepsilon \leq \alpha$. By the L^2-isometry property of stochastic integrals, it follows that $\varepsilon^{-1}\left(\mathscr{E}(\varepsilon\widetilde{\psi})_T - 1\right)$ converges in $L^2(W)$ to $\int_0^T u_s d\beta_s$.

The upshot of this convergence result is that

$$E\left[F(\beta)\int_0^T u_s d\beta_s\right] = E\left[\int_0^T F'(\beta, dt)\psi_t\right],$$

hence, using integration by parts in the right-hand side,

$$E\left[\int_0^T \phi_s u_s ds\right] = E\left[\int_0^T \phi_s d\beta_s \int_0^T u_s d\beta_s\right] = E\left[\int_0^T u_s F'(\beta,]s, T])ds\right].$$

The result now follows from the definition of a predictable projection.

It remains to prove the integrability property we have used twice above. But, by the DDS Theorem, there is a BM γ such that $\widetilde{\psi} = \gamma_{\langle\widetilde{\psi},\widetilde{\psi}\rangle}$ and since $\langle\widetilde{\psi}, \widetilde{\psi}\rangle_T \leq c$ for some constant c, we get

$$E\left[\exp\left(\alpha\left(\widetilde{\psi}_T^*\right)^2\right)\right] \leq E\left[\exp\left(\alpha\left(\gamma_c^*\right)^2\right)\right] < \infty,$$

for α sufficiently small. □

Remark. If one wants to avoid the use of predictable projections, one can state the above result by saying that ϕ is equal to the projection in the Hilbert space $L^2(dW \otimes dt)$ of the process $F'(\beta,]\cdot, T])$ on the subspace of equivalence classes of predictable processes (see Exercise (5.18) Chap. IV).

The remainder of this section will be devoted to asymptotic results on Brownian motion.

(2.5) Definition. *The functional I_T on* **W** *defined by*

$$\begin{aligned} I_T(\phi) &= \frac{1}{2}\int_0^T \phi'(s)^2 ds = \frac{1}{2}\|\phi\|_H^2 \quad \text{if } \phi \in H, \\ &= +\infty \quad \text{otherwise,} \end{aligned}$$

is called the action functional *of Brownian motion.*

We will often drop T from the notation. We begin with a few remarks about I_T.

(2.6) Lemma. *A real-valued function ϕ on $[0, T]$ is absolutely continuous with derivative ϕ' in $L^2([0, T])$ if and only if*

$$M(\phi) = \sup \sum_i \left(\phi(t_{i+1}) - \phi(t_i)\right)^2 / (t_{i+1} - t_i),$$

where the supremum is taken on all finite subdivisions of $[0, T]$, is finite and in that case

$$\int_0^T \phi'(s)^2 ds = M(\phi).$$

Proof. If $\phi' \in L^2$, a simple application of Cauchy-Schwarz inequality proves that

$$M(\phi) \le \int_0^T \phi'(s)^2 ds.$$

Suppose conversely that $M(\phi) < \infty$; for disjoint intervals (α_i, β_i) of $[0, T]$, Cauchy-Schwarz inequality again gives us

$$\begin{aligned}\sum_i |\phi(\beta_i) - \phi(\alpha_i)| &\le \left(\sum_i \frac{(\phi(\beta_i) - \phi(\alpha_i))^2}{|\beta_i - \alpha_i|}\right)^{1/2} \left(\sum_i |\beta_i - \alpha_i|\right)^{1/2} \\ &\le M(\phi)^{1/2} \left(\sum_i |\beta_i - \alpha_i|\right)^{1/2}\end{aligned}$$

which proves that ϕ is absolutely continuous. If (α_i^n) is the n-th dyadic partition of $[0, T]$, we know, by an application of Martingale theory (Exercise (2.13) Chap. II) that the derivative ϕ' of ϕ is the limit a.e. of the functions $\psi^n(s) = (2^n/T)\left(\phi\left(\alpha_{i+1}^n\right) - \phi\left(\alpha_i^n\right)\right) 1_{[\alpha_i^n, \alpha_{i+1}^n[}(s)$. By Fatou's lemma, it follows that

$$\int_0^T \phi'(s)^2 ds \le \underline{\lim}_n \int_0^T \left(\psi^n(s)\right)^2 ds \le M(\phi).$$

□

(2.7) Proposition. *i) The functional $\phi \to I_T(\phi)$ is lower semi-continuous on* $\mathbf{W}$.
ii) For any $\lambda \ge 0$, the set $K_\lambda = \{\phi : I_T(\phi) \le \lambda\}$ is compact in $\mathbf{W}$.

Proof. The result can be proved componentwise so it is enough to prove it for $d = 1$. Moreover, i) is an easy consequence of ii). To prove ii), we pick a sequence $\{\phi_n\}$ converging to ϕ in $\mathbf{W}$ and such that $I_T(\phi_n) \le \lambda$ for every n. It is easy to see that $M(\phi) \le \sup_n M(\phi_n) \le 2\lambda$, which, by the lemma, proves our claim. □

Our first asymptotic result is known as the theorem of *large deviations* for Brownian motion. We will need the following definition.

(2.8) Definition. *The* Cramer transform *of a set $A \subset \mathbf{W}$ is the number $\Lambda_T(A)$ ($\le \infty$) defined by*

$$\Lambda_T(A) = \inf\{I_T(\phi) : \phi \in A\}.$$

In what follows, T will be fixed and so will be dropped from the notation. We will be interested in the probability $W[\varepsilon\beta \in A]$, as ε goes to zero. Clearly, if A is closed and if the function 0 does not belong to A, it will go to zero and we shall determine the speed of this convergence. We will need the following lemmas known as the *Ventcell-Freidlin estimates* where $B(\phi, \delta)$ denotes the open ball centered in ϕ and with radius δ in $\mathbf{W}$.

(2.9) Lemma. *If $\phi \in H$, for any $\delta > 0$,*

$$\underline{\lim}_{\varepsilon \to 0} \varepsilon^2 \log W[\varepsilon\beta \in B(\phi, \delta)] \ge -I(\phi).$$

Proof. By Theorem (2.2)

$$\begin{aligned} W[\varepsilon\beta \in B(\phi,\delta)] &= W[\beta - \varepsilon^{-1}\phi \in B(0,\delta\varepsilon^{-1})] \\ &= \int_{B(0,\delta\varepsilon^{-1})} \exp\left\{-\frac{1}{\varepsilon}\int_0^T \phi'(s)d\beta_s - \frac{1}{2\varepsilon^2}\int_0^T \phi'(s)^2 ds\right\} dW \\ &= \exp\left(-\varepsilon^{-2}I(\phi)\right)\int_{B(0,\delta\varepsilon^{-1})} \exp\left(-\frac{1}{\varepsilon}\int_0^T \phi'(s)d\beta_s\right) dW. \end{aligned}$$

Now on the one hand $W[B(0,\delta\varepsilon^{-1})] \geq 3/4$ for ε sufficiently small; on the other hand, Tchebicheff's inequality implies that

$$W\left[\int_0^T \phi'(s)d\beta_s \geq 2\sqrt{2I(\phi)}\right] \leq \frac{1}{8I(\phi)}\left(\int_0^T \phi'(s)^2 ds\right) = \frac{1}{4},$$

which implies

$$W\left[\exp\left(-\varepsilon^{-1}\int_0^T \phi'(s)d\beta_s\right) \geq \exp\left(-2\varepsilon^{-1}\sqrt{2I(\phi)}\right)\right] \geq \frac{3}{4}.$$

As a result

$$W[\varepsilon\beta \in B(\phi,\delta)] \geq \frac{1}{2}\exp\left(-\varepsilon^2 I(\phi) - 2\varepsilon^{-1}\sqrt{2I(\phi)}\right)$$

for ε sufficiently small. The lemma follows immediately. □

(2.10) Lemma. *For any $\delta > 0$,*

$$\overline{\lim_{\varepsilon\to 0}}\, \varepsilon^2 \log W\left[\rho(\varepsilon\beta, K_\lambda) \geq \delta\right] \leq -\lambda$$

where ρ is the distance in **W**.

Proof. Set $\alpha = T/n$ and let $0 = t_0 < t_1 < \ldots < t_n = T$ be the subdivision of $[0,T]$ such that $t_{k+1} - t_k = \alpha$. Let L^ε be the function which componentwise is affine between t_k and t_{k+1} and equal to $\varepsilon\beta$ at these times. We have

$$\begin{aligned} W\left[\rho(\varepsilon\beta, L^\varepsilon) \geq \delta\right] &\leq \sum_{k=1}^{n} W\left[\max_{t_{k-1}\leq t\leq t_k} \left|\varepsilon\beta_t - L_t^\varepsilon\right| \geq \delta\right] \\ &\leq nW\left[\max_{0\leq t\leq\alpha} \left|\varepsilon\beta_t - L_t^\varepsilon\right| \geq \delta\right] \\ &= nW\left[\max_{0\leq t\leq\alpha} \left|\beta_t - \frac{t}{\alpha}\beta_\alpha\right| \geq \delta\varepsilon^{-1}\right] \\ &\leq nW\left[\max_{0\leq t\leq\alpha} |\beta_t| \geq (2\varepsilon)^{-1}\delta\right] \end{aligned}$$

because $\left\{\max_{0\le t\le\alpha}|\beta_t| > (2\varepsilon)^{-1}\delta\right\} \supset \left\{\max_{0\le t\le\alpha}\left|\beta_t - \frac{t}{\alpha}\beta_\alpha\right| \ge \delta\varepsilon^{-1}\right\}$, and by the exponential inequality (Proposition (1.8) Chap. II), this is still less than

$$2nd\exp\left(-n\delta^2/8\varepsilon^2 d\ T\right).$$

If n is sufficiently large, we thus get

$$\overline{\lim_{\varepsilon\to0}}\ \varepsilon^2\log W\left[\rho(\varepsilon\beta, L^\varepsilon)\ge\delta\right]\le-\lambda.$$

Let us fix such an n; then because

$$\{\rho(\varepsilon\beta, K_\lambda)\ge\delta\}\subset\{I(L^\varepsilon)>\lambda\}\cup\{\rho(\varepsilon\beta, L^\varepsilon)\ge\delta\},$$

it is enough, to complete the proof, to prove that

$$\overline{\lim}\ \varepsilon^2\log W\left[I(L^\varepsilon)>\lambda\right]\le-\lambda.$$

But

$$I(L^\varepsilon)=\frac{\varepsilon^2}{2}\sum_1^{nd}\eta_i^2$$

where the η_i are independent standard Gaussian variables. For every $\beta>0$, $E\left[\exp\left(\frac{1-\beta}{2}\eta_i^2\right)\right]=C_\beta<\infty$, and therefore, by Markov inequality,

$$\begin{aligned}W[I(L^\varepsilon)>\lambda] &= P\left[\exp\left(\frac{1-\beta}{2}\sum_1^{nd}\eta_i^2\right)>\exp\left(\frac{(1-\beta)\lambda}{\varepsilon^2}\right)\right]\\ &\le C_\beta^{nd}\exp\left(-\frac{(1-\beta)\lambda}{\varepsilon^2}\right),\end{aligned}$$

whence $\overline{\lim}\ \varepsilon^2\log W[I(L^\varepsilon)>\lambda]\le-(1-\beta)\lambda$ and, as β is arbitrary, the proof of the lemma is complete. □

We can now state

(2.11) Theorem (Large deviations). *For a Borel set $A\subset\mathbf{W}$,*

$$-\Lambda(\overset{\circ}{A})\le\underline{\lim}_{\varepsilon\to0}\ \varepsilon^2\log W[\varepsilon\beta\in A]\le\overline{\lim_{\varepsilon\to0}}\ \varepsilon^2\log W[\varepsilon\beta\in A]\le-\Lambda(\overline{A}).$$

Proof. If A is open, for any $\phi\in H\cap A$, and δ sufficiently small,

$$W[\varepsilon\beta\in A]\ge W[\rho(\varepsilon\beta,\phi)<\delta],$$

and so, by Lemma (2.9),

$$\underline{\lim}\ \varepsilon^2\log W[\varepsilon\beta\in A]\ge-\inf\{I(\phi),\phi\in A\}=-\Lambda(A).$$

Let now A be closed; we may suppose $\Lambda(A) > 0$ as the result is otherwise obvious. Assume first that $\Lambda(A) < \infty$. For $\Lambda(A) > \gamma > 0$, the sets A and $K = K_{\Lambda(A)-\gamma}$ are disjoint. Since K is compact, there is a number $\delta > 0$ such that $\rho(\psi, K) \geq \delta$ for every $\psi \in A$; by Lemma (2.10), we consequently get

$$\overline{\lim}\ \varepsilon^2 \log W[\varepsilon\beta \in A] \leq -\Lambda(A) + \gamma$$

and, since γ is arbitrary, the proof is complete. If $\Lambda(A) = +\infty$, the same reasoning applies with $K = K_M$ for arbitrary large M. □

We will now apply the preceding theorem to the proof of a beautiful result known as *Strassen's functional law of the iterated logarithm*. In what follows, we set $g(n) = (2n \log_2 n)^{-1/2}$, $n \geq 2$, and $X_n(t) = g(n)\beta_{nt}$, $0 \leq t \leq T$. For every ω, we have thus defined a sequence of points in $\mathbf{W}$, the asymptotic behavior of which is settled by the following theorem. The unit ball of the Hilbert space H which is equal to the set $K_{1/2}$ will be denoted by U.

(2.12) Theorem. *For W-almost every ω, the sequence $\{X_n(\cdot, \omega)\}$ is relatively compact in $\mathbf{W}$ and the set of its limit points is the set U.*

Proof. We first prove the relative compactness. For $\delta > 0$, let K^δ be the closed set of points ω in $\mathbf{W}$ such that $\rho(\omega, U) \leq \delta$. Using the semi-continuity of I_T, it may be seen that $\Lambda\left(\left(K^\delta\right)^c\right) > 1/2$, and thus, for fixed δ, we may choose γ such that $1 < \gamma < 2\Lambda\left(\left(K^\delta\right)^c\right)$. Pick $\lambda > 1$ and set $n(m) = [\lambda^m]$; by the scaling property of BM,

$$W\left[X_{n(m)} \notin K^\delta\right] = W\left[\sqrt{n(m)}g\left(n(m)\right)\beta_{.} \notin K^\delta\right],$$

thus by Theorem (2.11), for m sufficiently large,

$$W\left[X_{n(m)} \notin K^\delta\right] \leq \exp\left(-\gamma \log_2 n(m)\right) \leq ((m-1)\log\lambda)^{-\gamma}.$$

It follows that W-a.s., $X_{n(m)}$ belongs to K^δ for m sufficiently large. As this is true for every δ, it follows that the sequence $\left\{X_{n(m)}\right\}$ is a.s. relatively compact and that its limit points are in U. Clearly, there is a set B of full W-measure such that for $\omega \in B$, all the sequences $X_{n(m)}(\omega)$ where λ ranges through a sequence $S = \{\lambda_k\}$ decreasing to 1, are relatively compact and have their limit points in U. We will prove that the same is true for the whole sequence $\{X_n(\omega)\}$. This will involve no probability theory and we will drop the ω which is fixed throughout the proof.

Let $M = \sup_{h\in U} \|h\|_\infty$ and set $b(t) = \sqrt{t}$; observe that

$$\sup_{h\in U} |h(t) - h(s)| \leq b(|t-s|)$$

thanks to the Cauchy-Schwarz inequality.

Fix $\lambda \in S$; for any integer n, there is an m such that $n(m) \leq n < n(m+1)$ and we will write for short $N = n(m+1)$. We want to show that $\rho(X_n, U)$ tends to 0 as n tends to infinity which we will do by comparing X_n and X_N.

Pick $\delta > 0$ and then choose $\lambda > S$ sufficiently close to 1 so that $b(1 - (1/\lambda)T) < \delta$. These numbers being fixed, for n sufficiently large, we have $\rho(X_N, U) \leq \delta$. This entails that $\|X_N\|_\infty \leq M + \delta$ and that there is a function $k \in U$ such that $\|X_N - k\|_\infty < \delta$. We may now write

$$\begin{aligned}
\rho(X_n, U) &\leq \rho(X_N, U) + \rho(X_n, X_N) \\
&\leq \delta + \sup_{t \leq T} \left| \frac{g(n)}{g(N)} X_N\left(\frac{n}{N}t\right) - X_N(t) \right| \\
&\leq \delta + \left| \frac{g(n)}{g(N)} - 1 \right| \|X_N\|_\infty + \sup_{t \leq T} \left| X_N\left(\frac{n}{N}t\right) - X_N(t) \right| \\
&\leq \delta + \left| \frac{g(n)}{g(N)} - 1 \right| (M + \delta) + 2\delta + b\left(\left(1 - \frac{n}{N}\right)T\right).
\end{aligned}$$

But $\left| \frac{g(n)}{g(N)} - 1 \right| \leq \lambda - 1$ as the reader may check so that

$$\rho(X_n, U) \leq (\lambda - 1)(M + \delta) + 4\delta.$$

Since M is fixed and we could take λ and δ arbitrarily close to 1 and 0 respectively we have proved the relative compactness of the sequence $\{X_n\}$ and that the limit points are in U. It remains to prove that all the points of U are limit points of the sequence $\{X_n\}$.

We first observe that there is a countable subset contained in $\{h \in H; \|h\|_H < 1\}$ which is dense in U for the distance ρ. Therefore, it is enough to prove that if $h \in H$ and $\|h\|_H < 1$,

$$W\left[\varliminf_{n \to \infty} \rho(X_n, h) = 0\right] = 1.$$

To this end, we introduce, for every integer k, an operator L_k on $\mathbf{W}$ by

$$\begin{aligned}
&L_k\phi(t) = 0 \qquad \text{if } 0 \leq t \leq T/k, \\
&L_k\phi(t) = \phi(t) - \phi(T/k) \qquad \text{if } T/k \leq t \leq T.
\end{aligned}$$

For $k \geq 2$ and $m \geq 1$, we may write

$$\begin{aligned}
\rho\left(X_{k^m}, h\right) \leq & \sup_{t \leq T/k} |X_{k^m}(t)| + \sup_{t \leq T/k} |h(t)| + |X_{k^m}(T/k)| + |h(T/k)| \\
& + \|L_k\left(X_{k^m} - h\right)\|_\infty .
\end{aligned}$$

Since $\{X_n\}$ is a.s. relatively compact, by Ascoli's theorem, we may for a.s. every ω choose a $k(\omega)$ such that for $k \geq k(\omega)$ the first four terms of this inequality are less than any preassigned $\delta > 0$ for every m. It remains to prove that for a fixed k,

$$W\left[\varliminf_{m \to \infty} \|L_k\left(X_{k^m} - h\right)\|_\infty = 0\right] = 1.$$

But, on the one hand, the processes $L_k(X_{k^m})$ are independent as m varies; on the other hand, by the invariance properties of BM,

$$W\left[\|L_k(X_{k^m}-h)\|_\infty<\varepsilon\right]=W\left[\sup_{0\le t\le T(1-1/k)}\left|\sqrt{k^m}g(k^m)\beta_\cdot-\tilde{h}\right|<\varepsilon\right]$$

where $\tilde{h}(t)=h(t+T/k)$, $0\le t\le T(1-1/k)$. Since

$$\int_0^{T(1-1/k)}\tilde{h}'(s)^2ds\le\int_0^T h'(s)^2ds<1,$$

the large deviation theorem applied for open sets and on $[0,T(1-1/k)]$ instead of $[0,T]$ shows that for m sufficiently large

$$W\left[\|L_k(X_{k^m}-h)\|_\infty<\varepsilon\right]\ge\exp\left(-\gamma\log_2 k^m\right)=(m\log k)^{-\gamma}$$

for some $\gamma<1$. An application of Borel-Cantelli's Lemma ends the proof.

Remark. We have not used the classical law of the iterated logarithm of Chap. II which in fact may be derived from the above result (see Exercise (2.16)).

(2.13) Exercise. 1°) In the setting of Theorem (2.2), prove that for $t<T$,

$$\left.\frac{dW_h}{dW}\right|_{\mathscr{F}_t}=\exp\left\{\sum_i h_i'(t)\beta_t^i-\sum_i\int_0^t h_i''(s)\beta_s^i ds-\frac{1}{2}\int_0^t|h'(s)|^2ds\right\}.$$

It is noteworthy that this derivative may be written without stochastic integrals.

2°) Conversely, prove that W is the only measure on $\mathbf{W}$ which is quasi-invariant with respect to H and which admits the above derivatives.

[Hint: Consider functions h with compact support in $]0,t[$.]

(2.14) Exercise. 1°) Prove that for the BM^d, $d\ge 2$, absolutely continuous functions are polar in the sense of Exercise (1.20) of Chap. I (see also Exercise (2.26) in Chap. V).

2°) Prove that if B is a Brownian Bridge of dimension $d\ge 2$ between any two points of $\mathbb{R}^d$, then for any $x\in\mathbb{R}^d$,

$$P[\exists t\in]0,1[:B_t=x]=0.$$

[Hint: Use the transformations of Exercise (3.10) Chap. I.]

3°) If in 2°) we make $d=2$ and $B_0=B_1=0$, the index $I(x,B)$ of $x\ne 0$ with respect to the curve B_t, $0\le t\le 1$, is well defined. Prove that $P[I(x,B)=n]>0$ for every $n\in\mathbb{Z}$.

[Hint: Extend the result of Corollary (2.3) to the Brownian Bridge and use it together with the fact that x has the same index with respect to two curves which are homotopic in $\mathbb{R}^2\backslash\{x\}$.]

(2.15) Exercise. 1°) Recall from Exercise (3.26) Chap. III that there is a.s. a unique time σ such that $S_1 = B_\sigma$. Prove that

$$S_1 = \sqrt{2/\pi} + \int_0^1 \phi_s dB_s$$

where ϕ_s is the predictable projection of $1_{(\sigma>s)}$ (or its L^2-projection on the predictable σ-field).

[Hint: With the notation of Theorem (2.4), prove that $F'(\beta, \cdot) = \varepsilon_\sigma$.]

2°) Prove that there is a right-continuous version of $P\left[\sigma > t \mid \mathscr{F}_t\right]$ which is indistinguishable from ϕ_t and conclude that for $\Phi(x) = \int_x^\infty g_1(y)dy$,

$$\phi_t = 2\Phi\left((S_t - B_t)/\sqrt{1-t}\right).$$

[Hint: $P\left[\sigma > t \mid \mathscr{F}_t\right] = P\left[T_a < 1 - t\right]_{a=S_t-B_t}$.]

3°) (Alternative method). Compute $E\left[f(S_1) \mid \mathscr{F}_t\right]$ where f is a positive Borel function and deduce directly the formula of 1°).

(2.16) Exercise. 1°) In the setting of Theorem (2.12), let Φ be a real-valued continuous function on $\mathbf{W}$, and prove that

$$W\left[\overline{\lim_n}\, \Phi\left(X_n(\cdot)\right) = \sup_{h\in U} \Phi(h)\right] = 1.$$

2°) Derive therefrom the classical law of the iterated logarithm.

(2.17) Exercise. Let (ξ_n) be a sequence of independent identically distributed real random variables with mean 0 and variance 1 and set $S_n = \sum_1^n \xi_k$. Define a process S_t by

$$S_t = (1 - t + [t])S_{[t]} + (t - [t])S_{[t]+1}.$$

Prove that the sequence $X_n(t) = g(n)S_{nt}$, $0 \le t \le T$, has the same property as that of Theorem (2.12).

[Hint: Use the result in Exercise (5.10) of Chap. VI.]

§3. Functionals and Transformations of Diffusion Processes

In the study of diffusions and stochastic differential equations, Girsanov's theorem is used in particular to change the drift coefficient. One reduces SDE's to simpler ones by playing on the drift or, from another view-point, constructs new Markov processes by the addition of a drift. This will be used in Sect. 1 Chap. IX.

In this section, we will make a first use of this idea towards another goal, namely the computation of the laws of functionals of BM or other processes. We will give a general principle and then proceed to examples.

The situation we study is that of Sect. 2 Chap. VII. A field σ (resp. b) of $d \times d$ symmetric matrices (resp. vectors in $\mathbb{R}^d$) being given, we assume that for each $x \in \mathbb{R}^d$ there is a probability measure P_x on $\Omega = C(\mathbb{R}_+, \mathbb{R}^d)$ such that $(\Omega, \mathscr{F}_t^0, X_t, P_x)$ is a diffusion process in the sense of Definition (2.1) of Chap. VII with $a = \sigma\sigma'$. By Theorem (2.7) Chap. VII, for each P_x, there is a Brownian motion B such that

$$X_t = x + \int_0^t \sigma(X_s)dB_s + \int_0^t b(X_s)ds.$$

We moreover assume that P_x is, for each x, the unique solution to the martingale problem $\pi(x, \sigma\sigma', b)$.

Suppose now given a pair (f, F) of functions such that $D_t = \exp(f(X_t) - f(X_0) - \int_0^t F(X_s)ds)$ is a $(\mathscr{F}_t^0, P_x)$-continuous martingale for every x. By Proposition (1.13) we can define a new probability P_x^f on $\mathscr{F}_\infty^0$ by $P_x^f = D_t \cdot P_x$ on $\mathscr{F}_t^0$. If Z is an $\mathscr{F}_\infty^0$-measurable function on Ω, we will denote by $E_{P_x}[Z \mid X_t = \cdot]$ a Borel function ϕ on $\mathbb{R}^d$ such that $E_x[Z \mid X_t] = \phi(X_t)$.

(3.1) Proposition. *The term $(\Omega, \mathscr{F}_t^0, X_t, P_x^f)$ is a Markov process. For each x, and $t > 0$, the probability measures $P_t^f(x, dy)$ and $P_t(x, dy)$ are equivalent and, for each x, the Radon-Nikodym derivative is given by*

$$\frac{P_t^f(x, dy)}{P_t(x, dy)} = \exp\big(f(y) - f(x)\big) E_{P_x}\left[\exp\left(-\int_0^t F(X_s)ds\right) \mid X_t = y\right].$$

Proof. The measurability of the map $x \to P_x^f$ is obvious. Let g be a positive Borel function and Y a $\mathscr{F}_t^0$-measurable r.v. Because $D_{t+s} = D_t \cdot D_s \circ \theta_t$, we have, with obvious notation,

$$\begin{aligned} E_x^f\left[Yg(X_{t+s})\right] &= E_x\left[Yg(X_{t+s})D_{t+s}\right] \\ &= E_x\left[YD_t E_{X_t}\left[g(X_s)D_s\right]\right] = E_x^f\left[YE_{X_t}^f\left[g(X_s)\right]\right] \end{aligned}$$

which proves the first claim. The second follows from the identities

$$\begin{aligned} P_t^f g(x) &= E_x\left[D_t g(X_t)\right] = E_x\left[g(X_t)\exp\left(f(X_t) - f(X_0) - \int_0^t F(X_s)ds\right)\right] \\ &= E_x\left[g(X_t)\exp\left(f(X_t) - f(x)\right) E_x\left[\exp\left(-\int_0^t F(X_s)ds\right) \mid X_t\right]\right]. \end{aligned}$$

□

In the above Radon-Nikodym derivative, three terms intervene, the two semi-groups and the conditional expectation of the functional $\exp(-\int_0^t F(X_s)ds)$. This can be put to use in several ways, in particular to compute the conditional expectation when the two semi-groups are known. This is where Girsanov's theorem comes into play. Since $D_t = \mathscr{E}(M)_t$ for some local martingale M (for each P_x), then Girsanov's theorem permits to compute the infinitesimal generator of P_t^f, hence at least theoretically, the semi-group P_t^f itself. Conversely, the above formula gives P_t^f when the conditional expectation is known.

We now give a general method to find such pairs (f, F) and will afterwards take advantage of it to compute the laws of some Brownian functionals.

The extended generator L of X is equal on C^2-functions to

$$L = \frac{1}{2}\sum_{i,j=1}^{d} a_{ij}\frac{\partial^2}{\partial x_i \partial x_j} + \sum_{i=1}^{d} b_i \frac{\partial}{\partial x_i}$$

where $a = \sigma\sigma^t$. We recall that if $f \in C^2$ then

$$M_t^f = f(X_t) - f(X_0) - \int_0^t Lf(X_s)ds$$

is a continuous local martingale and we now show how to associate with f a function F satisfying the above hypothesis.

(3.2) Definition. *The* opérateur carré du champ Γ *is defined on* $C^2 \times C^2$ *by*

$$\Gamma(f, g) = L(fg) - fLg - gLf.$$

(3.3) Proposition. *If* $f, g \in C^2$, *then, under each* P_x,

$$\langle M^f, M^g\rangle_t = \int_0^t \Gamma(f, g)(X_s)ds.$$

Proof. Let us write $A_t \sim B_t$ if $A - B$ is a local martingale. Using the integration by parts formula, straightforward computations yield

$$\begin{aligned}\left(M_t^f\right)^2 &\sim \int_0^t L(f^2)(X_s)ds + \left(\int_0^t Lf(X_s)ds\right)^2 - 2f(X_t)\int_0^t Lf(X_s)ds \\ &\sim \int_0^t L(f^2)(X_s)ds - 2\int_0^t (fLf)(X_s)ds = \int_0^t \Gamma(f, f)(X_s)ds.\end{aligned}$$

The proof is completed by polarization. □

As a consequence of this proposition, if $f \in C^2$, then

$$\begin{aligned}\mathscr{E}(M^f)_t &= \exp\left\{f(X_t) - f(X_0) - \int_0^t Lf(X_s)ds - \frac{1}{2}\int_0^t \Gamma(f, f)(X_s)ds\right\} \\ &= (h(X_t)/h(X_0))\exp\left(\int_0^t (Lh(X_s)/h(X_s))\,ds\right),\end{aligned}$$

if $h = \exp(f)$. If this local martingale turns out to be a true martingale, then we may define the probability measures P_x^f as described at the beginning of the section, with $F = Lf + \frac{1}{2}\Gamma(f, f)$. In this setting, we get

(3.4) Proposition. *If* L *is the extended generator of the* P*-process, the extended generator of the* P^f*-process is equal on* C^2 *to* $L + \Gamma(f, \cdot) \equiv L + h^{-1}\Gamma(h, \cdot)$.

Proof. If $\phi \in C^2$, then $\phi(X_t) - \phi(X_0) - \int_0^t L\phi(X_s)ds$ is a P_x-local martingale and Girsanov's theorem implies that

$$\phi(X_t) - \phi(X_0) - \int_0^t L\phi(X_s)ds - \langle M^f, M^\phi\rangle_t$$

is a P_x^f-local martingale. The proof is completed by means of Proposition (3.3). □

We proceed by applying the above discussion to particular cases. Let us suppose that f is a solution to $Lf = 0$, which, thinking of the special case of BM, may be expressed by saying that f is "harmonic". Then $\Gamma(f, f) = L(f^2)$ and $F = \frac{1}{2}L(f^2)$. The extended generator of the P^f-process is equal on $\phi \in C^2$ to

$$L\phi + \langle \nabla f, \sigma\sigma^t \nabla\phi\rangle.$$

We see that the effect of the transformation is to change the drift of the process.

If the P-process is a BM^d and f is harmonic in the usual sense, then $F = |\nabla f|^2$ and the generator is given by $\frac{1}{2}\Delta\phi + \langle\nabla f, \nabla\phi\rangle$. We will carry through some computations for particular cases of harmonic functions. Let for instance δ be a vector in $\mathbb{R}^d$; then $f(x) = \langle\delta, x\rangle$ is a harmonic function, and plainly $\mathscr{E}(M^f)$ is a martingale. We get a Markov process with generator

$$A^f\phi = \frac{1}{2}\Delta\phi + \langle\delta, \nabla\phi\rangle,$$

which is the *Brownian motion with constant drift* δ, namely $B_t + t\delta$. Let us call P_x^δ instead of P_x^f the corresponding probability measures. By the above discussion

$$P_x^\delta = \exp\left\{\langle\delta, X_t - X_0\rangle - \frac{|\delta|^2 t}{2}\right\} \cdot P_x \qquad \text{on } \mathscr{F}_t^0$$

and the semi-group P_t^δ is given by

$$P_t^\delta(x, dy) = \exp\left\{\langle\delta, y - x\rangle - \frac{|\delta|^2 t}{2}\right\} \cdot P_t(x, dy),$$

where P_t is the Brownian semi-group. Of course in this simple case, the semi-group P_t^δ may be computed directly from P_t. Before proceeding to other examples, we shall study the probability measure P_0^δ.

We suppose that $d \geq 2$; since P_x^δ is absolutely continuous with respect to P_x on $\mathscr{F}_t^0$ and since the hitting time of a closed set is a $(\mathscr{F}_t^0)$-stopping time, it follows from the polarity of points for BM^2 that the hitting times of points are also a.s. infinite under P_x^δ. Thus, we may write a.s. $X_t = \rho_t\theta_t$ for all $t > 0$, where $\rho_t = |X_t|$ and the process θ takes its values in the unit sphere. We set $\mathscr{R}_t = \sigma(\rho_s, s \leq t)$ and $\mathscr{R}_\infty = \bigvee_t \mathscr{R}_t$.

(3.5) Lemma. *For each $t > 0$, the r.v. θ_t is, under P_0, independent of $\mathscr{R}_\infty$ and uniformly distributed on the unit sphere S^{d-1}.*

Proof. Suppose $d = 2$ and let Z be $\mathscr{R}_\infty$-measurable and ≥ 0 and G be a positive Borel function on S^{d-1}. Because of the invariance of P_0, i.e. the Wiener measure, by rotations, for every $\alpha \in [0, 2\pi]$,

$$E_0[ZG(\theta_t)] = E_0\left[ZG\left(e^{i\alpha}\theta_t\right)\right],$$

and integrating with respect to α,

$$E_0[ZG(\theta_t)] = E_0\left[Z\frac{1}{2\pi}\int_0^{2\pi} G\left(e^{i\alpha}\theta_t\right)d\alpha\right];$$

since the Lebesgue measure on S^1 is invariant by multiplication by a given point of S^1, we get

$$E_0[ZG(\theta_t)] = E_0[Z]\left(\frac{1}{2\pi}\int_0^{2\pi} G\left(e^{i\alpha}\right)d\alpha\right).$$

For $d > 2$ there is a slight difficulty which comes from the fact that S^{d-1} is only a homogeneous space of the rotation group. The details are left to the reader (see Exercise (1.17) Chap. III). □

We henceforth call μ^d the uniform distribution on S^{d-1} and if ϕ is a positive Borel function on $\mathbb{R}^d$, we set

$$M\phi(\rho) = \int_{S^{d-1}} \phi(\rho u)\mu^d(du).$$

(3.6) Corollary. *On the σ-algebra $\mathscr{R}_t$,*

$$P_0^\delta = M\phi(\rho_t)\exp\left(-\frac{|\delta|^2 t}{2}\right)\cdot P_0,$$

with $\phi(x) = \exp\{\langle\delta, x\rangle\}$.

Proof. Since $\mathscr{R}_t \subset \mathscr{F}_t^0$, we obviously have

$$P_0^\delta = E_0\left[\exp\left(\langle\delta, X_t\rangle - \frac{|\delta|^2 t}{2}\right) \mid \mathscr{R}_t\right]\cdot P_0 \quad \text{on } \mathscr{R}_t$$

and it is easily checked, taking the lemma into account, that the conditional expectation is equal to $\exp\left(-\frac{|\delta|^2 t}{2}\right)M\phi(\rho_t)$. □

We may now state

(3.7) Theorem. *Under P_0^δ, the process ρ_t is a Markov process with respect to $(\mathscr{R}_t)$. More precisely, there is a semi-group Q_t such that for any positive Borel function f on $\mathbb{R}_+$,*

$$E_0^\delta\left[f(\rho_{t+s}) \mid \mathscr{R}_t\right] = Q_s f(\rho_t).$$

Proof. Pick A in $\mathscr{R}_t$; we may write, using the notation of Corollary (3.6),

$$\int_A f(\rho_{t+s})\, dP_0^\delta = \exp\left(-\frac{|\delta|^2}{2}(t+s)\right) \int_A f(\rho_{t+s})\, M\phi(\rho_{t+s})\, dP_0,$$

and by the Markov property under P_0, this is equal to

$$\begin{aligned} &\exp\left(-\frac{|\delta|^2}{2}(t+s)\right) \int_A E_{X_t}\left[f(\rho_s) M\phi(\rho_s)\right] dP_0 \\ &= \exp\left(-\frac{|\delta|^2}{2}(t+s)\right) \int_A E_0\left[E_{X_t}\left[f(\rho_s)M\phi(\rho_s)\right] \mid \mathscr{R}_t\right] dP_0. \end{aligned}$$

By the same reasoning as in Corollary (3.6), this is further equal to

$$\exp\left(-\frac{|\delta|^2}{2}(t+s)\right) \int_A M\psi(\rho_t) dP_0$$

with $\psi(x) = E_x\left[f(\rho_s)M\phi(\rho_s)\right]$. Thus we finally have

$$\int_A f(\rho_{t+s}) dP_0^\delta = \int_A \exp\left(-\frac{|\delta|^2 s}{2}\right) (M\psi(\rho_t)/M\phi(\rho_t))\, dP_0^\delta.$$

This shows the first part of the statement and we now compute the semi-group Q_t.

Plainly, because of the geometrical invariance properties of BM, the function ψ depends only on $|x|$ and consequently $M\psi = \psi$. Thus, we may write

$$\begin{aligned} E_0^\delta\left[f(\rho_{t+s}) \mid \mathscr{R}_t\right] &= \exp\left(-\frac{|\delta|^2 s}{2}\right) \psi(\rho_t)/M\phi(\rho_t) \\ &= E_{\rho_t}\left[f(\rho_s)M\phi(\rho_s)\exp\left(-\frac{|\delta|^2 s}{2}\right)\right] \Big/ M\phi(\rho_t) \end{aligned}$$

where P_a is the law of the modulus of BM started at x with $|x| = a$. We will see in Chap. XI that this process is a Markov process whose transition semi-group has a density $p_s^d(a, \rho)$. Thus

$$E_0^\delta\left[f(\rho_{t+s}) \mid \mathscr{R}_t\right] = Q_s f(\rho_t)$$

where $Q_s(a, d\rho) = \frac{M\phi(\rho)}{M\phi(a)} \exp\left(-\frac{|\delta|^2 s}{2}\right) p_s^d(a, \rho) d\rho$. Since p_s^d is the density of a semi-group, it is readily checked that Q_t is a semi-group, which ends the proof. □

Remark. The process ρ_t is no longer a Markov process under the probability measure P_x^δ for $x \neq 0$.

We now turn to another example, still about Brownian motion, which follows the same pattern (see also Exercise (1.34)). Suppose that $d = 2$ and in complex notation take $f(z) = \alpha \log |z|$ with $\alpha \geq 0$. The function f is harmonic outside the polar set $\{0\}$. Moreover, for every t,

$$\sup_{s\leq t} \mathscr{E}(M^f)_s \leq \sup_{s\leq t} |Z_s|^\alpha;$$

since the last r.v. is integrable, it follows that the local martingale $\mathscr{E}(M^f)$ is actually a martingale so that our general scheme applies for P_a if $a \neq 0$. With the notation used above, which is that of Sect. 2 in Chap. V,

$$F(Z_s) = |\nabla f|^2 (Z_s) = \alpha^2 \rho_s^{-2}.$$

From Itô's formula, it easily follows that under P_a

$$\rho_t = \rho_0 + \widetilde{\beta}_t + \frac{1}{2}\int_0^t \rho_s^{-1} ds$$

where $\widetilde{\beta}_t = \int_0^t \rho_s^{-1} (X_s dX_s + Y_s dY_s)$ is a linear BM. But we also know that

$$\log \rho_t = \log \rho_0 + \int_0^t \rho_s^{-1} d\widetilde{\beta}_s,$$

hence $\langle \widetilde{\beta}, \log \rho\rangle_t = \int_0^t \rho_s^{-1} ds$. Thus, Girsanov's theorem implies that under P_a^f, the process $\overline{\beta}_t = \widetilde{\beta}_t - \alpha \int_0^t \rho_s^{-1} ds$ is a BM, and consequently $\rho_t = \rho_0 + \overline{\beta}_t + \frac{2\alpha+1}{2} \int_0^t \rho_s^{-1} ds$. The equations satisfied by ρ under P_a and P_a^f are of the same type. We will see in Sect. 1 Chap. XI, how to compute explicitly the density p_t^δ of the semi-group of the solution to

$$\rho_t = \rho_0 + \overline{\beta}_t + \frac{\delta - 1}{2}\int_0^t \rho_s^{-1} ds.$$

All this can be used to compute the law of θ_t, the "winding number" of Z_t around the origin. As $f(z)$ depends only on $|z|$, the discussion leading to Proposition (3.1) may as well be applied to ρ as to Z with the same function f. As a result, we may now compute the conditional Laplace transform of $C_t = \int_0^t \rho_s^{-2} ds$. We recall that I_ν is the modified Bessel function of index ν.

(3.8) Proposition. *For every α and $a \neq 0$,*

$$\begin{aligned} E_a\left[\exp(i\alpha(\theta_t - \theta_0)) \mid \rho_t = \rho\right] &= E_a\left[\exp\left(-\frac{\alpha^2}{2} C_t\right) \mid \rho_t = \rho\right] \\ &= I_{|\alpha|}\left(\frac{|a|\rho}{t}\right) \Big/ I_0\left(\frac{|a|\rho}{t}\right). \end{aligned}$$

Proof. The first equality follows from Theorem (2.12) in Chapter V and the second from Proposition (3.1) and the explicit formulas of Chap. XI, Sect. 1.

Remark. From this result, one may derive the asymptotic properties of θ_t proved in Theorem (4.1) Chap. X (see Exercise (4.9) in that chapter).

Our next example falls equally in the general set-up of Proposition (3.1). Suppose given a function F for which we can find a C^2-function f such that

$$F = Lf + \frac{1}{2}\Gamma(f, f);$$

then our general scheme may be applied to the computation of the conditional expectation of $\exp\left(-\int_0^t F(X_s)ds\right)$ given X_t.

Let us apply this to the linear BM with drift b, namely the process with generator

$$L\varphi = \frac{1}{2}\varphi'' + b\varphi'.$$

Again, it is easily seen that $\Gamma(f, f) = f'^2$, so that if we use the semi-group (P_t^f) associated with f, we can compute the conditional expectation of $\exp\left(-\int_0^t F(X_s)ds\right)$ with

$$F(x) = \frac{1}{2}f''(x) + b(x)f'(x) + \frac{1}{2}f'(x)^2.$$

By playing on b and f, one can thus get many explicit formulas. The best-known example, given in Exercise (3.14), is obtained for $b(x) = \lambda x$ and leads to the Cameron-Martin formula which is proved independently in Chap. XI.

Still with Brownian motion, we proceed with some other examples in which we consider not only f but its product νf by a constant ν and call P_x^ν instead of $P_x^{\nu f}$ the corresponding probability measures. Moreover, we take $b(x) = \alpha f'(x)$ for a constant α and assume that f satisfies the differential equation

Eq. (3.1) $$(f')^2 = -\frac{1}{2\alpha}f'' + \gamma$$

for some constant γ. The function F is then given by

$$F(x) = \frac{\left(\nu^2 + 2\alpha\nu\right)\gamma}{2} - \frac{\nu^2}{4\alpha}f''(x),$$

the general formula of Proposition (3.1) reads

$$E_x\left[\exp\left(-\frac{\nu^2}{4\alpha}\int_0^t f''(X_s)ds\right) \mid X_t = y\right] = \frac{P_t^\nu(x, dy)}{P_t(x, dy)}\exp\left\{\nu\int_y^x f'(u)du\right\},$$

and the infinitesimal generator of the P^ν-process is given by

$$L^\nu g = \frac{1}{2}g'' + (\alpha + \nu)f'g',$$

as follows from Proposition (3.4).

By solving Eq. (3.1) for f', we find for which drifts and functions the above discussion applies. This is done by setting $f' = -h'/2\alpha h$ and solving for h. Three cases occur.

Case 1. $\gamma = 0$. In that case, $f'(x) = \frac{1}{2\alpha}\frac{A}{Ax+B}$ where A and B are constants. If, in particular, we take $\alpha = 1/2$, $A = 1$, $B = 0$, the generator L^ν is then given by

$$L^\nu g(x) = \frac{1}{2}g''(x) + \left(\frac{1}{2} + \nu\right)\frac{g'(x)}{x}.$$

The P^ν-process is thus a *Bessel* process which will be studied in Chap. XI.

Case 2. $\gamma < 0$. Then $f'(x) = \sqrt{|\gamma|}\frac{A\cos mx + iB\sin mx}{-A\sin mx + B\cos mx}$ where $m = -2\alpha\sqrt{|\gamma|}$. In the special case $\gamma = -1$, $A = 1$, $B = 0$, we get as generator

$$L^\nu g(x) = \frac{1}{2}g''(x) + (\alpha + \nu)\cot(2\alpha x)g'(x)$$

which is the generator of the so-called *Legendre* process.

Case 3. $\gamma > 0$. Then $f'(x) = \sqrt{\gamma}\frac{A\cosh mx + B\sinh mx}{A\sinh mx + B\cosh mx}$ where $m = 2\alpha\sqrt{\gamma}$. For $\gamma = 1$, $A = 1$, $B = 0$, the generator we get is

$$L^\nu g(x) = \frac{1}{2}g''(x) + (\alpha + \nu)\coth(2\alpha x)g'(x).$$

The corresponding processes are the so-called *hyperbolic Bessel* processes.

We proceed to other important transformations of diffusions or, more generally, Markov processes.

Let h be "harmonic", that is, as already said, h is in the domain of the extended infinitesimal generator and $Lh = 0$. Suppose further that h is strictly positive. If we set $f = \log h$, our general scheme applies with $F = 0$ provided that h is $P_t(x, \cdot)$-integrable for every x and t. In that case one observes that h is invariant by the semi-group P_t, namely $P_t h = h$ for every t.

The semi-group obtained from P_t by using this particular function f, namely $\log h$, will be denoted by P_t^h and the corresponding probability measures by P_x^h. Plainly, $P_t^h\phi = h^{-1}P_t(h\phi)$. The process X under the probability measures P_x^h is called the *h-process* of the P_t-process and is very important in some questions which lie beyond the scope of this book. We will here content ourselves with the following remark for which we suppose that the P_t-process is a diffusion with generator L.

(3.9) Proposition. *The extended infinitesimal generator of the h-process is equal on the C^2-function ϕ to*

$$L^h\phi = h^{-1}L(h\phi).$$

Proof. If $\phi \in C_K^2$, then

$$P_t^h\phi(x) - \phi(x) - \int_0^t P_s^h\left(h^{-1}L(h\phi)\right) ds$$
$$= h^{-1}(x)\left[P_t(h\phi)(x) - h(x)\phi(x) - \int_0^t P_s(L(h\phi))ds\right] = 0,$$

and one concludes by the methods of Sect. 2, Chap. VII. □

In the case of Brownian motion, the above formula becomes

$$L^h\phi = \frac{1}{2}\Delta\phi + h^{-1}\langle\nabla h, \nabla\phi\rangle$$

which is again the generator of Brownian motion to which is added another kind of drift. Actually, we see that the h-process is pulled in the direction where $h^{-1}\nabla h$ is large. This is illustrated by the example of BES^3; we have already observed and used the fact that it is the h-process of the BM killed at 0 for $h(x) = x$ (see Exercise (1.15) in Chap. III, Sect. 3 in Chap. VI and Exercise (3.17) in this section).

Finally, we observe that in Proposition (3.1) we used only the multiplicative property of D_t. Therefore, given a positive Borel function g we may replace D_t by $N_t = \exp\left(-\int_0^t g(X_s)ds\right)$ which has the same multiplicative property. Thus, we define a new semi-group $P_t^{(g)}$ and probability measures $P_x^{(g)}$ by

$$P_t^{(g)}f(x) = E_x\left[N_t \cdot f(X_t)\right], \qquad P_x^{(g)} = N_t \cdot P_x \quad \text{on } \mathscr{F}_t^0.$$

Again, X is a Markov process for the probability measures $P_x^{(g)}$ and

$$P_t^{(g)}(x, dy) = E_x\left[\exp\left(-\int_0^t g(X_s)ds\right) \mid X_t = y\right] P_t(x, dy).$$

This transformation may be interpreted as curtailment of the life-time of X or *killing* of X and g appears as a *killing rate*. Evidence about this statement is also given by the form of the extended infinitesimal generator of the new process which is denoted by $L^{(g)}$.

(3.10) Proposition (Feynman-Kac formula). *If $\phi \in C^2$,*

$$L^{(g)}\phi = L\phi - g\phi.$$

Proof. Let $\phi \in C_K^\infty$; then, M^ϕ is a P_x-martingale and the integration by parts formula gives

$$N_t\phi(X_t) = \phi(X_0) - \int_0^t \phi(X_s)N_s g(X_s)ds + \int_0^t N_s L\phi(X_s)ds + \int_0^t N_s dM_s^\phi.$$

The last term is clearly a P_x-martingale and integrating with respect to P_x yields

$$E_x^{(g)}\left[\phi(X_t)\right] = E_x^{(g)}\left[\phi(X_0)\right] + E_x^{(g)}\left[\int_0^t (L\phi - g\phi)(X_s)ds\right].$$

Using Proposition (2.2) in Chap. VII, the proof is easily completed. □

Let us finally observe that these transformations are related to one another. Indeed, if $g = Lf + \frac{1}{2}\Gamma(f, f)$, we have, with the notation of Proposition (3.1),

$$P_t^{(g)}(x, dy) = \exp\big(-f(x)\big) P_t^f(x, dy) \exp\big(f(y)\big).$$

Thus, the semi-group $P_t^{(g)}$ appears as the h-transform of the semi-group P_t^f with $h = \exp(f)$.

(3.11) Exercise. In the situation of Theorem (3.7) but with $x \neq 0$ prove that the bidimensional process $(\rho_t, \int_0^t \rho_s^{-2} ds)$ is a Markov process with respect to $(\mathscr{R}_t)$ under P_x^δ.

* **(3.12) Exercise (Time-inversion).** If X is a process indexed by $t > 0$, we define $\widetilde{X}$ by

$$\widetilde{X}_t = t X_{1/t}, \qquad t > 0.$$

1°) With the notation of this section, prove that if P_x^δ is the law of X, then P_δ^x is the law of $\widetilde{X}$. In other words, for the BM with constant drift, time-inversion interchanges the drift and the starting point.

2°) Suppose that $d = 2$, $x = 0$ and $\delta \neq 0$, namely X is the complex BM with drift δ started at 0. Prove that $X_t = \rho_t \exp(i\gamma_{A_t})$ where $\rho_t = |X_t|$, γ is a linear BM independent of $\mathscr{R}_\infty$ and $A_t = \int_t^\infty \rho_s^{-2} ds$.

3°) If for $r > 0$ we set $T = \inf\{t : \rho_t = r\}$, then X_T is independent of $\mathscr{R}_T$. As a result, X_T and T are independent. Observe also that this holds equally for $\delta = 0$ (see also Exercise (1.17) Chap. III).

4°) For $r = 1$, prove that X_T follows the so-called von Mises distribution of density $C_\delta \exp(\langle \delta, \theta \rangle)$ with respect to the uniform distribution on S^1, where C_δ is a normalizing constant and $\langle \delta, \theta \rangle$ is the scalar product of δ and θ as vectors of $\mathbb{R}^2$.

(3.13) Exercise. Instead of Eq. (3.1) suppose that f satisfies the equation $f^2(x) = \beta f'(x) + \gamma$ where β and γ are constants independent of α, and carry the computations as far as possible.

* **(3.14) Exercise (O.U. processes and Lévy's formula).** 1°) Let X be the standard d-dimensional BM and $\rho = |X|$. Prove that Proposition (3.1) extends to

$$D_t = \exp\left\{\frac{\lambda}{2}\left(\rho_t^2 - dt\right) - \frac{\lambda^2}{2}\int_0^t \rho_s^2 ds\right\}$$

and prove that the infinitesimal generator of the transformed process is given on $\phi \in C^2$, by

$$\frac{1}{2}\Delta\phi(x) + \lambda\langle x, \nabla\phi(x)\rangle.$$

We call $\widetilde{P}_x$ the corresponding probability measures.

2°) Prove that under $\widetilde{P}_x$, the process X satisfies the SDE

$$X_t = x + B_t + \lambda \int_0^t X_s ds$$

where B is a d-dimensional BM. Deduce therefrom that it can be written $e^{\lambda t}\left(x+\widetilde{B}\left(\left(1-e^{-2\lambda t}\right)/2\lambda\right)\right)$, where $\widetilde{B}$ is a standard BM^d, and find its semi-group. [This kind of question will be solved in a general setting in the next chapter, but this particular case may be solved by using the method of Sect. 3 Chap. IV].

The process X may be called the d-dimensional OU process.

3°) For $d=1$ and $\lambda<0$ the process X is an OU process as defined in Exercise (1.13) Chap. III. Check that it can be made stationary by a suitable choice of the initial measure.

4°) Prove that

$$E_x\left[\exp\left(-\frac{\lambda^2}{2}\int_0^t \rho_s^2 ds\right) \mid \rho_t=\rho\right]$$
$$=\frac{\lambda t}{\sinh \lambda t}\exp\left(\frac{|x|^2+\rho^2}{2t}(1-\lambda t\coth \lambda t)\right) I_\nu\left(\frac{|x|\rho\lambda}{\sinh \lambda t}\right) \Big/ I_\nu\left(\frac{|x|\rho}{t}\right)$$

where $\nu=(d/2)-1$. The reader will observe that for $d=2$, this gives the law of the stochastic area S_t studied in Exercise (2.19) Chap. V. For $x=0$ or $\rho=0$, and $k^2=|x|^2+\rho^2$, the right-hand side becomes

$$\left(\frac{\lambda t}{\sinh \lambda t}\right)^{\nu+1}\exp\left(\frac{k^2}{2t}(1-\lambda t\coth \lambda t)\right).$$

5°) For $d=2$, prove that

$$E_0\left[\exp(i\lambda S_t) \mid B_t=z\right]=\frac{\lambda t}{\sinh \lambda t}\exp\left(-\frac{|z|^2}{2t}(\lambda t\coth \lambda t-1)\right).$$

(3.15) Exercise. Prove that the extended generator of the semi-group of BES^3 (Exercise (1.15) Chap. III and Sect. 3 Chap. VI) is equal on $C^2(]0,\infty[)$ to

$$\phi \longrightarrow \frac{1}{2}\phi''(x)+\frac{1}{x}\phi'(x).$$

[Hint: Use the form of the generator of BM killed at 0 found in Exercise (1.22) Chap. VII.]

(3.16) Exercise. If (P_t) and $(\widehat{P}_t)$ are in duality with respect to a measure ξ (see Sect. 4 Chap. VII), then the semi-group of the h-process of X_t and $(\widehat{P}_t)$ are in duality with respect to the measure $h\xi$. This was implicitly used in the time reversal results on BES^3 proved in Sect. 4 of Chap. VII.

(3.17) Exercise (Inverting Brownian motion in space). In $\mathbb{R}^d\backslash\{0\}$, $d\geq 3$, put $\phi(x)=x/|x|^2$.

1°) If B is a BM$^d(a)$ with $a\neq 0$, prove that $\phi(B)$ is a Markov process with transition function

$$Q_t(x,dy)=(2\pi t)^{-d/2}|y|^{-2d}\exp\left(-|x-y|^2/2t|x|^2|y|^2\right)dy.$$

2°) Call (τ_t) the time-change associated with $A_t = \int_0^t |B_s|^{-4} ds$ (see Lemma (3.12) Chap. VI) and prove that $Y_t = \phi(B_{\tau_t})$, $t < A_\infty$, is a solution to the SDE

$$Y_t = Y_0 + \beta_t - (d-2) \int_0^t \left(Y_s / |Y_s|^2 \right) ds,$$

where β is a BM stopped at time $A_\infty = \inf\{t : Y_t = 0\}$.

3°) Prove that the infinitesimal generator of Y is equal on $C^2(\mathbb{R}^d \backslash \{0\})$ to

$$\frac{1}{2}\Delta - (d-2)\langle y/|y|^2, \nabla \cdot \rangle,$$

and that Y is the h-process of BM^d associated with $h(x) = |x|^{2-d}$.

4°) Prove that under P_a, the law of A_∞ is given by the density

$$\left(\Gamma(\nu) \left(2|a|^2 \right)^\nu t^{\nu+1} \right)^{-1} \exp\left(-(2t|a|^2)^{-1} \right)$$

with $\nu = (d/2) - 1$.

5°) Conditionally on $A_\infty = u$, the process $(Y_t, t \leq u)$ is a Brownian Bridge between $a/|a|^2$ and 0 over the interval $[0, u]$.

This exercise may be loosely interpreted by saying that the BM^d looks like a Brownian Bridge on the Riemann sphere.

(3.18) Exercise. Let Y be a d-dimensional r.v. of law λ and set $B_t^\lambda = B_t + tY$ where B is a standard BM^d independent of Y. Call P^λ (resp. P^0) the law of B_t^λ (resp. B).

1°) Prove that $P^\lambda \lhd P^0$ with density $h_\lambda(B_t, t)$ where

$$h_\lambda(x, t) = \int \exp\left(\langle x, y \rangle - t|y|^2/2 \right) \lambda(dy).$$

2°) Prove that in the filtration $\left(\mathscr{F}_t^{B^\lambda} \right)$ the semimartingale decomposition of B^λ is given by

$$B_t^\lambda = B_t + \int_0^t \nabla_x (\log h_\lambda) \left(B_s^\lambda, s \right) ds.$$

Write down the explicit value of this decomposition in the following two cases:

i) $d = 1$ and $\lambda = \left(\delta_{-m} + \delta_m \right)/2$;

ii) $d = 3$ and λ is the rotation-invariant probability measure on the sphere of radius $r > 0$, in which case if R_t is the radial part of B_t^λ,

$$R_t = \gamma_t + r \int_0^t \coth(r R_s) ds$$

for a BM γ.

(3.19) Exercise (B.M. with drift and Lévy's equivalence). Let $\mu \geq 0$ and set $B_t^\mu = B_t + \mu t$ where B is a $\mathrm{BM}^1(0)$. Call L^μ the local time of B^μ at 0. In particular $B^0 = B$ and $L^0 = L$. Set $X^\mu = |B^\mu| + L^\mu$ and $X^0 = X$.

1°) Using the symmetry of the law of B, prove that for any bounded functional F

$$E\left[F\left(X_s^\mu, s \leq t\right)\right] = E\left[F\left(X_s, s \leq t\right)\cosh\left(\mu|B_t|\right)\exp\left(-\mu^2 t/2\right)\right].$$

2°) Using Lévy's identity and Corollary (3.6) of Chapter VI, prove that these expressions are yet equal to

$$E\left[F\left(X_s, s \leq t\right)\left(\sinh\left(\mu X_t\right)/\mu X_t\right)\exp\left(-\mu^2 t/2\right)\right].$$

3°) Put $S_t^\mu = \sup\left(B_s^\mu, s \leq t\right)$ and prove that the processes X^μ and $2S^\mu - B^\mu$ have the same law, namely, the law of the diffusion with infinitesimal generator

$$(1/2)d^2/dx^2 + \mu\coth(\mu x)d/dx.$$

4°) Prove that for $\mu \neq 0$ the processes $|B^\mu|$ and $S^\mu - B^\mu$ do not have the same law. [Hint: Their behaviors for large times are different.]

Prove more precisely that for every bounded functional F,

$$E\left[F\left(|B_s^\mu|, s \leq t\right)\right] = E\left[F\left(S_s^\mu - B_s^\mu, s \leq t\right)\Delta_t\right]$$

where $\Delta_t = \exp\left(-S_t^\mu\right)\left(1 + \exp\left(2\mu\left(S_t^\mu - B_t^\mu\right)\right)/2\right)$.

5°) Prove however that

$$\left(|B_s^\mu|, s \leq \tau_t \mid \tau_t < \infty\right) \overset{(d)}{=} \left(S_s^\mu - B_s^\mu, s \leq T_t\right),$$

where $\tau_t = \inf\left\{s \geq 0 : L_s^\mu > t\right\}$ and $T_t = \inf\left\{s \geq 0 : B_s^\mu = t\right\}$. Give closed form formulae for the densities of τ_t and T_t.

6°) Use the preceding absolute continuity relationship together with question 4°) in Exercise (4.9) Chapter VI to obtain a closed form for the Laplace transform of the law of

$$\inf\left\{s : S_s^\mu - B_s^\mu = a\right\}.$$

Notes and Comments

Sect. 1. What we call Girsanov's theorem – in agreement with most authors – has a long history beginning with Cameron-Martin ([1] and [2]), Maruyama ([1] and [2]), Girsanov [1], Van Schuppen-Wong [1]. Roughly speaking, the evolution has been from Gaussian processes to Markov processes, then to martingales. Cameron and Martin were interested in the transformation of the Brownian trajectory for which the old and new laws were equivalent; they first considered deterministic translations and afterwards random translations so as to deal with BM with non constant drift. The theory was extended to diffusions or more generally Markov processes by Maruyama and by Girsanov. Proposition (1.12) is typical of this stage. Finally with the advent of Martingale problems developed by Stroock and Varadhan, it became necessary to enlarge the scope of the results to martingales which was done by Van Schuppen and Wong; Theorem (1.4) is typically in the line of the latter. This already intricate picture must be completed by the relationship of Girsanov's theorem with the celebrated Feynman-Kac formula and the Doob's h-processes which are described in Sect. 3.

For the general results about changes of law for semimartingales let us mention Jacod-Mémin ([1], [2]) and Lenglart [3] and direct the reader to the book of Dellacherie and Meyer [1] vol. II. Other important papers in this respect are Yoeurp ([3], [4]) which stress the close connection between Girsanov's theorem and the theory of enlargements of filtrations. In fact the parenthood between the decomposition formulas of martingales in this theory and in the Girsanov set-up is obvious, but more precisely, Yoeurp [4] has shown that, by using Föllmer's measure, the enlargement decomposition formula could be interpreted as a special case of a wide-sense Girsanov's formula; further important developments continue to appear in this area.

The theory of enlargements of filtrations is one of the major omissions of this book. By turning a positive random variable into a stopping time (of the enlarged filtration), it provides alternative proofs for many results such as time reversal and path-decomposition theorems. The interested reader may look at Jeulin [2] and to *Grossissements de filtrations: exemples et applications*, Lecture Notes in Mathematics, vol. 1118, Springer (1985). A thorough exposition is also found in Dellacherie, Maisonneuve and Meyer [1]. Complementing Yoeurp's theoretical work mentioned above, enlargement of filtrations techniques have been used together with Girsanov's theorem; see for example Azéma-Yor [4], Föllmer-Imkeller [1] and Mortimer-Williams [1].

The theory of enlargements also allows to integrate anticipating processes with respect to a semimartingale; a different definition of the integral of such processes was proposed by Skorokhod [3] and continues to be the subject of many investigations (Buckdahn, Nualart, Pardoux, ...).

The Girsanov pair terminology is not standard and is introduced here for the first time. The proof of Kazamaki's criterion (Kazamaki [2]) presented here is due to Yan [1] (see also Lépingle-Mémin [1]).

Exercise (1.24) is taken from Liptser-Shiryaev [1] as well as Exercise (1.40) which appears also in the book of Friedman [1].

Sect. 2. The fundamental Theorem (2.2) is due to Cameron-Martin. We refer to Koval'chik [1] for an account of the subject and an extensive bibliography. Theorem (2.4) is from Clark [1], our presentation being borrowed from Rogers-Williams [1]. The result has been generalized to a large class of diffusions by Haussman [1], Ocone [2], and Bismut [2] for whom it is the starting point of his version of Malliavin's calculus.

Part of our exposition of the large deviations result for Brownian motion is borrowed from Friedman [1]. Theorem (2.11) is due to Schilder [1]. The theory of large deviations for Markov processes has been fully developed by Donsker and Varadhan in a series of papers [1]. We refer the reader to Azencott [1], Stroock [4] and Deuschel and Stroock [1]. The proof of Strassen's law of the iterated logarithm

(Strassen [1]) is borrowed from Stroock [4]. In connection with the result let us mention Chover [1] and Mueller [1].

Exercise (2.14) is from Yor [3]. The method displayed in Exercise (2.16) is standard in the theory of large deviations. The use of Skorokhod stopping times in Exercise (2.17) is the original idea of Strassen.

Sect. 3. The general principle embodied in Proposition (3.1) and which is the expression of Girsanov's theorem in the context of diffusions is found in more or less explicit ways in numerous papers such as Kunita [1], Yor [10], Priouret-Yor [1], Nagasawa [2], Elworthy [1], Elworthy-Truman [1], Ezawa et al. [2], Fukushima-Takeda [1], Oshima-Takeda [1], Ndumu [1], Truman [1], Gruet [3], among others.

The opérateur carré du champ was introduced by Kunita [1] and by Roth [1]. The reader is warned that some authors use 2Γ or $\Gamma/2$ instead of our Γ. Clearly, it can be defined only on sub-algebras of the domain of the (extended) infinitesimal generator and this raises the question of its existence when one studies a general situation (see Mokobodzki [1] who, in particular, corrects errors made earlier on this topic).

Proposition (3.4) may be found in Kunita [1] and Theorem (3.7) in Pitman-Yor [1]. For Exercise (3.12), let us mention Kent [1], Wendel [1] and [2], Pitman-Yor [1] and Watanabe ([1], [4]). Exercise (3.14) comes from Pitman-Yor [1] and Yor [10]; the formula of 4°) was the starting point for the decomposition of Bessel Bridges (see Pitman-Yor [2] and [3]). In connection with Exercise (3.14) let us also mention Gaveau [1] whose computation led him to an expression of the heat semigroup for the Heisenberg group. It is interesting to note that P. Lévy's formula for the stochastic area plays a central role in the probabilistic proof given by Bismut ([4] and [5]) of the Atiyah-Singer theorems.

Exercise (3.17) is taken from Yor [16], following previous work by L. Schwartz [1]. It is also related to Getoor [2], and the results have been further developed by Carne [3].

Renormalizations of the laws of Markov processes with a multiplicative functional and the corresponding limits in law have been considered in Roynette et al. [1].

Chapter IX. Stochastic Differential Equations

In previous chapters stochastic differential equations have been mentioned several times in an informal manner. For instance, if M is a continuous local martingale, its exponential $\mathscr{E}(M)$ satisfies the equality

$$\mathscr{E}(M)_t = 1 + \int_0^t \mathscr{E}(M)_s dM_s ;$$

this can be stated: $\mathscr{E}(M)$ is a solution to the stochastic differential equation

$$X_t = 1 + \int_0^t X_s dM_s,$$

which may be written in differential form

$$dX_t = X_t dM_t, \qquad X_0 = 1.$$

We have even seen (Exercise (3.10) Chap. IV) that $\mathscr{E}(M)$ is the only solution to this equation. Likewise we saw in Sect. 2 Chap. VII, that some Markov processes are solutions of what may be termed stochastic differential equations.

This chapter will be devoted to the formal definition and study of this notion.

§1. Formal Definitions and Uniqueness

Stochastic differential equations can be defined in several contexts of varying generality. For the purposes of this book, the following setting will be convenient.

As usual, the space $C(\mathbb{R}_+, \mathbb{R}^d)$ is denoted by $\mathbf{W}$. If $w(s)$, $s \geq 0$, denote the coordinate mappings, we set $\mathscr{B}_t = \sigma\big(w(s), s \leq t\big)$. A function f on $\mathbb{R}_+ \times \mathbf{W}$ taking values in $\mathbb{R}^r$ is predictable if it is predictable as a process defined on $\mathbf{W}$ with respect to the filtration $(\mathscr{B}_t)$. If X is a continuous process defined on a filtered space $(\Omega, \mathscr{F}_t, P)$, the map $s \to X_s(\omega)$ belongs to $\mathbf{W}$ and if f is predictable, we will write $f(s, X_.)$ or $f(s, X_.(\omega))$ for the value taken by f at time s on the path $t \to X_t(\omega)$. We insist that we write $X_.(\omega)$ here and not $X_s(\omega)$, because $f(s, X_.(\omega))$ may depend on the entire path $X_.(\omega)$ up to time s. The case where $f(s, w) = \sigma(s, w(s))$ for a function σ defined on $\mathbb{R}_+ \times \mathbb{R}^d$ is a particular, if important, case and we then have $f(s, X_.) = \sigma(s, X_s)$. In any case we have

(1.1) Proposition. *If X is $(\mathscr{F}_t)$-adapted, the process $f(t, X_.(\omega))$ is $(\mathscr{F}_t)$-predictable.*

Proof. Straightforward.

(1.2) Definition. *Given two predictable functions f and g with values in $d \times r$ matrices and d-vectors, a solution of the* stochastic differential equation *$e(f, g)$ is a pair (X, B) of adapted processes defined on a filtered probability space $(\Omega, \mathscr{F}_t, P)$ and such that*

i) B is a standard $(\mathscr{F}_t)$-Brownian motion in $\mathbb{R}^r$;
ii) for $i = 1, 2, \ldots, d$,

$$X_t^i = X_0^i + \sum_j \int_0^t f_{ij}(s, X_.)dB_s^j + \int_0^t g_i(s, X_.)ds.$$

Furthermore, we use the notation $e_x(f, g)$ if we impose the condition $X_0 = x$ a.s. on the solutions.

We will rather write ii) in vector form

$$X_t = X_0 + \int_0^t f(s, X_.)dB_s + \int_0^t g(s, X_.)ds$$

and will abbreviate "stochastic differential equation" to SDE. Of course, it is understood that all the integrals written are meaningful i.e., almost-surely,

$$\int_0^t \sum_{i,j} f_{ij}^2(s, X_.)ds < \infty, \qquad \int_0^t |g(s, X_.)|\, ds < \infty.$$

Consequently, a solution X is clearly a continuous semimartingale.

As we saw in Sect. 2 of Chap. VII, diffusions are solutions of SDE's of the simple kind where

$$f(s, X_.) = \sigma(s, X_s), \qquad g(s, X_.) = b(s, X_s)$$

for σ and b defined on $\mathbb{R}_+ \times \mathbb{R}^d$, or even in the homogeneous case

$$f(s, X_.) = \sigma(X_s), \qquad g(s, X_.) = b(X_s).$$

Such SDE's will be denoted $e(\sigma, b)$ rather than $e(f, g)$.

We also saw that to express a diffusion as a solution to an SDE we had to construct the necessary BM. That is why the pair (X, B) rather than X alone is considered to be the solution of the SDE.

We have already seen many examples of solutions of SDE's, which justifies the study of uniqueness, even though we haven't proved yet any general existence result. There are at least two natural definitions of uniqueness and this section will be devoted mainly to the study of their relationship.

(1.3) Definitions. 1°) *There is* pathwise uniqueness *for $e(f, g)$ if whenever (X, B) and (X', B') are two solutions defined on the same filtered space with $B = B'$ and $X_0 = X'_0$ a.s., then X and X' are indistinguishable.*

2°) *There is* uniqueness in law *for $e(f, g)$ if whenever (X, B) and (X', B') are two solutions with possibly different Brownian motions B and B' (in particular if (X, B) and (X', B') are defined on two different probability spaces $(\Omega, \mathscr{F}, \mathscr{F}_t, P)$ and $(\Omega', \mathscr{F}', \mathscr{F}'_t, P')$) and $X_0 \overset{(d)}{=} X'_0$, then the laws of X and X' are equal. In other words, X and X' are two versions of the same process.*

Uniqueness in law is actually equivalent to a seemingly weaker condition.

(1.4) Proposition. *There is uniquenes in law if, for every $x \in \mathbb{R}^d$, whenever (X, B) and (X', B') are two solutions such that $X_0 = x$ and $X'_0 = x$ a.s., then the laws of X and X' are equal.*

Proof. Let P be the law of (X, B) on the canonical space $C(\mathbb{R}_+, \mathbb{R}^{d+r})$. Since this is a Polish space, there is a regular conditional distribution $P(\omega, \cdot)$ for P with respect to $\mathscr{B}_0$. For almost every ω the last r coordinate mappings β^i still form a BM^r under $P(\omega, \cdot)$ and the integral

$$\int_0^t f(s, \xi_.)d\beta_s + \int_0^t g(s, \xi_.)ds,$$

where ξ stands for the vector of the first d coordinate mappings, makes sense. It is clear (see Exercise (5.16) Chap. IV) that, for almost every ω, the pair (ξ, β) is under $P(\omega, \cdot)$ a solution to $e(f, g)$ with $\xi_0 = \xi(\omega)$ $P(\omega, \cdot)$-a.s. If (X', B') is another solution we may likewise define $P'(\omega, \cdot)$ and the hypothesis implies that $P(\omega, \cdot) = P'(\omega, \cdot)$ for ω in a set of probability 1 for P and P'. If $X_0 \overset{(d)}{=} X'_0$ we get $P = P'$ and the proof is complete. □

Remark. The reader will find in Exercise (1.16) an example where uniqueness in law does not hold.

The relationship between the two kinds of uniqueness is not obvious. We will show that the first implies the second, but we start with another important

(1.5) Definition. *A solution (X, B) of $e(f, g)$ on $(\Omega, \mathscr{F}, \mathscr{F}_t, P)$ is said to be a* strong solution *if X is adapted to the filtration $(\mathscr{F}_t^B)$ i.e. the filtration of B completed with respect to P.*

By contrast, a solution which is not strong will be termed a weak solution.

For many problems, it is important to know whether the solutions to an SDE are strong. Strong solutions are "non-anticipative" functionals of the Brownian motion, that is, they are known up to time t as soon as B is known up to time t. This is important in as much as B is often the given data of the problem under consideration.

We now prepare for the main result of this section. Let $\mathbf{W}_1 = C(\mathbb{R}_+, \mathbb{R}^d)$ and $\mathbf{W}_2 = C(\mathbb{R}_+, \mathbb{R}^r)$. On $\mathbf{W}_1 \times \mathbf{W}_2$ we define, with obvious notation, the σ-algebras

$$\mathscr{B}_t^i = \sigma\left(w_s^i, s \le t\right), \qquad \mathscr{B}^i = \bigvee_t \mathscr{B}_t^i,$$

$$\widehat{\mathscr{B}}_t^i = \sigma\left(w_u^i - w_t^i, u \ge t\right), \quad i = 1, 2;$$

we observe that $\mathscr{B}^i = \mathscr{B}_t^i \vee \widehat{\mathscr{B}}_t^i$ for each t.

Let (X, B) be a solution to $e(f, g)$ and Q the image of P under the map $\phi : \omega \to (X_.(\omega), B_.(\omega))$ from Ω into $\mathbf{W}_1 \times \mathbf{W}_2$. The projection of Q on $\mathbf{W}_2$ is the Wiener measure and, as all the spaces involved are Polish spaces, we can consider a regular conditional distribution $Q(w_2, \cdot)$ with respect to this projection, that is, $Q(w_2, \cdot)$ is a probability measure on $\mathbf{W}_1 \times \mathbf{W}_2$ such that $Q(w_2, \mathbf{W}_1 \times \{w_2\}) = 1$ Q-a.s. and for every measurable set $A \subset \mathbf{W}_1 \times \mathbf{W}_2$, $Q(w_2, A) = E_Q\left[1_A \mid \mathscr{B}^2\right]$ Q-a.s.

(1.6) Lemma. *If $A \in \mathscr{B}_t^1$, the map $w_2 \to Q(w_2, A)$ is $\mathscr{B}_t^2$-measurable up to a negligible set.*

Proof. If $\mathscr{A}_1, \mathscr{A}_2, \widehat{\mathscr{A}}$ are three σ-algebras such that $\mathscr{A}_1 \vee \mathscr{A}_2$ is independent of $\widehat{\mathscr{A}}$ under a probability measure m, then for $A \in \mathscr{A}_1$,

$$m(A \mid \mathscr{A}_2) = m\left(A \mid \mathscr{A}_2 \vee \widehat{\mathscr{A}}\right) \qquad m\text{-a.s.}$$

Applying this to $\mathscr{B}_t^1, \mathscr{B}_t^2, \widehat{\mathscr{B}}_t^2$ and Q we get

$$Q(\cdot, A) = E_Q\left[1_A \mid \mathscr{B}^2\right] = E_Q\left[1_A \mid \mathscr{B}_t^2\right] \qquad Q\text{-a.s.},$$

which proves our claim.

(1.7) Theorem. *If pathwise uniqueness holds for $e(f, g)$, then*

i) uniqueness in law holds for $e(f, g)$;
ii) every solution to $e_x(f, g)$ is strong.

Proof. By Proposition (1.4), it is enough to prove i) to show that if (X, B) and (X', B') are two solutions defined respectively on (Ω, P) and (Ω', P') such that $X_0 = x$ and $X_0' = x$ a.s. for some $x \in \mathbb{R}^d$, then the laws of X and X' are equal.

Let $\mathbf{W}_1$ and $\mathbf{W}_1'$ be two copies of $C(\mathbb{R}_+, \mathbb{R}^d)$. With obvious notation derived from that in the previous lemma, we define a probability measure π on the product $\mathbf{W}_1 \times \mathbf{W}_1' \times \mathbf{W}_2$ by

$$\pi\left(dw_1, dw_1', dw_2\right) = Q(w_2, dw_1)\, Q'\left(w_2, dw_1'\right) W(dw_2)$$

where W is the Wiener measure on $\mathbf{W}_2$. If $\mathscr{F}_t = \sigma\left(w_1(s), w_1'(s), w_2(s), s \le t\right)$ then under π, the process $w_2(t)$ is an $(\mathscr{F}_t)$-BM^r. Indeed we need only prove that for any pair (s, t), with $s < t$, $w_2(t) - w_2(s)$ is independent of $\mathscr{F}_s$. Let $A \in \mathscr{B}_s^1$, $A' \in \mathscr{B}_s^{1'}$, $B \in \mathscr{B}_s^2$; by Lemma (1.6), for $\xi \in \mathbb{R}^r$,

$$
\begin{aligned}
E_\pi & \left[\exp(i\langle \xi, w_2(t) - w_2(s)\rangle) 1_A 1_{A'} 1_B\right] \\
& = \int_B \exp(i\langle \xi, w_2(t) - w_2(s)\rangle)\, Q(w_2, A)\, Q'\left(w_2, A'\right) W(dw_2) \\
& = \exp\left(-|\xi|^2(t-s)/2\right) \int_B Q(w_2, A)\, Q'\left(w_2, A'\right) W(dw_2) \\
& = \exp\left(-|\xi|^2(t-s)/2\right) \pi(A \times A' \times B)
\end{aligned}
$$

which is the desired result.

We now claim that (w_1, w_2) and (w_1', w_2) are two solutions to $e_x(f, g)$ on the same filtered space $\left(\mathbf{W}_1 \times \mathbf{W}_1' \times \mathbf{W}_2, \mathscr{F}_t, \pi\right)$. Indeed, if for instance

$$X_t = x + \int_0^t f(s, X_.)dB_s + \int_0^t g(s, X_.)ds$$

under P, then

$$w_1(t) = x + \int_0^t f(s, w_1)dw_2(s) + \int_0^t g(s, w_1)ds$$

under π because the joint law of $(f(s, X_.), g(s, X_.), B)$ under P is that of $(f(s, w_1), g(s, w_1), w_2)$ under π (see Exercise (5.16) in Chap. IV). Since moreover $w_1'(0) = x$ π-a.s., the property of path uniqueness implies that w_1 and w_1' are π-indistinguishable, hence $w_1(\pi) = w_1'(\pi)$, that is: $X(P) = X'(P')$ which proves i).

Furthermore, to say that w_1 and w_1' are π-indistinguishable is to say that π is carried by the set $\left\{(w_1, w_1', w_2) : w_1 = w_1'\right\}$. Therefore, for W-almost every w_2, under the probability measure $Q(w_2, dw_1) \otimes Q'\left(w_2, dw_1'\right)$ the variables w_1 and w_1' are simultaneously equal and independent; this is possible only if there is a measurable map F from $\mathbf{W}_2$ into $\mathbf{W}_1$ such that for W-almost every w_2,

$$Q(w_2, \cdot) = Q'(w_2, \cdot) = \varepsilon_{F(w_2)}.$$

But then the image of P by the map ϕ defined above Lemma (1.6) is carried by the set of pairs $(F(w_2), w_2)$, hence $X = F(B)$ a.s. By Lemma (1.6), X is adapted to the completion of the filtration of B. □

Remarks. 1) In the preceding proof, it is actually shown that the law of the pair (X, B) does not depend on the solution. Thus stated, the above result has a converse which is found in Exercise (1.20).

On the other hand, property i) alone does not entail pathwise uniqueness which is thus strictly stronger than uniqueness in law (see Exercise (1.19)). Likewise the existence of a strong solution is not enough to imply uniqueness, even uniqueness in law (see Exercise (1.16)).

2) We also saw in the preceding proof that for each x, if there is a solution to $e_x(f, g)$, then there is a function $F(x, \cdot)$ such that $F(x, B)$ is such a solution. It can be proved that this function may be chosen to be measurable in x as well, in which case for any random variable X_0, $F(X_0, B)$ is a solution to $e(f, g)$ with X_0 as initial value.

We now turn to some important consequences of uniqueness for the equations $e(\sigma, b)$ of the homogeneous type. In Proposition (2.6) of Chap. VII we saw that if (X, B) is a solution of $e_x(\sigma, b)$, then $X(P)$ is a solution to the martingale problem $\pi(x, a, b)$ with $a = \sigma\sigma^t$. By Proposition (1.4) it is now clear that the uniqueness of the solution to the martingale problem $\pi(x, a, b)$ for every $x \in \mathbb{R}^d$, implies uniqueness in law for $e(\sigma, b)$.

In what follows, we will therefore take as our basic data the locally bounded fields a and b and see what can be deduced from the uniqueness of the solution to $\pi(x, a, b)$ for every $x \in \mathbb{R}^d$. We will therefore be working on $\mathbf{W} = C(\mathbb{R}_+, \mathbb{R}^d)$ and with the filtration $(\mathscr{B}_t) = (\sigma(X_s, s \le t))$ where the X_t's are the coordinate mappings. If τ is a $(\mathscr{B}_t)$-stopping time, the σ-algebra $\mathscr{B}_\tau$ is countably generated (see Exercise (4.21), Chap. I) and for any probability measure P on $\mathbf{W}$, there is a regular conditional distribution $Q(w, \cdot)$ with respect to $\mathscr{B}_\tau$.

(1.8) Proposition. *If P is a solution to $\pi(x, a, b)$ and τ is a bounded stopping time, there is a P-null set N such that for $w \notin N$, the probability measure $\theta_\tau(Q(w, \cdot))$ is a solution to $\pi(X_\tau(w), a, b)$.*

Proof. For a fixed w, let $A = \{w' : X_0(w') = X_\tau(w)\}$. From the definition of a regular conditional distribution, it follows that

$$\theta_\tau\big(Q(w, \cdot)\big)(A) = Q\left(w, \{w' : X_\tau(w') = X_\tau(w)\}\right) = 1.$$

Thus, by Definition (2.3) in Chap. VII, we have to find a negligible set N such that for any $f \in C_K^\infty$, any $t > s$,

$$(+) \qquad E_{Q(w,\cdot)}\left[M_t^f \circ \theta_\tau \mid \theta_\tau^{-1}(\mathscr{B}_s)\right] = M_s^f \circ \theta_\tau \qquad Q(w, \cdot)\text{-a.s.}$$

for $w \notin N$. Equivalently, for $w \notin N$, we must have

$$(*) \qquad \int_A M_t^f \circ \theta_\tau(w') Q(w, dw') = \int_A M_s^f \circ \theta_\tau(w') Q(w, dw')$$

for any $A \in \theta_\tau^{-1}(\mathscr{B}_s)$ and $t > s$.

Recall that, by hypothesis, each M^f is a martingale. Let f be fixed and pick B in $\mathscr{B}_\tau$; by definition of Q, we have

$$E\left[1_B(w) \int_A M_t^f \circ \theta_\tau(w') Q(w, dw')\right] = E\left[1_B \cdot 1_A \cdot \left(M_{\tau+t}^f - M_\tau^f\right)\right].$$

Since $1_B \cdot 1_A$ is $\mathscr{B}_{\tau+s}$-measurable as well as M_τ^f and since, by the optional stopping theorem, $M_{\tau+t}^f$ is a $\mathscr{B}_{\tau+t}$-martingale, this is further equal to

$$E\left[1_B \cdot 1_A \left(M_{\tau+s}^f - M_\tau^f\right)\right] = E\left[1_B(w) \int_A M_s^f \circ \theta_\tau(w') Q(w, dw')\right].$$

As a result, there is a P-null set $N(A, f, s, t)$ such that $(*)$ holds for $w \notin N(A, f, s, t)$.

Now the equality (+) holds for every f in C_K^∞ if it holds for f in a countable dense subset $\mathscr{D}$ of C_K^∞; because of the continuity of X it holds for every s and t if it holds for s and t in $\mathbb{Q}$. Let $\mathscr{C}_s$ be a countable system of generators for $\theta_\tau^{-1}(\mathscr{B}_s)$; the set

$$N = \bigcup_{s,t\in\mathbb{Q}} \bigcup_{f\in\mathscr{D}} \bigcup_{A\in\mathscr{C}_s} N(A, f, s, t)$$

is P-negligible and is the set we were looking for.

(1.9) Theorem. *If for every $x \in \mathbb{R}^d$, there is one and only one solution P_x to the martingale problem $\pi(x, a, b)$ and, if for every $A \in \mathscr{B}(\mathbb{R}^d)$ and $t \geq 0$ the map $x \to P_x[X_t \in A]$ is measurable, then $(X_t, P_x, x \in \mathbb{R}^d)$ is a Markov process with transition function $P_t(x, A) = P_x[X_t \in A]$.*

Proof. For every event $\Gamma \in \mathscr{B}_\infty$, every bounded $\mathscr{B}_t$-stopping time τ and every $x \in \mathbb{R}^d$ we have, with obvious notation,

$$P_x\left[\theta_\tau^{-1}(\Gamma) \mid \mathscr{B}_\tau\right] = Q_x\left(\cdot, \theta_\tau^{-1}(\Gamma)\right);$$

the uniqueness in the statement together with the preceding result entails that

$$P_x\left[\theta_\tau^{-1}(\Gamma) \mid \mathscr{B}_\tau\right] = P_{X_\tau}[\Gamma] \qquad P_x\text{-a.s.}$$

Making $\tau = t$ and integrating, we get the semi-group property. □

Remark. With continuity assumptions on a and b, it may be shown that the semi-group just constructed is actually a Feller semi-group. This will be done in a special case in the following section.

Having thus described some of the consequences of uniqueness in law, we want to exhibit a class of SDE's for which the property holds. This will provide an opportunity of describing two important methods of reducing the study of SDE's to that of simpler ones, namely, the method of transformation of drift based on Girsanov's theorem and already alluded to in Sect. 3 of the preceding chapter and the method of time-change. We begin with the former which we treat both in the setting of martingale problems and of SDE's. For the first case, we keep on working with the notation of Proposition (1.8).

(1.10) Theorem. *Let a be a field of symmetric and non-negative matrices, b and c fields of vectors such that a, b and $\langle c, ac\rangle$ are bounded. There is a one-to-one and onto correspondence between the solutions to the martingale problems $\pi(x, a, b)$ and $\pi(x, a, b + ac)$. If P and Q are the corresponding solutions, then*

$$\left.\frac{dQ}{dP}\right|_{\mathscr{B}_t} = \exp\left\{\int_0^t \langle c(X_s), d\overline{X}_s\rangle - \frac{1}{2}\int_0^t \langle c, ac\rangle (X_s)ds\right\}$$

where $\overline{X}_t = X_t - \int_0^t b(X_s)ds$.

The displayed formula is the *Cameron-Martin formula*.

Proof. Let P be a solution to $\pi(x, a, b)$. By Proposition (2.4) in Chap. VII, we know that under P the process $\overline{X}$ is a vector local martingale with increasing process $\int_0^t a(X_s)ds$. If we set $Y_t = \int_0^t \langle c(X_s), d\overline{X}_s \rangle$, we have $\langle Y, Y \rangle_t = \int_0^t \langle c, ac \rangle (X_s)ds$ and, since $\langle c, ac \rangle$ is bounded, Novikov's criterion of Sect. 1 Chap. VIII asserts that $\mathscr{E}(Y)$ is a martingale. Thus one can define a probability measure Q by $Q = \mathscr{E}(Y)_t . P$ on $\mathscr{B}_t$, which is the formula in the statement. We now prove that Q is a solution to $\pi(x, a, b + ac)$ by means of Proposition (2.4) in Chap. VII.

For $\theta \in \mathbb{R}^d$, the process $M_t^\theta = \langle \theta, \overline{X}_t - x \rangle$ is a P-local martingale with increasing process $A_t = \int_0^t \langle \theta, a(X_s)\theta \rangle\, ds$. Thus by Theorem (1.4) in Chap. VIII, $M^\theta - \langle M^\theta, Y \rangle$ is a Q-local martingale with the same increasing process A_t. It is furthermore easily computed that

$$\langle M^\theta, Y \rangle_t = \left\langle \theta, \int_0^t ac(X_s)ds \right\rangle.$$

As a result, $\left\langle \theta, X_t - x - \int_0^t b(X_s)ds - \int_0^t ac(X_s)ds \right\rangle$ is a Q-local martingale with increasing process A_t which proves our claim.

The fact that the correspondence is one-to-one and onto follows from Proposition (1.10) in Chap. VIII applied on each subinterval $[0, t]$. □

The above result has an SDE version which we now state.

(1.11) Theorem. *Let f (resp. g, h) be predictable functions on* $\mathbf{W}$ *with values in the symmetric non-negative $d \times d$ matrices (resp. d-vectors) and assume that h is bounded. Then, there exist solutions to $e_x(f, g)$ if and only if there exist solutions to $e_x(f, g + fh)$. There is uniqueness in law for $e(f, g)$ if and only if there is uniqueness in law for $e(f, g + fh)$.*

Proof. If (X, B) is a solution to $e(f, g)$ on a space $(\Omega, \mathscr{F}, P)$, we define a probability measure Q by setting $Q = \mathscr{E}(M)_t . P$ on $\mathscr{F}_t$ where $M_t = \int_0^t h(s, X_.)dB_s$. The process $\widetilde{B}_t = B_t - \int_0^t h(s, X_.)ds$ is a BM and $(X, \widetilde{B})$ is a solution of $e(f, g + fh)$ under Q. The details are left to the reader as an exercise. □

The reader will observe that the density $\frac{dQ}{dP}\big|_{\mathscr{F}_t}$ is simpler than in the Cameron-Martin formula. This is due to the fact that we have changed the accompanying BM. One can also notice that the assumption on h may be replaced by

$$E\left[\exp\left(\frac{1}{2}\int_0^t \|h(s, X_.)\|^2\, ds\right)\right] < \infty \qquad \text{for every } t > 0.$$

(1.12) Corollary. *Assume that, for every s and x, the matrix $\sigma(s, x)$ is invertible and that the map $(s, x) \to \sigma(s, x)^{-1}$ is bounded; if $e(\sigma, 0)$ has a solution, then for any bounded measurable b, the equation $e(\sigma, b)$ has a solution. If uniqueness in law holds for $e(\sigma, 0)$, it holds for $e(\sigma, b)$.*

Proof. We apply the previous result with $f(s, X_.) = \sigma(s, X_s)$, $g(s, X_.) = b(s, X_s)$ and $h(s, X_.) = -\sigma(s, X_s)^{-1}b(s, X_s)$.

Remark. Even if the solutions of $e(f, 0)$ are strong, the solutions obtained for $e(f, g)$ by the above method of transformation of drift are not always strong as will be shown in Sect. 3.

We now turn to the method of time-change.

(1.13) Proposition. *Let γ be a real-valued function on $\mathbb{R}^d$ such that $0 < k \leq \gamma \leq K < \infty$; there is a one-to-one and onto correspondence between the solutions to the martingale problem $\pi(x, a, b)$ and the solutions to the martingale problem $\pi(x, \gamma a, \gamma b)$.*

Proof. With the notation of Proposition (1.8) define

$$A_t = \int_0^t \gamma(X_s)ds,$$

and let (τ_t) be the associated time-change (Sect. 1 Chap. V). We define a measurable transformation ϕ on $\mathbf{W}$ by setting $X(\phi(w))_t = X_{\tau_t}(w)$. Let P be a solution to the martingale problem $\pi(x, a, b)$; for any pair (s, t), $s < t$, and $A \in \mathscr{B}_s$, we have, for $f \in C_K^\infty$,

$$\begin{aligned} &\int_A \left(f(X_t) - f(X_s) - \int_s^t \gamma(X_u)Lf(X_u)du \right) d\phi(P) \\ &= \int_{\phi^{-1}(A)} \left(f(X_{\tau_t}) - f(X_{\tau_s}) - \int_s^t \gamma(X_{\tau_u})Lf(X_{\tau_u})du \right) dP \\ &= \int_{\phi^{-1}(A)} \left(f(X_{\tau_t}) - f(X_{\tau_s}) - \int_{\tau_s}^{\tau_t} Lf(X_u)du \right) dP \end{aligned}$$

thanks to the time-change formula of Sect. 1 Chap. V. Now since $\phi^{-1}(A) \in \mathscr{B}_{\tau_s}$ the last integral vanishes, which proves that $\phi(P)$ is a solution to $\pi(x, \gamma a, \gamma b)$. Using γ^{-1} instead of γ, we would define a map ψ such that $\psi(\phi(P)) = P$ which completes the proof. □

Together with the result on transformation of drift, the foregoing result yields the following important example of existence and uniqueness.

(1.14) Corollary. *If σ is a bounded function on the line such that $|\sigma| \geq \varepsilon > 0$ and b a bounded function on $\mathbb{R}_+ \times \mathbb{R}$, there is existence and uniquenss in law for the SDE $e(\sigma, b)$. Moreover, if P_x is the law of the solution such that $X_0 = x$, for any $A \in \mathscr{B}(\mathbb{R})$, the map $x \to P_x[X_t \in A]$ is measurable.*

Proof. By Corollary (1.12) it is enough to consider the equation $e(\sigma, 0)$ and since the BM started at x is obviously the only solution to $e_x(1, 0)$, the result follows from the previous Proposition. The measurability of $P_.[X_t \in A]$ follows from the fact that the P_x's are the images of W_x under the same map.

Remark. By Theorem (1.9) the solutions of $e(\sigma, b)$ form a homogeneous Markov process when b does not depend on s. Otherwise, the Markov process we would get would be non homogeneous. Finally it is worth recording that the above argument does not carry over to $d > 1$, where the corresponding result, namely that for uniformly elliptic matrices, is much more difficult to prove.

(1.15) Exercise. Let (Y, B) be a solution to $e = e_y(f, g)$ and suppose that f never vanishes. Set

$$A_t = \int_0^t (2 + Y_s/(1 + |Y_s|))\, ds$$

and call T_t the inverse of A_t. Prove that $X_t = B_{T_t}$ is a pure local martingale if and only if Y is a strong solution to e.

\# **(1.16) Exercise.** 1°) Let $a(x) = 1 \wedge |x|^\alpha$ with $0 < \alpha < 1/2$ and B be the standard linear BM. Prove that the process $\int_0^t a^{-2}(B_s)ds$ is well-defined for any $t > 0$; let τ_t be the time-change associated with it.

2°) Prove that the processes $X_t = B_{\tau_t}$ and $X_t = 0$ are two solutions for $e_0(a, 0)$ for which consequently, uniqueness in law does not hold. Observe that the second of these solutions is strong.

\# **(1.17) Exercise.** A family X^x of $\mathbb{R}^d$-valued processes with $X_0^x = x$ a.s. is said to have the *Brownian scaling property* if for any $c > 0$, the processes $c^{-1}X_{c^2t}^x$ and $X_t^{c^{-1}x}$ have the same law. If uniqueness in law holds for $e(\sigma, b)$ and if X^x is a solution to $e_x(\sigma, b)$, prove that if $\sigma(cx) = \sigma(x)$ and $cb(cx) = b(x)$ for every $c > 0$ and $x \in \mathbb{R}^d$, then X^x has the Brownian scaling property. In particular if ϕ is a function on the unit sphere and $b(x) = \|x\|^{-1}\phi(x/\|x\|)$, the solutions to $e_x(\alpha I_d, b)$ have the Brownian scaling property.

* **(1.18) Exercise.** In the situation of Corollary (1.14) let (X, B) be a solution to $e_x(\sigma, b)$ and set $Y_t = X_t - x - \int_0^t b(s, X_s)ds$.

1°) Let $\mathbf{W}^x$ be the space of continuous functions w on $\mathbb{R}_+$, such that $w(0) = x$. For $w \in \mathbf{W}^x$, set

$$\begin{aligned} \psi_n(t, w) &= \sigma^{-1}(x) \quad \text{for} \quad 0 \le t < 2^{-n}, \\ \psi_n(t, w) &= 2^n \int_{(k-1)2^{-n}}^{k2^{-n}} \sigma^{-1}(w_s)ds \quad \text{for} \quad k2^{-n} \le t < (k+1)2^{-n}. \end{aligned}$$

Prove that for every t,

$$E\left[\left(\int_0^t \left(\psi_n(s, X_.) - \sigma^{-1}(X_s)\right) dY_s\right)^2\right] \underset{n\to\infty}{\longrightarrow} 0.$$

2°) Prove that there is an adapted function Φ from $\mathbf{W}^x$ to $\mathbf{W}^0$ which depends only on the law of X and is such that $B = \Phi(X)$.

3°) Derive from 2°) that if there exists a strong solution, then there is pathwise uniqueness.

[Hint: Prove that if (X, B) and (X', B) are two solutions, then $(X, B) \overset{(d)}{=} (X', B)$.]

\# **(1.19) Exercise.** 1°) If β is a BM(0) and $B_t = \int_0 \operatorname{sgn}(\beta_s) d\beta_s$, prove that (β, B) and $(-\beta, B)$ are two solutions to $e_0(\operatorname{sgn}, 0)$.

More generally, prove that, if $(\varepsilon_u, u \geq 0)$ is predictable with respect to the natural filtration of β, and takes only values $+1$ and -1, then $\left(\varepsilon_{g_t}\beta_t, B_t; t \geq 0\right)$ is a solution to $e_0(\operatorname{sgn}, 0)$, where $g_t = \sup\{s < t : \beta_s = 0\}$.

2°) Prove that $e_0(\operatorname{sgn}, 0)$ cannot have a strong solution.

[Hint: If X is a solution, write Tanaka's formula for $|X|$.]

(1.20) Exercise. 1°) Retain the notation of Lemma (1.6) and prove that (X, B) is a strong solution to $e_x(f, g)$ if and only if there is an adapted map F from $\mathbf{W}_2$ into $\mathbf{W}_1$ such that

$$Q(w_2, \cdot) = \varepsilon_{F(w_2)} \qquad Q\text{-a.s.}$$

2°) Let (X, B) and (X', B) be two solutions to $e_x(f, g)$ with respect to the same BM. Prove that if

i) $(X, B) \stackrel{(d)}{=} (X', B)$,
ii) one of the two solutions is strong,

then $X = X'$.

§2. Existence and Uniqueness in the Case of Lipschitz Coefficients

In this section we assume that the functions f and g of Definition (1.2) satisfy the following Lipschitz condition: there exists a constant K such that for every t and w,

$$\left|f(t, w) - f(t, w')\right| + \left|g(t, w) - g(t, w')\right| \leq K \sup_{s \leq t} |w(s) - w'(s)|$$

where $|\ |$ stands for a norm in the suitable space. Under this condition, given a Brownian motion B in $\mathbb{R}^r$, we will prove that for every $x \in \mathbb{R}^d$, there is a unique process X such that (X, B) is a solution to $e_x(f, g)$; moreover this solution is strong. As the pair (B_t, t) may be viewed as a $r+1$-dimensional semimartingale we need only prove the more general

(2.1) Theorem. *Let $(\Omega, \mathscr{F}_t, P)$ be a filtered space such that $(\mathscr{F}_t)$ is right-continuous and complete and Z a continuous r-dimensional semimartingale. If f satisfies the above Lipschitz condition and if, for every y, $f(\cdot, \bar{y})$ is locally bounded where $\bar{y}(t) \equiv y$, then for every $x \in \mathbb{R}^d$, there is a unique (up to indistinguishability) process X such that*

$$X_t = x + \int_0^t f(s, X_.) dZ_s.$$

Moreover, X is $(\mathscr{F}_t^Z)$-adapted.

Proof. We deal only with the case $d = 1$, the added difficulties of the general case being merely notational. If $M + A$ is the canonical decomposition of Z we first suppose that the measures $d\langle M, M\rangle_t$ and $|dA|_t$ on the line are dominated by the Lebesgue measure dt.

Let x be a fixed real number. For any process U with the necessary measurability conditions, we set

$$(SU)_t = x + \int_0^t f(s, U_\cdot)dZ_s.$$

If V is another such process, we set

$$\Phi_t(U, V) = E\left[\sup_{s\le t}|U_s - V_s|^2\right].$$

Because any two real numbers h and k satisfy $(h+k)^2 \le 2(h^2+k^2)$, we have

$$\begin{aligned}\Phi_t(SU, SV) \;\le\; & 2E\left[\sup_{s\le t}\left(\int_0^s (f(r, U_\cdot) - f(r, V_\cdot))\,dM_r\right)^2\right.\\ & \left.+\sup_{s\le t}\left(\int_0^s |f(r, U_\cdot) - f(r, V_\cdot)|\;|dA|_r\right)^2\right]\end{aligned}$$

and by the Doob and Cauchy-Schwarz inequalities, it follows that

$$\begin{aligned}\Phi_t(SU, SV) \;\le\; & 8E\left[\left(\int_0^t (f(r, U_\cdot) - f(r, V_\cdot))\,dM_r\right)^2\right]\\ & +2E\left[\left(\int_0^t |dA|_s\right)\left(\int_0^t |f(r, U_\cdot) - f(r, V_\cdot)|^2\;|dA|_r\right)\right]\\ \le\; & 8E\left[\int_0^t (f(r, U_\cdot) - f(r, V_\cdot))^2\,d\langle M, M\rangle_r\right]\\ & +2tE\left[\int_0^t |f(r, U_\cdot) - f(r, V_\cdot)|^2\;|dA|_r\right]\\ \le\; & 2K^2(4+t)E\left[\int_0^t \sup_{s\le r}|U_s - V_s|^2 dr\right]\\ =\; & 2K^2(4+t)\int_0^t \Phi_r(U, V)dr.\end{aligned}$$

Let us now define inductively a sequence (X^n) of processes by setting $X^0 \equiv x$ and $X^n = S(X^{n-1})$; let us further pick a time T and set $C = 2K^2(4+T)$. Using the properties of f it is easy to check that $D = \Phi_T(X^0, X^1)$ is finite. It then follows from the above computation that for every $t \le T$ and every n,

$$\Phi_t\left(X^{n-1}, X^n\right) \le DC^nT^n/n!\,.$$

Consequently

$$\sum_{n=1}^{\infty} \left\| \sup_{s \le t} \left| X_s^n - X_s^{n-1} \right| \right\|_2 < \infty.$$

Thus, the series $\sum_{n=1}^{\infty} \sup_{s \le t} \left| X_s^n - X_s^{n-1} \right|$ converges a.s. and as a result, X^n converges a.s., uniformly on every bounded interval, to a continuous process X. By Theorem (2.12) Chap. IV, $X = SX$, in other words X is a solution to the given equation.

To prove the uniqueness, we consider two solutions X and Y and put $T_k = \inf\{t : |X_t| \text{ or } |Y_t| > k\}$. Let $\widetilde{S}$ be defined as S but with Z^{T_k} in lieu of Z. Then it is easily seen that $X^{T_k} = \widetilde{S}(X^{T_k})$ and likewise for Y, so that for $t < T$

$$\Phi_t \left(X^{T_k}, Y^{T_k} \right) = \Phi_t \left(\widetilde{S} X^{T_k}, \widetilde{S} Y^{T_k} \right) \le C \int_0^t \Phi_s \left(X^{T_k}, Y^{T_k} \right) ds.$$

Since by the properties of f, the function $\Phi_t(X^{T_k}, Y^{T_k})$ is locally bounded, Gronwall's lemma implies that $\Phi_.(X^{T_k}, Y^{T_k})$ is identically zero, whence $X = Y$ on $[0, T_k \wedge T]$ follows. Letting k and T go to infinity completes the proof in the particular case.

The general case can be reduced to the particular case just studied by a suitable time-change. The process $A'_t = t + \langle M, M \rangle_t + \int_0^t |dA|_s$ is continuous and strictly increasing. If we use the time-change C_t associated with A'_t, then $\widetilde{M}_t = M_{C_t}$ and $\widetilde{A}_t = A_{C_t}$ satisfy the hypothesis of the particular case dealt with above. Since $C_t \le t$, one has

$$|f(C_t, U_.) - f(C_t, V_.)| \le K \sup_{s \le t} |U_s - V_s|,$$

and this condition is sufficient for the validity of the reasoning in the first part of the proof, so that the equation $\widetilde{X}_t = x + \int_0^t f(C_s, \widetilde{X}_.) d\widetilde{Z}_s$ has a unique solution. By the results of Sect. 1 Chap. V, the process $X_t = \widetilde{X}_{A'_t}$ is the unique solution to the given equation. □

As in the case of ordinary differential equations, the above result does not provide any practical means of obtaining closed forms in concrete cases. It is however possible to do so for the class of equations defined below. The reader may also see Exercise (2.8) for a link between SDE's and ODE's (ordinary differential equations).

(2.2) Definition. *A stochastic equation is called* linear *if it can be written*

$$Y_t = H_t + \int_0^t Y_s dX_s$$

where H and X are two given continuous semimartingales.

It can also be written as

$$dY_t = dH_t + Y_t dX_t, \qquad Y_0 = H_0.$$

An important example is the *Langevin equation* $dV_t = dB_t - \beta V_t dt$, where B is a linear BM and β a real constant, which was already studied in Exercise (3.14) of Chap. VIII. Another example is the equation $Y_t = 1 + \int_0^t Y_s dX_s$ for which we know (Sect. 3 Chap. IV) that the unique solution is $Y = \mathscr{E}(X)$. Together with the formula for ordinary linear differential equations, this leads to the closed form for solutions of linear equations, the existence and uniqueness of which are ensured by Theorem (2.1).

(2.3) Proposition. *The solution to the linear equation of Definition (2.2) is*

$$Y_t = \mathscr{E}(X)_t \left(H_0 + \int_0^t \mathscr{E}(X)_s^{-1} \left(dH_s - d\langle H, X\rangle_s \right) \right);$$

in particular, if $\langle H, X\rangle = 0$, *then*

$$Y_t = \mathscr{E}(X)_t \left(H_0 + \int_0^t \mathscr{E}(X)_s^{-1} dH_s \right).$$

Proof. Let us compute $\int_0^t Y_s dX_s$ for Y given in the statement. Because of the equality $\mathscr{E}(X)_s dX_s = d\mathscr{E}(X)_s$ and the integration by parts formula, we get

$$\begin{aligned}
\int_0^t & Y_s dX_s \\
&= H_0 \int_0^t \mathscr{E}(X)_s dX_s + \int_0^t \mathscr{E}(X)_s dX_s \left(\int_0^s \mathscr{E}(X)_u^{-1} \left(dH_u - d\langle H, X\rangle_u \right) \right) \\
&= -H_0 + H_0 \mathscr{E}(X)_t + \mathscr{E}(X)_t \int_0^t \mathscr{E}(X)_s^{-1} \left(dH_s - d\langle H, X\rangle_s \right) \\
&\quad - \int_0^t \mathscr{E}(X)_s \left(\mathscr{E}(X)_s^{-1} dH_s - \mathscr{E}(X)_s^{-1} d\langle H, X\rangle_s \right) \\
&\quad - \left\langle \mathscr{E}(X), \int_0^{\cdot} \mathscr{E}(X)_s^{-1} \left(dH_s - d\langle H, X\rangle_s \right) \right\rangle_t \\
&= Y_t - H_t + \langle H, X\rangle_t - \left\langle \int_0^{\cdot} \mathscr{E}(X)_s dX_s, \int_0^{\cdot} \mathscr{E}(X)_s^{-1} dH_s \right\rangle_t \\
&= Y_t - H_t,
\end{aligned}$$

which is the desired result. □

Remark. This proof could also be written without prior knowledge of the form of the solution (see Exercise (2.6) 2°).

One may also prove that if H is a progressive process and not necessarily a semimart., the process

$$Y = H - \mathscr{E}(X) \int H \, d\left(\mathscr{E}(X)^{-1} \right)$$

is a solution of the linear Equation (2.2). The reader will check that this jibes with the formula of Proposition (2.3) if H is a semimart..

The solution to the Langevin equation starting at v is thus given by

$$V_t = e^{-\beta t}\left(v + \int_0^t e^{\beta s} dB_s\right).$$

For $\beta > 0$ this is the OU process with parameter β of Exercise (1.13) of Chap. III (see Exercise (2.16)). The integral $\int_0^t V_s ds$ is sometimes used by physicists as another mathematical model of physical Brownian motion. Because of this interpretation, the process V is also called the OU *velocity* process of parameter β. In physical interpretations, β is a strictly positive number. For $\beta = 0$, we get $V_t = v + B_t$. In all cases, it also follows from the above discussion that the infinitesimal generator of the OU process of parameter β is the differential operator $\frac{1}{2}\frac{\partial^2}{\partial x^2} - \beta x \frac{\partial}{\partial x}$.

We now go back to the general situation of the equation $e(f, g)$ with f and g satisfying the conditions stated at the beginning of the section. We now know that for a given Brownian motion B on a space $(\Omega, \mathscr{F}_t, P)$ and every $x \in \mathbb{R}^d$, there is a unique solution to $e_x(f, g)$ which we denote by X_t^x. We will prove that X^x may be chosen within its indistinguishability class so as to get continuity in x.

(2.4) Theorem. *If f and g are bounded, there exists a process X_t^x, $x \in \mathbb{R}^d$, $t \in \mathbb{R}_+$, with continuous paths with respect to both variables t and x such that, for every x,*

$$X_t^x = x + \int_0^t f(s, X_\cdot^x) dB_s + \int_0^t g(s, X_\cdot^x) ds \qquad P\text{-a.s.}$$

Proof. Pick $p \geq 2$ and $t > 0$. Retain the notation of the proof of Theorem (2.1) writing S_x, rather than S, to stress the starting point x.

Because $|a + b + c|^p \leq 3^{p-1}(|a|^p + |b|^p + |c|^p)$, we have

$$\begin{aligned}
&\sup_{s \leq t} \left|S_x(U)_s - S_y(V)_s\right|^p \\
&\leq 3^{p-1}\left\{|x - y|^p + \sup_{s \leq t}\left|\int_0^s (f(r, U_\cdot) - f(r, V_\cdot))\, dB_r\right|^p \right. \\
&\quad \left. + \sup_{s \leq t}\left|\int_0^s (g(r, U_\cdot) - g(r, V_\cdot))\, dr\right|^p\right\}.
\end{aligned}$$

Thanks to the BDG and Hölder inequalities

$$\begin{aligned}
&E\left[\sup_{s \leq t}\left|\int_0^s (f(r, U_\cdot) - f(r, V_\cdot))\, dB_r\right|^p\right] \\
&\leq C_p E\left[\left(\int_0^t (f(r, U_\cdot) - f(r, V_\cdot))^2\, dr\right)^{p/2}\right] \\
&\leq C_p t^{(p-2)/2} E\left[\int_0^t |f(r, U_\cdot) - f(r, V_\cdot)|^p\, dr\right] \\
&\leq K^p C_p t^{(p-2)/2} E\left[\int_0^t \sup_{s \leq r} |U_s - V_s|^p dr\right].
\end{aligned}$$

Likewise,

$$\sup_{s\le t}\left|\int_0^s (g(r,U_\cdot)-g(r,V_\cdot))\,dr\right|^p \le K^p t^{p-1}\int_0^t \sup_{s\le r}|U_s-V_s|^p dr.$$

Applying this to $U = X^x$ and $V = X^y$ where X^x (resp. X^y) is a solution to $e_x(f,g)$ (resp. $e_y(f,g)$) and setting $h(t) = E\left[\sup_{s\le t}\left|X_s^x - X_s^y\right|^p\right]$, it follows that

$$h(t) \le b_t^p|x-y|^p + c_t^p\int_0^t h(s)ds$$

for two constants b_t^p and c_t^p. Gronwall's lemma then implies that there is a constant a_t^p depending only on p and t such that

$$E\left[\sup_{s\le t}|X_s^x - X_s^y|^p\right] \le a_t^p|x-y|^p.$$

By Kolmogorov's criterion (Theorem (2.1) of Chap. I), we can get a bicontinuous modification of X and it is easily seen that for each x, the process X^x is still a solution to $e_x(f,g)$.

Remark. The hypothesis that f and g are bounded was used to insure that the function h of the proof is finite. In special cases, h may be finite with unbounded f and g and the result will still obtain.

From now on, we turn to the case of $e(\sigma,b)$ where σ and b are bounded Lipschitz functions of x alone. If we set

$$P_t(x,A) = P\left[X_t^x \in A\right]$$

we know from Theorem (1.9) that each X^x is a Markov process with transition function P_t. We actually have the

(2.5) Theorem. *The transition function P_t is a Feller transition function.*

Proof. Pick f in C_0. From the equality $P_t f(x) = E\left[f(X_t^x)\right]$, it is plain that the function $P_t f$ is continuous. It is moreover in C_0; indeed,

$$|P_t f(x)| \le \sup_{y:|y-x|\le r}|f(y)| + \|f\|P\left[|X_t^x - x| > r\right]$$

and

$$\begin{aligned} P\left[|X_t^x - x| > r\right] &\le r^{-2}E\left[|X_t^x - x|^2\right] \\ &\le 2r^{-2}E\left[\left|\int_0^t \sigma\left(X_s^x\right)dB_s\right|^2 + \left|\int_0^t b\left(X_s^x\right)ds\right|^2\right] \\ &\le 2k^2r^{-2}(t+t^2) \end{aligned}$$

where k is a uniform bound for σ and b. By letting x, then r, go to infinity, we get $\lim_{|x|\to\infty} P_t f(x) = 0$.

On the other hand, for each x, $t \to P_t f(x)$ is also clearly continuous which completes the proof.

With respect to the program outlined at the end of Sect. 1 in Chap. VII we see that the methods of stochastic integration have allowed us to construct Feller processes with generators equal on C_K^2 to

$$\frac{1}{2}\sum a_{ij}(x)\frac{\partial^2}{\partial x_i \partial x_j} + \sum b_i(x)\frac{\partial}{\partial x_i},$$

whenever σ and b are bounded and Lipschitz continuous and $a = \sigma\sigma^t$.

\# **(2.6) Exercise (More on Proposition (2.3)).** 1°) Denoting by $*$ the backward integral (Exercise (2.18) of Chap. IV), show that in Proposition (2.3) we can write

$$Y_t = \mathscr{E}(X)_t \left(y + \int_0^t \mathscr{E}(X)_s^{-1} * dH_s \right)$$

which looks even more strikingly like the formulas for ordinary linear differential equations.

2°) Write down the proof of Proposition (2.3) in the following way: set $Y = \mathscr{E}(X)Z$ and find the conditions which Z must satisfy in order that Y be a solution.

(2.6) Bis Exercise (Vector linear equations). If X is a $d \times d$ matrix of cont. semimarts. and H a $r \times d$ matrix of locally bounded predictable processes, we define the right stochastic integral $(H \cdot dX)_t = \int_0^t H_s dX_s$ as the $r \times d$ matrix whose general term is given by

$$\sum_k \int_0^t H_s^{ik} dX_s^{kj}.$$

Likewise, there is a left integral $dX \cdot H$ provided that H and X have matching dimensions.

1°) If Y is a $r \times d$ matrix of cont. semimarts., we define $\langle Y, X \rangle$ as the $r \times d$ matrix whose entries are the processes

$$\sum_k \langle Y^{ik}, X^{kj} \rangle.$$

Prove that

$$d(YX) = dY \cdot X + Y \cdot dX + d\langle Y, X \rangle,$$

and that, provided the dimensions match,

$$\langle H \cdot dY, X \rangle = H \cdot d\langle Y, X \rangle, \quad \langle Y, dX \cdot H \rangle = d\langle Y, X \rangle \cdot H.$$

2°) Given X, let us call $\mathscr{E}(X)$ the $d \times d$ matrix-valued process which is the unique solution of the SDE

$$U_t = I_d + \int_0^t U_s dX_s,$$

and set $\mathscr{E}'(X) = \mathscr{E}(X^t)^t$. Prove that $\mathscr{E}'(X)$ is the solution to a linear equation involving left integrals. Prove that $\mathscr{E}'(-X + \langle X, X \rangle)$ is the inverse of the matrix

$\mathscr{E}(X)$ (which thus is invertible, a fact that can also be proved by showing that its determinant is the solution of a linear equation in dimension 1).

[Hint: Compute $d\big(\mathscr{E}(X)\mathscr{E}'(-X + \langle X, X\rangle)\big)$.]

3°) If H is a $r \times d$ matrix of cont. semimarts., prove that the solution to the equation

$$Y_t = H_t + \int_0^t Y_s dX_s,$$

is equal to

$$\left(H_0 + \int_0^t (dH_s - d\langle H, X\rangle_s)\, \mathscr{E}(X_s)^{-1}\right) \mathscr{E}(X)_t.$$

State and prove the analogous result for the equation

$$Y_t = H_t + \int_0^t dX_s Y_s.$$

(2.7) Exercise. Let F be a real-valued continuous function on $\mathbb{R}$ and f the solution to the ODE

$$f'(s) = F(s)f(s); \qquad f(0) = 1.$$

1°) Let G be another continuous function and X a continuous semimartingale. Prove that

$$Z_t = f(t)\left[z + \int_0^t f(u)^{-1} G(u) dX_u\right]$$

is the unique solution to the SDE

$$Z_t = z + \int_0^t F(u) Z_u du + \int_0^t G(u) dX_u.$$

Moreover

$$Z_t = \frac{f(t)}{f(s)} Z_s + \int_s^t \frac{f(t)}{f(u)} G(u) dX_u.$$

2°) If X is the BM, prove that Z is a Gaussian Markov process.

3°) Write down the multidimensional version of questions 1°) and 2°). In particular solve the d-dimensional Langevin equation

$$dX_t = \sigma dB_t + \beta X_t dt$$

where σ and β are $d \times d$ matrices.

[Hint: use Exercise (2.6) bis.]

* **(2.8) Exercise (Doss-Süssman method).** Let σ be a C^2-function on the real line, with bounded derivatives σ' and σ'' and b be Lipschitz continuous. Call $h(x, s)$ the solution to the ODE

$$\frac{\partial h}{\partial s}(x, s) = \sigma\big(h(x, s)\big), \qquad h(x, 0) = x.$$

Let X be a continuous semimartingale such that $X_0 = 0$ and call D_t the solution of the ODE

$$\frac{dD_t}{dt} = b\left(h\left(D_t, X_t(\omega)\right)\right) \exp\left\{-\int_0^{X_t(\omega)} \sigma'\left(h(D_t, s)\right) ds\right\}; \; D_0 = y.$$

Prove that $h(D_t, X_t)$ is the unique solution to the equation

$$Y_t = y + \int_0^t \sigma(Y_s) \circ dX_s + \int_0^t b(Y_s) ds$$

where $\circ$ stands for the Stratonovich integral. Moreover $D_t = h(Y_t, -X_t)$.

(2.9) Exercise. For semimartingales H and X, call $\mathscr{E}_X(H)$ the solution to the linear equation in Definition (2.2).

1°) Prove that if K is another semimartingale,

$$\mathscr{E}_X(H + K \cdot X) = \mathscr{E}_X(H + K) - K.$$

2°) Let X and Y be two semimartingales and set $Z = X + Y + \langle X, Y\rangle$. For any two semimartingales H and K, there is a semimartingale L such that $\mathscr{E}_X(H) \cdot \mathscr{E}_Y(K) = \mathscr{E}_Z(L)$. For $H = K = 1$, one finds the particular case treated in Exercise (3.11) Chap. IV.

3°) If X, Y, H are three semimartingales, the equation $Z = H + \langle Z, Y\rangle \cdot X$ has a unique solution given by

$$Z = H + \mathscr{E}_{\langle X, Y\rangle}(\langle H, Y\rangle) \cdot X.$$

* **(2.10) Exercise (Explosions).** The functions σ and b on $\mathbb{R}^d$ are said to be locally Lipschitz if, for any $n \in \mathbb{N}$, there exists a constant C_n such that

$$\|\sigma(x) - \sigma(y)\| + \|b(x) - b(y)\| \leq C_n \|x - y\|$$

for x and y in the ball $B(0, n)$.

1°) If σ and b are locally Lipschitz, prove that for any $x \in \mathbb{R}^d$ and any BM B on a space $(\Omega, \mathscr{F}, P)$ there exists a unique $(\mathscr{F}^B_\cdot)$-adapted process X such that, if $e = \inf\{t : |X_t| = +\infty\}$, then X is continuous on $[0, e[$ and

$$X_t = x + \int_0^t \sigma(X_s) dB_s + \int_0^t b(X_s) ds$$

on $\{t < e\}$. The time e is called the *explosion* time of X.

[Hint: Use globally Lipschitz functions σ_n and b_n which agree with σ and b on $B(0, n)$.]

2°) If there is a constant K such that

$$\|\sigma(x)\|^2 + \|b(x)\|^2 \leq K(1 + \|x\|^2)$$

for every $x \in \mathbb{R}^d$, prove that $E\left[|X_t|^2\right] < \infty$ and conclude that $P[e = \infty] = 1$.

[Hint: If $T_n = \inf\{t : |X_t| > n\}$, develop $\left|X_{t \wedge T_n}\right|^2$ by means of Itô's formula.]

3°) Let $\widehat{\mathbf{W}}$ be the space of functions w from $\mathbb{R}_+$ to $\mathbb{R}^d \cup \{\Delta\}$ such that if $\zeta(w) = \inf\{t : w(t) = \Delta\}$, then $w(s) = \Delta$ for any $s \geq \zeta$ and w is continuous on $[0, \zeta(w)[$. The space $\widehat{\mathbf{W}}$ contains the space $C(\mathbb{R}_+, \mathbb{R}^d)$ on which ζ is identically $+\infty$. Call Y the coordinate process and, with obvious notation, let Q_x be the law on $\widehat{\mathbf{W}}$ of the process X of 1°) and P_x be the law of BM(x). For $d = 1$ and $\sigma = 1$ prove that

$$Q_x\big|_{\mathscr{F}_t \cap (t<\zeta)} = \mathscr{E}\left(\int_0^\cdot b(Y_s)dY_s\right)_t \cdot P_x\big|_{\mathscr{F}_t}.$$

As a result, $\mathscr{E}\left(\int_0^\cdot b(Y_s)dY_s\right)$ is a true martingale if and only if $Q_x(\zeta < \infty) = 0$.

4°) Prove that for any $\alpha \in \mathbb{R}$, $\mathscr{E}\left(\alpha \int_0^\cdot B_s dB_s\right)$ is a true martingale although Kazamaki's criterion does not apply for $t \geq \alpha^{-1}$.

\# **(2.11) Exercise (Zvonkin's method).** 1°) Suppose that σ is a locally Lipschitz (Exercise(2.10)) function on $\mathbb{R}$, bounded away from zero and b is Borel and locally bounded; prove that the non-constant solutions h to the differential equation $\frac{1}{2}\sigma^2 h'' + bh' = 0$ are either strictly increasing or strictly decreasing and that $g = (\sigma h') \circ h^{-1}$ is locally Lipschitz.

2°) In the situation of 1°) prove that pathwise uniqueness holds for the equation $e(\sigma, b)$.

[Hint: Set up a correspondence between the solutions to $e(\sigma, b)$ and those to $e(g, 0)$ and use the results in the preceding exercise.]

This method amounts to putting the solution on its natural scale (see Exercise (3.20) in Chap. VII).

3°) Prove that the solution to $e(\sigma, b)$ where σ is bounded and $\geq \varepsilon > 0$ and b is Lebesgue integrable, is recurrent.

\# **(2.12) Exercise (Brownian Bridges).** The reader is invited to look first at questions 1°) and 2°) in Exercise (3.18) of Chap. IV.

1°) Prove that the solution to the SDE

$$X_t^x = \beta_t + \int_0^t \frac{x - X_s^x}{1 - s}\, ds, \qquad t \in [0, 1[,\ x \in \mathbb{R},$$

is given by

$$X_t^x = xt + \beta_t - (1 - t)\int_0^t \frac{\beta_s ds}{(1 - s)^2} = xt + (1 - t)\int_0^t \frac{d\beta_s}{1 - s}.$$

2°) Prove that $\lim_{t\uparrow 1} X_t^x = x$ a.s. and that if we set $X_1^x = x$, then $X_t^x, t \in [0, 1]$, is a Brownian Bridge.

[Hint: If g is a positive continuous decreasing function on $[0, 1]$ such that $g(1) = 0$, then for any positive f such that $\int_0^1 f(u)g(u)du < +\infty$, $\lim_{t\uparrow 1} g(t)\int_0^t f(u)du = 0$; apply this to $g(t) = 1 - t$ and $f(t) = |\beta_1 - \beta_t|(1 - t)^{-2}$.]

3°) If P^x is the law of X^x on $C([0, 1], \mathbb{R})$, the above questions together with 1°) in Exercise (3.18) of Chap. IV give another proof of Exercise (3.16) in Chap. I, namely that P^x is a regular disintegration of the Wiener measure on $C([0,1],\mathbb{R})$ with respect to $\sigma(B_1)$ where B is the canonical process.

4°) We retain henceforth the situation and notation of Exercise (3.18) in Chap. IV. Prove that X^x and β have the same filtration.

5°) Prove that the $(\mathscr{G}_t)$-predictable processes are indistinguishable from the processes $K\big(B_1(\omega), s, \omega\big)$ where K is a map on $\mathbb{R}\times(\mathbb{R}_+\times\Omega)$ which is $\mathscr{B}(\mathbb{R})\times\mathscr{P}(\beta)$-measurable, $\mathscr{P}(\beta)$ being the σ-algebra of predictable sets with respect to the filtration of β. Prove that any square-integrable $(\mathscr{G}_t)$-martingale can be written

$$f(B_1) + \int_0^t K_s d\beta_s$$

where f and K satisfy some integrability conditions.

(2.13) Exercise. 1°) Let B be the linear BM, b a bounded Borel function and set

$$Y_t = B_t - \int_0^t b(B_s)ds.$$

Prove that $\mathscr{F}_t^Y = \mathscr{F}_t^B$ for every t.

[Hint: Use Girsanov's Theorem and Exercise (2.11) 2° above.]

** 2°) If b is not bounded, this is no longer true. For example if $Y_t = B_t - \int_0^t B_s^{-1} ds$ where the integral was defined in Exercise (1.19) of Chap. VI, then

$$E\left[B_1 \mid \sigma(Y_s, s\le 1)\right] = 0.$$

[Hint: For $f\in L^2([0,1])$, prove that $E[B_1\mathscr{E}(f\cdot Y)_1]=0$ and use the argument of Lemma (3.1) Chap. V. One may also look at 4°) in Exercise (1.29) Chap. VI.]

* **(2.14) Exercise. (Stochastic differential equation with reflection).** Let $\sigma(s,x)$ and $b(s,x)$ be two functions on $\mathbb{R}_+\times\mathbb{R}_+$ and B a BM. For $x_0\ge 0$, we call solution to the SDE with reflection $e_{x_0}(\sigma,b)$ a pair (X,K) of processes such that

i) the process X is continuous, positive, $\mathscr{F}^B$-adapted and

$$X_t = x_0 + \int_0^t \sigma(s,X_s)dB_s + \int_0^t b(s,X_s)ds + K_t.$$

ii) the process K is continuous, increasing, vanishing at 0, $\mathscr{F}^B$-adapted and

$$\int_0^\infty X_s dK_s = 0.$$

If σ and b are bounded and satisfy the global Lipschitz condition

$$|\sigma(s,x)-\sigma(s,y)| + |b(s,x)-b(s,y)| \le C|x-y|$$

for every s, x, y and for a constant C, prove that there is existence and uniqueness for the solutions to $e_{x_0}(\sigma,b)$ with reflection.

[Hint: Use a successive approximations method with Lemma (2.1) of Chap. VI as the means to go from one step to the following one.]

* **(2.15) Exercise (Criterion for explosions).** We retain the situation of Exercise (2.11) and set

$$\begin{aligned} s(x) &= \int_0^x \exp\left(-\int_0^y 2b(z)\sigma^{-2}(z)dz\right)dy, \\ m(x) &= 2\int_0^x \exp\left(\int_0^y 2b(z)\sigma^{-2}(z)dz\right)\sigma^{-2}(y)dy, \\ k(x) &= \int_0^x m(y)s(dy). \end{aligned}$$

The reader is invited to look at Exercise (3.20) in Chap. VII for the interpretations of s and m.

1°) If U is the unique solution of the differential equation $\frac{1}{2}\sigma^2 U'' + bU' = U$ such that $U(0) = 1$ and $U'(0) = 0$, check that

$$U(x) = 1 + \int_0^x ds(y) \int_0^y U(z)dm(z)$$

and prove that

$$1 + k \leq U \leq \exp(k).$$

2°) If e is the explosion time, prove that $\exp(-t \wedge e)U(X_{t\wedge e})$, where X is the solution to $e_x(\sigma, b)$, is a positive supermartingale for every x. Conclude that if $k(-\infty) = k(+\infty) = \infty$, then $P_x[e = \infty] = 1$ for every x.

3°) If as usual $T_0 = \inf\{t : X_t = 0\}$, prove that $\exp(-t \wedge T_0)U(X_{t\wedge T_0})$ is a bounded P_x-martingale for every x and conclude that if either $k(-\infty) < \infty$ or $k(\infty) < \infty$, then $P_x[e < \infty] > 0$ for every x.

4°) Prove that if $P_x[e < \infty] = 1$ for every x in $\mathbb{R}$, then one of the following three cases occurs

i) $k(-\infty) < \infty$ and $k(+\infty) < \infty$,
ii) $k(-\infty) < \infty$ and $s(+\infty) = \infty$,
iii) $k(+\infty) < \infty$ and $s(-\infty) = \infty$.

5°) Assume to be in case i) above and set

$$\begin{aligned} G(x, y) &= \big(s(x) - s(-\infty)\big)\big(s(+\infty) - s(y)\big)/\big(s(+\infty) - s(-\infty)\big) && \text{if } x \leq y \\ &= G(y, x) && \text{if } y \leq x, \end{aligned}$$

and set $U_1(x) = \int G(x, y)m(y)dy$ (again see Sect. 3 Chap. VII for the rationale). Prove that $U_1(X_{t\wedge e}) + t \wedge e$ is a local martingale and conclude that $E_x[e] < \infty$, hence $P_x[e < \infty] = 1$.

6°) Prove that in the cases ii) and iii), $P_.[e < \infty] = 1$.

(2.16) Exercise. 1°) Prove that for $\beta > 0$, the solution V_t to the Langevin equation is the OU process with parameter β and size $1/2\beta$.

2°) Prove that if X_0 is independent of B, the solution V such that $V_0 = X_0$ is stationary if $X_0 \sim \mathcal{N}(0, 1/2\beta)$. In that case

$$V_t = (2\beta)^{-1/2} e^{-\beta t} \widetilde{B}\left(e^{2\beta t}\right) = (2\beta)^{-1/2} e^{\beta t} \widetilde{\Gamma}\left(e^{-2\beta t}\right)$$

where $\widetilde{B}$ and $\widetilde{\Gamma}$ are two standard linear BM's such that $\widetilde{B}_u = u\widetilde{\Gamma}_{1/u}$.

** **(2.17) Exercise.** Exceptionally in this exercise we do not ask for condition i) of Definition (1.2) to be satisfied.

1°) Prove that the stochastic equation

$$X_t = B_t + \int_0^t \frac{X_s}{s}\, ds$$

has a solution for $t \in [0, 1]$.

[Hint: Use a time reversal in the solution to the equation of Exercise (2.12).]

2°) Prove that for $0 < \varepsilon < t \leq 1$

$$X_t = \frac{t}{\varepsilon} X_\varepsilon + t \int_\varepsilon^t \frac{dB_u}{u}.$$

3°) Prove that X is not $\mathscr{F}^B$-adapted.

[Hint: If the solution were strong, the two terms on the right-hand side of the equality in 2°) would be independent. By letting ε tend to zero this would entail that for $\lambda \neq 0$, $E[\exp(i\lambda X_t)]$ is identically zero.]

4°) By the same device, prove that if ϕ is a locally bounded Borel function on $]0, \infty[$, the equation

$$X_t = B_t + \int_0^t \phi(s) X_s ds$$

does not have an $\mathscr{F}^B$-adapted solution as soon as

$$\int_0^1 \exp\left(2 \int_u^1 \phi(s) ds\right) du = \infty.$$

(2.18) Exercise (Laws of exponentials of BM). Let β and γ be two independent standard linear BM's and for μ, ν in $\mathbb{R}$ set $\beta_t^{(\mu)} = \beta_t + \mu t$ and $\gamma_t^{(\nu)} = \gamma_t + \nu t$.

1°) Prove that the process

$$X_t = \exp\left(\beta_t^{(\mu)}\right)\left(x_0 + \int_0^t \exp\left(-\beta_s^{(\mu)}\right) d\gamma_s^{(\nu)}\right)$$

is for every real x_0, a diffusion with infinitesimal generator equal to

$$\left((1 + x^2)/2\right) d^2/dx^2 + \left(x(\mu + (1/2)) + \nu\right) d/dx.$$

2°) Prove that X has the same law as $\sinh(Y_t)$, $t \geq 0$, where Y is a diffusion with generator

$$(1/2) d^2/dy^2 + \left(\mu \tanh(y) + \nu \cosh(y)^{-1}\right) d/dy.$$

3°) Derive from 2°) that for fixed t,

$$\int_0^t \exp\left(\beta_s^{(\mu)}\right) d\gamma_s^{(\nu)} \stackrel{(d)}{=} \sinh(Y_t).$$

In particular we get (**Bougerol's identity**): for fixed t,

$$\int_0^t \exp(\beta_s) d\gamma_s \stackrel{(d)}{=} \sinh(\beta_t).$$

[Hint: For a Lévy process X vanishing at 0, the processes X_s, $s \leq t$, and $X_t - X_{t-s}$, $s \leq t$, are equivalent.]

4°) Prove that on $\mathscr{F}_t$, when $\nu = 0$ and $Y_0 = 0$, the law P^μ of Y has the density

$$\cosh(X_t)^\mu \exp\left(-\left((\mu - \mu^2)/2\right) \int_0^t \cosh(X_s)^{-2} ds\right) \exp\left(-(\mu^2/2)t\right)$$

with respect to the Wiener measure. Prove as a consequence that for fixed t,

$$\int_0^t \exp(\beta_s + s) d\gamma_s \stackrel{(d)}{=} \sinh(\beta_t + \varepsilon t),$$

where ε is a Bernoulli r.v. independent of β.

[Hint: See case i) in 2°) Exercise (3.18) Chapter VIII.]

(2.19) Exercise (An extension of Lévy's equivalence). 1°) Let $\lambda \in \mathbb{R}$. Prove that pathwise uniqueness holds for the one-dimensional equation $e(1, b_\lambda)$, where $b_\lambda(x) = -\lambda \operatorname{sgn}(x)$.

[Hint: Use Exercise (2.11)].

2°) We denote by X^λ a solution to $e(1, b_\lambda)$ starting from 0, and by B^λ a Brownian motion with constant drift λ, starting from 0, and $S_t^\lambda = \sup_{s \leq t} B_s^\lambda$.

Prove that:

$$(S_t^\lambda - B_t^\lambda, t \geq 0) \stackrel{(d)}{=} (|X_t^\lambda|, t \geq 0)$$

[Hint: Use Skorokhod's lemma (2.1) in Chap. VI].

§3. The Case of Hölder Coefficients in Dimension One

In this section, we prove a partial converse to Theorem (1.7) by describing a situation where uniqueness in law entails pathwise uniqueness, thus extending the scope of the latter property. We rely heavily on local times and, consequently, have to keep to dimension 1. We study the equation $e(\sigma, b)$ where σ is locally bounded on $\mathbb{R}_+ \times \mathbb{R}$ and consider pairs of solutions to this equation defined on the same space and with respect to the same BM. We recall that any solution X is a continuous semimartingale and we denote by $L^x(X)$ the right-continuous version of its local times.

What may seem surprising in the results we are about to prove is that we get uniqueness in cases where there is no uniqueness for ODE's (ordinary differential equations). This is due to the regularizing effect of the quadratic variation of BM as will appear in the proofs. One can also observe that in the case of ODE's, the supremum of two solutions is a solution; this is usually not so for SDE's because of the appearance of a local time. In fact we have the following

(3.1) Proposition. *If X^1 and X^2 are two solutions of $e(\sigma, b)$ such that $X_0^1 = X_0^2$ a.s., then $X^1 \vee X^2$ is a solution if and only if $L^0(X^1 - X^2)$ vanishes identically.*

Proof. By Tanaka's formula,

$$\begin{aligned} X_t^1 \vee X_t^2 &= X_t^1 + \left(X_t^2 - X_t^1\right)^+ = X_t^1 + \int_0^t 1_{(X_s^2 > X_s^1)} d\left(X^2 - X^1\right)_s \\ &\quad + \frac{1}{2} L_t^0\left(X^2 - X^1\right), \end{aligned}$$

and replacing X^i, $i = 1, 2$ by $X_0^i + \int_0^{\cdot} \sigma(s, X_s^i) dB_s + \int_0^{\cdot} b(s, X_s^i) ds$, it is easy to check that

$$\begin{aligned} X_t^1 \vee X_t^2 &= \left(X_0^1 \vee X_0^2\right) + \int_0^t \sigma\left(s, X_s^1 \vee X_s^2\right) dB_s + \int_0^t b\left(s, X_s^1 \vee X_s^2\right) ds \\ &\quad + \frac{1}{2} L_t^0\left(X^2 - X^1\right), \end{aligned}$$

which establishes our claim. □

The following result is the key to this section.

(3.2) Proposition. *If uniqueness in law holds for $e(\sigma, b)$ and if $L^0(X^1 - X^2) = 0$ for any pair (X^1, X^2) of solutions such that $X_0^1 = X_0^2$ a.s., then pathwise uniqueness holds for $e(\sigma, b)$.*

Proof. By the preceding proposition, if X^1 and X^2 are two solutions, $X^1 \vee X^2$ is also a solution; but X^1 and $X^1 \vee X^2$ cannot have the same law unless they are equal, which completes the proof. □

The next lemma is crucial to check the above condition on the local time. In the sequel ρ will always stand for a Borel function from $]0, \infty[$ into itself such that $\int_{0+} da/\rho(a) = +\infty$.

(3.3) Lemma. *If X is a continuous semimartingale such that, for some $\varepsilon > 0$ and every t*

$$A_t = \int_0^t 1_{(0 < X_s \leq \varepsilon)} \rho(X_s)^{-1} d\langle X, X\rangle_s < \infty \qquad \text{a.s.},$$

then $L^0(X) = 0$.

Proof. Fix $t > 0$; by the occupation times formula (Corollary (1.6) Chap. VI)

$$A_t = \int_0^\varepsilon \rho(a)^{-1} L_t^a(X) da.$$

If $L_t^0(X)$ did not vanish a.s., as $L_t^a(X)$ converges to $L_t^0(X)$ when a decreases to zero, we would get $A_t = \infty$ with positive probability, which is a contradiction.

(3.4) Corollary. *Let b_i, $i = 1, 2$ be two Borel functions; if*

$$|\sigma(s, x) - \sigma(s, y)|^2 \le \rho(|x - y|)$$

for every s, x, y and if X^i, $i = 1, 2$, are solutions to $e(\sigma, b_i)$ with respect to the same BM, *then $L^0(X^1 - X^2) = 0$.*

Proof. We have

$$\begin{aligned} X_t^1 - X_t^2 &= X_0^1 - X_0^2 + \int_0^t \left(\sigma(s, X_s^1) - \sigma(s, X_s^2)\right) dB_s \\ &\quad + \int_0^t \left(b_1(s, X_s^1) - b_2(s, X_s^2)\right) ds, \end{aligned}$$

and therefore

$$\begin{aligned} &\int_0^t \rho(X_s^1 - X_s^2)^{-1} 1_{(X_s^1 > X_s^2)} d\left\langle X^1 - X^2, X^1 - X^2 \right\rangle_s \\ &= \int_0^t \rho(X_s^1 - X_s^2)^{-1} \left(\sigma(s, X_s^1) - \sigma(s, X_s^2)\right)^2 1_{(X_s^1 > X_s^2)} ds \le t. \end{aligned}$$

□

We may now state

(3.5) Theorem. *Pathwise uniqueness holds for $e(\sigma, b)$ in each of the following cases:*

i) $|\sigma(x) - \sigma(y)|^2 \le \rho(|x - y|)$, $|\sigma| \ge \varepsilon > 0$ *and b and σ are bounded;*
ii) $|\sigma(s, x) - \sigma(s, y)|^2 \le \rho(|x - y|)$ *and b is Lipschitz continuous i.e., for each compact H and each t there is a constant K_t, such that for every x, y in H and $s \le t$*

$$|b(s, x) - b(s, y)| \le K_t |x - y|;$$

iii) $|\sigma(x) - \sigma(y)|^2 \le |f(x) - f(y)|$ *where f is increasing and bounded, $\sigma \ge \varepsilon > 0$ and b is bounded.*

Remark. We insist that in cases i) and iii) σ does not depend on s, whereas in ii) non-homogeneity is allowed.

Proof. i) By Corollary (1.14), since $|\sigma| \ge \varepsilon$, uniqueness in law holds for $e(\sigma, b)$. The result thus follows from Proposition (3.2) through Corollary (3.4).

ii) Let X^1 and X^2 be two solutions with respect to the same BM and such that $X_0^1 = X_0^2$ a.s. By the preceding corollary,

$$|X_t^1 - X_t^2| = \int_0^t \operatorname{sgn}(X_s^1 - X_s^2) d(X_s^1 - X_s^2).$$

The hypothesis made on σ and b entails that we can find a sequence (T_n) of stopping times converging to ∞, such that, if we set $Y^i = (X^i)^{T_n}$, $i = 1, 2$, for fixed n, then $\sigma\left(s, Y_s^i\right)$ is bounded and

$$\left|b\left(s, Y_s^1\right) - b\left(s, Y_s^2\right)\right| \leq C_t \left|Y_s^1 - Y_s^2\right|$$

for $s \leq t$ and some constant C_t. As a result,

$$\left|Y_t^1 - Y_t^2\right| - \int_0^t \operatorname{sgn}\left(Y_s^1 - Y_s^2\right)\left(b\left(s, Y_s^1\right) - b\left(s, Y_s^2\right)\right) ds$$

is a martingale vanishing at 0 and we have

$$E\left[\left|Y_t^1 - Y_t^2\right|\right] \leq C_t \int_0^t E\left[\left|Y_s^1 - Y_s^2\right|\right] ds.$$

Using Gronwall's lemma, the proof is now easily completed.

iii) By Corollary (1.14) again, the condition $\sigma \geq \varepsilon$ implies uniqueness in law for $e(\sigma, b)$; we will prove the statement by applying Corollary (3.4) with $\rho(x) = x$ and Proposition (3.2). To this end, we pick a $\delta > 0$ and consider

$$\begin{aligned} E&\left[\int_0^t (X_s^1 - X_s^2)^{-1} 1_{(X_s^1 - X_s^2 > \delta)} d\left\langle X^1 - X^2, X^1 - X^2\right\rangle_s\right] \\ &\leq E\left[\int_0^t \left(f(X_s^1) - f(X_s^2)\right)(X_s^1 - X_s^2)^{-1} 1_{(X_s^1 - X_s^2 > \delta)} ds\right] \stackrel{\text{def}}{=} K(f)_t. \end{aligned}$$

We now choose a sequence $\{f_n\}$ of uniformly bounded increasing C^1-functions such that $\lim_n f_n(x) = f(x)$ for any x which is not a discontinuity point for f. The set D of discontinuity points for f is countable; by the occupation times formula, the set of times s such that X_s^1 or X_s^2 belongs to D has a.s. zero Lebesgue measure, and consequently

$$\lim_n \left(f_n(X_s^1) - f_n(X_s^2)\right) = f(X_s^1) - f(X_s^2)$$

for almost all $s \leq t$. It follows that $K(f)_t = \lim_n K(f_n)_t$.

For $u \in [0, 1]$, set $Z^u = X^2 + u(X^1 - X^2)$; we have

$$\begin{aligned} K(f_n)_t &= E\left[\int_0^t \left(\int_0^1 f_n'(Z_s^u) du\right) 1_{(X_s^1 - X_s^2 > \delta)} ds\right] \\ &\leq \int_0^1 E\left[\int_0^t f_n'(Z_s^u) ds\right] du \\ &\leq \frac{1}{\varepsilon^2} \int_0^1 E\left[\int_{\mathbb{R}} f_n'(a) L_t^a(Z^u) da\right] du, \end{aligned}$$

because $Z_t^u(\omega) = Z_0^u(\omega) + \int_0^t \sigma^u(s, \omega) dB_s(\omega) + \int_0^t b^u(s, \omega) ds$ where $\sigma^u \geq \varepsilon$. Moreover, $|\sigma^u| + |b^u| \leq M$ for a constant M; by a simple application of Tanaka's formula

$$\sup_{a,u} E\left[L_t^a(Z^u)\right] = C < \infty,$$

and it follows that

$$K(f_n)_t \leq \varepsilon^{-2} C \sup_n \| f_n \| .$$

Hence $K(f)_t$ is bounded by a constant independent of δ; letting δ go to zero, we see that the hypothesis of Lemma (3.3) is satisfied for $\rho(x) = x$ which completes the proof.

Remarks. 1°) In iii) the hypothesis $\sigma \geq \varepsilon$ cannot be replaced by $|\sigma| \geq \varepsilon$; indeed, we know that there is no pathwise uniqueness for $e(\sigma, 0)$ when $\sigma(x) = \text{sgn}(x)$.

2°) The significance of iii) is that pathwise uniqueness holds when σ is of bounded quadratic variation. This is the best that one can obtain as is seen in Exercise (1.16).

3°) The hypothesis of Theorem (3.5) may be slightly weakened as is shown in Exercises (3.13) and (3.14).

At the beginning of the section, we alluded to the difference between SDE's and ODE's. We now see that if, for instance, $\sigma(x) = \sqrt{|x|}$, then the ODE $dX_t = \sigma(X_t)dt$ has several solutions whereas the SDE $e(\sigma, b)$ has only one. The point is that for SDE's majorations are performed by using the increasing processes of the martingale parts of the solutions and thus it is $(\sigma(x) - \sigma(y))^2$ and not $|\sigma(x) - \sigma(y)|$ which comes into play. We now turn to other questions.

A consequence of the above results together with Corollary (1.14) is that for bounded b, the equation $e(1, b)$ has always a solution and that this solution is strong (another proof is given in Exercise (2.11)). We now give an example, known as *Tsirel'son example*, which shows that this does not carry over to the case where b is replaced by a function depending on the entire past of X.

We define a bounded function τ on $\mathbb{R}_+ \times \mathbf{W}$ in the following way. For a strictly increasing sequence $(t_k, k \in -\mathbb{N})$, of numbers such that $0 < t_k < 1$ for $k < 0$, $t_0 = 1$, $\lim_{k \to -\infty} t_k = 0$, we set

$$\begin{aligned} \tau(t, \omega) &= \left[\frac{\omega(t_k) - \omega(t_{k-1})}{t_k - t_{k-1}} \right] \qquad \text{if } t_k < t \leq t_{k+1}, \\ &= 0 \qquad \text{if } t = 0 \text{ or } t > 1, \end{aligned}$$

where $[x]$ is the *fractional* part of the real number x. This is clearly a predictable function on $\mathbf{W}$. If (X, B) is a solution to $e(1, \tau)$, then on $]t_k, t_{k+1}]$

$$X_t - X_{t_k} = B_t - B_{t_k} + \left[\frac{X_{t_k} - X_{t_{k-1}}}{t_k - t_{k-1}} \right] (t - t_k);$$

if we set for $t_k < t \leq t_{k+1}$,

$$\eta_t = \left(X_t - X_{t_k}\right) / (t - t_k), \qquad \varepsilon_t = \left(B_t - B_{t_k}\right) / (t - t_k),$$

we have

$$(*) \qquad \eta_t = \varepsilon_t + \left[\frac{X_{t_k} - X_{t_{k-1}}}{t_k - t_{k-1}} \right] = \varepsilon_t + \left[\eta_{t_k} \right].$$

(3.6) Proposition. *The equation* $e(1, \tau)$ *has no strong solution. More precisely*

i) for every t *in* $[0, 1]$, *the r.v.* $[\eta_t]$ *is independent of* $\mathscr{F}_1^B$ *and uniformly distributed on* $[0, 1]$;

ii) for any $0 < s \leq t$, $\mathscr{F}_t^X = \sigma([\eta_s]) \vee \mathscr{F}_t^B$.

Proof. The first statement is an obvious consequence of properties i) and ii), and ii) follows easily from the definitions. We turn to proving i).

Let $p \in \mathbb{Z} - \{0\}$ and set $d_k = E\left[\exp\left\{2i\pi p\eta_{t_k}\right\}\right]$; by (*), we have

$$\begin{aligned} d_k &= E\left[\exp\left\{2i\pi p\left(\varepsilon_{t_k} + \left[\eta_{t_{k-1}}\right]\right)\right\}\right] \\ &= E\left[\exp\left\{2i\pi p\left(\varepsilon_{t_k} + \eta_{t_{k-1}}\right)\right\}\right] \\ &= d_{k-1} E\left[\exp\left(2i\pi p\varepsilon_{t_k}\right)\right] = d_{k-1}\exp\left\{-2\pi^2 p^2 \left(t_k - t_{k-1}\right)^{-1}\right\}, \end{aligned}$$

because ε_{t_k} is independent of $\mathscr{F}_{t_{k-1}}$, where $(\mathscr{F}_t)$ is the filtration with respect to which (X, B) is defined. It follows that

$$|d_k| \leq |d_{k-1}| \exp\left\{-2\pi^2 p^2\right\} \leq \ldots \leq |d_{k-n}| \exp\left\{-2n\pi^2 p^2\right\},$$

and consequently, $d_k = 0$ for every k. This proves that $\left[\eta_{t_k}\right]$ is uniformly distributed on $[0, 1]$.

Define further $\mathscr{B}_k^n = \sigma\left(B_u - B_v, t_n \leq u \leq v \leq t_k\right)$; then

$$\begin{aligned} &E\left[\exp\left\{2i\pi p\eta_{t_k}\right\} \middle| \mathscr{B}_k^{k-n}\right] \\ &\quad = E\left[\exp\left\{2i\pi p\left(\varepsilon_{t_k} + \varepsilon_{t_{k-1}} + \ldots + \varepsilon_{t_{k-n+1}} + \eta_{t_{k-n}}\right)\right\} \middle| \mathscr{B}_k^{k-n}\right] \\ &\quad = \exp\left\{2i\pi p\left(\varepsilon_{t_k} + \ldots + \varepsilon_{t_{k-n+1}}\right)\right\} d_{k-n} \end{aligned}$$

since $\eta_{t_{k-n}}$ is independent of $\mathscr{B}_k^{k-n}$. The above conditional expectation is thus zero; it follows easily that $E\left[\exp\left\{2i\pi p\eta_{t_k}\right\} \mid \mathscr{F}_{t_k}^B\right] = 0$ and since $\mathscr{F}_{t_k}$ is independent of $\left\{B_t - B_{t_k}, t \geq t_k\right\}$,

$$E\left[\exp\left\{2i\pi p\eta_{t_k}\right\} \mid \mathscr{F}_1^B\right] = 0.$$

Finally for $t_k < t \leq t_{k+1}$, we have

$$E\left[\exp\left\{2i\pi p\eta_t\right\} \mid \mathscr{F}_1^B\right] = \exp\left\{2i\pi p\varepsilon_t\right\} E\left[\exp\left\{2i\pi p\eta_{t_k}\right\} \mid \mathscr{F}_1^B\right] = 0$$

and this being true for every $p \neq 0$, proves our claim.

Remark. As a result, there does exist a Brownian motion B on a space $(\Omega, \mathscr{F}_t, P)$ and two processes X^1, X^2 such that (X^i, B), $i = 1, 2$, is a solution of $e(1, \tau)$. The reader will find in Exercise (3.17) some information on the relationship between X^1 and X^2. Moreover, other examples of non existence of strong solutions may be deduced from this one as is shown in Exercise (3.18).

We will now use the techniques of this section to prove *comparison theorems* for solutions of SDE's. Using the same notation as above, we assume that either $(\sigma(s, x) - \sigma(s, y))^2 \leq \rho(|x - y|)$ or that σ satisfies hypothesis iii) of Theorem (3.5).

(3.7) Theorem. *Let b^i, $i = 1, 2$, be two bounded Borel functions such that $b^1 \geq b^2$ everywhere and one of them at least satisfies a Lipschitz condition. If X^i, $i = 1, 2$, are solutions to $e(\sigma, b^i)$ defined on the same space with respect to the same* BM *and if $X_0^1 \geq X_0^2$ a.s., then*

$$P\left[X_t^1 \geq X_t^2 \textit{ for all } t \geq 0\right] = 1.$$

Proof. It was shown in Corollary (3.4) and in the proof of Theorem (3.5) that, in each case, $L^0(X^1 - X^2) = 0$, and therefore

$$\begin{aligned} \phi(t) &= E\left[(X_t^2 - X_t^1)^+\right] = E\left[\int_0^t 1_{(X_s^2 > X_s^1)}\left(b^2(s, X_s^2) - b^1(s, X_s^1)\right) ds\right] \\ &\leq E\left[\int_0^t 1_{(X_s^2 > X_s^1)}\left(b^1(s, X_s^2) - b^1(s, X_s^1)\right) ds\right]. \end{aligned}$$

Thus, if b^1 is Lipschitz with constants K_t,

$$\phi(t) \leq K_t E\left[\int_0^t 1_{(X_s^2 > X_s^1)}\left|X_s^2 - X_s^1\right| ds\right] = K_t \int_0^t \phi(s) ds,$$

and we conclude by using Gronwall's lemma and the usual continuity arguments.

If b^2 is Lipschitz, using the same identity again, we have

$$\begin{aligned} \phi(t) &\leq E\left[\int_0^t 1_{(X_s^2 > X_s^1)}\left|b^2(s, X_s^2) - b^2(s, X_s^1)\right| ds \right. \\ &\qquad \left. + \int_0^t 1_{(X_s^2 > X_s^1)}\left(b^2(s, X_s^1) - b^1(s, X_s^1)\right) ds\right] \\ &\leq E\left[\int_0^t 1_{(X_s^2 > X_s^1)}\left|b^2(s, X_s^2) - b^2(s, X_s^1)\right| ds\right] \end{aligned}$$

since $b^2 \leq b^1$, and we complete the proof as in the first case. □

With more stringent conditions, we can even get strict inequalities.

(3.8) Theorem. *Retain the hypothesis of Theorem (3.7). If the functions b^i do not depend on s and are continuous and σ is Lipschitz continuous and if one of the following conditions is in force:*

i) $b^1 > b^2$ *everywhere,*
ii) $|\sigma| \geq \varepsilon > 0$, *either b^1 or b^2 is Lipschitz and there exists a neighborhood $V(x)$ of a point x such that*

$$\int_{V(x)} 1_{(b^1(a) = b^2(a))} da = 0,$$

then, if (X^i, B) is a solution of $e_x(\sigma, b^i)$,

$$P\left[X_t^1 > X_t^2 \text{ for all } t > 0\right] = 1.$$

Proof. In case i) one can suppose that either b^1 or b^2 is Lipschitz continuous, because it is possible to find a Lipschitz function b^3 such that $b^1 > b^3 > b^2$.

We now suppose that b^1 is Lipschitz, the other case being treated in similar fashion. We may write

$$X_t^1 - X_t^2 = H_t + \int_0^t (X_s^1 - X_s^2) dM_s$$

where

$$H_t = \int_0^t \left(b^1(X_s^2) - b^2(X_s^2)\right) ds$$

and

$$M_t = \int_0^1 1_{(X_s^1 \neq X_s^2)} (X_s^1 - X_s^2)^{-1} \left[\left(\sigma(s, X_s^1) - \sigma(s, X_s^2)\right) dB_s + \left(b^1(X_s^1) - b^1(X_s^2)\right) ds\right].$$

The hypothesis made on σ and b^i entails that H and M are continuous semi-martingales; by Proposition (2.3) we consequently have

$$X_t^1 - X_t^2 = \mathscr{E}(M)_t \int_0^t \mathscr{E}(M)_s^{-1} dH_s,$$

and it is enough to prove that for every $t > 0$, $H_t > 0$ a.s. This property is obviously true in case i); under ii), by the occupation times formula,

$$H_t \geq \varepsilon^{-2} \int_{-\infty}^{+\infty} \left(b^1(a) - b^2(a)\right) L_t^a(X^2) da.$$

If $L_t^x(X^2) > 0$ for all $t > 0$, the result will follow from the right-continuity in a of L_t^a. Thus we will be finished once we have proved the following

(3.9) Lemma. *If X is a solution of $e_x(\sigma, b)$ and, moreover $|\sigma| \geq \varepsilon > 0$, then almost surely, $L_t^x(X) > 0$ for every $t > 0$.*

Proof. We may assume that X is defined on the canonical space. By Girsanov's theorem, there is a probability measure Q for which X is a solution to $e(\sigma, 0)$. The stochastic integrals being the same under P and Q, the formula

$$|X_t - x| = \int_0^t \operatorname{sgn}(X_s - x) dX_s + L_t^x$$

shows that the local time is the same under P and Q. But under Q we have

$$X_t = x + \int_0^t \sigma(s, X_s) dB_s,$$

hence $X = \beta_{A_t}$ where β is a BM(x) and A_t a strictly increasing process of time-changes. The result follows immediately. □

\# **(3.10) Exercise (Stochastic area).** 1°) Give another proof of 3°) in Exercise (2.19) Chap. V in the following way: using the results of this chapter, prove that $(\mathscr{F}_t^{\rho}) \subset (\mathscr{F}_t^{\widetilde{\beta}})$ where $\widetilde{\beta}$ is defined within the proof of Theorem (2.11) Chap. V and compute $\langle S, S\rangle$ and $\langle S, \widetilde{\beta}\rangle$.

2°) Prove that

$$E\left[e^{i\lambda S_t}\right] = \left(E\left[\exp\left(-\frac{\lambda^2}{8}\int_0^t \beta_s^2 ds\right)\right]\right)^2$$

where β is a linear BM. The exact value of this function of λ and t is computed in Sect. 1 Chap. XI.

* **(3.11) Exercise. (Continuation of Exercise (4.15) of Chap. V).** Let T_t^n be the time-change associated with $\langle M^n, M^n\rangle$.

1°) Let $Z_t^n = B_{T_t^n}^{n+1}$ and prove that

$$(Z_t^n)^2 = c_n \int_0^t Z_s^n d\beta_s^n + d_n t$$

where c_n and d_n are two constants and β^n is the DDS Brownian motion of M^n.

2°) If n is odd, prove that M^n is pure.

[Hint: Use Theorem (3.5) to show that T_t^n is $\mathscr{F}^{\beta^n}$-measurable.]

* **(3.12) Exercise.** 1°) Retain the notation of Exercise (1.27) Chap. VIII and suppose that $(\mathscr{F}_t) = (\mathscr{F}_t^B)$ where B is a BM under P. Let τ be the Tsirel'son drift and define

$$Q = \mathscr{E}\left(\int_0^{\cdot} \tau(s, B_{\cdot})dB_s\right)_1 \cdot P \qquad \text{on } \mathscr{F}_1.$$

Set

$$A_t = \int_0^t (2 + B_s/(1+|B_s|))\, ds,$$

and call T_t the inverse of A. Using Exercise (1.15) prove that $G_P^Q(B)_{T_{\cdot}}$ is not pure, hence that a Girsanov transform of a pure martingale may fail to be pure.

2°) In the situation of Proposition (3.6), let U_t be the inverse of the process $\int_0^t (1+\tau(s, X_{\cdot}))\, ds$ and set $M_t = B_{U_t}$. Prove that M is not pure although $\mathscr{F}_\infty^M = \sigma\left(\mathscr{F}_\varepsilon^M, \mathscr{F}_\infty^B\right)$ for every $\varepsilon > 0$. This question is independent of the first. The following question shows that the situation is different if we replace purity by extremality.

3°) Let M be a $(\mathscr{F}_t)$-loc. mart. with the following representation property: for every $\varepsilon > 0$ and every $X \in L^2(\mathscr{F}_\infty)$ there is a suitable predictable process Φ_ε such that

$$X = E\left[X \mid \mathscr{F}_\varepsilon\right] + \int_\varepsilon^\infty \Phi_\varepsilon(s) dM_s.$$

Prove that M has the $(\mathscr{F}_t)$-PRP.

(3.13) Exercise. Prove that Theorem (3.5) is still true if in iii) we drop the hypothesis that f is bounded or if we replace the hypothesis $\sigma \geq \varepsilon$ by : for every $r > 0$, there is a number $\varepsilon_r > 0$ such that $\sigma \geq \varepsilon_r$ on $[-r, r]$.

(3.14) Exercise. Prove that parts i) and ii) of Theorem (3.5) are still true if the hypothesis on σ reads: there are locally integrable functions g and c and a number $\delta > 0$ such that for every x, for every $y \in [x - \delta, x + \delta]$,

$$(\sigma(s, x) - \sigma(s, y))^2 \leq \left(c(s) + g(x)\sigma^2(s, x)\right) \rho(|x - y|).$$

(3.15) Exercise. Let γ be a predictable function on $\mathbf{W}$ and τ be the Tsirel'son drift; define

$$\overline{\tau}(s, w) = \tau(s, w) + \gamma\left(s, w - \int_0^{\cdot} \tau(u, w)du\right).$$

Let (X, B) be a solution to $e_0(1, \overline{\tau})$ on the space $(\Omega, \mathscr{F}_t, P)$.

1°) If $\beta = X - \int_0^{\cdot} \tau(u, X_{\cdot})du$, prove that $(\mathscr{F}_t^B) \subset (\mathscr{F}_t^{\beta})$.

2°) Find a probability measure Q on $(\Omega, \mathscr{F}_\infty)$, equivalent to P, for which β is a BM and (X, β) a solution to $e_0(1, \tau)$. Derive therefrom that (X, B) is not a strong solution to $e_0(1, \overline{\tau})$.

\# **(3.16) Exercise.** 1°) If B is a standard linear BM, prove that the process $Z_t = B_t^2$ satisfies the SDE

$$Z_t = 2\int_0^t \sqrt{Z_s}d\beta_s + t$$

and derive therefrom another proof of the equality $\mathscr{F}_t^{\beta} = \mathscr{F}_t^{|B|}$ of Corollary (2.2) in Chap. VI.

2°) More generally if B is a $\mathrm{BM}^d(0)$, show that $|B|$ and the linear BM

$$\beta_t = \sum_{i=1}^d \int_0^t |B_s|^{-1} B_s^i dB_s^i$$

have the same filtration (see Sect. 3 Chap. VI and Sect. 1 Chap. XI).

3°) If A is a symmetric $d \times d$-matrix and B a $\mathrm{BM}^d(0)$, prove that the local martingale $\int_0^{\cdot} \langle AB_s, dB_s\rangle$ has the same filtration as a BM^r where r is the number of distinct, non zero eigenvalues of A. In particular prove that a planar BM (B^1, B^2) has the same filtration as $(|B^1 + B^2|, |B^1 - B^2|)$.

[Hint: Use Exercises (1.36) and (3.20) in Chap. IV.]

* **(3.17) Exercise.** 1°) Suppose given a Gaussian vector local martingale B in $\mathbb{R}^d$ on a space $(\Omega, \mathscr{F}_t, P)$ such that $\langle B^i, B^j\rangle_t = \rho_{ij}t$ with $\rho_{i,i} = 1$. For each i, we suppose that there is an $(\mathscr{F}_t)$-adapted process X^i such that (X^i, B^i) is a solution of $e(1, \tau)$ with $X_0 = B_0 = 0$. With obvious notation derived from those in Proposition (3.6) prove that the law of the random vector $[\eta_t] = \left([\eta_t^i], i = 1, \ldots, d\right)$ is independent of t and is invariant by the translations $x_i \to x_i + u_i$ (mod. 1) if $\sum p_i u_i = 0$ for

any $(p_i) \in \mathbb{Z}^d$ such that $\sum \rho_{ij} p_i p_j = 0$. Prove further that this random variable is independent of $\mathscr{F}_1^B$.

2°) Suppose from now on that all the components B^i of B are equal to the same linear BM β, and let α be a vector random variable independent of $\mathscr{F}_1^\beta$, whose law is carried by $([0, 1[)^d$ and is invariant under translations $x_i \to x_i + u$ (mod. 1) for every $u \in \mathbb{R}$ (not $\mathbb{R}^d$!). Set $[\eta_{t_{-1}}] = \alpha$ and prove that one can define recursively a unique process η such that

$$\eta_t = \varepsilon_t + [\eta_{t_k}] \qquad \text{for } t \in]t_k, t_{k+1}].$$

For any t, the vector random variable $[\eta_t]$ is independent of $\mathscr{F}_1^\beta$.

3°) Prove that the family of σ-algebras $\mathscr{G}_t = \mathscr{F}_t^\beta \vee \sigma([\eta_t])$ is a filtration and that β is a $(\mathscr{G}_t)$-Brownian motion.

4°) If, for $t \in]t_l, t_{l+1}]$, we define

$$X_t = \beta_t + \sum_{k \le l} (t_k - t_{k-1})[\eta_{t_{k-1}}] + (t - t_l)[\eta_{t_l}]$$

the process X is $(\mathscr{G}_t)$-adapted and (X^i, β) is for each i a solution to $e(1, \tau)$.

5°) Prove that for any $Z \in L^2(\mathscr{F}_1^X, P)$, there is a $(\mathscr{F}_t^X)$-predictable process ϕ such that $E\left[\int_0^1 \phi_s^2 ds\right] < \infty$ and

$$Z = Z_0 + \int_0^1 \phi_s d\beta_s$$

where Z_0 is $\mathscr{F}_0^X$-measurable.

* **(3.18) Exercise (Tsirel'son type equation without a drift).** With the same sequence (t_k) as for the Tsirel'son drift we define a predictable function f on $\mathbf{W}$ by

$$f(t, w) = \operatorname{sgn}(w(t_k) - w(t_{k-1})) \qquad \text{if } t_k < t \le t_{k+1}.$$

1°) Let B be an $(\mathscr{F}_t)$-BM on $(\Omega, \mathscr{F}_t, P)$ and $(X^i, B), i = 1, 2, \dots, n$ be solutions to $e_0(f, 0)$; if we define η by

$$\eta_t^i = \operatorname{sgn}(X_t^i - X_{t_k}^i) \qquad \text{if } t_k < t \le t_{k+1},$$

then, for any $t \in]0, 1]$, the law of η_t is invariant by the symmetry $(x_1, \dots, x_n) \to (-x_1, \dots, -x_n)$, η_t is independent of $\mathscr{F}_1^B$ and almost-surely

$$\sigma(X_s, s \le t) = \mathscr{F}_t^B \vee \sigma(\eta_s) \qquad \text{for any } s \le t.$$

2°) Conversely, if α is a r.v. taking its values in $\{-1, 1\}^n$, which is $\mathscr{F}_\infty$-measurable, independent of $\mathscr{F}_1^B$, and such that its law is invariant by the above symmetry, then there exists a filtration $(\mathscr{G}_t)$ on Ω and a $(\mathscr{G}_t)$-adapted process X such that:

i) B is a $(\mathscr{G}_t)$-BM;
ii) for each i, (X^i, B) is a solution to $e_0(f, 0)$ and $\alpha^i = \operatorname{sgn}(X_{t_0}^i - X_{t_{-1}}^i)$.

(3.19) Exercise. (Continuation of Exercise (2.24) of Chap. VI). With obvious definition prove that there is path uniqueness for the equation $(*)$.

[Hint: One can use Exercise (1.21) Chapter VI.]

(3.20) Exercise. Retain the situation and notation of Theorem (3.8).

1°) For ρ as in Lemma (3.3), prove that, for every $t > 0$,

$$\int_0^t \rho(X_s^1 - X_s^2)^{-1} ds = \infty \qquad \text{a.s.}$$

[Hint: Use the expression of $X^1 - X^2$ as a function of M and H given in the proof of Theorem (3.8).]

2°) If in addition $b_1 - b_2 \geq a > 0$ and if now ρ is such that $\int_{0+} \rho(u)^{-1} du < \infty$, then

$$\int_0^t \rho(X_s^1 - X_s^2)^{-1} ds < \infty \qquad \text{a.s.}$$

** **(3.21) Exercise.** Retain the situation and notation of Theorem (3.7) and suppose that σ and b_1 satisfy some Lipschitz conditions. Prove that when a tends to zero, for every $t > 0$ and $\varepsilon \in]0, 1/2[$,

$$L_t^a(X^1 - X^2) = 0(a^{1/2-\varepsilon}) \qquad \text{a.s.}$$

[Hint: Use Corollary (1.9) in Chap. VI and the exponential formulas of the proof of Theorem (3.8).]

Notes and Comments

Sect. 1. The notion of stochastic differential equation originates with Itô (see Itô [2]). To write this section, we made use of Ikeda-Watanabe [2], Stroock-Varadhan [1] and Priouret [1]. For a more general exposition, see Jacod [2]. The important Theorem (1.7) is due to Yamada and Watanabe [1]. The result stated in Remark 2) after Theorem (1.7) was proved in Kallenberg [2].

Exercise (1.15) is taken from Stroock-Yor [1] and Exercise (1.16) from Girsanov [2]. The result in Exercise (1.18) is due to Perkins (see Knight [7]). Pathwise uniqueness is a property which concerns all probability spaces. Kallsen [1] has found a Tsirel'son-like example of an SDE which enjoys the existence and uniqueness properties on a particular space but not on all spaces.

Sect. 2. As for many results on SDE's, the results of this section originate with Itô. Theorem (2.1) was proved by Doléans-Dade [1] for general (i.e. non continuous) semimartingales. Proposition (2.3) comes from an unpublished paper of Yoeurp and Yor.

Theorem (2.4) and its corollaries are taken from Neveu [2] and Priouret [1], but of course, most ideas go back to Itô. Theorem (2.4) is the starting point for the theory of flows of SDE's in which, for instance, one proves, under appropriate hypothesis, the differentiability in x of the solutions. It also leads to some aspects of stochastic differential geometry. An introduction to these topics is provided by the lecture course of Kunita [4].

Exercise (2.6)bis is taken from Jacod [3] and Karandikar [1].

Exercise (2.8) is due to Doss [1] and Sussman ([1] and [2]). Exercise (2.9) is from Yoeurp [3] and Exercise (2.10) is taken in part from Ikeda-Watanabe [2]. Exercise (2.12), inspired by Jeulin-Yor [2] and Yor [10], originates with Itô [6] and provides a basic example for the theory of enlargements of filtrations. Exercise (2.14) is taken from El Karoui and Chaleyat-Maurel [1]. Exercise (2.15) describes results which are due to Feller; generally speaking, the contribution of Feller to the theory of diffusions is not sufficiently stressed in these Notes and Comments.

Exercise (2.17) is very close to Chitashvili-Toronjadze [1] and is further developed in Jeulin-Yor [4]. It would be interesting to connect the results in this exercise with those of Carlen [1] and Carlen-Elworthy [1]. For Exercise (2.18) see Bougerol [1] and Alili [1], Alili-Dufresne-Yor [1]; related results, for Lévy processes instead of BM with drift, are found in Carmona et al. [1].

Some explicit solutions to SDE's are exhibited in Kloeden-Platen [1] using Itô's formula.

The result of Exercise (2.19) may be found in Fitzsimmons [1] who studies in fact a converse to Lévy's equivalence.

Sect. 3. Our exposition of Theorem (3.5) is based on Le Gall [1] who improved earlier results of Nakao [1] and Perkins [5]. Problems of stability for solutions of such one-dimensional SDE's are studied in Kawabata-Yamada [1] and Le Gall [1].

The proof given here for the Tsirel'son example is taken from Stroock-Yor [1] and is inspired by a proof due to Krylov which is found in Liptser-Shiryayev [1]. Beněs [1] gives another proof, as well as some extensions. The exercises linked to this example are also mainly from Stroock-Yor [1] with the exception of Exercise (3.17) which is from Le Gall-Yor [1]. Further general results about Tsirel'son's equation in discrete time are developed in Yor [19].

Notice that in the Tsirel'son example, despite the fact that $(\mathscr{F}_t^B)$ is strictly coarser than $(\mathscr{F}_t^X)$, the latter filtration is still a strong Brownian filtration as was discussed in the Notes and Comments of Chap. V.

For the comparison theorems see Yamada [1], Ikeda-Watanabe ([1] and [2]) and Le Gall [1], but there are actually many other papers, too numerous to be listed here, devoted to this question.

Exercise (3.10) is from Williams [4] and Yor [11] and Exercise (3.11) is a result of Stroock-Yor [2]. Exercise (3.16) is taken from Yor [7]; with the notation of this exercise let us mention the following open

Question 1. If in 3°) the matrix A is no longer supposed to be symmetric, is the filtration of the martingale still that of a BM^r and, in the affirmative, what is r in terms of A?

A partial answer is found in Auerhan-Lépingle [1]; further progress on this question has been made by Malric [1].

Chapter X. Additive Functionals of Brownian Motion

§1. General Definitions

Although we want as usual to focus on the case of linear BM, we shall for a while consider a general Markov process for which we use the notation and results of Chap. III.

(1.1) Definition. *An* additive functional *of X is a $\overline{\mathbb{R}}_+$-valued, $(\mathscr{F}_t)$-adapted process $A = \{A_t, t \geq 0\}$ defined on Ω and such that*

i) it is a.s. non-decreasing, right-continuous, vanishing at zero and such that $A_t = A_{\zeta-}$ on $\{\zeta \leq t\}$;
ii) for each pair (s, t), $A_{s+t} = A_t + A_s \circ \theta_t$ a.s.

A continuous additive functional *(abbreviated CAF) is an additive functional such that the map $t \to A_t$ is continuous.*

Remark. In ii) the negligible set depends on s and t, but by using the right-continuity it can be made to depend only on t.

The condition $A_t = A_{\zeta-}$ on $\{\zeta \leq t\}$ means that the additive functional does not increase once the process has left the space. Since by convention $f(\Delta) = 0$ for any Borel function on E, if Γ is a Borel subset of E, this condition is satisfied by the occupation time of Γ, namely $A_t = \int_0^t 1_\Gamma(X_s)ds$, which is a simple but very important example of a CAF. In particular $A_t = t \wedge \zeta$, which corresponds to the special case $\Gamma = E$, is a CAF.

Let X be a Markov process with jumps and for $\varepsilon > 0$ put

$$T_\varepsilon = \inf\{t > 0 : d\,(X_t, X_{t^-}) > \varepsilon\}.$$

Then T_ε is an a.s. strictly positive stopping time and if we define inductively a sequence (T_n) by

$$T_1 = T_\varepsilon, \qquad T_n = T_{n-1} + T_\varepsilon \circ \theta_{T_{n-1}},$$

the reader will prove that $A_t = \sum_1^\infty 1_{(T_n \leq t)}$ is a purely discontinuous additive functional which counts the jumps of magnitude larger than ε occuring up to time t.

We shall now give the fundamental example of the local time of Brownian motion which was already defined in Chap. VI from the stochastic calculus point of view. Actually, all we are going to say is valid more generally for linear Markov processes X which are also continuous semimartingales such that $\langle X, X\rangle_t = \int_0^t \phi(X_s)ds$ for some function ϕ and may even be extended further by time-changes (Exercise (1.25) Chap. XI). This is in particular the case for the OU process or for the Bessel processes of dimension $d \geq 1$ or the squares of Bessel processes which we shall study in Chap. XI. The reader may keep track of the fact that the following discussion extends trivially to these cases.

We now consider the BM as a Markov process, that is to say we shall work with the canonical space $\mathbf{W} = C(\mathbb{R}_+, \mathbb{R})$ and with the entire family of probability measures P_a, $a \in \mathbb{R}$.

With each P_a, we may, by the discussion in Sect. 1, Chap. VI, associate a process L which is the local time of the martingale B at zero, namely, such that

$$|B_t| = |a| + \int_0^t \operatorname{sgn}(B_s)dB_s + L_t \qquad P_a\text{-a.s.}$$

Actually, L may be defined simultaneously for every P_a, since, thanks to Corollary (1.9) in Chap. VI,

$$L_t = \varlimsup_k \frac{1}{2\varepsilon_k} \int_0^t 1_{]-\varepsilon_k,\varepsilon_k[}(B_s)ds \qquad \text{a.s.},$$

where $\{\varepsilon_k\}$ is any sequence of real numbers decreasing to zero. The same discussion applies to the local time at a and yields a process L^a. By the results in Chap. VI, the map $(a, t) \to L_t^a$ is a.s. continuous.

Each of the processes L^a is an additive functional and even a *strong additive functional* which is the content of

(1.2) Proposition. *If T is a stopping time, then, for every a*

$$L^a_{T+S} = L^a_T + L^a_S(\theta_T) \qquad P_b\text{-a.s.}$$

for every b and every positive random variable S.

Proof. Set $I(\varepsilon) =]a - \varepsilon, a + \varepsilon[$; it follows from Corollary (1.9) in Chap. VI that if T is a stopping time

$$L^a_{T+S} = L^a_T + \lim_{\varepsilon \downarrow 0} \frac{1}{2\varepsilon} \int_0^S 1_{I(\varepsilon)}(B_u(\theta_T))\,du \qquad P_b\text{-a.s.}$$

for every b. By the Strong Markov property of BM

$$\begin{aligned} P_b\left[L^a_S(\theta_T) = \lim_{\varepsilon \downarrow 0} \frac{1}{2\varepsilon} \int_0^S 1_{I(\varepsilon)}(B_u(\theta_T))\,du\right] \\ = E_b\left[E_{B_T}\left[L^a_S = \lim_{\varepsilon \downarrow 0} \frac{1}{2\varepsilon} \int_0^S 1_{I(\varepsilon)}(B_u)du\right]\right] = 1. \end{aligned}$$

Consequently,

$$L^a_{T+S} = L^a_T + L^a_S \circ \theta_T \qquad \text{a.s.}$$

A case of particular interest is that of two stopping times T and S. We saw (Proposition (3.3) Chap. III) that $T + S \circ \theta_T$ is again a stopping time and the preceding result reads

$$L^a_{T+S\circ\theta_T} = L^a_T + L^a_S \circ \theta_T.$$

We draw the attention of the reader to the fact that $L^a_S \circ \theta_T$ is the map $\omega \to L^a_{S(\theta_T(\omega))}(\theta_T(\omega))$, whereas the $L^a_S(\theta_T)$ in the statement of the proposition is the map $\omega \to L^a_{S(\omega)}(\theta_T(\omega))$.

The upcoming consequence of the preceding result is very important in Chap. XII on Excursion theory as well as in other places. As in Chap. VI, we write τ_t for the time-change associated with $L_t = L^0_t$.

(1.3) Proposition. *For every t, there is a negligible set Γ_t such that for $\omega \notin \Gamma_t$ and for every $s > 0$,*

$$\tau_{t+s}(\omega) = \tau_t(\omega) + \tau_s\left(\theta_{\tau_t}(\omega)\right), \qquad \tau_{(t+s)-}(\omega) = \tau_t(\omega) + \tau_{s-}\left(\theta_{\tau_t}(\omega)\right).$$

Proof. Plainly, τ_t is a.s. finite and $\tau_{t+s} > \tau_t$. Therefore

$$\begin{aligned}\tau_{t+s} &= \inf\{u > 0 : L_u > t + s\} \\ &= \tau_t + \inf\left\{u > 0 : L_{\tau_t + u} > t + s\right\}.\end{aligned}$$

Using the strong additivity of L and the fact that $L_{\tau_t} = t$, we have that, almost-surely, for every s,

$$\begin{aligned}\tau_{t+s} &= \tau_t + \inf\left\{u > 0 : L_{\tau_t} + L_u(\theta_{\tau_t}) > t + s\right\} \\ &= \tau_t + \inf\left\{u > 0 : L_u(\theta_{\tau_t}) > s\right\} = \tau_t + \tau_s \circ \theta_{\tau_t}.\end{aligned}$$

The second claim follows at once from the first. □

Remarks. 1°) The same result is clearly true for L^a in place of L. It is in fact true for any finite CAF, since it is true, although outside the scope of this book, that every additive functional of a Markov process has the strong additivity property.

2°) Since the processes (L_t) and (S_t) have the same law under P_0 (Sect. 2 Chap. VI), it follows from Sect. 3 Chap. III that the process (τ_t) is a stable subordinator of index 1/2. This may also be proved using the above result and the strong Markov property as is outlined in Exercise (1.11).

3°) Together with Proposition (3.3) in Chap. III, the above result entails that $\theta_{\tau_{t+s}} = \theta_{\tau_t} \circ \theta_{\tau_s}$ a.s.

Our goal is to extend these properties to all continuous additive functionals of linear BM. This will be done by showing that those are actually integrals of local times, thus generalizing what is known for occupation times. To this end, we will need a few results which we now prove but which, otherwise, will be used sparingly in the sequel.

Again, we consider a general Markov process. By property i) of the definition of additive functionals, for every ω, we can look upon $A_t(\omega)$ as the distribution function of a measure on $\mathbb{R}_+$ just as we did for the increasing processes of Sect. 1

Chap. IV. If F_t is a process, we shall denote by $\int_0^t F_s dA_s$ its integral on $[0, t]$ with respect to this measure provided it is meaningful. Property ii) of the definition is precisely an invariance property of the measure dA_s.

(1.4) Proposition. *If f is a bounded positive Borel function and A_t is finite for every t, the process $(f \cdot A)$ defined by*

$$(f \cdot A)_t = \int_0^t f(X_s) dA_s$$

is an additive functional. If A is a CAF, *then $(f \cdot A)$ is a* CAF.

Proof. The process X is progressively measurable with respect to $(\mathscr{F}_t)$, thus so is $f(X)$ and as a result, $(f \cdot A)_t$ is $\mathscr{F}_t$-measurable for each t. Checking the other conditions of Definition (1.1) is easy and left to the reader. □

Remarks. 1°) The hypothesis that f is bounded serves only to ensure that $t \to (f \cdot A)_t$ is right-continuous. If f is merely positive and $(f \cdot A)$ is right-continuous, then it is an additive functional. The example of the uniform motion to the right and of $f(x) = (1/x)1_{(x>0)}$ shows that this is not always the case.

2°) If $A_t = t \wedge \zeta$ and $f = 1_\Gamma$, then $(f \cdot A)_t$ is the occupation time of Γ by X.

(1.5) Definition. *For $\alpha \geq 0$, the function*

$$U_A^\alpha(x) = E_x\left[\int_0^\infty e^{-\alpha t} dA_t\right]$$

*is called the α-*potential *of A. We also write $U_A^\alpha f(x)$ for the α-potential of $f \cdot A$, in other words*

$$U_A^\alpha f(x) = E_x\left[\int_0^\infty e^{-\alpha t} f(X_t) dA_t\right].$$

For $A_t = t \wedge \zeta$, we have $U_A^\alpha f = U^\alpha f$, and, more generally, it is easily checked that the map $(x, \Gamma) \to U_A^\alpha 1_\Gamma(x)$ is a kernel on $(E, \mathscr{E})$ which will be denoted $U_A^\alpha(x, \cdot)$. Moreover, these kernels satisfy a resolvent-type equation.

(1.6) Proposition. *For $\alpha, \beta \geq 0$, if $U_A^\alpha f(x)$ and $U_A^\beta f(x)$ are finite,*

$$U_A^\alpha f(x) - U_A^\beta f(x) = (\beta - \alpha) U^\alpha U_A^\beta f(x) = (\beta - \alpha) U^\beta U_A^\alpha f(x).$$

Proof. Using the definitions and the Markov property

$$\begin{aligned} U^\alpha U_A^\beta f(x) &= E_x\left[\int_0^\infty e^{-\alpha t} E_{X_t}\left[\int_0^\infty e^{-\beta s} f(X_s) dA_s\right] dt\right] \\ &= E_x\left[\int_0^\infty e^{-\alpha t} E_x\left[\int_0^\infty e^{-\beta s} f\left(X_s(\theta_t)\right) dA_s(\theta_t) \mid \mathscr{F}_t\right] dt\right] \\ &= E_x\left[\int_0^\infty e^{-\alpha t} dt \int_0^\infty e^{-\beta s} f(X_{s+t}) dA_s(\theta_t)\right]. \end{aligned}$$

By property ii) in Definition (1.1), we consequently have

$$\begin{aligned} U^\alpha U_A^\beta f(x) &= E_x\left[\int_0^\infty e^{-\alpha t}dt \int_0^\infty e^{-\beta s} f(X_{s+t})dA_{s+t}\right] \\ &= E_x\left[\int_0^\infty e^{-(\alpha-\beta)t}dt \int_t^\infty e^{-\beta u} f(X_u)dA_u\right] \end{aligned}$$

and Fubini's theorem yields the desired result. □

We will prove that the map which with A, associates the family of kernels $U_A^\alpha(x,\cdot)$ is one-to-one. Let us first mention that as usual $A = B$ will mean that A and B are indistinguishable i.e. in this context, $P_x[\exists t : A_t \neq B_t] = 0$ for every x. Because of the right-continuity, this is equivalent to $P_x[A_t = B_t] = 1$ for every x and t. The following proposition will be used in §2.

(1.7) Proposition. *If A and B are two additive functionals such that for some $\alpha \geq 0$, $U_A^\alpha = U_B^\alpha < \infty$ and $U_A^\alpha f = U_B^\alpha f$ for every $f \in C_K^+$, then $A = B$.*

Proof. The hypothesis extends easily to $U_A^\alpha f = U_B^\alpha f$ for every $f \in b\mathscr{E}_+$. In particular $U_A^\alpha U^\alpha f = U_B^\alpha U^\alpha f$. But computations similar to those made in the preceding proof show that

$$U_A^\alpha U^\alpha f(x) = E_x\left[\int_0^\infty e^{-\alpha t} f(X_t)A_t dt\right]$$

and likewise for B. As a result,

$$\int_0^\infty e^{-\alpha t} E_x[f(X_t)A_t]dt = \int_0^\infty e^{-\alpha t} E_x[f(X_t)B_t]dt.$$

By the resolvent equation of Proposition (1.6), the same result is true for each $\beta \geq \alpha$.

If f is continuous and bounded, the map $t \to E_x[f(X_t)A_t]$ is finite since

$$\infty > U_A^\alpha(x) \geq E_x\left[\int_0^t e^{-\alpha s}dA_s\right] \geq e^{-\alpha t}E_x[A_t],$$

hence right-continuous and the same is true with B in place of A. The injectivity of the Laplace transform implies that

$$E_x[f(X_t)A_t] = E_x[f(X_t)B_t]$$

for every t and for every continuous and bounded f. This extends at once to $f \in b\mathscr{E}_+$. Finally, an induction argument using the Markov property and the additivity property of A and B shows that for $0 \leq t_1 \leq t_2 \leq \ldots \leq t_n \leq t$ and $f_k \in b\mathscr{E}_+$,

$$E_x\left[\prod_{k=1}^n f_k(X_{t_k}) \cdot A_t\right] = E_x\left[\prod_{k=1}^n f_k(X_{t_k}) \cdot B_t\right].$$

Since A_t and B_t are $\mathscr{F}_t$-measurable and $\mathscr{F}_t$ is generated by X_s, $s \leq t$, it follows that $A = B$. □

(1.8) Exercise. If A and B are two additive functionals such that for every $t > 0$ and every $x \in E$

$$E_x[A_t] = E_x[B_t] < \infty,$$

then $M_t = A_t - B_t$ is a $(\mathscr{F}_t, P_x)$-martingale for every $x \in E$. (Continuation in Exercise (2.22)).

\# **(1.9) Exercise (Extremal process).** 1°) For the standard linear BM, set, with the usual notation, $X_a = L_{T_a}$ with $a \geq 0$. Prove that the process $a \to X_a$ has independent (non stationary) increments. Prove further that for $x_1 < x_2 < \ldots < x_p$ and $a_1 < a_2 < \ldots a_p$,

$$P_0\left[X_{a_1} > x_1, X_{a_2} > x_2, \ldots, X_{a_p} > x_p\right]$$
$$= \exp\left(-\frac{x_1}{2a_1} - \frac{x_2 - x_1}{2a_2} \cdots - \frac{x_p - x_{p-1}}{2a_p}\right).$$

2°) If as usual τ_t is the time-change associated with L, prove that the process $Y_t = S_{\tau_t}$ is the inverse of the process X_a. Deduce therefrom, that for $t_1 < t_2 < \ldots < t_p$ and $y_1 < y_2 < \ldots < y_p$,

$$P_0\left[Y_{t_1} < y_1, Y_{t_2} < y_2, \ldots, Y_{t_p} < y_p\right]$$
$$= \exp\left(-\frac{t_1}{2y_1} - \frac{t_2 - t_1}{2y_2} \cdots - \frac{t_p - t_{p-1}}{2y_p}\right).$$

Some further information on these processes may be found in Exercise (4.11) in Chap. XII.

(1.10) Exercise. For $0 < \alpha < 1$, define

$$Z_\alpha(t) = \int_0^{+\infty} L^x_{\tau_t} x^{(1-2\alpha)/\alpha} dx$$

where τ_t is the time-change associated with $L = L^0$.

1°) Prove that Z_α is a stable subordinator of index α.

[Hint: Use the scaling properties of Exercise (2.11) in Chap. VI.]

2°) Prove that

$$\lim_{\alpha \to 0} (Z_\alpha(t))^\alpha = S_{\tau_t} \qquad \text{a.s.}$$

(see the previous exercise for the study of the limit process).

*# **(1.11) Exercise.** Let L be the local time of BM^1 at zero and τ_t its inverse.

1°) Deduce from Proposition (1.3) and the strong Markov property that τ_t is under P_0, an increasing Lévy process.

2°) From the definition of L in Chap. VI and the integration by parts formula, deduce that the α-potential ($\alpha > 0$) of L at 0, namely $U^\alpha_L(0)$ is equal to $(2\alpha)^{-1/2}$. This will also follow from a general result in the next section.

3°) For $x \in \mathbb{R}$, derive from the equality

$$U_L^\alpha(x) = \int_0^\infty E_x\left[\exp(-\alpha\tau_t)\right]dt,$$

that

$$E_x\left[\exp(-\alpha\tau_t)\right] = E_x\left[\exp(-\alpha\tau_0)\right]\exp\left(-t\sqrt{2\alpha}\right),$$

and conclude that under P_0, the process (τ_t) is a stable subordinator of index 1/2. We thus get another proof of this result independent of the equivalence in law of L_t and S_t. See also 6°) in Exercise (4.9) of Chap. VI.

4°) If β is an independent BM^d, prove that β_{τ_t} is a d-dimensional Cauchy process, i.e. the Lévy process in $\mathbb{R}^d$ such that the increments have Cauchy laws in $\mathbb{R}^d$ (see Exercise (3.24) in Chap. III).

(1.12) Exercise. Let X be a Markov process on $\mathbb{R}$ with continuous paths such that its t.f. has a density $p_t(x, y)$ with respect to the Lebesgue measure which is continuous in each of the three variables. Assume moreover that for each P_x, X is a semimartingale such that $\langle X, X\rangle_t = t$. If L^x is the family of its local times, prove that

$$\frac{d}{dt}E_x\left[L_t^y\right] = p_t(x, y)$$

for every t, x, y in $]0, \infty[\times\mathbb{R}\times\mathbb{R}$.

The reader is invited to compare this exercise with Exercise (4.16) Chap. VII.

\# **(1.13) Exercise.** Let $X = \left(X_t, \mathscr{F}_t, P_x\right)$ be a Markov process with state space E and A an additive functional of X. Prove that, for any P_x, the pair (X, A) is a $(\mathscr{F}_t)$-Markov process on $E \times \mathbb{R}_+$ with t.f. given by

$$Q_t\phi(x, a) = E_x\left[\phi\left(X_t, a + A_t\right)\right].$$

[Hint: The reader may find it useful to use the Markov property of Chap. III Exercise (3.19).] A special case was studied in Exercise (2.18) of Chap. VI.

(1.14) Exercise. 1°) Let L^a be the family of local times of BM and set $L_t^* = \sup_a L_t^a$. Prove that for every $p \in]0, \infty[$ there are two constants c_p and C_p such that

$$c_pE\left[T^{p/2}\right] \le E\left[\left(L_T^*\right)^p\right] \le C_pE\left[T^{p/2}\right]$$

for every stopping time T.

[Hint: Use Theorems (4.10) and (4.11) in Chap. IV.]

2°) Extend adequately the results to all continuous local martingales. This result is proved in Sect. 2 Chap. XI.

\# **(1.15) Exercise.** 1°) Let A be an additive functional of the Markov process X and e an independent exponential r.v. with parameter γ. Let

$$\zeta = \inf\{t : A_t > e\}.$$

Prove that the process $\widetilde{X}$ defined by

$$\widetilde{X}_t = X_t \quad \text{on } \{t < \zeta\}, \qquad \widetilde{X}_t = \Delta \quad \text{on } \{t \geq \zeta\}$$

is a Markov process for the probability measure P_x of X. Prove that the semigroup Q_t of $\widetilde{X}$ is given by

$$Q_t f(x) = E_x\left[f(X_t)\exp(-\gamma A_t)\right].$$

If X is the linear BM and A its local time at 0, the process $\widetilde{X}$ is called the *elastic Brownian motion.*

2°) Call $\widetilde{U}_\alpha$ the resolvent of the elastic BM. If f is bounded and continuous prove that $u = \widetilde{U}_\alpha f$ satisfies the equations

$$\begin{cases} \alpha u - u''/2 = f, \\ u'_+(0) - u'_-(0) = 2\gamma u(0); \end{cases}$$

loosely speaking, this means that the infinitesimal generator A of X is given by $Au = \frac{1}{2}u''$ on the space $\{u \in C^2(\mathbb{R}) : u'_+(0) - u'_-(0) = 2\gamma u(0)\}$.

** **(1.16) Exercise. (Conditioning with respect to certain random times).** Let X be a continuous Markov process such that there exists a family $P^s_{x,y}$ of probability measures on $(\Omega, \mathscr{F}_\infty)$ with the following properties

i) for each $s > 0$, there is a family $\mathscr{C}_s$ of $\mathscr{F}_s$-measurable r.v.'s, closed under pointwise multiplication, generating $\mathscr{F}_s$ and such that the map $(t, y) \to E^t_{x,y}[C]$ is continuous on $]s, \infty[\times E$ for every $x \in E$ and $C \in \mathscr{C}_s$.

ii) for every $\Gamma \in \mathscr{F}_t$, the r.v. $P^t_{x,X_t}(\Gamma)$ is a version of $E_x[1_\Gamma \mid X_t]$.

The reader will find in Chap. XI a whole family of processes satisfying these hypotheses.

1°) Prove that for every additive functional A of X such that $E_x[A_t] < \infty$ for every t, and every positive process H,

$$E_x\left[\int_0^\infty H_s dA_s\right] = E_x\left[\int_0^\infty E^s_{x,X_s}[H_s] dA_s\right].$$

2°) Let L be a positive a.s. finite $\mathscr{F}_\infty$-measurable r.v., λ a predictable process and Λ an additive functional such that

$$E_x[H_L] = E_x\left[\int_0^\infty H_s \lambda_s d\Lambda_s\right]$$

for every positive predictable process H (see Exercise (4.16) Chap. VII). If $\Gamma_x = \{(s, y) : E^s_{x,y}[\lambda_s] = 0\}$, prove that $P_x[(L, X_L) \in \Gamma_x] = 0$ and that

$$E_x[H_L \mid L = s, X_L = y] = E^s_{x,y}[H_s\lambda_s] / E^s_{x,y}[\lambda_s].$$

3°) Assume that in addition, the process X satisfies the hypothesis of Exercise (1.12) and that there is a measure ν such that $\Lambda_t = \int L^x_t \nu(dx)$. Prove that the law of the pair (L, X_L) under P_x has density $E^t_{x,y}[\lambda_t] p_t(x, y)$ with respect to $dt\,\nu(dy)$.

Give another proof of the final result of Exercise (4.16) Chap. VII.

§2. Representation Theorem for Additive Functionals of Linear Brownian Motion

With each additive functional, we will associate a measure which in the case of BM, will serve to express the functional as an integral of local times. To this end, we need the following definitions.

If N is a kernel on $(E, \mathscr{E})$ and m a positive measure, one sets, for $A \in \mathscr{E}$,

$$mN(A) = \int m(dx)N(x, A).$$

We leave to the reader the easy task of showing that the map $A \to mN(A)$ is a measure on $\mathscr{E}$ which we denote by mN.

(2.1) Definition. *A positive σ-finite measure m on $(E, \mathscr{E})$ is said to be* invariant *(resp.* excessive*) for X, or for its semi-group, if $mP_t = m$ (resp. $mP_t \leq m$) for every $t \geq 0$.*

For a positive measure m, we may define a measure P_m on $(\Omega, \mathscr{F})$ by setting as usual

$$P_m[\Gamma] = \int_E P_x[\Gamma]m(dx).$$

This extends what was done in Chap. III with a starting probability measure ν, but here P_m is not a probability measure if $m(E) \neq 1$. Saying that m is invariant is equivalent to saying that P_m is invariant by the maps θ_t for every t. Moreover, if m is invariant and bounded and if we normalize m so that $m(E) = 1$ then, under P_m, the process is stationary (Sect. 3 Chap. I). For any process with independent increments on $\mathbb{R}^d$, in particular BM^d, the Lebesgue measure is invariant, as the reader will easily check. Examples of excessive measures are provided by *potential measures* $\nu U = \int_0^\infty \nu P_t dt$ when they are σ-finite as is the case for transient processes such as BM^d for $d \geq 3$.

The definition of invariance could have been given using the resolvent instead of the semi-group. Plainly, if m is invariant (resp. excessive) then $mU^1 = m$ (resp. $mU^1 \leq m$) and the converse may also be proved (see Exercise (2.21)).

From now on, we assume that there is an excessive measure m which will be fixed throughout the discussion. For all practical purposes, this assumption is always in force as, for transient processes, one can take for m a suitable potential measure whereas, in the recurrent case, it can be shown under mild conditions that there is an invariant measure. In the sequel, a function on E is said to be integrable if it is integrable with respect to m.

Let A be an additive functional; under the above assumption, we set, for $f \in b\mathscr{E}_+$,

$$\nu_A(f) = \sup_{t>0} \frac{1}{t} E_m\left[(f \cdot A)_t\right].$$

(2.2) Proposition. *For every f and A,*

$$\nu_A(f) = \lim_{t \downarrow 0} \frac{1}{t} E_m\left[(f \cdot A)_t\right] = \lim_{\alpha \to \infty} \alpha \int m(dx) U_A^\alpha f(x).$$

Moreover, the second limit is an increasing limit.

Proof. Since m is excessive, it follows easily from the Markov property that $t \to E_m\left[(f \cdot A)_t\right]$ is sub-additive; thus, the first equality follows from the well-known properties of such functions. The second equality is then a consequence of the abelian theorem for Laplace transform and finally the limit is increasing as a consequence of the resolvent equation. □

For every $\alpha \geq 0$, the map $f \to \alpha \int m(dx) U_A^\alpha f(x)$ is a positive measure, and since one can interchange the order of increasing limits, the map $f \to \nu_A(f)$ is a positive measure which is denoted by ν_A.

(2.3) Definition. *The measure ν_A is called the* measure associated *with the additive functional A. If ν_A is a bounded (σ-finite) measure, A is said to be* integrable *(σ-*integrable*).*

It is clear that the measures ν_A do not charge polar sets. One must also observe that the correspondence between A and ν_A depends on m, but there is usually a canonical choice for m, for instance Lebesgue measure in the case of BM. Let us further remark that the measure associated with $A_t = t \wedge \zeta$ is m itself and that $\nu_{f \cdot A} = f \cdot \nu_A$. In particular, if $A_t = \int_0^t f(X_s) ds$ is an additive functional, then $\nu_A = f \cdot m$ and A is integrable if and only if f is integrable.

Let us finally stress that if m is invariant then ν_A is defined by the simpler formula

$$\nu_A(f) = E_m\left[\int_0^1 f(X_s) dA_s\right].$$

As a fundamental example, we now compute the measure associated with the local time L^a of linear BM. As we just observed, here m is the Lebesgue measure which is invariant and is in fact the only excessive measure of BM (see Exercise (3.14)).

(2.4) Proposition. *The local time L^a is an integrable additive functional and the associated measure is the Dirac measure at a.*

Proof. Plainly, we can make $a = 0$ and we write L for L^0. Proposition (1.3) in Chap. VI shows that the measure ν_L has $\{0\}$ for support and the total mass of ν_L is equal to

$$\int_{-\infty}^{+\infty} dx\, E_x[L_1] = \int_{-\infty}^{+\infty} dx\, E_0[L_1^x] = E_0\left[\int_{-\infty}^{+\infty} L_1^x dx\right] = 1$$

by the occupation times formula. The proof is complete. □

The measure ν_A will be put to use in a moment, thus it is important to know how large the class of σ-integrable additive functionals is. We shall need the following

(2.5) Lemma. *If C is an integrable AF and $h \in b\mathscr{E}_+$ and if there is an $\alpha > 0$ such that $U_A^\alpha h \le U_C^\alpha$, then $\nu_A(h) < \infty$.*

Proof. For $\beta \ge \alpha$, the resolvent equation yields

$$\beta \int \left(U_C^\beta - U_A^\beta h\right) dm = \beta \int \left(U_C^\alpha - U_A^\alpha h\right) dm - (\beta - \alpha) \int \beta U^\beta \left(U_C^\alpha - U_A^\alpha h\right) dm,$$

and because m is excessive

$$\begin{aligned}\beta \int \left(U_C^\beta - U_A^\beta h\right) dm &\ge \beta \int \left(U_C^\alpha - U_A^\alpha h\right) dm - (\beta - \alpha) \int \left(U_C^\alpha - U_A^\alpha h\right) dm \\ &= \alpha \int \left(U_C^\alpha - U_A^\alpha h\right) dm \ge 0.\end{aligned}$$

Thus $\beta \int U_A^\beta h\, dm \le \int \beta U_C^\beta dm$ for every $\beta \ge \alpha$, which entails the desired result. □

We may now state

(2.6) Theorem. *Every CAF is σ-integrable. Moreover, E is the union of a sequence of universally measurable sets E_n such that the potentials $U_A^1(\cdot, E_n)$ are bounded and integrable.*

Proof. Let f be a bounded, integrable, strictly positive Borel function and set

$$\phi(x) = E_x\left[\int_0^\zeta e^{-t} f(X_t) e^{-A_t} dt\right].$$

Plainly, $0 < \phi \le \|f\|$; let us compute $U_A^1\phi$. We have

$$\begin{aligned}U_A^1\phi(x) &= E_x\left[\int_0^\zeta e^{-t} E_{X_t}\left[\int_0^\zeta e^{-s} f(X_s) e^{-A_s} ds\right] dA_t\right] \\ &= E_x\left[\int_0^\zeta e^{-t} E_x\left[\int_0^\zeta e^{-s} f(X_{s+t}) e^{-A_s \circ \theta_t} ds \,\middle|\, \mathscr{F}_t\right] dA_t\right].\end{aligned}$$

Using the result in 1°) of Exercise (1.13) of Chap. V and Proposition (4.7) Chap. 0, this is further equal to

$$\begin{aligned}&E_x\left[\int_0^\zeta e^{-t} dA_t \int_0^\zeta e^{-s} f(X_{s+t}) e^{-A_{s+t}} e^{A_t} ds\right] \\ &= E_x\left[\int_0^\zeta e^{-s} f(X_s) e^{-A_s} ds \int_0^s e^{A_t} dA_t\right] \\ &= E_x\left[\int_0^\zeta e^{-s} f(X_s)\left(1 - e^{-A_s}\right) ds\right] = U^1 f(x) - \phi(x).\end{aligned}$$

It remains to set $E_n = \{\phi \geq 1/n\}$ or $\{\phi > 1/n\}$ and to apply the preceding lemma with $C_t = \int_0^t f(X_s)ds$ and $h = n^{-1}1_{E_n}$. □

Now and for the remainder of this section, we specialize to the case of linear BM which we denote by B. We recall that the resolvent U^α is a convolution kernel which is given by a continuous and symmetric density. More precisely

$$U^\alpha f(x) = \int u^\alpha(x, y) f(y) dy$$

where $u^\alpha(x, y) = \left(\sqrt{2\alpha}\right)^{-1} \exp\left(-\sqrt{2\alpha}|y - x|\right)$. As a result, if m is the Lebesgue measure, then for two positive functions f and g,

$$(*) \qquad \int g(U^\alpha f)\, dm = \int (U^\alpha g) f\, dm.$$

Moreover, these kernels have the *strong Feller property*: if f is a bounded Borel function, then $U^\alpha f$ is continuous. This is a well-known property of convolution kernels.

Finally, we will use the following observation: if f is a Borel function on $\mathbb{R}$ and $f(B_t)$ is a.s. right-continuous at $t = 0$, then f is a continuous function; this follows at once from the law of the iterated logarithm. This observation applies in particular to the α-potentials of a CAF when they are finite. Indeed, in that case,

$$U_A^\alpha(B_t) = E_x\left[\int_0^\infty e^{-\alpha s} dA_s \circ \theta_t \,|\mathscr{F}_t\right] = e^{\alpha t} E_x\left[\int_t^\infty e^{-\alpha s} dA_s \,|\mathscr{F}_t\right] \quad P_x\text{-a.s.},$$

and this converges to $U_A^\alpha(x)$ when $t \downarrow 0$ thanks to Theorem (2.3) in Chap. II. □

We may now state

(2.7) Proposition. *In the case of linear BM, for every continuous additive functional A, the measure ν_A is a Radon measure.*

Proof. Thanks to Lemma (2.5) it is enough to prove that the sets $E_n = \{\phi > 1/n\}$ in the proof of Theorem (2.6) are open, hence that ϕ is continuous. Again, this follows from the right-continuity of $\phi(B_t)$ at $t = 0$ which is proved as above. We have P_x-a.s.,

$$\begin{aligned}
\phi(B_t) &= E_{B_t}\left[\int_0^\zeta e^{-u} f(B_u) e^{-A_u} du\right] \\
&= E_x\left[\int_0^\zeta e^{-u} f(B_{u+t}) e^{-A_{u+t}+A_t} du \,|\mathscr{F}_t\right] \\
&= e^{A_t} e^t E_x\left[\int_t^\zeta e^{-u} f(B_u) e^{-A_u} du \,|\mathscr{F}_t\right]
\end{aligned}$$

and this converges P_x-a.s. to $\phi(x)$ by Corollary (2.4) of Chap. II. □

We shall now work in the converse direction and show that, with each Radon measure ν, we can associate a CAF A such that $\nu = \nu_A$. Although it could be done in a more general setting, we keep with linear BM.

(2.8) Theorem. *If A is a* CAF, *then for every $\alpha > 0$ and $f \in \mathscr{E}_+$,*

$$U_A^\alpha f(x) = \int u^\alpha(x, y) f(y) \nu_A(dy).$$

In particular, $U_A^\alpha = \int u^\alpha(\cdot, y)\nu_A(dy)$.

Proof. Since $\nu_{f\cdot A} = f \cdot \nu_A$, it is enough to prove the particular case.

Supppose first that U_A^α is bounded and integrable and let ϕ be in C_K^+. By Proposition (2.2),

$$\int U^\alpha \phi(y) \nu_A(dy) = \lim_{\beta\to\infty} \beta E_m \left[\int_0^\infty e^{-\beta s} U^\alpha \phi(B_s) dA_s\right].$$

The function $s \to e^{-\beta s} U^\alpha \phi(B_s)$ is continuous, hence, is the limit of the sums

$$S_n = \sum_{k\geq 0} \exp\big(-\beta(k+1)/n\big) U^\alpha \phi(B_{k/n}) 1_{]k/n,(k+1)/n]};$$

as a function on $\Omega \times \mathbb{R}_+$, the sum S_n is smaller than $\alpha^{-1}\|\phi\| e^{-\beta s}$, which thanks to the hypothesis that ν_A is bounded, is integrable with respect to the measure $P_m \otimes dA_s$ defined on $\Omega \times \mathbb{R}_+$ by

$$P_m \otimes dA_s(\Gamma) = E_m\left[\int_0^\infty 1_\Gamma(\omega, s) dA_s\right].$$

Consequently, Lebesgue's dominated convergence theorem tells us that

$$E_m\left[\int_0^\infty e^{-\beta s} U^\alpha \phi(B_s) dA_s\right] = \lim_{n\to\infty} E_m\left[\int_0^\infty S_n(s) dA_s\right].$$

Using the Markov property and the fact that m is an invariant measure, this is further equal to

$$\begin{aligned}
&\lim_{n\to\infty} \sum_{k\geq 0} E_m\left[\exp\big(-\beta(k+1)/n\big) U^\alpha \phi(B_{k/n}) E_{B_{k/n}}\left[A_{1/n}\right]\right] \\
&= \lim_{n\to\infty} \left(\sum_{k\geq 0} \exp\big(-\beta(k+1)/n\big)\right) \int U^\alpha \phi(y) E_y\left[A_{1/n}\right] m(dy) \\
&= \lim_{n\to\infty} \exp(-\beta/n)\big(1 - \exp(-\beta/n)\big)^{-1} \int \phi(y) U^\alpha\left(E_\cdot\left[A_{1/n}\right]\right)(y) m(dy)
\end{aligned}$$

where we lastly used the duality relation $(*)$ above. As for every β we have $\lim_{n\to\infty} \beta \exp(-\beta/n)/n\big(1 - \exp(-\beta/n)\big) = 1$, it follows that

$$\begin{aligned}\int U^\alpha \phi(y)\nu_A(dy) &= \lim_{n\to\infty} n \int \phi(y) U^\alpha \left(E_.\left[A_{1/n}\right]\right)(y) m(dy) \\ &= \lim_{n\to\infty} \int_0^\infty n e^{-\alpha t} E_{\phi\cdot m}\left[E_{X_t}\left[A_{1/n}\right]\right] dt \\ &= \lim_{n\to\infty} \int_0^\infty n e^{-\alpha t} E_{\phi\cdot m}\left[A_{t+1/n} - A_t\right] dt \\ &= \lim_{n\to\infty} \left\{ n\left(e^{\alpha/n} - 1\right) \int_{1/n}^\infty e^{-\alpha s} E_{\phi\cdot m}[A_s] ds - n \int_0^{1/n} e^{-\alpha s} E_{\phi\cdot m}[A_s] ds \right\} \\ &= \alpha \int_0^\infty e^{-\alpha s} E_{\phi\cdot m}[A_s] ds = E_{\phi\cdot m}\left[\int_0^\infty e^{-\alpha s} dA_s\right],\end{aligned}$$

where, to obtain the last equality, we used the integration by parts formula. Thus, we have obtained the equality

$$\int U^\alpha \phi(y)\nu_A(dy) = \int U_A^\alpha(x)\phi(x) m(dx)$$

namely

$$\iint u^\alpha(x, y)\phi(x) m(dx)\nu_A(dy) = \int U_A^\alpha(x)\phi(x) m(dx),$$

which entails

$$\int u^\alpha(x, y)\nu_A(dy) = U_A^\alpha(x) \qquad m\text{-a.e.}.$$

Both sides are continuous, the right one by the observation made before (2.7) and the left one because of the continuity of u^α, so this equality holds actually everywhere and the proof is complete in that case. The general case where A is σ-integrable follows upon taking increasing limits. □

We now state our main result which generalizes the occupation times formula (see below Definition (2.3)).

(2.9) Theorem. *If A is a* CAF *of linear* BM, *then*

$$A_t = \int_{-\infty}^{+\infty} L_t^a \nu_A(da).$$

As a result, any CAF *of linear* BM *is a strong additive functional.*

Proof. Since ν_A is a Radon measure and $a \to L_t^a$ is for each t a continuous function with compact support, the integral

$$\widetilde{A}_t = \int_{-\infty}^{+\infty} L_t^a \nu_A(da)$$

is finite and obviously defines a CAF. Let us compute its associated measure. For $f \in \mathscr{E}_+$, and since m is invariant,

$$\begin{aligned}
\nu_{\widetilde{A}}(f) &= E_m\left[\int_0^1 f(B_s)d\widetilde{A}_s\right] \\
&= E_m\left[\int_{-\infty}^{+\infty}\left(\int_0^1 f(B_s)dL_s^a\right)\nu_A(da)\right] \\
&= \int_{-\infty}^{+\infty}\nu_A(da)E_m\left[\int_0^1 f(B_s)dL_s^a\right] = \nu_A(f).
\end{aligned}$$

Thus $\nu_{\widetilde{A}} = \nu_A$ and by Proposition (1.7) and the last result, it is easily seen that $A = \widetilde{A}$ (up to equivalence) which completes the proof. □

Remark. It is true that every additive functional of a Feller process is a strong additive functional, but the proof of this result lies outside the scope of this book.

We now list a few consequences of the above theorem.

(2.10) Corollary. *Every* CAF *of linear* BM *is finite and the map* $A \to \nu_A$ *is one-to-one.*

Proof. Left to the reader. □

Let us mention that for the right-translation on the line there exist non finite CAF's i.e. such that $A_t = \infty$ for finite t, for instance $\int_0^t f(X_s)ds$ with $f = |x|^{-1}1_{(x<0)}$.

(2.11) Corollary. *For every* CAF, *there is a modification such that* A *is* $(\mathscr{F}_t^0)$-*adapted.*

Proof. We recall that $(\mathscr{F}_t^0)$ is the uncompleted filtration of BM. The property is clear for local times by what was said in the last section, hence carries over to all CAF's. □

We also have the

(2.12) Corollary. *For any* CAF A *of linear* BM, *there exists a convex function* f *such that*

$$A_t = f(B_t) - f(B_0) - \int_0^t f'_-(B_s)dB_s.$$

Proof. Let f be a convex function whose second derivative is precisely $2\nu_A$ (Proposition (3.2) in the Appendix). If we apply Tanaka's formula to this function, we get exactly the formula announced in the statement. □

(2.13) Corollary. *If* A *is a* CAF *and* f *a positive Borel function, then*

$$\int_0^t f(B_s)dA_s = \int L_t^a f(a)\nu_A(da).$$

Proof. Obvious.

(2.14) Corollary. *For any non zero* CAF A *of linear* BM, $A_\infty = \infty$ *a.s.*

Proof. Because of Corollary (2.4) Chapter VI and the translation invariance of BM, $L^a_\infty = \infty$ for every a simultaneously almost-surely; this obviously entails the Corollary.

Remark. A more general result will be proved in Proposition (3.14).

Let us further mention that, although we are not going to prove it in this book, there are no additive functionals of BM other than the continuous ones; this is closely related to Theorem (3.4) in Chap. V.

We will now turn to the investigation of the time-change associated with CAF A, namely

$$\tau_t = \inf\{s : A_s > t\}, \qquad t \geq 0.$$

Because A is an additive functional and not merely an increasing process, we have the

(2.15) Proposition. *For every t, there is a negligible set outside which*

$$\tau_{t+s} = \tau_t + \tau_s \circ \theta_{\tau_t} \qquad \text{for every } s.$$

Proof. Same as for the local time in Sect. 1. □

In the sequel, A is fixed and we denote by $S(A)$ the support of the measure ν_A. We write T^A for $T_{S(A)}$. We will prove that A_t increases only when the Brownian motion is in $S(A)$. As a result, A is strictly increasing if and only if ν_A charges every open set.

(2.16) Lemma. *The times τ_0 and T^A are a.s. equal.*

For notational convenience, we shall write T for T^A, throughout the proof.

Proof. Pick a in $S(A)$; it was seen in the proof of Proposition (2.5) in Chap. VI that for $t > 0$, L^a_t is > 0 P_a-a.s. By the continuity of L^x_t in x, it follows that $L^{\cdot}_t$ is > 0 on a neighborhood of a P_a-a.s., which entails

$$A_t = \int_{-\infty}^{+\infty} L^x_t \nu_A(dx) > 0 \qquad P_a\text{-a.s.}$$

Let now x be an arbitrary point in $\mathbb{R}$; using the strong additivity of A and the strong Markov property, we get, for $t > 0$,

$$P_x\left[A_{T+t} > 0\right] \geq E_x\left[P_{B_T}\left[A_t > 0\right]\right] = 1$$

since, $S(A)$ being a closed set, $B_T \in S(A)$ a.s. This shows that $\tau_0 \leq T$ a.s.

On the other hand, by Corollary (2.13),

$$\int_0^t 1_{S(A)^c}(B_s)dA_s = \int_{S(A)^c} L^a_t \nu_A(da) = 0 \qquad \text{a.s.,}$$

hence

$$A_t = \int_0^t 1_{S(A)}(B_s)dA_s \qquad \text{a.s.,}$$

which entails that $A_T = 0$ and $\tau_0 \geq T$ a.s. □

(2.17) Proposition. *The following three sets are a.s. equal:*

i) the support of the measure dA_t*;*
ii) the complement $\sum_A$ *of* $\bigcup_s \left]\tau_{s-}, \tau_s\right[$*;*
iii) the set $\Gamma = \{t : B_t \in S(A)\}$.

Proof. The equality of the sets featured in i) and ii) is proved as in the case of the local time, hence $\sum_A$ is the support of dA_t.

Again, because $\int_0^t 1_{S(A)^c}(B_s)dA_s = 0$, we have $\Gamma^c \subset \bigcup_s \left]\tau_{s-}, \tau_s\right[$ since these intervals are the intervals of constancy of A; consequently, $\sum_A \subset \Gamma$. To prove the reverse inclusion, we observe that

$$\sum_A = \{t : A(t+\varepsilon) - A(t-\varepsilon) > 0 \text{ for all } \varepsilon > 0\}.$$

Consequently,

$$\left\{\omega : \Gamma(\omega) \not\subset \sum_A(\omega)\right\} \subset \bigcup \left\{A_s - A_r = 0, r + T^A \circ \theta_r < s\right\}$$

where the union is over all pairs (r, s) of rational numbers such that $0 \leq r < s$. But for each x,

$$P_x\left[A_s - A_r = 0, r + T^A \circ \theta_r < s\right] = E_x\left[P_{B_r}\left[A_{s-r} = 0, T^A < s - r\right]\right]$$

and this vanishes thanks to the preceding lemma. which completes the proof. □

We now study what becomes of the BM under the time-change associated with an additive functional A, thus substantiating in part the comment made in Sect. 3 Chap. VII about linear Markov processes on natural scale being time-changed BM's. We set $\widehat{B}_t = B_{\tau_t}$; the process $\widehat{B}$ is adapted to the filtration $(\widehat{\mathscr{F}_t}) = (\mathscr{F}_{\tau_t})$. Since $t \to \tau_t$ is right-continuous, the filtration $(\widehat{\mathscr{F}_t})$ is right-continuous; it is also complete for the probability measures (P_x) of BM. If T is a $(\widehat{\mathscr{F}_t})$-stopping time, then τ_T i.e. the map $\omega \to \tau_{T(\omega)}(\omega)$ is a $(\mathscr{F}_t)$-stopping time; indeed for any s

$$\{\tau_T < s\} = \bigcup_{q \in \mathbb{Q}_+} \{T \leq q\} \cap \{\tau_q < s\}$$

and since $\{T \leq q\} \in \mathscr{F}_{\tau_q}$, each of these sets is in $\mathscr{F}_s$. Moreover,

$$\tau_{T+t} = \tau_T + \tau_t \circ \theta_{\tau_T} \qquad \text{for every } t \quad \text{a.s.}$$

as can be deduced from Proposition (2.15) using an approximation of T by decreasing, countably valued stopping times. All these facts will be put to use to prove the

(2.18) Theorem. *If the support of* ν_A *is an interval* I*, the process* $(\widehat{B}_t, \widehat{\mathscr{F}_t}, P_x, x \in I)$ *is a continuous regular strong Markov process on* I *with natural scale and speed measure* $2\nu_A$*. Its semi-group* Q_t *is given by* $Q_t f(x) = E_x[f(B_{\tau_t})]$.

Proof. We first observe that B_{τ_t} has continuous paths. Indeed, by the foregoing proposition, if $\tau_{t-} \neq \tau_t$, then $B_{\tau_{t-}}$ must be a finite end-point of I and B must be leaving I at time τ_{t-}. But then τ_t is the following time at which B reenters I and it can obviously reenter it only at the same end-point. As a result $B_{\tau_t} = B_{\tau_{t-}}$ which proves our claim.

We now prove that $\widehat{B}$ has the strong Markov property. Let f be a positive Borel function; by the remarks preceding the statement, if T is a $(\widehat{\mathscr{F}_t})$-stopping time

$$E_x\left[f(\widehat{B}_{t+T}) \mid \widehat{\mathscr{F}_t}\right] = E_x\left[f(B_{\tau_t} \circ \theta_{\tau_T}) \mid \mathscr{F}_{\tau_T}\right] = Q_t f(\widehat{B}_T)$$

P_x-a.s. on $\{T < \infty\}$. The result follows from Exercise (3.16) Chap. III.

That $\widehat{B}$ is regular and on its natural scale is clear. It remains to compute its speed measure. Let $J =]a, b[$ be an open sub-interval of I. With the notation of Sect. 3 Chap. VII we recall that for $x \in J$, and a positive Borel function f,

$$E_x\left[\int_0^{T_a \wedge T_b} f(B_s) ds\right] = \int_J E_x\left[L^y_{T_a \wedge T_b}\right] f(y) dy = 2\int_J G_J(x, y) f(y) dy;$$

it follows that $E_x\left[L^{\cdot}_{T_a \wedge T_b}\right] = 2G_J(x, \cdot)$ a.e. and, by continuity, everywhere. We now compute the potential of $\widehat{B}$. Define $\widehat{T}_a = \inf\{t : \widehat{B}_t = a\} = \inf\{t : \tau_t = T_a\}$ and $\widehat{T}_b$ similarly. By the time-change formula in Sect. 1 Chap. V

$$\begin{aligned} E_x&\left[\int_0^{\widehat{T}_a \wedge \widehat{T}_b} f(\widehat{B}_s) ds\right] \\ &= E_x\left[\int_0^{\tau_{\widehat{T}_a \wedge \widehat{T}_b}} f(B_s) dA_s\right] = E_x\left[\int_0^{T_a \wedge T_b} f(B_s) dA_s\right] \\ &= \int_J E_x\left[L^y_{T_a \wedge T_b}\right] f(y) \nu_A(dy) = 2\int_J G_J(x, y) f(y) \nu_A(dy) \end{aligned}$$

which completes the proof. Actually, we ought to consider the particular case of end-points but the same reasoning applies. □

Using Corollary (2.11), it is easy to check that $Q_t f$ is a Borel function if f is a Borel function. One can also check, using the change of variable formula in Stieltjes integrals that the resolvent of $\widehat{B}$ is given by

$$V^\alpha f(x) = E_x\left[\int_0^\infty e^{-\alpha A_t} f(B_t) dA_t\right].$$

An unsatisfactory aspect of the above result is that if I has a finite end-point, this point must be a reflecting or slowly reflecting boundary and in particular belongs to E in the notation of Sect. 3 Chap. VII. We haven't shown how to obtain a process on $]0, \infty[$ as a time-changed BM. We recall from Exercise (3.23) Chap. VII that if the process is on its natural scale, then 0 must be a natural boundary. We will briefly sketch how such a process may be obtained from BM.

Let ν be a Radon measure on $]0, \infty[$ such that $\nu(]0, \varepsilon[) = \infty$ for every $\varepsilon > 0$; then the integral

$$\int L_t^a \nu_A(da)$$

no longer defines an additive functional because it is infinite P_0-a.s. for any $t > 0$. But it may be viewed as a CAF of the BM killed at time T_0. The associated time-changed process will then be a regular process on $]0, \infty[$ on natural scale and with 0 as natural boundary. We leave as an exercise for the reader the task of writing down the proof of these claims; he may also look at the exercises for particular cases and related results.

It is also important to note that Theorem (2.18) has a converse. Namely, starting with X, one can find a BM B and a CAF of B such that X is the associated time-changed process (see Exercise (2.34)).

(2.19) Exercise. If A is a CAF of BM and (τ_t) the inverse of the local time at 0, prove that the process A_{τ_t} has stationary independent increments. See also Proposition (2.7) Chap. XII.

(2.20) Exercise. Let B be the BM^d, $d > 1$, and v a unit vector. The local time at y of the linear BM $X = \langle v, B\rangle$ is an additive functional of B. Compute its associated measure.

(2.21) Exercise. If m is a σ-finite measure and $mU^1 = m$, then m is invariant.

[Hint: If $m(A) < \infty$, prove that $\alpha m U^\alpha(A) = m(A)$ for each α and use the properties of Laplace transforms.]

(2.22) Exercise (Signed additive functionals). If, in the definition of a CAF, we replace the requirement that it be increasing by that it merely be of finite variation on each bounded interval, we get the notion of signed CAF.

1°) Prove that if A is a signed CAF there exist two CAF's A^+ and A^- such that $A = A^+ - A^-$, this decomposition being minimal.

2°) In the case of BM, extend to signed CAF's the results of this section. In particular for a signed CAF A, there exists a function f which is locally the difference of two convex functions, such that $f''(dx) = 2\nu_A(dx)$ and

$$A_t = f(B_t) - f(B_0) - \int_0^t f'_-(B_s)dB_s.$$

(2.23) Exercise (Semimartingale functions of BM). 1°) If $f(B_t)$ is for each P_x a continuous semimartingale, prove that f is locally the difference of two convex functions.

[Hint: Use the preceding exercise and Exercise (3.13) in Chap. II.]

As it is known that all additive functionals of BM are continuous, one can actually remove the hypothesis that $f(B_t)$ is continuous and prove that it is so.

** 2°) If $f(B_t)$ is a continuous semimartingale under P_ν for one starting measure ν, prove that the result in 1°) still holds.

[Hint: Prove that the hypothesis of 1°) holds.]

\# **(2.24) Exercise (Skew Brownian motion).** Let $0 < \alpha < 1$ and define

$$g_\alpha(x) = (1-\alpha)^{-2} \quad \text{if } x \geq 0, \qquad g_\alpha(x) = \alpha^{-2} \quad \text{if } x < 0.$$

If B is the linear BM, we call Y^α the process obtained from B by the time-change associated with the additive functional

$$\int_0^t g_\alpha(B_s)ds.$$

Finally we set $r_\alpha(x) = x/(1-\alpha)$ if $x \geq 0$, $r_\alpha(x) = x/\alpha$ if $x < 0$ and $X_t^\alpha = r_\alpha(Y_t^\alpha)$. The process X^α is called the *Skew Brownian Motion with parameter* α. For completeness, the reader may investigate the limit cases $\alpha = 0$ and $\alpha = 1$.

1°) Compute the scale function, speed measure and infinitesimal generator of X^α. As a result X^α has the transition density of Exercise (1.16), Chap. III. Prove that $P_0[X_t > 0] = \alpha$ for every t.

2°) A skew BM is a semimartingale. Prove that for $\alpha \neq 1/2$, its local time is discontinuous in the space variable at level 0 and compute its jump.

3°) Prove that $X_t^\alpha = \beta_t + (2\alpha - 1)\widetilde{L}_t$ where β is a BM and $\widetilde{L}$ the symmetric local time of X^α at zero given by

$$\widetilde{L}_t = \lim_{\varepsilon \downarrow 0} \frac{1}{2\varepsilon} \int_0^t 1_{[-\varepsilon,\varepsilon]}(X_s^\alpha)ds.$$

* 4°) Let γ be a constant and let X be a solution to the equation $X_t = \beta_t + \gamma\widetilde{L}_t$ where $\widetilde{L}$ is the symmetric local time of X as defined in 3°). Compute $L^0(X)_t$ and $L^{0-}(X)_t$ as functions of $\widetilde{L}_t$. Derive therefrom that a solution X to the above equation exists if and only if $|\gamma| \leq 1$.

[Hint: Use Exercise (2.24) Chapter VI. See also Exercise (3.19) in Chapter IX.]

* **(2.25) Exercise (Additive local martingales).** 1°) Let A be a continuous additive functional of linear BM. Prove that a.s. the measures dA_t are absolutely continuous with rspect to dt if and only if there is a positive Borel function f such that

$$A_t = \int_0^t f(B_s)ds.$$

2°) Prove that M is a continuous process vanishing at 0 and such that

i) it is an $(\mathscr{F}_t)$-local martingale for every P_ν,
ii) for every pair (s, t),

$$M_{t+s} - M_t = M_s \circ \theta_t \qquad \text{a.s.},$$

if and only if there is a Borel function f such that

$$M_t = \int_0^t f(B_s)dB_s \qquad \text{a.s.}$$

[Hint: Use the representation result of Sect. 3 Chap. V and the fact that $\langle M, B\rangle$ is an additive functional.]

\# **(2.26) Exercise (Continuation of Exercise (1.13)).** 1°) Suppose that X is the linear BM and prove that the pair (X, A) has the strong Markov property for every $(\mathscr{F}_t)$-stopping time.

2°) If $A_t = \int_0^t f(X_s)ds$, then (X, A) is a diffusion in $\mathbb{R}^2$ with generator $\frac{1}{2}\frac{\partial^2}{\partial x_1^2} + f(x_1)\frac{\partial}{\partial x_2}$. In particular, the process of Exercise (1.12) of Chap. III has the generator $\frac{1}{2}\frac{\partial^2}{\partial x_1^2} + x_1\frac{\partial}{\partial x_2}$.

* **(2.27) Exercise.** 1°) In the setting of Theorem (2.18), check that ν_A is excessive for $\widetilde{B}$.

2°) If more generally X is a linear Markov process on a closed interval and is not necessarily on natural scale, prove that the speed measure is excessive.

3°) Extend this result to the general case.

[Hint: Use Exercise (3.18) in Chap. VII.]

(2.28) Exercise. In the setting of Theorem (2.18) suppose that $\nu_A(dx) = V(x)dx$ with $V > 0$ on $\overset{\circ}{I} = \mathrm{Int}(I)$. Prove that the extended infinitesimal generator of $\widehat{B}$ is given on $C^2(\overset{\circ}{I})$ by $\widehat{A}f = \frac{1}{2}V^{-1}f''$. Check the answer against the result in Theorem (3.12) Chap. VII and compare with Proposition (1.13) Chap. IX.

[Hint: Use the characterization of the extended generator in terms of martingales.]

* **(2.29) Exercise.** 1°) In the setting of Theorem (2.18), if $A_t = \int_0^t 1_{(B_s>0)}ds + \lambda L_t^0$ with $0 < \lambda < \infty$, prove that 0 is a slowly reflecting boundary for $\widehat{B}$.

2°) Prove that if $\widehat{B}$ is a Feller process, the domain $\mathscr{D}_A$ of the infinitesimal generator of $\widehat{B}$ is

$$\left\{f \in C^2(]0, \infty[), \quad f''(0) = \lim_{x\to 0} f''(x) \quad \text{exists and} \quad f'_+(0) = \lambda f''(0)\right\},$$

and that $Af(x) = \frac{1}{2}f''(x)$ for $x > 0$, $Af(0) = \lambda^{-1}f'_+(0)$.

\# **(2.30) Exercise.** Take up the skew BM X^α of Exercise (2.24) with $0 < \alpha < 1$ and let L be its local time at 0.

1°) Prove that there is a BM β such that

$$X_t^\alpha = \beta_t + \big((2\alpha - 1)/2\alpha\big)L_t.$$

2°) Let a and b be two positive numbers such that $\alpha = b/(a+b)$ and put $\sigma(x) = a1_{(x>0)} - b1_{(x\le 0)}$. Set $Y_t = a\left(X_t^\alpha\right)^+ - b\left(X_t^\alpha\right)^-$ and $B_t = \left|X_t^\alpha\right| - \frac{1}{2}L_t\left(|X^\alpha|\right)$ and prove that (Y, B) is a solution to $e_0(\sigma, 0)$.

3°) By considering another skew BM, say Z^α, such that $|X^\alpha| = |Z^\alpha|$ (see Exercise (2.16) Chap. XII), prove that pathwise uniqueness does not hold for $e_0(\sigma, 0)$.

\# **(2.31) Exercise.** Let f be a positive Borel function on $\mathbb{R}$ and B be the linear BM.

1°) If f is not locally integrable prove that there exists a point a in $\mathbb{R}$ such that for every $t > 0$,

$$\int_0^t f(B_s)ds = \infty \qquad P_a\text{-a.s.}$$

2°) Prove that the following three conditions are equivalent

(i) $P_0\left[\int_0^t f(B_s)ds < \infty \ \forall t \in [0, \infty[\right] > 0$;

(ii) $P_x\left[\int_0^t f(B_s)ds < \infty \ \forall t \in [0, \infty[\right] = 1$ for every $x \in \mathbb{R}$;

(iii) f is locally integrable.

(2.32) Exercise. 1°) In the situation of Theorem (2.18) prove that there is a bi-continuous family of r.v.'s L_t^a such that for every positive Borel function f,

$$\int_0^t f(\widehat{B}_s)ds = \int_I f(a)L_t^a \nu_A(da) \qquad \text{a.s.}$$

2°) If in addition $I = \mathbb{R}$, then $\widehat{B}$ is a loc. mart.; prove that $\langle \widehat{B}, \widehat{B} \rangle_t = \tau_t$.

(2.33) Exercise. Construct an example of a continuous regular strong Markov process X on $\mathbb{R}$ which "spends all its time on $\mathbb{Q}$", i.e. the set $\{t : X_t \in \mathbb{R}\backslash\mathbb{Q}\}$ has a.s. Lebesgue measure 0.

(2.34) Exercise. 1°) With the notation of Proposition (3.5) Chap. VII prove that

$$m_I\left(X_{t\wedge\sigma_I}\right) + t \wedge \sigma_I$$

is a uniformly integrable P_x-martingale for every x in $\mathring{I}$.

2°) Assume that X is on natural scale and that $E = \mathbb{R}$. Call m its speed measure and prove that if B is the DDS Brownian motion of X and τ_t is the inverse of $\langle X, X \rangle$, then

$$\tau_t = \int L_t^a(B)m(da)$$

(see also Exercise (2.32)). In particular X is a pure loc. mart.

[Hint: The measure m is twice the opposite of the second derivative of the concave function m_I; use Tanaka's formula.]

§3. Ergodic Theorems for Additive Functionals

In Sect. 1 Chap. II and Sect. 2 Chap. V, we proved some recurrence properties of BM in dimensions 1 and 2. We are now taking this up to prove an ergodic result for occupation times or more generally additive functionals. Since at no extra cost we can cover other cases, we will consider in this section a Markov process X for which we use the notation and results of Chap. III. We assume in addition that the resolvent U^α has the strong Feller property, namely $U^\alpha f$ is continuous for every α and every bounded Borel function f and also that $P_t 1 = 1$ for every $t \geq 0$ which is equivalent to $P_.[\zeta = \infty] \equiv 1$.

Our first definition makes sense for any Markov process and is fundamental in the description of probabilistic potential theory.

(3.1) Definition. *A positive universally measurable function f is* excessive *for the process X (or for its semi-group) if*

i) $P_t f \leq f$ *for every* $t > 0$;
ii) $\lim_{t \downarrow 0} P_t f = f$.

A finite universally measurable function h is said to be invariant *if $P_t h = h$ for every t.*

(3.2) Proposition. *If f is excessive, $f(X_t)$ is a $(\mathscr{F}_t)$-supermartingale for every P_ν. If h is invariant, $h(X_t)$ is a martingale.*

Proof. By the Markov property, and property i) above

$$E_\nu\left[f\left(X_{t+s}\right) \mid \mathscr{F}_s\right] = E_{X_s}\left[f(X_t)\right] = P_t f(X_s) \leq f(X_s) \qquad P_\nu\text{-a.s.}$$

In the case of invariant functions, the inequality is an equality. □

This proposition, which used only property i) in the definition above, does not say anything about the possibility of getting a good version for the super-martingale $f(X_t)$; the property ii) is precisely what is needed to ensure that $f(X_t)$ is a.s. right-continuous, but we are not going to prove it in this book. We merely observe that, if f is excessive, then $\alpha U^\alpha f \leq f$ for every α and $\lim_{\alpha\to\infty} \alpha U_\alpha f = f$ as the reader will easily show; moreover, the limit is increasing and it follows easily from the strong Feller property of U^α that an excessive function is lower-semi-continuous. If h is invariant and bounded, then $\alpha U^\alpha h = h$ hence h is continuous; the martingale $h(X_t)$ is then a.s. right-continuous, a fact which we will use below. Moreover if conversely h is bounded and $\alpha U^\alpha h = h$ for every α, the continuity of h hence the right-continuity of $P_t h$ in t, entails, by the uniqueness property of Laplace transform that h is invariant.

(3.3) Definition. *An event Γ of $\mathscr{F}_\infty$ is said to be* invariant *if $\theta_t^{-1}(\Gamma) = \Gamma$ for every t. The σ-field $\mathscr{J}$ of invariant events is called the* invariant σ-field *and an $\mathscr{J}$-measurable r.v. is also called* invariant. *Two invariant r.v.'s Z and Z' are said to be* equivalent *if $P_x[Z = Z'] = 1$ for every x.*

Invariant r.v.'s and invariant functions on the state space are related by the following

(3.4) Proposition. *The formula $h(x) = E_x[Z]$ sets up a one-to-one and onto correspondence between the bounded invariant functions and the equivalence classes of bounded invariant r.v.'s. Moreover*

$$Z = \lim_{t\to\infty} h(X_t) \qquad a.s.$$

Proof. If Z is invariant, a simple application of the Markov property shows that $h(\cdot) = E_{\cdot}[Z]$ is invariant (notice that if we did not have $\zeta = \infty$ a.s., we would only get $P_t h \leq h$).

Conversely, since $h(X_t)$ is a right-continuous bounded martingale, it converges a.s. to a bounded r.v. Z which may be chosen invariant. Moreover, by Lebesgue's dominated convergence theorem, $h(x) = E_x[Z]$ for every x in E. The correspondence thus obtained is clearly one-to-one. □

Let A be a Borel set; the set

$$R(A) = \left\{\overline{\lim}_{t\to\infty} 1_A(X_t) = 1\right\} = \bigcap_t \{t + T_A \circ \theta_t < \infty\}$$

is the set of paths ω which hit A infinitely often as $t \to \infty$; it is in $\mathscr{F}^0_\infty$ since it is equal to $\bigcap_n \{n + T_A \circ \theta_n < \infty\}$. It is then clear that it is an invariant event. The corresponding invariant function $h_A = P_{\cdot}[R(A)]$ is the probability that A is hit at arbitrarily large times and $\lim_{t\to\infty} h_A(X_t) = 1_{R(A)}$ a.s. by the above result.

(3.5) Definition. *A set A is said to be* transient *if* $h_A \equiv 0$ *and* recurrent *if* $h_A \equiv 1$.

In general, a set may be neither recurrent nor transient but we have the

(3.6) Proposition. *The following three statements are equivalent:*

i) the bounded invariant functions are constant;
ii) the σ-algebra $\mathscr{I}$ is a.s. trivial;
iii) every set is either recurrent or transient.

Proof. The equivalence of i) and ii) follows immediately from Proposition (3.4), and it is clear that ii) implies iii).

We prove that iii) implies ii). Let $\Gamma \in \mathscr{I}$ and put $A = \{x : P_x[\Gamma] > a\}$ for $0 < a < 1$. We know that $1_\Gamma = \lim_{t\to\infty} P_{X_t}[\Gamma]$ a.s.; if A is recurrent, then $\Gamma = \Omega$ a.s. and if A is transient then $\Gamma = \emptyset$ a.s. □

Although we are not going to develop the corresponding theory, Markov processes have roughly two basic behaviors. Either they converge to infinity in which case they are called *transient*, or they come back at arbitrarily large times to relatively small sets, for instance open balls of arbitrarily small radius, in which case they are called *recurrent*. After proving a result pertaining to the transient case, we will essentially study the recurrent case. Let us first observe that because of the right-continuity of paths, if A is an open set, $R(A) = \left\{\overline{\lim}_{q\to\infty} 1_A(X_q) = 1\right\}$ where q runs through the rational numbers.

The following result applies in particular to BM^d, $d > 2$, in which case however it was already proved in Sect. 2 Chap. V.

(3.7) Proposition. *If for every relatively compact set A, the potential $U(\cdot, A)$ is finite, then the process converges to infinity.*

Proof. We have, by the Markov property

$$P_t\big(U(\cdot, A)\big)(x) = E_x\left[E_{X_t}\left[\int_0^\infty 1_A(X_s)ds\right]\right] = E_x\left[\int_t^\infty 1_A(X_s)ds\right];$$

it follows on the one hand that $U(\cdot, A)$ is excessive, hence lower-continuous, on the other hand that $\lim_{t\to\infty} P_t\big(U(\cdot, A)\big) = 0$. From the first property we deduce that $U(X_q, A)$ is a positive supermartingale indexed by $\mathbb{Q}_+$, and by the second property and Fatou's lemma, its limit as $q \to \infty$ is zero a.s.

Let now Γ and Γ' be two relatively compact open sets such that $\overline{\Gamma} \subset \Gamma'$. The function $U(\cdot, \Gamma')$ is strictly positive on Γ', because of the right-continuity of paths; by the lower-semi-continuity of $U(\cdot, \Gamma')$, there is a constant $a > 0$ such that $U(\cdot, \Gamma') \geq a$ on $\overline{\Gamma}$. Thus on the paths which hit Γ at infinitely large times, we have $\overline{\lim}_{q\to\infty} U(X_q, \Gamma') \geq a$. By the first paragraph, the set of these paths is a.s. empty. Therefore, Γ is a.s. not visited from some finite time on, and the proof is now easily completed. □

We now study the opposite situation.

(3.8) Definition. *The process X is said to be* Harris-recurrent *or merely* Harris *if there is an invariant measure m such that $m(A) > 0$ implies that A is recurrent.*

In the sequel, when we deal with Harris processes, we will always assume that the support of m is the whole space. Indeed, the support of an invariant measure is an absorbing set, a fact which is proved in the following way. Let Γ be the complement of the support; since Γ is open, the right-continuity of paths entails that the set of points from which the process can reach Γ is precisely $\Gamma' = \{x : U^\alpha(x, \Gamma) > 0\}$ for some $\alpha > 0$. Clearly $\Gamma' \supset \Gamma$ and since m is invariant, $\alpha m U^\alpha(\Gamma) = m(\Gamma) = 0$ which proves that $m(\Gamma') = 0$; as a result $\Gamma' = \Gamma$ and Γ^c is absorbing. Thus, one loses little by assuming that Γ is empty and in fact this is naturally satisfied in most cases. This condition implies that every open set is recurrent.

Conversely we have the following result which shows that BM^d, $d = 1, 2$, the OU process and many linear Markov processes such as the Bessel processes of low dimensions are Harris-recurrent.

(3.9) Proposition. *If X has an invariant measure and if every open set is recurrent, then X is Harris.*

Proof. If $m(A) > 0$, since $P_m[X_t \in A] = m(A)$ for every t, there is a constant $a > 0$ such that the set $\Gamma = \{x : P_x[T_A < \infty] > a\}$ is not empty.

Now the function $f \equiv P_\cdot[T_A < \infty]$ is excessive because $P_t f(x) = P_x\big[t + T_A \circ \theta_t < \infty\big] \leq f(x)$ and one checks that $\lim_{t\downarrow 0}(t + T_A \circ \theta_t) = T_A$ which implies that $\lim_{t\downarrow 0} P_t f(x) = f(x)$. As a result the set Γ is open; furthermore, by Corollary (2.4) in Chap. II,

$$\lim_{q\to\infty} P_{X_q}[T_A < \infty] = \lim_{q\to\infty} P_\cdot\big[q + T_A \circ \theta_q < \infty \mid \mathscr{F}_q\big] = 1_{R(A)} \qquad \text{a.s.}$$

and since Γ is recurrent, we find that $1_{R(A)} \geq a$ a.s. hence $R(A) = \Omega$ a.s. which completes the proof. □

For a Harris process, the equivalent conditions of Proposition (3.6) are in force.

(3.10) Proposition. *If X is a Harris process, the excessive functions and the bounded invariant functions are constant.*

Proof. If the excessive function f were not constant, we could find two constants $a < b$ such that the sets $J = \{f > b\}$ and $J' = \{f \leq a\}$ are not empty. The set J is open, hence recurrent, and by Fatou's lemma, for each $x \in E$,

$$f(x) \geq \varliminf_{q \to \infty} P_q f(x) \geq E_x \left[\varliminf_{q \to \infty} f(X_q) \right] \geq b$$

and we get a contradiction.

For a bounded harmonic function h, we apply the result just proved to $h + \|h\|$. □

By the occupation times formula together with Corollary (2.4) Chapter IV (or Corollary (2.14)), we know that in the case of BM^1, if m is the Lebesgue measure, $m(A) > 0$ implies

$$\int_0^\infty 1_A(X_s)ds = \infty \qquad \text{a.s.},$$

which is apparently stronger than the Harris condition. We will prove that actually this property is shared by every Harris process, in particular by BM^2. We consider a *strong* additive functional A which, as we already observed, is in fact no restriction.

(3.11) Proposition. *If ν_A does not vanish, then $A_\infty = \infty$ a.s.*

Of course, here ν_A is computed with respect to the invariant measure m which is the only invariant measure for X (see Exercise (3.14)).

Proof. For $\varepsilon > 0$, we set $\tau_\varepsilon = \inf\{t : A_t > \varepsilon\}$. If ν_A does not vanish, we may find $\varepsilon > 0$ and $a > 0$ such that

$$m\left(\{x : P_x[\tau_\varepsilon < \infty] > a\}\right) > 0.$$

Therefore, $\overline{\lim}_{t\to\infty} P_{X_t}[\tau_\varepsilon < \infty] \geq a$ a.s. But on the other hand, for $x \in E$,

$$P_{X_t}[\tau_\varepsilon < \infty] = P_x\left[t + \tau_\varepsilon \circ \theta_t < \infty \mid \mathscr{F}_t\right] \qquad P_x\text{-a.s.},$$

and by Corollary (2.4) in Chap. II, this converges P_x-a.s. to $1_{[\bigcap_t \{t+\tau_\varepsilon \circ \theta_t < \infty\}]}$. It follows that $\bigcap_t \{t + \tau_\varepsilon \circ \theta_t < \infty\} = \Omega$ a.s. and a fortiori $P_\cdot[\tau_\varepsilon < \infty] = 1$.

If we define now inductively the stopping times T^n by $T^1 = \tau_\varepsilon$ and $T^n = T^{n-1} + \tau_\varepsilon \circ \theta_{T^{n-1}}$, a simple application of the strong Markov property shows that $P_\cdot[T^n < \infty] = 1$. By the strong additivity of A, we have $A_{T^n} \geq n\varepsilon$ for every n, which completes the proof.

Remarks. 1°) The function $P_{\cdot}[\tau_\varepsilon < \infty]$ can be shown to be excessive, hence constant and the proof could be based on these facts.

2°) This result shows that for $m(A) > 0$ we have $U(\cdot, A) \equiv \infty$ which is to be compared with Proposition (3.7).

We now turn to the *limit-quotient theorem* which is the main result of this section.

(3.12) Theorem. *If X is Harris, if A and C are two integrable additive functionals and if $\|\nu_C\| > 0$, then*

$$\lim_{t\to\infty} (A_t/C_t) = \|\nu_A\|/\|\nu_C\| \qquad a.s.$$

By the preceding result, the condition $\|\nu_C\| > 0$ ensures that the quotient on the left is meaningful at least for t sufficiently large.

Proof. By afterwards taking quotients, it is clearly enough to prove the result when $C_t = \int_0^t f(X_s)ds$ where f is a bounded, integrable and strictly positive Borel function.

We will use the Chacon-Ornstein theorem (see Appendix) for the operator θ_a, $a > 0$. Since m is invariant for the process, the measure P_m on $(\Omega, \mathscr{F})$ is invariant by θ_a, so that $Z \to Z\circ\theta_a$ is a positive contraction of $L^1(P_m)$. Moreover, by the preceding result, we have

$$\sum_{n=0}^{\infty} C_a \circ \theta_{na} = C_\infty = \infty \qquad \text{a.s.}$$

which proves that the set D in the Hopf's decomposition of Ω with respect to θ_a is empty; in other words, $Z \to Z \circ \theta_a$ is conservative. We may therefore apply the Chacon-Ornstein theorem; by hypothesis, A_a and C_a are in $L^1(P_m)$ so that the limit

$$\lambda_a = \lim_{n\to\infty} (A_{na}/C_{na})$$

exists P_m-a.s. As $\lim_{t\to\infty} C_t = \infty$, it is easily seen that $\lim_{n\to\infty} C_{n+1}/C_n = 1$; therefore the inequalities

$$\begin{aligned}&\left(A_{[t/a]a}/C_{[t/a]a}\right)\left(C_{[t/a]a}/C_{[t/a]a+1}\right)\\ &\qquad \le A_t/C_t \le \left(A_{[t/a]a+1}/C_{[t/a]a+1}\right)\left(C_{[t/a]a+1}/C_{[t/a]a}\right)\end{aligned}$$

imply that

$$\lim_{t\to\infty} (A_t/C_t) = \lambda_a \qquad P_m\text{-a.s.}$$

As a result, there is a r.v. λ such that

$$\lim_{t\to\infty} (A_t/C_t) = \lambda \qquad P_m\text{-a.s.}$$

and $\lambda = \lambda \circ \theta_a$ P_m-a.s. for every $a > 0$. It follows from Propositions (3.6) and (3.10) that λ is P_m-a.s. equal to a constant. From the Chacon-Ornstein theorem, it follows that this constant must be $E_m[A_a]/E_m[C_a]$ for an arbitrary a, that is $\|\nu_A\|/\|\nu_C\|$.

Set

$$F = \left\{\omega : \lim_{t\to\infty} A_t(\omega)/C_t(\omega) = \|\nu_A\|/\|\nu_C\|\right\}.$$

We have just proved that $P_m(F^c) = 0$; moreover if $\omega \in F$ then $\theta_s(\omega) \in F$ for every s or, in other words, $1_F \leq 1_F \circ \theta_s$ for every s. But since $\lim_{t\to\infty} C_t = +\infty$ a.s. if $\theta_s(\omega) \in F$, then $\omega \in F$. Thus $1_F = 1_F \circ \theta_s$ a.s. for every s which implies that $P_.[F^c]$ is a bounded invariant function, hence a constant function, which has to be identically zero. The proof is complete.

Remarks. 1°) In the case of BM, the end of the proof could also be based on the triviality of the asymptotic σ-field (see Exercise (2.28) Chap. III). The details are left to the reader as an exercise.

2°) In the case of BM^1, one can give a proof of the above result using only the law of large numbers (see Exercise (3.16)).

3°) If the invariant measure m is bounded, as for instance is the case for the OU process (see Exercise (1.13) in Chap. III), then constant functions are integrable and taking $C_t = t$ in the above result we get

$$\lim_{t\to\infty} (A_t/t) = \|\nu_A\|/m(E) \qquad \text{a.s.}$$

Thus in that case the additive functionals increase like t. When $m(E) = \infty$, it would be interesting to describe the speed at which additive functionals increase to infinity. This is tackled for BM^2 in the following section and treated in the case of BM^1 in Sect. 2 Chap. XIII. (See also the Notes and Comments).

4°) Some caution must be exercised when applying the above result to occupation times because there are integrable functions f such that $\int_0^t f(X_s)ds$ is not an additive functional; one may have $\int_0^t f(X_s)ds = \infty$ for every $t > 0$, P_x-a.s. for x in a polar set (see Exercise (3.17)). The reader will find in Exercise (2.6) of Chap. XI how to construct examples of such functions in the case of BM^2 (see also Exercise (2.31) above in the case of linear BM). The limit-quotient theorem will then be true P_x-a.s. for x outside a polar set. If f is bounded, the result is true for $\int_0^t f(X_s)ds$ without qualification.

(3.13) Exercise. Under the assumptions of this section, if X is Harris then for every $A \in \mathscr{E}$ either $U(\cdot, A) \equiv \infty$ or $U(\cdot, A) \equiv 0$. Prove moreover that all cooptional times are equal to 0 a.s.

[Hint: See Exercise (4.14) in Chap. VII].

(3.14) Exercise. Suppose X is Harris with invariant measure m and that the hypotheses of this section are in force.

1°) Prove that m is equivalent to $U^\alpha(x, \cdot)$ for every $\alpha > 0$ and $x \in E$.

2°) Prove that m is the unique (up to multiplication by a constant) excessive measure for X.

[Hint: Prove that an invariant measure is equivalent to m, then use the limit-quotient theorem.]

(3.15) Exercise. Let X be a Harris process, A an integrable additive functional of X and C a σ-integrable but not integrable additive functional, prove that

$$\lim_{t\to\infty} (A_t/C_t) = 0 \qquad \text{a.s.}$$

[Hint: Use the ergodic theorem for A and $1_F \cdot C$ where $\nu_C(F) < \infty$.]

(3.16) Exercise. For the linear BM and for $a < b$ define

$$T^1 = T_b + T_a \circ \theta_{T_b}, \ldots, T^n = T^{n-1} + T^1 \cdot \theta_{T^{n-1}}, \ldots$$

1°) Let A be a continuous additive functional. Prove that the r.v.'s $Z_n = A_{T^n} - A_{T^{n-1}}$, $n > 1$, are independent and identically distributed under every P_x, $x \in \mathbb{R}$. If $\|\nu_A\| < \infty$, prove that the Z_n's are P_x-integrable.

[Hint: For this last fact, one can consider the case of local times and use the results in Sect. 4 Chap. VI.]

2°) Applying the law of large numbers to the variables Z_n, prove Theorem (3.12).

[Hint: Prove that $A_t / \inf\{n : T^n \geq t\}$ converges as t goes to infinity, then use quotients.]

3°) Extend the above pattern of proof to recurrent linear Markov processes.

(3.17) Exercise. Let X be Harris and f be positive and m-integrable. Prove that $\int_0^t f(X_s)ds < \infty$ P_x-a.s. for every $t > 0$ and for every x outside a polar set. That this result cannot be improved is shown in Exercise (2.6) Chap. XI.

* **(3.18) Exercise.** In the setting of Theorem (3.12), prove that

$$\lim_{t\to\infty} E_x[A_t]/E_x[C_t] = \|\nu_A\|/\|\nu_C\|$$

for m-almost every x.

[Hint: Prove that for each $a > 0$, P_a is a conservative contraction of $L^1(m)$ and apply the Chacon-Ornstein theorem.]

(3.19) Exercise. We retain the situation of Exercise (2.22) 2°) and we put $\nu_A = \nu_{A^+} - \nu_{A^-}$.

1°) If ν_A is bounded, $\nu_A(1) = 0$ and $\int |x|\ |\nu_A|(dx) < \infty$, prove that f is bounded and f'_- is in $L^1 \cap L^2$ of the Lebesgue measure.

[Hint: This question is solved in Sect. 2 Chap. XIII.]

2°) Under the hypothesis of 1°), prove that there is a constant C such that

$$|E_x[A_T]| \leq C$$

for every point x and stopping time T such that $E_x[T] < \infty$.

3°) If A^i, $i = 1, 2$, are positive integrable additive functionals of BM^1 such that $\|\nu_{A^i}\| > 0$ and $\int |x| \nu_{A^i}(dx) < \infty$, then for any probability measure μ on $\mathbb{R}$,

$$\lim_{t\to\infty} E_\mu\left[A_t^1\right]/E_\mu\left[A_t^2\right] = \|\nu_{A^1}\| / \|\nu_{A^2}\| .$$

The results in 2°) and 3°) are strengthenings of the result in the preceding exercise.

*# **(3.20) Exercise.** 1°) Let c be a positive real number. On the Wiener space $\mathbf{W}^d$ the transformation $\omega \to \omega(c\cdot)/\sqrt{c}$ is measurable and leaves the Wiener measure W invariant. By applying Birkhoff's theorem to this transformation, prove that for $d \geq 3$,

$$\lim_{t\to\infty} \frac{1}{\log t} \int_1^t |B_s|^{-2} ds = \frac{1}{d-2} \qquad W\text{-a.s.}$$

[Hint: To prove that the limit provided by Birkhoff's theorem is constant use the $0-1$ law for processes with independent increments. The value of the constant may be computed by elementary means or derived from Exercise (4.23) in Chap. IV.]

2°) Prove the companion central-limit theorem to the above a.s. result, namely, that, in distribution,

$$\lim_{t\to\infty} \sqrt{\log t}\left((\log t)^{-1} \int_1^t |B_s|^{-2} ds - (d-2)^{-1}\right) = N,$$

where N is a Gaussian r.v.

[Hint: Use the methods of Exercise (4.23) in Chap. IV.]

§4. Asymptotic Results for the Planar Brownian Motion

This section is devoted to some asymptotic results for functionals of BM^2. In particular it gives a partial answer to the question raised in Remark 3°) at the end of the previous section. We use the skew-product representation of BM^2 described in Theorem (2.11) of Chap. V and the notation thereof and work with the probability measure P_z for $z \neq 0$.

(4.1) Theorem (Spitzer). *As t converges to infinity, $2\theta_t/\log t$ converges in distribution to a Cauchy variable with parameter* 1.

Proof. Because of the geometric and scaling invariance properties of BM, we may assume that $z = 1$. For $r > 1$, define $\sigma_r = \inf\{u : |Z_u| = r\}$ and for $a > 0$, $T_a = \inf\{t > 0 : \beta_t = a\}$. From the representation theorem recalled above, it follows that $C_{\sigma_r} = T_{\log r}$. As a result

$$\theta_{\sigma_r} = \gamma_{C_{\sigma_r}} = \gamma_{T_{\log r}} \stackrel{(d)}{=} (\log r)\gamma_{T_1},$$

the last equality being a consequence of the independence of ρ and γ and of the scaling properties of T_a and γ (Proposition (3.10), Chap. III). Therefore for every $r > 1$,

$$\frac{1}{\log r}\theta_{\sigma_r} \stackrel{(d)}{=} C,$$

where C is a Cauchy variable with parameter 1; this can alternatively be written as

$$\frac{2}{\log t}\theta_{\sigma_{\sqrt{t}}} \overset{(d)}{=} C.$$

We will be finished if we prove that $\left(\theta_t - \theta_{\sigma_{\sqrt{t}}}\right)/\log t$ converges to zero in probability.

We have

$$\theta_t - \theta_{\sigma_{\sqrt{t}}} = \operatorname{Im}\int_{\sigma_{\sqrt{t}}}^{t} \frac{dZ_s}{Z_s} = \operatorname{Im}\int_{\sigma_{\sqrt{t}}}^{t} \frac{d\widehat{Z}_s}{1+\widehat{Z}_s},$$

where $\widehat{Z}_s = Z_s - 1$ is a $\mathrm{BM}^2(0)$. Setting $\widetilde{Z}_s = \widehat{Z}_{ts}/\sqrt{t}$, we get

$$\theta_t - \theta_{\sigma_{\sqrt{t}}} = \operatorname{Im}\int_{t^{-1}\sigma_{\sqrt{t}}}^{1} \frac{d\widetilde{Z}_s}{(1/\sqrt{t}) + \widetilde{Z}_s}.$$

Let Z' be a $\mathrm{BM}^2(0)$ fixed once and for all. Since

$$\sigma_{\sqrt{t}} = \inf\left\{u : \left|1+\widehat{Z}_u\right| = \sqrt{t}\right\} \overset{(d)}{=} t\inf\left\{u : \left|(1/\sqrt{t}) + Z'_u\right| = 1\right\}$$

we have

$$\theta_t - \theta_{\sigma_{\sqrt{t}}} \overset{(d)}{=} \operatorname{Im}\int_{\bar{\nu}_t}^{1} \frac{dZ'_s}{(1/\sqrt{t}) + Z'_s}$$

where $\bar{\nu}_t = \inf\left\{u : \left|(1/\sqrt{t}) + Z'_u\right| = 1\right\}$.

We leave as an exercise to the reader the task of showing that $\bar{\nu}_t$ converges in probability to $\sigma'_1 = \inf\left\{u : \left|Z'_u\right| = 1\right\}$ as t goes to infinity. It then follows from Exercise (1.25) in Chap. VIII that the last displayed integral converges in probability to the a.s. finite r.v. $\int_{\sigma'_1}^{1} dZ'_s/Z'_s$, which completes the proof. □

We will now go further by studying the windings of Z according as Z is close to or far away from 0. More precisely, pick a real number $r > 0$ and set

$$\theta_t^0 = \int_0^t 1_{(|Z_s|\le r)}d\theta_s, \qquad \theta_t^\infty = \int_0^t 1_{(|Z_s|>r)}d\theta_s.$$

We want to study the asymptotic behavior of the pair $(\theta_t^0, \theta_t^\infty)$ and we will actually treat a more general problem. Let ϕ be a bounded and positive function on $\mathbb{R}$; we assume that $m_1(\phi) = \int_{\mathbb{R}} \phi(x)dx < \infty$. We set

$$A_t = \int_0^t |Z_s|^{-2}\phi(\log|Z_s|)ds.$$

This may fail to be an additive functional because A_t may be infinite P_0-a.s. for every $t > 0$, but the equality $A_{t+s} = A_t + A_s \circ \theta_t$ holds P_z-a.s. for $z \neq 0$ (see Remark 4 below Theorem (3.12)). It is moreover integrable in the sense that $E_m[A_1] = 2\pi m_1(\phi)$ as is easily checked using the skew-product representation.

(4.2) Theorem. *Under P_z, $z \neq 0$, the 3-dimensional family of r.v.'s*

$$\frac{2}{\log t}\left(\theta_t^0, \theta_t^\infty, A_t\right)$$

converges in distribution as t converges to ∞, to

$$\left(\int_0^{T_1} 1_{(\beta_s \le 0)} d\gamma_s, \int_0^{T_1} 1_{(\beta_s \ge 0)} d\gamma_s, m_1(\phi) L_{T_1}^0\right)$$

where (β, γ) is a standard planar BM *starting at* 0, *L^0 is the local time of β at* 0 *and $T_1 = \inf\{t : \beta_t = 1\}$.*

In the following proof as well as in similar questions treated in Chap. XIII we will make an extensive use of the scaling transformations. If B is a BM and a is > 0, we will denote by $B^{(a)}$ the BM $a^{-1}B_{a^2}$, and anything related to $B^{(a)}$ will sport the superscript (a); when $a = 1$, we have $B^{(1)} = B$ and we drop the (1). For instance

$$T_1^{(a)} = \inf\left\{t : B_t^{(a)} = 1\right\} = a^{-2}T_a.$$

Proof. Again we may assume that $z = 1$ and as in the above proof we look at the given process at time σ_t. By the time-changes and properties already used in the previous proof, we get

$$\theta_{\sigma_t}^0 = \int_0^{\sigma_t} 1_{(\log|Z_s| \le \log r)} d\gamma_{C_s} = \int_0^{C_{\sigma_t}} 1_{(\beta_s \le \log r)} d\gamma_s = \int_0^{T_{\log t}} 1_{(\beta_s \le \log r)} d\gamma_s;$$

setting $a = \log t$, we get

$$(\log t)^{-1}\theta_{\sigma_t}^0 = \int_0^{a^{-2}T_a} 1_{(\beta_{a^2 s} \le \log r)} d\gamma_s^{(a)} = \int_0^{T_1^{(a)}} 1_{(\beta_s^{(a)} \le a^{-1}\log r)} d\gamma_s^{(a)}.$$

The same computation will yield

$$(\log t)^{-1}\theta_{\sigma_t}^\infty = \int_0^{T_1^{(a)}} 1_{(\beta_s^{(a)} \ge a^{-1}\log r)} d\gamma_s^{(a)}.$$

We turn to the third term for which we have

$$\begin{aligned} A_{\sigma_t} &= \int_0^{\sigma_t} \phi(\beta_{C_s}) dC_s = \int_0^{C_{\sigma_t}} \phi(\beta_s) ds = \int_0^{T_{\log t}} \phi(\beta_s) ds \\ &= a^2 \int_0^{T_1^{(a)}} \phi\left(a\beta_s^{(a)}\right) ds. \end{aligned}$$

Consequently, $(\log t)^{-1}\left(\theta_{\sigma_t}^0, \theta_{\sigma_t}^\infty, A_{\sigma_t}\right)$ has the same law as

$$\left(\int_0^{T_1} 1_{(\beta_s \le a^{-1}\log r)} d\gamma_s, \int_0^{T_1} 1_{(\beta_s \ge a^{-1}\log r)} d\gamma_s, a\int_0^{T_1} \phi(a\beta_s) ds\right)$$

where (β, γ) is a planar standard BM and $T_1 = \inf\{t : \beta_t = 1\}$. The first two terms converge in probability thanks to Theorem (2.12) in Chap. IV; as to the third, introducing $L^x_{T_1}$, the local time of β at x up to time T_1, and using the occupation times formula, it is equal to

$$a \int \phi(ax) L^x_{T_1} dx = \int \phi(y) L^{y/a}_{T_1} dy$$

which converges a.s. to $m_1(\phi) L^0_{T_1}$ by dominated convergence.

Thus we have proved that $\frac{1}{\log t}\left(\theta^0_{\sigma_t}, \theta^\infty_{\sigma_t}, A_{\sigma_t}\right)$ converges in distribution to

$$\left(\int_0^{T_1} 1_{(\beta_s \le 0)} d\gamma_s, \int_0^{T_1} 1_{(\beta_s \ge 0)} d\gamma_s, m_1(\phi) L^0_{T_1}\right).$$

Furthermore, as in the preceding proof, we have

$$P\text{-}\lim_{t\to\infty} \frac{2}{\log t}\left(\theta^0_t - \theta^0_{\sigma_{\sqrt{t}}}\right) = P\text{-}\lim_{t\to\infty} \frac{2}{\log t}\left(\theta^\infty_t - \theta^\infty_{\sigma_{\sqrt{t}}}\right) = 0.$$

Also

$$\begin{aligned} \frac{2}{\log t}\left(A_t - A_{\sigma_{\sqrt{t}}}\right) &= \frac{2}{\log t}\int_{\sigma_{\sqrt{t}}}^t |Z_s|^{-2} \phi(\log|Z_s|) ds \\ &\le \frac{2\|\phi\|_\infty}{\log t}\int_{\sigma_{\sqrt{t}}}^t |Z_s|^{-2} ds = \frac{2\|\phi\|_\infty}{\log t}\int_{t^{-1}\sigma_{\sqrt{t}}}^1 |\widetilde{Z}_s|^{-2} ds \end{aligned}$$

where $\widetilde{Z}_s = t^{-1/2} Z_{ts}$ and this converges to zero in probability; indeed, as in the end of the proof of Theorem (4.1), the last integral converges in law to $\int_{\sigma'_1}^1 |Z'_s|^{-2} ds$. □

Remark. It is noteworthy that the limiting expression does not depend on r. If, in particular, we make $r = 1$ and if we put together the expressions for $\theta^0_{\sigma_t}$ and $\theta^\infty_{\sigma_t}$ given at the beginning of the proof and the fact that $\frac{1}{\log t}(\theta^0_t - \theta^0_{\sigma_{\sqrt{t}}}, \theta^\infty_t - \theta^\infty_{\sigma_{\sqrt{t}}})$ converges to zero in probability, we have proved that

$$\frac{2}{\log t}\left\{(\theta^0_t, \theta^\infty_t) - \left(\int_0^{T_{\log\sqrt{t}}} 1_{(\beta_s \le 0)} d\gamma_s, \int_0^{T_{\log\sqrt{t}}} 1_{(\beta_s \ge 0)} d\gamma_s\right)\right\}$$

converges to zero in probability, a fact which will be used in Sect. 3 Chap. XIII.

We now further analyse the foregoing result by computing the law of the limit which we will denote by (W^-, W^+, Λ). This triplet takes its values in $\mathbb{R}^2 \times \mathbb{R}_+$.

(4.3) Proposition. *If* $m_1(\phi) = 1$, *for* $a > 0$, $(b, c) \in \mathbb{R}^2$,

$$E\left[\exp\left(-a\Lambda + ibW^- + icW^+\right)\right] = f(2a + |b|, c)$$

where $f(u, c) = \left(\cosh c + \left(\frac{u}{c}\right)\sinh c\right)^{-1}$ *for* $c \ne 0$, $f(u, 0) = (1 + u)^{-1}$.

Proof. By conditioning with respect to $\mathscr{F}_\infty^\beta$, we get

$$E\left[\exp\left(-a\Lambda + ibW^- + icW^+\right)\right] = E\left[\exp\left(-H_{T_1}\right)\right]$$

where

$$H_t = aL_t^0 + \frac{b^2}{2}\int_0^t 1_{(\beta_s \le 0)} ds + \frac{c^2}{2}\int_0^t 1_{(\beta_s \ge 0)} ds.$$

The integrand $\exp\left(-H_{T_1}\right)$ now involves only β and the idea is to find a function F such that $F(\beta_t)\exp(-H_t)$ is a suitable local martingale.

With the help of Tanaka's formula, the problem is reduced to solving the equation

$$F'' = \left(2a\delta_0 + b^2 1_{(x\le 0)} + c^2 1_{(x\ge 0)}\right) F$$

in the sense of distributions. We take

$$F(x) = \exp(|b|x)1_{(x<0)} + \left(\cosh(cx) + \frac{2a+|b|}{c}\sinh(cx)\right)1_{(x\ge 0)}.$$

Stopped at T_1, the local martingale $F(\beta_t)\exp(-H_t)$ is bounded. Thus, we can apply the optional stopping theorem which yields the result.

(4.4) Corollary. *i) The r.v.* $\Lambda = L_{T_1}^0$ *is exponentially distributed with parameter* $1/2$.

ii) Conditionally on Λ*, the r.v.'s* W^- *and* W^+ *are independent,* W^- *is a Cauchy variable with parameter* $\Lambda/2$ *and the characteristic function of* W^+ *is equal to* $(c/\sinh c)\exp\left(-\frac{\Lambda}{2}(c\coth c - 1)\right)$.

iii) The density of W^+ *is equal to* $(2\cosh(\pi x/2))^{-1}$.

Proof. The proof of i) is straightforward. It is also proved independently in Proposition (4.6) of Chap. VI.

To prove ii) set $f_{b,c}(\lambda) = \left(\frac{c}{\sinh c}\right)\exp\left(-\frac{\lambda}{2}(c\coth c - 1 + |b|)\right)$ and compute $E\left[\exp(-a\Lambda)f_{b,c}(\Lambda)\right]$. Using the law of Λ found in i), this is easily seen to be equal to

$$\frac{c}{\sinh c}(1 + 2a + c\coth c - 1 + |b|)^{-1} = f(2a + |b|, c)$$

where f is the same as in Proposition (4.3). As a is an arbitrary positive number, it follows that

$$f_{b,c}(\Lambda) = E\left[\exp(ibW^- + icW^+) \mid \Lambda\right]$$

which proves ii).

Finally, the proof of iii) is a classical Fourier transform computation (see Sect. 6 Chap. 0).

Remark. The independence in ii) has the following intuitive meaning. The r.v. Λ accounts for the time spent on the boundary of the disk or for the number of times the process crosses this boundary. Once this is known, what occurs inside the disk is independent of what occurs outside. Moreover, the larger the number of these crossings, the larger in absolute value the winding number tends to be.

We finally observe that since we work with $z \neq 0$, by Remark 4 at the end of the preceding section, if G is an integrable additive functional and if $m_1(\phi) > 0$ we have $\lim(G_t/A_t) = \|\nu_G\|/2\pi m_1(\phi)$ P_z-a.s. As a result we may use G instead of A in the above results and get

(4.5) Corollary. *If G is any integrable additive functional,*

$$2\,(\log t)^{-1}\left(\theta_t^0, \theta_t^\infty, G_t\right)$$

converges in law under P_z to $\left(W^-, W^+, (2\pi)^{-1}\|\nu_G\|\Lambda\right)$ as t goes to infinity.

In Theorem (4.2) we were interested in the imaginary parts of

$$\frac{2}{\log t}\left(\int_0^t 1_{(|Z_s| \gtrless r)} \frac{dZ_s}{Z_s}\right).$$

For later needs, it is also worth recording the asymptotic behavior of the real parts. We set

$$N_t^0 = \operatorname{Re} \int_0^t 1_{(|Z_s| \le r)} \frac{dZ_s}{Z_s}, \qquad N_t^\infty = \operatorname{Re} \int_0^t 1_{(|Z_s| \ge r)} \frac{dZ_s}{Z_s}.$$

With the same notation as above, we have the

(4.6) Proposition. *As t converges to infinity, $2(\log t)^{-1}\left(N_t^0, N_t^\infty, A_t\right)$ converges in distribution to $\left(\frac{1}{2}\Lambda, \frac{1}{2}\Lambda - 1, m_1(\phi)\Lambda\right)$.*

Proof. The same pattern of proof as in Theorem (4.2) leads to the convergence of the law of $2(\log t)^{-1}\left(N_t^0, N_t^\infty, A_t\right)$ towards that of

$$\left(\int_0^{T_1} 1_{(\beta_s \le 0)} d\beta_s, \int_0^{T_1} 1_{(\beta_s \ge 0)} d\beta_s, m_1(\phi) L_{T_1}^0\right).$$

Thus the result follows immediately from Tanaka's formula.

(4.7) Exercise. Deduce Theorem (4.1) from Theorem (4.2).

(4.8) Exercise. With the notation of Theorem (4.1) prove that $X_u = \theta\left(\sigma_{\exp(u)}\right)$ and $Y_u = \theta\left(\sigma_{\exp(-u)}\right)$ are two Cauchy processes.

* **(4.9) Exercise.** Prove Theorem (4.1) as a Corollary to Proposition (3.8) in Chap. VIII.

[Hint: Use the explicit expressions for the density of ρ_t and make the change of variable $\rho = u\sqrt{t}$.]

* **(4.10) Exercise.** Let B be a $BM^2(0)$ and call $\widetilde{\theta}_t$, $t > 0$ a continuous determination of $\arg(B_t)$, $t > 0$. Prove that as t converges to 0, the r.v.'s $2\widetilde{\theta}_t/\log t$ converge in distribution to the Cauchy r.v. with parameter 1.

[Hint: By scaling or time-inversion, $\left(\widetilde{\theta}_1 - \widetilde{\theta}_t\right) \overset{(d)}{=} \left(\widetilde{\theta}_{1/t} - \widetilde{\theta}_1\right)$.]

* **(4.11) Exercise (Another proof of Theorem (4.1)).** 1°) With the notation of Theorems (4.1) and (4.2) prove that

$$a^{-2}C_t = \inf\left\{u : a^2 \int_0^u \exp\left(2a\beta_s^{(a)}\right) ds > t\right\}.$$

2°) Using the Laplace method prove that for a fixed BM, say B,

$$\lim_{a\to\infty} (2a)^{-1} \log\left(\int_0^u \exp(2aB_s)ds\right) - \sup_{0\le s\le u} B_s = 0$$

holds a.s. for every u (see Exercise (1.18) Chap. I).

3°) Prove that for $a = \log t/2$,

$$P\text{-}\lim_{t\to\infty} \left\{a^{-2}C_t - T_1^{(a)}\right\} = 0.$$

[Hint: The processes $\beta^{(a)}$ have all the law of B which shows that the convergence holds in law.]

4°) Give another proof of Theorem (4.1) based on the result in 3°).

* **(4.12) Exercise.** Let Z be a $\mathrm{BM}^2(1)$ and θ be the continuous determination of $\arg Z$ such that $\theta_0 = 0$. Set $T_n = \inf\{t : |Z_t| \ge n\}$.

1°) If $\tau = \inf\{t : \theta_t > 1\}$ prove that

$$\lim_{n\to\infty} (\log n) P[\tau > T_n]$$

exists.

[Hint: $P[\tau > T_n] = P[C_\tau > C_{T_n}]$.]

2°) If $\widetilde{\tau} = \inf\{t : |\theta_t| > 1\}$, prove that $P[\widetilde{\tau} > T_n] = 0(1/n)$.

Notes and Comments

Sect. 1. The basic reference for additive functionals is the book of Blumenthal and Getoor [1] from which most of our proofs are borrowed. There the reader will find, for instance, the proof of the strong additivity property of additive functionals and an account of the history of the subject. Our own exposition is kept to the minimum which is necessary for the asymptotic results of this chapter and of Chap. XIII. It gives no inkling of the present-day state of the art for which we recommend the book of Sharpe [3].

The extremal process of Exercise (1.9) is studied in Dwass [1] and Resnick [1]. It appears as the limit process in some asymptotic results as for instance in Watanabe [2] where one finds also the matter of Exercise (1.10).

Exercise (1.11) is actually valid in a much more general context as described in Chap. V of Blumenthal and Getoor [1]. If X is a general strong Markov process, and if a point x is regular for itself (i.e. x is regular for $\{x\}$ as defined in Exercise (2.24) of Chap. III), it can be proved that there exists an additive functional A

such that the measure dA_t is a.s. carried by the set $\{t : X_t = x\}$. This additive functional, which is unique up to multiplication by a constant, is called the local time of X at x. Thus, for a Markov process which is also a semimartingale, as is the case for BM, we have two possible definitions of local times. A profound study of the relationships between Markov processes and semi-martingales was undertaken by Çinlar et al. [1].

Exercise (1.14) is from Barlow-Yor ([1] and [2]), the method hinted at being from Bass [2] and B. Davis [5].

Exercise (1.16) is closely linked to Exercise (4.16) of Chap. VII. The interested reader shall find several applications of both exercises in Jeulin-Yor [3] and Yor [16]. For an up-date on the subject the reader is referred to Fitzsimmons et al. [1] who in particular work with less stringent hypotheses.

Sect. 2. This section is based on Revuz [1]. The representation theorem (2.9) is valid for every process having a local time at each point, for instance the linear Markov processes of Sect. 3 Chap. VII. It was originally proved in the case of BM in Tanaka [1]. Some of its corollaries are due to Wang [2]. The proof that all the additive functionals of BM are continuous may be found in Blumenthal-Getoor [1].

For BM^d, $d > 1$, there is no result as simple as (2.9), precisely because for $d > 1$, the one-point sets are polar and there are no local times. For what can nonetheless be said, the reader may consult Brosamler [1] (see also Meyer [7]) and Bass [1].

Exercise (2.23) is taken from Çinlar et al. [1]. The skew Brownian motion of Exercises (2.24) and (2.30) is studied in Harrison-Shepp [1], Walsh [3] and Barlow [4]. Walsh's multivariate generalization of the skew Brownian motion is studied by Barlow et al. [2]. Exercise (2.31) is due to Engelbert-Schmidt [1].

Sect. 3. Our exposition is based on Azéma et al. [1] (1967) and Revuz [2], but the limit quotient theorem had been known for a long time in the case of BM^1 (see Itô-McKean [1]) and BM^2 for which it was proved by Maruyama and Tanaka [1]. For the results of ergodic theory used in this section see for instance Krengel [1], Neveu [4] or Revuz [3].

Exercise (3.18) is from Azéma et al. [1] (1967) and Exercise (3.19) from Revuz [4]. Incidentally, let us mention the

Question 1. Can the result in Exercise (3.19) be extended to all Harris processes?

Exercise (3.20) is taken from Yor [17].

Sect. 4. Theorem (4.1) was proved by Spitzer [1] as a consequence of his explicit computation of the distribution of θ_t. The proof presented here, as well as the proof of Theorem (4.2) is taken from Messulam and Yor [1] who followed an idea of Williams [4] with an improvement of Pitman-Yor [5]. A variant of this proof based on Laplace's method is given in Exercise (4.11); this variant was used by Durrett [1] and Le Gall-Yor [2]. The almost-sure asymptotic behavior of winding numbers has been investigated by Bertoin-Werner ([1], [2]) and Shi [1].

The asymptotic property of additive functionals which is part of Theorem (4.2) was first proved by Kallianpur and Robbins [1]. This kind of result is proved for BM^1 in Sect. 2 Chap. XIII; for more general recurrent Markov processes we refer to Darling-Kac [1], Bingham [1] and the series of papers by Kasahara ([1], [2] and [3]).

The formula given in Proposition (4.3) may be found in the literature in various disguises; it is clearly linked to P. Lévy's formula for the stochastic area, and the reader is referred to Williams [5], Azéma-Yor [2] and Jeulin-Yor [3]. For more variations on Lévy's formula see Biane-Yor [2] and Duplantier [1] which contains many references.

Chapter XI. Bessel Processes and Ray-Knight Theorems

§1. Bessel Processes

In this section, we take up the study of Bessel processes which was begun in Sect. 3 of Chap. VI and we use the notation thereof. We first make the following remarks.

If B is a BM^δ and we set $\rho = |B|$, Itô's formula implies that

$$\rho_t^2 = \rho_0^2 + 2\sum_{i=1}^{\delta}\int_0^t B_s^i dB_s^i + \delta t.$$

For $\delta > 1$, ρ_t is a.s. > 0 for $t > 0$ and for $\delta = 1$ the set $\{s : \rho_s = 0\}$ has a.s. zero Lebesgue measure, so that in all cases we may consider the process

$$\beta_t = \sum_{i=1}^{\delta}\int_0^t \left(B_s^i/\rho_s\right) dB_s^i$$

which, since $\langle \beta, \beta\rangle_t = t$, is a linear BM; therefore ρ_t^2 satisfies the SDE

$$\rho_t^2 = \rho_0^2 + 2\int_0^t \rho_s d\beta_s + \delta t.$$

For any real $\delta \geq 0$, and $x \geq 0$, let us consider the SDE

$$Z_t = x + 2\int_0^t \sqrt{|Z_s|}d\beta_s + \delta t.$$

Since $\left|\sqrt{z} - \sqrt{z'}\right| < \sqrt{|z - z'|}$ for $z, z' \geq 0$, the results of Sect. 3 in Chap. IX apply. As a result, for every δ and x, this equation has a unique strong solution. Furthermore, as for $\delta = x = 0$, this solution is $Z_t \equiv 0$, the comparison theorems ensure that in all cases $Z_t \geq 0$ a.s. Thus the absolute value in the above SDE may be discarded.

(1.1) Definitions. *For every $\delta \geq 0$ and $x \geq 0$, the unique strong solution of the equation*

$$Z_t = x + 2\int_0^t \sqrt{Z_s}d\beta_s + \delta t$$

is called the square of δ-dimensional Bessel process *started at x and is denoted by $BESQ^\delta(x)$. The number δ is the* dimension *of $BESQ^\delta$.*

The law of $\mathrm{BES}Q^\delta(x)$ on $C(\mathbb{R}_+, \mathbb{R})$ is denoted by Q_x^δ. We will also use the number $\nu = (\delta/2) - 1$ which is called the *index* of the corresponding process, and write $\mathrm{BES}Q^{(\nu)}$ instead of $\mathrm{BES}Q^\delta$ if we want to use ν instead of δ and likewise $Q_x^{(\nu)}$. We will use ν and δ in the same statements, it being understood that they are related by the above equation.

We have thus defined a one-parameter family of processes which for integer dimensions, coincides with the squared modulus of BM^δ. For every t and every $a \geq 0$, the map $x \to Q_x^\delta[X_t \geq a]$ where X is the coordinate process is increasing, thanks to the comparison theorems, hence Borel measurable. By the monotone class theorem, it follows that $x \to Q_x^\delta[X_t \in A]$ is Borel measurable for every Borel set A. By Theorem (1.9) in Chap. IX, these processes are therefore Markov processes. They are actually Feller processes, which will be a corollary of the following *additivity* property of the family $\mathrm{BES}Q^\delta$. If P and Q are two probability measures on $C(\mathbb{R}_+, \mathbb{R})$, we shall denote by $P * Q$ the convolution of P and Q, that is, the image of $P \otimes Q$ on $C(\mathbb{R}_+, \mathbb{R})^2$ by the map $(\omega, \omega') \to \omega + \omega'$. With this notation, we have the following result which is obvious for integer dimensions.

(1.2) Theorem. *For every $\delta, \delta' \geq 0$ and $x, x' \geq 0$,*

$$Q_x^\delta * Q_{x'}^{\delta'} = Q_{x+x'}^{\delta+\delta'}.$$

Proof. For two independent linear BM's β and β', call Z and Z' the corresponding two solutions for (x, δ) and (x', δ'), and set $X = Z + Z'$. Then

$$X_t = x + x' + 2\int_0^t \left(\sqrt{Z_s}d\beta_s + \sqrt{Z_s'}d\beta_s'\right) + (\delta + \delta')t.$$

Let β'' be a third BM independent of β and β'. The process γ defined by

$$\gamma_t = \int_0^t 1_{(X_s>0)} \frac{\sqrt{Z_s}d\beta_s + \sqrt{Z_s'}d\beta_s'}{\sqrt{X_s}} + \int_0^t 1_{(X_s=0)} d\beta_s''$$

is a linear BM since $\langle \gamma, \gamma \rangle_t = t$ and we have

$$X_t = (x + x') + 2\int_0^t \sqrt{X_s}d\gamma_s + (\delta + \delta')t$$

which completes the proof.

Remark. The family Q_x^δ is not the only family with this property as is shown in Exercise (1.13).

(1.3) Corollary. *If μ is a measure on $\mathbb{R}_+$ such that $\int_0^\infty (1 + t)d\mu(t) < \infty$, there exist two numbers A_μ and $B_\mu > 0$ such that*

$$Q_x^\delta\left[\exp\left(-\int_0^\infty X_t d\mu(t)\right)\right] = A_\mu^x B_\mu^\delta,$$

where X is the coordinate process.

Proof. Let us call $\phi(x,\delta)$ the left-hand side. The hypothesis on μ entails that

$$\phi(x,\delta) \geq \exp\left(-Q_x^\delta\left(\int_0^\infty X_t d\mu(t)\right)\right) = \exp\left(-\int_0^\infty (x+\delta t)d\mu(t)\right) > 0.$$

Furthermore, from the theorem it follows easily that

$$\phi(x+x',\delta+\delta') = \phi(x,\delta)\phi(x',\delta'),$$

so that

$$\phi(x,\delta) = \phi(x,0)\phi(0,\delta).$$

Each of the functions $\phi(\cdot,0)$ and $\phi(0,\cdot)$ is multiplicative and equal to 1 at 0. Moreover, they are monotone, hence measurable. The result follows immediately. □

By making $\mu = \lambda\varepsilon_t$, we get the Laplace transform of the transition function of $\mathrm{BES}Q^\delta$. We need the corresponding values of A_μ and B_μ which we compute by taking $\delta = 1$. We then have for $\lambda > 0$,

$$Q_x^1\left[\exp(-\lambda X_t)\right] = Q_x^1\left[\exp\left(-\lambda\int_0^\infty X_s\varepsilon_t(ds)\right)\right] = E_{\sqrt{x}}\left[\exp(-\lambda B_t^2)\right]$$

where B is BM^1. This is easily computed and found equal to

$$(1+2\lambda t)^{-1/2}\exp\big(-\lambda x/(1+2\lambda t)\big).$$

As a result

$$Q_x^\delta\left[\exp(-\lambda X_t)\right] = (1+2\lambda t)^{-\delta/2}\exp\big(-\lambda x/(1+2\lambda t)\big).$$

By inverting this Laplace transform, we get the

(1.4) Corollary. *For $\delta > 0$, the semi-group of $\mathrm{BES}Q^\delta$ has a density in y equal to*

$$q_t^\delta(x,y) = \frac{1}{2}\left(\frac{y}{x}\right)^{\nu/2}\exp\big(-(x+y)/2t\big)I_\nu\left(\sqrt{xy}/t\right), \qquad t>0, x>0,$$

where ν is the index corresponding to δ and I_ν is the Bessel function of index ν.

For $x = 0$, this density becomes

$$q_t^\delta(0,y) = (2t)^{-\delta/2}\Gamma(\delta/2)^{-1}y^{\delta/2-1}\exp(-y/2t).$$

The semi-group of $\mathrm{BES}Q^0$ is given by, for $x > 0$,

$$Q_t^0(x,\cdot) = \exp(-x/2t)\varepsilon_0 + \Xi_t(x,\cdot)$$

where $\Xi_t(x,\cdot)$ has the density

$$q_t^0(x,y) = (2t)^{-1}(y/x)^{-1/2}\exp\big(-(x+y)/2t\big)I_1\left(\sqrt{xy}/t\right)$$

(recall that $I_1 = I_{-1}$).

A consequence of these results is, as announced, that BESQ^δ is a Feller process. This may be seen either by using the value of the density or by observing that for $f \in C_0([0,\infty[)$, $Q_x^\delta[f(X_t)]$ is continuous in both x and t; this follows from the special case $f(x) = \exp(-\lambda x)$ and the Stone-Weierstrass theorem. Thus we may apply to these processes all the results in Chap. III. We proceed to a few observations on their behavior.

The comparison theorems and the known facts about BM in the lower dimensions entail that:

(i) for $\delta \geq 3$, the process BESQ^δ is transient and, for $\delta \leq 2$, it is recurrent,
(ii) for $\delta \geq 2$, the set $\{0\}$ is polar and, for $\delta \leq 1$, it is reached a.s. Furthermore for $\delta = 0$, $\{0\}$ is an absorbing point, since the process $X \equiv 0$ is then clearly a solution of the SDE of Definitions (1.1).

These remarks leave some gaps about the behavior of BESQ^δ for small δ. But if we put

$$s_\nu(x) = -x^{-\nu} \quad \text{for } \nu > 0, \qquad s_0(x) = \log x, \qquad s_\nu(x) = x^{-\nu} \quad \text{for } \nu < 0$$

and if T is the hitting time of $\{0\}$, then by Itô's formula, $s_\nu(X)^T$ is a local martingale under Q_x^δ. In the language of Sect. 3 Chap. VII, the function s_ν is a scale function for BESQ^δ, and by the reasonings of Exercise (3.21) therein, it follows that for $0 \leq \delta < 2$ the point 0 is reached a.s.; likewise, the process is transient for $\delta > 2$. It is also clear that the hypotheses of Sect. 3 Chap. VII are in force for BESQ^δ with $E = [0,\infty[$ if $\delta < 2$ and $E =]0,\infty[$ if $\delta \geq 2$. In the latter case, 0 is an entrance boundary; in the former, we have the

(1.5) Proposition. *For $\delta = 0$, the point 0 is absorbing. For $0 < \delta < 2$, the point 0 is instantaneously reflecting.*

Proof. The case $\delta = 0$ is obvious. For $0 < \delta < 2$, if X is a BESQ^δ, it is a semimartingale and by Theorem (1.7) Chap. VI, we have, since obviously $L_t^{0-}(X) = 0$,

$$L_t^0(X) = 2\delta \int_0^t 1_{(X_s=0)} ds.$$

On the other hand, since $d\langle X, X\rangle_t = 4X_t dt$, the occupation times formula tells us that

$$\begin{aligned} t \geq \int_0^t 1_{(0<X_s)} ds &= \int_0^t 1_{(0<X_s)} (4X_s)^{-1} d\langle X, X\rangle_s \\ &= \int_0^\infty (4a)^{-1} L_t^a(X) da. \end{aligned}$$

If $L_t^0(X)$ were not $\equiv 0$, we would have a contradiction. As a result, the time spent by X in 0 has zero Lebesgue measure and by Corollary (3.13) Chap. VII, this proves our claim.

Remarks. i) A posteriori, we see that the term $\int_0^t 1_{(X_s=0)} d\beta_s''$ in the proof of Theorem (1.2) is actually zero with the exception of the case $\delta = \delta' = 0$.

ii) Of course, a $\mathrm{BES}Q^\delta$ is a semimartingale and, for $\delta \geq 2$, it is obvious that $L^0(X) = 0$ since 0 is polar.

This result also tells us that if we call m_ν the speed measure of $\mathrm{BES}Q^\delta$ then for $\delta > 0$, we have $m_\nu(\{0\}) = 0$. To find m_ν on $]0, \infty[$ let us observe that by Proposition (1.7) of Chap. VII, the infinitesimal generator of $\mathrm{BES}Q^\delta$ is equal on $C_K^2(\,]0,\infty[\,)$ to the operator

$$2x\frac{d^2}{dx^2} + \delta\frac{d}{dx}\,.$$

By Theorem (3.14) in Chap. VII, it follows that, for the above choice of the scale function, m_ν must be the measure with density with respect to the Lebesgue measure equal to

$$x^\nu/2\nu \quad \text{for } \nu > 0, \qquad 1/2 \quad \text{for } \nu = 0, \qquad -x^\nu/2\nu \quad \text{for } \nu < 0.$$

The reader can check these formulas by straightforward differentiations or by using Exercise (3.20) in Chap. VII.

Let us now mention the scaling properties of $\mathrm{BES}Q^\delta$. Recall that if B is a standard BM^δ and $B_t^x = x + B_t$ then for any real $c > 0$, the processes $B_{c^2t}^x$ and $cB_t^{x/c}$ have the same law. This property will be called the Brownian scaling property. The processes $\mathrm{BES}Q$ have a property of the same ilk.

(1.6) Proposition. *If X is a $\mathrm{BES}Q^\delta(x)$, then for any $c > 0$, the process $c^{-1}X_{ct}$ is a $\mathrm{BES}Q^\delta(x/c)$.*

Proof. By a straightforward change of variable in the stochastic integral, one sees that

$$c^{-1}X_{ct} = c^{-1}x + 2\int_0^t \left(c^{-1}X_{cs}\right)^{1/2} c^{-1/2}dB_{cs} + \delta t$$

and since $c^{-1/2}B_{ct}$ is a BM, the result follows from the uniqueness of the solution to this SDE. □

We now go back to Corollary (1.3) to show how to compute the constants A_μ and B_μ; this will lead to the computation of the exact laws of some Brownian functionals. Let us recall (see Appendix 8) that if μ is a Radon measure on $[0, \infty[$, the differential equation (in the distribution sense) $\phi'' = \phi\mu$ has a unique solution ϕ_μ which is positive, non increasing on $[0, \infty[$ and such that $\phi_\mu(0) = 1$. The function ϕ_μ is convex, so its right-hand side derivative ϕ_μ' exists and is ≤ 0. Moreover, since ϕ_μ is non increasing, the limit $\phi_\mu(\infty) = \lim_{x\to\infty}\phi_\mu(x)$ exists and belongs to $[0, 1]$. In fact, $\phi_\mu(\infty) < 1$ if we exclude the trivial case $\mu = 0$; indeed, if $\phi_\mu(\infty) = 1$, then ϕ_μ is identically 1 and $\mu = 0$.

We suppose henceforth that $\int(1 + x)d\mu(x) < \infty$; we will see in the proof below that this entails that $\phi_\mu(\infty)$ is > 0. We set

$$X_\mu = \int_0^\infty X_t d\mu(t).$$

In this setting we get the exact values of the constants A_μ and B_μ of Corollary (1.3).

(1.7) Theorem. *Under the preceding assumptions,*

$$Q_x^\delta\left[\exp\left(-\frac{1}{2}X_\mu\right)\right] = \phi_\mu(\infty)^{\delta/2}\exp\left(\frac{x}{2}\phi'_\mu(0)\right).$$

Proof. The function ϕ'_μ is right-continuous and increasing, hence $F_\mu(t) = \phi'_\mu(t)/\phi_\mu(t)$ is right-continuous and of finite variation. Thus we may apply the integration by parts formula (see Exercise (3.9) Chap. IV) to get

$$F_\mu(t)X_t = F_\mu(0)x + \int_0^t F_\mu(s)dX_s + \int_0^t X_s dF_\mu(s);$$

but

$$\begin{aligned}\int_0^t X_s dF_\mu(s) &= \int_0^t X_s \frac{d\phi'_\mu(s)}{\phi_\mu(s)} - \int_0^t X_s \frac{\phi'_\mu(s)}{\phi_\mu^2(s)} d\phi_\mu(s)\\ &= \int_0^t X_s d\mu(s) - \int_0^t X_s F_\mu(s)^2 ds.\end{aligned}$$

As a result, since $M_t = X_t - \delta t$ is a Q_x^δ-continuous local martingale, the process

$$\mathscr{E}\left(\frac{1}{2}\int_0^\cdot F_\mu(s)dM_s\right)_t = \exp\left(\frac{1}{2}\int_0^t F_\mu(s)dM_s - \frac{1}{2}\int_0^t X_s F_\mu(s)^2 ds\right)$$

is a continuous local martingale and is equal to

$$Z_t^\mu = \exp\left\{\frac{1}{2}\left[F_\mu(t)X_t - F_\mu(0)x - \delta\log\phi_\mu(t)\right] - \frac{1}{2}\int_0^t X_s d\mu(s)\right\}.$$

Since F_μ is negative and X positive, this local martingale is bounded on $[0, a]$ and we may write

(†) $$E\left[Z_a^\mu\right] = E\left[Z_0^\mu\right] = 1.$$

The theorem will follow upon letting a tend to $+\infty$.

Firstly, using Proposition (1.6), one easily sees that (X_a/a) converges in law as a tends to $+\infty$. Secondly,

$$\phi'_\mu(x) = -\left(\phi_\mu\mu\right)(]x,\infty[)$$

(see Appendix 3), and consequently,

$$0 > aF_\mu(a) \geq \int_{]a,\infty[} x\, d\mu(x)$$

which goes to 0 as a tends to $+\infty$. As a result, $F_\mu(a)X_a$ converges in probability to 0 as a tends to $+\infty$ and passing to the limit in (†) yields

$$\left(\lim_{a\to\infty} \exp((-\delta/2)\log\phi_\mu(a))\right)\exp(-xF_\mu(0)/2)\, Q_x^\delta\left(\exp(-X_\mu/2)\right) = 1.$$

Since $X_\mu < \infty$ a.s., this shows that $\phi_\mu(\infty) > 0$ and completes the proof.

Remarks. 1°) It can be seen directly that Z_t^μ is continuous by computing the jumps of the various processes involved and observing that they cancel. We leave the details as an exercise to the reader.

2°) Since $aF_\mu(a)$ converging to 0 was all important in the proof above, let us observe that, in fact, it is equivalent to $-a\mu([a,\infty[)$ as a goes to infinity.

This result allows us to give a proof of the *Cameron-Martin formula* namely

$$E\left[\exp\left(-\lambda\int_0^1 B_s^2 ds\right)\right] = \left(\cosh\sqrt{2\lambda}\right)^{-1/2},$$

where B is a standard linear BM. This was first proved by analytical methods but is obtained by making $x = 0$ and $\delta = 1$ in the following

(1.8) Corollary.

$$Q_x^\delta\left[\exp\left(-\frac{b^2}{2}\int_0^1 X_s ds\right)\right] = (\cosh b)^{-\delta/2}\exp\left(-\frac{1}{2}xb\tanh b\right).$$

Proof. We must compute ϕ_μ when $\mu(ds) = b^2 ds$ on $[0,1]$. It is easily seen that on $[0,1]$, we must have $\phi_\mu(t) = \alpha\cosh bt + \beta\sinh bt$ and the condition $\phi_\mu(0) = 1$ forces $\alpha = 1$. Next, since ϕ_μ is constant on $[1,\infty[$ and ϕ'_μ is continuous, we must have $\phi'_\mu(1) = 0$, namely

$$b\sinh b + \beta b\cosh b = 0$$

which yields $\beta = -\tanh b$. Thus $\phi_\mu(t) = \cosh bt - (\tanh b)\sinh bt$ on $[0,1]$ which permits to compute $\phi_\mu(\infty) = \phi_\mu(1) = (\cosh b)^{-1}$ and $\phi'_\mu(0) = -b\tanh b$.

Remark. This corollary may be applied to the stochastic area which was introduced in Exercises (2.19) Chap. V and (3.10) Chap. IX.

We have dealt so far only with the squares of Bessel processes; we now turn to Bessel processes themselves. The function $x \to \sqrt{x}$ is a homeomorphism of $\mathbb{R}_+$. Therefore if X is a Markov process on $\mathbb{R}_+$, $\sqrt{X}$ is also a Markov process. By applying this to the family BESQ^δ, we get the family of Bessel processes.

(1.9) Definition. *The square root of BES$Q^\delta(a^2)$, $\delta \geq 0$, $a \geq 0$, is called the* Bessel process of dimension δ *started at a and is denoted by BES$^\delta(a)$. Its law will be denoted by P_a^δ.*

For integer dimensions, these processes were already introduced in Sect. 3 of Chap. VI: they can be realized as the modulus of the corresponding BM^δ.

Some properties of $\mathrm{BES}Q^\delta$ translate to similar properties for BES^δ. The Bessel processes are Feller processes with continuous paths and satisfy the hypothesis of Sect. 3 Chap. VII. Using Exercise (3.20) or Exercise (3.18) Chap. VII, it is easily seen that the scale function of BES^δ may be chosen equal to

$$-x^{-2\nu} \quad \text{for } \nu > 0, \qquad 2\log x \quad \text{for } \nu = 0, \qquad x^{-2\nu} \quad \text{for } \nu < 0,$$

and with this choice of the scale function, the speed measure is given by the densities

$$x^{2\nu+1}/\nu \quad \text{for } \nu > 0, \qquad x \quad \text{for } \nu = 0, \qquad -x^{2\nu+1}/\nu \quad \text{for } \nu < 0.$$

Moreover, for $0 < \delta < 2$, the point 0 is instantaneously reflecting and for $\delta \geq 2$ it is polar. Using Theorem (3.12) Chap. VII, one can easily compute the infinitesimal generator of BES^δ.

The density of the semi-group is also obtained from that of $\mathrm{BES}Q^\delta$ by a straightforward change of variable and is found equal, for $\delta > 0$, to

$$p_t^\delta(x, y) = t^{-1}(y/x)^\nu y \exp\left(-(x^2 + y^2)/2t\right) I_\nu(xy/t) \quad \text{for } x > 0, t > 0$$

and

$$p_t^\delta(0, y) = 2^{-\nu} t^{-(\nu+1)} \Gamma(\nu + 1)^{-1} y^{2\nu+1} \exp(-y^2/2t).$$

We may also observe that, since for $\delta \geq 2$, the point 0 is polar for $\mathrm{BES}Q^\delta(x)$, $x > 0$, we may apply Itô's formula to this process which we denote by X and to the function $\sqrt{x}$. We get

$$X_t^{1/2} = \sqrt{x} + \beta_t + \frac{\delta - 1}{2} \int_0^t X_s^{-1/2} ds$$

where β is a BM. In other words, $\mathrm{BES}^\delta(a)$, $a > 0$, is a solution to the SDE

$$\rho_t = a + \beta_t + \frac{\delta - 1}{2} \int_0^t \rho_s^{-1} ds.$$

By Exercise (2.10) Chap. IX, it is the only solution to this equation. For $\delta < 2$ the situation is much less simple; for instance, because of the appearance of the local time, BES^1 is not the solution to an SDE in the sense of Chap. IX (see Exercise (2.14) Chap. IX).

Finally Proposition (1.6) translates to the following result, which, for $\delta \geq 2$, may also be derived from Exercise (1.17) Chap. IX.

(1.10) Proposition. BES^δ *has the Brownian scaling property.*

We will now study another invariance property of this family of processes. Let X^δ be a family of diffusions solutions to the SDE's

$$X_t^\delta = x + B_t + \int_0^t b_\delta(X_s^\delta) ds$$

where b_δ is a family of Borel functions. Let f be a positive strictly increasing C^2-function with inverse f^{-1}. We want to investigate the conditions under which the process $f(X_t^\delta)$ belongs to the same family up to a suitable time change.

Itô's formula yields

$$f(X_t^\delta) = f(x) + \int_0^t f'(X_s^\delta)dB_s + \int_0^t \left(f'(X_s^\delta)b_\delta(X_s^\delta) + \frac{1}{2}f''(X_s^\delta)\right) ds.$$

Setting $\tau_t = \inf\left\{u : \int_0^u f'^2(X_s^\delta)ds > t\right\}$ and $Y_t^\delta = f(X_{\tau_t}^\delta)$, the above equation may be rewritten

$$Y_t^\delta = f(x) + \beta_t + \int_0^t \left(\left(f'b_\delta + \frac{1}{2}f''\right)/f'^2\right) \circ f^{-1}(Y_s^\delta)ds$$

where β is a BM. Thus, if we can find $\widetilde{\delta}$ such that $\left(f'b_\delta + \frac{1}{2}f''\right)/f'^2 = b_{\widetilde{\delta}} \circ f$, then the process Y^δ will satisfy a SDE of the given family.

This leads us to the following result where ρ_ν is a $\mathrm{BES}^{(\nu)}$.

(1.11) Proposition. *Let p and q be two conjugate numbers $\left(p^{-1} + q^{-1} = 1\right)$. If $\nu > -1/q$, there is a* $\mathrm{BES}^{(\nu q)}$ *defined on the same probability space as ρ_ν such that*

$$q\rho_\nu^{1/q} = \rho_{\nu q}\left(\int_0^\cdot \rho_\nu^{-2/p}(s)ds\right).$$

Proof. For $\nu \geq 0$, the process ρ_ν lives on $]0, \infty[$ and $x^{1/q}$ is twice differentiable on $]0, \infty[$. Thus we may apply the above method with $f(x) = qx^{1/q}$ and since then, with $b_\nu(x) = \left(\nu + \frac{1}{2}\right)x^{-1}$, we have

$$\left(b_\nu f' + \frac{1}{2}f''\right)/f'^2 = b_{\nu q} \circ f,$$

the result follows from the uniqueness of the solutions satisfied by the Bessel processes for $\nu \geq 0$.

For $\nu < 0$, one can show that $q\rho_\nu^{1/q}(\tau_t)$, where τ_t is the time-change associated with $\int_0^\cdot \rho_\nu^{-2/p}(s)ds$, has on $]0, \infty[$ the same generator as $\rho_{\nu q}$; this is done as in Exercise (2.29) Chap. X. Since the time spent in 0 has zero Lebesgue measure, the boundary 0 is moreover instantaneously reflecting; as a result the generator of $q\rho_\nu^{1/q}(\tau_t)$ is that of $\rho_{\nu q}(t)$ and we use Proposition (3.14) Chap. VII to conclude. The details are left to the reader. □

This invariance principle can be put to use to give explicit expressions for the laws of some functionals of BM.

(1.12) Corollary. *In the setting of Proposition (1.11), if $\rho_\nu(0) = 0$ a.s., then*

$$\int_0^1 \rho_\nu^{-2/p}(s)ds \stackrel{(d)}{=} \left(\int_0^1 \rho_{\nu q}^{2q/p}(s)ds\right)^{-1/q}.$$

Proof. Let $C_t = \int_0^t \rho_\nu^{-2/p}(s)ds$ and τ_t the associated time-change. Within the proof of the proposition we saw that $q\rho_\nu^{1/q}(\tau_s) = \rho_{\nu q}(s)$; since $d\tau_t = \rho_\nu^{2/p}(\tau_t)dt$, it follows that

$$\tau_t = \int_0^t \left(q^{-1}\rho_{\nu q}(s)\right)^{2q/p} ds.$$

It remains to prove that C_1 has the same law as $\tau_1^{-1/q}$. To this end, we first remark that $\{C_1 > t\} = \{\tau_t < 1\}$; we then use the scaling property of ρ_ν to the effect that

$$\tau_t = \int_0^1 \left(q^{-1}\rho_{\nu q}(tu)\right)^{2q/p} t\, du \overset{(d)}{=} t^q \tau_1,$$

which yields

$$P\{C_1 > t\} = P\{t^q\tau_1 < 1\} = P\{\tau_1^{-1/q} > t\}.$$

□

The point of this corollary is that the left-hand side is, for suitable ν's, a Brownian functional whereas the Laplace transform of the right-hand side may be computed in some cases. For instance, making $p = q = 2$, we get

$$\int_0^1 \rho_\nu^{-1}(s)ds \overset{(d)}{=} 2\left(\int_0^1 \rho_{2\nu}^2(s)ds\right)^{-1/2},$$

and the law of the latter r.v. is known from Corollary (1.8).

(1.12bis) Exercise (Continuation of Corollary (1.12)).
1°) In the setting of Proposition (1.11) prove that

$$E\left[\left(\int_0^1 \rho_{\nu q}(s)^{2q/p}ds\right)^{-1/q}\right] = q\Gamma\left((1/q)+\nu\right)/\Gamma(1+\nu).$$

2°) Deduce that

$$E\left[\left(\sup_{s\le 1}\rho_{\nu q}(s)\right)^{-2}\right] = \left(2(1+\nu q)\right)^{-1}.$$

Explain this result in view of Exercise (1.18) 1°).

(1.13) Exercise. 1°) Let X^δ, $\delta \geq 2$, be a family of diffusions which are solutions to the SDE's

$$X_t^\delta = x + B_t + \int_0^t b_\delta(X_s^\delta)ds.$$

Prove that the laws of the processes $(X^\delta)^2$ satisfy the additivity property of Theorem (1.2) provided that

$$2xb_\delta(x) + 1 = \delta + bx^2$$

for a constant b. This applies in particular to the family of Euclidean norms of $\mathbb{R}^\delta$-valued OU processes, namely $X_t = |Y_t|$, where

$$Y_t = y + B_t + \frac{b}{2}\int_0^t Y_s ds.$$

(See Exercise (3.14) Chap. VIII.)

2°) Prove that the diffusion with infinitesimal generator given on $C_K^\infty(]0,\infty[)$ by

$$2x\, d^2/dx^2 + (2\beta x + \delta)d/dx$$

with $\beta \neq 0$ and $\delta \geq 0$, may be written

$$e^{2\beta t} X\left(\left(1 - e^{-2\beta t}\right)/2\beta\right) \quad \text{or} \quad e^{2\beta t} Y\left(\left|e^{-2\beta t} - 1\right|\right)/2|\beta|$$

where X and Y are suitable $\mathrm{BES}Q^\delta$ processes.

[Hint: Start with $\delta = 1$ in which case the process is the square of a OU process, then use 1°).]

(1.14) Exercise (Infinitely divisible laws on W). 1°) For every $\delta \geq 0$ and $x \geq 0$, prove that the law Q_x^δ is infinitely divisible, i.e. for every n, there is a law $P^{(n)}$ on $\mathbf{W} = C(\mathbb{R}_+, \mathbb{R})$ such that

$$Q_x^\delta = P^{(n)} * \ldots * P^{(n)} \qquad (n \text{ terms}).$$

2°) Let B and C be two independent Brownian motions starting from 0, and R be a $\mathrm{BES}^\delta(x)$ process independent from B. Prove that the laws of

a) BC; b) $\int_0^\cdot (B_s dC_s - C_s dB_s)$; c) $\int_0^\cdot R_s dB_s$;
d) $\int_0^\cdot dR_s^2 \varphi(s)$, where $\varphi : \mathbb{R}_+ \to \mathbb{R}_+$ is bounded, Borel

are infinitely divisible.

(1.15) Exercise. Compute A_μ and B_μ in Corollary (1.3) when μ is a linear combination of Dirac masses i.e. $\mu = \sum \lambda_i \varepsilon_{t_i}$.

[Hint: Use the Markov property.]

\# **(1.16) Exercise. (More on the counterexample of Exercise (2.13) in Chap. V).** Let X be a $\mathrm{BES}^\delta(x)$, and recall that if $\delta > 2$, $x > 0$, the process $X_t^{2-\delta}$ is a local martingale.

1°) Using the fact that $I_\nu(z)$ is equivalent to $\Gamma(\nu+1)^{-1}(z/2)^\nu$ as z tends to zero, prove that for $\alpha < \delta/(\delta - 2)$ and $\varepsilon > 0$,

$$\sup_{\varepsilon \leq t \leq 1} E\left[X_t^{\alpha(2-\delta)}\right] < +\infty.$$

[Hint: Use the scaling properties of BES^δ and the comparison theorems for SDE's.]

2°) For every $p \geq 1$, give an example of a positive continuous local martingale bounded in L^p and which is not a martingale.

[Hint: Show that $E\left[X_t^{2-\delta}\right]$ is not constant in t.]

* **(1.17) Exercise.** Let M be in $\mathcal{M}$ (Exercise (3.26) in Chap. IV) and such that $\lim_{t\downarrow 0} M_t = +\infty$. Prove that the continuous increasing process

$$A_t = \int_0^t \exp(-2M_s)\, d\langle M, M\rangle_s$$

is finite-valued and that there exists a BES^2, say X, such that $\exp(-M_t) = X_{A_t}$.

*# **(1.18) Exercise.** 1°) If X is a $\mathrm{BES}^\delta(0)$, prove that $T_1 = \inf\{t : X_t = 1\}$ and $\left(\sup_{t\le 1} X_t\right)^{-2}$ have the same law.

[Hint: Write $\{T_1 > t\} = \left\{\sup_{s\le t} X_s < 1\right\}$ and use the scaling invariance properties.]

2°) If moreover $\delta > 2$, prove that $L = \sup\{t : X_t = 1\}$ and $\left(\inf_{t\ge 1} X_t\right)^{-2}$ have the same law.

3°) Let Y be a $\mathrm{BES}^\delta(1)$ independent of X. Show, using the scaling invariance of X, that $\inf_{t\ge 1} X_t$ and $X_1 \cdot \inf_{t\ge 0} Y_t$ have the same law.

4°) Using the property that $Y^{2-\delta}$ is a local martingale, show that $\inf_{t\ge 0} Y_t$ has the same law as $U^{1/\delta-2}$, where U is uniformly distributed on $[0, 1]$.

[Hint: Use Exercise (3.12) Chap. II.]

5°) Show finally that the law of the r.v. L defined in 2°) above is that of Z_1^{-2} where Z is a $\mathrm{BES}^{\delta-2}(0)$. Compare this result with the result obtained for the distribution of the last passage time of a transient diffusion in Exercise (4.16) of Chap. VII.

(1.19) Exercise. For $p > 0$, let X be a $\mathrm{BES}^{2p+3}(0)$. Prove that there is a $\mathrm{BES}^3(0)$, say Y, such that

$$X_t^{2p+1} = Y\left(\int_0^t (2p+1)^2\, X_s^{4p} ds\right).$$

[Hint: Use Itô's formula and the time-change associated with $\int_0^t (2p+1)^2 X_s^{4p} ds$.]

(1.20) Exercise. For $\delta \ge 1$, prove that BES^δ satisfies the various laws of the iterated logarithm.

[Hint: Use Exercise (1.21) in Chap. II and the comparison theorems of Sect. 3, Chap. IX.]

(1.21) Exercise. For $0 < \delta < 1$, prove that the set $\{t : X_t = 0\}$ where X is a BES^δ is a perfect set with empty interior.

*# **(1.22) Exercise.** 1°) Let $(\mathscr{F}_t^0)$ be the filtration generated by the coordinate process X. For the indexes $\mu, \nu \ge 0$, T a bounded $(\mathscr{F}_t^0)$-stopping time, Y any $\mathscr{F}_{T+}^0$-measurable positive r.v. and any $a > 0$, prove that

$$P_a^{(\mu)}\left[Y \exp\left(-\frac{\nu^2}{2}\int_0^T X_s^{-2} ds\right)(X_T/a)^{-\mu}\right]$$
$$= P_a^{(\nu)}\left[Y \exp\left(-\frac{\mu^2}{2}\int_0^T X_s^{-2} ds\right)(X_T/a)^{-\nu}\right].$$

[Hint: Begin with $\mu = 0$, $\nu > 0$ and use the $P_a^{(0)}$-martingale $\mathscr{E}(\nu M)$ where $M = \log(X/a)$ as described in Exercise (1.34) of Chap. VIII.]

2°) Let W_a be the law of the linear BM started at a and define a probability measure R_a by

$$R_a = \left(X_{t\wedge T_0}/a\right) \cdot W_a \quad \text{on} \quad \mathscr{F}_t^0,$$

where $T_0 = \inf\{t : X_t = 0\}$. Show that R_a is the law of BES_a^3 i.e. $P_a^{(1/2)}$ in the notation of this exercise.

[Hint: Use Girsanov's theorem to prove that, under R_a, the process X satisfies the right SDE.]

3°) Conclude that for every $\nu \geq 0$,

$$P_a^{(\nu)} = \left(X_{t\wedge T_0}/a\right)^{\nu+1/2} \exp\left(-((\nu^2 - 1/4)/2) \int_0^t X_s^{-2} ds\right) \cdot W_a \quad \text{on } \mathscr{F}_t^0.$$

\# **(1.23) Exercise.** Prove that for $\nu > 0$, and $b > 0$, the law of $\{X_{L_b-t}, t < L_b\}$ under $P_0^{(\nu)}$ is the same as the law of $\{X_t, t < T_0\}$ under $P_b^{(-\nu)}$.

[Hint: Use Theorem (4.5) Chap. VII.]

In particular L_b under $P_0^{(\nu)}$ has the same law as T_0 under $P_b^{(-\nu)}$. [This generalizes Corollary (4.6) of Chap. VII.] Prove that this common law is that of $2/\gamma_\nu$.

[Hint: Use Exercise (1.18).]

(1.24) Exercise. Let Z be the planar BM. For $\alpha < 2$, prove that $\mathscr{E}(-\alpha \log|Z|)$ is not a martingale.

[Hint: Assume that it is a martingale, then follow the scheme described above Proposition (3.8) Chap. VIII and derive a contradiction.]

(1.25) Exercise. 1°) Prove that even though $\mathrm{BES}^{(\nu)}$ is not a semimartingale for $\nu < -1/2$, it has nonetheless a bicontinuous family of occupation densities namely a family l_t^x of processes such that

$$\int_0^t f(\rho_\nu(s))\, ds = \int_0^\infty f(x) l_t^x m_\nu(dx) \qquad \text{a.s.}$$

Obviously, this formula is also true for $\nu \geq -1/2$.

[Hint: Use Proposition (1.11) with $q = -2\nu$.]

2°) Prove that for $\nu \in]-1, 0[$ the inverse τ of the local time at 0 is a stable subordinator with index $(-\nu)$ (see Sect. 4 Chap. III), i.e.

$$E\left[\exp(-\lambda\tau_t)\right] = \exp(-ct\lambda^{-\nu}).$$

(1.26) Exercise. (Bessel processes with dimension δ in $]0, 1[$). Let $\delta > 0$ and ρ be a BES^δ process.

1°) Prove that for $\delta \geq 1$, ρ is a semimartingale which can be decomposed as

$$(1) \qquad \rho_t = \rho_0 + \beta_t + ((\delta - 1)/2) \int_0^t \rho_s^{-1} ds, \quad \text{if } \delta > 1,$$

and

$$(2) \qquad \rho_t = \rho_0 + \beta_t + \frac{1}{2} l_t^0, \quad \text{if } \delta = 1,$$

where l^0 is the local time of ρ at 0. This will serve as a key step in the study of ρ for $\delta < 1$.

2°) For $0 < \alpha < 1/2$ and the standard BM^1 B, prove that

$$\begin{aligned} |B_t|^{1-\alpha} &= |B_0|^{1-\alpha} + (1-\alpha)\int_0^t |B_s|^{-\alpha}\,\mathrm{sgn}(B_s)dB_s \\ &\quad + \big(-\alpha(1-\alpha)/2\big)\mathrm{P.V.}\int_0^t |B_s|^{-1-\alpha}\,ds, \end{aligned}$$

where P.V. is defined by

$$\mathrm{P.V.}\int_0^t |B_s|^{-1-\alpha}\,ds = \int_{-\infty}^{\infty} |b|^{-1-\alpha}\left(l_t^b - l_t^0\right)db.$$

[Hint: For the existence of the principal value, see Exercise (1.29) Chap. VI.]

3°) Let now ρ be a BES^δ with $\delta \in]0,1[$; denote by l^a the family of its local times defined by

$$\int_0^t \phi(\rho_s)ds = \int_0^\infty \phi(x)l_t^x x^{\delta-1}dx$$

in agreement with Exercise (1.25) above. Prove that

$$\rho_t = \rho_0 + \beta_t + \big((\delta-1)/2\big)k_t$$

where $k_t = \mathrm{P.V.}\ \int_0^t \rho_s^{-1}ds$ which, by definition, is equal to $\int_0^\infty a^{\delta-2}\left(l_t^a - l_t^0\right)da$.

[Hint: Use 2°) as well as Proposition (1.11) with $\nu = -1/2$.]

(1.27) Exercise. 1°) Prove that $Q_x^4\left(xX_t^{-1}\right) = 1 - \exp(-x/2t)$.

[Hint: $\xi^{-1} = \int_0^\infty \exp(-\lambda\xi)d\lambda$.]

2°) Comparing with the formulae given in Corollary (1.4) for Q_t^4 and Q_t^0, check, using the above result, that $Q_t^0(x, \mathbb{R}_+) = 1$.

* **(1.28) Exercise (Lamperti's relation).** 1°) Let B be a $\mathrm{BM}^1(0)$. For $\nu \geq 0$, prove that there exists a $\mathrm{BES}^{(\nu)}$, say $R^{(\nu)}$, such that

$$\exp(B_t + \nu t) = R^{(\nu)}\left(\int_0^t \exp\big(2(B_s + \nu s)\big)ds\right).$$

2°) Give an adequate extension of the previous result for any $\nu \in \mathbb{R}$.

3°) Prove that, for $a \in \mathbb{R}$, $a \neq 0$, and $b > 0$, the law of $\int_0^\infty ds\,\exp(aB_s - bs)$ is equal to the law of $2/(a^2\gamma_{(2b/a^2)})$, where γ_α denotes a Gamma (α) variable.

[Hint: Use Exercise (1.23).]

(1.29) Exercise (An extension of Pitman's theorem to transient Bessel processes). Let $(\rho_t, t \geq 0)$ be a $\mathrm{BES}^\delta(0)$ with $\delta > 2$, and set $J_t = \inf_{s \geq t}\rho_s$. Prove that there exists a Brownian motion $(\gamma_t, t \geq 0)$ such that

$$\rho_t = \gamma_t + 2J_t - \frac{\delta-3}{2}\int_0^t \frac{ds}{\rho_s}.$$

[Hint: Use Proposition (1.11).]

(1.30) Exercise (A property equivalent to Pitman's theorem). Let B be a BM^1 and S its supremum. Assume that for every t, conditionally on $\mathscr{F}_t^{2S-B}$, the r.v. S_t is uniformly distributed on $[0, 2S_t - B_t]$, then prove that the process $2S - B$ is a $BES^3(0)$. Compare with the results in Section VI.3, notably Corollary (3.6).

[Hint: Prove that $(2S_t - B_t)^2 - 3t$ is a local martingale.]

(1.31) Exercise (Seshadri's identities). In the notation of Sections 2 and 3 in Chapter VI, write

$$(2S_t - B_t)^2 = (S_t - B_t)^2 + r_t^2 = S_t^2 + \rho_t^2.$$

1°) For fixed $t > 0$, prove that the r.v.'s r_t^2 and ρ_t^2 are exponentially distributed and that $(S_t - B_t)^2$ and r_t^2 are independent; likewise S_t^2 and ρ_t^2 are independent.

[Hint: Use the results on Gamma and Beta variables from Sect. 6 Chap. 0.]

2°) Prove further that the processes r_t^2 and ρ_t^2 are not $BESQ^2(0)$, however tempting it might be to think so.

[Hint: Write down their semimartingale decompositions.]

(1.32) Exercise. (Asymptotic distributions for functionals X_μ whenever $\int_0^\infty (1+t)d\mu(t) = \infty$). 1°) Let X be a $BESQ^\delta(x)$ with $\delta > 0$; for $\alpha < 2$, prove that, as t tends to infinity, $t^{\alpha-2}\int_1^t u^{-\alpha} X_u\, du$ converges in distribution to $\int_0^1 u^{-\alpha} Y_u\, du$, where Y is a $BESQ^\delta(0)$.

2°) In the same notation, prove that $(\log t)^{-1}\int_1^t u^{-2} X_u\, du$ converges a.s. to $\delta = E[Y_1]$. Prove further that $(\log t)^{-1/2}\left(\delta(\log t) - \int_1^t u^{-2} X_u\, du\right)$ converges in distribution to $2\gamma_\delta$ where γ is a $BM^1(0)$.

[Hint: Develop $(X_t/t)\,, t \geq 1$, as a semimartingale.]

(1.33) Exercise. ("Square" Bessel processes with negative dimensions). 1°) Let $x, \delta > 0$, and β be a BM^1. Prove that there exists a unique strong solution to the equation

$$Z_t = x + 2\int_0^t \sqrt{|Z_s|}\, d\beta_s - \delta t.$$

Let $Q_x^{-\delta}$ denote the law of this process on $\mathbf{W} = C(\mathbb{R}_+, \mathbb{R})$.

2°) Show that $T_0 = \inf\{t : Z_t = 0\} < \infty$ a.s., and identify the process $\left\{-Z_{T_0+t}, t \geq 0\right\}$.

3°) Prove that the family $\left\{Q_x^\gamma, \gamma \in \mathbb{R}, x \geq 0\right\}$ does not satisfy the additivity property, as presented in Theorem (1.2).

(1.34) Exercise (Complements to Theorem 1.7). Let μ be a positive, diffuse, Radon measure on $\mathbb{R}_+$. Together with ϕ_μ, introduce the function $\psi_\mu(t) = \phi_\mu(t)\int_0^t \frac{ds}{\phi_\mu^2(s)}$.

1°) Prove that ψ_μ is a solution of the Sturm-Liouville equation $\phi'' = \mu\phi$, and that, moreover, $\psi_\mu(0) = 0$, $\psi_\mu'(0) = 1$.

Note the Wronskian relation

$$W(\phi_\mu, \psi_\mu) = \phi_\mu\psi_\mu' - \phi_\mu'\psi_\mu \equiv 1.$$

2°) Prove that, for every $t \geq 0$, one has

$$Q_x^\delta\left(\exp\left(-\frac{1}{2}\int_0^t X_s d\mu(s)\right)\right) = \frac{1}{\left(\psi'_\mu(t)\right)^{\delta/2}} \exp\left(\frac{x}{2}\left(\phi'_\mu(0) - \frac{\phi'_\mu(t)}{\psi'_\mu(t)}\right)\right).$$

3°) Check, by letting $t \to \infty$, that the previous formula agrees with the result given in Theorem 1.7.

[Hint: If $\int^\infty t\, d\mu(t) < \infty$, then: $\psi'_\mu(t) \to 1/\phi_\mu(\infty)$ (> 0).]

4°) Let B be a $\mathrm{BM}^1(0)$. Prove that, if ν is a second Radon measure on $\mathbb{R}_+$, then

$$E\left[\exp\left\{\int_0^t B_s \nu(ds) - \frac{1}{2}\int_0^t B_s^2 \mu(ds)\right\}\right]$$
$$= \frac{1}{\left(\psi'(t)\right)^{1/2}} \exp\left\{\frac{1}{2}\int_0^t du \left(\int_u^t \frac{\phi(s)}{\phi(u)}\nu(ds)\right)^2 - \theta(t)\left(\int_0^t \psi(s)\nu(ds)\right)^2\right\}$$

where $\theta(t) = \phi'_\mu(t)/2\psi'_\mu(t)$.

[Hint: Use the change of probability measure considered in the proof of Theorem (3.2) below.]

§2. Ray-Knight Theorems

Let B be the standard linear BM and $\left(L_t^a\right)$ the family of its local times. The Ray-Knight theorems stem from the desire to understand more thoroughly the dependence of $\left(L_t^a\right)$ in the space variable a. To this end, we will first study the process

$$Z_a = L_{T_1}^{1-a}, \qquad 0 \leq a \leq 1,$$

where $T_1 = \inf\{t : B_t = 1\}$. We will prove that this is a Markov process and in fact a $\mathrm{BES}Q^2$ restricted to the time-interval $[0, 1]$. We call $(\mathscr{Z}_a)_{a\in[0,1]}$, the complete and right-continuous filtration generated by Z. For this filtration, we have a result which is analogous to that of Sect. 3, Chap. V, for the Brownian filtration.

(2.1) Proposition. *Any r.v. H of $L^2(\mathscr{Z}_a)$, $0 \leq a \leq 1$, may be written*

$$H = E[H] + \int_0^{T_1} h_s 1_{(B_s > 1-a)} dB_s$$

where h is predictable with respect to the filtration of B and

$$E\left[\int_0^{T_1} h_s^2 1_{(B_s > 1-a)} ds\right] < \infty.$$

Proof. The subspace $\mathscr{H}$ of r.v.'s H having such a representation is closed in $L^2(\mathscr{Z}_a)$, because

$$E[H^2] = E[H]^2 + E\left[\int_0^{T_1} h_s^2 1_{(B_s>1-a)} ds\right]$$

and one can argue as in Sect. 3 of Chap. V.

We now consider the set of r.v.'s K which may be written

$$K = \exp\left\{-\int_0^a g(b) Z_b db\right\}$$

with g a positive C^1-function with compact support contained in $]0, a[$. The vector space generated by these variables is an algebra of bounded functions which, thanks to the continuity of Z, generates the σ-field $\mathscr{Z}_a$. It follows from the monotone class theorem that this vector space is dense in $L^2(\mathscr{Z}_a)$. As a result, it is enough to prove the representation property for K.

Set $U_t = \exp\left\{-\int_0^t g(1-B_s)ds\right\}$; thanks to the occupation times formula, since $g(1-x)$ vanishes on $]0, 1-a[$,

$$K = \exp\left\{-\int_{1-a}^1 g(1-x) L_{T_1}^x dx\right\} = U_{T_1}.$$

If $F \in C^2$, the semimartingale $M_t = F(B_t)U_t$ may be written

$$M_t = F(0) - \int_0^t F(B_s)U_s g(1-B_s)ds + \int_0^t U_s F'(B_s)dB_s + \frac{1}{2}\int_0^t U_s F''(B_s)ds.$$

We may pick F so as to have $F' \equiv 0$ on $]-\infty, 1-a]$, $F(1) \neq 0$ and $F''(x) = 2g(1-x)F(x)$. We then have, since $F' = F' 1_{]1-a,\infty[}$,

$$M_{T_1} = F(0) + \int_0^{T_1} U_s F'(B_s) 1_{(B_s>1-a)} dB_s$$

and, as $K = F(1)^{-1} M_{T_1}$, the proof is complete.

We may now state what we will call the *first Ray-Knight theorem.*

(2.2) Theorem. *The process* Z_a, $0 \leq a \leq 1$ *is a* BES$Q^2(0)$ *restricted to the time interval* $[0, 1]$.

Proof. From Tanaka's formula

$$(B_t - (1-a))^+ = \int_0^t 1_{(B_s>1-a)} dB_s + \frac{1}{2} L_t^{1-a},$$

it follows that

$$Z_a - 2a = -2\int_0^{T_1} 1_{(B_s>1-a)} dB_s.$$

It also follows that Z_a is integrable; indeed, for every t

$$E\left[L_{t\wedge T_1}^{1-a}\right] = 2E\left[\left(B_{t\wedge T_1} - (1-a)\right)^+\right]$$

and passing to the limit yields $E[Z_a] = 2a$.

Now, pick $b < a$ and H a bounded $\mathscr{Z}_b$-measurable r.v. Using the representation of the preceding proposition, we may write

$$\begin{aligned} E[(Z_a - 2a)H] &= E\left[-2\int_0^{T_1} h_s 1_{(B_s>1-b)} 1_{(B_s>1-a)} ds\right] \\ &= E\left[-2\int_0^{T_1} h_s 1_{(B_s>1-b)} ds\right] \\ &= E[(Z_b - 2b)H]. \end{aligned}$$

Therefore, $Z_a - 2a$ is a continuous martingale and by Corollary (1.13) of Chap. VI, its increasing process is equal to $4\int_0^a Z_u du$. Proposition (3.8) of Chap. V then asserts that there exists a BM β such that

$$Z_a = 2\int_0^a Z_u^{1/2} d\beta_u + 2a;$$

in other words, Z is a $\mathrm{BES}Q^2(0)$ on $[0, 1]$. □

Remarks. 1°) This result may be extended to the local times of some diffusions, by using for example the method of time-substitution as described in Sect. 3 Chap. X (see Exercise (2.5)).

2°) The process Z_a is a positive submartingale, which bears out the intuitive feeling that L_t^a has a tendency to decrease with a.

3°) In the course of the proof, we had to show that Z_a is integrable. As it turns out, Z_a has moments of all orders, and is actually an exponential r.v. which was proved in Sect. 4 Chap. VI.

4°) Using the scaling properties of BM and $\mathrm{BES}Q^2$, the result may be extended to any interval $[0, c]$.

We now turn to the *second Ray-Knight theorem*. For $x > 0$, we set

$$\tau_x = \inf\{t : L_t^0 > x\}.$$

(2.3) Theorem. *The process* $L_{\tau_x}^a$, $a \geq 0$, *is a* $\mathrm{BES}Q^0(x)$.

Proof. Let g be a positive C^1-function with compact support contained in $]0, \infty[$ and F_g the unique positive decreasing solution to the equation $F'' = gF$ such that $F_g(0) = 1$ (See the discussion before Theorem (1.7) and Appendix 8). If $f(\lambda, x) = \exp\left(-(\lambda/2)F_g'(0)\right) F_g(x)$, Itô's formula implies that, writing L for L^0,

$$\begin{aligned} f\left(L_t, B_t^+\right) &= 1 + \int_0^t f_x'\left(L_s, B_s^+\right) 1_{(B_s>0)} dB_s + \frac{1}{2}\int_0^t f_x'\left(L_s, B_s^+\right) dL_s \\ &\quad + \frac{1}{2}\int_0^t f_{x^2}''\left(L_s, B_s^+\right) 1_{(B_s>0)} ds + \int_0^t f_\lambda'\left(L_s, B_s^+\right) dL_s. \end{aligned}$$

In the integrals with respect to dL_s, one can replace B_s^+ by 0, and since $\frac{1}{2}f_x'(\lambda, 0) + f_\lambda'(\lambda, 0) = 0$, the corresponding terms cancel. Thus, using the integration by parts formula, and by the choice made of F_g, it is easily seen that

$f\left(L_t, B_t^+\right)\exp\left(-\frac{1}{2}\int_0^t g(B_s)ds\right)$ is a local martingale. This local martingale is moreover bounded on $[0, \tau_x]$, hence by optional stopping,

$$E\left[\exp\left(-\frac{1}{2}F_g'(0)L_{\tau_x}\right)F_g\left(B_{\tau_x}^+\right)\exp\left(-\frac{1}{2}\int_0^{\tau_x} g(B_s)ds\right)\right] = 1.$$

But $B_{\tau_x} = 0$ a.s. since τ_x is an increase time for L_t (see Sect. 2 Chap. VI) and of course $L_{\tau_x} = x$, so the above formula reads

$$E\left[\exp\left(-\frac{1}{2}\int_0^{\tau_x} g(B_s)ds\right)\right] = \exp\left(\frac{x}{2}F_g'(0)\right).$$

By the occupation times formula, this may also be written

$$E\left[\exp\left(-\frac{1}{2}\int_0^{\infty} g(u)L_{\tau_x}^u du\right)\right] = \exp\left(\frac{x}{2}F_g'(0)\right).$$

If we now compare with Theorem (1.7), since g is arbitrary, the proof is finished.

Remarks. 1°) The second Ray-Knight theorem could also have been proved by using the same pattern of proof as for the first (see Exercise (2.8)) and vice-versa (see Exercise (2.7)).

2°) The law of the r.v. $L_{\tau_x}^a$ has also been computed in Exercise (4.14) Chap. VI. The present result is much stronger since it gives the law of the process.

We will now use the first Ray-Knight theorem to give a useful BDG-type inequality for local times, a proof of which has already been hinted at in Exercise (1.14) of Chap. X.

(2.4) Theorem. *For every $p \in]0, \infty[$, there exist two constants $0 < c_p < C_p < \infty$ such that for every continuous local martingale vanishing at 0,*

$$c_p E\left[(M_\infty^*)^p\right] \leq E\left[(L_\infty^*)^p\right] \leq C_p E\left[(M_\infty^*)^p\right]$$

where L^a is the family of local times of M and $L_t^* = \sup_{a\in\mathbb{R}} L_t^a$.

Proof. One can of course, thanks to the BDG inequalities, use $\langle M, M\rangle_\infty^{1/2}$ instead of M_∞^* in the statement or in its proof. The occupation times formula yields

$$\langle M, M\rangle_\infty = \int_{-\infty}^{+\infty} L_\infty^a da = \int_{-M^*}^{M^*} L_\infty^a da \leq 2M^*L^*.$$

Therefore, there exists a constant d_p such that

$$E\left[(M_\infty^*)^p\right] \leq d_p E\left[(M_\infty^*)^p\right]^{1/2} E\left[(L_\infty^*)^p\right]^{1/2};$$

if $E\left[(M_\infty^*)^p\right]$ is finite, which can always be achieved by stopping, we may divide by $E\left[(M_\infty^*)^p\right]^{1/2}$ and get the left-hand side inequality.

We now turn to the right-hand side inequality. If T_t is the time-change associated with $\langle M, M\rangle$, we know from Sect. 1 Chap. V that $B_{\cdot} = M_{T_{\cdot}}$ is a BM, possibly stopped at $S = \langle M, M\rangle_\infty$. By a simple application of the occupation times formula, $\left(L^a_{T_t}\right)$ is the family of local times, say $\left(l^a_t\right)$, of $B_{\cdot\wedge S}$, and therefore $L^*_\infty = l^*_S$. Consequently, it is enough to prove the right-hand side inequality whenever M is a stopped BM and we now address ourselves to this situation.

We set $\xi_n = \inf\{t : |B_t| = 2^n\}$. For $n = 0$, we have, since $\xi_0 = T_1 \wedge T_{-1}$,

$$E\left[L^*_{\xi_0}\right] = E\left[\sup_{-1\leq a\leq 1} L^a_{\xi_0}\right] \leq E\left[\sup_{0\leq a\leq 1} L^a_{T_1}\right] + E\left[\sup_{-1\leq a\leq 0} L^a_{T_{-1}}\right].$$

By the first Ray-Knight theorem, $L^{1-a}_{T_1} - 2a$ is a martingale with moments of every order; thus, by Theorem (1.7) in Chap. II the above quantity is finite; in other words, there is a constant K such that $E\left[L^*_{\xi_0}\right] = K$ and using the invariance under scaling of (B_t, L_t) (Exercise (2.11) Chap. VI), $E\left[L^*_{\xi_n}\right] = 2^n K$.

We now prove that the right-hand side inequality of the statement holds for $p = 1$. We will use the stopping time

$$T = \inf\{\xi_n : \xi_n \geq S\}$$

for which $B^*_S \leq B^*_T \leq 2B^*_S + 1$ (the 1 on the right is necessary when $T = \xi_o$). Plainly, $E\left[L^*_S\right] \leq E\left[L^*_T\right]$ and we compute this last quantity by means of the strong Markov property. Let m be a fixed integer; we have

$$J = E\left[L^*_{T\wedge\xi_m}\right] = E\left[\sum_{n=0}^{m-1}\left(L^*_{T\wedge\xi_{n+1}} - L^*_{T\wedge\xi_n}\right)\right].$$

Obviously, $\xi_{n+1} = \xi_n + \xi_{n+1}\circ\theta_{\xi_n}$ which by the strong additivity of local times (Proposition (1.2) Chap. X), is easily seen to imply that

$$L^*_{\xi_{n+1}} \leq L^*_{\xi_n} + L^*_{\xi_{n+1}}\circ\theta_{\xi_n}.$$

Therefore,

$$J \leq E\left[\sum_{n=0}^{m-1} 1_{(T>\xi_n)} L^*_{\xi_{n+1}}\circ\theta_{\xi_n}\right] = E\left[\sum_{n=0}^{m-1} 1_{(T>\xi_n)} E_{B_{\xi_n}}\left[L^*_{\xi_{n+1}}\right]\right].$$

Furthermore, if $|a| = 2^n$, then, under the law P_a, $\xi_{n+1}(B) = \xi_{n+2}(B-a)$; since on the other hand $L^*_t(B-a) = L^*_t(B)$, we get

$$\begin{aligned} J &\leq E\left[\sum_{n=0}^{m-1} 1_{(T>\xi_n)} E_0\left[L^*_{\xi_{n+2}}\right]\right] \\ &= E\left[\sum_{n=0}^{m-1} 2^{n+2} K\, 1_{(T>\xi_n)}\right] \leq 4K E\left[B^*_{T\wedge\xi_m}\right]. \end{aligned}$$

By letting m tend to infinity, we finally get $E\left[L^*_T\right] \leq 4K E\left[B^*_T\right]$ and as a result,

$$E\left[L_S^*\right] \leq E\left[L_T^*\right] \leq 8KE\left[B_S^*+1\right].$$

By applying this inequality to the Brownian motion $c^{-1}B_{c^2\cdot}$, and to the time $c^{-2}S$, one can check that

$$E\left[L_S^*\right] \leq 8K\left(E\left[B_S^*\right]+c^2\right).$$

Letting c tend to zero we get our claim in the case $p=1$, namely, going back to M,

$$(*) \qquad E\left[L_\infty^*(M)\right] \leq 8KE\left[M_\infty^*\right].$$

To complete the proof, observe that by considering for a stopping time S, the loc.mart. M_{S+t}, $t \geq 0$, we get from (*):

$$E\left[L_\infty^*(M)-L_S^*(M)\right] \leq 8KE\left[M_\infty^*\right].$$

By applying the Garsia-Neveu lemma of Exercise (4.29) Chap. IV, with $A_t = L_t^*$, $X = 8KM_\infty^*$ and $F(\lambda)=\lambda^p$ we get the result for all p's. □

* **(2.5) Exercise (Local times of BES^d).** 1°) Let l^a, $a \geq 0$, be the family of local times of BES$^3(0)$. Prove that the process l_∞^a, $a \geq 0$, is a BES$Q^2(0)$.

[Hint: Use the time-reversal result of Corollary (4.6) of Chap. VII.]

2°) For $p>0$ let λ^a be the family of local times of BES$^{2p+3}(0)$. Prove that λ_∞^a, $a \geq 0$, has the same law as the process $\left|V_p(a)\right|^2$, $a \geq 0$, where

$$V_p(a) = a^{-p}\int_0^a s^p d\beta_s$$

with β a BM^2.

[Hint: Use Exercise (1.19) in this chapter and Exercise (1.23) Chap. VI.]

3°) The result in 2°) may also be expressed: if l^a is the family of local times of BES$^d(0)$ with $d \geq 3$, then

$$\left(l_\infty^a, a \geq 0\right) \overset{(d)}{=} \left((d-2)^{-1}a^{d-1}U(a^{2-d}), a \geq 0\right)$$

where U is a BES$Q^2(0)$.

[Hint: Use property iv) in Proposition (1.10) Chap. I.]

4°) Let f be a positive Borel function on $]0,\infty[$ which vanishes on $[b,\infty[$ for some $b>0$ and is bounded on $[c,\infty[$ for every $c>0$. If X is the BES$^d(0)$ with $d \geq 3$, prove that

$$\int_0^1 f(X_s)ds < \infty \quad \text{a.s.} \quad \text{iff} \quad \int_0^b rf(r)dr < \infty.$$

[Hint: Apply the following lemma: let μ be a positive Radon measure on $]0,1]$; let $(V_r, r \in]0,1])$ be a measurable, strictly positive process such that there exists a bounded Borel function ϕ from $]0,1]$ to $]0,\infty[$ for which the law of $\phi(r)^{-1}V_r$, does not depend on r and admits a moment of order 1. Then

$$\int_0^1 V_r d\mu(r) < \infty \quad \text{a.s.} \quad \text{iff} \quad \int_0^1 \phi(r)d\mu(r) < \infty. \;]$$

N.B. The case of dimension 2 is treated in the following exercise.

* **(2.6) Exercise.** 1°) Let X be a $\mathrm{BES}^2(0)$, λ^a the family of its local times and $T_1 = \inf\{t : X_t = 1\}$. Prove that the process $\lambda^a_{T_1}$, $0 < a < 1$, has the same law as $aU_{-\log a}$, $0 < a < 1$, where U is $\mathrm{BES}Q^2(0)$.

[Hint: Use 1°) of the preceding exercise and the result in Exercise (4.12) of Chap. VII.]

2°) With the same hypothesis and by the same device as in 4°) of the preceding exercise, prove that

$$\int_0^1 f(X_s)ds < \infty \quad \text{a.s.} \quad \text{iff} \quad \int_0^1 r|\log r| f(r)dr < \infty.$$

Conclude that for the planar BM, there exist functions f such that $\int_0^t f(B_s)ds = +\infty$ P_0-a.s. for every $t > 0$ although f is integrable for the two dimensional Lebesgue measure. The import of this fact was described in Remark 4 after Theorem (3.12) Chap. X.

* **(2.7) Exercise. (Another proof of the first Ray-Knight theorem).** 1°) Let Z be the unique positive solution to the SDE

$$Z_t = 2\int_0^t \sqrt{Z_s}d\beta_s + 2\int_0^t 1_{(0\le s\le 1)}ds, \qquad Z_0 = 0.$$

Prove that the stopping time $\sigma = \inf\{t : Z_t = 0\}$ is a.s. finite and > 1.

2°) Let g be a positive continuous function on $\mathbb{R}$ with compact support and f the strictly positive, increasing solution to the equation $f'' = 2fg$ such that $f'(-\infty) = 0$, $f(0) = 1$. With the notation of Theorem (2.2) prove that

$$E\left[\exp\left(-\int_{-\infty}^1 g(a)L^a_{T_1}da\right)\right] = f(1)^{-1}.$$

3°) Set $v(x) = f(1-x)$ for $x \ge 0$; check that

$$v(a\wedge 1)^{-1}\exp\left(Z_a\frac{v'(a)}{2v(a)} - \int_0^a g(1-b)Z_b db\right)$$

is a local martingale and conclude that

$$E\left[\exp\left(-\int_0^\infty g(1-b)Z_b db\right)\right] = f(1)^{-1}.$$

4°) Prove that $L^{1-a}_{T_1}$, $a \ge 0$, has the same law as Z_a, $a \ge 0$, which entails in particular Theorem (2.2).

* **(2.8) Exercise. (Another proof of the second Ray-Knight theorem).** 1°) In the situation of Theorem (2.3), call $(\mathscr{Z}'_a)$ the right-continuous and complete filtration of $L^a_{\tau_x}$, $a \ge 0$. Prove that any variable H in $L^2(\mathscr{Z}'_a)$ may be written

$$H = E[H] + \int_0^{\tau_x} h_s 1_{(0<B_s<a)}dB_s$$

for a suitable h.

2°) Prove that $L^a_{\tau_x} - x$ is a $(\mathscr{Z}'_a)$-martingale and derive therefrom another proof of Theorem (2.3).

* **(2.9) Exercise. (Proof by means of the filtration of excursions).** Let B be the standard linear BM. For $x \in \mathbb{R}$, call τ^x_t the time-change inverse of $\int_0^t 1_{(B_s \le x)} ds$ and set $\mathscr{E}_x = \sigma\left(B_{\tau^x_t}, t \ge 0\right)$.

1°) Prove that $\mathscr{E}_x \subset \mathscr{E}_y$ for $x \le y$.

2°) Prove that, for $H \in L^2(\mathscr{E}_x)$, there exists a $(\mathscr{F}_t)$-predictable process h, such that $E\left[\int_0^\infty h_s^2 1_{(B_s \le x)} ds\right] < \infty$ and

$$H = E[H] + \int_0^\infty h_s 1_{(B_s \le x)} dB_s.$$

3°) For $x \in \mathbb{R}$, define $\mathscr{G}^x_t = \bigcap_{\varepsilon > 0} \sigma\left(\mathscr{F}_{t+\varepsilon}, \mathscr{E}_x\right)$. Prove that if Y is a $(\mathscr{F}_t)$-local martingale, the process $\int_0^t 1_{(B_s < x)} dY_s$ is a local martingale with respect to the filtration $\left(\mathscr{G}^x_t\right)$.

4°) Let $a \in \mathbb{R}$, T be a $\left(\mathscr{G}^a_t\right)$-stopping time such that $B_T \le a$ a.s. and L^a_T is $\mathscr{E}_a$-measurable. For $a \le x < y$, set

$$V_t = \frac{1}{2}\left(L^y_t - L^x_t\right) + y^- - x^- - (B_t - y)^+ + (B_t - x)^+$$

and show that

$$E\left[\langle V, V\rangle_T \mid \mathscr{G}^x_0\right] < +\infty.$$

[Hint: For $p > 0$, set $F(z) = \cosh\sqrt{2p}\,(y - \sup(z, x))^+$ and $2c = \sqrt{2p} \tanh\sqrt{2p}(y - x)$. The process $U_t = F(B_{t\wedge T}) \exp(-c(L^x_T - L^x_{t\wedge T}))\times \exp(-p\int_0^{t\wedge T} 1_{(x<B_s\le y)} ds)$ is a bounded $(\mathscr{G}^x_t)$-martingale.]

5°) Prove that $L^y_T + 2y^-$, $y \ge a$ is an $\left(\mathscr{E}_y\right)_{y\ge a}$-continuous martingale with increasing process $4\int_a^y L^z_T dz$. Derive therefrom another proof of the Ray-Knight theorems.

* **(2.10) Exercise (Points of increase of Brownian motion).** Let B be the standard linear BM.

1°) Let r be a positive number and Γ the set of ω's for which there exists a time $U(\omega) < r$ such that

$$B_t(\omega) < B_{U(\omega)}(\omega) \quad \text{for } 0 \le t < U(\omega), \qquad B_t(\omega) > B_{U(\omega)}(\omega) \quad \text{for } U(\omega) < t \le r.$$

Let x be a rational number and set $S_t = \sup_{s\le t} B_s$; prove that a.s. on $\Gamma \cap \{B_U < x < S_r\}$, we have $L^{B_U}_U = L^{B_U}_{T_x} > 0$ and that consequently Γ is negligible.

2°) If f is a continuous function on $\mathbb{R}_+$, a point t_0 is called a point of increase of f if there is an $\varepsilon > 0$ such that

$$f(t) \le f(t_0) \quad \text{for } t_0 - \varepsilon \le t < t_0, \qquad f(t) \ge f(t_0) \quad \text{for } t_0 < t \le t_0 + \varepsilon.$$

Prove that almost all Brownian paths have no points of increase. This gives another proof of the non-differentiability of Brownian paths.

[Hint: Use Exercise (2.15) in Chap. VI to replace, in the case of the Brownian path, the above inequalities by strict inequalities.]

3°) Prove that for any $T > 0$ and any real number x the set

$$\Lambda_T = \{t \leq T : B_t = x\}$$

is a.s. of one of the following four kinds: $\emptyset$, a singleton, a perfect set, the union of a perfect set and an isolated singleton.

* **(2.11) Exercise.** 1°) Derive the first Ray-Knight theorem from the result in Exercise (4.17) Chap. VI by checking the equality of the relevant Laplace transforms.

2°) Similarly, prove that the process $L^a_{T_1}$, $a \leq 0$, has the law of $(1-a)^2 X(((1-a)^{-1} - (1-m)^{-1})^+)$ where X is a $\mathrm{BES}Q^4$ and m a r.v. on $]-\infty, 0[$ independent of X with density $(1-x)^{-2}$.

[Hint: For the law of m, see Proposition (3.13) i) in Chap. VI.]

* **(2.12) Exercise.** Let B be the standard linear BM, L^a the family of its local times and set $Y_t = L_t^{B_t}$.

1°) If $\mathscr{G}_s^{(t)} = \mathscr{F}_s^B \vee \sigma(B_t)$, prove that for each t

$$\frac{1}{2} Y_t = B_t \wedge 0 - \int_0^t 1_{(B_s > B_t)} dB_s$$

where the stochastic integral is taken in the filtration $\left(\mathscr{G}^{(t)}\right)$ (see Exercise (1.39) Chap. IV).

2°) Prove that for every t,

$$\sup_\tau E\left[\sum_\tau \left(Y_{s_{i+1}} - Y_{s_i}\right)^4\right] < \infty$$

where $\tau = (s_i)$ ranges through the finite subdivisions of $[0, t]$.

[Hint: Use the decomposition of B in the filtration $\left(\mathscr{G}^{(t)}\right)$ and the BDG inequalities for local martingales and for local times.]

(2.13) Exercise. Retain the notation of the preceding exercise and let S be a $\left(\mathscr{F}_t^B\right)$-stopping time such that the map $x \to L_S^x$ is a semimart.. Prove that if ϕ is continuous, then, in the notation of Exercise (1.35) Chapter VI,

$$P\text{-}\lim_{n\to\infty} \sum_{\Delta_n} \left(\int_0^S \phi(Y_u) dL_u^{a_{i+1}} - \int_0^S \phi(Y_u) dL_u^{a_i}\right)^2 = 4 \int_a^b L_S^x \phi\left(L_S^x\right)^2 dx.$$

By comparison with the result of Exercise (1.35) Chap. VI, prove that Y is not a $\left(\mathscr{F}_t^B\right)$-semimart..

[Hint: Use the result in 3°) Exercise (1.33) Chapter IV.]

(2.14) Exercise. (Time asymptotics via space asymptotics. Continuation to Exercise (1.32)) Prove the result of Exercise (3.20) Chapter X in the case $d = 3$, by considering the expression

$$\left(\log \sqrt{u}\right)^{-1} \int_0^\infty |B_s|^{-2} 1_{\{1 \leq |B_s| \leq \sqrt{u}\}} ds.$$

[Hint: Use the Ray-Knight theorems for the local times of $\mathrm{BES}^3(0)$ as described in Exercise (2.5).]

§3. Bessel Bridges

In this section, which will not be needed in the sequel save for some definitions, we shall extend some of the results of Sect. 1 to the so-called Bessel Bridges. We take $\delta > 0$ throughout.

For any $a > 0$, the space $\mathbf{W}_a = C([0,a],\mathbb{R})$ endowed with the topology of uniform convergence is a Polish space and the σ-algebra generated by the coordinate process X is the Borel σ-algebra (see Sect. 1 in Chap. XIII). As a result, there is a regular conditional distribution for $P_x^\delta[\cdot \mid X_a]$, namely a family $P_{x,y}^{\delta,a}$ of probability measures on $\mathbf{W}_a$ such that for any Borel set Γ

$$P_x^\delta[\Gamma] = \int P_{x,y}^{\delta,a}(\Gamma)\mu_a(dy)$$

where μ_a is the law of X_a under P_x^δ. Loosely speaking

$$P_{x,y}^{\delta,a}[\Gamma] = P_x^\delta[\Gamma \mid X_a = y].$$

For fixed x, δ and a, these transition probabilities are determined up to sets of measure 0 in y; but we can choose a version by using the explicit form found in Sect. 1 for the density p_t^δ. For $y > 0$, we may define $P_{x,y}^{\delta,a}$ by saying that for $0 < t_1 < \ldots < t_n < a$, the law of $\left(X_{t_1}, \ldots, X_{t_n}\right)$ under $P_{x,y}^{\delta,a}$ is given by the density

$$p_{t_1}^\delta(x, x_1)p_{t_2-t_1}^\delta(x_1, x_2)\ldots p_{a-t_n}^\delta(x_n, y)/p_a^\delta(x, y)$$

with respect to $dx_1dx_2\ldots dx_n$. This density is a continuous function of y on $\mathbb{R}_+\backslash\{0\}$. Moreover, since $I_\nu(z)$ is equivalent for small z to $c_\nu z^\nu$ where c_ν is a constant, it is not hard to see that these densities have limits as $y \to 0$ and that the limits themselves form a projective family of densities for a probability measure which we call $P_{x,y}^{\delta,a}$. From now on, $P_{x,y}^{\delta,a}$ will always stand for this canonical system of probability distributions. Notice that the map $(x, y) \to P_{x,y}^{\delta,a}$ is continuous in the weak topology on probability measures which is introduced in Chap. XIII.

The same analysis may be carried through with Q_x^δ instead of P_x^δ and leads to a family $Q_{x,y}^{\delta,a}$ of probability measures; thus, we lay down the

(3.1) Definition. *A continuous process, the law of which is equal to* $P_{x,y}^{\delta,a}$ *(resp.* $Q_{x,y}^{\delta,a}$*) is called the* Bessel Bridge *(resp.* Squared Bessel Bridge*) from x to y over* $[0,a]$ *and is denoted by* $\mathrm{BES}_a^\delta(x,y)$ *(resp.* $\mathrm{BES}Q_a^\delta(x,y)$*).*

All these processes are inhomogeneous Markov processes; one may also observe that the square of $\mathrm{BES}_a^\delta(x,y)$ is $\mathrm{BES}Q_a^\delta(x^2,y^2)$.

Of particular interest in the following chapter is the case of $\mathrm{BES}_a^3(0,0)$. In this case, since we have explicit expressions for the densities of BES^3 which are given in Sect. 3 of Chap. VI, we may compute the densities of $\mathrm{BES}_a^3(0,0)$ without having to refer to the properties of Bessel functions. Let us put $l_t(y) = (2\pi t^3)^{-1/2} y \exp\left(-(y^2/2t)\right)1_{(y>0)}$ and call q_t the density of the semigroup of BM killed at 0. If $0 < t_1 < t_2 < \ldots < t_n < a$, by the results in Sect. 3 Chap. VI, the density of $(X_{t_1}, \ldots, X_{t_n})$ under the law $P^{3,a}_{0,z}$ is equal to

$$l_{t_1}(y_1)q_{t_2-t_1}(y_1, y_2)\ldots q_{a-t_n}(y_n, z)/l_a(z).$$

Letting z converge to zero, we get the corresponding density for $P^{3,a}_{0,0}$, namely

$$2\left(2\pi a^3\right)^{1/2} l_{t_1}(y_1)q_{t_2-t_1}(y_1, y_2)\ldots q_{t_n-t_{n-1}}(y_{n-1}, y_n)l_{a-t_n}(y_n).$$

We aim at extending Theorem (1.7) to $\mathrm{BES}Q^\delta_a(x, y)$.

(3.2) Theorem. *Let μ be a measure with support in* $[0, 1]$. *There exist three constants A, $\widehat{A}$, B depending only on μ, such that*

$$Q^{\delta,1}_{x,y}\left[\exp\left(-\frac{1}{2}X_\mu\right)\right] = A^x \widehat{A}^y B^2 I_\nu\left(\sqrt{xy}B^2\right)/I_\nu\left(\sqrt{xy}\right).$$

Proof. We retain the notation used in the proof of Theorem (1.7) and define a probability measure $R^{\delta,\mu}_x$ on $\mathscr{F}_1 = \sigma(X_s, s \le 1)$ by $R^{\delta,\mu}_x = Z^\mu_1 \cdot Q^\delta_x$. The law of X_t under $R^{\delta,\mu}_x$ has a density $r^{\delta,\mu}_t(x, \cdot)$ which we propose to compute. By an application of Girsanov's theorem, under $R^{1,\mu}_x$, the coordinate process X is a solution to the SDE

$$\text{Eq. (3.1)} \qquad X_t = x + 2\int_0^t \sqrt{X_s}d\beta_s + 2\int_0^t F_\mu(s)X_s ds + t.$$

If H is a solution to

$$\text{Eq. (3.2)} \qquad H_t = u + B_t + \int_0^t F_\mu(s)H_s ds,$$

then H^2 is a solution to Eq. (3.1) with $x = u^2$ and $\beta_t = \int_0^t(\operatorname{sgn} H_s)dB_s$. But Eq. (3.2) is a linear equation the solution of which is given by Proposition (2.3) Chap. IX. Thus, H_t is a Gaussian r.v. with mean $um(t)$ and variance $\sigma^2(t)$ where

$$m(t) = \phi_\mu(t), \qquad \sigma^2(t) = \phi_\mu(t)^2\int_0^t \phi_\mu(s)^{-2}ds.$$

If we recall that $q^1_t(x, \cdot)$ is the density of the square of a Gaussian r.v. centered at $\sqrt{x}$ and with variance t, we see that we may write

$$r^{1,\mu}_t(x, \cdot) = q^1_{\sigma^2(t)}\left(xm^2(t), \cdot\right).$$

Furthermore, it follows from Theorem (1.2) that

$$R^{\delta,\mu}_x * R^{\delta',\mu}_{x'} = R^{\delta+\delta',\mu}_{x+x'};$$

as a result

$$r^{\delta,\mu}_t(x, \cdot) = q^\delta_{\sigma^2(t)}\left(xm^2(t), \cdot\right).$$

We turn to the proof of the theorem. For any Borel function $f \ge 0$,

$$\int Q^{\delta,1}_{x,y}\left[\exp\left(-\frac{1}{2}X_\mu\right)\right] f(y)q^\delta_1(x,y)dy$$
$$= R^{\delta,\mu}_x\left[\left(Z^\mu_1\right)^{-1}\exp\left(-\frac{1}{2}X_\mu\right)f(X_1)\right]$$
$$= \exp\left\{\frac{1}{2}\left(F_\mu(0)x+\delta\log\phi_\mu(1)\right)\right\}R^{\delta,\mu}_x\left[f(X_1)\right]$$

since $\phi'_\mu(1)=0$. Consequently, for Lebesgue almost every y,

$$Q^{\delta,1}_{x,y}\left[\exp\left(-\frac{1}{2}X_\mu\right)\right]$$
$$= \exp\left\{\frac{1}{2}\left(F_\mu(0)x+\delta\log\phi_\mu(1)\right)\right\}r^{\delta,\mu}_1(x,y)/q^\delta_1(x,y)$$
$$= \exp\left\{\frac{1}{2}\left(F_\mu(0)x+\delta\log\phi_\mu(1)\right)\right\}q^\delta_{\sigma^2(1)}\left(xm^2(1),y\right)/q^\delta_1(x,y).$$

Using the explicit expressions for $\sigma^2(1)$ and $m^2(1)$ and the value of q^δ_1 found in Sect. 1, we get the desired result for a.e. y and by continuity for every y. □

In some cases, one can compute the above constants. We thus get

(3.3) Corollary. *For every $b\geq 0$*

$$Q^{\delta,1}_{x,y}\left[\exp\left(-\frac{b^2}{2}\int_0^1 X_s ds\right)\right]$$
$$= (b/\sinh b)\exp\left\{\left(\frac{x+y}{2}\right)(1-b\coth b)\right\}I_\nu\left(b\sqrt{xy}/\sinh b\right)/I_\nu\left(\sqrt{xy}\right).$$

In particular,

$$Q^{\delta,1}_{x,0}\left[\exp\left(-\frac{b^2}{2}\int_0^1 X_s ds\right)\right] = Q^{\delta,1}_{0,x}\left[\exp\left(-\frac{b^2}{2}\int_0^1 X_s ds\right)\right]$$
$$= \left(\frac{b}{\sinh b}\right)^{\delta/2}\exp\left(\frac{x}{2}(1-b\coth b)\right).$$

Proof. The proof is patterned after Corollary (1.8). The details are left to the reader. □

We are now going to extend Corollary (1.12) to Bessel Bridges. We will need the following

(3.4) Lemma. *Let X be a real-valued Markov process, g a positive Borel function such that $\int_0^\infty g(X_s)^{-1}ds=\infty$. If we set*

$$C_t=\int_0^t g(X_s)^{-1}ds,\qquad \widehat{X}_t=X_{\widehat{C}_t},$$

where $\widehat{C}_t$ is the time-change associated with C, then $d\widehat{C}_t = g(\widehat{X}_t)dt$. Moreover, if we assume the existence of the following densities

$$p_t(x, y) = P_x[X_t \in dy]/dy, \quad \widehat{p}_t(x, y) = P_x[\widehat{X}_t \in dy]/dy,$$

$$h_{x,y}(t, u) = P_x[C_t \in du \mid X_t = y]/du, \quad \widehat{h}_{x,y}(t, u) = P_x[\widehat{C}_t \in du \mid \widehat{X}_t = y]/du,$$

then $dt\, du\, dy$-a.e.

$$\widehat{p}_u(x, y)g(y)\widehat{h}_{x,y}(u, t) = p_t(x, y)h_{x,y}(t, u).$$

Proof. The first sentence has already been proved several times. To prove the second, pick arbitrary positive Borel functions ϕ, $\widehat{\phi}$ and f; we have

$$E_x\left[\int_0^\infty dt\widehat{\phi}(t)f(X_t)\phi(C_t)\right]$$
$$= \int_0^\infty dt\widehat{\phi}(t)\int p_t(x, y)f(y)dy\int \phi(u)h_{x,y}(t, u)du,$$

but on the other hand, using the time-change formulas the same expression is equal to

$$E_x\left[\int_0^\infty dt\widehat{\phi}(\widehat{C}_t)f(\widehat{X}_t)g(\widehat{X}_t)\phi(t)\right]$$
$$= \int_0^\infty dt\phi(t)\int \widehat{p}_t(x, y)g(y)f(y)dy\int \widehat{\phi}(u)\widehat{h}_{x,y}(t, u)du.$$

Interchanging the roles played by t and u in the last expression and comparing with that given above yields the result.

We may now state

(3.5) Theorem. *For $\nu > 0$, p and q two conjugate numbers > 1 and $\rho_\nu(t)$, $0 \leq t \leq 1$ a $\mathrm{BES}_1^{(\nu)}(0, 0)$, we put*

$$x_{\nu,p} = \int_0^1 \rho_\nu(s)^{-2/p}ds, \qquad y_{\nu,p} = q^{2/p}\left(\int_0^1 \rho_{\nu q}^{2q/p}(s)ds\right)^{-1/q}.$$

Then, with $\lambda_\nu = (2^\nu\Gamma(\nu + 1))^{-1}$, for every positive Borel function f

$$\lambda_\nu E\left[f\left(x_{\nu,p}\right)\left(q^{-2}x_{\nu,p}\right)^{\nu q}\right] = \lambda_{\nu q}E\left[f\left(y_{\nu,p}\right)\right].$$

Proof. We use Lemma (3.4) with $g(x) = x^{2/p}$ and X a $\mathrm{BES}^{(\nu)}(0)$. Using Proposition (1.11), we see that

$$\widehat{X}_t = X_{\widehat{C}_t} = \left(q^{-1}R_{\nu q}(t)\right)^q$$

where $R_{\nu q}$ is a BES$^{(\nu q)}$. It follows from the first sentence of Lemma (3.4) that

$$\widehat{C}_t = \int_0^t g(\widehat{X}_s)ds = q^{-2q/p} \int_0^t R_{\nu q}(s)^{2q/p} ds.$$

By a straightforward change of variable, the density $\widehat{p}_t(0, y)$ of $\widehat{X}_t$ may be deduced from $p_t^{(\nu q)}(0, y)$ and is found to be equal to

$$\lambda_{\nu q} t^{-(\nu q+1)} q^{2\nu+1} y^{2\nu+(2/q)-1} \exp\left(-\left(q y^{1/q}\right)^2 / 2t\right).$$

If we write the equality of Lemma (3.4) for $u = 1$ and $x = 0$, we obtain

$$\lambda_{\nu q} q^{2\nu+1} \exp\left(-\left(q y^{1/q}\right)^2 / 2t\right) \widehat{h}_{0y}(1, t) = \lambda_\nu t^{-(\nu+1)} \exp\left(-y^2/2t\right) h_{0y}(t, 1)$$

because of the cancellation of the powers of y. If we further make $y = 0$, we get

$$\lambda_{\nu q} q^{2\nu+1} \widehat{h}_{00}(1, t) = \lambda_\nu t^{-(\nu+1)} h_{00}(t, 1).$$

Next, by the definition of h_{xy} and the scaling properties of X seen in Proposition (1.10), one can prove that

$$h_{00}(t, 1) = t^{-1/q} h_{00}(1, t^{-1/q});$$

this is left to the reader as an exercise. As a result

$$\lambda_{\nu q} q^{2\nu+1} \widehat{h}_{00}(1, t) = \lambda_\nu t^{-(\nu+1+(1/q))} h_{00}(1, t^{-1/q}).$$

Now, $\widehat{h}_{00}(t, 1)$ is the density of $y_{\nu,p}^{-q}$ and $h_{00}(1, t)$ is that of $x_{\nu,p}$. Therefore if f is a positive Borel function

$$\begin{aligned}
\lambda_{\nu q} E\left[f\left(y_{\nu,q}\right)\right] &= \lambda_{\nu q} \int f\left(t^{-1/q}\right) \widehat{h}_{00}(1, t)dt \\
&= \lambda_\nu q^{-(2\nu+1)} \int f\left(t^{-1/q}\right) t^{-(\nu+1+(1/q))} h_{00}\left(1, t^{-1/q}\right) dt.
\end{aligned}$$

Making the change of variable $t = u^{-q}$, this is further equal to

$$\begin{aligned}
\lambda_\nu q^{-(2\nu+1)} \int f(u) u^{\nu q} h_{00}(1, u) q \, du &= \lambda_\nu \int f(u) \left(u/q^2\right)^{\nu q} h_{00}(1, u)du \\
&= \lambda_\nu E\left[f\left(x_{\nu,p}\right)\left(x_{\nu,p}/q^2\right)^{\nu q}\right]
\end{aligned}$$

as was to be proved.

(3.6) Exercise. 1°) Prove that under Q_x^δ, the process

$$(1-u)^2 X_{u/(1-u)}, \qquad 0 \le u < 1,$$

has the law $Q_{x,0}^{\delta,1}$.

2°) Prove that

$$Q_{x,0}^{\delta,1} * Q_{x',0}^{\delta',1} = Q_{x+x',0}^{\delta+\delta',1}.$$

(3.7) Exercise. Let $\widetilde{P}$ be the law of X_{1-t}, $0 \le t \le 1$, when X has law P. Prove that

$$Q^{\delta,1}_{x,y} = \widetilde{Q}^{\delta,1}_{y,x}.$$

* **(3.8) Exercise.** State and prove a relationship between Bessel Bridges of integer dimensions and the modulus of Brownian Bridges. As a result, the corresponding Bessel Bridges are semimartingales.

(3.9) Exercise. If ρ is the $\mathrm{BES}^3_1(0,0)$, prove that

$$E\left[\int_0^1 \rho_s^{-1} ds\right] = \sqrt{2\pi}.$$

(3.10) Exercise. For a fixed $a > 0$ and any x, y, δ, t such that $t > a$, prove that $P^{\delta,t}_{x,y}$ has a density $Z^{(t)}_a$ on the σ-algebra $\mathscr{F}_a = \sigma(X_s, s \le a)$ with respect to P^δ_x. Show that as $t \to \infty$, $\left(Z^{(t)}_a\right)$ converges to 1 pointwise and in $L^1\left(P^\delta_x\right)$, a fact which has an obvious intuitive interpretation. The same result could be proved for the Brownian Bridge.

[Hint: Use the explicit form of the densities.]

* **(3.11) Exercise. (Stochastic differential equation satisfied by the Bessel Bridges).**

1°) Prove that for $t < 1$,

$$P^{\delta,1}_{0,0} = h(t, X_t) P^\delta_0 \quad \text{on } \mathscr{F}_t,$$

where $h(t,x) = \lim_{y\to 0} p^\delta_{1-t}(x,y)/p^\delta_1(0,y) = (1-t)^{-\delta/2}\exp\left(-x^2/2(1-t)\right)$.

[Hint: Compute $E_{P^\delta_0}\left[F(X_u, u \le t)\phi(X_1)\right]$ by conditioning with respect to $\sigma(X_1)$ and with respect to $\mathscr{F}_t$.]

2°) Prove that for $\delta \ge 2$, the Bessel Bridge between 0 and 0 over $[0,1]$ is the unique solution to the SDE

$$X_t = B_t + \int_0^t \left(\frac{\delta - 1}{2X_s} - \frac{X_s}{1-s}\right) ds, \qquad X_0 = 0.$$

[Hint: Use Girsanov's theorem.]

3°) Prove that the squared Bessel Bridge is the unique solution to the SDE

$$X_t = 2\int_0^t \sqrt{X_s} dB_s + \int_0^t \left(\delta - \frac{2X_s}{1-s}\right) ds.$$

(3.12) Exercise. 1°) Prove that the Bessel processes satisfy the hypothesis of Exercise (1.12) Chap. X.

2°) Prove that for $\delta > 2$,

$$P^{\delta,a}_{x,y} = \left(P^\delta_x\right)^{L_y}\left[\cdot \mid L_y = a\right],$$

where $L_y = \sup\{t : X_t = y\}$ and the notation P^{L_y} is defined in Sect. 4 of Chap. XII.

[Hint: Use Exercise (4.16) Chap. VII.]

Notes and Comments

Sect. 1. The systematic study of Bessel processes was initiated in Mc Kean [1]; beside their interest in the study of Brownian motion (see Chaps. VI and VII) they afford basic examples of diffusions and come in handy for testing general conjectures. Theorem (1.2) is due to Shiga-Watanabe [1]; actually, these authors characterize the family of diffusion processes which possess the additivity property of Theorem (1.2) (in this connection, see Exercise (1.13)). Corollaries (1.3) and (1.4) are from Molchanov [1]. An explicit Lévy-Khintchine formula for Q_x^δ is given in Exercise (4.21), Chap. XII. Exercise (1.14) presents further examples of infinitely divisible laws on $\mathbf{W} = C(\mathbb{R}_+, \mathbb{R})$. The following question arises naturally

Question 1. Under which condition on the covariance $\delta(s,t)$ of a continuous Gaussian process $(X_t, t \geq 0)$ is the distribution of $(X_t^2, t \geq 0)$ infinitely divisible?

Some answers are given by Eisenbaum [4].

Theorem (1.7) is taken from Pitman-Yor [3] and Corollary (1.8) is in Lévy [3]. Proposition (1.11) is in Biane-Yor [1] where these results are also discussed at the Itô's excursions level and lead to the computation of some of the laws associated with the Hilbert transform of Brownian local times, the definition of which was given in Exercise (1.29), Chap. VI.

Exercise (1.17) is due to Calais-Génin [1]. Exercise (1.22) is from Yor [10] and Pitman-Yor [1]; Gruet [3] develops similar results for the family of hyperbolic Bessel processes, mentioned in Chap. VIII, Sect. 3, Case 3.

The relationship described in this exercise together with the related Hartman-Watson distribution (see Yor [10]) have proven useful in several questions: the shape of random triangles (Kendall [1]), flows of Bessel processes (Hirsch-Song [1]), the study of SLE processes (Werner [1]).

For Exercise (1.25) see Molchanov-Ostrovski [1] who identify the distribution of the local time at 0 for a BES^δ, $\delta < 2$ as the Mittag-Leffler distribution; their result may be recovered from the result in question 2°).

Exercise (1.26) may serve as a starting point for the deep study of BES^δ, $0 < \delta < 1$, by Bertoin [6] who shows that $(0,0)$ is regular for the 2-dimensional process (ρ, k) and develops the corresponding excursion theory.

Exercise (1.28) is a particular example of the relationship between exponentials of Lévy processes and semi-stable processes studied by Lamperti ([1], [2]) who showed that (powers of) Bessel processes are the only semi-stable one-dimensional diffusions. Several applications of Lamperti's result to exponential functionals of Brownian motion are found in Yor ([27], [28]); different applications to Cauchy principal values are found in Bertoin [9].

The result presented in Exercise (1.29) is a particularly striking case of the extensions of Pitman's theorem to transient diffusions obtained by Saisho-Tanemura [1]; see also Rauscher [1], Takaoka [1] and Yor [23].

Exercise (1.33) is taken from A. Goïng's thesis [1]. Exercise (1.34) is taken partly from Pitman-Yor [3], partly from Föllmer-Wu-Yor [1].

Bessel processes and their squares have undergone a series of generalizations, especially with matrix-valued diffusion processes, leading for example to Wishart processes (Bru [1]–[3], Donati-Martin et al. [1]). Pitman's theorem has been generalized within this framework (Bougerol-Jeulin [1], O'Connell-Yor [1]) and shown to be closely connected with some important results of Littleman in Representation theory (Biane-Bougerol-O'Connell [1]).

Sect. 2. The Ray-Knight theorems were proved independently in Ray [1] and Knight [2]. The proof given here of the first Ray-Knight theorem comes from Jeulin-Yor [1] as well as the second given in Exercise (2.8). The proof in Exercise (2.7) is due to McGill [1] and provides the pattern for the proof of the second Ray-Knight theorem given in the text. Another proof is given in Exercise (2.9) which is based on Walsh [2] (see also Jeulin [3]). Exercise (2.11) is from Ray [1].

The proof of Theorem (2.4) is the original proof of Barlow-Yor [1], its difficulty stemming from the fact that, with this method, the integrability of L_1^* has to be taken care of. Actually, Bass [2] and Davis [5] proved that much less was needed and gave a shorter proof which is found in Exercise (1.14) of Chap. X. See Gundy [1] for some related developments in analysis.

Exercises (2.5) and (2.6) are from Williams [3] and Le Gall [3], the lemma in the Hint of (2.5) 4°) being from Jeulin ([2] and [4]) and the final result of (2.6) from Pitman-Yor [5]. The result of Exercise (2.10) is due to Dvoretsky et al [1] and Exercise (2.12) to Yor [12].

Although it is not discussed in this book, there is another approach, the so-called Dynkin's isomorphism theorem (Dynkin [3]), to Ray-Knight theorems (see Sheppard [1], Eisenbaum [2], [3], [5]). A number of consequences for local times of Lévy processes have been derived by Marcus and Rosen ([1], [2]).

Exercise (2.13) comes from discussions with B. Toth and W. Werner. Another proof, related to the discussion in Exercise (2.12), of the fact that $L_t^{B_t}$ is not a semimartingale has been given by Barlow [1]. The interest in this question stems from the desire to extend the Ray-Knight theorems to more general classes of processes (for instance diffusions) and possibly get processes in the space variables other than squares of Bessel or OU processes.

Sect. 3. For the results of this section, see Pitman-Yor ([1] and [2]) and Biane-Yor [1].

Chapter XII. Excursions

§1. Prerequisites on Poisson Point Processes

Throughout this section, we consider a measurable space $(U, \mathscr{U})$ to which is added a point δ and we set $U_\delta = U \cup \{\delta\}$, $\mathscr{U}_\delta = \sigma(\mathscr{U}, \{\delta\})$.

(1.1) Definition. *A process* $e = (e_t, t > 0)$ *defined on a probability space* $(\Omega, \mathscr{F}, P)$ *with values in* $(U_\delta, \mathscr{U}_\delta)$ *is said to be a* point process *if*

i) the map $(t, \omega) \to e_t(\omega)$ *is* $\mathscr{B}(]0, \infty[) \otimes \mathscr{F}$*-measurable;*
ii) the set $D_\omega = \{t : e_t(\omega) \neq \delta\}$ *is a.s. countable.*

The statement ii) means that the set $\{\omega : D_\omega$ is not countable$\}$ is contained in a P-negligible $\mathscr{F}$-measurable set.

Given a point process, with each set $\Gamma \in \mathscr{U}_\delta$, we may associate a new point process e^Γ by setting $e_t^\Gamma(\omega) = e_t(\omega)$ if $e_t(\omega) \in \Gamma$, $e_t^\Gamma(\omega) = \delta$ otherwise. The process e^Γ is the trace of e on Γ. For a measurable subset Λ of $]0, \infty[\times U$, we also set

$$N^\Lambda(\omega) = \sum_{t>0} 1_\Lambda(t, e_t(\omega)).$$

In particular, if $\Lambda =]0, t] \times \Gamma$, we will write N_t^Γ for N^Λ; likewise $N_{]s,t]}^\Gamma = \sum_{s<u\leq t} 1_\Gamma(e_u)$.

(1.2) Definition. *A point process is said to be* discrete *if* $N_t^U < \infty$ *a.s. for every* t. *The process* e *is* σ-discrete *if there is a sequence* (U_n) *of sets, the union of which is* U *and such that each* e^{U_n} *is discrete.*

If the process is σ*-discrete, one can prove, and we will assume, that all the* N^Λ*'s are random variables.*

Let us now observe that the Poisson process defined in Exercise (1.14) of Chap. II is the process N^U associated with the point process obtained by making $U = \mathbb{R}_+$ and $e_t(\omega) = t$ if there is an n such that $S_n(\omega) = t$, $e_t(\omega) = \delta$ otherwise. More generally we will set the

(1.3) Definition. *Let* $(\Omega, \mathscr{F}, \mathscr{F}_t, P)$ *be a filtered probability space. An* $(\mathscr{F}_t)$-Poisson process N *is a right-continuous adapted process, such that* $N_0 = 0$, *and for every* $s < t$, *and* $k \in \mathbb{N}$,

$$P\left[N_t - N_s = k \mid \mathscr{F}_s\right] = c^k \frac{(t-s)^k}{k!} \exp\left(-c(t-s)\right)$$

for some constant $c > 0$ called the parameter *of N. We set $\Delta N_t = N_t - N_{t-}$.*

We see that in particular, the Poisson process of Exercise (1.14) Chap. II is an $(\mathscr{F}_t)$-Poisson process for its natural filtration $(\mathscr{F}_t) = (\sigma(N_s, s \le t))$. We have moreover the

(1.4) Proposition. *A right-continuous adapted process is an $(\mathscr{F}_t)$-Poisson process if and only if it is a Lévy process which increases only by jumps a.s. equal to* 1.

Proof. Let N be an $(\mathscr{F}_t)$-Poisson process. It is clear from the definition that it is an integer-valued Lévy process and that the paths are increasing. As a result, for any fixed T, the set of jumps on $[0, T]$ is a.s. finite and consequently,

$$\sup_{0 \le t \le T} (N_t - N_{t-}) = \lim_n \max_{1 \le k \le n} \left(N_{kT/n} - N_{(k-1)T/n}\right) \quad \text{a.s.};$$

but

$$P\left[\max_{1 \le k \le n} \left(N_{kT/n} - N_{(k-1)T/n}\right) \le 1\right]$$
$$= P\left[N_{T/n} \le 1\right]^n = \left(e^{-cT/n}(1 + cT/n)\right)^n$$

which goes to 1 as n tends to infinity. Hence the jumps of N are a.s. of magnitude equal to 1.

Conversely, if N is a Lévy process which increases only by jumps of magnitude 1, the right-continuity of paths entails that each point is a holding point; moreover because of the space homogeneity, the times of jumps are independent exponential r.v.'s with the same parameter, hence the increments $N_t - N_s$ are Poisson r.v.'s. □

Here are a few more properties of the Poisson process, the proofs of which are left to the reader.

Firstly, recall from Exercise (1.14) Chapter II that $M_t = N_t - ct$ is a $(\mathscr{F}_t)$-martingale and $M_t^2 - ct$ is another one. More generally, the paths of N being increasing and right-continuous, the path by path Stieltjes integral with respect to $dN_s(\omega)$ is meaningful. Thus, if Z is a *predictable* process such that

$$E\left[\int_0^t |Z_s| ds\right] < \infty \qquad \text{for every } t,$$

then

$$\int_0^t Z_s dN_s - c \int_0^t Z_s ds = \int_0^t Z_s dM_s$$

is an $(\mathscr{F}_t)$-martingale. This may be seen by starting with elementary processes and passing to the limit. The property does not extend to *optional* processes (see

Exercise (1.14) below). Further, as for BM, we also get exponential martingales, namely, if $f \in L^1_{\text{loc}}(\mathbb{R}_+)$, then

$$L_t^f = \exp\left\{i\int_0^t f(s)dN_s - c\int_0^t \left(e^{if(s)} - 1\right)ds\right\}$$

is a $(\mathscr{F}_t)$-martingale. Again, this may be proved by starting with step functions and passing to the limit. It may also be proved as an application of the "chain rule" formula for Stieltjes integrals seen in Sect. 4 of Chap. 0.

Finally, because the laws of the jump times are diffuse, it is easily seen that for any Poisson process and every fixed time t, $\Delta N_t = N_t - N_{t-}$ is a.s. zero (this follows also from the general remark after Theorem (2.7) Chapter III).

(1.5) Proposition. *If N^1 and N^2 are two independent Poisson processes, then*

$$\sum_{s>0} \left(\Delta N_s^1\right)\left(\Delta N_s^2\right) = 0 \qquad \text{a.s.;}$$

in other words, the two processes almost surely do not jump simultaneously.

Proof. Let T_n, $n \geq 1$, be the successive jump times of N^1. Then

$$\sum_{s>0} \left(\Delta N_s^1\right)\left(\Delta N_s^2\right) = \sum_{n>0} \left(\Delta N_{T_n}^2\right) \qquad \text{a.s..}$$

Since $\Delta N_t^2 = 0$ a.s. for each t, by the independence of N^2 and the T_n's, we get $\Delta N_{T_n}^2 = 0$ a.s. for every n, which completes the proof. □

We now generalize the notion of Poisson process to higher dimensions.

(1.6) Definition. *A process $\left(N^1, \ldots, N^d\right)$ is a d-dimensional $(\mathscr{F}_t)$-Poisson Process if each N^i is a right-continuous adapted process such that $N_0^i = 0$ and if there exist constants c_i such that for every $t \geq s \geq 0$,*

$$P\left[\bigcap_{i=1}^d \left\{N_t^i - N_s^i = k_i\right\} \,\middle|\, \mathscr{F}_s\right] = \prod_{i=1}^d \exp\left(-c_i(t-s)\right)\frac{\left(c_i(t-s)\right)^{k_i}}{k_i!}.$$

By Proposition (1.5), no two components N^i and N^j jump simultaneously. We now work in the converse direction.

(1.7) Proposition. *An adapted process $N = \left(N^1, \ldots, N^d\right)$ is a d-dimensional $(\mathscr{F}_t)$-Poisson process if and only if*

i) each N^i is an $(\mathscr{F}_t)$-Poisson process,
ii) no two N^i's jump simultaneously.

Proof. We need only prove the sufficiency, i.e. that the r.v.'s $N_t^i - N_s^i$, $i = 1, \ldots, d$, are independent. For clarity's sake, we suppose that $d = 2$. For any pair (f_1, f_2) of simple functions on $\mathbb{R}_+$, the process

$$X_t = \exp\left\{i\left(\int_0^t f_1(s)dN_s^1 + \int_0^t f_2(s)dN_s^2\right)\right\}$$

changes only by jumps so that we may write

$$\begin{aligned} X_t &= 1 + \sum_{0<s\leq t} (X_s - X_{s-}) \\ &= 1 + \sum_{0<s\leq t} X_{s-}\left\{\exp\left(i\left(f_1(s)\Delta N_s^1 + f_2(s)\Delta N_s^2\right)\right) - 1\right\}. \end{aligned}$$

Condition ii) implies that, if $\Delta N_s^1 = 1$ then $\Delta N_s^2 = 0$ and vice-versa; as a result

$$X_t = 1 + \sum_{0<s\leq t} X_{s-}\left\{\left(e^{if_1(s)} - 1\right)\Delta N_s^1 + \left(e^{if_2(s)} - 1\right)\Delta N_s^2\right\}.$$

The process X_{s-} is predictable; using integration with respect to the martingales $M_t^i = N_t^i - c_i t$, we get

$$E[X_t] = 1 + E\left[\int_0^t X_{s-}\left\{\left(e^{if_1(s)} - 1\right)c_1 + \left(e^{if_2(s)} - 1\right)c_2\right\}ds\right].$$

But $\{s : X_s \neq X_{s-}\}$ is a.s. countable hence Lebesgue negligible, and consequently

$$\begin{aligned} E[X_t] &= 1 + E\left[\int_0^t X_s\left\{\left(e^{if_1(s)} - 1\right)c_1 + \left(e^{if_2(s)} - 1\right)c_2\right\}ds\right] \\ &= 1 + \int_0^t E[X_s]\left\{\left(e^{if_1(s)} - 1\right)c_1 + \left(e^{if_2(s)} - 1\right)c_2\right\}ds. \end{aligned}$$

As a result,

$$E[X_t] = \exp\left(\int_0^t c_1\left(e^{if_1(s)} - 1\right)ds\right)\exp\left(\int_0^t c_2\left(e^{if_2(s)} - 1\right)ds\right),$$

which completes the proof. □

We now turn to the most important definition of this section.

(1.8) Definition. *An* $(\mathscr{F}_t)$-Poisson point process *(in short:* $(\mathscr{F}_t)$*-PPP) is a* σ*-discrete point process* (e_t)*, such that*

i) the process e *is* $(\mathscr{F}_t)$*-adapted, that is, for any* $\Gamma \in \mathscr{U}_\delta$*, the process* N_t^Γ *is* $(\mathscr{F}_t)$*-adapted;*

ii) for any s *and* $t > 0$ *and any* $\Gamma \in \mathscr{U}_\delta$*, the law of* $N_{]s,s+t]}^\Gamma$ *conditioned on* $\mathscr{F}_s$ *is the same as the law of* N_t^Γ*.*

The property ii) may be stated by saying that the process is homogeneous in time and that the increments are independent of the past. By Propositions (1.4) and (1.7) each process N^Γ for which $N_t^\Gamma < \infty$ a.s. for every t, is a Poisson process and, if the sets Γ_i are pairwise disjoint and such that $N_t^{\Gamma_i} < \infty$ a.s., the process $\left(N_t^{\Gamma_i}, i = 1, 2, \ldots, d\right)$ is a d-dimensional Poisson process. Moreover, when $N_t^\Gamma < \infty$ a.s., then $E\left[N_t^\Gamma\right] < \infty$ and the map $t \to E\left[N_t^\Gamma\right]$ is additive, thus $\frac{1}{t}E\left[N_t^\Gamma\right]$ does not depend on $t > 0$.

(1.9) Definition. *The σ-finite measure n on $\mathscr{U}$ defined by*

$$n(\Gamma) = \frac{1}{t} E\left[N_t^\Gamma\right], \quad t > 0,$$

is called the characteristic measure *of e. The measure n is continued to U_δ by setting $n(\{\delta\}) = 0$.*

Thus, if $n(\Gamma) < \infty$, $n(\Gamma)$ is the parameter of the Poisson process N^Γ.

If $\Lambda \in \mathscr{B}(\mathbb{R}_+) \otimes \mathscr{U}_\delta$, the monotone class theorem implies easily that

$$E[N^\Lambda] = \int_0^\infty dt \int 1_\Lambda(t, u) n(du).$$

Let us also observe that, if $n(\Gamma) < \infty$, then $N_t^\Gamma - tn(\Gamma)$ is an $(\mathscr{F}_t)$-martingale and, more generally, we have the

(1.10) Proposition (Master Formula). *Let H be a positive process defined on $(\mathbb{R}_+ \times \Omega \times U_\delta)$, measurable with respect to $\mathscr{P}(\mathscr{F}_t) \otimes \mathscr{U}_\delta$ and vanishing at δ, then*

$$E\left[\sum_{s>0} H\big(s, \omega, e_s(\omega)\big)\right] = E\left[\int_0^\infty ds \int H(s, \omega, u) n(du)\right].$$

Proof. The set of processes H which satisfy this equality is a cone which is stable under increasing limits; thus, by the monotone class theorem, it is enough to check that the equality holds whenever $H(s, \omega, u) = K(s, \omega) 1_\Gamma(u)$ where K is a bounded positive $(\mathscr{F}_t)$-predictable process and $\Gamma \in \mathscr{U}$ with $n(\Gamma) < \infty$. In that case, since $N_t^\Gamma - tn(\Gamma)$ is a martingale, the left-hand side is equal to

$$\begin{aligned} E\left[\sum_{s>0} K(s, \omega) 1_\Gamma(e_s(\omega))\right] &= E\left[\int_0^\infty K(s, \omega) dN_s^\Gamma(\omega)\right] \\ &= n(\Gamma) E\left[\int_0^\infty K(s, \omega) ds\right] \end{aligned}$$

which completes the proof.

(1.11) Corollary. *If moreover*

$$E\left[\int_0^t ds \int H(s, \omega, u) n(du)\right] < \infty$$

for every t, the process

$$\sum_{0<s\leq t} H(s, \omega, e_s(\omega)) - \int_0^t ds \int H(s, \omega, u) n(du)$$

is a martingale.

Proof. Straightforward.

The following result shows that n characterizes e in the sense that two PPP with the same characteristic measure have the same law.

(1.12) Proposition (Exponential formulas). *If f is a $\mathscr{B}(\mathbb{R}_+) \otimes \mathscr{U}$-measurable function such that $\int_0^\infty ds \int |f(s,u)| n(du) < \infty$ then, for every $t \leq \infty$,*

$$E\left[\exp\left\{i \sum_{0<s\leq t} f(s, e_s)\right\}\right] = \exp\left\{\int_0^t ds \int \left(e^{if(s,u)} - 1\right) n(du)\right\}.$$

For $f \geq 0$, the following equality also holds,

$$E\left[\exp\left\{-\sum_{0<s\leq t} f(s, e_s)\right\}\right] = \exp\left\{-\int_0^t ds \int \left(1 - e^{-f(s,u)}\right) n(du)\right\}.$$

Proof. The process of bounded variation $X_t = \sum_{0\leq s\leq t} f(s, e_s)$ is purely discontinuous, so that, for $g \in \mathscr{B}(\mathbb{R}_+)$,

$$g(X_t) - g(X_0) = \sum_{s\leq t} (g(X_s) - g(X_{s-})).$$

If we put $\phi(t) = E\left[\exp(iX_t)\right]$, we consequently have

$$\phi(t) = 1 + E\left[\sum_{0\leq s\leq t} \exp\left\{i \sum_{0\leq s'<s} f(s', e_{s'})\right\} (\exp\{if(s, e_s)\} - 1)\right],$$

which, using the master formula is equal to

$$1 + E\left[\int_0^t ds \exp\left\{i \sum_{0<s'<s} f(s', e_{s'})\right\} \int \left(e^{if(s,u)} - 1\right) n(du)\right].$$

The function $\psi(\cdot) = \int \left(e^{if(\cdot,u)} - 1\right) n(du)$ is in $L^1(\mathbb{R}_+)$ thanks to the hypothesis made on f, and we have

$$\phi(t) = 1 + \int_0^t \psi(s)\phi(s-)ds.$$

It follows that ϕ is continuous, so that

$$\phi(t) = \exp\left\{\int_0^t \psi(s)ds\right\}$$

which is the first formula of the statement. The second formula is proved in much the same way. □

Remark. In the above proposition one cannot replace the function f by a non deterministic process H as in Proposition (1.10) (see Exercise (1.22)). Moreover Proposition (1.12) has a martingale version along the same line as in Corollary (1.11), namely

$$\exp\left\{\alpha \sum_{0<s\leq t} f(s, e_s) + \int_0^t ds \int \left(1 - e^{\alpha f(s,u)}\right) n(du)\right\}$$

where α is equal to i or -1, is a martingale.

We close this section with a lemma which will be useful later on. We set $S = \inf\left\{t : N_t^U > 0\right\}$.

(1.13) Lemma. *If $n(U) < \infty$, then S and e_S are independent and for any $\Gamma \in \mathscr{U}$*

$$P\left[e_S \in \Gamma\right] = n(\Gamma)/n(U).$$

Proof. Since $\Gamma \cap \Gamma^c = \emptyset$, the Poisson processes N^Γ and N^{Γ^c} are independent. Let T and T^c be their respective first jump times. For every t,

$$\{S > t; e_S \in \Gamma\} = \left\{t < T < T^c\right\}$$

and consequently, since T and T^c are independent exponential r.v.'s with parameters $n(\Gamma)$ and $n(\Gamma^c)$,

$$\begin{aligned} P\left[S > t; e_S \in \Gamma\right] &= \int_t^\infty n(\Gamma)e^{-n(\Gamma)s}ds \int_s^\infty n(\Gamma^c)e^{-n(\Gamma^c)s'}ds' \\ &= \left(n(\Gamma)/n(U)\right)e^{-n(U)t}; \end{aligned}$$

this shows in one stroke the independence of S and e_S and the formula in the statement and confirms the fact that S is exponentially distributed with parameter $n(U)$.

\# **(1.14) Exercise.** 1°) Let N be the Poisson process on $\mathbb{R}_+$ and $M_t = N_t - ct$. Prove that $\int_0^t N_{s-}dM_s$ is a martingale and that $\int_0^t N_s dM_s$ is not. Conclude that the process N is not predictable with respect to its natural filtration.

2°) If Z is predictable, then under a suitable integrability hypothesis,

$$\left(\int_0^t Z_s dM_s\right)^2 - c\int_0^t Z_s^2 ds$$

is a martingale. This generalizes the result in 3°) of Exercise (1.14) Chapter II.

\# **(1.15) Exercise.** 1°) Let $(\mathscr{F}_t)$ be the natural filtration of a Poisson process N_t and set $M_t = N_t - ct$. Prove that every $(\mathscr{F}_t)$-local martingale X_t may be written

$$X_0 + \int_0^t Z_s dM_s$$

for a $\mathscr{F}_t$-predictable process Z.

[Hint: Use the same pattern of proof as for Theorem (3.4) in Chap. V.]

2°) Let B_t be a BM independent of N_t and $(\mathscr{G}_t)$ the smallest right-continuous complete filtration for which B and N are adapted. Give a representation of $(\mathscr{G}_t)$-local martingales as stochastic integrals.

3°) Let N' be another $(\mathscr{F}_t)$-Poisson process. Deduce from 1°) that $N' = N$ a.s. The reader will observe the drastic difference with the Brownian filtration for which there exist many $(\mathscr{F}_t)$-BM's.

[Hint: Use 2°) of the previous exercise.]

4°) Describe all the $(\mathscr{F}_t)$-Poisson processes, when $(\mathscr{F}_t)$ is the natural filtration of a d-dimensional Poisson process.

[Hint: The discussion depends on whether the parameters of the components are equal or different.]

(1.16) Exercise. 1°) Let N be a Poisson process and $f \in L^1(\mathbb{R}_+)$. What is the necessary and sufficient condition under which the r.v. $N(f) = \int_0^\infty f(u)dN_u$ is a Poisson r.v.?

2°) If g is another function of $L^1(\mathbb{R}_+)$, find the necessary and sufficient condition under which $N(f)$ and $N(g)$ (not necessarily Poisson) are independent.

(1.17) Exercise. Let N be an $(\mathscr{F}_t)$-Poisson process and T a $(\mathscr{F}_t)$-stopping time. Prove that $N'_t = N_{T+t} - N_T$ is a Poisson process. [For the natural filtration of N, this is a consequence of the Strong Markov property of Sect. 3 Chap. III; the exercise is to prove this with the methods of this section.] Generalize this property to $(\mathscr{F}_t)$-Poisson point processes with values in $(U, \mathscr{U})$.

* **(1.18) Exercise.** Let $(E, \mathscr{E})$ be a measurable space and $\mathscr{M}$ the space of non-negative, possibly infinite, integer-valued measures on $(E, \mathscr{E})$. We endow $\mathscr{M}$ with the coarsest σ-field for which the maps $\mu \to \mu(A)$, $A \in \mathscr{E}$, are measurable. An $\mathscr{M}$-valued r.v. N is called a *Poisson random measure* if

i) for each $A \in \mathscr{E}$, the r.v. $N(A)$ is a Poisson r.v. (the map which is identically $+\infty$ being considered as a Poisson r.v.);
ii) if (A_i) is a finite sequence of pairwise disjoint sets of $\mathscr{E}$, the r.v.'s $N(A_i)$ are independent.

1°) Prove that if N is a Poisson random measure and if we set $\lambda(A) = E[N(A)]$, then λ is a measure on $\mathscr{E}$. Conversely, given a σ-finite measure λ on $(E, \mathscr{E})$ prove that there exist a probability space and a Poisson random measure N on this space such that $E[N(A)] = \lambda(A)$ for every $A \in \mathscr{E}$. The measure λ is called the *intensity* of N.

2°) Prove that an $\mathscr{M}$-valued r.v. N is a Poisson random measure with intensity λ, if and only if, for every $f \in \mathscr{E}_+$,

$$E\left[\exp\left(-\int f(x)N(dx)\right)\right] = \exp\left(-\int (1 - \exp(-f(x)))\lambda(dx)\right).$$

3°) Let n be a σ-finite measure on a measurable space $(U, \mathscr{U})$. Prove that there exists a PPP on U with characteristic measure n.

[Hint: Use 1°) with $d\lambda = dt\, dn$ on $\mathbb{R}_+ \times U$.]

\# **(1.19) Exercise.** Let N be a Poisson random measure on $(E, \mathscr{E})$ with intensity λ (see the preceding exercise) and f be a uniformly bounded positive function on $]0, \infty[\times E$ such that for every $s > 0$, the measure $f_s\lambda$ is a probability measure. Assume that there exist an exponential r.v. S with parameter 1 and an E-valued r.v. X such that, conditionally on $S = s$

i) N is a Poisson random measure with intensity $g_s\lambda$ where $g_s(x) = \int_s^\infty f_t(x)dt$;
ii) X has distribution $f_s\lambda$;
iii) X and N are independent.

Prove that $N^* = N + \varepsilon_X$ is a Poisson random measure with intensity $g_0\lambda$.
[Hint: Use 2°) in the preceding exercise.]

\# **(1.20) Exercise. (Chaotic representation for the Poisson process).** 1°) Retain the notation of Exercise (1.15) and let f be a positive locally bounded function on $\mathbb{R}_+$. Prove that

$$\mathscr{E}_t^f = \exp\left(-\int_0^t f(s)dN_s + c\int_0^t \left(1 - \exp(-f(s))\right)ds\right)$$

is a $(\mathscr{F}_t)$-martingale for which the process Z of Exercise (1.15) is equal to $\mathscr{E}_{s-}^f\left(\exp\left(-f(s)\right) - 1\right)$.

2°) Prove that every $Y \in L^2(\mathscr{F}_\infty)$ can be written

$$Y = E[Y] + \sum_{n=1}^\infty \int_0^\infty dM_{s_1}\int_{[0,s_1[} dM_{s_2}\dots\int_{[0,s_{n-1}[} dM_{s_n} f_n(s_1, \dots, s_n)$$

where (f_n) is a sequence of functions such that

$$\sum_{n=1}^\infty \int ds_1 \dots \int ds_n f_n^2(s_1, \dots, s_n) < \infty.$$

(1.21) Exercise. 1°) Let e be a PPP and in the notation of this section, let f be $\mathscr{B}(\mathbb{R}_+)\otimes\mathscr{U}_\delta$-measurable and positive. Assume that for each t, the r.v.

$$N_t^f = \sum_{0<s\le t} f(s, e_s)$$

is a.s. finite.

1°) Prove that the following two conditions are equivalent:

i) for each t, the r.v. N_t^f is a Poisson r.v.;
ii) there exists a set $\Gamma \in \mathscr{B}(\mathbb{R}_+)\otimes\mathscr{U}_\delta$ such that $f = 1_\Gamma \; ds \otimes dn$-a.e.

[Hint: See Exercise (1.16).]

2°) Prove that the process $\left(N_t^\Gamma; t \ge 0\right)$ satisfies the equality

$$N_t^\Gamma = \widehat{N}_{a(t)}$$

where $a(t) = \int_0^t n(\Gamma(s, \cdot))ds$ and $\widehat{N}$ is a Poisson process. Consequently, N^Γ is a Poisson process iff $n(\Gamma(s, \cdot))$ is Lebesgue a.e. constant in s.

(1.22) Exercise. 1°) Let N be a Poisson process with parameter c and call T_1 the time of its first jump. Prove that the equality

$$E\left[\exp\left(-\int f(s,\omega)dN_s(\omega)\right)\right] = E\left[\exp\left(-c\int (1-\exp(-f(s,\omega)))\,ds\right)\right]$$

fails to be true if $f(s,\omega) = \lambda 1_{\{0\le s\le T(\omega)\}}$, with $T = T_1$ and $\lambda > 0$.

2°) Prove that if g is not $ds \otimes dP$-negligible, the above equality cannot be true for all functions $f(s,\omega) = g(s,\omega)1_{\{0\le s\le T(\omega)\}}$ where T ranges through all the bounded $(\mathscr{F}_t)$-stopping times.

§2. The Excursion Process of Brownian Motion

In what follows, we work with the canonical version of BM. We denote by $\mathbf{W}$ the Wiener space, by P the Wiener measure, and by $\mathscr{F}$ the Borel σ-field of $\mathbf{W}$ completed with respect to P. We will use the notation of Sect. 2, Chap. VI.

Henceforth, we apply the results of the preceding section to a space $(U_\delta, \mathscr{U}_\delta)$ which we now define. For a $w \in \mathbf{W}$, we set

$$R(w) = \inf\{t > 0 : w(t) = 0\}.$$

The space U is the set of these functions w such that $0 < R(w) < \infty$ and $w(t) = 0$ for every $t \ge R(w)$. We observe that the graph of these functions lies entirely above or below the t-axis, and we shall call U_+ and U_- the corresponding subsets of U. The point δ is the function which is identically zero. Finally, $\mathscr{U}$ is the σ-algebra generated by the coordinate mappings. Notice that U_δ is a subset of the Wiener space $\mathbf{W}$ and that $\mathscr{U}_\delta$ is the trace of the Borel σ-field of $\mathbf{W}$. As a result, any Borel function on $\mathbf{W}$, as for instance the function R defined above, may be viewed as a function on U. This will often be used below without further comment. However, we stress that U_δ is negligible for P, and that we will define and study a measure n carried by U_δ, hence singular with respect to P.

(2.1) Definition. *The* excursion process *is the process* $e = (e_s, s > 0)$, *defined on* $(\mathbf{W}, \mathscr{F}, P)$ *with values in* $(U_\delta, \mathscr{U}_\delta)$ *by*

i) if $\tau_s(w) - \tau_{s-}(w) > 0$, *then* $e_s(w)$ *is the map*

$$r \longrightarrow 1_{[r\le \tau_s(w)-\tau_{s-}(w)]} B_{\tau_{s-}(w)+r}(w);$$

ii) if $\tau_s(w) - \tau_{s-}(w) = 0$, *then* $e_s(w) = \delta$.

We will sometimes write $e_s(r, w)$ *or* $e_s(r)$ *for the function* $e_s(w)$ *taken at time* r.

The process e is a point process. Indeed, to check condition i) in Definition (1.1), it is enough to consider the map $(t, w) \to X_r(e_t(w))$ where X_r is a fixed coordinate mapping on U, and it is easily seen that this map is measurable. Moreover, e_s is not equal to δ if, and only if, the local time L has a constant stretch at level s and e_s is then that part of the Brownian path which lies between the times τ_{s-} and τ_s at which B vanishes. It was already observed in Sect. 2 Chap. VI that there are almost-surely only countably many such times s, which ensures that condition ii) of Definition (1.1) holds.

(2.2) Proposition. *The process e is σ-discrete.*

Proof. The sets

$$U_n = \{u \in U;\ R(u) > 1/n\}$$

are in $\mathscr{U}$, and their union is equal to U. The functions

$$N_t^{U_n}(w) = \sum_{s \leq t} 1_{(e_s(w) \in U_n)}$$

are measurable. Indeed, the process $t \to \tau_t$ is increasing and right-continuous; if for $n \in \mathbb{N}$, we set

$$T_1 = \inf\{t > 0 : \tau_t - \tau_{t-} > 1/n\},$$

then $P[T_1 > 0] = 1$. If we define inductively

$$T_k = \inf\{t > T_{k-1} : \tau_t - \tau_{t-} > 1/n\}$$

then, the T_k's are random variables and

$$N_t^{U_n} = \sum_k 1_{(T_k \leq t)}$$

is a random variable. Moreover, $N_t^{U_n} < n\tau_t$, as is easily seen, which proves our claim.

We will also need the following

(2.3) Lemma. *For every $r > 0$, almost-surely, the equality*

$$e_{s+r}(w) = e_s\left(\theta_{\tau_r}(w)\right)$$

holds for all s.

Proof. This is a straightforward consequence of Proposition (1.3) in Chap. X.

We may now state the following important result.

(2.4) Theorem (Itô). *The excursion process (e_t) is an $\left(\mathscr{F}_{\tau_t}\right)$-Poisson point process.*

Proof. The variables N_t^{Γ} are plainly $\mathscr{F}_{\tau_t}$-measurable. Moreover, by the lemma and the Strong Markov property of Sect. 3 Chap. III, we have, using the notation of Definition (1.8),

$$\begin{aligned} P\left[N^{\Gamma}_{]r,t+r]} \in A \mid \mathscr{F}_{\tau_r}\right] &= P\left[N_t^{\Gamma} \circ \theta_{\tau_r} \in A \mid \mathscr{F}_{\tau_r}\right] \\ &= P_{B_{\tau_r}}\left[N_t^{\Gamma} \in A\right] = P\left[N_t^{\Gamma} \in A\right] \qquad \text{a.s.,} \end{aligned}$$

since $B_{\tau_r} = 0$ P-a.s. The proof is complete. □

Starting from B, we have defined the excursion process. Conversely, if the excursion process is known, we may recover B. More precisely

(2.5) Proposition. *We have*

$$\tau_t(w) = \sum_{s \le t} R\left(e_s(w)\right), \qquad \tau_{t-}(w) = \sum_{s<t} R\left(e_s(w)\right)$$

and

$$B_t(w) = \sum_{s \le L_t} e_s\left(t - \tau_{s-}(w), w\right)$$

where L_t can be recovered as the inverse of τ_t.

Proof. The first two formulas are consequences of the fact that $\tau_t = \sum_{s \le t}(\tau_s - \tau_{s-})$. For the third, we observe that if $\tau_{s-} < t < \tau_s$ for some s, then $L_t = s$, and for any $u < L_t$, $e_u(t - \tau_{u-}) = 0$; otherwise, t is an increase point for L, and then $e_{L_t}(t - \tau_{L_t}) = 0$ so that $B_t(\omega) = 0$ as it should be.

Remark. The formula for B can also be stated as $B_t = e_s\left(t - \tau_{s-}\right)$ if $\tau_{s-} \le t \le \tau_s$.

The characteristic measure of the excursion process is called the *Itô measure.* It will be denoted by n and its restrictions to U_+ and U_- by n_+ and n_-. It is carried by the sets of u's such that $u(0) = 0$. Our next task is to describe the measure n and see what consequences may be derived from the formulas of the foregoing section. We first introduce some notation.

If $w \in \mathbf{W}$, we denote by $i_0(w)$ the element u of U such that

$$u(t) = w(t) \quad \text{if } t < R(w), \qquad u(t) = 0 \quad \text{if } t \ge R(w).$$

If $s \in \mathbb{R}_+$, we may apply this to the path $\theta_s(w)$. We will put $i_s(w) = i_0(\theta_s(w))$. We observe that $R\left(\theta_s(w)\right) = R\left(i_s(w)\right)$.

We will also call G_w the set of strictly positive left ends of the intervals contiguous to the set $Z(w)$ of zeros of B, in other words, the set of the starting times of the excursions, or yet the set

$$\left\{\tau_{s-}(w) : \tau_{s-}(w) \ne \tau_s(w)\right\} = \left\{\tau_{s-}(w) : R\left(\theta_{\tau_{s-}}(w)\right) > 0\right\}.$$

Let H be a $\mathscr{P}(\mathscr{F}_t) \otimes \mathscr{U}_\delta$-measurable positive process vanishing at δ. The process $(s, w, u) \to H\left(\tau_{s-}(w), w; u\right)$ is then $\mathscr{P}\left(\mathscr{F}_{\tau_t}\right) \otimes \mathscr{U}_\delta$-measurable as is easily checked. The master formula of Sect. 1 applied to this process yields

$$E\left[\sum_s H\left(\tau_{s-}(w), w; e_s(w)\right)\right] = E\left[\int_0^\infty ds \int H\left(\tau_{s-}(w), w; u\right) n(du)\right].$$

The left-hand side may also be written $E\left[\sum_{\gamma \in G_w} H(\gamma, w, i_\gamma(w))\right]$ and in the right-hand side, owing to the fact that $\{s : \tau_{s-} \ne \tau_s\}$ is countable and that we integrate with respect to the Lebesgue measure, we may replace τ_{s-} by τ_s. We have finally the

(2.6) Proposition. *If H is as above,*

$$E\Big[\sum_{\gamma\in G_w} H\left(\gamma, w; i_\gamma(w)\right)\Big] = E\left[\int_0^\infty ds \int H\left(\tau_s(w), w; u\right) n(du)\right]$$
$$= E\left[\int_0^\infty dL_t(w) \int H(t, w; u) n(du)\right].$$

Proof. Only the second equality remains to be proved, but it is a direct consequence of the change of variables formula in Stieltjes integrals. □

The above result has a martingale version. Namely, provided the necessary integrability properties obtain,

$$\sum_{s\le t} H\left(\tau_{s-}(w), w; i_{\tau_{s-}}(w)\right) - \int_0^t ds \int H\left(\tau_s(w), w; u\right) n(du)$$

is a $\left(\mathscr{F}_{\tau_t}\right)$-martingale and

$$\sum_{\gamma\in G_w\cap[0,t]} H\left(\gamma, w; i_\gamma(w)\right) - \int_0^t dL_s(w) \int H(s, w; u) n(du)$$

is a $\left(\mathscr{F}_{d_t}\right)$-martingale, the proof of which is left to the reader as an exercise.

The exponential formula yields also some interesting results. In what follows, we consider an additive functional A_t such that $\nu_A(\{0\}) = 0$, or equivalently $\int 1_Z dA = 0$. In this setting, we have

(2.7) Proposition. *The random variable A_{τ_t} is infinitely divisible and the Laplace transform of its law is equal to*

$$\phi_t(\lambda) = \exp\left\{t \int \left(e^{-\lambda x} - 1\right) m_A(dx)\right\}$$

where m_A is the image of n under A_R, the random variable A_R being defined on U by restriction from $\mathbf{W}$ to U.

Proof. Since dA_t does not charge Z, we have

$$A_t = \int_0^t 1_{Z^c}(s) dA_s,$$

hence, thanks to the strong additivity of A,

$$A_{\tau_t} = \sum_{s\le t}\left(A_{\tau_s} - A_{\tau_{s-}}\right) = \sum_{s\le t} A_R \circ \theta_{\tau_{s-}}.$$

The exponential formula of Proposition (1.12) in Sect. 1 then implies that

$$\phi_t(\lambda) = E\left[\exp\left(-\lambda A_{\tau_t}\right)\right] = E\Big[\exp\Big(-\lambda \sum_{s\le t} A_R \circ \theta_{\tau_{s-}}\Big)\Big]$$
$$= \exp\left\{t \int n(du)\left(e^{-\lambda A_R(u)} - 1\right)\right\}$$

which is the announced result.

Remark. The process A_{τ_t} is a Lévy process (Exercise (2.19) Chap. X) and m_A is its Lévy measure; the proof could be based on this observation. Indeed, it is known in the theory of Lévy processes that the Lévy measure is the characteristic measure of the jump process which is a PPP. In the present case, because of the hypothesis made on ν_A, the jump process is the process itself, in other words $A_{\tau_t} = \sum_{s \le t} A_R(e_s)$; as a result

$$m_A(C) = E\Big[\sum_{s \le 1} 1_C\left(A_R(e_s)\right)\Big] = n\left\{A_R \in C\right\}.$$

The following result yields the "law" of R under n.

(2.8) Proposition. *For every* $x > 0$,

$$n(R > x) = (2/\pi x)^{1/2}.$$

Proof. The additive functional $A_t = t$ plainly satisfies the hypothesis of the previous result which thus yields

$$E\left[\exp\left(-\lambda\tau_t\right)\right] = \exp\left\{t\int m_A(dx)\left(e^{-\lambda x} - 1\right)\right\}.$$

By Sect. 2 in Chap. VI, the law of τ_a is that of T_a which was found in Sect. 3 of Chap. II. It follows that

$$\int_0^\infty m_A(dx)\left(1 - e^{-\lambda x}\right) = \sqrt{2\lambda}.$$

By the integration by parts formula for Stieltjes integrals, we further have

$$\lambda\int_0^\infty m_A(\,]x, \infty[\,)e^{-\lambda x}dx = \sqrt{2\lambda}.$$

Since it is easily checked that

$$\sqrt{2\lambda} = \lambda\int_0^\infty e^{-\lambda x}(2/\pi x)^{1/2}dx,$$

we get $m_A(]x, \infty[) = (2/\pi x)^{1/2}$; by the definition of m_A, the proof is complete.

Remarks. 1°) Another proof is hinted at in Exercise (4.13).

2°) Having thus obtained the law of R under n, the description of n will be complete if we identify the law of $u(t)$, $t < R$, conditionally on the value taken by R. This will be done in Sect. 4.

The foregoing proposition says that $R\left(e_s(w)\right)$ is a PPP on $\mathbb{R}_+$ with characteristic measure $\bar{n}$ given by $\bar{n}(]x, \infty[) = (2/\pi x)^{1/2}$. We will use this to prove another approximation result for the local time which supplements those in Chap. VI. For $\varepsilon > 0$, let us call $\eta_t(\varepsilon)$ the number of excursions with length $\ge \varepsilon$ which end at a time $s \le t$. If N is the counting measure associated with the PPP $R(e_s)$, one moment's reflection shows that $\eta_t(\varepsilon) = N^\varepsilon_{L_t}$, where $N^\varepsilon_u = N_u^{[\varepsilon, \infty[}$, and we have the

(2.9) Proposition. $P\left[\lim_{\varepsilon\downarrow 0}\sqrt{\frac{\pi\varepsilon}{2}}\eta_t(\varepsilon) = L_t \text{ for every } t\right] = 1.$

Proof. Let $\varepsilon_k = 2/\pi k^2$; then $\bar{n}([\varepsilon_k, \infty[) = k$ and the sequence $\left\{N_t^{\varepsilon_{k+1}} - N_t^{\varepsilon_k}\right\}$ is a sequence of independent Poisson r.v.'s with parameter t. Thus, for fixed t, the law of large numbers implies that a.s.

$$\lim_n \frac{1}{n} N_t^{\varepsilon_n} = \lim_n \sqrt{\frac{\pi\varepsilon_n}{2}} N_t^{\varepsilon_n} = t.$$

As N_t^ε increases when ε decreases, for $\varepsilon_{n+1} \le \varepsilon < \varepsilon_n$,

$$\sqrt{\frac{\pi\varepsilon_{n+1}}{2}} N_t^{\varepsilon_n} \le \sqrt{\frac{\pi\varepsilon}{2}} N_t^{\varepsilon} \le \sqrt{\frac{\pi\varepsilon_n}{2}} N_t^{\varepsilon_{n+1}}$$

and plainly

$$P\left[\lim_{\varepsilon\downarrow 0}\sqrt{\frac{\pi\varepsilon}{2}} N_t^\varepsilon = t\right] = 1.$$

We may find a set Σ of probability 1 such that for $w \in \Sigma$,

$$\lim_{\varepsilon\downarrow 0}\sqrt{\frac{\pi\varepsilon}{2}} N_t^\varepsilon(w) = t$$

for every rational t. Since N_t^ε increases with t, the convergence actually holds for all t's. For each $w \in \Sigma$, we may replace t by $L_t(w)$ which ends the proof.

Remarks. 1°) A remarkable feature of the above result is that $\eta_t(\varepsilon)$ depends only on the set of zeros of B up to t. Thus we have an approximation procedure for L_t, depending only on Z. This generalizes to the local time of regenerative sets (see Notes and Comments).

2°) The same kind of proof gives the approximation by downcrossings seen in Chap. VI (see Exercise (2.10)).

(2.10) Exercise. 1°) Prove that

$$n\left(\sup_{t<R}|u(t)| \ge x\right) = 1/x.$$

[Hint: If $\Lambda_x = \left\{u : \sup_{t<R} u(t) \ge x\right\}$, observe that L_{T_x} is the first jump time of the Poisson process $N_t^{\Lambda_x}$ and use the law of L_{T_x} found in Sect. 4 Chap. VI.]

2°) Using 1°) and the method of Proposition (2.9), prove, in the case of BM, the a.s. convergence in the approximation result of Theorem (1.10) Chap. VI.

3°) Let $a > 0$ and set $M_a(w) = \sup_{t\le g_{T_a}} w(t)$ where, as usual, g_{T_a} is the last zero of the Brownian path before the time T_a when it first reaches a. Prove that M_a is uniformly distributed on $[0, a]$. This is part of Williams' decomposition theorem (see Sect. 4 Chap. VII).

[Hint: If J_a (resp. J_y) is the first jump of the Poisson process N^{Λ_a} (resp. N^{Λ_y}), then $P[M_a < y] = P\left[J_y = J_a\right]$; use Lemma (1.13).]

(2.11) Exercise. Prove that the process X defined, in the notation of Proposition (2.5), by $X_t(w) = |e_s(t - \tau_{s-}(w), w)| - s$, if $\tau_{s-}(w) \leq t \leq \tau_s(w)$, is the BM

$$\beta_t = \int_0^t \operatorname{sgn}(B_s) dB_s$$

and that $Y_t(w) = s + |e_s(t - \tau_{s-}(w), w)|$ if $\tau_{s-}(w) \leq t \leq \tau_s(w)$, is a $\mathrm{BES}^3(0)$.

(2.12) Exercise. 1°) Let $A \in \mathscr{U}_\delta$ be such that $n(A) < \infty$. Observe that the number $C^A_{d_t}$ of excursions belonging to A in the interval $[0, d_t]$ (i.e. whose two end points lie between 0 and d_t) is defined unambiguously and prove that $E\left[C^A_{d_t}\right] = n(A)E[L_t]$.

2°) Prove that on $\{d_s < t\}$,

$$C^A_{d_t} - C^A_{d_s} = C^A_{d_t - d_s} \circ \theta_{d_s}, \quad L_{d_t} - L_{d_s} = L_{d_t - d_s} \circ \theta_{d_s},$$

and consequently that $C^A_{d_t} - n(A)L_t$ is a $(\mathscr{F}_{d_t})$-martingale.

[Hint: Use the strong Markov property of BM.]

\# **(2.13) Exercise (Scaling properties).** 1°) For any $c > 0$, define a map s_c on $\mathbf{W}$ or U_δ by

$$s_c(w)(t) = w(ct)/\sqrt{c}.$$

Prove that $e_t(s_c(w)) = s_c\left(e_{t\sqrt{c}}(w)\right)$ and that for $A \in \mathscr{U}_\delta$

$$n\left(s_c^{-1}(A)\right) = n(A)/\sqrt{c}.$$

[Hint: See Exercise (2.11) in Chap. VI.]

2°) **(Normalized excursions)** We say that $u \in U$ is normalized if $R(u) = 1$. Let U^1 be the subset of normalized excursions. We define a map ν from U to U^1 by

$$\nu(u) = s_{R(u)}(u).$$

Prove that for $\Gamma \subset U^1$, the quantity

$$\gamma(\Gamma) = n_+\left(\nu^{-1}(\Gamma) \cap (R \geq c)\right) / n_+(R \geq c)$$

is independent of $c > 0$. The probability measure γ may be called the *law of the normalized Brownian excursion.*

3°) Show that for any Borel subset S of $\mathbb{R}_+$,

$$n_+\left(\nu^{-1}(\Gamma) \cap (R \in S)\right) = \gamma(\Gamma) n_+(R \in S)$$

which may be seen as displaying the independence between the length of an excursion and its form.

4°) Let e^c be the first positive excursion e such that $R(e) \geq c$. Prove that

$$\gamma(\Gamma) = P\left[\nu(e^c) \in \Gamma\right].$$

[Hint: Use Lemma (1.13).]

\# **(2.14) Exercise.** Let $\Lambda_t(\varepsilon)$ be the total length of the excursions with length $< \varepsilon$, strictly contained in $[0, t[$. Prove that

$$P\left[\lim_{\varepsilon \downarrow 0} \sqrt{\frac{\pi}{2\varepsilon}} \Lambda_t(\varepsilon) = L_t \quad \text{for every } t\right] = 1.$$

[Hint: $\Lambda_t(\varepsilon) = -\int_0^\varepsilon x\eta_t(dx)$ where η_t is defined in Proposition (2.9).]

\# **(2.15) Exercise.** Let $S_t = \sup_{s \le t} B_s$ and $n_t(\varepsilon)$ the number of flat stretches of S of length $\ge \varepsilon$ contained in $[0, t]$. Prove that

$$P\left[\lim_{\varepsilon \downarrow 0} \sqrt{\frac{\pi\varepsilon}{2}} n_t(\varepsilon) = S_t \quad \text{for every } t\right] = 1.$$

(2.16) Exercise (Skew Brownian motion). Let (Y_n) be a sequence of independent r.v.'s taking the values 1 and -1 with probabilities α and $1-\alpha$ $(0 \le \alpha \le 1)$ and independent of B. For each w in the set on which B is defined, the set of excursions $e_t(w)$ is countable and may be given the ordering of $\mathbb{N}$. In a manner similar to Proposition (2.5), define a process X^α by putting

$$X_t^\alpha = Y_n \left| e_s(t - \tau_{s-}(w), w) \right|$$

if $\tau_{s-}(w) \le t < \tau_s(w)$ and e_s is the n-th excursion in the above ordering. Prove that the process thus obtained is a Markov process and that it is a skew BM by showing that its transition function is that of Exercise (1.16), Chap. I. Thus, we see that the skew BM may be obtained from the reflecting BM by changing the sign of each excursion with probability $1-\alpha$. As a result, a Markov process X is a skew BM if and only if $|X|$ is a reflecting BM.

* **(2.17) Exercise.** Let $A_t^+ = \int_0^t 1_{(B_s > 0)} ds$, $A_t^- = \int_0^t 1_{(B_s < 0)} ds$.

1°) Prove that the law of the pair $L_t^{-2}(A_t^+, A_t^-)$ is independent of t and that $A_{\tau_t}^+$ and $A_{\tau_t}^-$ are independent stable r.v.'s of index 1/2.

[Hint: $A_{\tau_t}^+ + A_{\tau_t}^- = \tau_t$ which is a stable r.v. of index 1/2.]

2°) Let a^+ and a^- be two positive real numbers, S an independent exponential r.v. with parameter 1. Prove that

$$E\left[\exp\left(-L_S^{-2}\left(a^+ A_S^+ + a^- A_S^-\right)\right)\right] = \int_0^\infty \exp(-\phi(s))\phi'(s)ds$$

where $\phi(s) = \frac{1}{\sqrt{2}}\left[\left(s^2 + a^+\right)^{1/2} + \left(s^2 + a^-\right)^{1/2}\right]$. Prove that, consequently, the pair $L_t^{-2}\left(A_t^+, A_t^-\right)$ has the same law as $\frac{1}{4}(T^+, T^-)$ where T^+ and T^- are two independent r.v.'s having the law of τ_1 and derive therefrom the arcsine law of Sect. 2 Chap. VI (see also Exercise (4.20) below). The reader may wish to compare this method with that of Exercise (2.33) Chap. VI.

(2.18) Exercise. Prove that the set of Brownian excursions can almost-surely be labeled by $\mathbb{Q}_+$ in such a way that $q < q'$ entails that e_q occurs before $e_{q'}$.

[Hint: Call e_1 the excursion straddling 1, then $e_{1/2}$ the excursion straddling $g_1/2, \dots$]

§3. Excursions Straddling a Given Time

From Sect. 4 in Chap. VI, we recall the notation

$$g_t = \sup\{s < t : B_s = 0\}, \qquad d_t = \inf\{s > t : B_s = 0\},$$

and set

$$A_t = t - g_t, \qquad \Lambda_t = d_t - g_t.$$

We say that an excursion *straddles* t if $g_t < t < d_t$. In that case, A_t is the age of the excursion at time t and Λ_t is its length. We have $\Lambda_t = R(i_{g_t})$.

(3.1) Lemma. *The map $t \to g_t$ is right-continuous.*

Proof. Let $t_n \downarrow t$; if there exists an n such that $g_{t_n} < t$, then $g_{t_m} = g_t$ for $m \geq n$; if $g_{t_n} \geq t$ for every n, then $t \leq g_{t_n} \leq t_n$ for every n, hence t is itself a zero of B and $g_t = t = \lim_n g_{t_n}$.

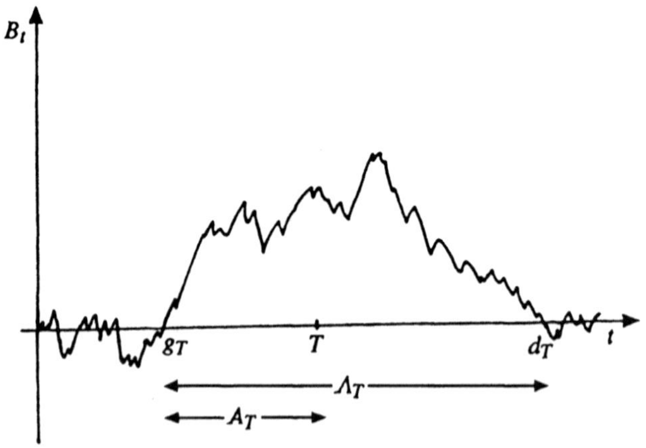

Fig. 8.

For a positive r.v. S we denote by $\mathscr{F}_S$ the σ-algebra generated by the variables H_S where H ranges through the optional processes. If S is a stopping time, this coincides with the usual σ-field $\mathscr{F}_S$. Let us observe that in general $S \leq S'$ does not entail $\mathscr{F}_S \subset \mathscr{F}_{S'}$, when S and S' are not stopping times; one can for instance find a r.v. $S \leq 1$ such that $\mathscr{F}_S = \mathscr{F}_\infty$.

Before we proceed, let us recall that by Corollary (3.3), Chap. V, since we are working in the Brownian filtration $(\mathscr{F}_t)$, there is no difference between optional and predictable processes.

(3.2) Lemma. *The family $(\check{\mathscr{F}}_t) = (\mathscr{F}_{g_t})$ is a subfiltration of $(\mathscr{F}_t)$ and if T is a $(\check{\mathscr{F}}_t)$-stopping time, then $\mathscr{F}_{g_T} \subset \check{\mathscr{F}}_T \subset \mathscr{F}_T$.*

Proof. As g_t is $\mathscr{F}_t$-measurable, the same reasoning as in Proposition (4.9) Chap. I shows that, for a predictable process Z, the r.v. Z_{g_t} is $\mathscr{F}_t$-measurable whence $\check{\mathscr{F}}_t \subset \mathscr{F}_t$ follows.

Now choose u in $\mathbb{R}_+$ and set $Z'_t = Z_{g_{t\wedge u}}$; thanks to Lemma (3.1) and to what we have just proved, Z' is $(\mathscr{F}_t)$-optional hence predictable. Pick $v > u$; if $g_v \le u$, then $g_u = g_v$ and since $g_{g_t} = g_t$ for every t,

$$Z'_{g_v} = Z_{g_v} = Z_{g_u},$$

and if $g_v > u$, then $Z'_{g_v} = Z_{g_u}$. As a result, each of the r.v.'s which generate $\check{\mathscr{F}}_u$ is among those which generate $\check{\mathscr{F}}_v$. It follows that $\check{\mathscr{F}}_u \subset \check{\mathscr{F}}_v$, that is, $(\check{\mathscr{F}}_t)$ is a filtration.

Let now T be a $(\check{\mathscr{F}}_t)$-stopping time. By definition, the σ-algebra $\mathscr{F}_{g_T}$ is generated by the variables Z_{g_T} with Z optional; but $Z_{g_T} = \left(Z_{g_\cdot}\right)_T$ and $Z_{g_\cdot}$ is $\mathscr{F}_{g_t}$-optional because of Lemma (3.1) which entails that Z_{g_T} is $\check{\mathscr{F}}_T$-measurable, hence that $\mathscr{F}_{g_T} \subseteq \check{\mathscr{F}}_T$. On the other hand, since $\check{\mathscr{F}}_t \subset \mathscr{F}_t$, the time T is also a $(\mathscr{F}_t)$-stopping time from which the inclusion $\check{\mathscr{F}}_T \subset \mathscr{F}_T$ is easily proved. □

We now come to one of the main results of this section which allows us to compute the laws of some particular excursions when n is known. If F is a positive $\mathscr{U}$-measurable function on U, for $s > 0$, we set

$$q(s, F) = n(R > s)^{-1} \int_{\{R>s\}} F\, dn \equiv n(F \mid R > s).$$

We recall that $0 < n(R > s) < \infty$ for every $s > 0$.

(3.3) Proposition. *For every fixed $t > 0$,*

$$E\left[F\left(i_{g_t}\right) \mid \check{\mathscr{F}}_t\right] = q(A_t, F) \qquad \text{a.s.},$$

and for a $(\check{\mathscr{F}}_t)$-stopping time T,

$$E\left[F\left(i_{g_T}\right) \mid \check{\mathscr{F}}_T\right] = q(A_T, F) \quad \text{a.s. on the set} \quad \{0 < g_T < T\}.$$

Proof. We know that, a.s., t is not a zero of B hence $0 < g_t < t$ and $q(A_t, F)$ is defined; also, if $s \in G_w$ and $s < t$, we have $s = g_t$ if and only if $s + R \circ \theta_s > t$. As a result, g_t is the only $s \in G_w$ such that $s < t$ and $s + R \circ \theta_s > t$. If Z is a positive $(\mathscr{F}_t)$-predictable process, we consequently have

$$E\left[Z_{g_t} F\left(i_{g_t}\right)\right] = E\Big[\sum_{s \in G_w} Z_s F(i_s) 1_{\{R\circ\theta_s > t-s>0\}}\Big].$$

We may replace $R\circ\theta_s$ by $R(i_s)$ and then apply the master formula to the right-hand side which yields

$$E\left[Z_{g_t} F\left(i_{g_t}\right)\right] = E\left[\int_0^\infty ds \int Z_{\tau_s}(w) F(u) 1_{\{R(u)>t-\tau_s(w)>0\}} n(du)\right].$$

Since by Proposition (2.8), for every $x > 0$, we have $n(R > x) > 0$, the right-hand side of the last displayed equality may be written

$$E\left[\int_0^\infty ds\ Z_{\tau_s}(w) n(R > t - \tau_s(w)) q(t - \tau_s(w), F)\right].$$

And, using the master formula in the reverse direction, this is equal to

$$E\left[\sum_{s \in G_w} Z_s q(t-s, F) 1_{\{R\circ\theta_s > t-s > 0\}}\right] = E\left[Z_{g_t} q(t - g_t, F)\right]$$

which yields the first formula in the statement.

To get the second one, we consider a sequence of countably valued $(\check{\mathscr{F}}_t)$-stopping times T_n decreasing to T. The formula is true for T_n since it is true for constant times. Moreover, on $\{0 < g_T < T\}$, one has $\{g_{T_n} = g_T\}$ from some n_0 onwards and $\lim_n 1_{\{g_{T_n} < T_n\}} = 1$; therefore, for bounded F,

$$\begin{aligned} E\left[F\left(i_{g_T}\right) \mid \check{\mathscr{F}}_T\right] &= \lim_n E\left[F\left(i_{g_{T_n}}\right) 1_{\{g_{T_n} < T_n\}} \mid \check{\mathscr{F}}_{T_n} \mid \check{\mathscr{F}}_T\right] \\ &= \lim_n E\left[q\left(A_{T_n}, F\right) 1_{\{g_{T_n} < T_n\}} \mid \check{\mathscr{F}}_T\right] = q(A_T, F) \end{aligned}$$

because $\lim_n A_{T_n} = A_T$, the function $q(\cdot, F)$ is continuous and A_T is $\check{\mathscr{F}}_T$-measurable. The extension to an unbounded F is easy. □

The foregoing result gives the conditional expectation of a function of the excursion straddling t with respect to the past of BM at the time when the excursion begins. This may be made still more precise by conditioning with respect to the length of this excursion as well. In the sequel, we write $E[\cdot \mid \check{\mathscr{F}}_t, \Lambda_t]$ for $E[\cdot \mid \check{\mathscr{F}}_t \vee \sigma(\Lambda_t)]$. Furthermore, we denote by $\nu(\cdot; F)$ a function such that $\nu(R; F)$ is a version of the conditional expectation $n(F \mid R)$. This is well defined since n is σ-finite on the σ-field generated by R and by Proposition (2.8), the function $r \to \nu(r; F)$ is unique up to Lebesgue equivalence. We may now state

(3.4) Proposition. *With the same hypothesis and notation as in the last proposition,*

$$E\left[F\left(i_{g_t}\right) \mid \check{\mathscr{F}}_t, \Lambda_t\right] = \nu(\Lambda_t; F),$$

and

$$E\left[F\left(i_{g_T}\right) \mid \check{\mathscr{F}}_T, \Lambda_T\right] = \nu(\Lambda_T; F) \quad \text{on } \{0 < g_T < T\}.$$

Proof. Let ϕ be a positive Borel function on $\mathbb{R}_+$; making use of the preceding result, we may write

$$\begin{aligned} E\left[Z_{g_t} \phi(\Lambda_t) F\left(i_{g_t}\right)\right] &= E\left[Z_{g_t} \phi\left(R\left(i_{g_t}\right)\right) F\left(i_{g_t}\right)\right] \\ &= E\left[Z_{g_t} q\left(A_t, \phi(R) F\right)\right]. \end{aligned}$$

But, looking back at the definition of q, we have

$$q(\cdot, \phi(R)F) = q(\cdot, \phi(R)\nu(R; F)),$$

so that using again the last proposition, but in the reverse direction, we get

$$E\left[Z_{g_t}\phi(\Lambda_t)F\left(i_{g_t}\right)\right] = E\left[Z_{g_t}\phi(\Lambda_t)\nu(\Lambda_t; F)\right]$$

which is the desired result. The generalization to stopping times is performed as in the preceding proof. □

We now prove an independence property between some particular excursions and the past of the BM up to the times when these excursions begin. A $\left(\mathscr{F}_t^0\right)$-stopping time T is said to be *terminal* if $T = t + T \circ \theta_t$ a.s. on the set $\{T > t\}$; hitting times, for instance, are terminal times. For such a time, $T = g_T + T \circ \theta_{g_T}$ a.s. on $\{g_T < T\}$. A time T may be viewed as defined on U by setting for $u = i_0(w)$,

$$T(u) = T(w) \quad \text{if } R(w) \geq T(w), \qquad T(u) = +\infty \quad \text{otherwise.}$$

By Galmarino's test of Exercise (4.21) in Chap. I, this definition is unambiguous. If $T(u) < \infty$, the length Λ_T of the excursion straddling T may then also be viewed as defined on U. Thanks to these conventions, the expressions in the next proposition make sense.

(3.5) Proposition. *If T is a terminal $\left(\mathscr{F}_t^0\right)$-stopping time, then on $\{0 < g_T < T\}$,*

$$E\left[F\left(i_{g_T}\right) \mid \mathscr{F}_{g_T}\right] = n\left(F 1_{(R>T)}\right)/n(R > T) \equiv n(F \mid R > T)$$

and

$$E\left[F\left(i_{g_T}\right) \mid \mathscr{F}_{g_T}, \Lambda_T\right] = \nu\left(\Lambda_T; F 1_{(R>T)}\right)/\nu\left(\Lambda_T; 1_{(R>T)}\right).$$

Proof. For a positive predictable process Z, the same arguments as in Proposition (3.3) show that

$$\begin{aligned}
&E\left[Z_{g_T} F\left(i_{g_T}\right) 1_{(0<g_T<T)}\right] \\
&\quad = E\left[\sum_{s \in G_w} 1_{(s<T)} Z_s F(i_s) 1_{(R(i_s)>T(i_s))}\right] \\
&\quad = E\left[\int_0^\infty ds\ Z_{\tau_s} 1_{(\tau_s<T)} \int F(u) 1_{(R(u)>T(u))} n(du)\right] \\
&\quad = E\left[\int_0^\infty ds\ Z_{\tau_s} 1_{(\tau_s<T)} n(R > T)\right] n\left(F 1_{(R>T)}\right)/n(R > T) \\
&\quad = E\left[\, Z_{g_T} 1_{(0<g_T<T)}\right] n\left(F 1_{(R>T)}\right)/n(R > T)
\end{aligned}$$

which proves the first half of the statement. To prove the second half we use the first one and use the same pattern as in Proposition (3.4).

Remark. As the right-hand side of the first equality in the statement is a constant, it follows that any excursion which straddles a terminal time T is independent of the past of the BM up to time g_T. This is the independence property we had announced. We may observe that, by Proposition (3.3), this property does not hold with a fixed time t in lieu of the terminal time T (see however Exercise (3.11)).

We close this section with an interesting application of the above results which will be used in the following section. As usual, we set

$$T_\varepsilon(w) = \inf\{t > 0 : w(t) > \varepsilon\}.$$

On U, we have $\{T_\varepsilon < \infty\} = \{T_\varepsilon < R\}$, and moreover

(3.6) Proposition. $n\left(\sup_{s \le R(u)} u(s) > \varepsilon\right) = n(T_\varepsilon < \infty) = 1/2\varepsilon.$

Proof. Let $0 < x < y$. The time T_x is a terminal time to which we may apply the preceding proposition with $F = 1_{(T_y < R)}$; it follows that

$$P\left[T_y\left(\theta_{g_{T_x}}\right) < \infty\right] = n(T_y < \infty)/n(T_x < \infty).$$

The left-hand side of this equality is also equal to $P_x\left[T_y < T_0\right] = x/y$; as a result, $n(T_\varepsilon < \infty) = c/\varepsilon$ for a constant c which we now determine.

Proposition (2.6) applied to $H(s, \cdot; u) = 1_{(T_\varepsilon < \infty)}(u) 1_{(s \le t)}$ yields

$$n(T_\varepsilon < \infty) E[L_t] = E\Big[\sum_{s \in G_w \cap [0,t]} 1_{(T_\varepsilon < \infty)}(i_s)\Big],$$

which, in the notation of Sect. 1 Chap. VI, implies that

$$cE[L_t] = E\left[\varepsilon\left(d_\varepsilon(t) \pm 1\right)\right];$$

letting ε tend to 0, by Theorem (1.10) of Chap. VI, we get $c = 1/2$.

Remark. This was also proved in Exercise (2.10).

(3.7) Exercise. Use the results in this and the preceding section to give the conditional law of d_1 with respect to g_1. Deduce therefrom another proof of 4°) and 5°) in Exercise (3.20) of Chap. III.

*# **(3.8) Exercise.** 1°) Prove that conditionally on $g_1 = u$, the process $(B_t, t \le u)$ is a Brownian Bridge over $[0, u]$. Derive therefrom that $B_{tg_1}/\sqrt{g_1}$, $0 \le t \le 1$ is a Brownian Bridge over $[0, 1]$ which is independent of g_1 and of $\left\{B_{g_1+u}, u \ge 0\right\}$ and that the law of g_1 is the arcsine law. See also Exercise (2.30) Chap. VI.

2°) The Brownian Bridge over $[0, 1]$ is a semimartingale. Let l^a be the family of its local times up to time 1. Prove that $L^a_{g_1}$ has the same law as $\sqrt{g_1}\, l^{a/\sqrt{g_1}}$ where g_1 is independent of the Bridge. In particular, $L^0_1 = \sqrt{g_1}\, l^0$; derive therefrom that l^0 has the same law as $\sqrt{2\mathbf{e}}$, where $\mathbf{e}$ is an exponential r.v. with parameter 1.

[Hint: See Sect. 6 Chap. 0.]

3°) Prove that the process $M_u = \left|B_{g_1+u(1-g_1)}\right| / \sqrt{1-g_1}$, $0 \le u \le 1$, is independent of the σ-algebra $\check{\mathscr{F}}_1$; M is called the *Brownian Meander* of length 1. Prove that M_1 has the law of $\sqrt{2\mathrm{e}}$ just as l^0 above.

[Hint: Use the scaling properties of n described in Exercise (2.13).]

4°) Prove that the joint law of (g_t, L_t, B_t) is

$$1_{(l\ge 0)} 1_{(s\le t)} \frac{l}{\sqrt{2\pi s^3}} \exp\left(-\frac{l^2}{2s}\right) \frac{|x|}{\sqrt{2\pi(t-s)^3}} \exp\left(-\frac{x^2}{2(t-s)}\right) ds\, dl\, dx.$$

[Hint: $(g_1, L_1, B_1) \overset{(d)}{=} \left(g_1, \sqrt{g_1} l^0, \sqrt{1-g_1} M_1\right)$ where g_1, l^0, M_1 are independent.]

* **(3.9) Exercise.** We retain the notation of the preceding exercise and put moreover $A_t = \int_0^t 1_{(B_s>0)} ds$ and $U = \int_0^1 1_{(\beta_s>0)} ds$ where β is a Brownian Bridge. We recall from Sect. 2 Chap. VI that the law of A_1 is the Arcsine law; we aim at proving that U is uniformly distributed on [0, 1].

1°) Let T be an exponential r.v. with parameter 1 independent of B. Prove that A_{g_T} and $\left(A_T - A_{g_T}\right)$ are independent. As a result,

$$A_T \overset{(d)}{=} TA_1 \overset{(d)}{=} Tg_1 U + T\varepsilon(1-g_1)$$

where ε is a Bernoulli r.v. and T, g_1, U, ε are independent.

2°) Using Laplace transform, deduce from the above result that

$$\frac{1}{2}N^2 U \overset{(d)}{=} \frac{1}{2}N^2 V$$

where N is a centered Gaussian r.v. with variance 1 which is assumed to be independent of U on one hand, and of V on the other hand, and where V is uniformly distributed on [0, 1]. Prove that this entails the desired result.

(3.10) Exercise. Prove that the natural filtration $\left(\mathscr{F}_t^g\right)$ of the process $g_.$ is strictly coarser than the filtration $\left(\check{\mathscr{F}}_t\right)$ and is equal to $\left(\mathscr{F}_t^L\right)$.

(3.11) Exercise. For $a > 0$ let $T = \inf\{t : t - g_t = a\}$. Prove the independence between the excursion which straddles T and the past of the BM up to time g_T.

§4. Descriptions of Itô's Measure and Applications

In Sects. 2 and 3, we have defined the Itô measure n and shown how it can be used in the statements or proofs of many results. In this section, we shall give several precise descriptions of n which will lead to other applications.

Let us first observe that when a σ-finite measure is given on a function space, as is the case for n and U, a property of the measure is a property of the "law" of the coordinate process when governed by this measure. Moreover, the measure is the unique extension of its restriction to the semi-algebra of measurable rectangles; in other words, the measure is known as soon as are known the finite-dimensional distributions of the coordinate process.

Furthermore, the notion of homogeneous Markov process makes perfect sense if the time set is $]0, \infty[$ instead of $[0, \infty[$ as in Chap. III. The only difference is that we cannot speak of "initial measures" any longer and if, given the transition semi-group P_t, we want to write down finite-dimensional distributions $P\left[X_{t_1} \in A_1, \ldots, X_{t_k} \in A_k\right]$ for k-uples $0 < t_1 < \ldots < t_k$, we have to know the measures $\mu_t = X_t(P)$. The above distribution is then equal to

$$\int_{A_1} \mu_{t_1}(dx_1) \int_{A_2} P_{t_2-t_1}(x_1, dx_2) \ldots \int_{A_k} P_{t_k-t_{k-1}}(x_{k-1}, dx_k).$$

The family of measures μ_t is known as the *entrance law*. To be an entrance law, (μ_t) has to satisfy the equality $\mu_t P_s = \mu_{t+s}$ for every s and $t > 0$. Conversely, given (μ_t) and a t.f. (P_t) satisfying this equation, one can construct a measure on the canonical space such that the coordinate process has the above marginals and therefore is a Markov process. Notice that the μ_t's may be σ-finite measures and that everything still makes sense; if μ is an invariant measure for P_t, the family $\mu_t = \mu$ for every t is an entrance law. In the situation of Chap. III, if the process is governed by P_ν, the entrance law is (νP_t). Finally, we may observe that in this situation, the semi-group needs to be defined only for $t > 0$.

We now recall some notation from Chap. III. We denote by Q_t the semi-group of the BM killed when it reaches 0 (see Exercises (1.15) and (3.29) in Chap. III). We recall that it is given by the density

$$q_t(x, y) = (2\pi t)^{-1/2}\left(\exp\left(-\frac{1}{2t}(y-x)^2\right) - \exp\left(-\frac{1}{2t}(y+x)^2\right)\right) 1_{(xy>0)}.$$

We will denote by $\lambda_t(dy)$ the measure on $\mathbb{R}\backslash\{0\}$ which has the density

$$m_t(y) = \left(2\pi t^3\right)^{-1/2} |y| \exp(-y^2/2t)$$

with respect to the Lebesgue measure dy. For fixed y, this is the density in t of the hitting time T_y as was shown in Sect. 3 Chap. III.

Let us observe that on $]0, \infty[$

$$m_t = -\frac{\partial g_t}{\partial y} = \lim_{x \to 0} \frac{1}{2x} q_t(x, \cdot).$$

Our first result deals with the coordinate process w restricted to the interval $]0, R[$, or to use the devices of Chap. III we will consider – in this first result only – that $w(t)$ is equal to the fictitious point δ on $[R, \infty[$. With this convention we may now state

(4.1) Theorem. *Under n, the coordinate process $w(t)$, $t > 0$, is a homogeneous strong Markov process with Q_t as transition semi-group and λ_t, $t > 0$, as entrance law.*

Proof. Everything being symmetric with respect to 0, it is enough to prove the result for n_+ or n_-. We will prove it for n_+, but will keep n in the notation for the sake of simplicity.

The space $(U_\delta, \mathscr{U}_\delta)$ may serve as the canonical space for the homogeneous Markov process (in the sense of Chap. III) associated with Q_t, in other words Brownian motion killed when it first hits $\{0\}$; we call (Q_x) the corresponding probability measures on $(U_\delta, \mathscr{U}_\delta)$. As usual, we call θ_r the shift operators on U_δ.

Our first task will be to prove the equality

eq. (4.1) $$n\left((u(r) \in A) \cap \theta_r^{-1}(\Gamma)\right) = n\left(1_A(u(r)) Q_{u(r)}(\Gamma)\right)$$

for $\Gamma \in \mathscr{U}$, $A \in \mathscr{B}(\mathbb{R}_+ - \{0\})$ and $r > 0$. Suppose that $n(u(r) \in A) > 0$ failing which the equality is plainly true. For $r > 0$, we have $\{u(r) \in A\} \subset \{r < R\}$ hence $n(u(r) \in A) < \infty$ and the expressions we are about to write will make sense. Using Lemma (1.13) for the process $e^{\{u(r) \in A\}}$, we get

$$n\left(1_A(u(r))\left(1_\Gamma \circ \theta_r\right)\right) / n(u(r) \in A) = P\left[e_S^{\{u(r) \in A\}}(w) \in \theta_r^{-1}(\Gamma)\right]$$

where P is the Wiener measure.

The time S which is the first jump time of the Poisson process $N^{\{u(r) \in A\}}$ is a $(\mathscr{F}_{\tau_t})$-stopping time; the times τ_{S-} and τ_S are therefore $(\mathscr{F}_t)$-stopping times. We set $T = \tau_{S-} + r$. The last displayed expression may be rewritten

$$P\left[\{B_T \in A\} \cap \{\widehat{B}_. \circ \theta_T \in \Gamma\}\right]$$

where $\widehat{B}_.$ stands for the BM killed when it hits $\{0\}$. By the strong Markov property for the $(\mathscr{F}_t)$-stopping time T, this is equal to

$$E\left[1_{\{B_T \in A\}} Q_{B_T}[\Gamma]\right].$$

As a result

$$n\left((u(r) \in A) \cap \theta_r^{-1}(\Gamma)\right) = n(u(r) \in A) \int \gamma(dx) Q_x[\Gamma]$$

where γ is the law of B_T under the restriction of P to $\{B_T \in A\}$. For a Borel subset C of $\mathbb{R}$, make $\Gamma = \{u(0) \in C\}$ in the above formula, which, since then $Q_x[\Gamma] = 1_C(x)$, becomes

$$n(u(r) \in A \cap C) = n(u(r) \in A)\gamma(C);$$

it follows that $\gamma(\cdot) = n(u(r) \in A \cap \cdot)/n(u(r) \in A)$ which proves eq. (4.1).

Let now $0 < t_1 < t_2 < \ldots < t_k < t$ be real numbers and let $f_1, \ldots, f_k, f$ be positive Borel functions on $\mathbb{R}$. Since $\widehat{B}$, the BM killed at 0, is a Markov process we have, for every x

eq. (4.2) $$Q_x\left[\prod_{i=1}^k f_i\left(u(t_i)\right) f(u(t))\right] = Q_x\left[\prod_{i=1}^k f_i\left(u(t_i)\right) Q_{t-t_k} f(u(t_k))\right].$$

Set $F = \prod_{i=2}^k f_i\left(u(t_i - t_1)\right)$. By repeated applications of equations (4.1) and (4.2), we may now write

$$\begin{aligned} n\left[\left(\prod_{i=1}^k f_i\left(u(t_i)\right)\right) f(u(t))\right] &= n\left[f_1\left(u(t_1)\right)\left(F \cdot f(u(t-t_1))\right) \circ \theta_{t_1}\right] \\ &= n\left[f_1\left(u(t_1)\right) Q_{u(t_1)}\left[F \cdot f\left(u(t-t_1)\right)\right]\right] \\ &= n\left[f_1\left(u(t_1)\right) Q_{u(t_1)}\left[F Q_{t-t_k} f\left((u(t_k - t_1)\right)\right]\right] \\ &= n\left[f_1\left(u(t_1)\right) F \circ \theta_{t_1} Q_{t-t_k} f\left(u(t_k)\right)\right] \\ &= n\left[\prod_{i=1}^k f_i\left(u(t_i)\right) Q_{t-t_k} f\left(u(t_k)\right)\right] \end{aligned}$$

which shows that the coordinate process u is, under n, a homogeneous Markov process with transition semi-group Q_t. By what we have seen in Chap. III, $\widehat{B}$ has the strong Markov property; using this in eq. (4.2) instead of the ordinary Markov property, we get the analogous property for n; we leave the details as an exercise for the reader.

The entrance law is given by $\lambda_t(A) = n(u(t) \in A)$ and it remains to prove that those measures have the density announced in the statement. It is enough to compute $\lambda_t([y, \infty[\,)$ for $y > 0$. For $0 < \varepsilon < y$,

$$\lambda_t([y, \infty[\,) = n(u(t) \geq y) = n(u(t) \geq y;\, T_\varepsilon < t)$$

where $T_\varepsilon = \inf\{t > 0 : u(t) > \varepsilon\}$. Using the strong Markov property for n just alluded to, we get

$$\begin{aligned} \lambda_t([y, \infty[\,) &= n\left(T_\varepsilon < t;\, Q_{u(T_\varepsilon)}(u(t - T_\varepsilon) \geq y)\right) \\ &= n\left(T_\varepsilon < t;\, Q_{t-T_\varepsilon}(\varepsilon, [y, \infty[\,)\right). \end{aligned}$$

Applying Proposition (3.5) with $F(u) = 1_{(T_\varepsilon(u) < t)} Q_{t - T_\varepsilon(u)}(\varepsilon, [y, \infty[\,)$ yields

$$\lambda_t([y, \infty[\,) = E\left[1_{(\widetilde{T}_\varepsilon < t)} Q_{t-\widetilde{T}_\varepsilon}(\varepsilon, [y, \infty[\,)\right] n(T_\varepsilon < R)$$

where $\widetilde{T}_\varepsilon = T_\varepsilon\left(i_{g_{T_\varepsilon}}\right)$. Using Proposition (3.6) and the known value of Q_t, this is further equal to

$$E\left[1_{(\widetilde{T}_\varepsilon < t)}\left(\Phi_{t-\widetilde{T}_\varepsilon}(y+\varepsilon) - \Phi_{t-\widetilde{T}_\varepsilon}(y-\varepsilon)\right)/2\varepsilon\right]$$

with $\Phi_t(y) = \int_{-\infty}^y g_t(z)dz$. If we let ε tend to zero, then $\widetilde{T}_\varepsilon$ converges to zero P-a.s. and we get $\lambda_t([y, \infty[\,) = g_t(y)$ which completes the proof.

Remarks. 1°) That (λ_t) is an entrance law for Q_t can be checked by elementary computations but is of course a consequence of the above proof.

2°) Another derivation of the value of (λ_t) is given in Exercise (4.9).

The above result permits to give *Itô's description of n* which was hinted at in the remark after Proposition (2.8). Let us recall that according to this proposition the density of R under n_+ is $(2\sqrt{2\pi r^3})^{-1}$. In the following result, we deal with the law of the Bessel Bridge of dimension 3 over $[0, r]$ namely $P^{3,r}_{0,0}$ which we will abbreviate to π_r. The following result shows in particular that the law of the normalized excursion (Exercise (2.13)) is the probability measure π_1.

(4.2) Theorem. *Under n_+, and conditionally on $R = r$, the coordinate process w has the law π_r. In other words, if $\Gamma \in \mathscr{U}_\delta^+$,*

$$n_+(\Gamma) = \int_0^\infty \pi_r(\Gamma \cap \{R = r\}) \frac{dr}{2\sqrt{2\pi r^3}} .$$

Proof. The result of Theorem (4.1) may be stated by saying that for $0 < t_1 < t_2 < \ldots < t_n$ and Borel sets $A_i \subset]0, \infty[$, if we set

$$\Gamma = \bigcap_{i=1}^{n} \{u(t_i) \in A_i\},$$

then

$$n_+(\Gamma) = \int_{A_1} m_{t_1}(x_1)dx_1 \int_{A_2} q_{t_2-t_1}(x_1, x_2)dx_2 \ldots \int_{A_n} q_{t_n-t_{n-1}}(x_{n-1}, x_n)dx_n .$$

On the other hand, using the explicit value for π_r given in Sect. 3 of Chap. XI, and taking into account the fact that $\Gamma \cap \{R < t_n\} = \emptyset$, the formula in the statement reads

$$\begin{aligned} n_+(\Gamma) \quad &= \quad \int_{t_n}^\infty \frac{dr}{2\sqrt{2\pi r^3}} \int_{A_1} 2\sqrt{2\pi r^3} m_{t_1}(x_1)dx_1 \int_{A_2} q_{t_2-t_1}(x_1, x_2)dx_2 \ldots \\ &\quad \ldots \int_{A_n} q_{t_n-t_{n-1}}(x_{n-1}, x_n) m_{r-t_n}(x_n)dx_n . \end{aligned}$$

But

$$\int_{t_n}^\infty \frac{dr}{2\sqrt{2\pi r^3}} 2\sqrt{2\pi r^3} m_{r-t_n}(x_n) = 1$$

as was seen already several times. Thus the two expressions for $n_+(\Gamma)$ are equal and the proof is complete.

This result has the following important

(4.3) Corollary. *The measure n is invariant under time-reversal; in other words, it is invariant under the map $u \to \hat{u}$ where*

$$\hat{u}(t) = u(R(u) - t) 1_{(R(u) \geq t)} .$$

Proof. By Exercise (3.7) of Chap. XI, this follows at once from the previous result.

This can be used to give another proof of the time-reversal result of Corollary (4.6) in Chap. VII.

(4.4) Corollary. *If B is a* BM(0) *and for $a > 0$, $T_a = \inf\{t : B_t = a\}$, if Z is a* $\mathrm{BES}^3(0)$ *and $\sigma_a = \sup\{t : Z_t = a\}$, then the processes $Y_t = a - B_{T_a - t}$, $t < T_a$ and Z_t, $t < \sigma_a$ are equivalent.*

Proof. We retain the notation of Proposition (2.5) and set $\beta_t = L_t - |B_t| = s - |e_s(t - \tau_{s-})|$ if $\tau_{s-} \le t \le \tau_s$. We know from Sect. 2 Chap. VI that β is a standard BM.

If we set $Z_t = L_t + |B_t| = s + |e_s(t - \tau_{s-})|$ if $\tau_{s-} \le t \le \tau_s$, Pitman's theorem (see Corollary (3.8) of Chap. VI) asserts that Z is a $\mathrm{BES}^3(0)$.

For $a > 0$ it is easily seen that

$$\tau_a = \inf\{t : L_t = a\} = \inf\{t : \beta_t = a\};$$

moreover

$$\tau_a = \sup\{t : Z_t = a\}$$

since $Z_{\tau_a} = L_{\tau_a} + |B_{\tau_a}| = a$ and for $t > \tau_a$ one has $L_t > a$.

We now define another Poisson point process with values in $(U, \mathscr{U})$ by setting

$$\begin{aligned} \widetilde{e}_s &= e_s \quad \text{if } s > a, \\ \widetilde{e}_s(t) &= e_{a-s}\left(R(e_{a-s}) - t\right), \qquad 0 \le t \le R(e_{a-s}), \quad \text{if } s \le a. \end{aligned}$$

In other words, for $s \le a$, $\widetilde{e}_s = \hat{e}_{a-s}$ in the notation of Corollary (4.3). Thus, for a positive $\mathscr{B}(\mathbb{R}_+) \times \mathscr{U}_\delta$-measurable function f,

$$\sum_{0<s\le a} f(s, \widetilde{e}_s) = \sum_{0<s\le a} f(a - s, \hat{e}_s),$$

and the master formula yields

$$E\left[\sum_{0<s\le a} f(s, \widetilde{e}_s)\right] = E\left[\int_0^a ds \int f(a - s, \hat{u}) n(du)\right];$$

by Corollary (4.3), this is further equal to

$$E\left[\int_0^a ds \int f(s, u) n(du)\right].$$

This shows that the PPP $\widetilde{e}$ has the same characteristic measure, hence the same law as e. Consequently the process $\widetilde{Z}$ defined by

$$\widetilde{Z}_t = s + |\widetilde{e}_s(t - \widetilde{\tau}_{s-})| \qquad \text{if } \widetilde{\tau}_{s-} \le t \le \widetilde{\tau}_s,$$

has the same law as Z. Moreover, one moment's reflection shows that

$$\widetilde{Z}_t = a - \beta(\tau_a - t) \qquad \text{for } 0 \le t \le \tau_a$$

which ends the proof. □

Let us recall that in Sect. 4 of Chap. X we have derived Williams' path decomposition theorem from the above corollary (and the reversal result in Proposition (4.8) of Chap. VII). We will now use this decomposition theorem to give another description of n and several applications to BM. We denote by M the maximum of positive excursions, in other words M is a r.v. defined on U_δ^+ by

$$M(u) = \sup_{s \le R(u)} u(s).$$

The law of M under n_+ has been found in Exercise (2.10) and Proposition (3.6) and is given by $n_+(M \ge x) = 1/2x$.

We now give *Williams' description* of n. Pick two independent $\mathrm{BES}^3(0)$ processes ρ and $\widetilde{\rho}$ and call T_c and $\widetilde{T}_c$ the corresponding hitting times of $c > 0$. We define a process Z^c by setting

$$Z_t^c = \begin{cases} \rho_t, & 0 \le t \le T_c, \\ c - \widetilde{\rho}(t - T_c), & T_c \le t \le T_c + \widetilde{T}_c, \\ 0, & t \ge T_c + \widetilde{T}_c. \end{cases}$$

For $\Gamma \in \mathscr{U}_\delta^+$, we put $N(c, \Gamma) = P\left[Z^c \in \Gamma\right]$. The map N is a kernel; indeed $\left(Z_t^c\right)_{t \ge 0} \stackrel{(d)}{=} \left(c Z^1_{t/c^2}\right)_{t \ge 0}$, thanks to the scaling properties of $\mathrm{BES}^3(0)$, so that N maps continuous functions into continuous functions on $\mathbb{R}$ and the result follows by a monotone class argument. By Proposition (4.8) in Chap. VII, the second part of Z^c might as well have been taken equal to $\widetilde{\rho}\left(T_c + \widetilde{T}_c - t\right)$, $T_c \le t \le T_c + \widetilde{T}_c$.

(4.5) Theorem. *For any* $\Gamma \in \mathscr{U}_\delta^+$

$$n_+(\Gamma) = \frac{1}{2} \int_0^\infty N(x, \Gamma) x^{-2} dx.$$

In other words, conditionally on its height being equal to c, the Brownian excursion has the law of Z^c.

Proof. Let $U_c = \{u : M(u) \ge c\}$; by Lemma (1.13), for $\Gamma \in \mathscr{U}_\delta^+$

$$n_+(\Gamma \cap U_c) = n_+(U_c) P[e^c \in \Gamma] = \frac{1}{2c} P[e^c \in \Gamma],$$

where e^c is the first excursion the height of which is $\ge c$. The law of this excursion is the law of the excursion which straddles T_c, i.e. the law of the process

$$Y_t = B_{g_{T_c}+t}, \qquad 0 \le t \le d_{T_c} - g_{T_c}.$$

By applying the strong Markov property to B at time T_c, we see that the process Y may be split up into two independent parts Y^1 and Y^2, with

$$Y_t^1 = B_{g_{T_c}+t}, \quad 0 \le t \le T_c - g_{T_c}; \qquad Y_t^2 = B_{T_c+t}, \quad 0 \le t \le d_{T_c} - T_c.$$

By the strong Markov property again, the part Y^2 has the law of $\widetilde{B}_t$, $0 \le t \le \widetilde{T}_0$, where $\widetilde{B}$ is a BM(c). Thus by Proposition (3.13) in Chap. VI, Y^2 may be described as follows: conditionally on the value M of the maximum, it may be further split up into two independent parts V^1 and V^2, with

$$V_t^1 = B_{T_c+t}, \quad 0 \le t \le T_M - T_c, \qquad V_t^2 = B_{T_M+t}, \quad 0 \le t \le d_{T_c} - T_M.$$

Moreover V^1 is a $\mathrm{BES}^3(c)$ run until it first hits M and V^2 has the law of $M - \rho_t$ where ρ is a $\mathrm{BES}^3(0)$ run until it hits M.

Furthermore, by Williams decomposition theorem (Theorem (4.9) Chap. VII), the process Y^1 is a $\mathrm{BES}^3(0)$ run until it hits c. By the strong Markov property for BES^3, if we piece together Y^1 and V^1, the process we obtain, namely

$$B_{g_{T_c}+t}, \qquad 0 \le t \le T_M - g_{T_c},$$

is a $\mathrm{BES}^3(0)$ run until it hits M.

As a result, we see that the law of e^c conditional on the value M of the maximum is that of Z^M. Since the law of this maximum has the density c/M^2 on $[c, \infty[$ as was seen in Proposition (3.13) of Chap. VI we get

$$P[e^c \in \Gamma] = c \int_c^\infty x^{-2} N(x, \Gamma) dx$$

which by the first sentence in the proof, is the desired result. □

To state and prove our next result, we will introduce some new notation. We will call $\mathscr{X}$ the space of real-valued continuous functions ω defined on an interval $[0, \zeta(\omega)] \subset [0, \infty[$. We endow it with the usual σ-fields $\mathscr{F}_t^0$ and $\mathscr{F}_\infty^0$ generated by the coordinates. The Itô measure n and the law of BM restricted to a compact interval may be seen as measures on $(\mathscr{X}, \mathscr{F}_\infty^0)$.

If μ and μ' are two such measures, we define $\mu \circ \mu'$ as the image of $\mu \otimes \mu'$ under the map $(\omega, \omega') \to \omega \circ \omega'$ where

$$\begin{aligned} \zeta(\omega \circ \omega') &= \zeta(\omega) + \zeta(\omega'), \\ \omega \circ \omega'(s) &= \omega(s), \quad \text{if } 0 \le s \le \zeta(\omega) \\ &= \omega(\zeta(\omega)) + \omega'(s - \zeta(\omega)) - \omega'(0) \quad \text{if } \zeta(\omega) \le s \le \zeta(\omega) + \zeta(\omega'). \end{aligned}$$

We denote by ${}^\vee\mu$ the image of μ under the time-reversal map $\omega \to \check{\omega}$ where

$$\zeta(\check{\omega}) = \zeta(\omega), \qquad \check{\omega}(s) = \omega(\zeta(\omega) - s), \qquad 0 \le s \le \zeta(\omega).$$

Finally, if T is a measurable map from $\mathscr{X}$ to $[0, \infty]$, we denote by μ^T the image of μ by the map $\omega \to k_T(\omega)$ where

$$\zeta(k_T(\omega)) = \zeta(\omega) \wedge T(\omega), \qquad k_T(\omega)(s) = \omega(s) \qquad \text{if } 0 \le s \le \zeta(\omega) \wedge T(\omega).$$

We also define, as usual:

$$T_a(\omega) = \inf\{t : \omega(t) = a\}, \qquad L_a(\omega) = \sup\{t : \omega(t) = a\}.$$

Although the law P_a of BM(a) cannot be considered as a measure on $\mathscr{X}^\cdot$, we will use the notation $P_a^{T_0}$ for the law of BM(a) killed at 0 which we may consider as a measure on $\mathscr{X}^\cdot$ carried by the set of paths ω such that $\zeta(\omega) = T_0(\omega)$. If S_3 is the law of BES$^3(0)$, we may likewise consider $S_3^{L_a}$ and the time-reversal result of Corollary (4.6) in Chap. VII then reads

$$\overset{\vee}{\left(P_a^{T_0}\right)} = S_3^{L_a}.$$

In the same way, the last result may be stated

$$n_+ = \frac{1}{2}\int_0^\infty a^{-2}\left(S_3^{T_a} \circ \overset{\vee}{\left(S_3^{T_a}\right)}\right) da.$$

The space U of excursions is contained in $\mathscr{X}^\cdot$ and carries n; on this subspace, we will write R instead of ζ in keeping with the notation used so far and also use w or ω indifferently. We may now state

(4.6) Proposition.

$$\int_0^\infty n^r(\cdot \cap \{r < R\})dr = \int_{-\infty}^{+\infty} \overset{\vee}{\left(P_a^{T_0}\right)} da.$$

Proof. Let (θ_t) be the usual translation operators and as above put

$$g_t(\omega) = \sup\{s < t : \omega(s) = 0\}, \qquad d_t(\omega) = \inf\{s > t : \omega(s) = 0\},$$

$$G_\omega = \{g_t(\omega), t \in \mathbb{R}_+\}.$$

We denote by E_0 the expectation with respect to the Wiener measure.

The equality in the statement will be obtained by comparing two expressions of

$$J = E_m\left[\int_0^\infty 1_{[T_0<t]} e^{-\lambda g_t}\, Y \circ k_{t-g_t} \circ \theta_{g_t} dt\right]$$

where $P_m = \int_{-\infty}^{+\infty} P_a da$, Y is a positive $\mathscr{F}_\infty^0$-measurable function and λ is > 0.

From Proposition (2.6) it follows that

$$\begin{aligned}
E_0&\left[\int_0^\infty e^{-\lambda g_t} Y \circ k_{t-g_t} \circ \theta_{g_t} dt\right] \\
&= E_0\left[\sum_{s \in G_\omega} e^{-\lambda s}\int_s^{d_s(\omega)} Y \circ k_{t-s} \circ \theta_s(\omega) dt\right] \\
&= E_0\left[\int_0^\infty e^{-\lambda s} dL_s\right] n\left(\int_0^R Y \circ k_r dr\right).
\end{aligned}$$

Using the strong Markov property for P_a we get

$$J = \int_{-\infty}^{+\infty} da\, E_a\left[e^{-\lambda T_0}\right] E_0\left[\int_0^\infty e^{-\lambda s} dL_s\right] n\left(\int_0^R Y \circ k_r dr\right).$$

But $E_a\left[e^{-\lambda T_0}\right] = e^{-|a|\sqrt{2\lambda}}$ so that $\int_{-\infty}^{+\infty} da\, E_a\left[e^{-\lambda T_0}\right] = \sqrt{2/\lambda}$ and by the results in Sect. 2 of Chap. X, $E_0\left[\int_0^\infty e^{-\lambda s} dL_s\right] = 1/\sqrt{2\lambda}$. As a result

$$J = \lambda^{-1}\int_0^\infty n(Y \circ k_r 1_{(r<R)})dr.$$

On the other hand, because $\zeta\left(k_{t-g_t} \circ \theta_{g_t}\right) = t$,

$$J = \int_0^\infty E_m^t\left[1_{[T_0<t]} e^{-\lambda g_t} Y \circ k_{t-g_t} \circ \theta_{g_t}\right] dt$$

and since obviously $\overset{\vee}{}(P_m^t) = P_m^t$,

$$J = \int_0^\infty E_m^t\left[1_{[T_0<t]} e^{-\lambda(t-T_0)} Z\right] dt$$

where $Z(\omega) = \left(Y \circ k_{T_0}\right)(\overset{\vee}{\omega})$; indeed $t - g_t$ is the hitting time of zero for the process reversed at t. Now the integrand under E_m^t depends only on what occurs before t and therefore we may replace E_m^t by E_m. Consequently

$$\begin{aligned} J &= E_m\left[Z\int_0^\infty 1_{[T_0<t]} e^{-\lambda(t-T_0)} dt\right] \\ &= \lambda^{-1} E_m[Z] = \lambda^{-1}\int_{-\infty}^{+\infty} E_a[Z] da. \end{aligned}$$

Comparing the two values found for J ends the proof.

Remark. Of course this result has a one-sided version, namely

$$\int_0^\infty n_+^r(\cdot \cap \{r < R\})dr = \int_0^\infty \overset{\vee}{}\left(P_a^{T_0}\right) da.$$

We may now turn to *Bismut's description* of n.

(4.7) Theorem. *Let $\bar{n}_+$ be the measure defined on $\mathbb{R}_+ \times U_\delta$ by*

$$\bar{n}_+(dt, du) = 1_{(0\le t\le R(u))} dt\; n_+(du).$$

Then, under $\bar{n}_+$ the law of the r.v. $(t, u) \to u(t)$ is the Lebesgue measure da and conditionally on $u(t) = a$, the processes $\{u(s), 0 \le s \le t\}$ and $\{u(R(u) - s), t \le s \le R(u)\}$ are two independent BES^3 *processes run until they last hit a.*

The above result may be seen, by looking upon U_δ as a subset of $\mathscr{X}$, as an equality between two measures on $\mathbb{R}_+ \times \mathscr{X}$. By the monotone class theorem, it is enough to prove the equality for the functions of the form $f(t)H(\omega)$ where H belongs to a class stable under pointwise multiplication and generating the σ-field on $\mathscr{X}$. Such a class is provided by the r.v.'s which may be written

$$H = \prod_{i=1}^{n} \int_0^\infty e^{-\lambda_i s} f_i(\omega(s)) ds$$

where λ_i is a positive real number and f_i a bounded Borel function on $\mathbb{R}$. It is clear that for each t, these r.v.'s may be written $Z_t \cdot Y_t \circ \theta_t$ where Z_t is $\mathscr{F}_t^0$-measurable. We will use this in the following

Proof. Using the Markov description of n proved in Theorem (4.1), we have

$$\begin{aligned}
&\int_{\mathbb{R}_+ \times U_\delta} f(t) Z_t(u) Y_t\,(\theta_t(u))\, \bar{n}_+(dt, du) \\
&= \int_0^\infty dt\ f(t) \int_{U_\delta} 1_{[t<R(u)]} Z_t(u) Y_t\,(\theta_t(u))\, n_+(du) \\
&= \int_0^\infty dt\ f(t) \int_{U_\delta} 1_{[1<R(u)]} Z_t(u) E^{T_0}_{u(t)}[Y_t] n_+(du) \\
&= \int_0^\infty dt\ f(t) \int_{U_\delta} Z_t(u) E^{T_0}_{u(t)}[Y_t] dn^t_+(\cdot \cap (t < R))
\end{aligned}$$

where $E^{T_0}_x$ is the expectation taken with respect to the law of BM(x) killed at 0. Using the one-sided version of Proposition (4.6) (see the remark after it), this is further equal to

$$\int_0^\infty da\ \overset{\vee}{E}{}^{T_0}_a \left[f(\zeta(\omega)) Z_\zeta(\omega) E^{T_0}_{\omega(\zeta)}[Y_\zeta] \right].$$

But for $\overset{\vee}{P}{}^{T_0}_a$, we have $\zeta = L_a$ a.s. hence in particular $\omega(\zeta) = a$ a.s. and for $P^{T_0}_a$, we have $\zeta = T_0$ a.s. so that we finally get

$$\int_0^\infty da\ \overset{\vee}{E}{}^{T_0}_a \left[f(L_a) Z_{L_a} \right] E^{T_0}_a \left[Y_{T_0} \right].$$

Using the time-reversal result of Corollary (4.4), this is precisely what was to be proved.

\# **(4.8) Exercise. (Another proof of the explicit value of λ_t in Theorem (4.1)).** Let f be a positive Borel function on $\mathbb{R}_+$ and U_p the resolvent of BM. Using the formula of Proposition (2.6), prove that

$$U_p f(0) = E\left[\int_0^\infty e^{-p\tau_s} ds \right] \int_0^\infty e^{-pu} \lambda_u(f) du.$$

Compute $\lambda_u(f)$ from the explicit form of U_p and the law of τ_s.

* **(4.9) Exercise.** Conditionally on $(R = r)$, prove that the Brownian excursion is a semimartingale over $[0, r]$ and, as such, has a family l^a, $a > 0$, of local times up to time r for which the occupation times formula obtains.

(4.10) Exercise. For $x \in \mathbb{R}_+$, let s_x be the time such that e_{s_x} is the first excursion for which $R(e_s) > x$. Let L be the length of the longest excursion e_u, $u < s_x$. Prove that

$$P[L < y] = (y/x)^{1/2} \qquad \text{for } y \leq x.$$

(4.11) Exercise. (Watanabe's process and Knight's identity). 1°) Retaining the usual notation, prove that the process $Y_t = S_{\tau_t}$ already studied in Exercise (1.9) of Chap. X, is a homogeneous Markov process on $[0, \infty[$ with semi-group T_t given by

$$T_0 = I, T_t f(x) = e^{-t/2x} f(x) + \int_x^\infty e^{-t/2y} \frac{t}{2y^2} f(y) dy.$$

[Hint: Use the description of BM by means of the excursion process given in Proposition (2.5)]. In particular,

$$P\left[S_{\tau_t} \leq a\right] = \exp(-t/2a).$$

Check the answers given in Exercise (1.27) Chap. VII.

* 2°) More generally, prove that

$$E\left[\exp\left(-\lambda^2 \tau_t^+/2\right) 1_{(S_{\tau_t} \leq a)}\right] = \exp(-\lambda t \coth(a\lambda)/2)$$

where $\tau_t^+ = \int_0^{\tau_t} 1_{(B_s > 0)} ds$.

3°) Deduce therefrom *Knight's identity*, i.e.

$$E\left[\exp\left(-\lambda^2 \tau_t^+ / 2S_{\tau_t}^2\right)\right] = 2\lambda / \sinh(2\lambda).$$

Prove that consequently,

$$\tau_t^+ / S_{\tau_t}^2 \overset{(d)}{=} \inf\{s : U_s = 2\},$$

where U is a $\mathrm{BES}^3(0)$.

[Hint: Prove and use the formula

$$\int \left(1 - \exp(-R/2) 1_{(M \leq x)}\right) dn^+ = (\coth x)/2.$$

where $M = \sup_{t<R} w(t)$.]

4°) Give another proof using time reversal.

(4.12) Exercise. (Continuation of Exercise (2.13) on normalized excursions). Let p be the density of the law of M (see Exercise (4.13)) under γ. Prove that

$$\int_0^\infty x p(x) dx = \sqrt{\pi/2},$$

that is: the mean height of the normalized excursion is $\sqrt{\pi/2}$.

[Hint: Use 3°) in Exercise (2.13) to write down the joint law of R and M under n as a function of p, then compare the marginal distribution of R with the distribution given in Proposition (2.8).]

(4.13) Exercise. 1°) Set $M(w) = \sup_{t<R} w(t)$; using the description of n given in Theorem (4.1), prove that

$$n_+(M \geq x) = \lim_{s \to 0} \left(\int_0^x \lambda_s(dy) Q_y [T_x < T_0] + \int_x^\infty \lambda_s(dy) \right)$$

and derive anew the law of M under n_+, which was already found in Exercise (2.10) and Proposition (3.6).

2°) Prove that $M_x = \sup \{B_t, t < g_{T_x}\}$ is uniformly distributed on $[0, x]$ (a part of Williams' decomposition theorem).

[Hint: If M_x is less than y, for $y \leq x$, the first excursion which goes over x is also the first to go over y.]

3°) By the same method as in 1°), give another proof of Proposition (2.8).

\# **(4.14) Exercise. (An excursion approach to Skorokhod problem).** We use the notation of Sect. 5 Chap. VI; we suppose that ψ_μ is continuous and strictly increasing and call ϕ its inverse.

1°) The stopping times

$$T = \inf\{t : S_t \geq \psi_\mu(B_t)\} \quad \text{and} \quad T' = \inf\{t : |B_t| \geq L_t - \phi(L_t)\}$$

have the same law.

2°) Prove that, in the notation of this chapter, the process $\{s, e_s\}$ is a PPP with values in $\mathbb{R}_+ \times U_\delta$ and characteristic measure $ds\, dn(u)$.

3°) Let $\Gamma_x = \{(s, u) \in \mathbb{R}_+ \times U : 0 \leq s \leq x \text{ and } M(u) \geq s - \phi(s)\}$ and $N_x = \sum_s 1_{\Gamma_x}(s, e_s)$. Prove that $P[L_{T'} \geq x] = P[N_x = 0]$ and derive therefrom that $\phi(S_T) = B_T$ has the law μ.

4°) Extend the method to the case where ψ_μ is merely right-continuous.

(4.15) Exercise. If $A = L^z$, $z > 0$, prove that the Lévy measure m_A of A_{τ_t} defined in Proposition (2.7) is given by

$$m_A(\,]x, \infty[\,) = (2z)^{-1} \exp(-x/2z), \quad x > 0.$$

* **(4.16) Exercise (Azéma martingale).** 1°) Using Proposition (3.3) prove, in the notation thereof, that if f is a function such that $f(|B_t|)$ is integrable,

$$E\left[f(|B_t|) \mid \check{\mathscr{F}}_t \right] = A_t^{-1} \int_0^\infty \exp\left(-y^2/2A_t\right) y f(y) dy.$$

[Hint: Write $B_t = B_{g_t + t - g_t} = B_{A_t}\left(i_{g_t}\right)$.]

2°) By applying 1°) to the functions $f(y) = y^2$ and $|y|$, prove that $t - 2g_t$ and $\sqrt{\frac{\pi}{2}(t - g_t)} - L_t$ are $(\check{\mathscr{F}}_t)$-martingales.

3°) If f is a bounded function with a bounded first derivative f' then

$$f(L_t) - \sqrt{\frac{\pi}{2}(t - g_t)} f'(L_t)$$

is a $(\check{\mathscr{F}}_t)$-martingale.

(4.17) Exercise. Let $\tau_a = \inf\{t : L_t = a\}$. Prove that the processes $\{B_t, t \leq \tau_a\}$ and $\left\{B_{\tau_a - t}, t \leq \tau_a\right\}$ are equivalent.

[Hint: Use Proposition (2.5) and Corollary (4.3).]

*# **(4.18) Exercise.** In the notation of Proposition (4.6) and Theorem (4.7), for P_0-almost every path, we may define the local time at 0 and its inverse τ_t; thus $P_0^{\tau_s}$ makes sense and is a probability measure on $\mathscr{X}$.

1°) Prove that

$$\int_0^\infty P_0^t dt = \int_0^\infty P_0^{\tau_s} ds \circ \int_0^\infty n^r(\cdot \cap (r < R)) dr.$$

This formula in another guise may also be derived without using excursion theory as may be seen in Exercise (4.26). We recall that it was proved in Exercise (2.29) of Chap. VI that

$$\int_0^\infty P_0^{\tau_s} ds = \int_0^\infty Q_t \frac{dt}{\sqrt{2\pi t}}$$

where Q_t is the law of the Brownian Bridge over the interval $[0, t]$.

2°) Call M^t the law of the Brownian meander of length t defined by scaling from the meander of length 1 (Exercise (3.8)) and prove that

$$n^t(\cdot \cap (t < R)) = M^t / \sqrt{2\pi t}.$$

As a result

$$(*) \qquad \int_0^\infty M^t \frac{dt}{\sqrt{2\pi t}} = \int_0^\infty S_3^{L_a} da.$$

3°) Derive from (*) *Imhof's relation*, i.e., for every t

$$(+) \qquad M^t = (\pi t/2)^{1/2} X_t^{-1} S_3^t$$

where $X_t(\omega) = \omega(t)$ is the coordinate process.

[Hint: In the left-hand side of $(*)$ use the conditioning given $L_a = t$, then use the result of Exercise (3.12) Chap. XI.]

By writing down the law of (ζ, X_ζ) under the two sides of $(*)$, one also finds the law of X_t under M^t which was already given in Exercise (3.8).

4°) Prove that (+) is equivalent to the following property: for any bounded continuous functional F on $C([0,1], \mathbb{R})$,

$$M^1(F) = \lim_{r \downarrow 0} (\pi/2)^{1/2} E_r \left[F 1_{[T_0 > 1]}\right] / r$$

where P_r is the probability measure of BM(r) and T_0 is the first hitting time of 0. This question is not needed for the sequel.

[Hint: Use Exercise (1.22) Chap. XI.]

5°) On $C([0,1], \mathbb{R})$, set $\mathscr{B}_t = \sigma(X_s, s \leq t)$. Prove that for $0 \leq t \leq 1$,

$$S_3^1\left[(\pi/2)^{1/2} X_1^{-1} \mid \mathscr{B}_t\right] = X_t^{-1} \phi\left((1-t)^{-1/2} X_t\right)$$

where $\phi(a) = \int_0^a \exp(-y^2/2)dy$. Observe that this shows, in the fundamental counterexample of Exercise (2.13) Chap. V, how much $1/X_t$ differs from a martingale.

[Hint: Use the Markov property of BES^3.]

6°) Prove that, under M^1, there is a Brownian motion β such that

$$X_t = \beta_t + \int_0^t \left(\frac{\phi'}{\phi}\right)\left(\frac{X_s}{\sqrt{1-s}}\right)\frac{ds}{\sqrt{1-s}}, \qquad 0 \le t \le 1,$$

which shows that the meander is a semimartingale and gives its decomposition in its natural filtration.

[Hint: Apply Girsanov's theorem with the martingale of 5°).]

** **(4.19) Exercise (Longest excursions).** If $B_s = 0$, call $D(s)$ the length of the longest excursion which occured before time s. The aim of this exercise is to find the law of $D(g_t)$ for a fixed t. For $\beta > 0$, we set

$$c_\beta = \int_0^\infty (1 - e^{-\beta t})(2\pi t^3)^{-1/2}dt, \qquad d_\beta(x) = \int_x^\infty e^{-\beta t}(2\pi t^3)^{-1/2}dt$$

and

$$\phi_s(x, \beta) = E\left[1_{(D(\tau_s)>x)} \exp(-\beta\tau_s)\right].$$

1°) If $L_\beta(x) = E\left[\int_0^\infty \exp(-\beta t) 1_{(D(g_t)>x)}dt\right]$, prove that

$$\beta L_\beta(x) = c_\beta \int_0^\infty \phi_s(x, \beta)ds.$$

2°) By writing

$$\phi_t(x, \beta) = E\left[\sum_{s\le t}\left\{1_{(D(\tau_s)>x)}\exp(-\beta\tau_s) - 1_{(D(\tau_{s-})>x)}\exp(-\beta\tau_{s-})\right\}\right],$$

prove that ϕ satisfies the equation

$$\phi_t(x, \beta) = -\left(c_\beta + d_\beta(x)\right)\int_0^t \phi_s(x, \beta)ds + d_\beta(x)\int_0^t e^{-c_\beta s}ds.$$

3°) Prove that

$$\beta L_\beta(x) = d_\beta(x)/\left(c_\beta + d_\beta(x)\right).$$

[Hint: $\{D(\tau_s) > x\} = \{D(\tau_{s-}) > x\} \cup \{\tau_s - \tau_{s-} > x\}$.]

4°) Solve the same problem with $D(d_t)$ in lieu of $D(g_t)$.

5°) Use the scaling property of BM to compute the Laplace transforms of $(D(g_1))^{-1}$ and $(D(d_1))^{-1}$.

** **(4.20) Exercise.** Let A be an additive functional of BM with associated measure μ and S_θ an independent exponential r.v. with parameter $\theta^2/2$.

1°) Use Exercise (4.18) 1°) to prove that for $\lambda > 0$,

$$E_0\left[\exp\left(-\lambda A_{S_\theta}\right)\right]$$
$$= \frac{\theta^2}{2}\int_0^\infty E_0\left[\exp\left(-\lambda A_{\tau_s} - \frac{\theta^2}{2}\tau_s\right)\right]ds \int_{-\infty}^{+\infty} E_a\left[\exp\left(-\lambda A_{T_0} - \frac{\theta^2}{2}T_0\right)\right]da.$$

2°) If ϕ and ψ are suitable solutions of the Sturm-Liouville equation $\phi'' = 2\left(\lambda\mu + \frac{\theta^2}{2}\right)\phi$, then

$$E\left[\exp(-\lambda A_{S_\theta})\right] = \left(\theta^2/2\phi'(0+)\right)\int_{-\infty}^{+\infty}\psi(a)da.$$

3°) With the notation of Theorem (2.7) Chap. VI find the explicit values of the expressions in 1°) for $A_t = A_t^+$ and derive therefrom another proof of the arcsine law. This question is independent of 2°).

[Hint: Use the independence of $A_{\tau_s}^+$ and $A_{\tau_s}^-$, the fact that $\tau_s = A_{\tau_s}^+ + A_{\tau_s}^-$ and the results in Propositions (2.7) and (2.8).]

** **(4.21) Exercise (Lévy-Khintchine formula for BESQ^δ).** If l^a is a family of local times of the Brownian excursion (see Exercise (4.9)), call M the image of n^+ under the map $u \to (a \to l^a_{R(u)}(u))$. The measure M is a measure on $\mathbf{W}_+ = C(\mathbb{R}_+, \mathbb{R}_+)$. If $f \in \mathbf{W}_+$ and X is a process we set

$$X_f = \int_0^\infty f(t)X_t dt.$$

1°) With the notation of Sect. 1 Chap. XI prove that for $x \geq 0$

$$Q_x^0\left[\exp(-X_f)\right] = \exp\left\{-x\int(1-\exp(-\langle f,\phi\rangle))M(d\phi)\right\}$$

where $\langle f,\phi\rangle = \int_0^\infty f(t)\phi(t)dt$.

[Hint: Use the second Ray-Knight theorem and Proposition (1.12).]

2°) For $\phi \in \mathbf{W}_+$ call ϕ^s the function defined by $\phi^s(t) = \phi\left((t-s)^+\right)$ and put $N = \int_0^\infty M_s ds$ where M_s is the image of M by the map $\phi \to \phi^s$. Prove that

$$Q_0^2\left[\exp(-X_f)\right] = \exp\left\{-2\int(1-\exp(-\langle f,\phi\rangle))N(d\phi)\right\}.$$

[Hint: Use 1°) in Exercise (2.7) of Chap. XI and the fact that for a BM the process $|B| + L$ is a $\mathrm{BES}^3(0)$.]

The reader is warned that M_s has nothing to do with the law M^s of the meander in Exercise (4.18).

3°) Conclude that

$$Q_x^\delta\left[\exp(-X_f)\right] = \exp\left\{-\int(1-\exp(-\langle f,\phi\rangle))(xM+\delta N)(d\phi)\right\}.$$

4°) Likewise prove a similar Lévy-Khintchine representation for the laws $Q^{\delta,\cdot}_{x,0}$ of the squares of Bessel bridges ending at 0; denote by M_0 and N_0 the corresponding measures, which are now defined on $C([0,1];\mathbb{R}_+)$.

5°) For a subinterval I of $\mathbb{R}_+$, and $x, y \in I$, with $x < y$, let $P_{x,y}$ be the probability distribution on $C(I,\mathbb{R}_+)$ of a process $X_{x,y}$ which vanishes off the interval (x, y), and on (x, y), is a $\mathrm{BES}Q^4_{y-x}(0,0)$ that is

$$X_{x,y}(v) = (y-x)Z\left(\frac{v-x}{y-x}\right)1_{(x\le v\le y)} \qquad (v \in I)$$

where Z has distribution $Q^{4,1}_{0,0}$.

Prove that the Lévy measures encountered above may be represented by the following integrals:

$$M = \frac{1}{2}\int_0^\infty y^{-2}P_{0,y}\,dy; \quad N = \frac{1}{2}\int_0^\infty dx\int_x^\infty (y-x)^{-2}P_{x,y}\,dy;$$

$$M_0 = \frac{1}{2}\int_0^1 y^{-2}P_{0,y}\,dy; \quad N_0 = \frac{1}{2}\int_0^1 dx\int_x^1 (y-x)^{-2}P_{x,y}\,dy.$$

\# **(4.22) Exercise.** Let ϕ and f be positive Borel functions on the appropriate spaces. Prove that

$$\int n_+(de)\int_0^{R(e)} \phi(s)f(e_s)ds = 2\int n_+(de)\phi(R(e))\int_0^{R(e)} f(2e_s)ds.$$

[Hint: Compute the left member with the help of Exercise (4.17) 2°) and the right one by using Theorem (4.1).]

* **(4.23) Exercise.** Prove that Theorem (4.7) is equivalent to the following result. Let ξ be the measure on $\mathbb{R}_+ \times \mathbf{W} \times \mathbf{W}$ given by

$$\xi(dt, dw, dw') = 1_{(t>0)}dt\ S_3(dw)S_3(dw')$$

and set $L(t, w) = \sup\{s : w(s) = t\}$. If we define an U_δ-valued variable e by

$$e_s(t,w,w') = \begin{cases} w(s) & \text{if } 0 \le s \le L(t,w) \\ w'(L(t,w)+L(t,w')-s) & \text{if } L(t,w) \le s \le L(t,w)+L(t,w') \\ 0 & \text{if } s \ge L(t,w)+L(t,w'), \end{cases}$$

then the law of (L, e) under ξ is equal to $\bar{n}_+$.

** **(4.24) Exercise (Chung-Jacobi-Riemann identity).** Let B be the standard BM and T an exponential r.v. with parameter $1/2$, independent of B.

1° Prove that for every positive measurable functional F,

$$E\left[F(B_u; u \le g_T) \mid L_T = s\right] = e^s E\left[F(B_u; u \le \tau_s)\exp(-\tau_s/2)\right],$$

and consequently that

$$E\left[F\left(B_u; u \leq g_T\right)\right] = \int_0^\infty E\left[F\left(B_u; u \leq \tau_s\right)\exp\left(-\tau_s/2\right)\right] ds.$$

2°) Let S^0, I^0 and l^0 denote respectively the supremum, the opposite of the infimum and the local time at 0 of the standard Brownian bridge $(b(t); t \leq 1)$. Given a $\mathcal{N}(0,1)$ Gaussian r.v. N independent of b, prove the three variate formula

$$P\left[|N|S^0 \leq x; |N|I^0 \leq y; |N|l^0 \in dl\right] = \exp(-l(\coth x + \coth y)/2)dl.$$

3°) Prove as a result that

$$P\left[|N|S^0 \leq x; |N|I^0 \leq y\right] = 2/(\coth x + \coth y)$$

and that, if $M^0 = \sup\{|b(s)|; s \leq 1\}$,

$$P\left[|N|M^0 \leq x\right] = \tanh x.$$

Prove *Csáki's formula*:

$$P\{S^0/S^0 + I^0 \leq v\} = (1-v)(1-\pi v\cot(\pi v)) \quad (0 < v < 1)$$

[Hint: Use the identity:

$$2v^2\int_0^\infty d\lambda\left(\frac{\sinh(v\lambda)}{v\sinh(\lambda)}\right)^2 = 1 - \pi v\cot(\pi v). \quad].$$

4°) Prove the Chung-Jacobi-Riemann identity:

$$\left(S^0 + I^0\right)^2 \overset{(d)}{=} \left(M^0\right)^2 + \left(\widetilde{M}^0\right)^2$$

where $\widetilde{M}^0$ is an independent copy of M^0.

5°) Characterize the pairs (S, I) of positive r.v.'s such that

i) $P[|N|S \leq x; |N|I \leq y] = 2/(h(x) + h(y))$ for a certain function h,
ii) $(S+I)^2 \overset{(d)}{=} M^2 + \widetilde{M}^2$,

where M and $\widetilde{M}$ are two independent copies of $S \vee I$.

(4.25) Exercise. (Brownian meander and Brownian bridges). Let $a \in \mathbb{R}$, and let Π_a be the law of the Brownian bridge $(B_t, t \leq 1)$, with $B_0 = 0$ and $B_1 = a$. Prove that, under Π_a, both processes $(2S_t - B_t, t \leq 1)$ and $(|B_t| + L_t, t \leq 1)$ have the same distribution as the Brownian meander $(m_t, t \leq 1)$ conditioned on $(m_1 \geq |a|)$.

[Hint: Use the relation (+) in Exercise (4.18) together with Exercise (3.20) in Chap. VI.]

In particular, the preceding description for $a = 0$ shows that, if $(b_t, t \leq 1)$ is a standard Brownian bridge, with $\sigma_t = \sup_{s \leq t} b_s$, and $(l_t, t \leq 1)$ its local time at 0, then

$$(m_t, t \leq 1) \overset{(d)}{=} (2\sigma_t - b_t, t \leq 1) \overset{(d)}{=} (|b_t| + l_t, t \leq 1).$$

Prove that under the probability measure $(\sigma_1/c) \cdot \Pi_0$ (resp. $(l_1/c) \cdot \Pi_0$), c being the normalization constant, the process $(2\sigma_t - b_t, t \leq 1)$ (resp. $(|b_t| + l_t, t \leq 1)$) is a BES^3.

(4.26) Exercise. 1°) With the notation of this section set $J = \int_0^\infty P_0^t\,dt$ and prove that

$$J = \int_{-\infty}^{\infty} da \int_0^\infty P_0^{\tau_s^a}\,ds = \int_{-\infty}^{\infty} da \int_0^\infty \left(P_0^{T_a} \circ P_0^{\tau_s^0}\right) ds,$$

where $\tau_s^a = \inf\{t : L_t^a > s\}$.

[Hint: Use the generalized occupation times formula of Exercise (1.13) Chap. VI.]

2°) Define a map $\omega \to \tilde{\omega}$ on $\mathscr{X}$ by

$$\zeta(\tilde{\omega}) = \zeta(\omega) \quad \text{and} \quad \tilde{\omega}(t) = \omega(0) + \omega(\zeta(\omega)) - \check{\omega}(t),$$

and call $\tilde{\mu}$ the image by this map of the measure μ. Prove that $\tilde{J} = J$ and that

$$(\mu \circ \mu')^{\sim} = \tilde{\mu}' \circ \tilde{\mu}$$

for any pair (μ, μ') of measures on $\mathscr{X}$.

3°) Prove that $P_0^{\tau_s^0} = \widetilde{\left(P_0^{\tau_s^0}\right)}$ and conclude that

$$J = \left(\int_0^\infty P_0^{\tau_s^0}\,ds\right) \circ \left(\int_{-\infty}^{\infty} \check{\left(P_0^{T_a}\right)} da\right).$$

[Hint: See Exercise (2.29) Chap. VI.]

Notes and Comments

Sect. 1. This section is taken mainly from Itô [5] and Meyer [4].

Exercise (1.19) comes from Pitman-Yor [8].

Sect. 2. The first breakthrough in the description of Brownian motion in terms of excursions and Poisson point processes was the paper of Itô [5]. Although some ideas were already, at an intuitive level, in the work of Lévy, it was Itô who put the subject on a firm mathematical basis, thus supplying another cornerstone to Probability Theory. Admittedly, once the characteristic measure is known all sorts of computations can be carried through as, we hope, is clear from the exercises of the following sections. For the results of this section we also refer to Maisonneuve [6] and Pitman [4].

The approximation results such as Proposition (2.9), Exercise (2.14) and those already given in Chap. VI were proved or conjectured by Lévy. The proofs were given and gradually simplified in Itô-McKean [1], Williams [6], Chung-Durrett [1] and Maisonneuve [4].

Exercise (2.17) may be extended to the computation of the distribution of the multidimensional time spent in the different rays by a Walsh Brownian motion (see Barlow et al. [1] (1989)).

Sect. 3. In this section, it is shown how the *global excursion* theory, presented in Section 2, can be applied to describe the laws of *individual excursions*, i.e. excursions straddling a given random time T. We have presented the discussion only for stopping times T w.r.t. the filtration $(\check{\mathscr{F}}_t) = (\mathscr{F}_{g_t})$, and terminal $(\mathscr{F}_t)$ stopping times. See Maisonneuve [7] for a general discussion. The *canevas* for this section is Getoor-Sharpe [5] which is actually written in a much more general setting. We also refer to Chung [1]. The filtration $(\mathscr{F}_{g_t})$ was introduced and studied in Maisonneuve [6].

The Brownian Meander of Exercise (3.8) has recently been much studied (see Imhof ([1] and [2]), Durrett et al [1], Denisov [1] and Biane-Yor [3]). It has found many applications in the study of Azéma's martingale (see Exercise (4.16) taken from Azéma-Yor [3]).

Sect. 4. Theorems (4.1) and (4.2) are fundamental results of Itô [5]. The proof of Corollary (4.4) is taken from Ikeda-Watanabe [2].

Williams' description of the Itô measure is found in Williams [7] and Rogers-Williams [1] (see also Rogers [1]) and Bismut's description appeared in Bismut [3]. The formalism used in the proof of the latter as well as in Exercise (4.18) was first used in Biane-Yor [1]. The paper of Bismut contains further information which was used by Biane [1] to investigate the relationship between the Brownian Bridge and the Brownian excursion and complement the result of Vervaat [1].

Exercise (4.8) is due to Rogers [3]. Knight's identity (Knight [8]) derived in Exercise (4.11) has been explained in Biane [2] and Vallois [3] using a pathwise decomposition of the pseudo-Brownian bridge (cf. Exercise (2.29) Chap. VI); generalizations to Bessel processes (resp. perturbed Brownian motions) have been given by Pitman-Yor [9] (resp. [23]). The Watanabe process appears in Watanabe [2]. Exercise (4.14) is from Rogers [1]. Exercise (4.16) originates with Azéma [2] and Exercise (4.17) with Biane et al. [1]. Exercise (4.18) is taken partly from Azéma-Yor [3] and partly from Biane-Yor ([1] and [3]) and Exercise (4.19) from Knight [6]. Exercise (4.20) is in Biane-Yor [4] and Exercise (4.21) in Pitman-Yor [2]; further results connecting the Brownian bridge, excursion and meander are presented in Bertoin-Pitman [1] and Bertoin et al. [1].

With the help of the explicit Lévy-Khintchine representation of Q_x^δ obtained in Exercise (4.21), Le Gall-Yor [5] extend the Ray-Knight theorems on Brownian local times by showing that, for any $\delta > 0$, Q_0^δ is the law of certain local times processes in the space variable. In the same Exercise (4.21), the integral representations of M, N, M_0 and N_0 in terms of squares of BES^4 bridges are taken from Pitman [5]. Exercise (4.22) is in Azéma-Yor [3], and Exercise (4.23) originates from Bismut [3].

The joint law of the supremum, infimum and local time of the Brownian bridge is characterized in Exercise (4.24), taken from work in progress by Pitman and Yor. The presentation which involves an independent Gaussian random variable, differs from classical formulae, in terms of theta functions, found in the literature (see e.g. Borodin and Salminen [1]). Csáki's formula in question 3°) comes from Csáki [1] and is further discussed in Pitman-Yor [13]. Chung's identity of question 4°) remains rather mysterious, although Biane-Yor [1] and Williams [9] explain partly its relation to the functional equation of the Riemann zeta function. See also

Smith and Diaconis [1] for a random walk approach to the functional equation, and Biane-Pitman-Yor [1] for further developments.

Exercise (4.25) is a development and an improvement of the corresponding result found in Biane-Yor [3] for $a = 0$, and of the remark following Theorem 4.3 in Bertoin-Pitman [1]. The simple proof of Exercise (4.26) is taken from Leuridan [1].

Chapter XIII. Limit Theorems in Distribution

§1. Convergence in Distribution

In this section, we will specialize the notions of Sect. 5 Chap. 0 to the Wiener space $\mathbf{W}^d$. This space is a Polish space when endowed with the topology of uniform convergence on compact subsets of $\mathbb{R}_+$. This topology is associated with the metric

$$d(\omega, \omega') = \sum_1^\infty 2^{-n} \frac{\sup_{t\leq n} |\omega(t) - \omega'(t)|}{1 + \sup_{t\leq n} |\omega(t) - \omega'(t)|} .$$

The relatively compact subsets in this topology are given by Ascoli's theorem. Let

$$V^N(\omega, \delta) = \sup\left\{|\omega(t) - \omega(t')|;\ |t - t'| \leq \delta \text{ and } t, t' \leq N\right\}.$$

With this notation, we have

(1.1) Proposition. *A subset Γ of $\mathbf{W}^d$ is relatively compact if and only if*

(i) the set $\{\omega(0), \omega \in \Gamma\}$ is bounded in $\mathbb{R}^d$;
(ii) for every N,

$$\lim_{\delta\downarrow 0} \sup_{\omega\in\Gamma} V^N(\omega, \delta) = 0.$$

In Sect. 5 Chap. 0, we have defined a notion of weak convergence for probability measures on the Borel σ-algebra of $\mathbf{W}^d$; the latter is described in the following

(1.2) Proposition. *The Borel σ-algebra on $\mathbf{W}^d$ is equal to the σ-algebra $\mathscr{F}$ generated by the coordinate mappings.*

Proof. The coordinate mappings are clearly continuous, hence $\mathscr{F}$ is contained in the Borel σ-algebra. To prove the reverse inclusion, we observe that by the definition of d, the map $\omega \to d(\omega, \omega')$ where ω' is fixed, is $\mathscr{F}$-measurable. As a result, every ball, hence every Borel set, is in $\mathscr{F}$.

Before we proceed, let us observe that the same notions take on a simpler form when the time range is reduced to a compact interval, but we will generally work with the whole half-line.

(1.3) Definition. *A sequence (X^n) of $\mathbb{R}^d$-valued continuous processes defined on probability spaces $(\Omega^n, \mathscr{F}^n, P^n)$ is said to* converge in distribution to a process *X if the sequence $(X^n(P^n))$ of their laws converges weakly on $\mathbf{W}^d$ to the law of X. We will write $X^n \xrightarrow{(d)} X$.*

In this definition, we have considered processes globally as $\mathbf{W}^d$-valued random variables. If we consider processes taken at some fixed times, we get a weaker notion of convergence.

(1.4) Definition. *A sequence (X^n) of (not necessarily continuous) $\mathbb{R}^d$-valued processes is said to converge to the process X in the* sense of finite distributions *if for any finite collection $(t_1, \dots, t_k)$ of times, the $\mathbb{R}^{dk}$-valued r.v.'s $\left(X^n_{t_1}, \dots, X^n_{t_k}\right)$ converge in law to $\left(X_{t_1}, \dots, X_{t_k}\right)$. We will write $X^n \xrightarrow{\text{f.d.}} X$.*

Since the map $\omega \to (\omega(t_1), \dots, \omega(t_k))$ is continuous on $\mathbf{W}^d$, it is easy to see that, if $X_n \xrightarrow{(d)} X$, then $X_n \xrightarrow{\text{f.d.}} X$. The converse is not true, and in fact continuous processes may converge in the sense of finite distributions to discontinuous processes as was seen in Sect. 4 of Chap. X and will be seen again in Sect. 3.

The above notions make sense for multi-indexed processes or in other words for $C\left((\mathbb{R}_+)^l, \mathbb{R}^d\right)$ in lieu of the Wiener space. We leave to the reader the task of writing down the extensions to this case (see Exercise (1.12)).

Convergence in distribution of a sequence of probability measures on $\mathbf{W}^d$ is fairly often obtained in two steps:

i) the sequence is proved to be weakly relatively compact;
ii) all the limit points are shown to have the same set of finite-dimensional distributions.

In many cases, one gets ii) by showing directly that the finite dimensional distributions converge, or in other words that there is convergence in the sense of finite distributions. To prove the first step above, it is usually necessary to use Prokhorov's criterion which we will now translate in the present context. Let us first observe that the function $V^N(\cdot, \delta)$ is a random variable on $\mathbf{W}^d$.

(1.5) Proposition. *A sequence (P_n) of probability measures on $\mathbf{W}^d$ is weakly relatively compact if and only if the following two conditions hold:*

i) for every $\varepsilon > 0$, there exist a number A and an integer n_0 such that

$$P_n[\,|\,\omega(0)| > A] \leq \varepsilon, \qquad \textit{for every } n \geq n_0;$$

ii) for every η, $\varepsilon > 0$ and $N \in \mathbb{N}$, there exist a number δ and an integer n_0 such that

$$P_n\left[V^N(\cdot, \delta) > \eta\right] \leq \varepsilon \qquad \textit{for every } n \geq n_0.$$

Remark. We will see in the course of the proof that we can actually take $n_0 = 0$.

Proof. The necessity with $n_0 = 0$ follows readily from Proposition (1.1) and Prokhorov's criterion of Sect. 5 Chap. 0.

Let us turn to the sufficiency. We assume that conditions i) and ii) hold. For every n_0, the finite family $(P_n)_{n\le n_0}$ is tight, hence satisfies i) and ii) for numbers A' and δ'. Therefore, by replacing A by $A \vee A'$ and δ by $\delta \wedge \delta'$, we may as well assume that conditions i) and ii) hold with $n_0 = 0$. This being so, for $\varepsilon > 0$ and $N \in \mathbb{N}$, let us pick $A_{N,\varepsilon}$ and $\delta_{N,k,\varepsilon}$ such that

$$\begin{aligned}\sup_n P_n\left[\,|\,\omega(0)| > A_{N,\varepsilon}\right] &\le 2^{-N-1}\varepsilon,\\ \sup_n P_n\left[V^N(\cdot,\delta_{N,k,\varepsilon}) > 1/k\right] &\le 2^{-N-k-1}\varepsilon,\end{aligned}$$

and set $K_{N,\varepsilon} = \left\{\omega : |\,\omega(0)| \le A_{N,\varepsilon},\quad V^N(\omega,\delta_{N,k,\varepsilon}) \le 1/k \text{ for every } k \ge 1\right\}$. By Proposition (1.1), the set $K_\varepsilon = \bigcap_N K_{N,\varepsilon}$ is relatively compact in $\mathbf{W}^d$ and we have $P_n\left(K_\varepsilon^c\right) \le \sum_N P_N\left(K_{N,\varepsilon}^c\right) \le \varepsilon$, which completes the proof. □

We will use the following

(1.6) Corollary. *If $X^n = \left(X_1^n,\dots,X_d^n\right)$ is a sequence of d-dimensional continuous processes, the set $(X^n(P^n))$ of their laws is weakly relatively compact if and only if, for each j, the set of laws $X_j^n(P^n)$ is weakly relatively compact.*

Hereafter, we will need a condition which is slightly stronger than condition ii) in Proposition (1.5).

(1.7) Lemma. *Condition ii) in Proposition (1.5) is implied by the following condition: for any N and $\varepsilon, \eta > 0$, there exist a number δ, $0 < \delta < 1$, and an integer n_0, such that*

$$\delta^{-1}P_n\left[\left\{\omega:\ \sup_{t\le s\le t+\delta}|\,\omega(s)-\omega(t)| \ge \eta\right\}\right] \le \varepsilon \text{ for } n \ge n_0 \text{ and for all } t \le N.$$

Proof. Let N be fixed, pick $\varepsilon, \eta > 0$ and let n_0 and δ be such that the condition in the statement holds. For every integer i such that $0 \le i < N\delta^{-1}$, define

$$A_i = \left\{\sup_{i\delta\le s\le (i+1)\delta\wedge N}|\,\omega(i\delta)-\omega(s)| \ge \eta\right\}.$$

As is easily seen $\left\{V^N(\cdot,\delta) < 3\eta\right\} \supset \bigcap_i A_i^c$, and consequently for every $n \ge n_0$, we get

$$P_n\left[V^N(\cdot,\delta) \ge 3\eta\right] \le P_n\Big[\bigcup_i A_i\Big] \le (1+[N\delta^{-1}])\delta\varepsilon < (N+1)\varepsilon$$

which proves our claim. □

The following result is very useful.

(1.8) Theorem (Kolmogorov's criterion for weak compactness). *Let (X^n) be a sequence of $\mathbb{R}^d$-valued continuous processes such that*

i) the family $\{X_0^n(P^n)\}$ of initial laws is tight in $\mathbb{R}^d$,
ii) there exist three strictly positive constants α, β, γ such that for every $s, t \in \mathbb{R}_+$ and every n,

$$E_n\left[\left|X_s^n - X_t^n\right|^\alpha\right] \leq \beta|s-t|^{\gamma+1};$$

then, the set $\{X^n(P^n)\}$ of the laws of the X_n's is weakly relatively compact.

Proof. Condition i) implies condition i) of Proposition (1.5), while condition ii) of Proposition (1.5) follows at once from Markov inequality and the result of Theorem (2.1) (or its extension in Exercise (2.10)) of Chap. I. □

We now turn to a first application to Brownian motion. We will see that the Wiener measure is the weak limit of the laws of suitably interpolated random walks. Let us mention that the existence of Wiener measure itself can be proved by a simple application of the above ideas.

In what follows, we consider a sequence of independent and identically distributed, centered random variables ξ_k such that $E\left[\xi_k^2\right] = \sigma^2 < \infty$. We set $S_0 = 0$, $S_n = \sum_{k=1}^n \xi_k$. If $[x]$ denotes the integer part of the real number x, we define the continuous process X^n by

$$X_t^n = \left(\sigma\sqrt{n}\right)^{-1}\left(S_{[nt]} + (nt - [nt])\xi_{[nt]+1}\right).$$

(1.9) Theorem (Donsker). *The processes X^n converge in distribution to the standard linear Brownian motion.*

Proof. We first prove the convergence of finite-dimensional distributions. Let $t_1 < t_2 < \ldots < t_k$; by the classical central limit theorem and the fact that $[nt]/n$ converges to t as n goes to $+\infty$, it is easily seen that $\left(X_{t_1}^n, X_{t_2}^n - X_{t_1}^n, \ldots, X_{t_k}^n - X_{t_{k-1}}^n\right)$ converges in law to $\left(B_{t_1}, B_{t_2} - B_{t_1}, \ldots, B_{t_k} - B_{t_{k-1}}\right)$ where B is a standard linear BM. The convergence of finite-dimensional distributions follows readily.

Therefore, it is sufficient to prove that the set of the laws of the X_n's is weakly relatively compact. Condition i) of Proposition (1.5) being obviously in force, it is enough to show that the condition of Lemma (1.7) is satisfied.

Assume first that the ξ_k's are bounded. The sequence $|S_k|^4$ is a submartingale and therefore for fixed n

$$P\left[\max_{i\leq n}|S_i| > \lambda\sigma\sqrt{n}\right] \leq E\left[|S_n|^4\right]\left(\lambda\sigma\sqrt{n}\right)^{-4}.$$

One computes easily that $E\left[S_n^4\right] = nE\left[\xi_1^4\right] + 3n(n-1)\sigma^4$. As a result, there is a constant K independent of the law of ξ_k such that

$$\overline{\lim_{n\to\infty}}\, P\left[\max_{i\leq n}|S_i| > \lambda\sigma\sqrt{n}\right] \leq K\lambda^{-4}.$$

By truncating and passing to the limit, it may be proved that this is still true if we remove the assumption that ξ_k is bounded. For every $k \geq 1$, the sequence

$\{S_{n+k} - S_k\}$ has the same law as the sequence $\{S_n\}$ so that finally, there exists an integer n_1 such that

$$P\left[\max_{i \le n} |S_{i+k} - S_k| > \lambda\sigma\sqrt{n}\right] \le K\lambda^{-4}$$

for every $k \ge 1$ and $n \ge n_1$. Pick ε and η such that $0 < \varepsilon, \eta < 1$ and then choose λ such that $K\lambda^{-2} < \eta\varepsilon^2$; set further $\delta = \varepsilon^2\lambda^{-2}$ and choose $n_0 > n_1\delta^{-1}$. If $n \ge n_0$, then $[n\delta] \ge n_1$, and the last displayed inequality may be rewritten as

$$P\left[\max_{i \le [n\delta]} |S_{i+k} - S_k| \ge \lambda\sigma\sqrt{[n\delta]}\right] \le \eta\varepsilon^2\lambda^{-2}.$$

Since $\lambda\sqrt{[n\delta]} \le \varepsilon\sqrt{n}$, we get

$$\delta^{-1}P\left[\max_{i \le [n\delta]} |S_{i+k} - S_k| \ge \varepsilon\sigma\sqrt{n}\right] \le \eta$$

for every $k \ge 1$ and $n \ge n_0$. Because the X_n's are linear interpolations of the random walk (S_n), it is now easy to see that the condition in Lemma (1.7) is satisfied for every N and we are done. □

To illustrate the use of weak convergence as a tool to prove existence results, we will close this section with a result on solutions to martingale problems. At no extra cost, we will do it in the setting of Itô processes (Definition (2.5), Chap. VII).

We consider functions a and b defined on $\mathbb{R}_+ \times \mathbf{W}^d$ with values respectively in the sets of symmetric non-negative $d \times d$-matrices and $\mathbb{R}^d$-vectors. We assume these functions to be progressively measurable with respect to the filtration $(\mathscr{F}_t^0)$ generated by the coordinate mappings $\omega(t)$. The reader is referred to the beginning of Sect. 1 Chap. IX. With the notation of Sect. 2 Chap. VII, we may state

(1.10) Theorem. *If a and b are continuous on $\mathbb{R}_+ \times \mathbf{W}^d$, then for any probability measure μ on $\mathbb{R}^d$, there exists a probability measure P on $\mathbf{W}^d$ such that*

i) $P[\omega(0) \in A] = \mu(A)$;
ii) for any $f \in C_K^2$, *the process* $f(\omega(t)) - f(\omega(0)) - \int_0^t L_s f(\omega(s))ds$ *is a* $(\mathscr{F}_t^0, P)$*-martingale, where*

$$L_s f(\omega(s)) = \frac{1}{2}\sum a_{ij}(s,\omega)\frac{\partial^2 f}{\partial x_i \partial x_k}(\omega(s)) + \sum_i b_i(s,\omega)\frac{\partial f}{\partial x_i}(\omega(s)).$$

Proof. For each integer n, we define functions a_n and b_n by

$$a_n(t,\omega) = a([nt]/n, \omega), \qquad b_n(t,\omega) = b([nt]/n, \omega).$$

These functions are obviously progressively measurable and we call L_s^n the corresponding differential operators.

Pick a probability space $(\Omega, \mathscr{F}, P)$ on which a r.v. X_0 of law μ and a $\mathrm{BM}^d(0)$ independent of X_0, say B, are defined. Let σ_n be a square root of a_n. We define inductively a process X^n in the following way; we set $X_0^n = X_0$ and if X^n is defined up to time k/n, we set for $k/n < t \leq (k+1)/n$,

$$X_t^n = X_{k/n}^n + \sigma_n(k/n, X_\cdot^n)(B_t - B_{k/n}) + b_n(k/n, X_\cdot^n)(t - k/n).$$

Plainly, X^n satisfies the SDE

$$X_t^n = \int_0^t \sigma_n(s, X_\cdot^n) dB_s + \int_0^t b_n(s, X_\cdot^n) ds$$

and if we call P^n the law of X^n on $\mathbf{W}^d$, then $P^n[\omega(0) \in A] = \mu(A)$ and $f(\omega(t)) - f(\omega(0)) - \int_0^t L_s^n f(\omega(s)) ds$ is a P^n-martingale for every $f \in C_K^2$.

The set (P^n) is weakly relatively compact because condition i) in Theorem (1.8) is obviously satisfied and condition ii) follows from the boundedness of a and b and the Burkholder-Davis-Gundy inequalities applied on the space Ω.

Let P be a limit point of (P^n) and $(P^{n'})$ be a subsequence converging to P. We leave as an exercise to the reader the task of showing that, since for fixed t the functions $\int_0^t L_s^n f(\omega(s)) ds$ are equi-continuous on $\mathbf{W}^d$ and converge to $\int_0^t L_s f(\omega(s)) ds$, then

$$E_P\left[\left(f(\omega(t)) - \int_0^t L_s f(\omega(s)) ds\right)\phi\right] =$$

$$\lim_{n' \to \infty} E_{P^{n'}}\left[\left(f(\omega(t)) - \int_0^t L_s^{n'} f(\omega(s)) ds\right)\phi\right]$$

for every continuous bounded function ϕ. If $t_1 < t_2$ and ϕ is $\mathscr{F}_{t_1}^0$-measurable it follows that

$$E_P\left[\left(f(\omega(t_2)) - f(\omega(t_1)) - \int_{t_1}^{t_2} L_s f(\omega(s)) ds\right)\phi\right] = 0$$

since the corresponding equality holds for $P^{n'}$ and $L_s^{n'}$. By the monotone class theorem, this equality still holds if ϕ is merely bounded and $\mathscr{F}_{t_1}^0$-measurable; as a result, $f(\omega(t)) - f(\omega(0)) - \int_0^t L_s f(\omega(s)) ds$ is a P-martingale and the proof is complete. □

Remarks. With respect to the results in Sect. 2 Chap. IX, we see that we have dropped the Lipschitz conditions. In fact, the hypothesis may be further weakened by assuming only the continuity in ω of a and b for each fixed t. On the other hand, the existence result we just proved is not of much use without a uniqueness result which is a much deeper theorem.

\# **(1.11) Exercise.** 1°) If (X^n) converges in distribution to X, prove that $(X^n)^*$ converges in distribution to X^* where, as usual, $X_t^* = \sup_{s \leq t} |X_s|$.

2°) Prove the reflection principle for BM (Sect. 3 Chap. III) by means of the analogous reflection principle for random walks. The latter is easily proved in the case of the simple random walk, namely with the notation of Theorem (1.9), $P[\xi_k = 1] = P[\xi_k = -1] = 1/2$.

* **(1.12) Exercise.** Prove that a family (P_λ) of probability measures on $C\left((\mathbb{R}_+)^k, \mathbb{R}\right)$ is weakly relatively compact if there exist constants α, β, γ, $p > 0$ such that $\sup_\lambda E_\lambda[\,|X_0|^p] < \infty$, and for every pair (s, t) of points in $(\mathbb{R}_+)^k$

$$\sup_\lambda E_\lambda\left[\,|X_s - X_t|^\alpha\right] \le \beta|s - t|^{k+\gamma}$$

where X is the coordinate process.

* **(1.13) Exercise.** Let β_s^n, $s \in [0, 1]$ and γ_t^n, $t \in [0, 1]$ be two independent sequences of independent standard BM's. Prove that the sequence of doubly indexed processes

$$X_{(s,t)}^n = n^{-1/2} \sum_{i=1}^n \beta_s^i \gamma_t^i$$

converges in distribution to the Brownian sheet. This is obviously an infinite-dimensional central-limit theorem.

(1.14) Exercise. In the setting of Donsker's theorem, prove that the processes

$$\left(\sigma\sqrt{n}\right)^{-1}\left(S_{[nt]} + (nt - [nt])\xi_{[nt]+1} - tS_n\right), \qquad 0 \le t \le 1,$$

converge in distribution to the Brownian Bridge.

(1.15) Exercise. Let (M^n) be a sequence of (super) martingales defined on the same filtered space and such that

i) the sequence (M^n) converges in distribution to a process M;
ii) for each t, the sequence $\left(M_t^n\right)$ is uniformly integrable.

Prove that M is a (super) martingale for its natural filtration.

* **(1.16) Exercise.** Let (M^n) be a sequence of continuous local martingales vanishing at 0 and such that $(\langle M^n, M^n\rangle)$ converges in distribution to a deterministic function a. Let P_n be the law of M^n.

1°) Prove that the set (P_n) is weakly relatively compact.

[Hint: One can use Lemma (4.6) Chap. IV.]

2°) If, in addition, the M^n's are defined on the same filtered space and if, for each t, there is a constant $\alpha(t)$ such that $\langle M^n, M^n\rangle_t \le \alpha(t)$ for each n, show that (P_n) converges weakly to the law W_a of the gaussian martingale with increasing process $a(t)$ (see Exercise (1.14) Chap. V).

[Hint: Use the preceding exercise and the ideas of Proposition (1.23) Chap. IV.]

3°) Let $(M^n) = \left(M_i^n, i = 1, \dots, k\right)$ be a sequence of multidimensional local martingales such that (M_i^n) satisfies for each i all the above hypotheses and, in addition, for $i \neq j$, the processes $\langle M_i^n, M_j^n\rangle$ converge to zero in distribution. Prove that the laws of M^n converge weakly to $W_{a_1} \otimes \dots \otimes W_{a_k}$.

[Hint: One may consider the linear combinations $\sum u_i M_i^n$.]

The two following exercises may be solved by using only elementary properties of BM.

* **(1.17) Exercise (Scaling and asymptotic independence).** 1°) Using the notation of the following section, prove that if β is a BM, the processes β and $\beta^{(c)}$ are asymptotically independent as c goes to 0.

[Hint: For every $A > 0$, $(\beta_{c^2 t}, t \leq A)$ and $(\beta_{c^2 A+u} - \beta_{c^2 A}, u \geq 0)$ are independent.]

2°) Deduce from 1°) that the same property holds as c goes to infinity. (See also Exercise (2.9).)

Prove that for $c \neq 1$, the transformation $x \to X^{(c)}$ which preserves the Wiener measure, is ergodic. This ergodic property is the key point in the proof of Exercise (3.20), 1°), Chap. X.

3°) Prove that if $(\gamma_t, t \leq 1)$ is a process whose law P^γ on $C([0, 1], \mathbb{R})$ satisfies

$$P^\gamma_{|\mathscr{F}_t} \ll W_{|\mathscr{F}_t} \quad \text{for every } t < 1,$$

then the two-dimensional process $V_t^{(c)} = ((\gamma_t^{(c)}, \gamma_t), t \leq 1)$ converges in law as c goes to 0 towards $((\beta_t, \gamma_t), t \leq 1)$, where β is a BM which is independent of γ.

[Hint: Use Lemma (5.7) Chap. 0.]

4°) Prove that the law of $\gamma^{(c)}$ converges in total variation to the law of β i.e. the Wiener measure. Can the convergence in 3°) be strengthened into a convergence in total variation?

5°) Prove that $V^{(c)}$ converges in law as c goes to 0 whenever γ is a BB, a Bessel bridge or the Brownian meander and identify the limit in each case.

* **(1.18) Exercise. (A Bessel process looks eventually like a BM).** Let R be a $\mathrm{BES}^\delta(r)$ with $\delta > 1$ and $r \geq 0$. Prove that as t goes to infinity, the process $(R_{t+s} - R_t, s \geq 0)$ converges in law to a BM^1.

[Hint: Use the canonical decomposition of R as a semimartingale. It may be necessary to write separate proofs for different dimensions.]

§2. Asymptotic Behavior of Additive Functionals of Brownian Motion

This section is devoted to the proof of a limit theorem for stochastic integrals with respect to BM. As a corollary, we will get (roughly speaking) the growth rate of occupation times of BM.

In what follows, B is a standard linear BM and L^a the family of its local times. As usual, we write L for L^0. The Lebesgue measure is denoted by m.

(2.1) Proposition. *If f is integrable,*

$$\lim_{n\to\infty} n \int_0^{\cdot} f(nB_s)ds = m(f)L \qquad a.s.,$$

and, for each t, the convergence of $n\int_0^t f(nB_s)ds$ to $m(f)L_t$ holds in L^p for every $p \geq 1$. Both convergences are uniform in t on compact intervals.

Proof. By the occupation times formula

$$n\int_0^t f(nB_s)ds = \int_{-\infty}^{+\infty} f(a)L_t^{a/n}da.$$

For fixed t, the map $a \to L_t^a$ is a.s. continuous and has compact support; thus, the r.v. $\sup_a L_t^a$ is a.s. finite and by the continuity of $L_t^{\cdot}$ and the dominated convergence theorem,

$$\lim_n n\int_0^t f(nB_s)ds = m(f)L_t \qquad \text{a.s.}$$

Hence, this is true simultaneously for every rational t; moreover, it is enough to prove the result for $f \geq 0$ in which case all the processes involved are increasing and the proof of the first assertion is easily completed.

For the second assertion, we observe that

$$\left|\int_{-\infty}^{+\infty} f(a)L_t^{a/n}da\right| \leq \|f\|_1 \left(\sup_a L_t^a\right)$$

and, since by Theorem (2.4) in Chap. XI, $\sup_a L_t^a$ is in L^p for every p, the result follows from the dominated convergence theorem.

The uniformity follows easily from the continuity of L_t^a in both variables. □

The following is a statement about the asymptotic behavior of additive functionals, in particular occupation times. The convergence in distribution involved is that of processes (see Sect. 1), not merely of individual r.v.'s.

(2.2) Proposition. *If A is an integrable* CAF,

$$\lim_{n\to\infty} \frac{1}{\sqrt{n}} A_{n\cdot} = \nu_A(1)L_{\cdot} \qquad \textit{in distribution.}$$

Proof. Since (see Exercise (2.11) Chap. VI) $L_{n\cdot}^a \stackrel{(d)}{=} \sqrt{n}L_{\cdot}^{a/\sqrt{n}}$, it follows that

$$\frac{1}{\sqrt{n}}A_{n\cdot} = \frac{1}{\sqrt{n}}\int L_{n\cdot}^a \nu_A(da) \stackrel{(d)}{=} \int L_{\cdot}^{a/\sqrt{n}}\nu_A(da)$$

and the latter expression converges a.s. to $\nu_A(1)L_{\cdot}$ by the same reasoning as in the previous proposition. □

The above result is satisfactory for $\nu_A(1) \neq 0$; it says that a positive integrable additive functional increases roughly like $\nu_A(1)\sqrt{t}$. On the contrary, the case $\nu_A(1) = 0$ must be further investigated and will lead to a central-limit type theorem with interesting consequences. Moreover, measures with zero integral are important when one wants to associate a potential theory with linear BM.

If we refer to Corollary (2.12) in Chap. X, we see that we might as well work with stochastic integrals and that is what we are going to do. To this end, we need a result which will be equally very useful in the following section. It is an *asymptotic version of Knight's theorem* (see Sect. 1 Chap. V).

In what follows, $(M_j^n, 1 \leq j \leq k)$ will be a sequence of k-tuples of continuous local martingales vanishing at 0 and such that $\langle M_j^n, M_j^n\rangle_\infty = \infty$ for every n and j. We call $\tau_j^n(t)$ the time-change associated with $\langle M_j^n, M_j^n\rangle$ and β_j^n the DDS Brownian motion of M_j^n.

(2.3) Theorem. *If, for every t, and every pair (i, j) with $i \neq j$*

$$\lim_{n\to\infty} \langle M_i^n, M_j^n\rangle_{\tau_i^n(t)} = \lim_{n\to\infty} \langle M_i^n, M_j^n\rangle_{\tau_j^n(t)} = 0$$

in probability, then the k-dimensional process $\beta^n = \left(\beta_j^n, 1 \leq j \leq k\right)$ *converges in distribution to a* BM^k.

Proof. The laws of the processes β_j^n are all equal to the one-dimensional Wiener measure. Therefore, the sequence $\{\beta^n\}$ is weakly relatively compact and we must prove that, for any limit process, the components, which are obviously linear BM's, are independent.

It is no more difficult to prove the results in the general case than in the case $k = 2$ for which we introduce the following handier notation. We consider two sequences of continuous local martingales (M^n) and (N^n). We call $\mu^n(t)$ and $\eta^n(t)$ the time-changes associated with $\langle M^n, M^n\rangle$ and $\langle N^n, N^n\rangle$ respectively and β^n and γ^n the corresponding DDS Brownian motions.

If $0 = t_0 < t_1 < \ldots < t_p = t$ and if we are given scalars $f_0, \ldots, f_{p-1}$ and $g_0, \ldots, g_{p-1}$, we set

$$f = \sum_j f_j 1_{]t_j, t_{j+1}]}, \qquad \beta^n(f) = \sum_j f_j \left(\beta_{t_{j+1}}^n - \beta_{t_j}^n\right),$$

$$g = \sum_j g_j 1_{]t_j, t_{j+1}]}, \qquad \gamma^n(g) = \sum_j g_j \left(\gamma_{t_{j+1}}^n - \gamma_{t_j}^n\right).$$

Let us first observe that if we set

$$U_s^n = \int_0^s f\left(\langle M^n, M^n\rangle_u\right) dM_u^n, \qquad V_s^n = \int_0^s g\left(\langle N^n, N^n\rangle_u\right) dN_u^n,$$

then $\beta(f) = U_\infty^n$ and $\gamma^n(g) = V_\infty^n$. Therefore writing $E\left[\mathscr{E}\left(i\left(U^n + V^n\right)\right)_\infty\right] = 1$ yields

$$E[\exp(i(\beta^n(f) + \gamma^n(g))) \cdot H^n] = \exp\left(-\frac{1}{2}\int \left(f^2 + g^2\right)(t)dt\right)$$

where

$$\begin{aligned} H^n &= \exp\left(\int_0^\infty f\left(\langle M^n, M^n\rangle_s\right) g\left(\langle N^n, N^n\rangle_s\right) d\langle M^n, N^n\rangle_s\right) \\ &= \exp\left(\sum_{i,j} f_i g_j \int d\langle M^n, N^n\rangle_s 1_{[\mu^n(t_i), \mu^n(t_{i+1})]}(s) 1_{[\eta^n(t_j), \eta^n(t_{j+1})]}(s)\right). \end{aligned}$$

The hypothesis entails plainly that H^n converges to 1 in probability; thus the proof will be finished if we can apply the dominated convergence theorem. But Kunita-Watanabe's inequality (Proposition (1.15) Chap. IV) and the time-change formulas yield

$$H^n \leq \exp(\|f\|_2\|g\|_2)$$

and we are done. □

We will make a great use of a corollary to the foregoing result which we now describe. For any process X and for a fixed real number $h > 0$, we define the *scaled* process $X^{(h)}$ by

$$X_t^{(h)} = h^{-1}X(h^2 t).$$

The importance of the scaling operation has already been seen in the case of BM. If M is a continuous local martingale and β its DDS Brownian motion, then $\beta^{(h)}$ is the DDS Brownian motion of $h^{-1}M$ as is stated in Exercise (1.17) of Chap. V.

We now consider a family M_i, $i = 1, 2, \ldots, k$ of continuous local martingales such that $\langle M_i, M_i\rangle_\infty = \infty$ for every i and call β_i their DDS Brownian motions. We set $M_i^n = M_i/\sqrt{n}$ and call β_i^n the DDS Brownian motion of M_i^n. As observed above, $\beta_i^n(t) = \beta_i(nt)/\sqrt{n}$.

(2.4) Corollary. *The k-dimensional process* $\beta^n = \left(\beta_i^n, i = 1, \ldots, k\right)$ *converges in distribution to a* BM^k *as soon as*

$$\lim_{t\to\infty} \langle M_i, M_j\rangle_t / \langle M_i, M_i\rangle_t = 0$$

almost surely for every $i, j \leq k$ *with* $i \neq j$.

Proof. If $\tau_i(t)$ (resp. $\tau_i^n(t)$) is the time-change associated with $\langle M_i, M_i\rangle$ (resp. $\langle M_i^n, M_i^n\rangle$), then $\tau_i^n(t) = \tau_i(nt)$ and consequently $\langle M_i^n, M_j^n\rangle_{\tau_i^n(t)} = n^{-1}\langle M_i, M_j\rangle_{\tau_i(nt)}$. The hypothesis entails that $t^{-1}\langle M_i, M_j\rangle_{\tau_i(t)}$ converges a.s. to 0 as t goes to $+\infty$, so that the result follows from Theorem (2.3). □

The foregoing corollary has a variant which often comes in handy.

(2.5) Corollary. *If there is a positive continuous strictly increasing function* ϕ *on* $\mathbb{R}_+$ *such that*

i) $\phi(t)^{-1}\langle M_i, M_i\rangle_t \xrightarrow[t\to\infty]{(d)} U_i$, $i = 1, 2, \ldots, k$ *where* U_i *is a strictly positive r.v.,*

ii) $\phi(t)^{-1}\sup_{s\leq t}\left|\langle M_i, M_j\rangle_s\right| \xrightarrow[t\to\infty]{} 0$ *in probability for every* $i, j \leq k$ *with* $i \neq j$,

then the conclusion of Corollary (2.4) holds.

Proof. Again it is enough to prove that, for $i \neq j$, $t^{-1}\langle M_i, M_j\rangle_{\tau_i(t)}$ converges to 0 in probability and in fact we shall show that $t^{-1}X_t$ converges to 0 in probability where $X_t = \sup\left\{\left|\langle M_i, M_j\rangle_s\right|; s \leq \tau_i(t)\right\}$.

Hypothesis i) implies that $t^{-1}\phi(\tau_i(t))$ converges to U_i^{-1} in distribution. For $\lambda > 0$ and $x > 0$, we have, using the fact that $X_.$ is increasing,

$$\begin{aligned} P\left[X_{\tau_i(t)} > \lambda t\right] &\leq P\left[\phi(\tau_i(t)) \geq tx\right] + P\left[X_{\tau_i(t)} > \lambda t; \tau_i(t) < \phi^{-1}(tx)\right] \\ &\leq P\left[\phi(\tau_i(t)) \geq tx\right] + P\left[X_{\phi^{-1}(tx)} > \lambda t\right]. \end{aligned}$$

Pick $\varepsilon > 0$; since U_i is strictly positive we may choose x sufficiently large and $T > 0$, such that, for every $t \geq T$,

$$P\left[\phi(\tau_i(t)) \geq tx\right] \leq \varepsilon.$$

Hypothesis ii) implies that there exists $T' \geq T$ such that for every $t \geq T'$,

$$P\left[X_{\phi^{-1}(tx)} > \lambda t\right] \leq \varepsilon.$$

It follows that for $t \geq T'$,

$$P\left[X_{\tau_i(t)} > \lambda t\right] \leq 2\varepsilon,$$

which is the desired result. □

We now return to the problem raised after Proposition (2.2). We consider Borel functions f_i, $i = 1, 2, \ldots, k$ in $L^1(m) \cap L^2(m)$ which we assume to be pairwise orthogonal in L^2, i.e. $\int f_i f_j dm = 0$ for $i \neq j$. We set

$$M_i^n = \sqrt{n} \int_0^{\cdot} f_i(nB_s)dB_s.$$

(2.6) Theorem (Papanicolaou-Stroock-Varadhan). *The $(k+1)$-dimensional process $\left(B, M_1^n, \ldots, M_k^n\right)$ converges in distribution to $(\beta, \|f_i\|_2\gamma_l^i, i = 1, 2, \ldots, k)$ where $\left(\beta, \gamma^1, \ldots, \gamma^k\right)$ is a BM^{k+1} and l is the local time of β at zero.*

Proof. We have

$$\langle M_i^n, M_j^n \rangle_t = n \int_0^t (f_i f_j)(nB_s)ds,$$

so that by Proposition (2.1), we have a.s.

$$\lim_{n\to\infty} \langle M_i^n, M_i^n \rangle_t = \|f_i\|_2^2 L_t; \text{ for } i \neq j, \lim_{n\to\infty} \langle M_i^n, M_j^n \rangle_t = \lim_{n\to\infty} \langle M_i^n, B \rangle_t = 0$$

uniformly in t on compact intervals. Thus it is not difficult to see that the hypotheses of Theorem (2.3) obtain; as a result, if we call B_i^n the DDS Brownian motion of M_i^n, the process $\left(B, B_1^n, \ldots, B_k^n\right)$ converges in distribution to $\left(\beta, \gamma^1, \ldots, \gamma^k\right)$. Now $\left(B, M_1^n, \ldots, M_k^n\right)$ is equal to $\left(B, B_i^n\left(\langle M_i^n, M_i^n \rangle\right), i = 1, \ldots, k\right)$ and it is plain that $\left(B, \langle M_i^n, M_i^n \rangle, i = 1, \ldots, k, B_i^n, i = 1, \ldots, k\right)$ converges in distribution to $\left(\beta, \|f_i\|_2^2 l, i = 1, \ldots, k, \gamma^i, i = 1, \ldots, k\right)$. The result follows. □

Remark. Instead of n, we could use any sequence (a_n) converging to $+\infty$ or, for that matter, consider the real-indexed family

$$M_i^\lambda = \sqrt{\lambda} \int_0^{\cdot} f_i(\lambda B_s)dB_s$$

and let λ tend to $+\infty$. The proof would go through just the same.

We will now draw the consequences we have announced for an additive functional A which is the difference of two integrable positive continuous additive functionals (see Exercise (2.22) Chap. X) and such that $\nu_A(1) = 0$. In order to be able to apply the representation result given in Theorem (2.9) of Chap. X, we will have to make the additional assumption that

$$\int |x||\nu_A|(dx) < \infty.$$

As in Sect. 3 of the Appendix, we set

$$F(x) = \int |x - y|\nu_A(dy).$$

The function F is the difference of two convex functions and its second derivative in the sense of distributions is $2\nu_A$. Let F'_- be its left derivative which is equal to $\nu_A(]-\infty, \cdot[)$. We have the

(2.7) Lemma. *The function F is bounded and F'_- is in $\mathscr{L}^1(m) \cap \mathscr{L}^2(m)$. Moreover*

$$\|F'_-\|_2^2 = I(\nu_A)$$

where $I(\nu_A) = -(1/2)\int\int |x - y|\nu_A(dx)\nu_A(dy)$ is called the energy of ν_A.

Proof. Since

$$\int |x - y||\nu_A|(dy) \le |x|\|\nu_A\| + \int |y||\nu_A|(dy),$$

the integral $I(\nu_A)$ is finite and we may apply Fubini's theorem to the effect that

$$\begin{aligned} I(\nu_A) &= -\iint_{x>y} (x - y)\nu_A(dx)\nu_A(dy) \\ &= -\iiint_{x>z>y} \nu_A(dx)\nu_A(dy)dz \\ &= -\int_{-\infty}^{+\infty} dz \left(\int_z^{\infty} \nu_A(dx)\right)\left(\int_{-\infty}^{z} \nu_A(dy)\right). \end{aligned}$$

The set of z's such that $\nu_A(\{z\}) \neq 0$ is countable and for the other z's, it follows from the hypothesis $\nu_A(1) = 0$ that

$$\nu_A(]-\infty, z[) = -\nu_A(]z, \infty[).$$

Thus the proof of the equality in the statement is complete.

By the same token

$$\begin{aligned} \int_0^{\infty} |F'_-(x)|dx &= \int_0^{\infty} |\nu_A(\,]x, \infty[\,)|dx \le \int_0^{\infty} |\nu_A|(\,]x, \infty[\,)dx \\ &= \int_0^{\infty} x|\nu_A|(dx) < \infty, \end{aligned}$$

and likewise

$$\int_{-\infty}^{0} |F'_-(x)|dx \le \int_{-\infty}^{0} |x||\nu_A|(dx) < \infty.$$

Consequently, F'_- is in $\mathscr{L}^1$ and it follows that F is bounded. □

We may now prove that additive functionals satisfying the above set of hypotheses are, roughly speaking, of the order of $t^{1/4}$ as t goes to infinity.

(2.8) Proposition. *If $\nu_A(1) = 0$ and $\int |x||\nu_A|(dx) < \infty$, the 2-dimensional process $\left(n^{-1/2}B_{n.}, n^{-1/4}A_{n.}\right)$ converges in distribution to $\left(\beta, I(\nu_A)^{1/2}\gamma_l\right)$, where (β, γ) is a* BM^2 *and l the local time of β at* 0.

Proof. By the representation result in Theorem (2.9) of Chap. X and Tanaka's formula,

$$n^{-1/4}A_{n.} = n^{-1/4}[F(B_{n.}) - F(0)] - n^{-1/4}\int_0^{n\cdot} F'_-(B_s)dB_s.$$

Since F is bounded, the first term on the right goes to zero as n goes to infinity and, therefore, it is enough to study the stochastic integral part.

Setting $s = nu$, we see that we might as well study the limit of

$$\left(n^{-1/2}B_{n.}, n^{-1/4}\int_0^{\cdot} F'_-(B_{nu})dB_{nu}\right),$$

and since $B_{n.} \overset{(d)}{=} \sqrt{n}B$, this process has the same law as

$$\left(B_{.}, n^{1/4}\int_0^{\cdot} F'_-\left(\sqrt{n}B_u\right)dB_u\right).$$

Because F'_- is in $\mathscr{L}^1(m) \cap \mathscr{L}^2(m)$, it remains to apply the remark following Theorem (2.6). □

Remark. Propositions (2.2) and (2.8) are statements about the speed at which additive functionals of linear BM tend to infinity. In dimension $d > 2$, there is no such question as integrable additive functionals are finite at infinity but, for the planar BM, the same question arises and it was shown in Sect. 4 Chap. X that integrable additive functionals are of the order of $\log t$. However, as the limiting process is not continuous, one has to use other notions of convergence.

* **(2.9) Exercise.** 1°) In the situation of Theorem (2.3), if there is a sequence of positive random variables L_n such that

i) $\lim_n \langle M_i^n, M_i^n\rangle_{L_n} = +\infty$ in probability for each i;

ii) $\lim_n \sup_{s\le L_n} \left|\langle M_i^n, M_j^n\rangle_s\right| = 0$ in probability for each pair i, j with $i \ne j$,

prove that the conclusion of the Theorem holds.

2°) Assume now that there are only two indexes and write M^n for M_1^n and N^n for M_2^n. Prove that if there is a sequence (L_n) of positive random variables such that

i') $\lim_n \langle M^n, M^n\rangle_{L_n} = \infty$ in probability,
ii) $\lim_n \sup_{s\le L_n} |\langle M^n, N^n\rangle_s| = 0$,

then the conclusion of the Theorem holds.

3°) Deduce from the previous question that if $\beta_.$ is a BM, and if c converges to $+\infty$, then β and $\beta^{(c)}$ are asymptotically independent.

Remark however that the criterion given in Corollary (2.4) does not apply in the particular case of a pair $\left(M, \frac{1}{c}M\right)$ as $c \to \infty$. Give a more direct proof of the asymptotic independence of β and $\beta^{(c)}$.

* **(2.10) Exercise.** For f in $L^2 \cap L^1$, prove that for fixed t, the random variables $\sqrt{n}\int_0^t f(nB_s)dB_s$ converge weakly to zero in L^2 as n goes to infinity. As a result, the convergence in Theorem (2.6) cannot be improved to convergence in probability.

* **(2.11) Exercise.** Let $0 = a_0 < a_1 < \ldots < a_k < \infty$ be a finite sequence of real numbers. Prove that the $(k+1)$-dimensional process

$$\left(B_t, \frac{\sqrt{n}}{2}\left(L_t^{a_i/n} - L_t^{a_{i-1}/n}\right), i = 1, 2, \ldots, k\right)$$

converges in distribution to

$$\left(\beta, \sqrt{a_i - a_{i-1}}\gamma_l^i, i = 1, 2, \ldots, k\right)$$

where $(\gamma^i, i = 1, 2, \ldots, k)$ is a k-dimensional BM independent of β and l is the local time of β at 0.

* **(2.12) Exercise.** 1°) Let

$$X(t, a) = \int_0^t 1_{[0,a]}(B_s)dB_s.$$

Prove that for $p \geq 2$, there exists a constant C_p such that for $0 \leq s \leq t \leq 1$ and $0 \leq a \leq b \leq 1$,

$$E\left[\,|X(t,b) - X(s,a)|^p\right] \leq C_p\left((t-s)^{p/2} + (b-a)^{p/2}\right).$$

2°) Prove that the family of the laws P_λ of the doubly indexed processes

$$\left(B_t, \lambda^{1/2}\int_0^t 1_{[0,a]}(\lambda B_s)dB_s\right)$$

is weakly relatively compact.

[Hint. Use Exercise (1.12).]

3°) Prove that, as λ goes to infinity, the doubly-indexed processes

$$\left(B_t, \lambda^{1/2}\left(L_t^{a/\lambda} - L_t^0\right)/2\right)$$

converge in distribution to $\left(B_t, \mathbb{B}\left(L_t^0, a\right)\right)$, where $\mathbb{B}$ is a Brownian sheet independent of B.

[Hint: Use the preceding Exercise (2.11).]

4°) For $\nu > 0$, prove that

$$\varepsilon^{\nu} \int_{\varepsilon}^{\infty} a^{-(3/2+\nu)} \left(L_t^a - L_t^0\right) da \xrightarrow[\varepsilon \to 0]{(d)} 2 \int_0^{\infty} e^{-(\nu+1/2)u} \mathbb{B}\left(L_t^0, e^u\right) du.$$

5°) Let $\tau_x = \inf\left\{u : L_u^0 > x\right\}$; the processes $\lambda^{1/2}\left(L_{\tau_x}^{a/\lambda} - x\right)/2$ converge in distribution, as λ tends to $+\infty$, to the process $\sqrt{x}\gamma_a$ where γ_a is a standard BM. This may be derived from 3°) but may also be proved as a consequence of the second Ray-Knight theorem (Sect. 2 Chap. XI).

* **(2.13) Exercise.** With the notation of Theorem (1.10) in Chap. VI, prove that

$$\lim_{\varepsilon \to 0} \frac{1}{\sqrt{\varepsilon}} \left(\varepsilon d_{\varepsilon}(\cdot) - \frac{1}{2} L\right) = \gamma_l$$

in the sense of finite distributions, where as usual, l is the local time at 0 of a BM independent of γ.

[Hint: If $M_t^{\varepsilon} = \frac{1}{\sqrt{\varepsilon}} \int_0^t \theta_s^{\varepsilon} dB_s$ and P^{ε} is the law of $(B_t, L_t, M_t^{\varepsilon})$, prove that the set $(P^{\varepsilon}, \varepsilon > 0)$ is relatively compact.]

* **(2.14) Exercise.** In the notation of this section, if (x_i), $i = 1, \ldots, k$ is a sequence of real numbers, prove that $\left(B, \varepsilon^{-1/2}\left(L^{x_i+\varepsilon} - L^{x_i}\right), i = 1, \ldots, k\right)$ converges in distribution as $\varepsilon \to 0$, to $\left(B, 2\beta_{L^{x_i}}^i, i = 1, \ldots, k\right)$, where $\left(B, \beta^1, \ldots, \beta^k\right)$ is a BM^{k+1}.

** **(2.15) Exercise.** Prove, in the notation of this section, that for any $x \in \mathbb{R}$, $\varepsilon^{-1/2}\left[\varepsilon^{-1}\int_0^{\cdot} 1_{[x,x+\varepsilon]}(B_s)ds - L^x\right]$ converges in distribution to $\left(2/\sqrt{3}\right)\beta_{L^x}$, as ε tends to 0. The reader will notice that this is the "central-limit" theorem associated with the a.s. result of Corollary (1.9) in Chap. VI.

[Hint: Extend the result of the preceeding exercise to $\left(L^{x_i+\varepsilon z} - L^{x_i}\right)$ and get a doubly indexed limiting process.]

* **(2.16) Exercise (A limit theorem for the Brownian motion on the unit sphere).** Let Z be a $\mathrm{BM}^d(a)$ with $a \neq 0$ and $d \geq 2$; set $\rho = |Z|$. Let V be the process with values in the unit sphere of $\mathbb{R}^d$ defined by

$$Z_t = \rho_t V_{C_t}$$

where $C_t = \int_0^t \rho_s^{-2} ds$. This is the *skew-product* decomposition of BM^d.

1°) Prove that there is a BM^d, say B, independent of ρ and such that

$$V_t = V_0 + \int_0^t \sigma(V_s) dB_s - \frac{d-1}{2} \int_0^t V_s ds$$

where σ is the field of matrices given by

$$\sigma_{ij}(x) = \delta_{ij} - x_i x_j.$$

2°) If $X_t = \int_0^t \sigma(V_s)dB_s$, prove that $\langle X^i, X^j\rangle_t = \langle X^i, B^j\rangle_t$.
[Hint: Observe that $\sigma(x)x = 0$, $\sigma(x)y = y$ if $\langle x, y\rangle = 0$, hence $\sigma^2(x) = \sigma(x)$.]
3°) Show that

$$\lim_{t\to\infty} t^{-1}\langle X^i, B^j\rangle_t = \delta_{ij}(1 - d^{-1}) \qquad \text{a.s.}$$

4°) Prove that the $2d$-dimensional process $\left(c^{-1}B_{c^2t}, (2c)^{-1}\int_0^{c^2t} V_s ds\right)$ converges in distribution, as c tends to ∞, to the process

$$\left(B_t, d^{-1}\left(B_t + (d-1)^{-1/2}B'_t\right)\right)$$

where (B, B') is a BM^{2d}.

§3. Asymptotic Properties of Planar Brownian Motion

In this section, we take up the study of some asymptotic properties of complex BM which was initiated in Sect. 4 of Chap. X. We will use the asymptotic version of Knight's theorem (see the preceding section) which gives a sufficient condition for the DDS Brownian motions of two sequences of local martingales to be asymptotically independent. We will also have to envisage below the opposite situation in which these BM's are asymptotically equal. Thus, we start this section with a sufficient condition to this effect.

(3.1) Theorem. *Let (M_i^n), $i = 1, 2$, be two sequences of continuous local martingales and β_i^n their associated DDS Brownian motions. If $R_n(t)$ is a sequence of processes of time-changes such that the following limits exist in probability for every t,*

i) $\lim_n \langle M_1^n, M_1^n\rangle_{R_n(t)} = \lim_n \langle M_2^n, M_2^n\rangle_{R_n(t)} = t$,
ii) $\lim_n \langle M_1^n - M_2^n, M_1^n - M_2^n\rangle_{R_n(t)} = 0$,

then, $\lim_n \sup_{s\le t}\left|\beta_1^n(s) - \beta_2^n(s)\right| = 0$ *in probability for every t.*

Proof. If T_i^n is the time-change associated with $\langle M_i^n, M_i^n\rangle$,

$$\begin{aligned}\left|\beta_1^n(t) - \beta_2^n(t)\right| \le\ & \left|M_1^n\left(T_1^n(t)\right) - M_1^n\left(R_n(t)\right)\right| + \left|M_1^n\left(R_n(t)\right) - M_2^n\left(R_n(t)\right)\right| \\ & + \left|M_2^n\left(R_n(t)\right) - M_2^n\left(T_2^n(t)\right)\right|.\end{aligned}$$

By Exercise (4.14) Chap. IV, for fixed t, the left-hand side converges in probability to zero if each of the terms

$$\langle M_1^n, M_1^n\rangle_{R_n(t)}^{T_1^n(t)}, \quad \langle M_1^n - M_2^n, M_1^n - M_2^n\rangle_{R_n(t)}, \quad \langle M_2^n, M_2^n\rangle_{R_n(t)}^{T_2^n(t)}$$

converges in probability to zero. Since $\langle M_1^n, M_1^n\rangle_{T_1^n(t)} = t$, this follows readily from the hypothesis.

As a result, $\left|\beta_1^n - \beta_2^n\right| \xrightarrow{\text{f.d.}} 0$. On the other hand, Kolmogorov's criterion (1.8) entails that the set of laws of the processes $\beta_1^n - \beta_2^n$ is weakly relatively compact; thus, $\beta_1^n - \beta_2^n \xrightarrow{(d)} 0$. This implies (Exercise (1.11)) that $\sup_{s\le t}\left|\beta_1^n(s) - \beta_2^n(s)\right|$ converges in distribution, hence in probability, to zero. □

The following results are to be compared with Corollary (2.4). We now look for conditions under which the DDS Brownian motions are asymptotically equal.

(3.2) Corollary. *If M_i, $i = 1, 2$, are continuous local martingales and R is a process of time-changes such that the following limits exist in probability*

i)
$$\lim_{u\to\infty} \frac{1}{u}\langle M_i, M_i\rangle_{R(u)} = 1 \textit{ for } i = 1, 2,$$

ii)
$$\lim_{u\to\infty} \frac{1}{u}\langle M_1 - M_2, M_1 - M_2\rangle_{R(u)} = 0,$$

then $\frac{1}{\sqrt{u}}(\beta_1(u\cdot) - \beta_2(u\cdot))$ converges in distribution to the zero process as u tends to infinity.

Proof. By the remarks in Sect. 5 Chap. 0, it is equivalent to show that the convergence holds in probability uniformly on every bounded interval. Moreover, by Exercise (1.17) Chap. V (see the remarks before Corollary (2.4)), $\frac{1}{\sqrt{u}}\beta_i(u\cdot)$ is the DDS Brownian motion of $\frac{1}{\sqrt{u}}M_i$. Thus, we need only apply Theorem (3.1) to $\left(\frac{1}{\sqrt{u}}M_1, \frac{1}{\sqrt{u}}M_2\right)$. □

The above corollary will be useful later on. The most likely candidates for R are mixtures of the time-changes μ_t^i associated with $\langle M^i, M^i\rangle$ and actually the following result shows that $\mu_t^1 \vee \mu_t^2$ will do.

(3.3) Proposition. *The following two assertions are equivalent:*

(i)
$$\lim_{t\to\infty} \frac{1}{t}\langle M_1 - M_2, M_1 - M_2\rangle_{\mu_t^1\vee\mu_t^2} = 0 \textit{ in probability;}$$

(ii)
$$\lim_{t\to\infty} \frac{1}{t}\langle M_1, M_1\rangle_{\mu_t^2} = \lim_{t\to\infty} \frac{1}{t}\langle M_2, M_2\rangle_{\mu_t^1} = 1 \textit{ in probability,}$$

and
$$\lim_{t\to\infty} \frac{1}{t}\langle M_1 - M_2, M_1 - M_2\rangle_{\mu_t^1\wedge\mu_t^2} = 0 \textit{ in probability.}$$

Under these conditions, the convergence stated in Corollary (3.2) holds.

Proof. From the "Minkowski" inequality of Exercise (1.47) in Chap. IV, we conclude that

$$\left|\left(t^{-1}\langle M_1, M_1\rangle_{\mu_t^2}\right)^{1/2} - 1\right| \le \left(t^{-1}\langle M_1 - M_2, M_1 - M_2\rangle_{\mu_t^2}\right)^{1/2}.$$

By means of this inequality, the proof that i) implies ii) is easily completed.

To prove the converse, notice that

$$\begin{aligned}\langle M_1 - M_2, M_1 - M_2\rangle_{\mu_t^1\wedge\mu_t^2}^{\mu_t^1\vee\mu_t^2} &= \left|\langle M_1 - M_2, M_1 - M_2\rangle_{\mu_t^1} - \langle M_1 - M_2, M_1 - M_2\rangle_{\mu_t^2}\right| \\ &= \left|\langle M_1, M_1\rangle_{\mu_t^2} - \langle M_2, M_2\rangle_{\mu_t^1} + 2\left(\langle M_1, M_2\rangle_{\mu_t^1} - \langle M_1, M_2\rangle_{\mu_t^2}\right)\right|.\end{aligned}$$

Since by Kunita-Watanabe inequality

$$\left|\langle M_1, M_2\rangle_{\mu_t^1} - \langle M_1, M_2\rangle_{\mu_t^2}\right| \leq \left|t - \langle M_1, M_1\rangle_{\mu_t^2}\right|^{1/2} \left|t - \langle M_2, M_2\rangle_{\mu_t^1}\right|^{1/2},$$

the equivalence of ii) and i) follows easily. □

The foregoing proposition will be used under the following guise.

(3.4) Corollary. *If* $\langle M_1, M_1\rangle_\infty = \langle M_2, M_2\rangle_\infty = \infty$ *and*

$$\lim_{t\to\infty} \langle M_1 - M_2, M_1 - M_2\rangle_t / \langle M_i, M_i\rangle_t = 0 \textit{ almost-surely},$$

for $i = 1, 2$, *then the conclusion of Proposition (3.3) holds.*

Proof. The hypothesis implies that μ_t^i is finite and increases to $+\infty$ as t goes to infinity. Moreover

$$\langle M_1 - M_2, M_1 - M_2\rangle_{\mu_t^i} / \langle M_i, M_i\rangle_{\mu_t^i} = \frac{1}{t}\langle M_1 - M_2, M_1 - M_2\rangle_{\mu_t^i},$$

so that condition i) in the Proposition is easily seen to be satisfied. □

From now on, we consider a complex BM Z such that $Z_0 = z_0$ a.s. and pick points $z_1, \dots, z_p$ in $\mathbb{C}$ which differ from z_0. For each j, we set

$$\chi_t^j = \int_0^t \frac{dZ_s}{Z_s - z_j} = \log\left|\frac{Z_t - z_j}{z_0 - z_j}\right| + i\theta_t^j,$$

where θ_t^j is the continuous determination of $\arg\left(\frac{Z_t - z_j}{z_0 - z_j}\right)$ which vanishes for $t = 0$. The process $(2\pi)^{-1}\theta_t^j$ is the "winding number" of Z around z_j up to time t; we want to further the results of Sect. 4 in Chap. X by studying the simultaneous asymptotic properties of the θ_t^j, $j = 1, \dots, p$.

Let us set

$$C_t^j = \int_0^t \left|Z_s - z_j\right|^{-2} ds$$

and denote by T_t^j the time-change process which is the inverse of C^j. As was shown in Sect. 2 Chap. V, for each j, there is a complex BM $\zeta^j = \beta^j + i\gamma^j$ such that

$$\chi_t^j = \zeta_{C_t^j}^j, \qquad \zeta^j = \chi_{T^j}^j.$$

We observe that up to a time-change, β^j is ≤ 0 when Z is inside the disk $D_j = D\left(z_j, |z_0 - z_j|\right)$ and ≥ 0 when Z is outside D_j.

We now recall the notation introduced in Chap. VI before Theorem (2.7). Let β be a standard linear BM and l its local time at 0. We put

$$M_t^+ = \int_0^t 1_{(\beta_s > 0)} d\beta_s, \qquad M_t^- = \int_0^t 1_{(\beta_s < 0)} d\beta_s,$$

and call $\alpha^\pm$ the time-changes associated with $\langle M^\pm, M^\pm \rangle$. Let β^+ and β^- be the positive and negative parts of β and put $\rho_t^\pm = \beta^\pm_{\alpha_t^\pm}$. By the results in Chapter VI, $(\delta^+, \delta^-) = \left(M^+_{\alpha^+}, M^-_{\alpha^-}\right)$ is a planar BM such that

$$\rho_t^\pm = \pm\delta_t^\pm + \frac{1}{2} l_{\alpha_t^\pm}.$$

The process $\delta^\pm$ is the DDS Brownian motion of $M^\pm$. Moreover, $\rho^\pm$ are reflecting BM's, $\left(\rho^\pm, \frac{1}{2} l_{\alpha_t^\pm}\right)$ have the same law as $(|\beta|, l)$ and

$$\frac{1}{2} l_{\alpha_t^+} = \sup_{s \le t} \left(-\delta_s^+\right), \qquad \frac{1}{2} l_{\alpha_t^-} = \sup_{s \le t} \left(\delta_s^-\right).$$

The processes $l_{\alpha_t^\pm}$ are the local times at 0 of $\rho^\pm$ (Exercise (2.14) Chap. VI).

(3.5) Proposition. *The processes ρ^+ and ρ^- are independent. Moreover, there are measurable functions f and g from* **W** $\times$ **W** *to* **W** *such that*

$$\beta = f(\rho^+, \rho^-) = g(\delta^+, \delta^-).$$

Proof. The first part follows from the independence of δ^+ and δ^-. To prove the second part, we observe that $\beta_t = \rho^+\left(\langle M^+, M^+\rangle_t\right) + \rho^-\left(\langle M^-, M^-\rangle_t\right)$; thus, it is enough to prove that $\langle M^\pm, M^\pm\rangle$ are measurable functions of $\rho^\pm$. Calling $L^\pm$ the local time of $\rho^\pm$ at zero, we have

$$l = L^+\left(\langle M^+, M^+\rangle\right) = L^-\left(\langle M^-, M^-\rangle\right).$$

Moreover, as $\langle M^+, M^+\rangle_t + \langle M^-, M^-\rangle_t = t$, one can guess that

$$\langle M^+, M^+\rangle_t = \inf\left\{s : L_s^+ > L_{t-s}^-\right\}$$

which is readily checked. Since ρ^+ and ρ^- are functions of δ^+ and δ^-, the proof is complete.

Remark. To some extent, this is another proof of the fact that Brownian motion may be recovered from its excursion process (Proposition (2.5) Chap. XII), as ρ^+ and ρ^- may be seen as accounting respectively for the positive and negative excursions.

In the sequel, we are going to use simultaneously the above $\pm$ notational pattern for several BM's which will be distinguished by superscripts; the superscripts will be added to the $\pm$. For instance, if β^j is the real part of the process ζ^j defined above

$$\beta^j = g\left(\delta^{j+}, \delta^{j-}\right).$$

The following remark will be important.

(3.6) Lemma. *The process δ^{j+} is the* DDS *Brownian motion of the local martingale*

$$N_t^{j+} = \mathrm{Re}\int_0^t 1_{D_j^c}(Z_s)\frac{dZ_s}{Z_s - z_j}.$$

The same result holds for δ^{j-} *with* D_j *instead of* D_j^c *and, naturally, the corresponding local martingale will be called* N^{j-}.

Proof. It is easily seen that $N_t^{j+} = M_{T_t^j}^{j+}$ and since δ^{j+} is the DDS Brownian motion of M^{j+}, by Exercise (1.17) in Chap. V, it is also the DDS Brownian motion of N^{j+}. □

We now introduce some more notation pertaining to the imaginary part γ^j of ζ^j. We call γ_t^{j+} and γ_t^{j-} the DDS Brownian motions of the local martingales

$$\theta_t^{j+} = \int_0^t 1_{D_j^c}(Z_s)d\theta_s^j, \qquad \theta_t^{j-} = \int_0^t 1_{D_j}(Z_s)d\theta_s^j.$$

As in the previous proof, it is seen that

$$\langle\theta^{j+}, \theta^{j+}\rangle_t = \int_0^{C_t^j} 1_{(\beta_s^j \geq 0)}ds,$$

and, by the same reasoning, γ_t^{j+} is also the DDS Brownian motion of $\int_0^t 1_{(\beta_s^j\geq 0)}d\gamma_s^j$, namely

$$\gamma_t^{j+} = \int_0^{\alpha_t^{j+}} 1_{(\beta_s^j \geq 0)}d\gamma_s^j.$$

The same result holds for γ^{j-} with the obvious changes.

Moreover, it is plain that $\gamma^j = \gamma^{j+}(\langle M^{j+}, M^{j+}\rangle) + \gamma^{j-}(\langle M^{j-}, M^{j-}\rangle)$ so that, by Proposition (3.5), the knowledge of the four processes $(\rho^{j+}, \rho^{j-}, \gamma^{j+}, \gamma^{j-})$ is equivalent to the knowledge of (ρ^j, γ^j).

Our next result will make essential use of the scaling operation. Let us insist that for $h > 0$,

$$X^{(h)}(t) = h^{-1}X(h^2 t).$$

In particular, we denote by $\zeta^{j(h)}$ the Brownian motion

$$\left(\zeta^j\right)^{(h)}(t) = h^{-1}\zeta^j(h^2 t).$$

We must observe that the family $\left(\beta^{j\pm}, M^{j\pm}, \delta^{j\pm}, \rho^{j\pm}\right)$ of processes associated with the planar BM $\zeta^{j(h)}$ by the above scheme is actually equal to

$$\left(\beta^{j\pm(h)}, M^{j\pm(h)}, \delta^{j\pm(h)}, \rho^{j\pm(h)}\right).$$

Indeed, it is obvious for $\beta^{\pm}$ and we have

$$\begin{aligned}\int_0^t 1_{(\beta_s^{j(h)}>0)}d\beta_s^{j(h)} &= \frac{1}{h}\int_0^t 1_{(\beta_{h^2 s}^j>0)}d\beta_{h^2 s}^j \\ &= \frac{1}{h}\int_0^{h^2 t} 1_{(\beta_s^j>0)}d\beta_s^j = M_t^{j+(h)}.\end{aligned}$$

As δ^{j+} is the DDS Brownian motion of M^{j+}, Exercise (1.17) in Chap. V tells us that $\delta^{j+(h)}$ is the DDS Brownian motion of $M_t^{j+(h)}$ which entails our claim in the case of $\delta^{\pm}$. Finally, the claim is also true for $\rho^{\pm}$ since it is a function of $\delta^{\pm}$.

We may now state

(3.7) Theorem. *The $2p$-dimensional process $\left(\zeta^{1(h)}, \dots, \zeta^{p(h)}\right)$ converges in distribution as h tends to infinity to a process $\left(\zeta^{1\infty}, \dots, \zeta^{p\infty}\right)$, the law of which is characterized by the following three properties:*

i) each $\zeta^{j\infty}$ is a complex BM;

ii) if we keep the same notational device as above with the obvious changes, then the processes $\rho^{j+\infty} + i\gamma^{j+\infty}$ are all identical;

iii) if we call $\rho^{+\infty} + i\gamma^{+\infty}$ the common value of the processes $\rho^{j+\infty} + i\gamma^{j+\infty}$, then the processes $\rho^{+\infty} + i\gamma^{+\infty}, \rho^{1-\infty} + i\gamma^{1-\infty}, \dots, \rho^{p-\infty} + i\gamma^{p-\infty}$, are independent.

Proof. By Corollary (1.6), the set of laws under consideration is weakly relatively compact. Therefore, all we have to prove is that every limit law satisfies i) through iii) of the statement.

We first observe that property i) is obvious. Next, to prove that the $\rho^{j+\infty}$ are identical, we may as well prove that the $\delta^{j+\infty}$ are identical. Now, by Corollary (5.8) Chap. 0, the processes $\delta^{j+\infty}$ are the limits in distribution of the processes $\delta^{j+(h)}$. Furthermore, by Lemma (3.6), the processes $\delta^{j+(h)}$ are the scaled DDS Brownian motions of the local martingales N^{j+}. Thus, it remains to prove that we can apply Corollary (3.4) to the local martingales N^{j+}. But by Sect. 2 Chap. V

$$\langle N^{j+}, N^{j+}\rangle_t = \int_0^t f^j(Z_s)ds$$

with $f^j(z) = \left|z - z_j\right|^{-2} 1_{D_j^c}(z)$ and likewise

$$\langle N^{j+} - N^{k+}, N^{j+} - N^{k+}\rangle_t = \int_0^t f^{jk}(Z_s)ds$$

with $f^{jk}(z) = \left|\frac{1}{z-z_j} 1_{D_j^c}(z) - \frac{1}{z-z_k} 1_{D_k^c}(z)\right|^2$. As the functions f^{jk} are integrable with respect to the Lebesgue measure in the plane, whereas the functions f^j are not, the ergodic theorem of Sect. 3 Chap. X (see Exercise (3.15) Chap. X) shows that the hypotheses of Corollary (3.4) are satisfied. As a result, the processes $\rho^{j+\infty}$ are identical. The same pattern of proof applies to the processes $\gamma^{j+\infty}$ without any changes. This proves ii).

We now turn to the proof of iii). By the same reasoning as in the proof of ii), it is enough to prove that

$$\left(\delta^{1+}, \gamma^{1+}, \delta^{1-}, \gamma^{1-}, \dots, \delta^{p-}, \gamma^{p-}\right)^{(h)}$$

converges in distribution to a BM^{2p+2}. By Lemma (3.6) and Exercise (1.17) in Chap. V, $\delta^{1+(h)}$ (resp. $\delta^{j-(h)}$) is the DDS Brownian motion of $\frac{1}{h}N^{1+}$ (resp. $\frac{1}{h}N^{j-}$) and likewise $\gamma^{1+(h)}$ (resp. $\gamma^{j-(h)}$) is the DDS Brownian motion of $\frac{1}{h}\theta^{1+}$ (resp. $\frac{1}{h}\theta^{j-}$). Thus, we need only apply Corollary (2.4) to the local martingales

$$\left(N^{1+}, \theta^{1+}, N^{1-}, \theta^{1-}, \ldots, N^{p-}, \theta^{p-}\right).$$

Let M be any of these martingales; then, as in the first part of the proof

$$\langle M, M\rangle_t = \int_0^t f(Z_s)ds$$

for a function f which is not integrable with respect to the Lebesgue measure. On the other hand, if M, N are two local martingales of the above list

$$\langle M, N\rangle_t = \int_0^t f(Z_s)ds$$

where, this time, f is integrable. For instance, for N^{j-} and N^{k-}, we get

$$f(z) = \left(z - z_j, z - z_k\right) \left|z - z_j\right|^{-2} |z - z_k|^{-2} 1_{D_j \cap D_k}(z)$$

which is integrable since

$$|f(z)| \leq \left|z - z_j\right|^{-1} |z - z_k|^{-1} 1_{D_j \cap D_k}(z);$$

the other cases are either trivial or similar. In any case, it is easily deduced from the ergodic theorem (see Exercise (3.15) Chap. X) that the hypotheses of Corollary (2.4) are satisfied. This completes the proof. □

The foregoing theorem allows to generalize Corollary (4.5) of Chap. X to several points. As in there, A_t will be an additive functional and we will assume that $\|\nu_A\| = 2\pi$.

(3.8) Theorem. *As t goes to infinity,*

$$\frac{2}{\log t}\left(\left(\theta_t^{j+}, \theta_t^{j-}\right), j = 1, \ldots, p, A_t\right)$$

converges in distribution to $\left((W^+, W^{j-}), j = 1, \ldots, p, \Lambda\right)$ where, for each j, the triple $\left(W^+, W^{j-}, \Lambda\right)$ has the law described in Corollary (4.4) of Chap. X, and, conditionally on Λ, the $p+1$ variables $\left(W^+, W^{j-}, j = 1, \ldots, p\right)$ are independent.

Proof. From Theorem (4.2) in Chap. X, we know that for each j, $\frac{2}{\log t}(\theta_t^{j+}, \theta_t^{j-}, A_t)$ converges to $(W^{j+}, W^{j-}, \Lambda)$; thus what we have to prove is the relationship between these triples when j varies, that is between W^{j+} and W^{j-}, $j = 1, \ldots, p$, given Λ. In the remark after Theorem (4.2) Chap. X, we pointed out that

$$\frac{2}{\log t}\left(\theta_t^{j+}, \theta_t^{j-}\right) - \left(\int_0^{T^j_{(\log t/2)}} 1_{(\beta_s^j \geq 0)} d\gamma_s^j, \int_0^{T^j_{(\log t/2)}} 1_{(\beta_s^j \leq 0)} d\gamma_s^j\right)$$

where $T_a^j = \inf\{t : \beta_t^j = a\}$, converges in probability to zero. With each planar BM $Z = X + iY$ we associate a bidimensional r.v. $W(Z)$ by setting

$$W(Z) = \left(\int_0^{T_1} 1_{(X_s \geq 0)} dY_s, \int_0^{T_1} 1_{(X_s \leq 0)} dY_s \right)$$

where $T_1 = \inf\{t : X_t = 1\}$. Thanks to the scaling properties of the family T_a^j, it is not hard to see that

$$\frac{2}{\log t} \left(\int_0^{T^j_{(\log t/2)}} 1_{(\beta_s^j \geq 0)} d\gamma_s^j, \int_0^{T^j_{(\log t/2)}} 1_{(\beta_s^j \leq 0)} d\gamma_s^j \right) = W\left(\zeta^{j(h)}\right)$$

if $h = \frac{1}{2}\log t$. By another application of Corollary (5.8) Chap. 0, it follows that $\frac{2}{\log t}((\theta_t^{j+}, \theta_t^{j-}), j = 1, \ldots, p)$ converges in distribution to $(W(\zeta^{j\infty}), j = 1, \ldots, p)$.

As a result, the r.v.'s W^{j+} which depend on $\beta^{j+\infty}$ alone are all equal to the same variable W^+. For the same reason, $T_1^{j\infty}$ does not depend on j. Furthermore, conditionally on Λ, each W^{j-} is independent of W^+, hence of $T_1^{j\infty}$, and becomes a function of $\rho^{j-\infty} + i\gamma^{j-\infty}$ alone. The independence follows from Theorem (3.7). □

We now record the asymptotic distribution for the windings θ^j themselves.

(3.9) Corollary. *The limiting distribution of* $\left(\frac{2}{\log t}\theta_t^j, j = 1, 2, \ldots, p\right)$ *is the law of* $\left(W_j = W^+ + W^{j-}, j = 1, \ldots, p\right)$ *which may be described as follows:*

i) $W^{j-} = HY_j$, *where*

ii) the r.v.'s Y_j *are independent Cauchy variables with parameter* 1 *which are also independent of the pair* (W^+, H)*;*

iii) the Laplace-Fourier transform of the pair (W^+, H) *is given by*

$$E\left[\exp\left(-aH + ivW_+\right)\right] = \left[\cosh v + (a/v)\sinh v\right]^{-1}.$$

Proof. This is a reformulation of Corollary (4.4) in Chap. X with $H = \Lambda/2$. □

In Theorem (4.2) of Chap. X, we saw that the result is independent of the radius of the disk used to distinguish between "small" and "big" windings. In this section, we have used, for convenience sake, the disks D_j of radius $|z_0 - z_j|$, but it is, likewise, inessential. This is implied by the next result which will also be used in the proof of the last theorem of this section.

(3.10) Proposition. *If* f *is locally bounded and square-integrable with respect to the* 2*-dimensional Lebesgue measure, then*

$$\frac{1}{\log t} M_t = \frac{1}{\log t} \int_0^t f(Z_s) dZ_s$$

converges in probability to zero as t *goes to infinity.*

Proof. Since M is conformal,

$$\langle \mathrm{Re}M, \mathrm{Re}M\rangle_t = \langle \mathrm{Im}M, \mathrm{Im}M\rangle_t = \int_0^t |f|^2(Z_s)ds$$

and we know that $\frac{1}{\log t}\int_0^t |f|^2(Z_s)ds$ converges in distribution to a finite r.v. It follows that $\langle \mathrm{Re}M, \mathrm{Re}M\rangle_t/(\log t)^2$ converges in probability to zero and by Exercise (4.14) in Chap. IV, $\mathrm{Re}M_t/(\log t)$ and $\mathrm{Im}M_t/(\log t)$ converge in probability to zero. □

Remark. The assumption that f is locally bounded is only made to ensure that $\int_0^t |f|^2(Z_s)ds$ is P_0-a.s. finite for every $t > 0$.

The foregoing discussion entails further asymptotic results. We keep the same setting and notation and we write $\mathrm{Res}(f, a)$ for the residue of f at a.

(3.11) Theorem. *Let f be holomorphic in $\mathbb{C}\setminus\{z_1, \dots, z_p\}$ and Γ an open, relatively compact set such that $\{z_1, \dots, z_p\} \subset \Gamma$; then*

$$\frac{2}{\log t}\int_0^t f(Z_s)1_\Gamma(Z_s)dZ_s \xrightarrow[t\to\infty]{(d)} \sum_{j=1}^p \mathrm{Res}(f, z_j)\left\{\frac{\Lambda}{2} + iW^{j-}\right\}.$$

If f is moreover holomorphic at infinity with $\lim_{z\to\infty} f(z) = 0$, then

$$\frac{2}{\log t}\int_0^t f(Z_s)dZ_s$$

converges in distribution as $t \to \infty$, to

$$\sum_{j=1}^p \mathrm{Res}(f, z_j)\left\{\frac{\Lambda}{2} + iW^{j-}\right\} + \mathrm{Res}(f, \infty)\left\{\frac{\Lambda}{2} - 1 + iW^+\right\}.$$

Proof. By the preceding Proposition, we may as well suppose that Γ is the union of disjoint disks $\Gamma_j = D(z_j, \varepsilon_j)$ with $\varepsilon_j > 0$ and sufficiently small, so that we look for the limit of $\sum_{j=1}^p F_t^j$ with $F_t^j = \frac{2}{\log t}\int_0^t f(Z_s)1_{\Gamma_j}(Z_s)dZ_s$. Within Γ_j we may write $f(z) = h_j(z) + g_j\left(\frac{1}{z-z_j}\right)$ with h_j holomorphic in a neighborhood of Γ_j and

$$g_j(z) = \mathrm{Res}(f, z_j)z + \tilde{g}_j(z)z^2$$

for an entire function $\tilde{g}_j$. We set

$$H_t^j = \frac{2}{\log t}\int_0^t 1_{\Gamma_j}(Z_s)\tilde{g}_j\left(\frac{1}{Z_s - z_j}\right)\frac{dZ_s}{(Z_s - z_j)^2}.$$

By Proposition (3.10),

$$F_t^j - \mathrm{Res}(f, z_j)\left\{\frac{2}{\log t}\int_0^t 1_{\Gamma_j}(Z_s)\frac{dZ_s}{Z_s - z_j}\right\} - H_t^j$$

converges to zero in probability. We moreover claim that H_t^j converges to zero in probability.

Let $\widetilde{G}_j$ be the antiderivative of $\tilde{g}_j$ vanishing at 0. By Itô's formula for conformal martingales,

$$\widetilde{G}_j\left(\frac{1}{Z_t - z_j}\right) = \widetilde{G}_j\left(\frac{1}{z_0 - z_j}\right) - \int_0^t \tilde{g}_j\left(\frac{1}{Z_s - z_j}\right)\frac{dZ_s}{(Z_s - z_j)^2}.$$

Since Z_t converges to infinity in probability, the left hand side converges to $\widetilde{G}_j(0) = 0$ in probability. As a result

$$\frac{2}{\log t}\int_0^t \tilde{g}_j\left(\frac{1}{Z_s - z_j}\right)\frac{dZ_s}{(Z_s - z_j)^2}$$

converges to 0 in probability. But the real part of the conformal martingale

$$\int_0^t \tilde{g}_j\left(\frac{1}{Z_s - z_j}\right)1_{\Gamma_j^c}(Z_s)\frac{dZ_s}{(Z_s - z_j)^2}$$

has a bracket equal to $\int_0^t \phi(Z_s)ds$ where

$$\phi(z) = \left|\tilde{g}_j\left((z - z_j)^{-1}\right)\right|^2 \left|z - z_j\right|^{-4} 1_{\Gamma_j^c}(z)$$

is integrable. By the same reasoning as in Proposition (3.10), our claim is proved.

The first statement is then an easy consequence of Theorem (3.8) and of Proposition (4.6) of Chap. X.

For the second statement, we write $f(z) = -\mathrm{Res}(f,\infty)\frac{1}{z} + \frac{1}{z^2}g(1/z)$ for $|z| \geq \eta$ and g holomorphic in a neighborhood of $\{|z| \leq 1/\eta\}$ where η has been chosen sufficiently large. We have to add to the previous limit that of

$$\frac{2}{\log t}\left\{-\mathrm{Res}(f,\infty)\int_0^t 1_{(Z_s \geq \eta)}\frac{dZ_s}{Z_s} + \int_0^t 1_{(Z_s \geq \eta)}g\left(\frac{1}{Z_s}\right)\frac{dZ_s}{Z_s^2}\right\}.$$

This first part converges in distribution to $-\mathrm{Res}(f,\infty)\left(\frac{A}{2} - 1 + iW^+\right)$ thanks to Theorem (3.8) and to Proposition (4.6) of Chap. X and the second part converges in probability to zero by the same reasoning as for $\tilde{g}_j$ above. □

* **(3.12) Exercise.** Let n be an integer and let τ_t be the time-change inverse of

$$\int_0^t |nZ_s|^{2((1/n)-1)}ds.$$

Prove that, with the notation of this section

$$\frac{2}{\log t}\left(\theta_{\tau_t}^1, \ldots, \theta_{\tau_t}^p\right)$$

converges in law to $n\left(W^{1-} + W^+, \ldots, W^{p-} + W^+\right)$.

[Hint: Use Theorem (3.7).]

** **(3.13) Exercise (Mutual windings).** Let $B^1, \dots, B^p$ be p complex BM's on a filtered probability space $(\Omega, \mathscr{F}, \mathscr{F}_t, P)$ which are correlated as follows: for every k and l, $k \neq l$, there exists a matrix $A_{k,l}$ such that for every u and v in $\mathbb{R}^2(\simeq \mathbb{C})$,

$$\left(u, B_t^k\right)\left(v, B_t^l\right) - \left(u, A_{k,l}v\right)t, \quad t \geq 0,$$

is a martingale.

1°) Show that, if for every $k \neq l$, the matrix $A_{k,l}$ is not an orthogonal matrix and if $B_0^i \neq 0$ a.s. for every i, then

$$\frac{2}{\log t}\left(\theta_t^i, i \leq p\right) \xrightarrow[t\to\infty]{(d)} (C_i, i \leq p)$$

where the C_i's are independent Cauchy r.v.'s with parameter 1.

[Hint: Show that for $i \neq j$, $\int_0^t \left|d\langle\theta^i, \theta^j\rangle_s\right| / \log t$, $\int_0^t \left|d\langle\log|B^i|, \theta^j\rangle_s\right| / \log t$ and $\int_0^t \left|d\langle\log|B^i|, \log|B^j|\rangle_s\right| / \log t$ are bounded in probability as $t \to \infty$.]

2°) Let B be a BM^3 and $D^1, \dots, D^p$, p different straight lines which intersect at zero. Assume that B_0 is a.s. not in D^i for every i. Define the winding numbers θ_t^i, $i \leq p$, of B around D^i, $i \leq p$. Show that as a consequence of the previous question, the same convergence in law as in 1°) holds.

3°) Let $B^1, \dots, B^n$, be n independent planar BM's such that $B_0^i \neq B_0^j$ a.s. Call $\theta_t^{i,j}$ the winding number of $B^i - B^j$ around 0. Show that

$$\frac{2}{\log t}\left(\theta_t^{i,j}; 1 \leq i < j \leq n\right) \xrightarrow[t\to\infty]{(d)} \left(C^{i,j}; 1 \leq i < j \leq n\right)$$

where the $C^{i,j}$'s are independent Cauchy r.v.'s with parameter 1.

Notes and Comments

Sect. 1. For the basic definitions and results, as well as for those in Sect. 5 Chap. 0, we refer to the books of Billingsley [1] and Parthasarathy [1]. A more recent exposition is found in the book of Jacod and Shiryaev [1]. Our proof of Donsker's theorem (Donsker [1]) as well as Exercise (1.14) are borrowed from the former. This theorem is constantly being used as a tool to obtain properties of Brownian motion which have first been remarked on its random walk skeletons. This method, which we have refrained from using in this book, is, for instance, found in Pitman [1] and Le Gall [5]. It is interesting to note that conversely some original limit laws on random walks can only be understood in terms of Brownian motion as is seen in the work of Le Gall [7] completing former work of Jain and Pruitt.

Proposition (1.10) is taken from Stroock-Varadhan [1]; although it is of marginal importance in our development it is fundamental in theirs and, as is more generally the case with Martingale problems, has been extended to many situations.

For Exercise (1.13), see Nualart [1] and Yor [14]. Exercise (1.15) is due to Pagès [1] and Exercise (1.16) is inspired by Rebolledo ([1] and 2]).

Exercises (1.17) and (1.18) were suggested respectively by J. Pitman and L. Dubins. Exercise (1.17) allows to simplify the proofs of some limit theorems found in Getoor-Sharpe [4] (see also Jeulin [2] page 128).

Sect. 2. The main result of this section is due to Papanicolaou et al. [1], but their proof was different. The asymptotic version of Knight's theorem comes from Pitman-Yor [5]; another proof is found in Le Gall-Yor [2] and Exercise (2.9) is a variation on the same theme.

Kasahara and Kotani [1] have studied the same problem as Papanicolaou et al. [1] in the case of BM^2. We also refer to Kasahara [4] to whom Exercise (2.13) is due. Biane [3] unifies the asymptotic limit theorems for (multiple) additive functionals of several Brownian motions in $\mathbb{R}^d$.

A number of extensions of Exercises (2.11) through (2.15) have been obtained in recent years by Berman, Borodin [1] and in particular Rosen in the case of stable Lévy processes.

The SDE presentation of the Brownian motion on the sphere found in Exercise (2.16), is a very particular case of that given in Lewis [1] and Van den Berg-Lewis [1]; more generally, see Rogers-Williams [1] and Elworthy [1] for constructions of Brownian motions on surfaces.

Sect. 3. This section is entirely taken from Pitman and Yor ([4], [5] and [7]). Exercise (3.13) is taken from Yor [22] who answers a question of Mitchell Berger. The result in question 2°) of this exercise was originally obtained in a different manner in Le Gall-Yor [3]. More general asymptotic studies for the windings of BM^3 around curves in $\mathbb{R}^3$ are obtained in Le Gall-Yor [4]; the computation of the characteristic functions of the limit laws led the authors to some extension of the Ray-Knight theorems for Brownian local times, presented in Le Gall-Yor [5]; see also the Notes and Comments on Sect. 4 of Chap. XII.

Knight [11] and Yamazaki [1] give convergence results in the sense of fdd's which are closely related to what is called "log-scaling laws", namely limit theorems such as

$$\theta(\exp \lambda u)/u \xrightarrow[u\to\infty]{f\cdot d} F(\lambda)$$

found in Pitman-Yor [7].

Another extension of these results is provided by Watanabe [5] who studies asymptotics of Abelian differentials along Brownian paths on a Riemann surface.

Supplementing these multidimensional limits in law, there are also deep investigations of the pathwise behavior of multiwindings, such as for example of their speed of transience originating with Lyons-McKean [1] and continuing with Gruet [1], Gruet-Mountford [1] and Mountford [1].

We also mention that limit theorems for a large class of diffusions, including the Jacobi processes (see, e.g., Warren-Yor [1]) are developed in Hu-Shi-Yor [1]. These limit theorems are closely related to the asymptotics of diffusion processes in random environments (Kawazu-Tanaka [1], Tanaka [4]).

Intensive discussions of recent studies on the geometry of the planar Brownian curve are found in Le Gall [9] and Duplantier et al. [1].

Appendix

§1. Gronwall's Lemma

Theorem. *If ϕ is a positive locally bounded Borel function on $\mathbb{R}_+$ such that*

$$\phi(t) \leq a + b\int_0^t \phi(s)ds$$

for every t and two constants a and b with $b \geq 0$, then $\phi(t) \leq ae^{bt}$. If in particular $a = 0$ then $\phi \equiv 0$.

Proof. Plainly,

$$\begin{aligned}\phi(t) &\leq a + b\left(\int_0^t \left(a + b\int_0^s \phi(u)du\right)ds\right)\\ &= a + abt + b^2\int_0^t (t-u)\phi(u)du \leq a + abt + b^2 t\int_0^t \phi(u)du.\end{aligned}$$

Proceeding inductively one gets in this fashion

$$\phi(t) \leq a + abt + \ldots + ab^n\frac{t^n}{n!} + \frac{b^{n+1}t^n}{n!}\int_0^t \phi(u)du.$$

Since ϕ is locally bounded, the last term on the right converges to zero as n tends to infinity and the result follows.

§2. Distributions

Let U be a fixed open set in $\mathbb{R}^d$. We denote by C_K^∞ the space of infinitely differentiable functions on U which have a compact support contained in U.

(2.1) Definition. *A sequence (ϕ_n) in C_K^∞ is said to converge to an element ϕ of C_K^∞ if the supports of the ϕ_n's are contained in a fixed compact subset of U and if the k-th derivatives of $\phi_n - \phi$ converge uniformly to zero for every $k \geq 0$.*

(2.2) Definition. *A distribution T on U is a linear form on C_K^∞ such that $T(\phi_n)$ converges to 0 whenever (ϕ_n) is a sequence in C_K^∞ which converges to zero as n tends to infinity.*

We will also write $\langle T, \phi \rangle$ for the value taken by the distribution T on the function ϕ of C_K^∞. With every Radon measure μ on U, we associate a distribution T_μ by setting

$$\langle T_\mu, \phi \rangle = \int \phi d\mu.$$

Likewise, if f is a locally integrable Borel function we write T_f for T_μ where $\mu(dx) = f(x)dx$; in other words

$$\langle T_f, \phi \rangle = \int \phi(x) f(x) dx.$$

(2.3) Definition. *If T is a distribution and $\partial^\alpha / \partial x_1^{\alpha_1} \dots \partial x_d^{\alpha_d}$ a partial derivation operator, we define the corresponding* partial derivative *of T by setting for $\phi \in C_K^\infty$*

$$\left\langle \frac{\partial^\alpha T}{\partial x_1^{\alpha_1} \dots \partial x_d^{\alpha_d}}, \phi \right\rangle = (-1)^{|\alpha|} \left\langle T, \frac{\partial^\alpha \phi}{\partial x_1^{\alpha_1} \dots \partial x_d^{\alpha_d}} \right\rangle$$

where $|\alpha| = \alpha_1 + \dots \alpha_d$.

This obviously defines another distribution and in the case of T_f above, if f is $|\alpha|$ times continuously differentiable, then

$$\frac{\partial^\alpha T_f}{\partial x_1^{\alpha_1} \dots \partial x_d^{\alpha_d}} = T_g,$$

where

$$g = \frac{\partial^\alpha f}{\partial x_1^{\alpha_1} \dots \partial x_d^{\alpha_d}}.$$

§3. Convex Functions

We recall that a real-valued function f defined on an open interval I of $\mathbb{R}$ is *convex* if

$$f(tx + (1-t)y) \leq tf(x) + (1-t)f(y),$$

for every $0 \leq t \leq 1$ and $x, y \in I$. It follows from this definition that for fixed x, the ratio $(f(y) - f(x))/(y - x)$ increases with y. This, in turn, entails immediately that in each point x the function f has a left-hand derivative $f'_-(x)$ and a right-hand derivative $f'_+(x)$ and that, for $y > x$

$$f'_+(x) \leq \frac{f(y) - f(x)}{y - x} \leq f'_-(y).$$

We moreover have the

(3.1) Proposition. *The functions f'_- and f'_+ are increasing, respectively left and right-continuous and the set $\{x : f'_-(x) \neq f'_+(x)\}$ is at most countable.*

Proof. Since $f'_-(x) \le f'_+(x)$, the first property follows at once from the above inequality. To prove that f'_+ is right-continuous, we interchange increasing limits to the effect that if $a_n \downarrow 0$ and $b_m \downarrow 0$,

$$\begin{aligned} \lim_{n\to\infty} f'_+(x+a_n) &= \lim_{n\to\infty}\left(\lim_{m\to\infty}\frac{f(x+a_n+b_m)-f(x+a_n)}{b_m}\right) \\ &= \lim_{m\to\infty}\frac{f(x+b_m)-f(x)}{b_m} = f'_+(x). \end{aligned}$$

Finally, f'_- and f'_+ have only countably many discontinuities and where f'_- is continuous, we have $f'_- = f'_+$ thanks to the above inequalities. □

We now study the second derivative of f. If f is C^2, then f'' is positive, as is easily seen. More generally, we have the

(3.2) Proposition. *The second derivative f'' of f in the sense of distributions is a positive Radon measure; conversely, for any Radon measure μ on $\mathbb{R}$, there is a convex function f such that $f'' = \mu$ and for any interval I and $x \in \mathring{I}$,*

$$\begin{aligned} f(x) &= \frac{1}{2}\int_I |x-a|\mu(da) + \alpha_I x + \beta_I \\ f'_-(x) &= \frac{1}{2}\int_I \operatorname{sgn}(x-a)\mu(da) + \alpha_I \end{aligned}$$

where α_I and β_I are constants and $\operatorname{sgn} x = 1$ *if $x > 0$ and -1 if $x \le 0$.*

Proof. Let $\phi \in C_K^\infty$; the derivative Df of f in the sense of distributions is given by

$$\begin{aligned} \langle Df, \phi\rangle &= -\int f(x)\phi'(x)dx = -\int\left(\lim_{\varepsilon\to 0}\frac{\phi(x+\varepsilon)-\phi(x)}{\varepsilon}\right) f(x)dx \\ &= -\lim_{\varepsilon\to 0}\int \phi(x)\left(\frac{f(x-\varepsilon)-f(x)}{\varepsilon}\right)dx = \int \phi(x) f'_-(x)dx. \end{aligned}$$

By the integration by parts formula for Stieltjes integrals, the second derivative is the measure associated with the increasing function f'_-. Of course by the above results, we could have used f'_+ instead of f'_- without altering the result.

Conversely, if $I \subset J$ the integrals

$$\frac{1}{2}\int_I |x-a|\mu(da) \qquad \text{and} \qquad \frac{1}{2}\int_J |x-a|\mu(da)$$

are convex on $\mathring{I}$ and differ by an affine function. As a result one can define a convex function f on the whole line such that on $\mathring{I}$

$$f(x) = \frac{1}{2}\int_I |x-a|\mu(da) + \alpha_I x + \beta_I.$$

An application of Lebesgue's theorem yields, for $x \in \overset{\circ}{I}$,

$$f'_-(x) = \frac{1}{2}\int_I \operatorname{sgn}(x-a)\mu(da) + \alpha_I$$

and if ϕ is a test function with support in $\overset{\circ}{I}$, then

$$\int f'_-(x)\phi'(x)dx = \int_I \mu(da)\left(\frac{1}{2}\int_{-\infty}^{+\infty}\phi'(x)\operatorname{sgn}(x-a)dx\right) = -\int \phi(a)\mu(da)$$

which proves that the second derivative of f is μ.

The convex function determined by μ is of course unique only up to addition of an affine function. If the measure μ is such that $\int |x-a|\mu(da)$ is finite for every x, which will in particular be the case if μ has compact support, then one can globally state that

$$f(x) = \frac{1}{2}\int |x-a|\mu(da) + \alpha x + \beta.$$

The constants α and β can be fixed by specifying special values for f in two points. If in particular for $a < b$ we demand that $f(a) = f(b) = 0$ then one can give for f a more compact expression which we now describe in a slightly more general setting.

Let s be a continuous, strictly increasing function on $I = [a, b]$. We will say that f is s-convex if for $a < c_1 < x < c_2 < b$,

$$(s(c_2) - s(c_1))\, f(x) \leq (s(c_2) - s(x))\, f(c_1) + (s(x) - s(c_1))\, f(c_2).$$

Exactly as above one can define the right and left s-derivatives $df_\pm/ds$ by taking the appropriate limits of the ratios $(f(y)-f(x))/(s(y)-s(x))$. At the points where they are equal we say that f has an s-derivative. The functions thus defined are increasing and determine as above a measure μ.

If for $x \leq y$ we set

$$G(x, y) = G(y, x) = (s(x) - s(a))(s(b) - s(y))/(s(b) - s(a)),$$

then if f is s-convex and if $f(a) = f(b) = 0$,

$$f(x) = -\int_a^b G(x, y)\mu(dy).$$

Indeed, using the integration by parts formula for Stieltjes integrals,

$$\int_a^b G(x, y)\mu(dy) = \frac{s(b) - s(x)}{s(b) - s(a)}\int_{]a,x]}(s(y) - s(a))\mu(dy)$$
$$+ \frac{s(x) - s(a)}{s(b) - s(a)}\int_{]x,b[} s(b) - s(y))\mu(dy)$$

$$= \frac{s(b)-s(x)}{s(b)-s(a)}\left[(s(x)-s(a))\frac{df_+}{ds}(x) - \int_a^x \frac{df_+}{ds}(b)ds(y)\right]$$
$$+ \frac{s(x)-s(a)}{s(b)-s(a)}\left[-(s(b)-s(x))\frac{df_+}{ds}(x) + \int_x^b \frac{df_+}{ds}(b)ds(y)\right]$$
$$= -f(x).$$

Naturally, all we have said is valid for concave functions with the obvious changes.

§4. Hausdorff Measures and Dimension

Let h be a strictly increasing continuous function on $\mathbb{R}_+$ such that $h(0) = 0$ and $h(\infty) = \infty$. Let B be a Borel subset of a metric space E. The *Hausdorff h-measure* of B is the number

$$\Lambda^h(B) = \liminf_{\varepsilon \downarrow 0}\left(\sum_n h(|I_n|)\right)$$

where the infimum is over all coverings $\bigcup I_n$ of B where I_n is a closed set in E with diameter $|I_n| \leq \varepsilon$. Of special interest is the case where $h(t) = t^\alpha$, $\alpha > 0$, in which case we will write Λ^α and speak of α-measure. If $E = \mathbb{R}^d$, Λ^d is the ordinary Lebesgue measure.

(4.1) Lemma. *If* $h(t) = \nu(t)k(t)$ *with* $\lim_{t\to 0}\nu(t) = 0$, *then* $\Lambda^h(F) > 0$ *implies* $\Lambda^k(F) = \infty$.

Proof. Pick $\varepsilon > 0$; there is an $\eta > 0$ such that $v \leq \eta$ implies $\nu(v) \leq \varepsilon$. Let $\bigcup I_n$ be a covering of F with $|I_n| \leq v \leq \eta$. Then

$$\sum_n k(|I_n|) = \sum_n h(|I_n|)/\nu(|I_n|) \geq \frac{1}{\varepsilon}\sum_n h(|I_n|);$$

it follows that

$$\sum_n k(|I_n|) \geq \frac{1}{\varepsilon}\Lambda^h(F),$$

hence

$$\Lambda^k(F) \geq \frac{1}{\varepsilon}\Lambda^h(F),$$

and since ε is arbitrary, the proof is complete. □

A consequence of this lemma is that there is a number α_0 such that $\Lambda^\alpha(F) = +\infty$ if $\alpha < \alpha_0$ and $\Lambda^\alpha(F) = 0$ if $\alpha > \alpha_0$ (the number $\Lambda^{\alpha_0}(F)$ itself may be zero, non zero and finite or infinite). The number α_0 is called the *Hausdorff dimension* of F. For instance, one can prove that the dimension of the Cantor "middle third" set is $\log 2/\log 3$.

§5. Ergodic Theory

Let $(E, \mathscr{E}, m)$ be a σ-finite measure space. A positive contraction T of $L^1(m)$ is a linear operator on $L^1(M)$ with norm ≤ 1 and mapping positive (classes of) functions into positive (classes of) functions. A basic example of such a contraction is the map $f \to f \circ \theta$ where θ is a measurable transformation of $(E, \mathscr{E})$ which leaves m invariant.

(5.1) Theorem. (Hopf's decomposition theorem). *There is an m-essentially unique partition $C \cup D$ of E such that for any $f \in L^1_+(m)$*

i) $\sum_{k=0}^{\infty} T^k f = 0$ *or* $+\infty$ *on* C,

ii) $\sum_{k=0}^{\infty} T^k f < \infty$ *on* D.

If $D = \emptyset$, the contraction T is said to be *conservative*. In that case, the sums $\sum_{k=0}^{\infty} T^k f$ for $f \in L^1_+(m)$ take on only the values 0 and $+\infty$. The sets $\{\sum_{k=0}^{\infty} T^k f = \infty\}$ where f runs through $L^1_+(m)$ form a σ-algebra denoted by $\mathscr{C}$ and called the *invariant* σ-algebra. If all these sets are either $\emptyset$ or E (up to equivalence) or in other words if $\mathscr{C}$ is m-a.e. trivial then T is called *ergodic*.

We now state the basic *Chacon-Ornstein theorem*.

(5.2) Theorem. *If T is conservative and g is an element of $L^1_+(m)$ such that $m(g) > 0$, then for every $f \in L^1(m)$,*

$$\lim_{n\to\infty} \left(\sum_{k=0}^{n} T^k f \Big/ \sum_{k=0}^{n} T^k g \right) = E[f \mid \mathscr{C}]/E[g \mid \mathscr{C}] \qquad m\text{-a.e.}$$

The conditional expectations on the right are taken with respect to m. If m is unbounded this means that the quotient is equal to $E[(f/h) \mid \mathscr{C}]/E[(g/h) \mid \mathscr{C}]$ where h is a strictly positive element in $L^1(m)$ and the conditional expectations are taken with respect to the bounded measure $h \cdot m$; it can be shown that the result does not depend on h. If T is ergodic the quotient on the right is simply $m(f)/m(g)$.

The reader is referred to Revuz [3] for the proof of these results.

§6. Probabilities on Function Spaces

Let E be a Polish space and set $\Omega = C(\mathbb{R}_+, E)$. Let us call X the canonical process and set $\mathscr{F}_t = \sigma(X_s, s \leq t)$ and $\mathscr{F}_\infty = \sigma(X_s, s \geq 0)$.

(6.1) Theorem. *If for every $t \geq 0$, there exists a probability measure P^t on $\mathscr{F}_t$ such that for every $s < t$, P^t coincides with P^s on $\mathscr{F}_s$, then there exists a probability measure P on $\mathscr{F}_\infty$ which for every t coincides with P^t on $\mathscr{F}_t$.*

For the proof of this result the reader can refer to the book of Stroock and Varadhan [1] p. 34; see also Azéma-Jeulin [1].

§7. Bessel Functions

The modified Bessel function I_ν is defined for $\nu \geq -1$ and $x > 0$, by

$$I_\nu(x) = \sum_{k=0}^{\infty} (x/2)^{2k+\nu}/k!\Gamma(\nu + k + 1).$$

Observe that for $\nu = -1$ and $k = 0$ the term $\Gamma(\nu + k + 1)$ is infinite, and therefore the first term in the above series vanishes. By using the relationship $\Gamma(z + 1) = z\Gamma(z)$, one thus sees that $I_1 = I_{-1}$. For some details about these functions we refer the reader to Lebedev [1], pages 108–111.

This family of functions occurs in many computations of probability laws. Call for instance $d_x^{(\nu)}$ the density of a random variable with conditional law $\gamma_{\nu+k+1}$ where k is random with a Poisson law of parameter $x > 0$ and $\nu > -1$. Then

$$\begin{aligned} d_x^{(\nu)}(y) &= \sum_{k=0}^{\infty} e^{-x}\left(x^k/k!\right) y^{\nu+k} e^{-y}/\Gamma(\nu + k + 1) \\ &= e^{-(x+y)}(y/x)^{\nu/2} I_\nu\left(2\sqrt{xy}\right). \end{aligned}$$

Replacing x and y by $x/2t$ and $y/2t$ we find that, for $\nu > -1$,

$$q_t^{(\nu)}(x, y) = (1/2t)\exp(-(x + y)/2t)(y/x)^{\nu/2} I_\nu\left(\sqrt{xy}/t\right)$$

where $t > 0$, $x > 0$, $y > 0$, is also a probability density, in fact the density of $\text{BESQ}^{(\nu)}$ as found in section 1, Chapter IX. At that point we needed to know the Laplace transform of this density which is easily found from the above. Indeed, the Laplace transform of γ_k is equal to $(\lambda + 1)^{-k}$ and therefore the Laplace transform of $d_x^{(\nu)}$ is equal to

$$\sum_{k=0}^{\infty} e^{-x}(x^k/k!)(\lambda + 1)^{-(\nu+k+1)} = (\lambda + 1)^{-(\nu+1)} \exp(-\lambda x/(\lambda + 1)).$$

From this, using the same change of variables as before, one gets that the Laplace transform of $q_t^{(\nu)}(x, \cdot)$ is equal to

$$(2\lambda t + 1)^{-(\nu+1)} \exp(-\lambda x/(2\lambda t + 1)).$$

Another formula involving Bessel functions and which was of interest in Sect.3, Chap. VIII, is the following. If $x \in \mathbb{R}^d$ we call $\xi(x)$ the angle of Ox with a fixed axis, and if μ^d is the uniform probability measure on the unit sphere S^{d-1}, then

$$\int_{S^{d-1}} \exp(\rho \cos \xi(x))\mu^d(dx) = (2/\rho)^\nu \Gamma(\nu + 1) I_\nu(\rho/2)$$

where $\nu = (d/2) - 1$. This can be proved directly from the definition of I_ν by writing the exponential as a series and computing $\int_{S^{d-1}} \cos \xi(x)^p \mu^d(dx)$; to this end, it is helpful to use the duplication formula

$$\Gamma(2z) = (2\pi)^{-1/2} 2^{2z-(1/2)} \Gamma(z)\Gamma(z + (1/2)).$$

§8. Sturm-Liouville Equation

Let μ be a positive Radon measure on $\mathbb{R}$. Then there exists a unique, positive, decreasing function ϕ_μ such that

$$(*) \qquad \phi_\mu(0) = 1, \qquad \phi''_\mu = \phi_\mu \mu,$$

where ϕ''_μ is the second derivative in the sense of distributions (Appendix 2). Observe from (*) that since ϕ_μ is positive, it is convex, and ϕ''_μ is equal to $d\phi'_\mu$ where ϕ'_μ is the right derivative of ϕ_μ (Appendix 3).

To prove this existence and uniqueness result we transform (*) into the Riccati equation

$$(+) \qquad g(x) = 1 + \mu(]a, x]) - \int_a^x g^2(y)dy,$$

where $a \in \mathbb{R}$. We claim that this equation has a unique solution g on $[a, \infty[$ which satisfies the inequality $g(x) \geq 1/(1 + x - a)$, for $x \geq a$.

Indeed, since the function $x \to x^2$ is locally Lipschitz, there is a unique maximal solution to (+) on an interval $[a, \alpha[$ with $\alpha > a$. Obviously g is of finite variation on every bounded interval. We will first prove that g is > 0 on $[a, \alpha[$. Indeed, suppose that $g(x-) \leq 0$ for some $x \in [a, \alpha[$ and set $\gamma = \inf\{x \in [a, \alpha[: g(x-) \leq 0\}$. For $a \leq x < \gamma$, we have

$$(\ddagger) \qquad g(x) = g(x-) + \mu(\{x\}) > 0.$$

On the other hand, by Proposition (4.6) Chap. 0, we may write

$$\begin{aligned} 1/g(x) &= 1 - \int_{]a,x]} (g(y)g(y-))^{-1} dg(y) \\ &= 1 + (x - a) - \int_{]a,x]} (g(y)g(y-))^{-1} d\mu(y) \leq 1 + (x - a). \end{aligned}$$

As a result, $g(\gamma) \geq 1/(1 + \gamma - a) + \mu(\{\gamma\}) > 0$, and since g is right-continuous this contradicts the definition of γ. That g is > 0 then follows from ($\ddagger$).

Now, since g is > 0 on $[a, \alpha[$, if α is finite, rewriting (+) as

$$g(x) + \int_a^x g^2(y)dy = 1 + \mu(]a, x]),$$

we see that g is bounded on $]a, \alpha[$ and by letting x increase to α we get

$$g(\alpha-) + \int_a^\alpha g^2(y)dy = 1 + \mu(]a, \alpha[).$$

If we set $g(\alpha) = g(\alpha-) + \mu(\{\alpha\})$ and solve the equation (+) for $x > \alpha$, we see that α cannot be maximal. As a result $\alpha = \infty$ and we have proved our claim.

Next, if g is the solution to (+), we set, for $x \geq a$,

$$\psi(x) = \exp\left(\int_a^x g(y)dy\right).$$

One sees rapidly that $\psi(x) \geq 1 + x - a$ on $[a, \infty[$ and that $\psi'' = \psi\mu$. We further set

$$\phi(x) = \psi(x) \int_x^\infty \psi(y)^{-2} dy.$$

The function ϕ is > 0 and is another solution to the equation $\phi'' = \phi\mu$. Moreover, because ψ' is increasing, we have

$$\begin{aligned} \phi'(x) &= \psi'(x) \int_x^\infty \psi(y)^{-2} dy - (1/\psi(x)) \\ &\leq \int_x^\infty \left(\psi'(y)/\psi(y)^2\right) dy - (1/\psi(x)) = 0, \end{aligned}$$

which shows that ϕ is decreasing.

The space of solutions to the equation $\phi'' = \phi\mu$ is the space of functions $u\psi + v\phi$ with $u, v \in \mathbb{R}$. Since ψ increases to $+\infty$ at infinity, the only positive bounded solutions are of the form $v\phi$ with $v \geq 0$. If for $a < 0$ we put $\phi_\mu = \phi/\phi(0)$ we get the unique solution to (*) that we were looking for.

Bibliography

Airault, H. and Föllmer, H.

[1] Relative densities of semimartingales. Invent. Math. **27** (1974) 299–327.

Albeverio, S., Fenstad, J.E., Hoegh-Krohn, R., and Lindstrom, T.

[1] Non standard methods in stochastic analysis and mathematical physics. Academic Press, New York 1986.

Aldous, D.

[1] The continuous random tree II: an overview. In: M.T. Barlow and N.H. Bingham (eds.) Stochastic analysis. Camb. Univ. Press 1991.

Alili, L.

[1] Fonctionnelles exponentielles et certaines valeurs principales des temps locaux browniens. Thèse de doctorat de l'université de Paris VI. 1995.

Alili, L., Dufresne, D., and Yor, M.

[1] Sur l'identité de Bougerol pour les fonctionnelles exponentielles du mouvement brownien avec drift. In: Exponential functionals and principal values related to Brownian motion. Biblioteca de la Revista Matematica Iberoamericana, 1997, pp. 3–14.

Aronszajn, N.

[1] Theory of reproducing kernels. Trans. Amer. Math. Soc. **68** (1950) 337–404.

Atsuji, A.

[1] Some inequalities for some increasing additive functionals of planar Brownian motion and an application to Nevanlinna theory. J.F. Sci. Univ. Tokyo Sect. I-A, **37** (1990) 171–187.

[2] Nevanlinna theory via stochastic calculus. J. Funct. Anal. **132**, 2 (1995) 437–510.

[3] On the growth of meromorphic functions on the unit disc and conformal martingales. J. Math. Sci. Univ. Tokyo **3** (1996) 45–56.

Attal, S., Burdzy, K., Emery, M., and Hu, Y.

[1] Sur quelques filtrations et transformations browniennes. Sém. Prob. XXIX. Lect. Notes in Mathematics, vol. 1613. Springer, Berlin Heidelberg New York 1995, 56–69.

Auerhan, J. and Lépingle, D.

[1] Les filtrations de certaines martingales du mouvement brownien dans $\mathbb{R}^n$ (II). Sém. Prob. XV. Lect. Notes in Mathematics, vol. 850. Springer, Berlin Heidelberg New York 1981, pp. 643–668.

Azéma, J.

[1] Quelques applications de la théorie générale des processus I. Invent. Math. **18** (1972) 293–336.

[2] Représentation multiplicative d'une surmartingale bornée. Z.W. **45** (1978) 191–212.

[3] Sur les fermés aléatoires. Sém. Prob. XIX. Lect. Notes in Mathematics, vol. 1123. Springer, Berlin Heidelberg New York 1985, pp. 397–495.

Azéma, J., Duflo, M., and Revuz, D.
[1] Mesure invariante sur les classes récurrentes des processus de Markov. Z.W. **8** (1967) 157–181.

Azéma, J., Gundy, R.F., and Yor, M.
[1] Sur l'intégrabilité uniforme des martingales continues. Sém. Prob. XIV. Lect. Notes in Mathematics, vol. 784. Springer, Berlin Heidelberg New York 1980, pp. 53–61.

Azéma, J., and Jeulin, T.
[1] Précisions sur la mesure de Föllmer. Ann. I.H.P. **22** (3) (1976) 257–283.

Azéma, J., and Rainer, C.
[1] Sur l'équation de structure $d[X, X]_t = dt - X_{t-}^{+} dX_t$. Sém. Prob. XXVIII, Lect. Notes in Mathematics, vol. 1583. Springer, Berlin Heidelberg New York 1994, 236–255.

Azéma, J., and Yor, M.
[1] En guise d'introduction. Astérisque 52–53, Temps Locaux (1978) 3–16
[2] Une solution simple au problème de Skorokhod. Sém. Prob. XIII. Lect. Notes in Mathematics, vol. 721. Springer, Berlin Heidelberg New York 1979, pp. 90–115 and 625–633
[3] Etude d'une martingale remarquable. Sém. Prob. XXIII. Lect. Notes in Mathematics, vol. 1372. Springer, Berlin Heidelberg New York 1989, pp. 88–130.
[4] Sur les zéros des martingales continues. Sém. Prob. XXVI. Lect. Notes in Mathematics, vol. 1526. Springer, Berlin Heidelberg New York 1992, pp. 248–306.

Azencott, R.
[1] Grandes déviations et applications. Ecole d'Eté de Probabilités de Saint-Flour VIII. Lect. Notes in Mathematics, vol. 774. Springer, Berlin Heidelberg New York 1980, pp. 1–176.

Barlow, M.T.
[1] $L(B_t, t)$ is not a semi-martingale. Sém. Prob. XVI. Lect. Notes in Mathematics, vol. 920. Springer, Berlin Heidelberg New York 1982, pp. 209–211.
[2] One-dimensional stochastic differential equation with no strong solution. J. London Math. Soc. **26** (1982) 335–345.
[3] Continuity of local times for Lévy processes. Z.W. **69** (1985) 23–35.
[4] Skew Brownian motion and a one dimensional stochastic differential equation. Stochastics **25** (1988) 1–2.
[5] Necessary and sufficient conditions for the continuity of local times of Lévy processes. Ann. Prob. **16** (1988) 1389–1427.
[6] Study of a filtration expanded to include an honest time. Z. für Wahr. **44** (1978) 307–323.
[7] Construction of a martingale with given absolute value. Ann. Prob. **9** (1981) 314–320.

Barlow, M.T., Emery, M., Knight, F.B., Song, S., Yor, M.
[1] Autour d'un théorème de Tsirel'son sur des filtrations browniennes et non-browniennes. Sém. Probab. XXXII, Lect. Notes in Mathematics, vol. 1686. Springer, Berlin Heidelberg New York 1998, pp. 264–305.

Barlow, M.T., Jacka, S.D., and Yor, M.
[1] Inequalities for a pair of processes stopped at a random time. Proc. London Math. Soc. **52** (1986) 142–172.

Barlow, M.T., and Perkins, E.
[1] One-dimensional stochastic differential equations involving a singular increasing process. Stochastics **12** (1984) 229–249.
[2] Strong existence and non-uniqueness in an equation involving local time. Sém. Prob. XVII. Lect. Notes in Mathematics, vol. 986. Springer, Berlin Heidelberg New York 1986, pp. 32–66.

Barlow, M.T., Pitman, J.W., and Yor, M.
[1] Une extension multidimensionnelle de la loi de l'arc sinus. Sém. Prob. XXIII. Lect. Notes in Mathematics, vol. 1372. Springer, Berlin Heidelberg New York 1989, pp. 294–314.
[2] On Walsh's Brownian motions. Sém. Prob. XXIII. Lect. Notes in Mathematics, vol. 1372. Springer, Berlin Heidelberg New York 1989, pp. 275–293.

Barlow, M.T. and Yor, M.
[1] (Semi-)martingale inequalities and local times. Z.W. **55** (1981) 237–254.
[2] Semi-martingale inequalities via the Garsia-Rodemich-Rumsey lemma and applications to local times. J. Funct. Anal. **49** (1982) 198–229.
[3] Sur la construction d'une martingale continue de valeur absolue donnée. Sém. Prob. XIV, Lect. Notes in Mathematics 704, p. 62–75, Springer (1980).

Bass, R.F.
[1] Joint continuity and representation of additive functionals of d-dimensional Brownian motion. Stoch. Proc. Appl. **17** (1984) 211–228.
[2] L_p-inequalities for functionals of Brownian motion. Sém. Prob. XXI. Lect. Notes in Mathematics, vol. 1247. Springer, Berlin Heidelberg New York 1987, pp. 206–217.
[3] Probabilistic techniques in Analysis. Prob. and its App. Springer, Berlin Heidelberg New York 1995.
[4] Diffusions and Elliptic Operators. Springer, Berlin Heidelberg New York 1997.
[5] Skorokhod embedding via stochastic integrals. Sem. Prob. XVII, Lect. Notes in Mathematics 986, p. 221–224, Springer (1983).

Bass, R.F., and Burdzy, K.
[1] Stochastic Bifurcation Models. Ann. Prob. **27** (1999) 50–108.

Bass, R.F., and Griffin, P.S.
[1] The most visited site of Brownian motion and simple random walk. Z.W. **70** (1985) 417–436.

Beghdadi-Sakrani, S.
[1] Une martingale non-pure, dont la filtration est brownienne. Sém. Prob. XXXVI, Lect. Notes in Mathematics 1801, p. 348–359, Springer (2003).
[2] Some remarkable pure martingales. Ann. Inst. H. Poincaré **39** (2003) 287–299.
[3] The uniqueness class of continuous local martingales. Bernoulli **8** (2002) 207–217.
[4] Calcul stochastique pour des mesures signées. Sém. Prob. XXXVI, Lect. Notes in Mathematics 1801, p. 366–382, Springer (2003).
[5] On pathwise uniqueness of stochastic differential equations without drift. Jour. Th. Prob., Vol. 6, n° 4, p. 789–812 (2003).

Beghdadi-Sakrani, S. and Emery, M.
[1] On certain probabilities equivalent to coin-tossing, d'après Schachermayer. Sém. Prob. XXXIII, p. 240–256, Lect. Notes in Mathematics 1709, Springer (1999).

Beněs, V.
[1] Non existence of strong non-anticipating solutions to SDE's; Implications for functional DE's, filtering and control. Stoch. Proc. Appl. **5** (1977) 243–263.

[2] Realizing a weak solution on a Probability space. Stoch. Proc. Appl. **7** (1978) 205–225.

Bertoin, J.
[1] Sur une intégrale pour les processus à α-variation bornée. Ann. Prob. **17** (1989) 1521–1535.
[2] Applications de la théorie spectrale des cordes vibrantes aux fonctionnelles additives principales d'un brownien réfléchi. Ann. I.H.P. **25**, 3 (1989) 307–323.
[3] Complements on the Hilbert transform and the fractional derivatives of Brownian local times. J. Math. Kyoto Univ. **30** (4) (1990) 651–670.
[4] On the Hilbert transform of the local times of a Lévy process. Bull. Sci. Math. **119** (2) (1995) 147–156.
[5] An extension of Pitman's theorem for spectrally positive Lévy processes. Ann. Prob. **20** (3) (1993) 1464–1483.
[6] Excursions of a $\mathrm{BES}_0(d)$ and its drift term $(0 < d < 1)$. Prob. Th. Rel. F. **84** (1990) 231–250.
[7] Lévy processes. Cambridge Univ. Press 1996.
[8] Subordinators: examples and applications. XXVIIe Ecole d'Eté de St. Flour: Summer 1997. Lect. Notes in Mathematics. Springer, Berlin Heidelberg New York 1998.
[9] Cauchy's principal value of local times of Lévy processes with no negative jumps via continuous branching processes. Elec. J. of Prob. **2** 1997, Paper 6.

Bertoin, J., Chaumont, L., and Pitman, J.
[1] Path transformation of first passage bridges. Elec. Comm. in Prob. **8** (2003) 155–166.

Bertoin, J., and Le Jan, Y.
[1] Representation of measures by balayage from a regular recurrent point. Ann. Prob. **20** (1992) 538–548.

Bertoin, J. and Pitman, J.
[1] Path transformations connecting Brownian bridge, excursion and meander. Bull. Sci. Math. **118** (1994) 147–166.

Bertoin, J. and Werner, W.
[1] Comportement asymptotique du nombre de tours effectués par la trajectoire brownienne plane. Sém. Prob. XXVIII, Lect. Notes in Mathematics, vol. 1583. Springer, Berlin Heidelberg New York 1994, pp. 164–171.
[2] Asymptotic windings of planar Brownian motion revisited via the Ornstein-Uhlenbeck process. Sém. Prob. XXVIII, Lect. Notes in Mathematics, vol. 1583. Springer, Berlin Heidelberg New York 1994, pp. 138–152.
[3] Stable windings. Ann. Prob. **24**, 3, July 1996, pp. 1269–1279.

Bernard, A. and Maisonneuve, B.
[1] Décomposition atomique de martingales de la classe H^1. Sém. Prob. XI. Lect. Notes in Mathematics, vol. 581. Springer, Berlin Heidelberg New York 1977, pp. 303–323.

Besicovitch, A.S., and Taylor, S.J.
[1] On the complementary intervals of a linear closed set of zero Lebesgue measure. J. London Math. Soc. (1954) 449–459.

Biane, P.
[1] Relations entre pont brownien et excursion normalisée du mouvement brownien. Ann. I.H.P. **22**, 1 (1986) 1–7.
[2] Sur un calcul de F. Knight. Sém. Prob. XXII, Lect. Notes in Mathematics, vol. 1321. Springer, Berlin Heidelberg New York 1988, pp. 190–196.
[3] Comportement asymptotique de certaines fonctionnelles additives de plusieurs mouvements browniens. Sém. Prob. XXIII, Lect. Notes in Mathematics, vol. 1372. Springer, Berlin Heidelberg New York 1989, pp. 198–233.

[4] Decomposition of Brownian trajectories and some applications. In: A. Badrikian, P.A. Meyer, J.A. Yan (eds.) Prob. and Statistics; rencontres franco-chinoises en Probabilités et Statistiques. Proceedings of WuHan meeting, pp. 51–76. World Scientific, 1993.

Biane, P., Bougerol, P., and O'Connell, N.
[1] Littleman paths and Brownian paths. Preprint, March 2004.

Biane, P., Le Gall, J.F., and Yor, M.
[1] Un processus qui ressemble au pont brownien. Sém. Prob. XXI, Lect. Notes in Mathematics, vol. 1247. Springer, Berlin Heidelberg New York 1987, pp. 270–275.

Biane, P., Pitman, J., and Yor, M.
[1] Probabilistic interpretation of the Jacobi and the Riemann zeta functions via Brownian excursions. Bull. AMS **38** (2001) 435–465.

Biane, P., and Yor, M.
[1] Valeurs principales associées aux temps locaux browniens. Bull. Sci. Math. **111** (1987) 23–101.
[2] Variations sur une formule de P. Lévy. Ann. I.H.P. **23** (1987) 359–377.
[3] Quelques précisions sur le méandre brownien. Bull. Sci. Math. **112** (1988) 101–109.
[4] Sur la loi des temps locaux browniens pris en un temps exponentiel. Sém. Prob. XXII, Lect. Notes in Mathematics, vol. 1321. Springer, Berlin Heidelberg New York 1988, pp. 454–466.

Billingsley, P.
[1] Convergence of probability measures. Wiley and Sons, New York 1979. Second edition (1999).

Bingham, N.H.
[1] Limit theorem for occupation times of Markov processes. Z.W. **17** (1971) 1–22.
[2] The strong arc sine law in higher dimensions. In: Bergelson, March, Rosenblatt (eds.) Convergence in Ergodic Theory and Probability. de Gruyter 1996, pp. 111–116.

Bismut, J.M.
[1] Martingales, the Malliavin calculus and Hypoellipticity under general Hörmander's conditions. Z.W. **56** (1981) 469–506.
[2] On the set of zeros of certain semi-martingales. Proc. London Math. Soc. (3) **49** (1984) 73–86.
[3] Last exit decomposition and regularity at the boundary of transition probabilities. Z.W. **69** (1985) 65–98.
[4] The Atiyah-Singer theorems. J. Funct. Anal. **57** (1984) 56–99 and 329–348.
[5] Formules de localisation et formules de Paul Lévy. Astérisque 157–158, Colloque Paul Lévy sur les processus stochastiques (1988) 37–58.

Blumenthal, R.M.
[1] Excursions of Markov processes. Probability and its applications. Birkhäuser 1992.

Blumenthal, R.M., and Getoor, R.K.
[1] Markov processes and potential theory. Academic Press, New York 1968.

Borodin, A.N.
[1] On the character of convergence to Brownian local time I and II. Z.W. **72** (1986) 231–250 and 251–277.

Borodin, A.N., and Salminen, P.
[1] Handbook of Brownian motion – Facts and formulae. Birkhäuser 2002 (second edn.).

Bougerol, P.
[1] Exemples de théorèmes locaux sur les groupes résolubles. Ann. IHP **29** (4) (1983) 369–391.

Bougerol, P. and Jeulin, Th.
[1] Paths in Weyl chambers and random matrices. Prob. Theory Rel. Fields **124** (2002) 517–543.

Bouleau, N.
[1] Sur la variation quadratique de certaines mesures vectorielles. Zeit. für Wahr. **61** (1982) 261–270.

Bouleau, N., and Yor, M.
[1] Sur la variation quadratique des temps locaux de certaines semimartingales. C.R. Acad. Sci. Paris, Série I **292** (1981) 491–494.

Boylan, E.S.
[1] Local times of a class of Markov processes. Ill. J. Math. **8** (1964) 19–39.

Brassesco, S.
[1] A note on planar Brownian motion. Ann. Prob. **20** (3) (1992) 1498–1503.

Breiman, L.
[1] Probability. Addison-Wesley Publ. Co., Reading, Mass. 1968.

Brosamler, G.
[1] Quadratic variation of potentials and harmonic functions. Trans. Amer. Math. Soc. **149** (1970) 243–257.

Brossard, I., and Chevalier, L.
[1] Classe $L \log L$ et temps local. C.R. Acad. Sci. Paris, Ser. A Math. **305** (1987) 135–137.

Brown, T.C., and Nair, M.G.
[1] A simple proof of the multivariate random time change theorem for point processes. J. Appl. Prob. **25** (1988) 210–214.

Bru, M.F.
[1] Processus de Wishart. CRAS, Série I **308** (1989) 29–32.
[2] Diffusions of perturbed principal component analysis. J. Multi. Anal. **29** (1989) 127–136.
[3] Wishart processes. J. Th. Prob. **4** (1999) 725–751.

Burdzy, K.
[1] Brownian paths and cones. Ann. Prob. **13** (1985) 1006–1010.
[2] Cut points and Brownian paths. Ann. Prob. **17** (1989) 1012–1036.
[3] Geometric properties of two-dimensional Brownian paths. Prob. Th. Rel. F. **81** (1989) 485–505.

Burkholder, D.L.
[1] Martingale transforms. Ann. Math. Stat. **37** (1966) 1494–1504.
[2] Distribution function inequalities for martingales. Ann. Prob. **1** (1973) 19–42.
[3] Exit times of Brownian motion, harmonic majorization and Hardy spaces. Adv. Math. **26** (1977) 182–205.

Burkholder, D.L., and Gundy, R.F.
[1] Extrapolation and interpolation of quasi-linear operators on martingales. Acta. Math. **124** (1970) 249–304.

Calais, J.Y., and Génin, M.
[1] Sur les martingales locales continues indexées par]0, ∞[. Sém. Proba. XXII, Lect. Notes in Mathematics, vol. 986. Springer, Berlin Heidelberg New York 1988, pp. 454–466.

Cameron, R.H., and Martin, W.T.
[1] Transformation of Wiener integrals under translations. Ann. Math. **45** (1944) 386–396.
[2] Evaluations of various Wiener integrals by use of certain Sturm-Liouville differential equations. Bull. Amer. Math. Soc. **51** (1945) 73–90.
[3] Transformation of Wiener integrals under a general class of linear transformations. Trans. Amer. Math. Soc. **18** (1945) 184–219.

Carlen, E.
[1] The pathwise description of quantum scattering in stochastic mechanics. In: S. Albeverio et al. (eds.) Stochastic processes in quantum physics, Lect. Notes in Physics, vol. 262. Springer, Berlin Heidelberg New York 1986, pp. 139–147.

Carlen, E., and Elworthy, D.
[1] Stochastic and quantum mechanical scattering on hyperbolic spaces. In: D. Elworthy and N. Ikeda (eds.) Asymptotic problems in probability theory: stochastic models and diffusions on fractals. Pitman Research Notes (1993) vol. 283.

Carlen, E., and Krée, P.
[1] Sharp L^p-estimates on multiple stochastic integrals. Ann. Prob. **19** (1) (1991) 354–368.

Carmona, P., Petit, F., and Yor, M.
[1] Some extensions of the Arcsine law as partial consequences of the scaling property of Brownian motion. Prob. Th. Rel. Fields **100** (1994) 1–29.
[2] Beta variables as times spent in $[0, \infty)$ by certain perturbed Brownian motions. J. London Math. Soc. (2) **58** (1998) 239–256.
[3] An identity in law involving reflecting Brownian motion, derived from generalized arc-sine laws for perturbed Brownian motions. Stoch. Proc. and their App. (1999) 323–334.
[4] Beta-gamma random variables and intertwining relations between certain Markov processes. Revista Ibero Americana **14** (2) (1998) 311–367.
[5] A trivariate law for certain processes related to perturbed Brownian motions. Ann. IHP (2004).

Carne, T.K.
[1] The algebra of bounded holomorphic functions. J. Funct. Anal. **45** (1982) 95–108.
[2] Brownian motion and Nevanlinna Theory. Proc. London Math. Soc. (3) **52** (1986) 349–368.
[3] Brownian motion and stereographic projection. Ann. I.H.P. **21** (1985) 187–196.

Centsov, N.V.
[1] Limit theorems for some classes of random functions. Selected Translations in Mathematics. Statistics and Probability **9** (1971) 37–42.

Chacon, R., and Walsh, J.B.
[1] One-dimensional potential embedding. Sém. Prob. X, Lect. Notes in Mathematics, vol. 511. Springer, Berlin Heidelberg New York 1976, pp. 19–23.

Chaleyat-Maurel, M., and Yor, M.
[1] Les filtrations de X et X^+, lorsque X est une semimartingale continue. Astérisque 52–53 (1978) 193–196.

Chan, J., Dean, D.S., Jansons, K.M., and Rogers, L.C.G.
[1] On polymer conformations in elongational flows. Comm. Math. Phys. **160** (2) (1994) 239–257.

Chaumont, L. and Yor, M.
[1] Exercises in Probability: from measure theory to random processes via conditioning. Camb. Univ. Press 2003.

Chen, L.H.Y.
[1] Poincaré-type inequalities via stochastic integrals. Z.W. **69** (1985) 251–277.

Chevalier, L.
[1] Un nouveau type d'inégalités pour les martingales discrètes. Z.W. **49** (1979) 249–256.

Chitashvili, R., and Mania, M.
[1] On functions transforming a Wiener process into a semimartingale. Prob. Th. Rel. Fields **109** (1997) 57–76.

Chitashvili, R.J., and Toronjadze, T.A.
[1] On one dimensional stochastic differential equations with unit diffusion coefficient; structure of solutions. Stochastics **4** (1981) 281–315.

Chou, S.
[1] Sur certaines généralisations de l'inégalité de Fefferman. Sém. Prob. XVIII, Lect. Notes in Mathematics, vol. 1059. Springer, Berlin Heidelberg New York 1984, pp. 219–222.

Chover, J.
[1] On Strassen's version of the log log law. Z.W. **8** (1967) 83–90.

Chung, K.L.
[1] Excursions in Brownian motion. Ark. för Math. **14** (1976) 155–177.
[2] Lectures from Markov processes to Brownian motion. Springer, Berlin Heidelberg New York 1982.
[3] Green, Brown and Probability. World Scientific, Singapore (1995).

Chung, K.L., and Durrett, R.
[1] Downcrossings and local time. Z.W. **35** (1976) 147–149.

Chung, K.L., and Williams, R.T.
[1] Introduction to stochastic integration. Second edition. Birkhäuser, Boston 1989.

Chung, K.L., and Zhao, Z.
[1] From Schrödinger's equation to Brownian motion. Grundlehren 312. Springer, Berlin Heidelberg New York 1995.

Cinlar, E., Jacod, J., Protter, P., and Sharpe, M.J.
[1] Semi-martingales and Markov processes. Z.W. **54** (1980) 161–219.

Clark, J.M.C.
[1] The representation of functionals of Brownian motion by stochastic integrals. Ann. Math. Stat. **41** (1970) 1282–1295; **42** (1971) 1778.

Cocozza, C. and Yor, M.
[1] Démonstration d'un théorème de Knight à l'aide de martingales exponentielles. Sém. Prob. XIV, Lect. Notes in Mathematics, vol. 784. Springer, Berlin Heidelberg New York 1980, pp. 496–499.

Csáki, E.
[1] On some distributions concerning maximum and minimum of a Wiener process. In B. Gyires, editor, Analytic Function Methods in Probability Theory, number 21 in Colloquia Mathematica Societatis János Bolyai, p. 43–52. North-Holland, 1980 (1977, Debrecen, Hungary).

Csáki, E., Shi, Z., and Yor, M.
[1] Fractional Brownian motions as "higher order" fractional derivatives of Brownian local times. J. Bolyai Society (eds. I. Berkes, E. Csaki, M. Csörgö). Proceedings of a Conference on Limit Theorems (July 1999, Balatonlelle), vol. I, p. 365–388, 2002, Budapest.

Dambis, K.E.
[1] On the decomposition of continuous submartingales. Theor. Prob. Appl. **10** (1965) 401–410.

Darling, D.A., and Kac, M.
[1] Occupation times for Markov processes. Trans. Amer. Math. Soc. **84** (1957) 444–458.

Davis, B.
[1] Picard's theorem and Brownian motion. Trans. Amer. Math. Soc. **213** (1975) 353–362.
[2] On Kolmogorov's inequality $\|\tilde{f}\|_p \leq c_p \|f\|_1$, $0 < p < 1$. Trans. Amer. Math. Soc. **222** (1976) 179–192.
[3] Brownian motion and analytic functions. Ann. Prob. **7** (1979) 913–932.
[4] On Brownian slow points. Z.W. **64** (1983) 359–367.
[5] On the Barlow-Yor inequalities for local time. Sém. Prob. XXI, Lect. Notes in Mathematics, vol. 1247. Springer, Berlin Heidelberg New York 1987, pp. 218–220.
[6] Weak limits of perturbed random walks and the equation $Y_t = B_t + \alpha \sup_{s \leq t} Y_s + \beta \inf_{s \leq t} Y_s$. Ann. Prob. **24** (1996) 2007–2017.
[7] Brownian motion and random walk perturbed at extrema. Prob. Th. Rel. F. **113** (1999) 501–518.

Davis, M.H.A., and Varaiya, P.
[1] The multiplicity of an increasing family of σ-fields. Ann. Prob. **2** (1974) 958–963.

Dean, D.S., and Jansons, K.M.
[1] A note on the integral of the Brownian Bridge. Proc. R. Soc. London A **437** (1992) 792–730.

De Blassie, R.D.
[1] Exit times from cones in $\mathbb{R}^n$ of Brownian motion. Prob. Th. Rel. F. **74** (1989) 1–29.
[2] Remark on: Exit times from cones in $\mathbb{R}^n$ of Brownian motion. Prob. Th. Rel. F. **79** (1989) 95–97.

Deheuvels, P.
[1] Invariance of Wiener processes and Brownian Bridges by integral transforms and applications. Stoch. Proc. Appl. **13**, 3, (1982) 311–318.

Delbaen, F., and Schachermayer, W.
[1] Arbitrage possibilities in Bessel processes and their relations to local martingales. Prob. Th. Rel. F. **102** (3) (1995) 357–366.

Dellacherie, C.
[1] Intégrales stochastiques par rapport aux processus de Wiener et de Poisson. Sém. Prob. VIII, Lect. Notes in Mathematics, vol. 381. Springer, Berlin Heidelberg New York 1974, pp. 25–26.

Dellacherie, C., Maisonneuve, B., and Meyer, P.A.
[1] Probabilités et Potentiel. Hermann, Paris, vol. V. Processus de Markov (fin). Complements de Calcul stochastique. 1992.

Dellacherie, C., and Meyer, P.A.
[1] Probabilités et potentiel. Hermann, Paris, vol. I 1976, vol. II 1980, vol. III 1983, vol. IV 1987.

Dellacherie, C., Meyer, P.A., and Yor, M.
[1] Sur certaines propriétés des espaces H^1 et BMO. Sém. Prob. XII, Lect. Notes in Mathematics, vol. 649. Springer, Berlin Heidelberg New York 1978, pp. 98–113.

De Meyer, B.
[1] Une simplification de l'argument de Tsirel'son sur le caractère non-brownien des processus de Walsh. Sém. Prob. XXXIII, Lect. Notes in Mathematics, vol. 1709. Springer, Berlin Heidelberg New York (1999), pp. 217–220.

Denisov, I.V.
[1] A random walk and a Wiener process near a maximum. Theor. Prob. Appl. **28** (1984) 821–824.

Deuschel, J.D., and Stroock, D.W.
[1] An introduction to the theory of large deviations. Academic Press, New York 1988.

Doléans-Dade, C.
[1] Quelques applications de la formule du changement de variables pour les semimartingales. Z.W. **16** (1970) 181–194.
[2] On the existence and unicity of solutions of stochastic integral equations. Z.W. **36** (1976) 93–101.

Doléans-Dade, C., and Meyer, P.A.
[1] Intégrales stochastiques par rapport aux martingales locales. Sém. Prob. IV, Lect. Notes in Mathematics, vol. 124. Springer, Berlin Heidelberg New York 1970, pp. 77–107.

Donati-Martin, C.
[1] Transformation de Fourier et temps d'occupation browniens. C.R. Acad. Sci. Paris, Série I **308** (1989) 515–517.

Donati-Martin, C., and Yor, M.
[1] Mouvement brownien et inégalité de Hardy dans L^2. Sém. Prob. XXIII, Lect. Notes in Mathematics, vol. 1372. Springer, Berlin Heidelberg New York 1989, pp. 315–323.
[2] Fubini's theorem for double Wiener integrals and the variance of the Brownian path. Ann. I.H.P. **27** (1991) 181–200.
[3] Some Brownian Functionals and their laws. Ann. Prob. **25**, 3 (1997) 1011–1058.

Donati-Martin, C., Doumerc, Y., Matsumoto, H., and Yor, M.
[1] Some properties of the Wishart processes and a Matrix extension of the Hartman-Watson laws. Publ. RIMS Kyoto 2004.

Donsker, M.D.
[1] An invariance principle for certain probability limit theorems. Mem. Amer. Math. Soc. **6** (1951) 1–12.

Donsker, M.D., and Varadhan, S.R.S.
[1] Asymptotic evaluation of certain Markov processes expectations for large time I, II and III. Comm. Pure Appl. Math. **28** (1975) 1–47; **29** (1976) 279–301 and 389–461.

Doob, J.L.
[1] Stochastic processes. Wiley, New York 1953.
[2] Classical potential Theory and its probabilistic counterpart. Springer, New York Berlin 1984.

Doss, H.
[1] Liens entre équations différentielles stochastiques et ordinaires. C.R. Acad. Sci. Paris, Sér. A **233** (1976) 939–942 et Ann. I.H.P. **13** (1977) 99–125.

Doss, H., and Lenglart, E.
[1] Sur l'existence, l'unicité et le comportement asymptotique des solutions d'équations différentielles stochastiques. Ann. I.H.P. **14** (1978) 189–214.

Douglas, R.
[1] On extremal measures and subspace density. Michigan Math. J. **11** (1964) 644–652.

Dubins, L.
[1] Rises and upcrossings of non-negative martingales, III. J. Math. **6** (1962) 226–241.
[2] A note on upcrossings of martingales. Ann. Math. Stat. **37** (1966) 728.
[3] On a theorem of Skorokhod. Ann. Math. Stat. **39** (1968) 2094–2097.

Dubins, L., Emery, M., and Yor, M.
[1] On the Lévy transformation of Brownian motions and continuous martingales. Sém. Prob. XXVII, Lect. Notes in Mathematics, vol. 1577. Springer, Berlin Heidelberg New York 1993, pp. 122–132.

Dubins, L., Feldman, J., Smorodinsky, M., and Tsirel'son, B.
[1] Decreasing sequences of σ-fields and a measure change for Brownian motion. Ann. Prob. **24** (1996) 882–904.

Dubins, L., and Gilat, D.
[1] On the distribution of maxima of martingales. Proc. Amer. Math. Soc. **68** (1978) 337–338.

Dubins, L., and Schwarz, G.
[1] On continuous martingales. Proc. Nat. Acad. Sci. USA **53** (1965) 913–916.
[2] On extremal martingales distributions. Proc. Fifth Berkeley Symp. **2** (1) (1967) 295–297.
[3] A sharp inequality for sub-martingales and stopping times. Astérisque **157-158** (1988) 129–145. Colloque Paul Lévy sur les processus stochastiques. (Palaiseau, 1987).

Dubins, L., and Smorodinsky, M.
[1] The modified discrete Lévy transformation is Bernoulli. Sém. Prob. XXVI, Lect. Notes in Mathematics, vol. 1528. Springer, Berlin Heidelberg New York 1992, pp. 157–161.

Dudley, R.M.
[1] Wiener functionals as Itô integrals. Ann. Prob. **5** (1) (1977) 140–141.

Duplantier, B.
[1] Areas of planar Brownian curves. J. Phys. A. Math. Gen. **22** (13) (1989) 3033–3048.

Duplantier, B., Lawler, G.F., LeGall, J.F., and Lyons, T.J.
[1] The geometry of the Brownian curve. Bull. Sci. Math. **117** (1993) 91–106.

Durbin, J.
[1] The first passage density of the Brownian motion process to a curved boundary (with an appendix by D. Williams). J. Appl. Prob. **29** (2) (1992) 291–304.

[2] A reconciliation of two different expressions for the first passage density of Brownian motion to a curved boundary. J. Appl. Prob. **25** (1988) 829–832.

Durrett, R.
[1] A new proof of Spitzer's result on the winding of 2-dimensional Brownian motion. Ann. Prob. **10** (1982) 244–246.
[2] Brownian motion and martingales in analysis. Wadsworth, Belmont, Calif. 1984.
[3] Probability: Theory and examples. Duxbury Press, Wadsworth, Pacific Grove, Cal. 2nd ed. 1996.

Durrett, R., and Iglehart, D.L.
[1] Functionals of Brownian meander and Brownian excursion. Ann. Prob. **5** (1977) 130–135.

Durrett, R.T., Iglehart, D.L., and Miller, D.R.
[1] Weak convergence to Brownian meander and Brownian excursion. Ann. Prob. **5** (1977) 117–129.

Dvoretsky, A., Erdös, P., and Kakutani, S.
[1] Non increase everywhere of the Brownian motion process. Proc. Fourth Berkeley Symposium **2** (1961) 103–116.

Dwass, H.
[1] On extremal processes I and II. Ann. Math. Stat. **35** (1964) 1718–1725 and Ill. J. Math. **10** (1966) 381–391.

Dynkin, E.B.
[1] Markov processes. Springer, Berlin Heidelberg New York 1965.
[2] Self-intersection gauge for random walks and Brownian motion. Ann. Prob. **16** (1988) 1–57.
[3] Local times and random fields. Seminar on Stoch. Proc. 1983. Birkhäuser 1984, pp. 69–84.

Eisenbaum, N.
[1] Un théorème de Ray-Knight lié au supremum des temps locaux browniens. PTRF **87** (1) (1990) 79–95.
[2] Dynkin's isomorphism theorem and the Ray-Knight theorems. PTRF **99** (2) (1994) 321–355.
[3] Une version sans conditionnement du théorème d'isomorphisme de Dynkin. Sém. Prob. XXIX, Lect. Notes in Mathematics 1613, p. 266–289, Springer (1995).
[4] On the infinite divisibility of squared Gaussian processes. PTRF **125** (3) (2003) 381–392.
[5] Isomorphism theorems for Markov chains. Stoch. Proc. App. **110** (2004) 247–258.

Eisenbaum, N., Kaspi, H., Marcus, M.B., Rosen, J., and Shi, Z.
[1] A Ray-Knight theorem for symmetric Markov processes. Ann. Prob. **28** (4) (2000) 1781–1796.

El Karoui, N.
[1] Sur les montées des semi-martingales. Astérisque 52–53, Temps Locaux (1978) 63–88.

El Karoui, N., and Chaleyat-Maurel, M.
[1] Un problème de réflexion et ses applications au temps local et aux équations différentielles stochastiques sur $\mathbb{R}$-cas continu. Astérisque 52–53, Temps Locaux (1978) 117–144.

El Karoui, N., and Weidenfeld, G.
[1] Théorie générale et changement de temps. Sém. Prob. XI, Lect. Notes in Mathematics, vol. 581. Springer, Berlin Heidelberg New York 1977, pp. 79–108.

Elworthy, K.D.
[1] Stochastic differential equations on manifolds. London Math. Soc. Lecture Notes Series 70. Cambridge University Press (1982).

Elworthy, K.D., Li, X.M., and Yor, M.
[1] On the tails of the supremum and the quadratic variation of strictly local martingales. Sém. Prob. XXXI, Lect. Notes in Mathematics, vol. 1655. Springer, Berlin Heidelberg New York 1997, pp. 113–125.
[2] The importance of strictly local martingales: applications to radial Ornstein-Uhlenbeck processes. PTRF **115** (1999) 325–355.

Elworthy, K.D., and Truman, A.
[1] The diffusion equation: an elementary formula. In: S. Albeverio et al. (eds.) Stochastic processes in quantum physics. Lect. Notes in Physics, vol. 173. Springer, Berlin Heidelberg New York 1982, pp. 136–146.

Emery, M.
[1] Une définition faible de BMO. Ann. I.H.P. **21** (1) (1985) 59–71.
[2] On the Azéma martingales. Sém. Prob. XXIII, XXIV. Lect. Notes in Mathematics, vols. 1372, 1426. Springer, Berlin Heidelberg New York (1989), (1990), pp. 66–87; pp. 442–447.
[3] Quelques cas de représentation chaotique. Sém. Prob. XXV. Lect. Notes in Mathematics, vol. 1485. Springer, Berlin Heidelberg New York 1991, pp. 10–23.
[4] On certain probabilities equivalent to coin-tossing, d'après Schachermayer. Sém. Prob. XXXIII, Lect. Notes in Mathematics, vol. 1709. Springer, Berlin Heidelberg New York 1999, pp. 240–256
[5] Espaces probabilisés filtrés: de la théorie de Vershik au mouvement brownien, via des idées de Tsirelson. Séminaire Bourbaki, Novembre 2000, n° 882, 20 p.
[6] Stochastic Calculus in Manifolds, with an Appendix by P.A. Meyer. Universitext, Springer 1989.
[7] Classical probability theory: an outline of stochastic integrals and diffusions. In. Quantum Prob. Comm. **XI** (2003) 87–121.

Emery, M., and Perkins, E.
[1] La filtration de $B + L$. Z.W. **59** (1982) 383–390.

Emery, M., and Schachermayer, W.
[1] Brownian filtrations are not stable under equivalent time changes. Sém. Prob. XXXIII, Lect. Notes in Mathematics, vol. 1709. Springer Berlin Heidelberg New York 1999, pp. 267–276.
[2] A remark on Tsirel'son's stochastic differential equation. Sém. Prob. XXXIII, Lect. Notes in Mathematics, vol. 1709. Springer Berlin Heidelberg New York 1999, pp. 291–303.

Emery, M., Stricker, C., and Yan, J.A.
[1] Valeurs prises par les martingales locales continues à un instant donné. Ann. Prob. **11** (1983) 635–641.

Emery, M., and Yor, M.
[1] Sur un théorème de Tsirel'son relatif à des mouvements browniens corrélés et à la nullité de certains temps locaux. Sém. Prob. XXXII, Lect. Notes in Mathematics, vol. 1686. Springer, Berlin Heidelberg New York 1998, pp. 306–312.

Engelbert, H.J., and Hess, J.
[1] Integral representation with respect to stopped continuous local martingales. Stochastics **4** (1980) 121–142.

Engelbert, H.J., and Schmidt, W.
[1] On solutions of stochastic differential equations without drift. Z.W. **68** (1985) 287–317.

Ethier, S.N., and Kurtz, T.G.
[1] Markov processes. Characterization and convergence. Wiley and Sons, New York 1986.

Evans, S.
[1] On the Hausdorff dimension of Brownian cone points. Math. Proc. Camb. Philos. Soc. **98** (1985) 343–353.

Ezawa, H., Klauder, J.R., and Shepp, L.A.
[1] Vestigial effects of singular potentials in diffusion theory and quantum mechanics. J. Math. Phys. **16**, 4 (1975) 783–799.
[2] A path-space picture for Feynman-Kac averages. Ann. Phys. **88**, 2 (1974) 588–620.
[3] On the divergence of certain integrals of the Wiener process. Ann. Inst. Fourier, XXIV, 2 (1974) 189–193

Feldman, J., and Smorodinsky, M.
[1] Simple examples of non-generating Girsanov processes. Sém. Prob. XXXI, Lect. Notes in Mathematics, vol. 1655. Springer, Berlin Heidelberg New York 1997, pp. 247–251.

Feldman, J., and Tsirel'son, B.
[1] Decreasing sequences of σ-fields and a measure change for Brownian motion, II. Ann. Prob. **24** (1996) 905–911.

Feller, W.
[1] Diffusion process in one dimension. Trans. Amer. Math. Soc. **77** (1954) 1–31.
[2] On second order differential operators. Ann. Math. **61** (1955) 90–105.
[3] Differential operators with the positive maximum property. Ill. J. Math. **3** (1959) 182–186.

Fisk, D.L.
[1] Sample quadratic variation of sample continuous, second order martingales. Z.W. **6** (1966) 273–278.

Fitzsimmons, P.J.
[1] A converse to a theorem of P. Lévy. Ann. Prob. **15** (1987) 1515–1523.

Fitzsimmons, P.J., and Getoor, R.K.
[1] Limit theorems and variation properties for fractional derivatives of the local time of a stable process. Ann. I.H.P. **28** (1992) 311–333.
[2] On the distribution of the Hilbert transform of the local time of a symmetric Lévy process. Ann. Prob. **20** (1992) 1484–1497.

Fitzsimmons, P.J., Pitman, J.W., and Yor, M.
[1] Markovian bridges: Construction, Palm interpretation, and splicing. Seminar on Stochastic Processes 1992. Birkhäuser, pp. 101–134.

Föllmer, H.
[1] Stochastic holomorphy. Math. Ann. **207** (1974) 245–265.
[2] Calcul d'Itô sans probabilités. Sém. Prob. XV, Lect. Notes in Mathematics, vol. 850. Springer, Berlin Heidelberg New York 1981, pp. 143–150.

[3] Von der Brownsche Bewegung zum Brownschen Blatt: einige neuere Richtungen in der Theorie der stochastischen Prozesse. In: W. Jäger, J. Moser, R. Remmert (eds.) Perspectives in Mathematics, Anniversary of Oberwolfach, 1984. Birkhäuser, Basel 1984.
[4] The exit measure of a supermartingale. Z.W. **21** (1972) 154–166.
[5] On the representation of semimartingales. Ann. Prob. **1** (4) (1973) 580–589.

Föllmer, H., and Imkeller, P.
[1] Anticipation cancelled by a Girsanov transformation: a paradox on Wiener space. Ann. I.H.P. **29** (4) (1993) 569–586.

Föllmer, H., and Protter, P.
[1] On Itô's formula for d-dimensional Brownian motion. Prob. Th. Rel. Fields **116** (2000) 1–20.

Föllmer, H., Protter, P., and Shiryaev, A.N.
[1] Quadratic covariation and an extension of Itô's formula. Bernoulli **1** (1/2) (1995) 149–169.

Föllmer, H., Wu, C.T., and Yor, M.
[1] Canonical decomposition of linear transformations of two independent Brownian motions. Stoch. Proc. and their App. **84** (1) (1999) 137–164.
[2] Weak Brownian motions of arbitrary order. Ann. IHP **36** (4) (2000) 447–487.

Freedman, D.
[1] Brownian motion and diffusion. Holden-Day, San Francisco 1971.

Freidlin, M.I.
[1] Action functionals for a class of stochastic processes. Theor. Prob. Appl. **17** (1972) 511–515.

Friedman, A.
[1] Stochastic differential equations and applications I and II. Acadamic Press, New York 1975 and 1976.

Fujisaki, M., Kallianpur, G., and Kunita, H.
[1] Stochastic differential equations for the non-linear filtering problem. Osaka J. Math. **9** (1972) 19–40.

Fujita, T., Petit, F., and Yor, M.
[1] Pricing path dependent options in a Black-Scholes market from the distribution of homogeneous Brownian functionals. J. App. Prob. **41** (2004) 1–18.

Fukushima, M. and Takeda, M.
[1] A transformation of symmetric Markov processes and the Donsker-Varadhan theory. Osaka J. Math. **21** (1984) 311–326.

Fukushima, M. and Tanaka, H.
[1] Poisson point processes attached to symmetric diffusions. To appear. Sém. Prob. XXXVIII, Springer 2005.

Galmarino, A.R.
[1] Representation of an isotropic diffusion as a skew product. Z.W. **1** (1963) 359–378.

Galtchouk, L., and Novikov, A.A.
[1] On Wald's equation. Discrete time case. Sém. Prob. XXXI, Lect. Notes in Mathematics, vol. 1655. Springer, Berlin Heidelberg New York 1997, pp. 126–135.

Garsia, A., Rodemich, E., and Rumsey, J., Jr.
[1] A real variable lemma and the continuity of paths of some Gaussian processes. Indiana Univ. Math. J. **20** (1970/1971) 565–578.

Gaveau, B.
[1] Principe de moindre action, propagation de la chaleur et estimées sous-elliptiques sur certains groupes nilpotents. Acta. Math. **139** (1977) 96–153.

Geiger, J., and Kersting, G.
[1] Winding numbers for 2-dimensional positive recurrent diffusions. Potential Analysis **3**, 2 (1994) 189–201.

Geman, D., and Horowitz, J.
[1] Occupation densities. Ann. Prob. **8** (1980) 1–67.

Geman, D., Horowitz, J., and Rosen, J.
[1] A local time analysis of intersections of Brownian paths in the plane. Ann. Prob. **12** (1985) 86–107.

Getoor, R.K.
[1] Another limit theorem for local time. Z.W. **34** (1976) 1–10.
[2] The Brownian escape process. Ann. Prob. **7** (1979) 864–867.
[3] Infinitely divisible probabilities on the hyperbolic plane. Pacific J. Math. **11** (1961) 1287–1308.
[4] Excursions of a Markov process. Ann. Prob. **7** (2) (1979) 244–266.

Getoor, R.K., and Sharpe, M.J.
[1] Conformal martingales. Invent. Math. **16** (1972) 271–308.
[2] Last exit times and additive functionals. Ann. Prob. **1** (1973) 550–569.
[3] Last exit decompositions and distributions. Indiana Univ. Math. J. **23** (1973) 377–404.
[4] Excursions of Brownian motion and Bessel processes. Z.W. **47** (1979) 83–106.
[5] Excursions of dual processes. Adv. Math. **45** (1982) 259–309.
[6] Naturality, standardness and weak duality for Markov processes. Z.W. **67** (1984) 1–62.

Gikhman, I.I., and Skorokhod, A.V.
[1] Theory of stochastic processes I, II and III. Springer, Berlin Heidelberg New York 1974, 1975, and 1979.

Gilat, D.
[1] Every non-negative submartingale is the absolute value of a martingale. Ann. Prob. **5** (1977) 475–481.

Girsanov, I.V.
[1] On transforming a certain class of stochastic processes by absolutely continuous substitution of measures. Theor. Prob. Appl. **5** (1960) 285–301.
[2] An example of non-uniqueness of the solution to the stochastic differential equation of K. Itô. Theor. Prob. Appl. **7** (1962) 325–331.

Göing, A.
[1] Some generalizations of Bessel processes. Ph. D. Thesis, ETH Zürich (January 1998).

Göing-Jaeschke, A. and Yor, M.
[1] A survey and some generalizations of Bessel processes. Bernoulli **9** (2003) 313–350.
[2] A clarification about Hitting times densities for Ornstein-Uhlenbeck processes. Finance and Stochastics **7/3** (2003) 413–415.

Goswami, A., and Rao, B.V.
[1] Conditional expectation of odd chaos given even. Stochastics and Stochastics Reports **35** (1991) 213–224.

Graversen, S.E.
[1] "Polar"-functions for Brownian motion. Z.W. **61** (1981) 261–270.

Graversen, S.E. and Shyriaev, A.N.
[1] An extension of P. Lévy's distributional properties to the case of a Brownian motion with drift. Bernoulli **6** (4) (2000) 615–620.

Graversen, S.E. and Vuolle-Apiala, J.
[1] On Paul Lévy's arc sine law and Shiga-Watanabe's time inversion result. Prob. Math. Stoch. **20** (1) (2000) 63–73.

Greenwood, P., and Perkins, E.
[1] A conditional limit theorem for random walks and Brownian local times on square root boundaries. Ann. Prob. **11** (1983) 227–261.

Gruet, J.C.
[1] Sur la transience et la récurrence de certaines martingales locales continues à valeurs dans le plan. Stochastics **28** (1989) 189–207.
[2] Semi-groupe du mouvement brownien hyperbolique. Stochastics and Stochastic Reports **56** (1996) 53–61.
[3] Windings of hyperbolic Brownian motion. In: M. Yor (ed.) Exponential Functionals and Principal Values related to Brownian Motion. Bib. Rev. Ibero-Amer. (1997)
[4] On the length of the homotopic Brownian word in the thrice punctured sphere. Prob. Th. Rel. Fields **111** (4) (1998) 489–516.

Gruet, J.C., and Mountford, T.S.
[1] The rate of escape for pairs of windings on the Riemann sphere. J. London Math. Soc. **48** (2) (1993) 552–564.

Gundy, R.F.
[1] Some topics in probability and analysis. Conf. Board of the Math. Sciences. Regional Conf. Series in Maths. **70** (1989)

Hardin, C.D.
[1] A spurious Brownian motion. Proc. Amer. Math. Soc. **93** (1985) 350.

Harrison, J.M., and Shepp, L.A.
[1] On skew Brownian motion. Ann. Prob. **9** (1981) 309–313.

Hartman, P., and Wintner, A.
[1] On the law of the iterated logarithm. Amer. J. Math. **63** (1941) 169–176.

Haussmann, U.G.
[1] Functionals of Itô processes as stochastic integrals. Siam J. Contr. Opt. **16**, 2 (1978) 252–269.

Hawkes, J., and Barlow, M.T.
[1] Application de l'entropie métrique à la continuité des temps locaux des processus de Lévy. C.R. Acad. Sci. Paris **301**, Série I, 5 (1985) 237–239.

Hawkes, J., and Truman, A.
[1] Statistics of local time and excursions for the Ornstein-Uhlenbeck process. In: M.T. Barlow, N. Bingham (eds.) Stochastic Analysis. London Math. Soc. Lect. Notes, vol. 167, 1991.

Hirsch, F. and Song, S.
[1] Two parameter Bessel processes. Stoch. Proc. App. **83** (1999) 187–209.

Hobson, D.
[1] The maximum maximum of a martingale. Sém. Prob. XXXII, Lect. Notes in Mathematics, vol. 1686. Springer, Berlin Heidelberg New York 1998, pp. 250–263.

Hu, Y., Shi, Z., and Yor, M.
[1] Rates of convergence of diffusions with drifted Brownian potentials. Trans. Amer. Math. Soc. **351** (1999) 3915–3934.

Hu, Y., and Warren, J.
[1] Ray-Knight theorems related to a stochastic flow. Stoch. Prob. and their App. **86** (2) (2000) 287–306.

Hu, Y., and Yor, M.
[1] Asymptotic studies of Brownian functionals. Bolyai Society Math. Studies **9** (1999) 187–217.

Hunt, G.A.
[1] Markov processes and potentials I, II and III. Ill. J. of Math. **1** (19657) 44–93 and 316–369; Ill. J. Math. **2** (1958) 151–213.

Ibragimov, I.
[1] Sur la régularité des trajectoires des fonctions aléatoires. C.R. Acad. Sci. Paris, Série A 289 (1979) 545–547.

Idrissi Khamlichi, A.
[1] Problèmes de polarité pour des semimartingales bidimensionnelles continues. Thèse de 3° Cycle, Université de Paris VII, 1986.

Ikeda, N., and Manabe, S.
[1] Integral of differential forms along the paths of diffusion processes. Publ. R.I.M.S., Kyoto Univ. **15** (1978) 827–852.

Ikeda, N., and Watanabe, S.
[1] A comparison theorem for solutions of stochastic differential equations and its applications. Osaka J. Math. **14** (1977) 619–633.
[2] Stochastic differential equations and diffusion processes. Second edition North-Holland Publ. Co., Amsterdam Oxford New York; Kodansha Ltd., Tokyo 1989.

Imhof, J.P.
[1] Density factorizations for Brownian motion and the three-dimensional Bessel Processes and applications. J. Appl. Prob. **21** (1984) 500–510.
[2] On Brownian bridge and excursion. Studia Scient. Math. Hungaria **20** (1985) 1–10.
[3] A simple proof of Pitman's $2M - X$ theorem. Adv. Appl. Prob. **24** (1992) 499–501.

Isaacson, D.
[1] Stochastic integrals and derivatives. Ann. Math. Sci. **40** (1969) 1610–1616.

Itô, K.
[1] Stochastic integral. Proc. Imp. Acad. Tokyo **20** (1944) 519–524.
[2] Stochastic differential equations. Memoirs A.M.S. **4**, 1951.
[3] Multiple Wiener integrals. J. Math. Soc. Japan **3** (1951) 157–169.
[4] Lectures on stochastic processes. Tata Institute, Bombay 1961.
[5] Poisson point processes attached to Markov processes. Proc. Sixth Berkeley Symp. Math. Stat. Prob., vol. 3. University of California, Berkeley, 1970, pp. 225–239.
[6] Extension of stochastic integrals. Proc. Internat. Symp. SDE Kyoto 1976, pp. 95–109.

[Most of K. Itô's papers are reprinted in: D. Stroock, S. Varadhan (eds.) *Kiyosi Itô Selected Papers*. Springer (1987)]

Itô, K., and McKean, H.P.
[1] Diffusion processes and their sample paths. Springer, Berlin Heidelberg New York 1965.

Itô, K., and Watanabe, S.
[1] Transformation of Markov processes by multiplicative functionals. Ann. Inst. Fourier **15** (1965) 15–30.

Jacka, S.D.
[1] Optimal stopping and best constants for Doob-like inequalities. I. The case $p = 1$. Ann. Prob. **19** (4) (1991) 1798–1821.

Jacod, J.
[1] A general theorem of representation for martingales. Proc. AMS Prob. Symp. Urbana 1976, 37–53.
[2] Calcul stochastique et problèmes de martingales. Lect. Notes in Mathematics, vol. 714. Springer, Berlin Heidelberg New York 1979.
[3] Equations différentielles linéaires: la méthode de variation des constantes. Sém. Prob. XVI, Lect. Notes in Mathematics, vol. 920. Springer, Berlin Heidelberg New York 1982, pp. 442–446.

Jacod, J. and Mémin, J.
[1] Caractéristiques locales et conditions de continuité absolue pour les semimartingales. Z.W. **35** (1976) 1–37.
[2] Sur l'intégrabilité uniforme des martingales exponentielles. Z.W. **42** (1978) 175–204.

Jacod, J. and Protter, P.
[1] Time reversal on Lévy processes. Ann. Prob. **16** (1988) 620–641.

Jacod, J. and Shiryaev, A.N.
[1] Limit theorems for stochastic processes. Springer, Berlin Heidelberg New York 1987.

Jacod, J. and Yor, M.
[1] Étude des solutions extrémales et représentation intégrale des solutions pour certains problèmes de martingales. Z.W. **38** (1977) 83–125.

Janson, S.
[1] On complex hypercontractivity. J. Funct. Anal. **151** (1997) 270–280.
[2] Gaussian Hilbert spaces. Cambridge Univ. Press (1997)

Jeulin, T.
[1] Un théorème de J. Pitman. Sém. Prob. XIII, Lect. Notes in Mathematics, vol. 721. Springer, Berlin Heidelberg New York 1979, pp. 332–359.
[2] Semi-martingales et grossissement d'une filtration. Lect. Notes in Mathematics, vol. 833. Springer, Berlin Heidelberg New York 1980.
[3] Application de la théorie du grossissement à l'étude des temps locaux browniens. In: Grossissements de filtrations: Exemples et applications. Lect. Notes in Mathematics, vol. 1118. Springer, Berlin Heidelberg New York 1985, pp. 197–304.
[4] Sur la convergence absolue de certaines intégrales. Sém. Prob. XVI, Lect. Notes in Mathematics, vol. 920. Springer, Berlin Heidelberg New York 1982, pp. 248–256.
[5] Ray-Knight's theorems on Brownian local times and Tanaka formula. Seminar on Stoch. Proc. 1983, Birkhäuser 1984, pp. 131–142.

Jeulin, T., and Yor, M.
[1] Autour d'un théorème de Ray. Astérisque 52–53, Temps locaux (1978) 145–158.
[2] Inégalité de Hardy, semimartingales et faux-amis. Sém. Prob. XIII, Lect. Notes in Mathematics, vol. 721. Springer, Berlin Heidelberg New York 1979, pp. 332–359.
[3] Sur les distributions de certaines fonctionnelles du mouvement brownien. Sém. Prob. XV, Lect. Notes in Mathematics, vol. 850. Springer, Berlin Heidelberg New York 1981, pp. 210–226.
[4] Filtration des ponts browniens et équations différentielles linéaires. Sém. Prob. XXIV, Lect. Notes in Mathematics, vol. 1426. Springer, Berlin Heidelberg New York 1990, pp. 227–265.
[5] Une décomposition non canonique du drap brownien. Sém. Prob. XXVI, Lect. Notes in Mathematics, vol. 1528. Springer, Berlin Heidelberg New York 1992, pp. 322–347.
[6] Moyennes mobiles et semimartingales. Sém. Prob. XXVII, Lect. Notes in Mathematics, vol. 1557. Springer, Berlin Heidelberg New York 1993, pp. 53–77.

Johnson, G., and Helms, L.L.
[1] Class (D) supermartingales. Bull. Amer. Math. Soc. **69** (1963) 59–62.

Kac, M.
[1] On distribution of certain Wiener functionals. Trans. Amer. Math. Soc. **65** (1949) 1–13
[2] On some connections between probability theory and differential and integral equations. Proc. Second Berkeley Symp. Math. Stat. Prob. Univ. California Press 1951, pp. 189–251.
[3] Integration in Function spaces and some of its applications. Lezioni Fermiane. Acc. Nat. dei Lincei. Scuola Normale Sup. Pisa (1980).

Kahane, J.P.
[1] Some random series of functions. Second edition, Cambridge studies in Advanced Mathematics, **5**, 1985.
[2] Brownian motion and classical analysis. Bull. London Math. Soc. **7** (1976) 145–155.
[3] Le mouvement brownien. In: Matériaux pour l'histoire des Mathématiques au XXe siècle. Soc. Math. France (Dec. 1997).

Kallenberg, O.
[1] Some time change representation of stable integrals, via predictable transformations of local martingales. Stoch. Proc. and their App. **40** (1992) 199–223.
[2] On an independence criterion for multiple stochastic integrals. Ann. Prob. **19**, 2 (1991) 483–485.
[3] On the existence of universal Functional Solutions to classical SDE's. Ann. Prob. **24**, 1 (1996) 196–205.
[4] Foundations of Modern Probability. Springer 1997. Second edn. 2002.

Kallianpur, G.
[1] Stochastic filtering theory. Springer, Berlin Heidelberg New York 1980.
[2] Some recent developments in nonlinear filtering theory. In: N. Ikeda, S. Watanabe, M. Fukushima, H. Kunita (eds.) Itô's Stochastic Calculus and Probability Theory. Springer, Berlin Heidelberg New York 1996, pp. 157–170.

Kallianpur, G., and Robbins, H.
[1] Ergodic property of Brownian motion process. Proc. Nat. Acad. Sci. USA **39** (1953) 525–533.

Kallssen, J.
[1] An up to indistinguishability unique solution to a stochastic differential equation that is not strong. Sém. Prob. XXXIII, Lect. Notes in Mathematics, vol. 1709. Springer, Berlin Heidelberg New York 1999, pp. 315–326.

Karandikar, R.L.
[1] A.s. approximation results for multiplicative stochastic integrals. Sém. Prob. XVI, Lect. Notes in Mathematics, vol. 920. Springer, Berlin Heidelberg New York 1982, pp. 384–391.

Karatzas, I. and Shreve, S.E.
[1] Brownian motion and stochastic calculus. Springer, Berlin Heidelberg New York 1988.

Karlin, S. and Taylor, H.M.
[1] A second course in stochastic processes. Academic Press 1981.

Kasahara, Y.
[1] Limit theorems of occupation times for Markov processes. R.I.M.S. **3** (1977) 801–818.
[2] Limit theorems for occupation times of Markov processes. Japan J. Math. **7** (1981) 291–300.
[3] A limit theorem for slowly increasing occupation times. Ann. Prob. **10** (1982) 728–736.
[4] On Lévy's downcrossing theorem. Proc. Japan Acad. **56** A (10) (1980) 455–458.

Kasahara, Y., and Kotani, S.
[1] On limit processes for a class of additive functionals of recurrent diffusion processes. Z.W. **49** (1979) 133–143.

Kawabata, S., and Yamada, T.
[1] On some limit theorems for solutions of stochastic differential equations. Sém. Prob. XVI, Lect. Notes in Mathematics, vol. 920. Springer, Berlin Heidelberg New York 1982, pp. 412–441.

Kawazu, K., and Tanaka, H.
[1] A diffusion process in a Brownian environment with drift. J. Math. Soc. Japan **49** (1997) 189–211.

Kazamaki, N.
[1] Change of time, stochastic integrals and weak martingales. Z.W. **22** (1972) 25–32.
[2] On a problem of Girsanov. Tohoku Math. J. **29** (1977) 597–600.
[3] A remark on a problem of Girsanov. Sém. Prob. XII, Lect. Notes in Mathematics, vol. 649. Springer, Berlin Heidelberg New York 1978, pp. 47–50.
[4] Continuous Exponential martingales and BMO. Lect. Notes in Mathematics, vol. 1579. Springer, Berlin Heidelberg New York 1994.

Kazamaki, N. and Sekiguchi, T.
[1] On the transformation of some classes of martingales by a change of law. Tohoku Math. J. **31** (1979) 261–279.
[2] Uniform integrability of continuous exponential martingales. Tohoku Math. J. **35** (1983) 289–301.

Kendall, D.G.
[1] The Mardia-Dryden shape distribution for triangles: a stochastic calculus approach. J. App. Prob. **28** (1991) 225–230.

Kennedy, D.
[1] Some martingales related to cumulative sum tests and single-server queues. Stoch. Proc. Appl. **4** (1976) 261–269.

Kent, J.
[1] The infinite divisibility of the Von Mises-Fischer distribution for all values of the parameter in all dimensions. Proc. London Math. Soc. **35** (1977) 359–384.
[2] Some probabilistic properties of Bessel functions. Ann. Prob. **6** (1978) 760–770.

Kertz, R.P., and Rösler, U.
[1] Martingales with given maxima and terminal distributions. Israel J. of Maths. **69**, 2 (1990) 173–192.

Khintchine, A.Y.
[1] Asymptotische Gesetze der Wahrscheinlichkeitsrechnung. (Ergebnisse der Mathematik, Bd. 2). Springer, Berlin Heidelberg 1933, pp. 72–75.

Knight, F.B.
[1] Brownian local time and taboo processes. Trans. Amer. Math. Soc. **143** (1969) 173–185.
[2] Random walks and a sojourn density process of Brownian motion. Trans. Amer. Math. Soc. **107** (1963) 56–86
[3] A reduction of continuous square-integrable martingales to Brownian motion. Lect. Notes in Mathematics, vol. 190. Springer, Berlin Heidelberg New York 1970, pp. 19–31.
[4] An infinitesimal decomposition for a class of Markov processes. Ann. Math. Stat. **41** (1970) 1510–1529.
[5] Essentials of Brownian motion and diffusion. Math. Surv. **18** Amer. Math. Soc., Providence, Rhode-Island 1981.
[6] On the duration of the longest excursion. Sem. Stoch. Prob. 1985. Birkhäuser, Basel 1986, pp. 117–147.
[7] On invertibility of Martingale time changes. Sem. Stoch. Proc. 1987. Birkhäuser, Basel 1988, pp. 193–221.
[8] Inverse local times, positive sojourns, and maxima for Brownian motion. Astérisque **157-158** (1988) 233–247.
[9] Calculating the compensator: methods and examples. Sem. Stoch. Proc. 1990. Birkhäuser, Basel 1991, pp. 241–252.
[10] Foundations of the prediction process. Clarendon Press, Oxford 1992.
[11] Some remarks on mutual windings. Sem. Prob. XXVII, Lect. Notes in Mathematics, vol. 1557. Springer, Berlin Heidelberg New York 1993, pp. 36–43.
[12] A remark about Walsh's Brownian motion. Colloque in honor of J.P. Kahane (Orsay, June 1993). In: The Journal of Fourier Analysis and Applications, Special Issue, 1995, pp. 1600–1606.
[13] On the upcrossing chains of stopped Brownian motion. Sém. Prob. XXXII, Lect. Notes in Mathematics, vol. 1686. Springer, Berlin Heidelberg New York 1998, pp. 343–375.

Knight, F.B. and Maisonneuve, B.
[1] A characterization of stopping times. Ann. Prob. **22** (1994) 1600–1606.

Kolmogorov, A.N.
[1] Über das Gesetz des itierten Logarithmus. Math. Ann. **101** (1929) 126–135.

Koval'chik, I.M.
[1] The Wiener integral. Russ. Math. Surv. **18** (1963) 97–135.

Kôno, N.
[1] Démonstration probabiliste du théorème de d'Alembert. Sém. Prob. XIX, Lect. Notes in Mathematics, vol. 1123. Springer, Berlin Heidelberg New York 1985, pp. 207–208.

Khoshnevisan, D.
[1] Exact rates of convergence to Brownian local time. Ann. Prob. 22,3 (1994) 1295–1330.

Krengel, U.
[1] Ergodic theorems. de Gruyter, Berlin New York 1985.

Küchler, U.
[1] On sojourn times, excursions and spectral measures connected with quasi-diffusions. Jour. Math. Kyoto U. **26**, 3 (1986) 403–421.

Kunita, H.
[1] Absolute continuity of Markov processes and generators. Nagoya Math. J. **36** (1969) 1–26.
[2] Absolute continuity of Markov processes. Sém. Prob. X, Lect. Notes in Mathematics, vol. 511. Springer, Berlin Heidelberg New York 1976, pp. 44–77.
[3] On backward stochastic differential equations. Stochastics **6** (1982) 293–313.
[4] Stochastic differential equations and stochastic flows of diffeomorphisms. Ecole d'Eté de Probabilités de Saint-Flour XII, Lect. Notes in Mathematics, vol. 1097. Springer, Berlin Heidelberg New York 1984, pp. 143–303.
[5] Some extensions of Itô's formula. Sém. Prob. XV, Lect. Notes in Mathematics, vol. 850. Springer, Berlin Heidelberg New York 1981, pp. 118–141.
[6] Stochastic flows and stochastic differential equations. Cambridge University Press 1990.

Kunita, H., and Watanabe, S.
[1] On square-integrable martingales. Nagoya J. Math. **30** (1967) 209–245.

Kurtz, T.G.
[1] Representations of Markov processes as multiparameter time changes. Ann. Prob. **8** (1980) 682–715.

Lamperti, J.
[1] Continuous state branching processes. Bull. A.M.S. **73** (1967) 382–386.
[2] Semi-stable Markov processes I. Z.W. **22** (1972) 205–255.
[3] Semistable stochastic processes. Trans. Amer. Math. Soc. **104** (1962) 62–78.

Lane, D.A.
[1] On the fields of some Brownian martingales. Ann. Prob. **6** (1978) 499–508.

Lebedev, N.N.
[1] Special functions and their applications. Dover Publications, 1972.

Ledoux, M.
[1] Inégalités isopérimétriques et calcul stochastique. Sém. Prob. XXII, Lect. Notes in Mathematics, vol. 1321. Springer, Berlin Heidelberg New York 1988, pp. 249–259.

Ledoux, M., and Talagrand, M.
[1] Probability on Banach spaces. Springer, Berlin Heidelberg New York 1991.

Le Gall, J.F.
[1] Applications du temps local aux équations différentielles stochastiques unidimensionnelles. Sém. Prob. XVII, Lect. Notes in Mathematics, vol. 986. Springer, Berlin Heidelberg New York 1983, pp. 15–31.
[2] Sur la saucisse de Wiener et les points multiples du mouvement Brownien. Ann. Prob. **14** (1986) 1219–1244.
[3] Sur la mesure de Hausdorff de la courbe brownienne. Sém. Prob. XIX, Lect. Notes in Mathematics, vol. 1123. Springer, Berlin Heidelberg New York 1985, pp. 297–313.

[4] Sur le temps local d'intersection du mouvement brownien plan et la méthode de renormalisation de Varadhan. Sém. Prob. XIX, Lect. Notes in Mathematics, vol. 1123. Springer, Berlin Heidelberg New York 1985, pp. 314–331.

[5] Une approche élémentaire des théorèmes de décomposition de Williams. Sém. Prob. XX, Lect. Notes in Mathematics, vol. 1204. Springer, Berlin Heidelberg New York 1986, pp. 447–464.

[6] Sur les fonctions polaires pour le mouvement brownien. Sém. Prob. XXII, Lect. Notes in Mathematics, vol. 1321. Springer, Berlin Heidelberg New York 1988, pp. 186–189.

[7] Propriétés d'intersection des marches aléatoires, I. Convergence vers le temps local d'intersection, II. Etude des cas critiques. Comm. Math. Phys. **104** (1986) 471–507 and 509–528.

[8] Introduction au mouvement brownien. Gazette des Mathématiciens 40, Soc. Math. France (1989) 43–64.

[9] Some properties of planar Brownian motion. Ecole d'été de Saint-Flour XX, 1990, Lect. Notes in Mathematics, vol. 1527. Springer, Berlin Heidelberg New York 1992, pp. 112–234.

[10] The uniform random tree in a Brownian excursion. Prob. Th. Rel. Fields **96** (1993) 369–383.

[11] Mouvement brownien et calcul stochastique. Cours de troisième cycle. Université Paris VI, 1994.

Le Gall, J.F., and Meyre, T.

[1] Points cônes du mouvement brownien plan, le cas critique. Prob. Th. Rel. Fields **93** (1992) 231–247.

Le Gall. J.F., and Yor, M.

[1] Sur l'équation stochastique de Tsirel'son. Sém. Prob. XVII, Lect. Notes in Mathematics, vol. 986. Springer, Berlin Heidelberg New York 1983, pp. 81–88.

[2] Etude asymptotique de certains mouvements browniens complexes avec drift. Z.W. **71** (1986) 183–229.

[3] Etude asymptotique des enlacements du mouvement brownien autour des droites de l'espace. Prob. Th. Rel. Fields **74** (1987) 617–635.

[4] Enlacements du mouvement brownien autour des courbes de l'espace. Trans. Amer. Math. Soc. **317** (1990) 687–722.

[5] Excursions browniennes et carrés de processus de Bessel. C.R. Acad. Sci. Paris, Série I **303** (1986) 73–76.

Lehoczky, J.

[1] Formulas for stopped diffusion processes with stopping times based on the maximum. Ann. Prob. **5** (1977) 601–608.

Le Jan, Y.

[1] Martingales et changements de temps. Sém. Prob. XIII, Lect. Notes in Mathematics, vol. 721. Springer, Berlin Heidelberg New York 1979, pp. 385–389.

Lenglart, E.

[1] Sur la convergence p.s. des martingales locales. C.R. Acad. Sci. Paris **284** (1977) 1085–1088.

[2] Relation de domination entre deux processus. Ann. I.H.P. **13** (1977) 171–179.

[3] Transformation de martingales locales par changement absolument continu de probabilités. Z.W. **39** (1977) 65–70.

Lenglart, E., Lépingle, D., and Pratelli, M.

[1] Une présentation unifiée des inégalités en théorie des martingales. Sém. Prob. XIV, Lect. Notes in Mathematics, vol. 784. Springer, Berlin Heidelberg New York 1980, pp. 26–48.

Lépingle, D.
[1] Sur le comportement asymptotique des martingales locales. Sém. Prob. XII, Lect. Notes in Mathematics, vol. 649. Springer, Berlin Heidelberg New York 1978, pp. 148–161.
[2] Une remarque sur les lois de certains temps d'atteinte. Sém. Prob. XV, Lect. Notes in Mathematics, vol. 850. Springer, Berlin Heidelberg New York 1981, pp. 669–670.

Lépingle, D. and Mémin, J.
[1] Sur l'intégrabilité des martingales exponentielles. Z.W. **42** (1978) 175–203.

Leuridan, C.
[1] Une démonstration élémentaire d'une identité de Biane et Yor. Sém. Prob. XXX, Lect. Notes in Mathematics, vol. 1626. Springer, Berlin Heidelberg New York 1996, pp. 255–260.
[2] Le théorème de Ray-Knight à temps fixe. Sém. Prob. XXXII, Lect. Notes in Mathematics, vol. 1686. Springer, Berlin Heidelberg New York 1998, pp. 376–406.
[3] Les théorèmes de Ray-Knight et la mesure d'Itô pour le mouvement brownien sur le tore $\mathbb{R}/\mathbb{Z}$. Stochastics and Stochastic Reports **55** (1995) 109–128.

Lévy, P.
[1] Le mouvement brownien plan. Amer. J. Math. **62** (1940) 487–550.
[2] Processus stochastiques et mouvement brownien. Gauthier-Villars, Paris 1948.
[3] Wiener random functions and other Laplacian random functions. Proc. Second Berkeley Symp., vol. II, 1950, pp. 171–186.
[4] Sur un problème de M. Marcinkiewicz. C.R. Acad. Sci. Paris **208** (1939) 318–321. Errata p. 776.
[5] Sur certains processus stochastiques homogènes. Compositio Math. **7** (1939) 283–339.

Lewis, J.T.
[1] Brownian motion on a submanifold of Euclidean space. Bull. London Math. Soc. (1986) 616–620.

Liptser, R.S., and Shiryaev, A.N.
[1] Statistics of random processes, I and II. Springer Verlag, Berlin, 1977 and 1978. Second edition (2000).

Lorang, G., and Roynette, B.
[1] Etude d'une fonctionnelle liée au pont de Bessel. Ann. Inst. H. Poincaré **32** (1996) 107–133.

Lyons, T.J., and McKean, H.P., Jr.
[1] Windings of the plane Brownian motion. Adv. Math. **51** (1984) 212–225.

Lyons, T.J., and Zhang, T.S.
[1] Decomposition of Dirichlet processes and its application. Ann. Prob. **22** (1994) 494–524.

McGill, P.
[1] A direct proof of the Ray-Knight theorem. Sém. Prob. XV, Lect. Notes in Mathematics, vol. 850. Springer, Berlin Heidelberg New York 1981, pp. 206–209.
[2] Markov properties of diffusion local time: a martingale approach. Adv. Appl. Prob. **14** (1982) 789–810.
[3] Integral representation of martingales in the Brownian excursion filtration. Sém. Prob. XX, Lect. Notes in Mathematics, vol. 1204. Springer, Berlin Heidelberg New York 1986, pp. 465–502.

McGill, P., Rajeev, B., and Rao, B.V.

[1] Extending Lévy's characterization of Brownian motion. Sém. Prob. XXII, Lect. Notes in Mathematics, vol. 1321. Springer, Berlin Heidelberg New York 1988, pp. 163–165.

McKean, H.P., Jr.

[1] The Bessel motion and a singular integral equation. Mem. Coll. Sci. Univ. Kyoto, Ser. A, Math. **33** (1960) 317–322.

[2] Stochastic integrals. Academic Press, New York 1969.

[3] Brownian local time. Adv. Math. **15** (1975) 91–111.

[4] Brownian motion and the general diffusion: scale and clock. Proceedings of the First Bachelier Conference (July 2000). eds: H. Geman, D. Madan, S.R. Pliska, T. Vorst. Springer (2002), p. 75–92.

Maisonneuve, B.

[1] Systèmes régénératifs. Astérisque 15. Société Mathématique de France 1974.

[2] Exit systems. Ann. Prob. **3** (1975) 399–411.

[3] Une mise au point sur les martingales locales continues définies sur un intervalle stochastique. Sém. Prob. XI, Lect. Notes in Mathematics, vol. 528. Springer, Berlin heidelberg New York 1977, pp. 435–445.

[4] On Lévy's downcrossing theorem and various extensions. Sém. Prob. XV, Lect. Notes in Mathematics, vol. 850. Springer, Berlin Heidelberg New York 1981, pp. 191–205.

[5] Ensembles régénératifs de la droite. Z.W. **63** (1983) 501–510.

[6] On the structure of certain excursions of a Markov process. Z.W. **47** (1979) 61–67.

[7] Excursions chevauchant un temps aléatoire quelconque. In: Hommage à P.A. Meyer et J. Neveu. Astérisque **236** (1996) 215–226.

[8] Martingales de valeur absolue donnée, d'après Protter-Sharpe. Ann. Prob. **5** (1977) 475–481.

Malliavin, P.

[1] Stochastic Analysis. Springer, Berlin Heidelberg New York 1997.

Malric, M.

[1] Filtrations browniennes et balayage. Ann. I.H.P. **26** (1990) 507–540.

[2] Densité des zéros des transformés de Lévy itérés d'un mouvement brownien. C.R. Acad. Sci. Paris; Ser. I **336** (2003) 499–504.

Mandelbrot, B.B., and Van Ness, J.W.

[1] Fractional Brownian motions, fractional noises and applications. SIAM Rev. **10** (1968) 422–437.

Mandl, P.

[1] Analytical treatment of one-dimensional Markov process. Springer, Berlin Heidelberg New York 1968.

March, P. and Sznitman, A.S.

[1] Some connections between excursion theory and the discrete random Schrödinger equation with random potentials. Prob. Th. Rel. Fields **67** (1987) 11–54.

Marcus, M.B. and Rosen, J.

[1] Sample path properties of the local times of strongly symmetric Markov processes via Gaussian processes. Ann. Prob. **20**, 4 (1992) 1603–1684.

[2] Gaussian chaos and sample path properties of additive functionals of symmetric Markov processes. Ann. Prob. **24**, 3 (1996) 1130–1177.

[3] Gaussian processes and local times of symmetric Lévy processes. In: Lévy Processes, p. 67–88, eds. O. Barndorff-Nielsen, T. Mikosch, S. Resnick, Birkhäuser (2001).

[4] New perspectives on Ray's theorem for the local times of diffusions. Ann. Prob. **312** (2003) 882–913.

Maruyama, G.
[1] On the transition probability functions of Markov processes. Nat. Sci. Rep. Ochanomizu Univ. **5** (1954) 10–20.
[2] Continuous Markov processes and stochastic equations. Rend. Circ. Palermo **10** (1955) 48–90.

Maruyama, G., and Tanaka, H.
[1] Ergodic property of n-dimensional Markov processes. Mem. Fac. Sci. Kyushu Univ. **13** (1959) 157–172.

Meilijson, I.
[1] There exists no ultimate solution to Skorokhod's problem. Sém. Prob. XVI, Lect. Notes in Mathematics, vol. 920. Springer, Berlin Heidelberg New York 1982, pp. 392–399.
[2] On the Azéma-Yor stopping time. Sém. Prob. XVII, Lect. Notes in Mathematics, vol. 986. Springer, Berlin Heidelberg New York 1983, pp. 225–226.

Méléard, S.
[1] Application du calcul stochastique à l'étude des processus de Markov réguliers sur [0, 1]. Stochastics **19** (1986) 41–82.

Messulam, P., and Yor, M.
[1] On D. Williams' "pinching method" and some applications. J. London Math. Soc. **26** (1982) 348–364.

Métivier, M.
[1] Semimartingales: a course on stochastic processes. de Gruyter, Berlin New York 1982.

Meyer, P.A.
[1] Processus de Markov. Lect. Notes in Mathematics, vol. 26. Springer, Berlin Heidelberg New York 1967.
[2] Processus de Markov: La frontière de Martin. Lect. Notes in Mathematics, vol. 77. Springer, Berlin Heidelberg New York 1979.
[3] Démonstration simplifiée d'un théorème de Knight. Sém. Prob. V, Lect. Notes in Mathematics, vol. 191. Springer, Berlin New York 1971, pp. 191–195.
[4] Processus de Poisson ponctuels, d'après K. Itô. Sém. Prob. V, Lect. Notes in Mathematics, vol. 191. Springer, Berlin Heidelberg New York 1971, pp. 177–190.
[5] Un cours sur les intégrales stochastiques. Sém. Prob. X, Lect. Notes in Mathematics, vol. 511. Springer, Berlin Heidelberg New York 1976, pp. 245–400.
[6] Démonstration probabiliste de certaines inégalités de Littlewood-Paley. Sém. Prob. XII, Lect. Notes in Mathematics, vol. 511. Springer, Berlin Heidelberg New York 1976, pp. 125–183.
[7] La formule d'Itô pour le mouvement brownien, d'après Brosamler. Sém. Prob. XII, Lect. Notes in Mathematics, vol. 649. Springer, Berlin Heidelberg New York 1978, pp. 763–769.
[8] Flot d'une équation différentielle stochastique. Sém. Prob. XV, Lect. Notes in Mathematics, vol. 850. Springer, Berlin Heidelberg New York 1981, pp. 103–117.
[9] Sur l'existence de l'opérateur carré du champ. Sém. Prob. XX, Lect. Notes in Mathematics, vol. 1204. Springer, Berlin Heidelberg New York 1986, pp. 30–33.
[10] Formule d'Itô généralisée pour le mouvement Brownien linéaire, d'après Föllmer, Protter, Shiryaev. Sém. Prob. XXXI, Lect. Notes in Mathematics, vol. 1655. Springer, Berlin Heidelberg New York 1997, pp. 252–255.

[11] Sur une transformation du mouvement brownien due à Jeulin et Yor. Sém. Prob. XXVIII, Lect. Notes in Mathematics 1583, p. 98–101, Springer (1994).
[12] Les processus stochastiques de 1950 à nos jours. In: Developments of Mathematics 1950–2000, ed. J.P. Pier, Birkhäuser (2000), p. 815–818.

Meyer, P.A., Smythe, R.T., and Walsh, J.B.
[1] Birth and death of Markov processes. Proc. Sixth Berkeley Symposium III. University of California Press, 1971, pp. 295–305.

Millar, P.W.
[1] Martingale integrals. Trans. Amer. Math. Soc. **133** (1968) 145–166.
[2] Stochastic integrals and processes with stationary independent increments. Proc. Sixth Berkeley Symp. **3** (1972) 307–332.

Mokobodzki, G.
[1] Opérateur carré du champ: un contre-exemple. Sém. Prob. XXIII, Lect. Notes in Mathematics, vol. 1372. Springer, Berlin Heidelberg New York 1989, pp. 324–325.

Molchanov, S.
[1] Martin boundaries for invariant Markov processes on a solvable group. Theor. Prob. Appl. **12** (1967) 310–314.

Molchanov, S.A., and Ostrovski, E.
[1] Symmetric stable processes as traces of degenerate diffusion processes. Theor. Prob. Appl. **14** (1) (1969) 128–131.

Monroe, I.
[1] Processes that can be imbedded in Brownian motion. Ann. Prob. **6**, 1 (1978) 42–56.
[2] Embedding right-continuous martingales in Brownian motion. Ann. Math. Stat. **43** (1972) 1293–1311.

Mortimer, T.M. and Williams, D.
[1] Change of measure up to a random time: theory. J. Appl. Prob. **28** (1991) 914–918.

Mountford, T.S.
[1] Transience of a pair of local martingales. Proc. A.M.S. **103** (3) (1988) 933–938.
[2] Limiting behavior of the occupation of wedges by complex Brownian motion. Prob. Th. Rel. Fields **84** (1990) 55–65.
[3] The asymptotic distribution of the number of crossings between tangential circles by planar Brownian motion. J. London Math. Soc. **44** (1991) 184–192.

Mueller, C.
[1] A unification of Strassen's law and Lévy's modulus of continuity. Z.W. **56** (1981) 163–179.

Nagasawa, M.
[1] Time reversions of Markov processes. Nagoya Math. J. **24** (1964) 177–204.
[2] Transformations of diffusions and Schrödinger processes. Prob. Th. Rel. Fields **82** (1989) 106–136.
[3] Schrödinger equations and diffusion theory. Birkhäuser 1993.
[4] Stochastic processes in quantum physics. Birkhäuser 2000.

Nakao, S.
[1] On the pathwise uniqueness of solutions of one-dimensional stochastic diffferential equations. Osaka J. Math. **9** (1972) 513–518.
[2] Stochastic calculus for continuous additive functionals of zero energy. Zeit. für Wahr. **68** (1985) 557–578.

Neveu, J.
[1] Processus aléatoires gaussiens. Les Presses de l'Univ. de Montréal, 1968.
[2] Intégrales stochastiques et application. Cours 3° Cycle. Laboratoire de Probabilites, Université de Paris VI, 1972.
[3] Sur l'espérance conditionnelle par rapport à un mouvement Brownien. Ann. I.H.P. **12**, 2 (1976) 105–110.
[4] Bases mathématiques du calcul des probabilités, 2éme édition. Masson, Paris 1971.

Ndumu, N.M.
[1] The heat kernel formula in a geodesic chart and some applications to the eigenvalue problem of the 3-sphere. Prob. Th. Rel. F. **88** (3) (1991) 343–361.

Norris, J.R., Rogers, L.C.G., and Williams, D.
[1] Self-avoiding random walk: A Brownian model with local time drift. Prob. Th. Rel. F. **74** (1987) 271–287.

Novikov, A.A.
[1] On moment inequalities for stochastic integrals. Theor. Prob. Appl. **16** (1971) 538–541.
[2] On an identity for stochastic integrals. Theor. Prob. Appl. **17** (1972) 717–720.

Nualart, D.
[1] Weak convergence to the law of two-parameter continuous processes. Z.W. **55** (1981) 255–269.
[2] The Malliavin calculus and related topics. Springer, Berlin Heidelberg New York 1995.

Obloj, J.
[1] The Skorokhod problem and its offsprings. Preprint Arkiv. (2004).

Obloj, J. and Yor, M.
[1] An explicit solution to Skorokhod problem for the age of Brownian excursions and Azéma martingale. Stoch. Proc. App. **110** (2003) 93–110.

Ocone, D.L.
[1] A symmetry characterization of conditionally independent increment martingales. In: D. Nualart and M. Sanz (eds.) Proceedings of the San Felice workshop on Stochastic Analysis (1991). Birkhäuser 1993, 147–167.
[2] Malliavin's calculus and stochastic integral representations of functionals of diffusion processes. Stochastics and Stochastic Reports **12** (1984) 161–185.

O'Connell, N. and Yor, M.
[1] A representation for non-colliding random walks. Elec. Comm. Prob. **7** (2002) 1–12.

Orey, S.
[1] Two strong laws for shrinking Brownian tubes. Z.W. **63** (1983) 281–288.

Orey, S., and Pruitt, W.
[1] Sample functions of the N-parameter Wiener process. Ann. Prob. **1** (1973) 138–163.

Orey, S., and Taylor, S.J.
[1] How often on a Brownian path does the law of the iterated logarithm fail? Proc. London Math. Soc. **28** (1974) 174–192.

Oshima, Y., and Takeda, M.
[1] On a transformation of symmetric Markov processes and recurrence property. Lect. Notes in Mathematics, vol. 1250. Proceedings Bielefeld. Springer, Berlin Heidelberg New York 1987.

Ouknine, Y., and Rutkowski, M.
[1] Local times of functions of continuous semimartingales. Stochastic Anal. and Applications **13** (1995) 211–232.

Pagès, H.
[1] Personal communication.

Paley, R., Wiener, N., and Zygmund, A.
[1] Note on random functions. Math. Z. **37** (1933) 647-668.

Papanicolaou, G., Stroock, D.W., and Varadhan, S.R.S.
[1] Martingale approach to some limit theorems. Proc. 1976. Duke Conf. On Turbulence. Duke Univ. Math. Series III, 1977.

Parthasarathy, K.R.
[1] Probability measures on metric spaces. Academic Press, New York 1967.

Pauwels, E.J., and Rogers, L.C.G.
[1] Skew-product decompositions of Brownian motions. Contemp. Math. **73** (1988) 237–262.

Pazy, A.
[1] Semi-groups of linear operators and applications to partial differential equations. (Applied Mathematical Sciences, vol. 44). Springer, Berlin Heidelberg New York 1983.

Perkins, E.
[1] A global intrinsic characterization of Brownian local time. Ann. Prob. **9** (1981) 800–817.
[2] The exact Hausdorff measure of the level sets of Brownian motion. Z.W. **58** (1981) 373–388.
[3] Local time is a semimartingale. Z.W. **60** (1982) 79–117.
[4] Weak invariance principles for local time. Z.W. **60** (1982) 437–451.
[5] Local time and pathwise uniqueness for stochastic differential equations. Sém. Prob. XVI, Lect. Notes in Mathematics, vol. 920. Springer, Berlin Heidelberg New York 1982, pp. 201–208.
[6] On the Hausdorff dimension of Brownian slow points. Z.W. **64** (1983) 369–399.
[7] The Cereteli-Davis solution to the H^1-embedding problem and an optimal embedding in Brownian motion. Seminar on stochastic processes. Birkhäuser, Basel 1985, pp. 172–223.

Petit, F.
[1] Quelques extensions de la loi de l'arcsinus. C.R. Acad. Sci. Paris 315 Série I (1992) 855–858.

Pierre, M.
[1] Le problème de Skorokhod: une remarque sur la démonstration d'Azéma-Yor. Sém. Prob. XIV, Lect. Notes in Mathematics, vol. 784. Springer, Berlin Heidelberg New York 1980, pp. 392–396.

Pitman, J.W.
[1] One-dimensional Brownian motion and the three-dimensional Bessel process. Adv. Appl. Prob. **7** (1975) 511–526.
[2] A note on L_2 maximal inequalities. Sém. Prob. XV, Lect. Notes in Mathematics, vol. 850. Springer, Berlin Heidelberg New York 1981, pp. 251–258.
[3] Stationary excursions. Sém. Prob. XVI, Lect. Notes in Mathematics, vol. 1247. Springer, Berlin Heidelberg New York 1987, pp. 289–302.
[4] Lévy systems and path decompositions. Seminar on stochastic processes. Birkhäuser, Basel 1981, pp. 79–110.

[5] Cyclically stationary Brownian local time processes. Prob. Th. Rel. Fields **106** (1996) 299–329.
[6] The SDE solved by local times of a Brownian excursion or bridge derived from the height profile of a random tree or forest. Ann. Prob. **27** (1999) 261–283.
[7] The distribution of local times of a Brownian bridge. Sém. Prob. XXXIII, Lect. Notes in Mathematics, vol. 1709. Springer, Berlin Heidelberg New York 1999, pp. 388–394.

Pitman, J.W., and Rogers, L.C.G.
[1] Markov functions. Ann. Prob. **9** (1981) 573–582.

Pitman, J.W., and Yor, M.
[1] Bessel processes and infinitely divisible laws. In: D. Williams (ed.) Stochastic integrals. Lect. Notes in Mathematics, vol. 851. Springer, Berlin Heidelberg New York 1981.
[2] A decomposition of Bessel bridges. Z.W. **59** (1982) 425–457.
[3] Sur une décomposition des ponts de Bessel. In: Functional Analysis in Markov processes. Lect. Notes in Mathematics, vol. 923. Springer, Berlin Heidelberg New York 1982, pp. 276–285.
[4] The asymptotic joint distribution of windings of planar Brownian motion. Bull. Amer. Math. Soc. **10** (1984) 109–111.
[5] Asymptotic laws of planar Brownian motion. Ann. Prob. **14** (1986) 733–779.
[6] Some divergent integrals of Brownian motion. In: D. Kendall (ed.) Analysis and Geometric Stochastics, Supplement to Adv. Appl. Prob. (1986) 109–116.
[7] Further asymptotic laws of planar Brownian motion. Ann. Prob. **17** (3) (1989) 965–1011.
[8] Arcsine laws and interval partitions derived from a stable subordinator. Proc. London Math. Soc. **65** (3) (1992) 326–356.
[9] Dilatations d'espace-temps, réarrangements des trajectoires browniennes, et quelques extensions d'une identité de Knight. C.R. Acad. Sci. Paris, t 316, Série I (1993) 723–726.
[10] Decomposition at the maximum for excursions and bridges of one dimensional diffusions. In: N. Ikeda, S. Watanabe, M. Fukushima, H. Kunita (eds.) Itô's stochastic calculus and Probability Theory. Springer, Berlin Heidelberg New York 1996, pp. 293–310.
[11] Quelques identités en loi pour les processus de Bessel. In: Hommage à P.A. Meyer et J. Neveu. Astérisque **236**, Soc. Math. France 1996.
[12] Ranked Functionals of Brownian excursions. Comptes Rendus Acad. Sci. Paris, t 326, Série I, January 1998, 93–97.
[13] Path decompositions of a Brownian bridge related to the ratio of its maximum and amplitude. Studia Sci. Math. Hung. **35** (1999) 457–474.

Port, S.C. and Stone, C.J.
[1] Brownian motion and classical potential theory. Academic Press, New York 1978.

Pratelli, M.
[1] Le support exact du temps local d'une martingale continue. Sém. Prob. XIII, Lect. Notes in Mathematics, vol. 721. Springer, Berlin Heidelberg New York 1979, pp. 126–131.

Priouret, P.
[1] Processus de diffusion et équations différentielles stochastiques. Ecole d'Eté de Probabilités de Saint-Flour III, Lect. Notes in Mathematics, vol. 390. Springer, Berlin Heidelberg New York 1974, pp. 38–113.

Priouret, P. and Yor, M.

[1] Processus de diffusion à valeurs dans $\mathbb{R}$ et mesures quasi-invariantes sur $C(\mathbb{R}, \mathbb{R})$. Astérisque 22–23 (1975) 247–290.

Protter, Ph.

[1] Stochastic Integration and Differential Equations. In: App. of Maths. vol. 21, 2nd edition (2003), Springer.

[2] A partial introduction to financial asset pricing. Stoch. Proc. App. **91** (2001) 169–203.

Protter, Ph. and Sharpe, M.J.

[1] Martingales with given absolute value. Ann. Prob. **7** (1979) 1056–1058.

Rao, M.

[1] Brownian Motion and Classical Potential Theory. Aarhus University, Lect. Notes Series 47, February 1977.

Rauscher, B.

[1] Some remarks on Pitman's theorem. Sém. Prob. XXXI, Lect. Notes in Mathematics, vol. 1655. Springer, Berlin Heidelberg New York 1997, pp. 266–271.

Ray, D.B.

[1] Sojourn times of a diffusion process. Ill. J. Math. **7** (1963) 615–630.

Rebolledo, R.

[1] La méthode des martingales appliquée à l'étude de la convergence en loi des processus. Mémoire de la S.M.F. **62** (1979)

[2] Central limit theorems for local martingales. Z.W. **51** (1980) 269–286.

Resnick, S.I.

[1] Inverses of extremal processes. J. Appl. Prob. (1974) 392–405.

Revuz, D.

[1] Mesures associées aux fonctionnelles additives de Markov I. Trans. Amer. Math. Soc. **148** (1970) 501–531.

[2] Lois du tout ou rien et comportement asymptotique pour les probabilités de transition des processus de Markov. Ann. I.H.P. **19** (1983) 9–24.

[3] Markov chains. North-Holland, Amsterdam New York 1984.

[4] Une propriété ergodique des chaines de Harris et du mouvement Brownien linéaire (unpublished).

Robbins, H., and Siegmund, D.

[1] Boundary crossing probabilities for the Wiener process and sample sums. Ann. Math. Stat. **41** (1970) 1410–1429.

Rogers, L.C.G.

[1] Williams characterization of the Brownian excursion law: proof and applications. Sém. Prob. XV, Lect. Notes in Mathematics, vol. 850. Springer, Berlin Heidelberg New York 1981, pp. 227–250.

[2] Characterizing all diffusions with the $2M - X$ property. Ann. Prob. **9** (1981) 561-572.

[3] Itô excursions via resolvents. Z.W. **63** (1983) 237–255.

[4] A guided tour through excursions. Bull. London Math. Soc. **21** (1989) 305–341.

[5] The joint law of the maximum and terminal value of a martingale. Prob. Th. Rel. F. **95** (4) (1993) 451–466.

[6] Continuity of martingales in the Brownian excursion filtration. Prob. Th. Rel. F. **76** (1987) 291–298.

Rogers, L.C.G., and Walsh, J.B.
[1] The intrinsic local time sheet of Brownian motion. Prob. Th. Rel. F. **88** (1991) 363–379.
[2] The exact $\frac{4}{3}$-variation of a process arising from Brownian motion. Stochastics and Stoch. Reports **51** (1994) 267–291.
[3] $A(t, B_t)$ is not a semi-martingale. Seminar on Stoch. processes 1990. Prog. in Prob. **24** (1991) 275–283. Birkhäuser.
[4] Local time and stochastic area integrals. Ann. Prob. **19** (1991) 457–482.

Rogers, L.C.G., and Williams, D.
[1] Diffusions, Markov processes and Martingales, vol. 2: Itô calculus. Wiley and Sons, New York 1987.
[2] Diffusions, Markov processes and Martingales, vol. 1: Foundations. Wiley and sons, New York 1994.

Root, D.H.
[1] The existence of certain stopping times on Brownian motion. Ann. Math. Stat. **40** (1969) 715–718.

Rosen, J.
[1] A local time approach to the self-intersection of Brownian paths in space. Comm. Math. Phys. **88** (1983) 327–338.

Rosen, J. and Yor, M.
[1] Tanaka formulae and renormalization for triple intersections of Brownian motion in the plane. Ann. Prob. **19** (1991) 142–159.

Roth, J.P.
[1] Opérateurs dissipatifs et semi-groupes dans les espaces de fonctions continues. Ann. Inst. Fourier **26** (1976) 1–97.

Roynette, B., Vallois, P., and Yor, M.
[1] Limiting laws associated with Brownian motion perturbated by normalized exponential weights. Studia Math. Hung, 2004.

Ruiz de Chavez, J.
[1] Le théorème de Paul Lévy pour des mesures signées. Sém. Prob. XVIII, Lect. Notes in Mathematics, vol. 1059. Springer, Berlin Heidelberg New York 1984, pp. 245–255.

Saisho, Y., and Tanemura, H.
[1] Pitman type theorem for one-dimensional diffusion processes. Tokyo J. Math. **18** (2) (1990) 429–440.

Sato, K.I.
[1] Lévy processes and infinitely divisible distributions. Camb. Univ. Press (1999).

Schilder, M.
[1] Asymptotic formulas for Wiener integrals. Trans. Amer. Math. Soc. **125** (1966) 63–85.

Schwartz, L.
[1] Le mouvement brownien sur $\mathbb{R}^N$, en tant que semi-martingale dans S_N. Ann. I.H.P. **21**, 1 (1985) 15–25.

Seshadri, V.
[1] Exponential models, Brownian motion and independence. Can. J. of Stat. **16** (1988) 209–221.
[2] The inverse Gaussian distribution. Clarendon Press, Oxford 1993.

Sharpe, M.J.
[1] Local times and singularities of continuous local martingales. Sém. Prob. XIV, Lect. Notes in Mathematics, vol. 784. Springer, Berlin Heidelberg New York 1980, pp. 76–101.
[2] Some transformation of diffusions by time reversal. Ann. Prob. **8** (1980) 6, 1157–1162.
[3] General theory of Markov processes. Academic Press, New York 1989.

Shepp, L.A.
[1] Radon-Nikodym derivatives of Gaussian measures. Ann. Math. Stat. **37** (1966) 321–354.
[2] On the integral of the absolute value of the pinned Wiener process. Ann. Prob. **10** (1982) 234–239.
[3] The joint density of the maximum and its location for a Wiener process with drift. J. Appl. Prob. **16** (1979) 423–427.

Sheppard, P.
[1] On the Ray-Knight Markov properties of local times. J. London Math. Soc. **31** (1985) 377–384.

Shi, Z.
[1] Lim inf behaviours of the windings and Lévy's stochastic areas of planar Brownian motion. Sém. Prob. XXVIII, Lect. Notes in Mathematics, vol. 1583. Springer, Berlin Heidelberg New York 1994, pp. 122–137.
[2] Sinaï's walk via stochastic calculus. In: Milieux Aléatoires. Panoramas et Synthèses n° 12, 2001, p. 53–74.

Shiga, T., and Watanabe, S.
[1] Bessel diffusion as a one-parameter family of diffusion processes. Z.W. **27** (1973) 37–46.

Shyriaev, A.N., and Cherny, A.
[1] Some distributional properties of a Brownian motion with drift and an extension of P. Lévy's theorem. Theory Prob. App. **44** (2) (1999) 412–418.

Skorokhod, A.
[1] Stochastic equation for diffusion processes in a bounded region, I and II. Theor. Prob. Appl. **6** (1961) 264–274; **7** (1962) 3–23.
[2] Studies in the theory of random processes. Addison-Wesley, Reading, Mass. 1965.
[3] On a generalization of a stochastic integral. Theor. Prob. and Appl. **20** (1975) 219–233.

Simon, B.
[1] Functional integration and quantum physics. Academic Press, New York 1979.

Smith, L., and Diaconis, P.
[1] Honest Bernoulli excursions. J. Appl. Prob. **25** (3) (1988) 464–477.

Song, S.Q., and Yor, M.
[1] Inégalités pour les processus self-similaires arrètés à un temps quelconque. Sém. Prob. XXI, Lect. Notes in Mathematics, vol. 1247. Springer, Berlin Heidelberg New York 1987, pp. 230–245.

Spitzer, F.
[1] Some theorems concerning 2-dimensional Brownian motion. Trans. Amer. Math. Soc. **87** (1958) 187–197.
[2] Recurrent random walk and logarithmic potential. Proc. Fourth Berkeley Symp. on Math. Stat. and Prob. II. University of California 1961, pp. 515–534.
[3] Electrostatic capacity in heat flow and Brownian motion. Z.W. **3** (1964) 110–121.

[These papers are reprinted in: Durrett, R. and Kesten, H. (eds.), Random walks, Brownian motion and interacting particle systems. Birkhäuser 1992.]

Stoll, A.
[1] Self-repellent random walks and polymer measures in two dimensions. Doctoral dissertation. Bochum, 1985.

Strassen, V.
[1] An invariance principle for the law of the iterated logarithm. Z.W. **3** (1964) 211–226.

Stricker, C.
[1] Quasi-martingales, martingales locales, semi-martingales et filtrations. Z.W. **39** (1977) 55–63.
[2] Sur un théorème de H.J. Engelbert et J. Hess. Stochastics **6** (1981) 73–77.

Stricker, C. and Yor, M.
[1] Calcul stochastique dépendant d'un paramètre. Z.W. **45** (1978) 109–133.

Stroock, D.W.
[1] On the growth of stochastic integrals. Z.W. **18** (1971) 340–344.
[2] Topics in stochastic differential equations. Tata Institute, Bombay 1982.
[3] Some applications of stochastic calculus to partial differential equations. Ecole d'Eté de Probabilités de Saint-Flour XI, Lect. Notes in Mathematics, vol. 976. Springer, Berlin Heidelberg New York 1983, pp. 268–382.
[4] An introduction to the theory of large deviations. (Universitext). Springer, Berlin Heidelberg New York 1984.
[5] Probability Theory. An analytic view. Cambridge Univ. Press 1994.
[6] Markov Processes from K. Itô's perspective. Princeton Univ. Press 2003.

Stroock, D.W. and Varadhan, S.R.S.
[1] Multidimensional diffusion processes. Springer, Berlin Heidelberg New York 1979.

Stroock, D.W. and Yor, M.
[1] On extremal solutions of martingale problems. Ann. Sci. Ecole Norm. Sup. **13** (1980) 95–164.
[2] Some remarkable martingales. Sém. Prob. XV, Lect. Notes in Mathematics, vol. 850. Springer, Berlin Heidelberg New York 1981, pp. 590–603.

Sussman, H.J.
[1] An interpretation of stochastic differential equations as ordinary differential equations which depend on the sample point. Bull. Amer. Math. Soc. **83** (1977) 296–298.
[2] On the gap between deterministic and stochastic ordinary differential equations. Ann. Prob. **6** (1978) 19–41.

Sznitman, A.S. and Varadhan, S.R.S.
[1] A multidimensional process involving local time. Proc. Th. Rel. F. **71** (1986) 553–579.

Takaoka, K.
[1] On the martingales obtained by an extension due to Saisho, Tanemura and Yor of Pitman's theorem. Sém. Prob. XXXI, Lect. Notes in Mathematics, vol. 1655. Springer, Berlin Heidelberg New York 1997, pp. 256–265.

Tanaka, H.
[1] Note on continuous additive functionals of the 1-dimensional Brownian path. Z.W. **1** (1963) 251–257.
[2] Time reversal of random walks in one dimension. Tokyo J. Math. **12** (1989) 159–174.
[3] Time reversal of random walks in $\mathbb{R}^d$. Tokyo J. Math. **13** (1990) 375–389.

[4] Diffusion processes in random environments. Proc. ICM (S.D. Chatterji, ed.). Birkhäuser, Basel 1995, pp. 1047–1054.

[Selected papers of H. Tanaka have appeared in *Stochastic Processes*, World Scientific (2002), eds. M. Maejima, T. Shiga.]

Taylor, S.J.

[1] The α-dimensional measure of the graph and the set of zeros of a Brownian path. Proc. Cambridge Phil. Soc. **51** (1955) 265–274.

Toby, E. and Werner, W.

[1] On windings of multidimensional reflected Brownian motion. Stoch. and Stoch. Reports, vol. 55 (1995), pp. 315–327.

Trotter, H.F.

[1] A property of Brownian motion paths. Ill. J. Math. **2** (1958) 425–433.

Truman, A.

[1] Classical mechanics, the diffusion heat equation and Schrödinger equation. J. Math. Phys. **18** (1977) 2308–2315.

Truman, A. and Williams, D.

[1] A generalized Arcsine law and Nelson's mechanics of one-dimensional time-homogeneous diffusions. In: M. Pinsky (ed.) Diffusion processes and related problems in Analysis I. Birkhäuser, Boston 1990, pp. 117–135.

Tsirel'son, B.

[1] An example of a stochastic differential equation having no strong solution. Theor. Prob. Appl. **20** (1975) 427–430.

[2] Triple points: from non-brownian filtrations to harmonic measures. Geom. Funct. Anal. (GAFA) **7** (1997) 1071–1118.

[3] Within and beyond the reach of Brownian innovations. Proceedings of the International Congress of Mathematicians, Doc. Math., extra vol. III, 1998, pp. 311–320.

[4] Non classical Brownian motions, stochastic flows, continuous products I. Arkiv / PR / 0402431 / (Feb. 2004).

Uppman, A.

[1] Sur le flot d'une équation différentielle stochastique. Sém. Prob. XVI, Lect. Notes in Mathematics, vol. 920. Springer, Berlin Heidelberg New York 1982, pp. 268–284.

Vallois, P.

[1] Le problème de Skorokhod sur $\mathbb{R}$: une approche avec le temps local. Sém. Prob. XVII, Lect. Notes in Mathematics, vol. 986. Springer, Berlin Heidelberg New York 1983, pp. 227–239.

[2] Sur la loi du maximum et du temps local d'une martingale continue uniformément intégrable. Proc. London Math. Soc. **3** (69) (1994) 399–427.

[3] Sur la loi conjointe du maximum et de l'inverse du temps local du mouvement brownien: application à un théorème de Knight. Stochastics and Stochastics Reports **35** (1991) 175–186.

[4] Une extension des théorèmes de Ray-Knight sur les temps locaux browniens. Prob. Th. Rel. Fields **88** (1991) 443–482.

[5] Diffusion arrêtée au premier instant où l'amplitude atteint un niveau donné. Stoch. and Stoch. Reports **43** (1993) 93–115.

[6] Decomposing the Brownian path via the range process. Stoch. Proc. and their Appl. **55** (1995) 211–226.

Van den Berg, M. and Lewis, J.T.
[1] Brownian motion on a hypersurface. Bull. London Math. Soc. **17** (1985) 144–150.

Van Schuppen, J.H. and Wong, E.
[1] Transformations of local martingales under a change of law. Ann. Prob. **2** (1974) 879–888.

Ventsel, A.D.
[1] Rough limit theorems on large deviations for Markov processes 1 and 2. Theory Prob. Appl. **21** (1976) 227–242 and 499–512.

Ventsel, A.D. and Freidlin, M.I.
[1] On small random pertubations of dynamical systems. Russ. Math. Surv. **25** (1970) 1–55.
[2] Some problems concerning stability under small random perturbations. Theor. Prob. Appl. **17** (1972) 269–283.

Vershik, A.
[1] Decreasing sequence of measurable partitions and their applications. Soviet Math. Dokl. 11, 4 (1970) 1007–1011.
[2] The theory of decreasing sequences of measurable partitions. Saint-Petersburg Math. J. **6** (1994) 705–761.

Vervaat, W.
[1] A relation between Brownian bridge and Brownian excursion. Ann. Prob. **7** (1) (1979) 141–149.

Volkonski, V.A.
[1] Random time changes in strong Markov processes. Theor. Prob. Appl. **3** (1958) 310–326.

von Weizsäcker, H.
[1] Exchanging the order of taking suprema and countable intersection of σ-algebras. Ann. I.H.P. **19** (1983) 91–100.

Vostrikova, L. and Yor, M.
[1] Some invariance properties (of the laws) of Ocone's martingales. Sém. Prob. XXXIV, Lect. Notes in Mathematics 1729, pp. 417–431, Springer (2000).

Vuolle-Apiala, J. and Graversen, S.E.
[1] Duality theory for self-similar processes. Ann. I.H.P. **23**, 4 (1989) 323–332.

Walsh, J.B.
[1] A property of conformal martingales. Sém. Prob. XI, Lect. Notes in Mathematics, vol. 581. Springer, Berlin Heidelberg New York 1977, pp. 490–492.
[2] Excursions and local time. Astérisque **52-53** , Temps locaux (1978) 159–192.
[3] A diffusion with a discontinuous local time. Astérisque **52-53** , Temps locaux (1978) 37–45.
[4] The local time of the Brownian sheet. Astérisque **52-53** , Temps locaux (1978) 47–62.
[5] Downcrossings and the Markov property of local time. Astérisque **52-53** , Temps locaux (1978) 89–116.
[6] Propagation of singularities in the Brownian sheet. Ann. Prob. **10** (1982) 279–288.
[7] Stochastic integration with respect to local time. Seminar on Stochastic Processes 1989. Birkhäuser 1983, pp. 237–302.
[8] Some remarks on $A(t, B_t)$. Sém. Prob. XXVII, Lect. Notes in Mathematics, vol. 1557. Springer, Berlin Heidelberg New York 1993, pp. 173–176.

Wang, A.T.
[1] Quadratic variation of functionals of Brownian motion. Ann. Prob. **5** (1977) 756–769.
[2] Generalized Itô's formula and additive functionals of the Brownian path. Z.W. **41** (1977) 153–159.

Warren, J.
[1] Branching processes, the Ray-Knight theorem and sticky Brownian motion. Sém. Prob. XXXI, Lect. Notes in Mathematics, vol. 1655. Springer, Berlin Heidelberg New York 1997, pp. 1–15.
[2] On the joining of sticky Brownian motion. Sém. Prob. XXXIII, Lect. Notes in Mathematics, vol. 1709. Springer, Berlin Heidelberg New York 1999, pp. 257–266.

Warren, J. and Yor, M.
[1] The Brownian burglar: conditioning Brownian motion by its local time process. Sém. Prob. XXXII, Lect. Notes in Mathematics, vol. 1686. Springer, Berlin Heidelberg New York 1998, pp. 328–342.

Watanabe, S.
[1] On time-inversion of one-dimensional diffusion processes. Z.W. **31** (1975) 115–124.
[2] A limit theorem for sums of i.i.d. random variables with slowly varying tail probability. P.R. Krishnaia (ed.) Multivariate analysis, vol. 5. North-Holland, Amsterdam 1980, pp. 249–261.
[3] Generalized Arcsine laws for one-dimensional diffusion processes and random walks. In: M. Cranston and M. Pinsky (eds.) Proceedings of Symposia in pure mathematics, vol. 57. Amer. Math. Soc., Providence, Rhode Island (1995) 157–172.
[4] Bilateral Bessel diffusion processes with drift and time inversion. To appear (1998).
[5] Asymptotic windings of Brownian motion paths on Riemann surfaces. Acta Appl. Math. **63** (2000), n° 1–3, 441–464
[6] The existence of a multiple spider martingale in the natural filtration of a certain diffusion in the plane. Sém. Prob. XXXIII, Lect. Notes in Mathematics, vol. 1709. Springer, Berlin Heidelberg New York 1999, pp. 277–290.
[7] Invariants of one-dimensional diffusion processes and applications. Korean Math. Soc. **35** (1998), n° 3, 637–658.

Weber, M.
[1] Analyse infinitésimale de fonctions aléatoires. Ecole d'été de Saint-Flour XI-1981, Lect. Notes in Mathematics, vol. 976. Springer, Berlin Heidelberg New York 1983, pp. 383–465.

Weinryb, S.
[1] Etude d'une équation différentielle stochastique avec temps local. Sém. Prob. XVII, Lect. Notes in Mathematics, vol. 986. Springer, Berlin Heidelberg New York 1983, pp. 72–77.

Weinryb, S. and Yor, M.
[1] Le mouvement brownien de Lévy indexé par $\mathbb{R}^3$ comme limite centrale de temps locaux d'intersection. Sém. Prob. XXII, Lect. Notes in Mathematics, vol. 1321. Springer, Berlin Heidelberg New York 1988, pp. 225–248.

Wendel, J.W.
[1] Hitting spheres with Brownian motion. Ann. Prob. **8** (1980) 164–169.
[2] An independence property of Brownian motion with drift. Ann. Prob. **8** (1980) 600–601.

Werner, W.
[1] Girsanov's transformation for SLE (k, ρ) processes, intersection exponents and hiding exponents. Ann. Sci. Toulouse (2004).

Widder, D.V.
[1] The Heat Equation. Academic Press 1975.

Wiener, N.
[1] Differential space. J. Math. Phys. **2** (1923) 132–174.
[2] The homogeneous chaos. Amer. J. Math. **60** (1930) 897–936.

Williams, D.
[1] Markov properties of Brownian local time. Bull. Amer. Math. Soc. **76** (1969) 1035–1036.
[2] Decomposing the Brownian path. Bull. Amer. Math. Soc. **76** (1970) 871–873.
[3] Path decomposition and continuity of local time for one dimensional diffusions I. Proc. London Math. Soc. (3) **28** (1974) 738–768.
[4] A simple geometric proof of Spitzer's winding number formula for 2-dimensional Brownian motion. University College, Swansea (1974).
[5] On a stopped Brownian motion formula of H.M. Taylor. Sém. Prob. X, Lect. Notes in Mathematics, vol. 511. Springer, Berlin Heidelberg New York 1976, pp. 235–239.
[6] On Lévy's downcrossing theorem. Z.W. **40** (1977) 157–158.
[7] Diffusions, Markov processes and Martingales, vol. 1: Foundations. Wiley and Sons, New York 1979.
[8] Conditional excursion theory. Sém. Prob. XIII, Lect. Notes in Mathematics, vol. 721. Springer, Berlin Heidelberg New York 1979, pp. 490–494.
[9] Brownian motion and the Riemann zeta function. In: D. Welsh and G. Grimmett (eds.) Disorder in Physical Systems. Festschrift for J. Hammersley. Oxford 1990, pp. 361–372.
[10] Probability with Martingales. Cambridge Univ. Press 1990.
[11] Weighing the Odds. Camb. Univ. Press (2001).
[12] Some basic theorems on Harnesses. In: Stoch. Analysis (A Tribute to the memory of Rollo Davidson), p. 349–363, Wiley (1973).
[13] Brownian motion as a Harness. Unpublished manuscript 1980.
[14] A "non-stopping" time with the optional stopping property. Bull. London Math. Soc. **34** (2002) 610–612.

Wong, E.
[1] The construction of a class of stationary Markoff processes. In: Stoch. Processes in Math. Physics and Engineering. Proc. of Symposia in Applied Maths., vol. XVI. Am. Math. Soc. 1964, pp. 264–276.

Wong, E. and Zakai, M.
[1] The oscillation of stochastic integrals. Z.W. **4** (1965) 103–112.

Yamada, T.
[1] On a comparison theorem for solutions of stochastic differential equations. J. Math. Tokyo Univ. **13** (1973) 497–512.
[2] On some representations concerning the stochastic integrals. Proba. and Math. Statistics **4**, 2 (1984) 153–166.
[3] On the fractional derivative of Brownian local times. J. Math. Kyoto Univ. **21**, 1 (1985) 49–58.
[4] On some limit theorems for occupation times of one-dimensional Brownian motion and its continuous additive functionals locally of zero energy. J. Math. Kyoto Univ. **26**, 2 (1986) 309–222.
[5] Representations of continuous additive functionals of zero energy via convolution type transforms of Brownian local time and the Radon transform. Stochastics and Stochastics Reports **46**, 1 (1994) 1–15.

[6] Principal values of Brownian local times and their related topics. In: N. Ikeda, S. Watanabe, M. Fukushima, H. Kunita (eds.) Itô's stochastic calculus and Probability theory. Springer, Berlin Heidelberg New York 1996, pp. 413–422.

Yamada, T. and Ogura, Y.
[1] On the strong comparison theorem for the solutions of stochastic differential equation. Z.W. **56** (1981) 3–19.

Yamada, T. and Watanabe, S.
[1] On the uniqueness of solutions of stochastic differential equations. J. Math. Kyoto Univ. **11** (1971) 155–167.

Yamazaki, Y.
[1] On limit theorems related to a class of "winding-type" additive functionals of complex Brownian motion. J. Math. Kyoto Univ. **32** (4) (1992) 809–841.

Yan, J.A.
[1] A propos de l'intégrabilité uniforme des martingales exponentielles. Sém. Prob. XVI, Lect. Notes in Mathematics, vol. 920. Springer, Berlin Heidelberg New York 1982, pp. 338–347.
[2] Sur un théorème de Kazamaki-Sekiguchi. Sém. Prob. XVII, Lect. Notes in Mathematics, vol. 986. Springer, Berlin Heidelberg New York 1983, pp. 121–122.

Yan, J.A. and Yoeurp, C.
[1] Représentation des martingales comme intégrales stochastiques de processus optionnels. Sém. Prob. X, Lect. Notes in Mathematics, vol. 511. Springer, Berlin Heidelberg New York 1976, pp. 422–431.

Yoeurp, C.
[1] Compléments sur les temps locaux et les quasi-martingales. Astérisque **52-53** , Temps locaux (1978) 197-218.
[2] Sur la dérivation des intégrales stochastiques. Sém. Prob. XIV, Lect. Notes in Mathematics, vol. 784. Springer, Berlin Heidelberg New York 1980, pp. 249–253.
[3] Contribution au calcul stochastique. Thèse de doctorat d'état, Université de Paris VI, 1982.
[4] Théorème de Girsanov généralisé et grossissement de filtrations. In: Grossissement de filtrations: exemples et applications. Lect. Notes in Mathematics, vol. 1118. Springer, Berlin Heidelberg New York 1985, pp. 172–196.

Yor, M. (ed.)
[1] Exponential functionals and principal values related to Brownian motion. Bib. Revista Mat. Ibero-Americana, 1997.

Yor, M.
[1] Sur les intégrales stochastiques optionnelles et une suite remarquable de formules exponentielles. Sém. Prob. X, Lect. Notes in Mathematics, vol. 511. Springer, Berlin Heidelberg New York 1976, pp. 481–500.
[2] Sur quelques approximations d'intégrales stochastiques. Sém. Prob. XI, Lect. Notes in Mathematics, vol. 528. Springer, Berlin Heidelberg New York 1977, pp. 518–528.
[3] Formule de Cauchy relative à certains lacets browniens. Bull. Soc. Math. France **105** (1977) 3–31.
[4] Sur la continuité des temps locaux associés à certaines semi-martingales. Astérisque **52-53** , Temps locaux (1978) 23–36.
[5] Un exemple de processus qui n'est pas une semi-martingale. Astérisque **52-53** (1978) 219–221.

[6] Sous-espaces denses dans L^1 et H^1 et représentation des martingales. Sém. Prob. XII, Lect. Notes in Mathematics, vol. 649. Springer, Berlin Heidelberg New York 1978, pp. 264–309.

[7] Les filtrations de certaines martingales du mouvement brownien dans $\mathbb{R}^n$. Sém. Prob. XIII, Lect. Notes in Mathematics, vol. 721. Springer, Berlin Heidelberg New York 1979, pp. 427–440.

[8] Sur le balayage des semi-martingales continues. Sém. Prob. XIII, Lect. Notes in Mathematics, vol. 721. Springer, Berlin Heidelberg New York 1979, pp. 453–471.

[9] Sur l'étude des martingales continues extrémales. Stochastics **2** (1979) 191–196.

[10] Loi de l'indice du lacet brownien et distribution de Hartman-Watson. Z.W. **53** (1980) 71–95.

[11] Remarques sur une formule de P. Lévy. Sém. Prob. XIV, Lect. Notes in Mathematics, vol. 784. Springer, Berlin heidelberg New York 1980, pp. 343–346.

[12] Sur un processus associé aux temps locaux browniens. Ann. Sci. Univ. Clermont-Ferrand II, **20** (1982) 140–148.

[13] Sur la transformée de Hilbert des temps locaux browniens et une extension de la formule d'Itô. Sém. Prob. XVI, Lect. Notes in Mathematics, vol. 920. Springer, Berlin Heidelberg New York 1982, pp. 238–247.

[14] Le drap brownien comme limite en loi de temps locaux linéaires. Sém. Prob. XVII, Lect. Notes in Mathematics, vol. 986. Springer, Berlin Heidelberg New York 1983, pp. 89–105.

[15] Une inégalité optimale pour le mouvement brownien arrèté à un temps quelconque. C.R. Acad. Sci. Paris, Sér. A **296** (1983) 407–409.

[16] A propos de l'inverse du mouvement brownien dans $\mathbb{R}^n$ ($n \geq 3$). Ann. I.H.P. **21**, 1 (1985) 27–38.

[17] Renormalisation et convergence en loi pour les temps locaux d'intersection du mouvement brownien dans $\mathbb{R}^3$. Sém. Prob. XIX, Lect. Notes in Mathematics, vol. 1123. Springer, Berlin Heidelberg New York 1985, pp. 350–365.

[18] Sur la représentation comme intégrale stochastique des temps locaux du mouvement brownien dans R^n. Sém. Prob. XX, Lect. Notes in Mathematics, vol. 1204. Springer, Berlin Heidelberg New York 1986, pp. 543–552.

[19] De nouveaux résultats sur l'équation de Tsirel'son. C.R. Acad. Sci. Paris, Série I **309**, (1989) 511–514.

[20] Remarques sur certaines constructions des mouvements browniens fractionnaires. Sém. Prob. XXII, Lect. Notes in Mathematics, vol. 1321. Springer, Berlin Heidelberg New York 1988, pp. 217–224.

[21] On stochastic areas and averages of planar Brownian motion. J. Phys. A. Math. Gen. **22** (1989) 3049–3057.

[22] Etude asymptotique des nombres de tours de plusieurs mouvements browniens complexes corrélés. In: Durrett, R. and Kesten, H. (eds.), Random walks, Brownian motion and interacting particle systems. Birkhäuser, Basel 1992, pp. 441–455.

[23] Some aspects of Brownian motion. Part I: Some special functionals. Lectures in Mathematics ETH Zürich. Birkhäuser 1992. Part II: Some recent martingale problems. Lect. Notes in Maths. ETH Zürich. Birkhäuser 1997.

[24] The distribution of Brownian quantiles. J. Appl. Prob. **32** (1995) 405–416.

[25] Random Brownian scaling and some absolute continuity relationships. In: E. Bolthausen, M. Dozzi, F. Russo (eds.) Progress in probability, vol. 36. Birkhäuser (1995) 243–252.

[26] Inégalités de martingales continues arrètées à un temps quelconque, I, II. In: Grossissements de filtrations: exemples et applications. Lect. Notes in Mathematics, vol. 1118. Springer, Berlin Heidelberg New York 1985, pp. 110–171.

[27] Sur certaines fonctionnelles exponentielles du mouvement brownien réel. J. Appl. Prob. **29** (1992) 202–208.

[28] On some exponential functionals of Brownian motion. Adv. App. Prob. **24** (1992) 509–531.

Zaremba, P.
[1] Skorokhod problem-elementary proof of the Azéma-Yor formula. Prob. Math. Stat. **6** (1985) 11–17.

Zvonkin, A.K.
[1] A transformation of the phase space of a process that removes the drift. Math. USSR Sbornik **2** (1974) 129–149.

Index of Notation

$\mathscr{A}$, $\mathscr{A}^+$	Space of finite variation processes 119
a.s., a.e.	Almost sure, almost surely, almost everywhere
$\mathbb{B}$	Brownian sheet 39
BM	Brownian motion 19
BM^d	d-dimensional Brownian motion 20
$\mathrm{BM}^d(x)$	d-dimensional Brownian motion started at x 20
BB	Brownian Bridge 37
BES^δ, $\mathrm{BES}^\delta(x)$, $\mathrm{BES}^{(\nu)}$	Bessel processes of dimension δ, of index ν 445
BESQ^δ, $\mathrm{BESQ}^{(\nu)}$	Squares of Bessel processes 439, 440
$\mathrm{BES}^\delta_a(x, y)$	Bessel Bridge 463
$\mathrm{BESQ}^\delta_a(x, y)$	Square of Bessel Bridge 463
BMO, BMO_p	Space of martingales with bounded mean oscillation 75
CAF	Continuous additive functional 401
Cont. semi. mart.	Continuous semimartingale 127
$C_k(E)$	Space of continuous functions with compact support on the space E
$C_0(E)$	Space of continuous functions with limit 0 at infinity 88
$C^{p,q}$	Space of differentiable functions on a product space
DDS	Dambis-Dubins-Schwarz 181
$\mathscr{E}$	σ-field of Borel sets and space of Borel functions
$\mathscr{E}_+$	Space of positive Borel functions
$b\mathscr{E}$	Space of bounded Borel functions
$\mathscr{E}^*$	Space of universally measurable functions
$\mathscr{E}(M)$, $\mathscr{E}^f$, $\mathscr{E}^\lambda(M)$	Exponential martingales 148, 149
$e(f, g)$, $e_x(\sigma, b)$	Stochastic differential equations 366
F.V.	Finite variation 5, 119
$\mathscr{F}_T$	σ-field of the stopping time T 44

$\mathscr{F}_t^X$	Right-continuous and complete filtration generated by X 98
$\mathscr{F}_t^\nu$	Completion of $\mathscr{F}_t$ with respect to ν 93
g_t, $g_t(x)$	Density of centered Gaussian variables with variance t
$g_t(\omega)$, $d_t(\omega)$	Last zero before t, first zero after t 239, 260
$\mathbb{H}^2$	Space of L^2-bounded martingales 129
H^2	Space of L^2-bounded continuous martingales 129
H_0^2	Space of L^2-bounded continuous martingales vanishing at 0 129
H^p, H^1	Spaces of martingales 59
$\mathscr{J}$	σ-field of invariant events 423
$K \cdot M$, $K \cdot X$ $\int_0^t K_s dX_s$	Stochastic integrals 138, 140
$L^2(M)$, $\mathscr{L}^2(M)$,	Spaces of processes 137
$L^2_{loc}(M)$	140
$L_t^a(X)$	Family of local times of X 222
$L \log L$	Class of martingales 58
loc. mart.	Local martingale 123
$\log_2 = \log\log$	Iterated logarithm 56
M^f	Martingale additive functional 149, 284
M^*, $M_t^* = \sup_{s\le t} \|M_s\|$	Bilateral supremum 54
N, Nf, N, M, mN	Kernel notation 80
$\mathscr{O}$	Optional σ-field 172
ODE	Ordinary differential equation 382
OU	Ornstein-Uhlenbeck process 37
$\mathscr{P}$	Predictable σ-field 171
PRP	Predictable representation property 209
P_t	Semi-group 80
P_T	Law of X at time T 104
$p_t^\delta(x,y)$	Density of Bessel semigroup 446
$P_{x,y}^{\delta,a}$	Law of Bessel bridge 463
P_x^δ	Law of Bessel process 445
PPP	Poisson point process 474
Q_x^δ	Law of Square of Bessel process 440
$Q_{x,y}^{\delta,a}$	Law of Square of Bessel Bridge 463
$q_t^\delta(x,y)$	Density of Square of Bessel semigroup 441

R	Life-time of an excursion 480
r.v.	Random variable 2
SDE.	Stochastic differential equation 366
S	Supremum process of BM: $S_t = \sup_{s \le t} B_s$, 54
sgn	The sign function 222
t.f.	Transition function 80
U, U_δ	Spaces of excursions 480
$\mathscr{U}, \mathscr{U}_\delta$	σ-fields of spaces of excursions 480
W	Wiener measure 35
$\mathbf{W}, \mathbf{W}^d$	Wiener space 35
$\mathscr{X}$	Space of paths 500
$X^{(c)}$	Scaled process 535
Z	Set of zeros of BM 109
θ_t, θ_T	Shift operators 36
$\sigma(X_t, t \in T)$	σ-field generated by the random variables $X_t, t \in T$ 1
Δ, δ	Cemetery 84
$\|\Delta\|$	Modulus of subdivision Δ 4
$\pi(x, a, b)$	Martingale problem 296
ν_A	Measure associated with the additive functional A 410
γ_a	Gamma r.v. of parameter a 12
$\beta_{a,b}$	Beta r.v. of parameters a and b 12
τ_s	Time-change associated with the local time at 0 241
$\langle M, M\rangle, \langle X, X\rangle$ $\langle M, N\rangle, \langle X, Y\rangle$	Brackets 120, 124, 125, 128
$\overset{(d)}{=}$	Equality in law 10
$\xrightarrow{\text{f.d.}}$	Convergence in the sense of finite distributions 516
P-lim	Convergence in probability 10
1_A	Characteristic function of the set A 1
$\lhd$	Absolute continuity on $\mathscr{F}_t$ 325

Index of Terms

Absorbing point 84, 97
Action functional 342
Adapted process 42
Additive functional 401
  continuous – 401
  strong – 402
  integrable – 410
  σ-integrable – 410
  signed – 419
Arcsine law
  First – 112
  Second – 242
Area
  Stochastic – 196, 396
Associativity of stochastic integrals 139
Asymptotic σ-field 99
Atom 76
Augmentation
  usual – of a filtration 45
Azéma's martingale 505

Bachelier's equation 268
Backward
  equation 282
  integral 144
Bernstein's inequality 153
Bessel
  bridges 463
  processes 445
  squared – process 440
Bismut description of Itô's measure 502
Blumenthal zero one law 95
BMO-martingales 75
Bougerol's identity 388
Boundary
  entrance – 306
  natural – 306
Bracket
  of two continuous local martingales 125
  of two semimartingales 128
Bridge
  Brownian – 37, 154, 384
  Bessel – 463
  Squared Bessel – 463
  Brownian – as conditioned Brownian motion 41
  pseudo-Brownian – 248
Brownian
  motion 19
  d-dimensional – motion 20
  standard linear – motion 19
  Bridge 37, 154, 384
  filtrations
  motion with drift 73, 352
  sheet 39
  $(\mathscr{F}_t)$– – motion 97
  skew – motion 87, 292
  excursion 480
  reflected – motion 86, 238
  killed – motion 87
  meander 493
  motion on the sphere 530
Burkholder-Davis-Gundy inequalities 160

Cadlag 34
Cameron-Martin
  formula 371, 445
  space 339
Canonical
  process 34
  realization 92
Cauchy
  r.v. 13
  process 116
Cemetery 84
Chacon-Ornstein theorem 548
Chain rule 6
Chaos, Wiener 201, 207
Chapman-Kolmogorov equation 80
Characteristic measure of a Poisson point process 475

Chernov inequality 292
Chung-Jacobi-Riemann identity 509
Clark's formula 341
Comparison theorems 393
Conformal
  local martingale 189
  invariance of Brownian motion 190
Continuous process 17
Convergence
  weak – 10
  in distribution 516
  in the sense of finite distributions 516
  in law 10
Cooptional time 313
Covariance 294
Cramer transform 343

Dambis 181
Debut 46
Deviations of Brownian motion, large 345
Differentiability of BM, non – 32
Diffusion
  process 294
  coefficient 294
Discrete
  P.P. 471
  σ-discrete P.P. 471
Distribution, convergence in 516
Dominated process 162
Donsker's theorem 518
Doob's inequality 54
Downcrossings 60
Doss-Süssmann method 382
Drift, Brownian motion with – 73, 352
  coefficient 294
Dubins-Schwarz 181
Dubins inequality 66
Dynkin's operator 310

Elastic BM 408
Energy 527
Enlargement
  of a probability space 182
  of a filtration 363
Entrance
  boundary 306
  law 494
Equivalent, – processes 18
Ergodic theorem 427, 548
  for BM 548
Excessive
  measure 409
  function 423
Excursions
  intervals 109
  process 480
  normalized – 486
Explosions, criterion for – 383
Exponential
  formulas 476
  local martingales 148
  inequality 153
  holding point 97
Extension
  Kolmogorov – theorem 34
Extremal
  martingales 213
  probability measures 210
  process 406

Fefferman's inequality 76
Feller
  process 90
  semi-groups 88
  property, strong 423
Feynman-Kac formula 358
Filtered space 41
Filtering 175, 207
Filtration 41
  Brownian – 98
  complete – 45
  natural – 42
  right-continuous – 42
  usual augmentation of a – 45
Fine topology 98
Finite dimensional distributions 18
  of BM 23
  convergence in the sense of – 516
Fokker-Planck equation 282
Forward equation 282
Fractional Brownian motion 38
Fubini's theorem for stochastic integrals 175

Galmarino's test 47
Garsia-Neveu lemma 170
Gaussian
  martingales 133, 186
  random variable 11
  process 36
  measure 16
  space 12
  Markov process 86
Gebelein's inequality 205
Generator
  infinitesimal – 281
  extended infinitesimal – 285

Girsanov
  theorem 327
  transformation 329
  pair 329
Good λ inequality 164
Gronwall's lemma 543

Hardy's inequality 75, 155
Heat process 20
Hermite polynomials 151
Hölder properties
  of BM 28, 30
  of semimarts. 187
  of local times 237
Holding
  point 97
  exponential – time 97
Homogeneous
  Markov process 81
  transition function 80
Hypercontractivity 206

Image of a Markov process 87
Independent increments 96
Indistinguishable 19
Index
  of a stable law 115
  of a Bessel process 440
Inequality
  Burkholder-Davis-Gundy – 160
  exponential – 153
  Chernov – 292
  Dubins – 66
  Bernstein – 153
  Fefferman – 76
  Gebelein – 205
  Doob's – 54
  Good λ – 164
  Hardy's – 155
  Kunita-Watanabe – 127
  maximal – 53
Infinitely divisible 115
Infinitesimal generator 281
  extended – 285
Initial distribution 81
Innovation process 175
Instantaneously reflecting 307
Integral, stochastic – 138, 140, 141
Integration by parts formula 146
Invariant
  measure 409
  function 423
  σ-field 423
  events 423

Irregular points 98
Iterated logarithm, law of the – 56
Itô
  integral 138
  formula 147
  –Tanaka formula 223
  processes 298
  measure of excursions 482

Kailath-Segal identity 159
Kazamaki's criterion 331
Kernel 79
Killed 100
  Brownian motion 87
Knight's theorem 183
  asymptotic version of – 524
  identity 504
Kolmogorov's continuity criterion 19
Kunita-Watanabe inequality 127

Lamperti's relation 452
Langevin's equation 378
Large numbers for local martingales
  law of – 186
Large deviations of Brownian motion 345
Last exit times 408
Law
  of a process 34
  convergence in – 10
  uniqueness in – 367
Lebesgue theorem for stochastic integrals
  142
Lévy
  characterization theorem 150, 158
  measure 115
  Khintchine formula 115
  process 96
$L \log L$ class 58
Limit-quotient theorem 427
Linear continuous Markov processes 300
Linear stochastic equation 377
Localization 123
Local extrema of BM 113
Local martingales 123
  exponential – 148
  pure – 212
  standard – 213
  conformal – 189
Local time
  of a continuous semimartingale 222
  of Brownian motion 238
Locally bounded process 140

Markov
process 81
property 83, 94
property, strong – 202
property of BM, strong 102, 156
processes, linear 300
Markovian 84
Martingale 51
problem 296
Master formula 475
Maximum principle, positive 283
Meander, Brownian – 493
Measure
associated with an additive functional 410
characteristic – of a Poisson point process 475
Itô – 482
Lévy – 115
Wiener – 35
Measurable process 126
Modification 19
Modulus of continuity, Lévy's – 30
Monotone class theorem 2

Natural boundary 306
Newtonian potential 100
Normalized excursion 486
Novikov's criterion 332

Occupation
time 20, 401
times formula 224
Opérateur carré du champ 351
Optional
σ-field 172
process 172
projection 173
stopping theorem 69
Ornstein-Uhlenbeck process 37
d-dimensional – 360
Orthogonal martingales 145

Papanicolaou-Stroock-Varadhan theorem 526
Parameter
of a OU process 37
of a Poisson process 58, 471
Path decomposition 255, 318
Pathwise uniqueness 367
Pitman's theorem 253
Point process 471
discrete – 471
σ-discrete – 471
Points of increase of Brownian motion 461
Poisson
process 471
point process 474
$(\mathscr{F}_t)$- – point process 474
Polar
functions for BM 24
sets 191
Polarization 124
Potential kernel 100
Newtonian – 100
Predictable 47
σ-field 47, 171
process 171
projection 173
stopping 76
stopping time 172
representation property 209
Process
adapted – 42
cadlag – 34
continuous – 17
increasing – 119
finite-variation – 119
locally bounded – 140
measurable – 126
optional – 172
predictable – 171
progressively measurable – 44
Prokhorov's criterion 10, 516
Pseudo-Brownian bridge 248
Pure martingales 212

Quadratic variation 28
of Brownian motion 29
of local mart. 124
of semimarts. 128
Quasi-invariance of Wiener measure 339
Quasi-left continuity 101
Quasi-martingales 134

Ray-Knight theorems 454
Recurrence of Brownian motion 58, 192
Recurrent
Harris – 425
process 424
set 424
Reduced Gaussian random variable 11
Reflecting
instantaneously – 307
slowly – 307
Brownian motion 86, 239

Reflection
  principle 105
  stochastic differential equation with – 385
Regular
  linear Markov process 300
  points 98
Reproducing kernel Hilbert space 21, 39, 339
Resolvent equation 89

Scale function 302
  of Bessel processes 442, 446
Scaling invariance
  of Brownian motion 21
  of local time 244
  of Bessel process 446
Section theorem 172
Semimartingale 127
Sheet, Brownian 39
Size of a OU process 37
Skew Brownian motion 292
Skew-product 194
Skorokod's lemma 239
Slowly reflecting 308
Space-time Markov process 85
Speed measure 305
Stable
  random variable 115
  process 116
  process, symmetric – 116
  subspace of H_0^2 174
Standard
  Brownian motion 19, 97
  local martingale 213
State space 15
Stationary
  process 36
  independent increments 96
Stochastic
  integral 138, 140, 141
  differential equation 366
  area 196, 396
  process 15
Stopped process 44
Stopping
  time 42
  optional – 69
Strassen's law of the iterated logarithm 346
Stratonovitch integral 144
Strong
  Markov property 102
  Markov property, extended – 111
  solution 367
Submarkovian 84
Submartingales 51
Subordinator 117
Supermartingale 51
Support of local times 235
Symmetric
  local times 234
  stable processes 116

Tanaka, Itô-Tanaka formula 223
Terminal time 491
Time
  change 180
  entry – 43
  hitting – 43
  inversion 359
  reversal 312
  stopping – 42
Transient
  set 424
  process 424
Transition
  function 80
  probability 80
Trap 97
Tsirel'son's example 392

Upcrossings 60

Variation 5
  finite – process 119
  quadratic –
  p-variation of BM 33
Vector
  local martingale, semimartingale 147
Ventcell-Freidlin estimates 343
Version 18

Watanabe's process 504
Weak convergence 10
Weak solution to a stochastic differential equation 367
Wiener
  measure 35
  space 35
  chaos 201, 207
Williams
  path decomposition theorem 318
  description of Itô's measure 499

Zvonkin method 384

Catalogue

Additivity property of squared Bessel processes 440

Approximations of L_t and S_t 227, 233

Asymptotic properties for the transition function of the Brownian motion 100

Central-limit theorem for stochastic integrals 160

Conditioning with respect to last exit times 408

Continuous local martingales on $]0, \infty[$ 135, 157

Continuous local martingales on stochastic intervals 136

Criterions for an exponential local martingale to be a martingale 331, 332, 338

Deterministic semimartingales are functions of bounded variation 133, 145

Differentiation of stochastic integrals 143

Discontinuous local time for
- i) a semimartingale 238
- ii) a diffusion 420

Exceptional points for the law of the iterated logarithm 59

Exponentials of semimartingales 149

An exponential local martingale which is not a martingale 335

An extremal martingale which is not pure 214

Functions of the BM which are martingales 74

A Fourier transform proof of the existence of occupation densities 166

A Girsanov transform of a BM with a strictly smaller filtration 397

A Girsanov transform of a pure martingale which is not pure 393

Hausdorff dimension of the set of zeros of Brownian motion 247

The intervals of constancy are the same for a continuous local martingale and for its bracket 125

Joint law of $(\widetilde{T}_a, L_{\widetilde{T}_a})$ 265

L^p-inequalities for local times 265

Law of large numbers for local martingales 186

A local martingale with strong integrability properties is not necessarily a martingale 194

A local martingale is Gaussian iff its bracket is deterministic 186

Minkowski inequality for local martingales 136

A $0-1$ law for additive functionals of BM 422

The planar Brownian curve: its Lebesgue measure is zero 23, 196

Polar functions for the BM^2: they include the functions of bounded variation 197

Polarity for the Brownian sheet 198

Principal values for Brownian local times 236

A progressive set which is not optional 175

Ray-Knight theorems for local times of Bessel processes 459

A uniformly integrable martingale which is not in H^1 75

A semimartingale $X = M + A$ such that $\mathscr{F}^X \subset \mathscr{F}^M$ strictly 259

Semimartingale functions of BM 419

Skew-product representation of Brownian motion 194

A slowly reflecting boundary 421

Solutions to linear SDE's 378, 381

Support of local times 235

Times at which the BM is equal to its supremum 113

A bounded $(\mathscr{F}_t^B)$-martingale which is the stochastic integral of an unbounded process of $L^2(B)$ 269

A Brownian filtration sandwiched between $\mathscr{F}^B$ and $\mathscr{F}^{|B|}$ 208

Kazamaki's criterion is not a necessary condition 384

The time spent by the Brownian Bridge on the positive half-line is uniformly distributed 493

Printing: Strauss GmbH, Mörlenbach
Binding: Schäffer, Grünstadt

Printed in the United States
131982LV00002B/1-18/A

9 783540 643258